An Ordinary Man

An Ordinary Man

The Surprising Life and Historic Presidency
of Gerald R. Ford

RICHARD NORTON SMITH

An Imprint of HarperCollins*Publishers*

HarperCollins books may be purchased for educational, business, or sales promotional use. For information, please email the Special Markets Department at SPsales@harpercollins.com.

FIRST EDITION

All images are courtesy of the Gerald R. Ford Library unless otherwise noted.

Library of Congress Cataloging-in-Publication Data has been applied for.

ISBN 978-0-06-268416-5

23 24 25 26 27 LBC 5 4 3 2 1

For Oliver Sipple

Nothing astonishes men so much as common sense and plain dealing.

—RALPH WALDO EMERSON

CONTENTS

Part IV: Riding the Tiger, October 1974–December 1976

Part V: When Is This Retirement Going to Start? 1976–2006

A NOTE TO THE READER

THE CIRCUMSTANCES SURROUNDING Gerald Ford's formative years, and the several boyhood names by which he was identified, could hardly be more muddled. Born Leslie King Jr. in July 1913, he was two weeks old when his mother, fearing for both their lives, abandoned her brutish husband. Eventually she and her infant son settled in Grand Rapids, Michigan. There she met and in time married a local businessman named Gerald Ford (Sr.). Throughout his childhood the future president was variously known as Junior King, Junior Ford, plain Junior or its diminutive Junie. I have employed these names as appropriate, while making every effort to avoid confusion with either his blood father or the stepfather he considered his true parent. The reader is both forewarned and reassured that the problem solves itself by the time Junie Ford reaches his fifteenth birthday and is rebranded Gerald R. Ford Junior, the name by which he was known until the 1962 death of his stepfather.

An Ordinary Man

A CAPACITY FOR SURPRISE

—◆—

Bob Woodward (to Ford speechwriter Robert Hartmann, seventeen
years after publication of Hartmann's tell-all White House memoir):
Is there anything that stayed hidden that I should know about?

Hartmann: Ford was hidden.

NO ONE WAS more surprised to find himself invited to a White House stag dinner in the spring of 1975 than Senator George McGovern, the prairie populist crushed by Richard Nixon in the 1972 presidential election. McGovern's fierce opposition to the Vietnam War had made him persona non grata to presidents of both parties. "Lyndon Johnson never invited me to dinner," the South Dakota Democrat told Nixon's successor. "And you can be sure that Richard Nixon didn't."

"I know, George," replied Gerald Ford. "That's why I asked you." Policy disagreements notwithstanding, "the house belongs to everyone. Now more than ever."

Ford's penchant for the unexpected was not always so gracefully executed. On a Sunday morning in September 1974, a disbelieving nation woke to the news that he was pardoning Nixon for any and all offenses relating to the June 1972 Watergate break-in and subsequent cover-up that led to the president's resignation. Carl Bernstein of the *Washington Post* spoke for millions when he told his journalistic partner Bob Woodward, "The son of a bitch pardoned the son of a bitch." Ford used even harsher language in explaining his action. "Nixon really fucked up the pea patch," he told a White House visitor. "It will cost me the '76 election, but I had to do it."*

---

* The friend was shocked, less by Ford's justification of the pardon than by his use of the F-word, something he had never heard him employ before. Ford's use of profanity was sparing and picturesque. The harshest insult in his lexicon—"He doesn't know his ass from Page Eight"—was an expression of mysterious origin, apparently unique to the thirty-eighth president.

Overnight Ford's approval rating dropped twenty-two points, a descent unmatched in the history of public opinion polling, and a fair measure of how profoundly he had misread the popular mood. Privately Ford countered that he wasn't forgiving Nixon so much as he was trying to forget him; more precisely, to redirect his energies, and the country's attention, toward a rapidly deteriorating economy as well as escalating tensions in Asia, Africa and the Middle East. Whatever his motives, Ford's methods of deciding the fate of his disgraced predecessor shredded any stitches of trust woven during his first month in office. With the stroke of a pen the nation's only unelected president had smudged his claims to legitimacy. He also wrote the opening sentence of his historical obituary.

Nearly half a century later, Ford remains the president with an asterisk beside his name. The least self-dramatizing of men, long before "No Drama Obama," he guided the nation through its worst constitutional crisis since the Civil War. Later he broke the back of the harshest economic downturn since the Great Depression. And he accomplished both with so little fanfare that he got scant credit for either. To the contrary: in popular memory Ford is wedged between three Shakespearean predecessors—the idealized JFK, tormented LBJ and self-destructive Nixon—and the transformative figure of Ronald Reagan, a sunny revolutionary who remade his party as he replaced the existing New Deal–ish consensus with one suspicious of, if not hostile to, government solutions.

In such company Ford appears out of his depth, the genial embodiment of a decade recalled more for polyester and platform shoes than as an incubator of meaningful change. It was his misfortune to take the helm in a time of cultural upheaval, during which irony curdled into snark and authority was flogged like a medieval penitent by *Saturday Night Live*, the satirical game changer that cut its teeth on the new president's alleged clumsiness and intellectual shortcomings. Admittedly no Great Communicator, Ford's verbal gaffes inspired a *New York Magazine* cover depicting him as Bozo the Clown.* A commemorative marker in his hometown of Grand Rapids allegedly read, GERALD FORD SLIPPED HERE.

That Ford should be the last American president capable of conducting his own press briefings on the annual federal budget testifies to an expertise

---

* Richard Reeves, author of the accompanying article, apologized in 1995 for its venomous tone. "We pay presidents not by the hour, but for their judgment—a word Ford routinely mispronounced as 'judge-a-ment,'" Reeves wrote in *American Heritage*. "In retrospect Ford's judgment turned out to be better than his pronunciation."

gained during a quarter century as a congressional appropriator. But it was no substitute for Rooseveltian eloquence or Reaganesque wit. This presents a special challenge to the Ford biographer, one neatly summarized in an exchange between the present author and a writer friend whose familiarity with my 2014 life of Nelson Rockefeller evoked a tart parallel. From Rockefeller to Gerald Ford, he mused. "That's the difference between a peacock and a brown sparrow." With the perspective of almost five decades and eight post-Ford presidencies, the subsequent cooling of partisan emotions and access to papers and oral histories formerly off-limits, the time is right for a comprehensive Ford biography that combines scholarly rigor with popular accessibility.

When Jimmy Carter in his 1977 inaugural address complimented his predecessor for all he had done to restore public trust during his brief White House tenure, he fostered an image at once flattering and curiously stilted. In the years since, most Ford portraiture has been of the two-dimensional variety. There is Ford the Healer, saddled with inherited failures (the economy, Vietnam, a runaway CIA) the resolution of which evokes his response when the family's golden retriever soiled the presidential carpet. Waving off some White House stewards, Ford said that "no man should have to clean up after another man's dog." Yet that is exactly what he was forced to do, repeatedly, as president. In the shrewd observation of biographer John Robert Greene, Ford the Healer becomes "the man who saved the country from Nixon, prepared the country for Carter and Reagan, and did little of substance on his own."

A second, even less complimentary scenario replaces Ford the Healer with Good Old Jerry, congressional warhorse–turned–presidential caretaker, a man whose personal decency could neither compensate for his limited vision nor overcome the stigma of his association with Richard Nixon. Notwithstanding his bold assertion on August 9, 1974, America's long national nightmare was *not* over. Far from it. Balancing continuity against change, having no wish to tar innocent staffers with the guilt of their White House superiors, Ford waited too long to expel Nixon holdovers whose allegiance would always be suspect. "Nobody leaves the plane without a parachute," he announced. On another occasion he declared, "My whole philosophy of life is—I don't assume somebody is trying to screw me." Such comments prompted a query rarely heard around 1600 Pennsylvania Avenue: Can a president be *too nice* to succeed?

Much as Ford's public life deserves fresh appraisal, it is his personal story that most calls into question the familiar formulas: *An ordinary man summoned by extraordinary events. A politician without guile. What you see is what you get.* The reality—at least the one I have been able to piece together—is considerably more nuanced. Ford's "nice guy" reputation masked an ambition intense even by modern standards. In 1960 he mounted a vigorous, if futile, campaign to be Richard Nixon's running mate. That he learned of Vice President Spiro Agnew's legal problems at least six months earlier than he ever acknowledged is revealed here for the first time, as is his cultivation of Democratic support to fill the vacancy left by Agnew's resignation in October 1973.

Regarded as the ultimate Nixon loyalist, Ford as vice president actually chafed in the role of presidential defender (he later confided to Dick Cheney the eight months he spent in the job were the worst of his life). When Nixon's old nemesis Earl Warren died in July 1974, Ford went out of his way to honor the liberal chief justice by pausing before his bier in the Supreme Court building, well aware of the offense it would give his boss. That same month Warren's successor as chief justice, Nixon appointee Warren Burger, joined a unanimous court in ruling that the embattled president must turn over to the Watergate prosecutor hitherto secret White House tape recordings. The decision effectively sealed Nixon's fate, yet even then efforts to spare him a courtroom trial and possible conviction put Ford's presidency at risk before it began.

Ford had his wife, Betty, to thank for questioning the motives of White House chief of staff Alexander Haig after Haig mentioned a presidential pardon in the same breath as a putative Nixon resignation. As president, Ford repaid the debt by embracing Betty's work on behalf of the Equal Rights Amendment, refusing to attend the 1975 Gridiron Dinner unless it was opened to women, and making Carla Hills only the third woman to occupy a seat around the Cabinet table. "Do you know who would make a great Supreme Court justice?" Ford asked White House lawyers when a high court vacancy presented itself in December 1975. "Barbara Jordan." It's hard to imagine anyone else in his administration identifying the decidedly left-of-center congresswoman from Texas as a suitable replacement for the retiring William O. Douglas. Clearly, Gerald Ford could be a surprising man.

WAS THERE EVER a Ford Administration?" inquired novelist John Updike. "Evidence for its existence seems to be scanty." For once the writer's powers of observation failed him. As the tectonic plates of consensus inched away from the Washington-centric outlook prevailing since the 1930s, Ford established himself as the first true post–New Deal president. Less coda than curtain raiser, his administration serves as a bridge between Nixonian pragmatism and the more doctrinaire conservatism of Ronald Reagan. Case in point: economic deregulation (which Ford preferred to call "regulatory reform"). It began on his watch with the railroad and financial services industries, gathered bipartisan momentum under Carter and Reagan and achieved global respectability and a new emphasis on privatization in Margaret Thatcher's Britain.

Ford's free-market approach represented a decisive break from his predecessor's wage and price controls. His economic policies anticipated the monetarist strategies of Paul Volcker and Alan Greenspan. Meeting economist Arthur Laffer in a hotel bar where the father of supply-side economics drew his famous curve for them, Ford agents Donald Rumsfeld and Dick Cheney were inducted into Reaganomics before its namesake was.* Hoping to reduce American dependence on Middle East oil, Ford proposed a variety of incentives to increase domestic energy production. Unable to legislate the rapid decontrol of oil or natural gas prices, he succeeded in establishing the Strategic Petroleum Reserve, a fifty-five-mile-per-hour speed limit and the first CAFE standards to make American cars more energy efficient.

The Ford domestic agenda stretched beyond economics and energy policy. In Boston, court-ordered school busing ignited riots reminiscent of the mobs outside Central High School in Little Rock a generation earlier. The resulting crisis competed for attention with a botched campaign to combat swine flu; a pair of presidential assassination attempts; IRAs and ATMs; the near insolvency of America's greatest city; the unanimous confirmation of a Supreme Court nominee selected for his intellect, not his ideology, after a congressional hearing in which no one raised the subject of abortion; and a bicentennial observance that joyously affirmed how much the country's mood had improved since the bleak summer of 1974.[†]

Foreign policy afforded a still greater contrast with the status quo as Ford, a career legislator, battled his former colleagues to uphold what he called "the imperiled presidency." As memorable as the image of Richard Nixon's helicopter carrying a disgraced leader into exile are the helicopters of Saigon evacuating Americans and thousands of their South Vietnamese

---

* Reagan himself would praise the Earned Income Tax Credit, signed into law by Ford in 1975, as the best anti-poverty, pro-family program ever enacted by Washington.

† One offshoot of the national birthday party: in February 1976, Ford became the first American president to proclaim Black History Month.

collaborators from rooftop landing pads in April 1975. Practically speaking, there was little Ford could do to prevent a final humiliation as North Vietnamese forces overran the South's capital. Yet in the days immediately after the fall of Saigon he averted an even greater surrender, by shaming a resistant Congress into funding the resettlement on American soil of thousands of Vietnamese refugees. To do less, Ford argued, was to add moral bankruptcy to military defeat.

The Helsinki Accords, signed in August 1975, traded recognition of existing postwar boundaries in Europe, as favored by the Soviet Union, for the mutual acceptance and enforcement of human rights on both sides of the Iron Curtain. Widely denounced at the time, Helsinki is now regarded as an important milestone on the road to European liberation. In the volatile Middle East, the American president took steps to separate the warring parties as a prelude to substantive peace talks. Ford might have achieved more in foreign affairs but for domestic politics. Fears of a Reagan challenge from the right caused the administration to slow-walk potentially historic deals restricting nuclear weapons and returning the Panama Canal to the nation whose sovereignty was affronted every day it remained in Yankee possession.

Ford displayed greater intestinal fortitude in revamping US policy toward Africa, dispatching Secretary of State Henry Kissinger on the eve of a crucial Republican primary in Texas to proclaim support for Black majority rule in the former Rhodesia (now Zimbabwe). This in turn sent an unmistakable signal to South Africa's apartheid regime that its days were numbered. So bold a departure, however admirable in the abstract, was unlikely to advance the president's chances among Lone Star Republicans sympathetic to Ronald Reagan's harsh criticism of détente with the Soviet Union.

No one grasped this more instinctively than James Baker, then a junior Commerce Department official, soon to distinguish himself as the president's shrewdest delegate counter and manager of his fall campaign against Jimmy Carter. Baker registered his unhappiness with Chief of Staff Dick Cheney, who shared his concerns, but had been unable to convince the president of the political risks entailed in Kissinger's mission. Cheney urged Baker, if he felt so strongly about the issue, to make his own appeal directly to "the old man." An appointment was scheduled, allowing Baker to reiterate for the president the dangers inherent in the administration's African U-turn. If Kissinger's travels couldn't be postponed, pleaded Baker, might not the secretary at least forgo a "big splashy press conference" in advance of his departure? Through it all, Ford sat quietly—everyone agrees he excelled at listening—puffing on the pipe he sometimes used to control the pace of decision-making.*

When his turn came, Ford thanked Baker warmly for sharing his views. He praised Kissinger for the job he was doing in Africa. Moreover, he added, "I think he ought to tell the American people about it. And, anyway, the thinking Republicans in Texas will understand."

"Mr. President," replied Baker, "with respect to this issue, there are no thinking Republicans in Texas right now."

Ford's ability to laugh at his own expense was severely tested that spring, as he struggled to put down Reagan's conservative uprising. The immediate problem was his campaign style, best described as plodding, and made worse by speeches that had the opposite of their intended effect. When he walked onto a stage, Ford once confessed, "I'm never quite who I want to be." Pitted against the mediagenic Reagan, whose Hollywood training and natural charisma put him in a class with FDR and JFK, Ford evoked a high school coach exhorting his underdog players before the big game. From Reagan's California the legendary strategist Stu Spencer was recruited to convince a sitting president, slow to grasp the theater of modern politics, that he risked squandering the advantage of incumbency unless he stayed off the campaign trail altogether.

Armed with the latest discouraging poll data, Spencer was admitted to the Oval Office, where an enthusiastic Ford anticipated a heavy travel schedule for the fall campaign. After several failed attempts at candor, Spencer abandoned euphemism in favor of the direct approach.

"Forgive me, Mr. President, as a campaigner, you're no fucking good."

It is hard to imagine an ordinary man—still less any modern American president—calmly accepting this evaluation. Yet Ford gave no sign of offense. Cut to the sequel, in which veteran journalist Jules Witcover included the incident in *Marathon*, his book on the 1976 campaign. Spencer's reaction can easily be imagined. Appalled to see his crude putdown in print, the campaign strategist vented his anger on Dick Cheney, the chief of staff who had sat in on the Ford-Spencer meeting. The stoical Cheney listened for a while, then took advantage of a brief pause in the verbal abuse to remind Spencer they had not been alone in the Oval Office. Indeed, a

---

* For a man said to lack guile, Ford had an unusual appreciation of the pipe as theater prop. "I learned that if you want a little time to think about something, you can go through the process of filling a pipe, lighting it, and it gives you . . . time to digest, to analyze, and to make a decision," Ford confessed to an interviewer in 1990. For good measure the ex-president offered his assessment of pipe versus cigar smokers. The former, he guessed, were "a little more deliberate," the latter "more flamboyant" and extroverted.

third person had been present as Spencer offered his blunt estimate of the president's alleged shortcomings on the stump.

Suddenly it dawned on Stu Spencer. Improbable as it seemed, it was Ford himself, quietly amused by his adviser's act of lèse-majesté, who had leaked the story to Witcover. He may have lost an election, but not his capacity for surprise.

SEVEN DAYS IN JUNE

———◆———

I was standing this far away from John Mitchell and he looked me
right in the eye and he lied to me. That damn guy lied to me.

—GERALD R. FORD

JERRY FORD IS nothing if not self-disciplined. A month shy of his fifty-ninth birthday, the congressman from Grand Rapids exudes the vitality of a much younger man. A lanky six feet in height, Ford may be losing his hair, as the sandy-blond thatch of his youth retreats to a darkening corona. But he retains his waistline. The punishing round of calisthenics with which he starts each day, followed by twenty minutes of laps in a backyard heated pool, is little changed from the regimen long ago prescribed a square-jawed high school football hero. The sound of her husband splashing in the olive light of predawn serves as an alarm clock of sorts for Ford's wife, Betty, at least on mornings when he is home. Since he became House minority leader in 1965, there have been years when he devotes more nights to the rubber chicken circuit than to 514 Crown View Drive in suburban Alexandria, Virginia.

Today, June 17, 1972, is such a day. Why would Jerry Ford spend Father's Day weekend attending to the needs of a Michigan district whose voters have returned him to office twelve times since 1948? Chalk it up to personal ambition, and a will to win undiminished in the quarter century since Ford took on the local political establishment and ousted an entrenched GOP congressman who thought America's global responsibilities ended somewhere around the Indiana state line. A friend privy to Ford's grueling travel schedule once prescribed time on a California golf course, golf being a Ford passion second only to politics. "I wish I could do this more often," Ford told his friend in the clubhouse after their game. But "I've got to go out there and win those seats so I can be Speaker."

His single-minded pursuit of the gavel once wielded by Henry Clay and Sam Rayburn exacts a heavy toll. After seven years as minority leader, even Jerry has grown weary of the endless road shows, the indistinguishable hotel rooms and smoky banquet halls that separate him from his wife and children. He feels guilt over the additional burdens imposed on Betty Ford, forced in his absence to function as mother-father, house disciplinarian and congressional surrogate. "She holds down the fort," says her oldest son, Mike. "She's the anchor for the whole family." A stylish former dancer and model, Betty gamely battles a pinched nerve and the grinding pain of arthritis, chronic conditions dulled by pills and a vodka and tonic or two, which help fill the hours between adolescent crises and Jerry's phone calls from the road.

Lately the Fords have pursued their own version of détente. Should November fail to deliver Jerry's long-sought goal of a Republican House, he will retire after one more campaign, at an age when he may reasonably hope to earn enough as a lawyer and part-time lobbyist to provide his family some comforts unaffordable on a congressman's salary. And Betty and her children can reclaim the man neatly summarized in the career advice he gives to anyone contemplating a life in politics. "*Don't*," Ford advises potential candidates, "if you

aren't willing to work 70 hours per week,
have family or marital problems,
expect to make a great deal of money,
do not like people and working on their problems,
are thinned skinned [*sic*] and can't take public criticism,
are only interested in the glamour of the title or the responsibility.

Over the years Ford's attention to the residents of the Fifth District has bordered on the pastoral. When, early in his congressional career, a visiting member of the Daughters of the American Revolution fell on a Washington street corner and broke her ankle, no one knew how she was going to get back to Michigan, until Ford offered to drive her there himself. A quarter century later, he still insists that every letter addressed to him receive a personal response, within twenty-four hours if possible. This includes high school debaters researching their topic, candidates for a small business loan and the female traveler who desires an introduction to officials at the US embassy in London so "she won't be lonely on Thanksgiving Day." Following a rash of UFO sightings in Southern Michigan, Ford was showered with letters and telegrams demanding a federal investigation. He duly complied, even while acknowledging doubts about "planet people" possessed of the antigravity secret roaming the universe at fifty thousand miles per hour.

Ford's Capitol Hill office opens at seven a.m., two hours ahead of his colleagues. "We campaign 365 days a year," he reminds his staff. As a result, scarcely a birthday, wedding, obituary, civic award or graduating class in West Michigan goes unrecognized by the United States Congress. An elderly couple, otherwise unknown to their congressman, is nevertheless touched to receive anniversary greetings under his signature. Years later, on learning that the wife is in a nursing home, close to death, Ford drops by for a consoling visit. "The strongest weapon in a political campaign is the good credited to you by word of mouth"—this Ford credo goes a long way toward explaining him and the congressional mindset he personifies. By stressing individual contacts over ideological mandates, Ford defines leadership in transactional terms, constituent service on a grand scale. His is a vision of government suspicious of visionaries. When asked the secret of his political success, Ford reveals more than he perhaps intends by replying, "I made everyone else's problems my problems."

To those who know him merely as a partisan scrapper and indifferent orator, Ford is an easy man to underestimate. At the White House, domestic policy chief John Ehrlichman imagines that in another life he might make a successful Grand Rapids insurance salesman. Detractors liken him to the furniture, solid but unexciting, for which his hometown is nationally recognized. "I keep reading that I'm a plodder, that I'm not one who shakes up the establishment," says Ford. It is an assessment he is quick to dispute, citing not only the youthful insurgency that first landed him, a Republican, in Congress in the Democratic year of 1948, but his subsequent willingness to cross party lines and substitute constructive opposition for outworn dogma. His press secretary, Paul Miltich, struggles to find an accurate label to encapsulate Ford's politics. He is "both a liberal and conservative," concludes Miltich, "a kind of young Eisenhower." Twice in recent months Ford has signed a discharge petition to pry legislation out of the House Rules Committee—once for the proposed Equal Rights Amendment to the Constitution, which would assure women legal equality with their male counterparts; and once for legislation halting the practice of court-ordered busing to achieve racial balance in the nation's schools.

A Nixon stalwart since their first encounter on the floor of the House in January 1949, Ford is a cosponsor of the president's Family Assistance Program who nevertheless objects to its proposed $2,400 annual income as inadequate for a family of four. Uncomfortable with his party's racially insensitive mantra of "law and order," he much prefers the phrase "order with justice under the law." A child of the Depression, Ford has been known to roam the halls of Congress switching off lights to save money.

Office workers testify to his habit of wearing pencils down to their stubs. Yet he also supports limits on campaign spending, abolition of the Electoral College and the extension of voting rights to eighteen-year-olds. His executive secretary, Mildred Leonard, says she has never known a more even-tempered man. True to form, Ford admonishes a newcomer to his Washington staff, "You have to be nice, even if you have a headache."

Obscured from all but a few intimates is the anger inherited from his blood father, a disreputable wife abuser whose violent behavior drove Ford's mother, less than a month after the birth of their only child, to seek safety under the roof of Chicago relations. Those who know him best liken Ford to a well-mannered member of the family Ursidae—"98% teddy bear and 2% grizzly." He is at his best in small groups, where his evident integrity and command of subject offset any charisma deficit. Along with a near-photographic memory for names and numbers, Ford has a functional turn of mind, impatient with abstractions. As an eighth grader his son Steve aspired to a place on his school's freshman football team. The boy shared this ambition with his father, who noted the long odds against such a promotion. "But I'll tell you what," the elder Ford interjected. "If you can do one thing that nobody else can do, I can get you on the team."

It didn't take long for him to identify such a talent. "There's nobody that can snap for punts," he told Steve, "and I'm going to teach you how." True to his word, in seemingly endless backyard drills, Michigan's onetime star center showed the boy how he had snapped the ball for the Wolverines. Evidently the lessons took, for Steve Ford made the freshman squad, and went on to earn his letter, all "because he"—his father the congressman—"came up with a plan to get me on the team."

H AS SUCCESS SPOILED Jerry Ford? Hometown critics point to his national profile and incessant travel as evidence that he has outgrown the people who first sent him to Washington. It is to counter such arguments that he returns as often as he does. A classic early bird who thrives on six hours of sleep a night, Ford begins this Saturday with a ninety-minute skull session courtesy of Vern Ehlers, a Berkeley-trained nuclear physicist lured away from the California school to build a physics department at Grand Rapids' Calvin College. Inspired by comments made at a meeting of the American Physical Society, Ehlers wrote his congressman offering to recruit other scientists to informally advise him on matters of relevance to their field. Expecting a form response, Ehlers was surprised to receive, two days later, a phone call from Ford's administrative assistant Frank Meyer.

"Jerry really thinks it's a great idea and he'd like to have you put that together."

Soon all three men were discussing the panel's organization and membership. Were they interested in hearing from Republican scientists exclusively? Ehlers asked. "Of course," said Meyer.

"Well, I don't see why," Ford interrupted. "We want scientific advice, not political advice."

Ever since he has consulted Ehlers and his brethren on issues ranging from surface coal mining to the proposed SST supersonic airplane (Ford supports the revolutionary aircraft despite objections from his brain trust). This morning he proves yet again that scientists have no monopoly on deductive reasoning. By nine forty-five, having wrung out all the useful information he can, Ford thanks the group's members for their time and expertise. Ehlers stays behind to pose a question of his own. "Here you are, an extremely busy congressman," he tells Ford. "You come back on the weekend; you meet with constituents; and then you come and meet with us and spend an hour and a half or two hours . . . and you seem to enjoy it." Why?

Ford slips an arm around the professor's shoulders. "Well, Vern, you've got to recognize one thing. You're the only people who meet with me for the purpose of giving me something instead of asking me for something. All day long I sit in that office and people come in and ask me for favors. And you come along and you're here to help me."*

A few minutes later Ford strides onto the stage of Welsh Auditorium, an art deco relic named for a Depression-era mayor of Grand Rapids. Nearly four thousand members of the Michigan VFW greet him as a kindred spirit. In his speech Ford combines praise of Nixon for his outreach to historic adversaries in Moscow and Beijing with a hard-liner's call for the development of new weapons systems, if only to enable future presidents to negotiate from a position of strength. Ford mentions a forthcoming trip to the People's Republic of China, on which he is to depart at the end of the week. Afterward a meager lunch of cottage cheese doused in A.1. sauce is followed by the dedication of a government housing project in the nearby village of Rockford. No doubt Jean McKee, his personable Democratic opponent in this fall's contest to represent the Fifth District, will contrast Ford's support of the Rogue Valley Towers with his earlier votes to cut Section 8 funding. She has a point, though Ford is far from the first

---

* An observation that would serve Ehlers well during his seventeen-year career in the House of Representatives (1993–2010), where he represented Ford's old district with distinction.

fiscal conservative to oppose a Washington-funded initiative, only to lobby its administrators for a seat on the gravy train.

For proof one need look no further than Vandenberg Center, a bland compound of boxy government office buildings in downtown Grand Rapids replacing a much-loved Victorian-era city hall. In the windblown plaza a rawboned abstract sculpture, all curves and sharp edges, evokes the rushing river from which the city derives its name. Forty-three feet of plated steel, painted industrial orange, *La Grande Vitesse* ("the great swiftness") resembles nothing so much as a derrick-less oil well pumping up crude from the Michigan tundra. The work of celebrated artist Alexander Calder, it is the brainchild of a local arts enthusiast named Nancy Mulnix. In the spring of 1967, prompted by a visiting curator from the Metropolitan Museum of Art in New York, Mulnix wrote Congressman Ford to determine whether the federal government might be willing, through the recently established National Endowment for the Arts, to help underwrite a civic artwork dignifying the featureless expanse of Vandenberg Center. Ford just as promptly endorsed her request with a call to Roger L. Stevens, founding chairman of the endowment.

A legendary theatrical producer, Stevens was to prove the wiliest of bureaucrats. By authorizing the nation's first publicly funded work of art for Grand Rapids—self-proclaimed Furniture City and Calvinist bastion—Stevens hoped to allay conservative suspicions that his was an elitist enterprise contemptuous of flyover country. Enlisting the House Republican leader as his cosponsor was strategic icing on the cake. One week after her appeal to Ford, Nancy Mulnix received an encouraging phone call from Stevens. Ultimately the NEA pledged $45,000 to the project. When at last the components were shipped across the Atlantic and bolted into place—Aaron Copland composed a fanfare for its June 1969 dedication—*La Grande Vitesse* transformed more than a barren patch of downtown Grand Rapids.

Equally dramatic was the change visible in the congressman representing the Fifth District. At the outset, Ford readily admitted to House colleagues, "I did not know what a Calder was. But I can assure the members that a Calder in the center of the city . . . has really helped rejuvenate Grand Rapids." As president, Ford would award Calder the Presidential Medal of Freedom and, at a time of budgetary austerity, authorize a significant hike in NEA spending. His personal taste in art remained defiantly traditional. When the builders of his presidential museum came to him seeking artistic guidance regarding a work to grace the building's entrance, Ford made two requests. He did not want a statue of himself greeting visitors, he told them. And whatever art the committee came up with, he hoped it would be representational and not abstract.*

Sunday, June 18

It was late last night when Ford returned to the Pantlind Hotel from Lansing and an appearance before the American Legion Boys State convention. Since nothing he said there was likely to stop the presses, the news that commands his attention on this quiet Sabbath is not to be found in the *Grand Rapids Press* but in a fragmentary report broadcast on his car radio en route to Grace Episcopal Church, where he frequently ushers with his brother Dick. Five men with apparent links to the Central Intelligence Agency were arrested early Saturday morning while attempting to bug Democratic National Committee offices located in the Watergate complex in Northwest Washington.

"Flabbergasted" by what he hears, Ford is naturally curious about the origins of the break-in. It is one more distraction in a week crowded with preparations for his China trip and an escalating battle to enact a federal revenue-sharing program, cornerstone of President Nixon's so-called New Federalism. Under the administration's plan, financially beleaguered state and local governments would receive $5.3 billion as a down payment on a five-year commitment totaling $30 billion. For Grand Rapids and its environs, this windfall translates into an additional $20 million to construct sewerage systems and provide summer jobs for inner-city youth. Because the money comes virtually without strings, replacing categorical grants that can ensnare federal aid in onerous red tape, revenue sharing poses a radical departure from the centralization of power in Washington that began with the New Deal and reached new heights under Lyndon Johnson's Great Society.

Ford has another reason to champion revenue sharing. Breaking with Senator Robert Taft's classic admonition that "the duty of the opposition is to oppose," Republicans under Ford have adopted a new battle cry: "We have a better way." Replacing a less aggressive GOP leader in January 1965, Ford appointed more than a dozen task forces to develop *Republican* pro-

---

* The museum planners neatly solved their problem by commissioning a lifelike figure of an astronaut strolling in space, thereby recognizing Ford's role in the creation and subsequent congressional oversight of NASA. In 2013 the spaceman made way for a splendid bronze likeness of the president by sculptor J Brett Grill duplicating the larger-than-life-sized figure of Ford that Grill fashioned for the rotunda of the United States Capitol.

posals on health insurance, voting rights, aid to education, environmental protection and the war on poverty. On his own he pitched a Human Investment Act designed to lure private dollars into the nation's worst pockets of unemployment. After a sluggish start, Ford and his legislative allies began to punch above their weight, their morale boosted by Ford's tireless campaigning for "constructive conservatives" like George H. W. Bush of Texas. "It is so important . . . that we offer solutions to problems and that we dispel the image of Republicans as aginners," wrote Bush, now Republican National Chairman but in 1967 a freshman congressman singled out for advancement by Leader Ford.

That said, nothing promoted by Ford has roused such fervent opposition as revenue sharing. Conservatives in both parties look askance at any initiative adding to the federal deficit. Liberals fear the slow unraveling of national standards and a diminished role for Washington in guaranteeing social and economic justice. Only recently the bill's sponsors were forced to postpone final action for two weeks, in the hope that Speaker Carl Albert would round up additional votes among his Democratic brethren. Even now the outcome is in doubt.

Monday, June 19

Ford begins his legislative workweek in the pastel blue Congressional Prayer Room just off the Capitol Rotunda. Here he and a quartet of Republican colleagues—collectively dubbed the Five Sisters—meet for spiritual reflection every Monday morning at seven thirty. By then it's a safe bet the supplicants will have seen this morning's *Washington Post*, the contents of which might drive less devout legislators to their knees. According to the *Post*, four of the five burglars arrested at the Watergate two days ago have ties to the Cuban exile community in South Florida. The fifth, James Mc-Cord, is reportedly security director for the Nixon reelection campaign.

The news prompts some waggish commentary from Democratic lawmaker Lionel Van Deerlin.

Mother is merely a shoplifter,
Sister picks pockets with me.
Brother prowls only in pawn shops—
But Dad's with the GOP.

Whether Ford is laughing may be doubted. As it happens, he has on his calendar a previously planned meeting with John Mitchell, the Nixon attorney general, who, earlier this year, left the Justice Department to oversee the president's reelection effort. First, however, Ford visits his longtime friend Jack Marsh, a former Democratic congressman from Virginia whose office is eight floors above the Committee for the Reelection of the President (CREEP) at 1701 Pennsylvania Avenue. "Jerry, I know this fellow McCord," Marsh greets him, "and I know he works closely with John Mitchell." A few weeks earlier McCord, then in the act of establishing a nearby office for Martha Mitchell, had asked if he could borrow Marsh's telephone and copying machine. The same visitor had identified himself as security chief of the Nixon campaign.

McCord's apparent involvement in the recent break-in rattles Marsh's nerves. It doesn't help that McCord has known ties to the CIA. Ford offers reassurance. "Dick Nixon is much too smart a politician to have been involved in anything like that." His subsequent encounter with Mitchell tests his confidence. Sufficiently curious to arrive a few minutes early at the offices of CREEP, Ford asks Mitchell directly:

"John, did you or anyone at the Nixon campaign committee have anything to do with the break-in at Democratic National Headquarters?"

"Absolutely not," says Mitchell. And the president? Mitchell repeats himself, fervently. Accepting what he is told, Ford returns to the Capitol. There the full House is going into session in an atmosphere conducive to election-year pandering. Before it adjourns for the day, members will unanimously approve an increase in benefits received by two million disabled veterans, not forgetting Spanish-American War widows in their largesse. Only after its regular business is concluded does Speaker Albert recognize a handful of lawmakers for prearranged special order speeches: orators like Florida representative Robert Sikes, who berates the Weather Bureau for its inflated forecasts about an early season hurricane named Agnes. When the storm falls short of its hype, inflicting no damage in his Panhandle district except to the tourist trade, the vigilant Mr. Sikes threatens a full-scale investigation of the bureau.

Ford misses Sikes's presentation, having promised an office interview to consumer advocate Ralph Nader as part of his Congress Project to survey the voting records of all 535 members.* In the day's most pleasant surprise, the minority leader finds his evening schedule uncluttered by the interest

---

* Ford's profile as crafted by Nader and his volunteer investigators said he "combines a calm, forbearing manner with a strong pragmatic sense and lack of arm twisting." The same document was harsher in its assessment of Ford's legislative priorities, disputing his commitment to environmental and consumer protection.

group receptions and candidate fundraisers that often prevent him from enjoying a family dinner at 514 Crown View Drive. En route to Alexandria in the chauffeur-driven black Cadillac that is one of the perks of his job, Ford scans the afternoon papers. A news junkie accustomed to reading half a dozen journals a day, his practiced eye is drawn to a quote in the *Washington Star* attributed to White House press secretary Ron Ziegler, and mocking the recent break-in at the DNC as "a third rate burglary."

Ford has no way of knowing that last night at his Watergate apartment John Mitchell presided over an alcohol-fueled discussion of how to destroy evidence of criminal wrongdoing. To his campaign deputy, a baby-faced Brutus named Jeb Stuart Magruder, Mitchell hinted slyly, "Maybe you ought to have a little fire at your house tonight."

Tuesday, June 20

"It is not far from the truth to say that Congress in session is Congress on public exhibition," Woodrow Wilson wrote in 1885, "while Congress in its committee rooms is Congress at work." A century later Wilson's claim is validated each morning the House is in session, as bills and budgets are incubated in some twenty-three standing committees. Most of what Ford knows about government he learned in committee. As minority leader, however, he is excused from such deliberations. His Tuesdays generally begin at the other end of Pennsylvania Avenue, in the White House Cabinet Room, where the portraits of Eisenhower, Theodore Roosevelt and the aforementioned Wilson may be bipartisan, but the guest list is limited to members of the GOP. Here Ford and Senate Republican Leader Hugh Scott flank President Nixon around an oval mahogany table donated to the room by Nixon himself.

Proximity should not be confused with intimacy. Little about the weekly strategic review is spontaneous, the agenda having been prepared in advance by Nixon chief of staff H. R. Haldeman and domestic policy czar John Ehrlichman. That both men hold Congress in minimum high regard is among the capital's worst-kept secrets. This morning the Cabinet Room is silent, owing to the president's late return from an extended weekend in Florida and the Bahamas. At the hour customarily reserved for his legislative allies, Nixon is instead discussing with Haldeman and John Mitchell how to prevent any investigation of the Watergate break-in from exposing a host of Nixon campaign "horrors," as Mitchell labels them.

This frees Ford to concentrate on the House debate over revenue sharing—more precisely, over the closed rule adopted on an 8–7 vote by the Rules Committee that will govern the debate *if* the full House goes along. Adoption of the rule will vastly expedite the approval process by precluding amendments and limiting overall debate to eight hours. At once parliamentarian, traffic cop, amateur psychologist and father confessor, Ford treats the members of his caucus as self-respecting adults. "He never cajoles," says Maryland representative Larry Hogan. "He is persuasive. He works hard. He's the kind of guy who, when he comes to you, you listen." A notable exception to this consensus is the famously prickly gentleman from Iowa, H. R. Gross. Looking as if he might have just stepped out of a Grant Wood painting, Gross is a self-appointed scourge of the bureaucracy. His indignant cry, "Who dreamed up this boondoggle?" is an integral part of the House's daily routine, like the chaplain's prayer.

Thoroughly bipartisan in his nitpicking, Gross opposed restoring former president Dwight Eisenhower to his five-star status if it meant he would draw a general's salary on top of his presidential pension. In November 1963 he demanded to know the gas bill for the eternal flame marking the grave of John F. Kennedy. "There are three parties in the House," Ford remarks in bemused exasperation. "Democrats, Republicans and H. R. Gross." After twenty-three years in Washington—he and Ford were sworn into office on the same day in 1949—Gross retains an evergreen capacity for outrage. It will surprise no one that he is a staunch opponent of the State and Local Fiscal Assistance Act of 1972, "this euphoniously labeled revenue sharing bill," which he insists in view of last year's record $26 billion federal deficit should more accurately be labeled a debt- and interest-sharing bill. His argument is taken up by other critics, loath to admit why Congress is so reluctant to loosen its grip on spending—the idea, long advanced by Ford, that in eliminating the bureaucratic middleman, Washington will surrender to grassroots Americans more than the nation's purse strings.

At the leader's desk he keeps tabs on the argument roiling his colleagues, scribbling notes while entertaining requests from favor-seeking members of his caucus. They reciprocate his loyalty, encapsulated in a trademark colloquialism. "I'll be with you sled's length," he tells them, and while they may scratch their heads over Ford's meaning, they know they can take his pledge to the bank. His reluctance to crack a whip should not be mistaken for timidity. Good Old Jerry can play hardball if the stakes are high enough. Case in point: when a vacancy opened on the Federal Communications Commission, activists in the White House Office of Telecommunications Policy urged the president to select Ted Ledbetter, a Washington, DC–based communications consultant. If nominated and confirmed, Ledbetter would be the first African-American FCC commissioner. Unfortunately for his chances, a rival candidate, veteran Detroit broadcaster James Quello, had enlisted the support of his good friend Gerald Ford.

The ensuing contest, uneven to begin with, was abruptly terminated through a phone call from White House congressional liaison Max Friedersdorf to Ledbetter's sponsors in the OTP. Minority Leader Ford wished them to know that, unless they dropped their promotional efforts on Ledbetter's behalf, he would see to it that their entire office was zeroed out in the forthcoming budget. Quello, a Democrat, got the job, and kept it for twenty-four years.*

The clock over the Speaker's rostrum is approaching six. Still the House drones on. Unable to wait any longer, Ford springs up the aisle and outside to a waiting car. He is due at Blair House for a dinner in honor of retiring members of Congress and hosted by Vice President Spiro Agnew. The two men have a relationship that is less rivalry than cultural misunderstanding. A quiet dissenter from Agnew's alliterative attacks on a press corps the vice president considers biased, Ford routinely supplies his private home phone number to scribes in search of a story (the main number at 514 Crown View Drive is already listed in the phone book).

Anyone questioning the power of the press to dictate Washington dinner table dialogue would change their tone after eavesdropping on the low murmur of conversation at Blair House. Again it is the *Washington Post* that spoils the party. Earlier today it exposed E. Howard Hunt, a CIA operative turned $100-a-day consultant, whose name and White House phone number have turned up in an address book belonging to one of the Watergate burglars. Hunt's cover blown, John Ehrlichman directs him to "get out of the country." As it happens, the messenger entrusted with these instructions by Ehrlichman is himself general counsel to the Nixon reelection campaign. More than that, he is Hunt's coconspirator in planning the Watergate break-in. And he has a history with Gerald Ford.

Wednesday, June 21

At the White House this morning, H. R. Haldeman succinctly identifies G. Gordon Liddy for the president as "the guy who did this." Nixon quickly falls in with Ehrlichman's plan to have Liddy take the rap for Watergate. The news of Liddy's involvement jolts Ford, and for good reason. Three years have

---

* Ironically, two of Ledbetter's strongest supporters, OTP chairman Clay Whitehead and policy assistant Brian Lamb, were instrumental in a presidential transition group instigated without Ford's knowledge early in 1974 by his onetime law partner Phil Buchen.

passed since Ford shared Liddy's résumé with the Treasury Department in hopes of securing a political appointment for this gun-loving, self-dramatizing soldier of fortune. As an assistant prosecutor in New York's Dutchess County, Liddy earned a reputation for lock-'em-up flamboyance sufficient to win him second place in a hard-fought 1968 Republican congressional primary. The winner was Hamilton Fish IV, liberal heir to the lawmaker immortalized by his Hudson Valley neighbor FDR as one-third of the incurably obstructionist trio of Martin, Barton and Fish. The younger Fish's chances in November looked good, complicated only by Liddy's continued presence on the ballot under the banner of the New York State Conservative Party.

That fall, Ford was guest of honor at a GOP picnic and rally where Fish and others sought his help in finding employment for the potential spoiler. "I said sure, I'd see what I could do," Ford recalled later. At the time he gave the matter little thought; over the years, he had entertained countless similar requests. Following Fish's victory in November, it became clear that Liddy's best chance for political preferment lay with the incoming Nixon Administration. After the required endorsements from local and state Republican organizations were produced, Ford made a phone call to Eugene Rossides, assistant secretary of the Treasury for Enforcement and Operations. In April 1969 Liddy received his political appointment as a special assistant for Organized Crime. A month elapsed before he thanked Ford for his assistance. He had intended to write earlier, said Liddy, "but as Ham Fish may have mentioned I interrupted a robbery a few blocks from the White House a short while ago, and in the course of losing a fight with four of them broke a bone or two."

Liddy's flair for publicity, no less than his appetite for clandestine action, made him ill-suited for the bureaucratic backwaters of Treasury. Originally assigned to Operation Intercept, an anti-drug campaign aimed at the porous border with Mexico, Liddy had been transferred out of Treasury in July 1971, after he delivered a fiery attack on the administration's gun control policy before the National Rifle Association. Far more to his liking was Liddy's new position, a $26,000-a-year gig beside Howard Hunt in the White House Special Investigations Unit formed to plug government leaks following publication of the purloined Pentagon Papers (thus the group's colloquial designation as the Plumbers).

THAT WEDNESDAY AFTERNOON is zero hour in the fierce campaign to enact revenue sharing. Debate over the proposed closed rule pits Democratic whip Thomas P. "Tip" O'Neill for the yes side against H. Allen Smith, a former FBI agent who, like Ronald Reagan, left a boyhood home in Dixon, Illinois, for the greener pastures of California. Each man is allotted thirty minutes in which to argue his side's case, the time to be apportioned among like-minded members anxious to see their names in print (not for another seven years will C-SPAN broadcast House proceedings to 3.5 million cable TV subscribers). A robust argument ensues as opponents claim the federal government has no revenue to share, and that such a program will make local and state governments more, not less, dependent on Washington.

To the contrary, counter supporters of the bill, most of the challenges of the coming decade—water and air pollution, transportation, the disposal of solid wastes—will demand local solutions. Passage of the legislation before them will kindle a revitalized federalism to coincide with America's bicentennial.

Amid the volleys of claim and counterclaim, Ford prowls his party's side of the chamber like a coach pacing the sidelines in the fourth quarter. In a close call wavering or uncommitted members can expect to hear from the minority leader or one of his lieutenants. Hesitant lawmakers are then lined up at the back of the House chamber, ready to be summoned in the event their votes spell the difference between an administration victory or defeat. "I need three," Ford calls out from his frontline position, and a sheepish trio of party loyalists break ranks to announce a change in their vote. Today he is relieved of such anxieties, so confident is he about the bill's chances of passage. As the time for debate expires, Mr. O'Neill of Massachusetts moves the previous question. Mr. Smith of California demands the yeas and nays. The closed rule is approved by a vote of 223 to 185. A motion is entertained to remake the House into a Committee of the Whole, prerequisite for any final action on the bill.

The arcane language billows like smoke, obscuring the fact that most important House business does not transpire beneath a watchful public in the galleries. Rather, it unfolds behind thick doors, around the dinner table and the punch bowl. Capitol Hill receptions constitute the fourth branch of government. A cottage industry for Washington caterers, they are the lifeblood of hungry staffers, and a blight on their employers. After more than twenty years of small talk and canapés, Ford avoids most such affairs, requiring as a condition of his attendance the presence of at least one representative from the Fifth District.

Even then, his participation is fleeting. Encircled by lobbyists and flatterers, he has learned to make his exit as inconspicuous as possible. "Keep your feet shuffling all the time," he advises less experienced colleagues. "You get to your destination in that way without offending anyone."

Tonight his calendar lists no fewer than six receptions, culminating in a big bash at the Sheraton Park Hotel courtesy of the National Education Association. Complicating his party-hopping efforts are torrential rains from the former Hurricane Agnes. Reportedly hundreds of motorists are trapped by floodwaters inundating Rock Creek Parkway and scenic Canal Road in the city's northwest quadrant. Mr. Sikes of Florida may wish to revise and extend his earlier remarks about the Weather Bureau.

Thursday, June 22

Though downgraded to tropical storm status, Agnes has overnight brought disaster to the mid-Atlantic region. The governor of Pennsylvania is driven from his Harrisburg mansion by rising waters. In the nation's capital, residents of the Watergate are observed on their balconies sipping champagne and gawking at refrigerators and caskets bobbing atop a furious Potomac River. Eventually sixteen deaths are attributed to the storm in the District alone. Defying the elements, Ford attends the monthly breakfast meeting of Michigan's Republican congressmen at the Capitol Hill Club. The weather that concerns him is political, and the breakfast affords a late opportunity to buttonhole hesitant colleagues before the final roll call on revenue sharing.

At midmorning, Ford stops by the White House at the invitation of General Alexander Haig, Henry Kissinger's deputy on the National Security Council. Too busy to see Ford, Nixon is preparing for an afternoon press conference, the first public test of the nascent Watergate cover-up. For all that they agree on, Ford finds it easier to defend Nixonian policies than the president's exclusionary approach to GOP dissidents. "You have to have some people way out in front," says Ford. He has never been one to read people out of their party.

"I can get tears in my eyes when I think about Jerry Ford," gushes Paul "Pete" McCloskey, an early, outspoken critic of the Vietnam War who infuriated the White House earlier this year by waging a quixotic campaign to deny Nixon renomination. A poor showing in the New Hampshire primary caused McCloskey to drop out of presidential politics and file for reelection from the Northern California district he has represented since 1967. His anti-Nixon apostasy guaranteed the rebel a hotly contested primary and likely defeat, until Ford flew to the West Coast to personally vouch for McCloskey's character as one of those "way out in front" without whom the political process would be diminished. On primary day, McCloskey eked out a win by 867 votes. Not surprisingly, he regards Jerry Ford as "the most decent man I ever knew in the political arena."

THIS EVENING IT'S a safe bet that Pete McCloskey won't be among one hundred Republican lawmakers descending on the White House to have their pictures taken with a president whose reelection they regard as certain. Also MIA is Minority Leader Ford, who is too immersed in last-minute wrangling over revenue sharing to break away. An atmosphere of anticlimax pervades the House chamber as arguments made familiar through repetition merely postpone what they cannot prevent. Only once does Ford personally intervene, moving swiftly to defeat an eleventh-hour attempt to push back implementation of revenue sharing by six months. Proponents say this would save $2.7 billion. Ford counters that any such delay would break faith with elected officials at the state and local levels.

In short, a deal is a deal. The amendment is rejected, and the final bill passed by a vote of 274 to 122. It is after eleven p.m. when Speaker Albert brings down his gavel. Tonight Ford's sleep will be even shorter than usual as he and Betty are leaving for China in the morning. Before then, however, he has one more obligation to fulfill.

Friday, June 23

A waterlogged capital is still recovering from its brush with a hurricane when, a few minutes after eight, Ford is shown into the Old Family Dining Room of the White House. Its name derives from Jacqueline Kennedy's decision to have her family take their meals in more private quarters on the second floor. Accompanying Ford is Congressman Hale Boggs of Louisiana, his Democratic counterpart in the House, and soon to be his China traveling companion. The two men have an easy camaraderie dating to their service on the Warren Commission, which investigated the assassination of John F. Kennedy. Frequent debaters at the National Press Club, they will sometimes choose their topic while riding down Pennsylvania Avenue to the club building at Fourteenth and F Streets. After a vigorous airing of differences before a lunch crowd of journalists and their guests, the minority and majority leaders retrace their steps in time to resume jousting in the well of the House.

Cementing their friendship is that rarest of attributes in official Washington, the ability to laugh at oneself. This was put to the test during a recent trip to the National Zoo. Intended to publicize their forthcoming China adventure, the visit produced the week's unlikeliest photo op. In the zoo's reconverted rhino house, Ford and Boggs were pictured greeting Ling-

Ling and Hsing-Hsing, two giant pandas given to the people of the United States by the Chinese government in recognition of Nixon's opening to the People's Republic. Accustomed as the two politicians may be to playing second fiddle to more electrifying performers, this was the first time a rival for public notice had urinated in their direction.

The incident might stand as a metaphor for their White House breakfast. A year has passed since Boggs's withering criticism of the aging J. Edgar Hoover and his FBI "gestapo" prompted a lively telephone exchange between Nixon and Ford in which the latter, telling the president what he wanted to hear, declared, "He's nuts!" (Nixon's parting admonition to Ford: "Keep kicking them in you-know-where.") The tone of this morning's dialogue is more civil. The president recycles memories of his February trip, describing meetings with the Chinese leadership in between spoonfuls of cottage cheese, wheat germ and yogurt flown in daily from California. An hour passes. The men rise from the table, shake hands and go their separate ways. Ford and Boggs board a White House helicopter for the short flight to Andrews Air Force Base, where they will meet up with their wives and the rest of their party.

At 10:04 a.m., even as the military jet carrying the congressional travelers taxis down the runway at Andrews, the door to the Oval Office opens to admit H. R. Haldeman. He has new and disturbing information to impart on "the Democratic break-in thing" being probed by the FBI. Unfortunately, he tells Nixon, "their investigation is now leading into some productive areas . . . it goes in some directions we don't want it to go."

The ensuing conversation will change American history. It will also make Gerald Ford president.

Part I

JUNIOR

1912–1950

SECRETS

—=◆◆=—

*So few people knew that I was not Jerry Ford's son in Grand
Rapids. It wasn't a kept secret—but nobody talked about
it. And we sure never talked about it in our house.*

—GERALD R. FORD

THE GRAND RIVER defines the city it divides, with a ragged symmetry little changed in the two centuries since rival entrepreneurs of faith and commerce staked their claims to opposite shores. In 1825 a Baptist missionary named Isaac McCoy established an outpost on the west side of the stream called Owashtanong by a long-vanished tribe of Indian mound builders.* McCoy's efforts to convert Ottawas and Potawatomies to the white man's religion were undercut by the appearance, a year later, of Louis Campau, a War of 1812 veteran who was as eager to sell the natives whiskey as the Reverend McCoy was to save their souls. Campau's crude log dwelling and trading post formed the nucleus of modern Grand Rapids, named for the turbulent waters that had once sustained pole-driven fishermen, and more recently challenged sidewinders churning the final thirty miles of Michigan's longest river.

The same waterway tested lumberjacks in red flannel and spiked boots who shepherded huge log drives culled from the state's virgin forests. The mere sight of these buoyant woodpiles prompted craftsmen like William "Deacon" Haldane to forsake his backyard cabinet shop in favor of a modern, water-powered factory on the river's east bank. Through the daily prayers and Bible readings he shared with his employees, Haldane united the competing priorities of the west and east sides, of God versus Mammon, immigrant cottage versus lumber baron's house on the hill.

In 1847 the first wave of Dutch émigrés reached West Michigan, Zeelanders and Frisians fleeing their homeland to practice agricultural innovation

---

* Today the site of McCoy's mission adjoins the latter-day burial mound of Gerald and Betty Ford.

and spiritual orthodoxy in a New Holland on the shores of Lake Michigan. Like their English Puritan predecessors, the Dutch professed a more rigorous interpretation of biblical truth than their state church allowed. Hardly a decade had passed, however, before purists in the Reformed Protestant Dutch Church (renamed in 1867 the Reformed Church in America) were being criticized for such heresies as hymn singing and Christmas trees. A midcentury schism led to establishment of the rival Christian Reformed Church in North America (CRC).* However they interpreted Calvinist tenets, the newcomers were to have a profound influence on the area's civic culture. Frugal and industrious, they placed strict limits on government and the taxes that underwrote it. Formally incorporated as a city in 1850, Grand Rapids boasted three thousand inhabitants, board sidewalks and a mayor paid $1 a year.

"The Dutch had a thing that if your father was a baker, you were a baker," recalls veteran Grand Rapids journalist Maury DeJonge. "I remember . . . my sociology professor said the most clannish people in the world are Jews. Next, right behind, are the Dutch." Dutch insularity expressed itself in a separate, faith-based school system, and fierce resistance to cultural amalgamation. "If you aren't Dutch, you aren't much," it was said, more in pride than jest. Reformed Church services were conducted in the mother tongue as late as the 1930s. Long before then, however, Grand Rapids had become a virtual babel of tongues, as the booming furniture industry attracted a continent's worth of nationalities. Poles, Germans and Lithuanians predominated on the West Side, while smaller enclaves of Italians, Greeks, Irish, Scandinavians and Syrians toiled in factories on both sides of the Grand. Mill owners recruited ball teams and marching bands from their multicultural workforce, a third of it composed of children occupied ten hours a day, six days a week, in the manufacture of coffins and church pews, school desks and hat racks, table clocks and parlor organs.

Singled out for praise at the 1876 Centennial Exposition in Philadelphia, Grand Rapids began calling itself America's Furniture City. Prosperity exacted its price as a growing population leveled existing hills and Indian mounds, dumping their contents into the Grand, until the river lost two-thirds of its former bed. The narrowed channel invited logjams like the 1883 disaster in which a floating forest of six hundred thousand logs swept

---

* In time each sect founded its own institution of higher learning to carry on the feud— Calvin College vs. Hope College. "Boy do they have an athletic rivalry," Gerald Ford reflected late in life. "When they get into a basketball game, all Christian ethics disappear."

away railroad and traffic bridges spanning the waterway. Spring was the most dangerous season along the Grand. Then heavy rains and fluctuating temperatures shattered the river's icy crust, with the resulting debris backing up for miles. When these artificial dams, menacing as blood clots, crumbled, torrents of frigid water reclaimed whole neighborhoods.

Gerald Ford Sr., the future president's stepfather, never forgot his boyhood experience dodging water holes while delivering newspapers during the Great Flood of March 1904. The resulting hardship was as unevenly distributed as the city's wealth in dry season. For while the east side escaped with a few flooded basements, half of those living on the opposite shore found themselves underwater. Eight thousand workers lost their jobs, though not for long, given the demand for Grand Rapids' signature product. Theodore Roosevelt turned to the Furniture City when decorating the Red Room of the White House. The same source supplied Herbert Hoover an executive desk to replace one lost in a 1929 Oval Office fire. Grand Rapids crafted seats for Boston's Fenway Park and office fittings for Frank Lloyd Wright's Johnson Wax building in Racine, Wisconsin.

At its peak, early in the twentieth century, 85 percent of the local workforce was employed in cutting, gluing, painting and polishing the nation's most sought-after furnishings for home and office. No union monitored working conditions or advocated for higher wages, though many on the West Side were receptive to labor organizers and their demands for a nine-hour workday and 10 percent raise. In April 1911, workers in sixty factories walked off the job. Order was restored, and the status quo reaffirmed, when the Christian Reformed Church declared union membership incompatible with the true faith. Henceforth a city fractured by geography and ethnic divisions was further polarized between industrial workers and the propertied establishment. For the future Gerald Ford Sr., erstwhile newspaper delivery boy, railroad clerk and apprentice paint salesman, labor solidarity was a concept best left to West Side Democrats. His ambitions pointed in a different direction, to the city's fashionable hill district, and beyond it the well-shaded lots and stout brick dwellings of East Grand Rapids.

Ford harbored deeply personal motivations toward middle-class respectability. He had been forced to drop out of school after the eighth grade, the only man in the family following a tragic train accident that took the life of his father, George R. Ford, in 1909. That, at least, is what he told his namesake stepson. In fact, the first Gerald Ford left school five years *before* his father's death, for reasons he was understandably reluctant to divulge. The parent in question, Ohio native George Ford, had quit the Buckeye

State for employment as a Grand Rapids cabinetmaker. In July 1872, aged nineteen, he married fifteen-year-old Zana Frances Pixley. The couple had five children, among them a boy who died in infancy, followed by three daughters and a second son, born December 9, 1889, to whom the Fords gave the name Gerald Rudolff.

Contemporary accounts depict George Ford as a gregarious joiner and vagabond salesman, highly esteemed among his Freemason brethren. He was also a gifted tinkerer whose invention of a cleaning device for locomotive boilers subjected him to prolonged legal bickering reminiscent of Dickens's *Bleak House*. It was to defend his patent claims, Ford asserted, that he moved to St. Louis in 1898. There he chartered the Ford Automatic Boiler Cleaning Company, with himself as president. His family remained in Grand Rapids. Gerald, nine years old at the time, may well have puzzled over his father's long absences, broken only by an occasional letter and infrequent visit. The last of these coincided with Christmas 1906. Two years passed. A message from St. Louis notified Frances Ford that her errant husband, having finally secured legal title to his invention, hoped to return to Grand Rapids for good.

How much credibility Mrs. Ford attached to such promises may be doubted, given the front page of the *Grand Rapids Herald* for January 24, 1909. The paper reported the accidental death of former resident George R. Ford, run down by a Cleveland, Cincinnati, Chicago and St. Louis train in Granite City, Illinois. Scandal quickly supplanted tragedy, for Ford was not alone at the time of his death. As readers throughout the Midwest soon learned, the fifty-six-year-old businessman had been accompanied by Mrs. Emma Tutton Loheide, half his age, who since Christmas had been posing as his wife. The counterfeit widow interrupted her vigil over Ford's battered body long enough to answer police inquiries. At length these exacted a concession that her marital claims rested on nothing more than the "soulmating" process, under which the late Mr. Ford had slipped a plain gold band on her ring finger, legitimizing their union without recourse to religious or civil authority. According to neighbors in Granite City, Illinois, Ford and the former Mrs. Loheide had been cohabiting for several months. Whatever her motive, she hadn't "married" Ford for his money; his estate consisted exclusively of patent rights to his boiler-cleaning device.[*]

---

[*] I am indebted to Don Holloway, the enterprising former curator of the Gerald R. Ford Museum in Grand Rapids, for uncovering George Ford's bigamous life during his research for the Ford Centennial in 2013.

Only the last-minute intervention of the Ford family prevented the bogus bride from burying her soul mate in a local cemetery more than four hundred miles from Grand Rapids. The dead man's children strained to dignify the long separation of George and Frances Ford, spinning for the press a story of maternal invalidism and regular financial support from a dutiful, if distant, spouse. For nineteen-year-old Gerald Ford, his father's double life was not just a subject to be avoided in polite conversation. Inevitably, it spurred development of his own, radically different, character as what one early Ford biographer called "a strong, self-made and very inner-directed man." In time the first Gerald Ford would famously admonish his own children to "Work hard, tell the truth, and come to dinner on time." Yet some truths were too painful to share. The awkward facts of George Ford's adulterous conduct would be withheld from his grandchildren— none of whom, apparently, including a future American president, thought it strange that George and Frances Ford should be buried in Michigan plots separated by over a hundred miles.

The taint surrounding his father lodged in the elder Gerald Ford's memory like a bone in the throat. More immediately, the shame and hardship inflicted on his mother by a faithless mate sensitized Ford to the risks assumed by any young woman in the marital lottery. He could scarcely imagine it at the time, but the ironic consequence of his father's betrayal was to involve Gerald with a divorcée reminded of her own dubious past with every childish cry for attention from her infant son.

E VEN AS YOUNG Gerald Ford set about distinguishing himself from his wayward father, Miss Dorothy Gardner relished life atop the social pyramid of Harvard, Illinois, a small town in the Fox River Valley, some sixty miles northwest of Chicago. Her father, Levi Gardner, a dry goods merchant and funeral home operator, had left his mark on Harvard as a member of the city council, as president of the board of education and, briefly, as mayor. His 1884 marriage to Adele Ayer, whose family name adorned Harvard's main street, cemented his lofty status. She may have narrowly missed out on direct *Mayflower* kinship, but Adele Gardner's penchant for ancestor worship was passed intact to her daughter Dorothy. So was a belief, by no means universally shared at the time, in the value of educating women. Dorothy attended St. Mary's College in Galesburg, Illinois, described by its Episcopal sponsor as "a school home where girls would become better daughters, where they would be systematically trained for the duties of wifehood and motherhood." Dorothy thrived at the school. Outgoing by nature, she made friends easily, impressing classmates with the unusual energy that was her hallmark. It surprised no one when they elected her class president.

Sometime prior to her June 1912 graduation, Dorothy was introduced to Leslie Lynch King, the brother of her college roommate. King made a compelling first impression. At twenty-eight, he was eight years older than Dorothy. Standing nearly six feet in height, with an athletic build and easy grin, he wore a straw hat at a rakish angle over his blue eyes and sandy-blond hair. A superb dancer, his coiled energy and volatile charm anticipated the character of Billy Bigelow, the charismatic but morally flawed barker in *Carousel*. An unmistakable aura of success clung to Leslie King, who boasted to Dorothy of his $6,000 salary as general manager of the Omaha Wool and Storage Company. The latter was part of the commercial empire assembled by Leslie's father, Charles Henry King. A Pennsylvania native, the elder King had moved with his wife, Martha, to the remote Sand Hills of northwest Nebraska in 1884. His arrival in a young state (neighboring Wyoming was still a territory) coincided with the advent of the railroad that was to enrich King and his heirs through a commercial trail of general stores, banks, lumberyards, and a stage coach freight line— each strategically placed to benefit from the development of a region only recently ceded to the white man.

By the time "C. H." shifted his base of operations to Omaha in 1905, he was a millionaire several times over. As befitting his exalted status, King purchased a fifteen-room mansion at 3202 Woolworth Avenue. To this address Leslie King brought Dorothy Gardner following their wedding on September 7, 1912, in the little Episcopal church in Harvard where Dorothy had been christened. The mother of the groom expressed displeasure with her new daughter-in-law even before the newlyweds embarked on a West Coast honeymoon. "I don't want to speak to her," Martha King announced as the young couple prepared to depart. "I hope that I never see her again."*

This sour send-off set the tone for a painful voyage of discovery, in the course of which Dorothy encountered a Leslie King—suspicious, foulmouthed and physically abusive, especially when under the influence of alcohol—drastically unlike her attentive suitor. Most of all her new

---

* Contained in the file documenting his parents' divorce, held back from researchers during President Ford's life, is a notation that helps explain the frosty reception accorded Dorothy Gardner. Quite simply, she wasn't the "rich widow from California" Leslie King's parents hoped their son would marry. Arthur C. Pancoast–Dorothy A. King, November 10, 1913, DGDP.

husband was jealous, at times irrationally so. In a Portland hotel on September 27 he struck Dorothy, hard, after misreading her nod to a female acquaintance leaving an elevator as a come-on to some young men sharing the ride. Rejecting her claims of innocence, once back in their room he swore at her. Then he packed his suitcase and stormed out, only to return several hours later. King's temper flared a second time a few nights later in a sleeping car bound for Los Angeles. Once again Dorothy was accused of speaking to a strange man. During the night she woke with a start, dimly aware that her husband "was sitting up in bed and striking me. I turned over and he then accused me of hitting him with a pillow and kicked me."

Not all their time was spent in recriminations, to be sure. Dorothy returned to her new home in Omaha pregnant, which raises the delicate, and admittedly speculative, question of whether Leslie forced himself upon a bride who could hardly wait to leave her husband and seek refuge under her parents' roof. If Dorothy's pregnancy was not altogether voluntary—something broadly hinted at in her formal divorce petition accusing King "of extreme cruelty towards the defendant in that he was very abusive of his marital relations"—it would help account for her reluctance in later years to discuss her wretchedly unhappy first marriage, other than a cryptic acknowledgment that "it didn't work out."

It could also explain why this fervently Christian woman, credited by her famous son with implanting his belief in the basic goodness of people, should nurse a lifelong bitterness toward the deadbeat father she pursued in the courts long after Jerry reached maturity. Given the temper of the times, Dorothy King had few options, none of them attractive. In November 1912 a penitent Leslie persuaded her to rejoin him in Omaha, which only intensified her feelings of betrayal when Leslie broke his promise to provide a home of their own, away from Dorothy's hateful mother-in-law. Awash in debts, some of which he reputedly covered by turning over property belonging to his young wife, King supplemented his salary by leeching funds from his father's business.

Adding to the domestic tensions in the King household was Dorothy's mother, a frequent visitor who, according to Leslie, let no occasion pass without poisoning her daughter's mind against him. In December 1912 Dorothy again quit Omaha for Oak Park, the Chicago suburb that her married sister, Tannisse, called home. It is altogether possible, as alleged by King, that his wife's mother urged Dorothy to walk away from the marriage and "a dog's life." Equally credible is Dorothy's claim that a hostile Martha King refused to acknowledge her presence under the same roof, even on Christmas Day, and that on at least one occasion Martha deliberately inflamed her son's suspicions by whispering to him of a (nonexistent) male stranger who had invited Dorothy to lunch.

That Christmas Dorothy attempted a reconciliation, on condition that she and Leslie live somewhere besides the big house on Woolworth Avenue belonging to his parents. For a while the couple occupied a basement apartment in a nearby building, which became the scene of renewed hostilities on New Year's Day 1913. Returning from a movie matinee with her visiting sister, Dorothy found Leslie in a rage, and everything of value in their apartment stacked in the kitchen. Sparks flew again in the King household as the time for Dorothy's delivery drew near, and her mother became a near-constant presence. That summer the Kings moved back into the Woolworth Avenue house, temporarily vacated by its owners so that their grandchild might at least begin life in reputable surroundings.

Monday, July 14, 1913, was the hottest day of the Omaha summer. Shortly after midnight Dorothy King gave birth to a baby boy weighing eight and a quarter pounds and named Leslie King Jr. Unfortunately, the appearance of his first child did nothing to moderate Leslie's behavior. If anything, the added responsibility of parenthood acted as a trip wire to his volatile disposition. His verbal outbursts caused pedestrians to stop and crowds to gather outside the big house. Things took a decided turn for the worse on July 28. With Dorothy still in bed suffering from an inflamed gallbladder, Leslie appeared in her room waving a butcher knife and making threats against mother and child. An attending nurse summoned two policemen to calm the agitated King and eventually remove him from the house.

He fought back, obtaining a court order requiring his meddlesome in-laws to vacate the Woolworth Avenue premises by the first week of August. In the event, King got more than he bargained for. Two days after his confrontation with the police, sixteen days after the birth of Leslie Lynch King Jr., the boy's father came home to find him gone, along with Dorothy and her parents.

O F ALL THE entries lovingly inscribed in the baby book of Leslie King Jr., chronicling every milestone from first spoken words—"Bye, bye," uttered at ten months—to first game of patty-cake, none is more poignant than the page headed BABY'S OUTINGS. Here, registered without comment or context, is baby's first automobile ride, a short journey by taxi-cab across the Mississippi River to Council Bluffs, Iowa, from which a train carried him safely to his maternal aunt's home in the Chicago suburbs. The date was July 30, 1913. Dorothy King, fearing the worst from her volatile spouse, had absconded with the baby and put Omaha behind her. On August 21, alleging cruelty and desertion on the part of his wife, Leslie King filed for divorce in district court. Two weeks later Dorothy countersued, seeking monthly alimony of $500 plus legal fees. When the hearing got underway at the end of November, King hardly bothered to argue his claims. A parade of witnesses, led by the police officers who had witnessed Leslie's knife-wielding meltdown, tipped the scales in Dorothy's favor even before she took the witness stand.

King and his lawyers instead concentrated on disproving their client's earlier boasts of financial independence. Leslie, they contended, was actually in debt to his father, who chose this moment to abandon Omaha and move to Southern California, far beyond the jurisdiction of any judge in Douglas County, Nebraska. At that C. H. King accepted more responsibility than his son. For as long as he lived, the elder King paid the $25 monthly child support, raised to $100 in 1931, assessed against Leslie by the court. Emulating his nomadic parents, Leslie changed his residence to Wyoming, where he might also enjoy legal immunity, as he rebuilt his fortunes through the King Investment Company. In 1919 he eloped to Reno, Nevada, with a woman he met at a New Year's Eve party in Los Angeles. On the subject of his earlier marriage, King observed a strict silence. As for the second Mrs. King, she was naturally sorry that the man she called "Daddy" did not live to see his son become president of the United States. Beyond that, she expressed regret that Leslie Jr. never got to know Leslie Sr.

"He was quite a man—very dashing, a bit spoiled but—but quite a fellow."

Her assessment was not shared by Judge A. C. Troupe. The divorce settlement he granted Dorothy in December 1913 included custody of the infant son whom she, loath to pronounce the name of her former husband, styled Junior. The impersonal term would stick, along with its detested diminutive, Junie (in naming his own sons, the adult Gerald Ford decreed "anything but Junior"). Under the circumstances, Dorothy probably welcomed an appeal to join her parents in Grand Rapids, where Levi Gardner hoped to strike it rich through real estate development. Unfortunately, the Gardners had scarcely moved into their Michigan home when Levi was diagnosed with a fatal kidney ailment known as Bright's disease. Unable to care for him alone, Adele Gardner reached out to Dorothy. The single

mother and her baby were invited to share the Gardners' modest bungalow at 1960 Terrace (later Prospect) Avenue. Dorothy listed herself in the
Grand Rapids city directory as a young widow, a white lie easily rationalized in a community of Dutch propriety and 134 churches.

Junior King attempted his first uncertain steps in the house on Terrace
Avenue. Perhaps in frustration at the halting rate of his progress, Junior one
day tossed his grandfather's shoes and stockings in the toilet. What might
have been thought mischievous, even endearing, as an isolated incident
quickly lost its charm when part of a behavioral pattern. In a kindergarten
class he attended following his mother's remarriage, the little boy's short
fuse earned him a reputation as "naughty Junior Ford." "If I inherited anything," Gerald Ford said of his biological father, "I inherited his temper."
This, he admitted, was "terrible." His playmates would have agreed. "A
very headstrong little boy," recalled one, still marveling after half a century
at the stubborn belligerence with which Junie single-handedly prevented a
group of older children from scaling a backyard cherry tree. "*My* tree," he
declared. And when a girl twice his age made the attempt, "he *stood* on her
hand, until she screamed. Then he took his foot off."

Junie compensated for his fierce possessiveness by offering children
rides in a wagon he pulled like a dray horse. Some of these same youngsters he pushed around during the week, often under the noses of their
teachers. In his memoirs, Ford described the shaming tactics employed
by his mother, who reminded him in front of a mirror how foolish he
looked when red-faced with anger. Where ridicule fell short, religion was
invoked, in particular Proverbs 16:32: "He that is slow to anger is better than the mighty . . ." (the same passage cited by Ida Eisenhower of
Abilene, Kansas, in addressing anger management issues with her son
Dwight). One day Dorothy handed Junie a copy of Rudyard Kipling's *If*,
as if exposure to British stoicism might cure the boy's tendency to fly off
the handle. Whatever inspiration Ford drew from Kipling's verse, privately
he attributed his hard-won self-control to another, more conventional approach taken by Dorothy Ford: "She gave me unshirted hell every time I
would blow up."

With the passage of time, Ford, like Ike, learned to govern himself before he governed others. His animus was of the impersonal kind. "I tend
to get more angry at things than people," he explained. To be sure, the
character-building virtues of Scouting taught the boy self-restraint as well
as self-reliance. Competitive athletics helped to channel his aggressive feelings. Even here, however, the will to win occasionally got the better of

him. High school classmates remember Ford's eviction from one game for excessive roughness, the same charge made after he "accidently" kneed a player in a game with rival school Davis. Brother Dick Ford chuckled over the golf game in which Jerry, then enrolled in Yale Law School, wrapped his club around a tree in frustration after requiring three shots to escape a sand trap.

Accounts by White House insiders invariably portray Ford as the calmest man in the room. Few guessed at the effort required to maintain such equanimity. Fewer still chanced to observe the dark red shading that sometimes crept above the presidential shirt collar, evidence of anger unexpressed.

THOUGH HER DIVORCE precluded Dorothy King from becoming a member of St. Mark's Episcopal Church, it didn't prevent her from being a regular attendant. It was at a church social in 1915 that she first met Gerald Ford, another congregant, who immediately caught, and returned, her fancy. The antithesis of Leslie King, twenty-five-year-old "Jerry" Ford was gainfully employed as a sales agent for the Alabastine Company, a local firm marketing a gypsum-based "sanitary wall finish" said to be immune to germs or disease-bearing insects. That he entertained higher ambitions was no secret. At six feet one and a little over two hundred pounds, with a ramrod-straight posture and character traits to match, Ford exhibited none of the blustery self-regard Dorothy associated with her first husband.

A practical joker and enthusiastic outdoorsman, and a co-owner with some friends of a modest cabin on the Pere Marquette River, a hundred miles north of Grand Rapids, Ford was as comfortable gutting a brown speckled trout as he was donning his Sunday best and performing the duties of church vestryman. A muscular Christian who neither smoked nor drank, Ford regarded his civic obligations as seriously as his spiritual ones. Mature and, above all, stable, Ford reached out to Dorothy's infant son, with whom he established an emotional bond second only to his feelings for the boy's mother. Levi Gardner's death in May 1916 heightened Dorothy's sense of vulnerability. No doubt it also prompted her to reflect on the odds of finding another mate as attentive or conscientious as Gerald Ford. Her suitor's obvious affection for Dorothy's little boy cinched the deal. A wedding date was set for February 1, 1917. Denied a church ceremony on account of her divorced status, Dorothy and Gerald exchanged vows at the house on Terrace Avenue. Following a wedding breakfast for one hundred guests, the newlyweds embarked on an extended trip east. They returned at the end of the month, an instant family comfortably ensconced in the working-class neighborhood around their rented duplex at 716 Madison Avenue.

For Junior King, this was to be his fourth home in as many years (not including the Omaha house of his birth). At least he would not be an only child much longer. In July 1918, Thomas Ford was born, to be followed in 1924 by Richard, and James, who came along three years later.* As an infant Tom contracted scarlet fever, causing the whole family to be quarantined. No sooner had he been pronounced healthy than Junie's doctors misdiagnosed a stomachache as appendicitis, and operated on him accordingly. The boy recovered from the unnecessary procedure faster than did his parents' faith in the medical profession. Otherwise he suffered nothing worse than the usual scrapes and bruises inflicted on a precocious athlete whose versatility qualified him for football, baseball, basketball, track and the swimming pool. He discovered golf before his tenth birthday, though he turned down an invitation to caddy at a local course, a decision he would regret as an adult whose enthusiasm for the game exceeded his finesse.

Gerald Ford Sr. encouraged Junie in his athletic pursuits, teaching him to swim at the local YMCA and inviting him on fishing expeditions to his Northern Michigan cabin. Sunday-afternoon excursions to a Lake Michigan swimming beach were another treat, with the Ford children making the trip in their father's open-roofed touring car.

By now Mr. Ford had graduated to a management position with the home improvement firm of Heystek and Canfield. His financial position seemingly assured, the young husband and father drew plans for a house of his own, on Rosewood Avenue in prosperous East Grand Rapids. George Ford's son had arrived. Yet even as his growing family enjoyed the fruits of his success, they could never entirely escape Dorothy's past. Though out of sight, Leslie King made certain that he was not out of mind. In February 1921, he rejected Dorothy's request for an increase in monthly child support. "With my present salary," the remarried King told her lawyer, "and with a *family* to care for," he was in no position to do more. He didn't stop there. "I have been thinking of getting permission from the court to have

---

* Admitted to the hospital maternity ward at the age of fourteen, Junie sat quietly in one corner of his mother's room as everyone fussed over newly born James Ford. Finally Dorothy observed that her eldest didn't seem very happy about his new brother. "Oh, he's all right, I guess," Junie replied, "I'm just thinking of all the work he's going to be for us." JCI, Thomas Ford, GRFL.

the boy half the time," King wrote with casual cruelty, "since he has grown up to realize that he has a father."[*]

Since nothing came of the threat, it might be passed off as more of Leslie's bravado, except for its coincidence with the worst of Junie's childish tantrums, compounded by a severe case of stuttering that plagued the boy for several years. The adult Ford attributed the latter to his ambidextrous nature—"Left handed when I've been sitting down and right handed standing up"—and persistent efforts by his parents and teachers at Madison Elementary School to make him exclusively right-handed. It is tempting to speculate how this condition, reported in less than 1 percent of the population, may have affected the adult Ford. Tempting, but in all likelihood fruitless, for while Ford exhibited traits—a quickness to anger and occasional awkwardness, both physical and verbal—commonly associated with so-called symmetrical brains, the same might be said of many people who favor one hand exclusively. In any event, it is hard to formulate a common denominator applicable to Benjamin Franklin, LeBron James, and Mahatma Gandhi.

From Madison Elementary, Ford transferred to a school near his family's East Grand Rapids residence. He wasn't there long. The recession that followed World War I cost Gerald Sr. his managerial job and the heavily mortgaged house it sustained. This setback, fraught with childish humiliation, may well have inspired the future president's lifelong aversion to mortgages, credit cards and deficit spending alike. In 1923 the Fords rented a three-story frame house at 649 Union Street, a much grittier area than East Grand Rapids. They lived there for seven years, or until Junie was a senior in high school. Both his parents took an active part in shaping their children's education, supplementing the usual classroom disciplines with weekly church attendance and the character molding of competitive athletics. In addition, Dorothy and Gerald Sr. encouraged their sons to be readers. When they were young, Dorothy read to them from the Oz books. On his own Junie devoured the dime-store novels of Horatio Alger in which poor boys defied the odds to achieve great and tangible rewards. Savored under the bedcovers long after the rest of the household had retired for the

---

[*] If, as seems obvious, Leslie is referring to himself, then Junior King knew more about his origins, and knew it earlier, than he later acknowledged. On the subject of his parents' turbulent relationship, Ford told biographer James Cannon in 1990, "Until I got through college, I really never knew the fundamental causes of their divorce." Only after he got to Yale did Dorothy Ford "somehow" bring up the subject—"I don't recall the circumstances—she might have shown me some clippings from the Omaha paper of the trial," which depicted King as a brutal wife-beating husband.

night, titles like *Strive and Succeed* and *Jed the Poor House Boy* distilled a formula for success neatly summarized in the credo Jerry learned from his stepfather—"The harder you work the luckier you are."

Life on Union Street routinely tested this theory. Most days began at six, as the boy remade the bed reluctantly vacated on sunless winter mornings when temperatures hovered well below zero. The Fords' sole defense against Michigan's merciless winter, a coal-driven furnace in the basement, demanded constant attention. Each morning there were ashes to remove and a fresh supply of coal to load in their place. At night Junie banked the temperamental device, adjusting the vent to feed a slow-burning fire. In between these vital operations he shoveled snow-covered sidewalks and driveways. As the days lengthened there were lawns to mow and a garage to clean out. Year-round, Junie took his turn washing dishes after dinner and tidying up the kitchen. Only then could he turn to his schoolwork, though not without a radio playing in the background, a habit his parents found exasperating.

Generally speaking, where domestic matters were concerned, Dorothy Ford ruled the roost at 649 Union Street. That his mother was the well-spring of Ford's drive and unflagging energy seems beyond dispute. "She was a doer," recalled her firstborn, bestowing the ultimate compliment in his lexicon. Other members of the family testify to a trademark intensity, whether playing her baby grand piano, stringing Christmas decorations, or canning fruit and vegetables to stock the Ford pantry against seasonal privation. A faithful member of the Grand Rapids Collectors Club, proud of her heirlooms and the status they implied, Dorothy had charisma and a force of character that impressed nearly everyone who met her. "She was strong physically, she was strong emotionally. She was exceptionally strong spiritually," Betty Ford observed about her future mother-in-law.

"Mother had a lot of guts," Gerald Ford said of the woman whose early defiance of convention gave him the chance to emulate Alger's boyish strivers. At the same time it left her vulnerable to neighborhood gossip about "Mrs. Ford and Mr. King." Football teammate Art Brown learned of his buddy's true relationship to Gerald Ford Sr. through the tattling of a rival businessman angered by Mr. Ford's decision to go into the paint and varnish business for himself. Presumably he was not alone. Spontaneous and empathetic, Dorothy Ford set an example of broad-mindedness, implanting the generosity of spirit—critics called it naïveté—that was to find ultimate expression in her son's pardon of his White House predecessor. (Conceding that forgiveness was a quality he learned from his mother, Ford noted, "Some of my friends, especially in the political arena, believe I over-did it.")*

The ultimate source of this attitude is not hard to trace. The Fords were regular churchgoers, praying to a loving God whose intervention in human affairs did not end on a Jerusalem hillside two thousand years ago. Gerald Ford's reluctance as president to advertise his faith would be contrasted with Jimmy Carter's overt appeals to his spiritual brethren (in 1976 Ford overruled exploitative campaign advisers who wanted to promote his long-ago stint as a Grand Rapids Sunday-school teacher). On the Sunday morning he announced the Nixon pardon, Ford slipped out of the White House to attend a service at nearby St. John's Church. It was characteristic at such a time that he should say a prayer rather than take a poll. Dorothy Ford would have expected nothing less. Mothering the Union Street neighborhood as an extended family, she made sure no visitor to the Ford home left without a slice of pie, some mouthwatering angel food cake or her locally famous molasses cookies. The latter were a particular favorite of young Jerry and his friend Burt Garel, the son of an African-American chauffeur and a frequent visitor to the Fords'. In wintertime the boys navigated the steep, icy slope of Union Street on homemade bobsleds.

Christmas was Dorothy's favorite season. Then the Ford residence, as described by a family friend, "was like an old-fashioned Christmas card come true." Mrs. Ford's doll collection, assembled in anticipation of future granddaughters, received a holiday infusion of new outfits made by its owner. On Christmas Eve, following services at Grace Episcopal Church, she and Jerry Sr. invited everyone to their place for steaming cocoa and Christmas cookies. Dorothy's culinary skills, never more appreciated than on such occasions, had their downside. Unfortunately the cook couldn't resist her own chocolate fudge or homemade caramels. As a result she fought a lifelong battle to control her weight, and to ward off heart disease, diabetes and other ailments associated with obesity. Together with her husband, Dorothy attended every University of Michigan football game as long as Jerry was on the squad, including two years when he mostly sat on the bench, and this despite a three-hour trip either way. Still, for all her pride

---

* "I always felt a person had some good quality in him or her," Ford explained. If it was not evident on the surface, he was usually willing to make the effort to confirm its presence elsewhere. "And so I never really ended up hating somebody." The author only heard Ford, as ex-president, speak disparagingly of two individuals, and even then the most damning epithet he could come up with was "He's a bad man." The two in question were John Dean and G. Gordon Liddy.

in his accomplishments, she questioned his decision to enter politics, which she regarded as an "abnormal" way to live or raise a family.

Clearly, Dorothy Ford was not a simple woman. Even now, it is not easy to reconcile the devout churchgoer constantly writing notes of encouragement or sympathy, baking or sewing for needy Grand Rapidians, with the "very stern, stoic woman" recalled by her granddaughter Linda Burba. A grandmother who conspicuously shied away from babysitting, she is remembered as "a very busy woman . . . very direct," according to Susan Ford Bales. "She never stayed with us; she stayed in a hotel. She stopped by and would spend several hours with the children; and then she would leave," trailing in her wake the impression of someone devoted to her family, but not defined by it, living for her children, not through them. To her eldest son, Dorothy Gardner Ford was "the most selfless person I have ever known." She made everyone else's problems her own, he asserted, employing the same verbal formula he would use to explain his success in politics.

To HIS FATHER, says Dick Ford, "character and integrity were the most important things in the whole world . . . If you did something wrong and it came to light, if you didn't lie about it, there was no problem." As illustration he cites his own youthful transgression of pouring a quart of oil into a car's radiator instead of the engine. His father told him not to do it again. Beyond that, "he appreciated my efforts, though wrongly directed." Junie didn't escape parental correction. Together with some neighborhood friends, he requisitioned the second floor of a barnlike garage for proscribed games of penny-ante poker. They hadn't reckoned with Mr. Ford's curiosity, or his athleticism. Inevitably the gamblers were discovered and their den shuttered. More embarrassing was the time young Jerry backed the family's six-passenger Chandler Sedan into a telephone pole, losing a spare tire in the process. The cost of its replacement came out of his pocket, a memory that still rankled fifty years later.

Scorning the self-satisfaction of some who profess themselves self-made, Mr. Ford's involvement with the Elks, Kent County Republicans, and local civil defense in World War II was all part of giving back to the community. He was a cofounder of Youth Commonwealth, an early attempt to assist city boys, many of them African-American, living in economically deprived neighborhoods of Grand Rapids. "There was always politics in the house," claims one family member. The elder Ford counted among his friends as many Democrats as Republicans. The lesson would not be lost on his namesake, any more than Senior's talent for establishing seemingly instant rapport with potential customers. "He went in with a true salesman's philosophy," says his son Dick. "How can I help you?" The same skill set sustained Congressman Ford on the campaign trail and in the party caucus. It was to prove invaluable in the Oval Office, where Ford's ability to forge bonds of trust with foreign leaders transcended differences of outlook or temperament.

For as long as he lived, Gerald Jr. would ask himself how Gerald Sr. might handle a given situation. Yet just as with his mother, there were limits to his stepfather's influence. The son took vigorous exception to the older man's view of lawyers as "an unnecessary commodity." On the other hand, in reaching the most important decision of his life Jerry emulated his step-father in marrying a divorced woman, while making sure to schedule the politically sensitive ceremony *after* a hotly contested Republican primary for Congress. Finally, there were less tangible qualities imbibed from growing up around Dorothy and Gerald Ford Sr. "Jerry is not demonstrative," claimed his sister-in-law Janet. In this, as in so much else, he took after his parents. "They were not touching people."

Which only made Gerald Ford's public reaction, on the final morning of his uphill 1976 campaign to win the presidency in his own right, all the more affecting. As the first of eighty million Americans began streaming to the polls to render a verdict on his two years in office, Grand Rapids friends and admirers crowded into the local airport for the unveiling of a mural depicting key scenes in Ford's life. Ford saluted the artist for his talent, and his neighbors for their generosity. But it was the portrayal of his parents, neither of whom had lived to share the moment, that brought the president of the United States to tears. Whatever he might have accomplished, he told a clearly moved audience, was due to "the training, the love, the leadership" of Gerald and Dorothy Ford. He paused, overcome by emotion and what Edith Wharton called "the poignancy of vanished things."

2

LUCK AND PLUCK

Not merely to exist, but to amount to something in life.
—South High School Class of 1931 motto

MINIMIZING ANY CHILDHOOD traumas he might have experienced, Gerald Ford preferred to dwell instead on the happy accidents of his life. These included a mother of sufficient courage to flee an abusive spouse, risking her reputation and, possibly, her safety, in atonement for a disastrous marriage; and a stepfather who compensated in character for whatever he might lack in consanguinity. Ford the optimist even detected silver linings in the loss of his family's East Grand Rapids home and their subsequent relocation to the slightly scruffy environs of 649 Union Street. Through a quirk of geography this gave Jerry the choice of three high schools, each with its own distinctive profile. Central was the favored option for aspiring college students. Ottawa Hills, less elitist than its more established rival, exerted nearly as powerful an appeal among the upwardly mobile set.

Finally there was South High, a staunch redbrick structure whose classrooms resembled, for the time, a patchwork quilt of ethnic and racial diversity. It was Ralph Conger, Central's longtime basketball coach, who urged Dorothy and Gerald Ford Sr. to enroll their son at South High, explaining "that's where he will learn more about living." In accepting this advice, Jerry's parents ensured that he would rub elbows with working- and middle-class youngsters from Italian, Polish, Syrian and African-American backgrounds, as well as the predominant Dutch. Conger's advice later bore fruit in a revealing exchange between Ford and Don Penny, an irreverent, tousle-haired Hollywood writer and humorist brought onto the White House staff to enliven the presidential speaking style. "You're like me in a way," Ford told Penny. "I can sense that you're a street kid."

Penny confirmed that he had grown up in Brooklyn, playing stickball in gritty tenement blocks.

"Well, I grew up in Grand Rapids and I prefer street kids to academics." That Ford should take an immediate liking to the wisecracking Penny did not surprise anyone who knew of his affinity, nurtured at South High, for those of seemingly exotic background. To some in the Class of 1931, the school's smattering of Jewish pupils appeared alien. Ford took a different view, ascribing the anti-Semitism around him to ignorance and resentment of classmates whose academic gifts, no less than their study habits, inspired a particularly virulent form of envy. "I never felt jealous," he would recall. "I just wish I was as smart, worked as hard." This was modesty speaking; none of his instructors, then or later, doubted Ford's work ethic. Lucy Reed, who taught him English in the seventh and tenth grades, remembered Ford as "quiet and attentive, yet always prepared" to recite required passages from memory.*

History, his favorite subject, was also the source of his best grades. The first captain of the South High football Trojans to earn a spot in the National Honor Society, young Ford did better in math than science. In six years his sole D came in French. By his own admission devoid of musical talent—"Betty says that I can't even listen in tune"—Ford was discouraged from joining the Glee Club or playing the clarinet. He fared better in spiritual pursuits. Enrolling in a YMCA offshoot called the Junior Hi-Y Club, Ford was chosen vice president of the group sworn to maintain "high standards of Christian life" through weekly discussions and "inspirational addresses" from local businessmen.

True to Ralph Conger's prediction, Ford received his most useful education outside the classroom. Inspired by its British counterpart, the Boy Scouts of America had been in existence little more than a decade when Junie Ford pledged himself to Grand Rapids Troop 15 in December 1924. It is hard to exaggerate Scouting's impact on Ford. As an adult he credited his five years in Troop 15 for instilling in him qualities of self-discipline and the need for preparation. Half that time was spent in compiling the prerequisite twenty-six merit badges, in subjects ranging from blacksmithing to first aid to animals, that entitled him to Scouting's highest honor. His 1927 induction as an Eagle Scout was, by Ford's own acknowledgment, among the proudest moments of his life.

It was also something of a turning point in the boy's evolving sense of self. Although his mother identified him on an accompanying award as J. K. (for Jr. King) Ford, it wasn't long before the fourteen-year-old Junior Ford was publicly reintroduced as Gerald R. Ford Jr. Though no formal adoption ever took place, and a legal change of name didn't occur until

---

* "Why are Jews so funny?" Ford once inquired of the freewheeling Penny, who replied, in character, "Well, it's hard to kill us when you're laughing." AI, Don Penny.

1935, Dorothy Ford's eldest son would never question the authority or example of the second man with whom he'd shared a name. Working as an unpaid assistant at Camp Shawondossee, not far from Muskegon, Ford won the admiration of younger boys even as he impressed elders with his dedication and mentoring skills.

In 1929 Governor Fred Green named him to the inaugural Honor Guard at Fort Mackinac, on the island of the same name in the chilly waters between Lakes Huron and Michigan. Ford and seven other Eagle Scouts spent that August conducting tours of the historic fort for summer visitors. In their free time they swam in the Straits of Mackinac, clambered over geological formations like Arch Rock and took in silent movies at the waterfront Orpheum Theater. At day's end there was the twilight cannon to fire before supper prepared alfresco. Sequestered in their hilltop bastion, with a moaning wind stirring tree branches during the short summer night, it wasn't difficult to conjure up shades of distant redcoats and Yankee fur traders. For the little occupying force of Scouts, Mackinac was the ultimate campout.

THE CITY TO which Ford returned in the fall of 1929 was entering its own twilight. Out-of-town buyers still crowded the Pantlind Hotel for the semiannual Furnishing Market, but Grand Rapids was bleeding trade to Southern venues where non-union labor was as cheaply abundant as the Carolina hardwood. A craving for mobility and mass entertainment had begun to weaken the grip of Dutch Calvinism, prompting one CRC congregation, enraged that a streetcar line passing close to its sanctuary would transport Sabbath revelers to a nearby amusement park, to tear up the tracks in the middle of the night.

The automobile craze achieved the same result with far less commotion. By the time Jerry Ford entered South High, three-quarters of Grand Rapidians owned a car. Ford himself paid $75 for a battered 1924 Model T used to commute between South High and the latest Ford residence, a fixer-upper on Lake Drive that the family occupied in 1930. His automotive knowledge was on par with his musical gifts, as Ford was the first to concede. One wintry night, too impoverished to buy antifreeze, he threw a blanket over his overheated engine instead of the radiator. The resulting fire ensured that Jerry was without wheels for the rest of the school year.

Raised in a politically active household, Ford presumably noted Governor Green's 1928 appointment of *Grand Rapids Herald* publisher Arthur Vandenberg to a vacant seat in the United States Senate. The new senator looked askance at US involvement in European affairs, a position Ford was to emulate as a Yale Law student. By the same token, Vandenberg's later conversion to an internationalist outlook foreshadowed Ford's entry into national politics as a champion of bipartisan foreign policy. It was a subject to which Jerry had been drawn, even as a high school student too young to vote. Assigned the imaginary task of selecting a Cabinet for newly elected President Herbert Hoover, Ford said he would retain Secretary of the Treasury Andrew Mellon, a conservative favorite, while entrusting the war department to victorious World War I general John J. Pershing. Half a century later he would get mostly high marks when called on to repeat the exercise for himself.

Ford's youthful political activity was dwarfed by his interest in athletics, especially football. After his parents, probably no one had more influence on the adolescent Jerry Ford than South High football coach Clifford Gettings. Their relationship began casually enough, on a grassless school playground one spring day in 1927. A corporal's guard of eighth grade boys in ill-fitting uniforms competed for Gettings's notice, first step to a place on his junior football squad. In Jerry's case, it wasn't his play that impressed the coach but his hair—uncombed and corn-colored, hinting at a Dutch pedigree he never claimed but didn't go out of his way to deny, either, at least not when campaigning for votes in West Michigan.

"Hey, Whitey," Ford quoted Gettings, "you're a center." The coach remembered things a bit differently. He asked Jerry what position he played. He didn't know, the boy replied. "I never played any position before." Gettings took a closer look at Ford's "long, lanky" form and designated him a center. The game as played by Ford and his teammates would be hardly recognizable to modern fans. Because each of the eleven men on Gettings's roster was expected to handle defense as well as offense it wasn't enough for the center to master a variety of intricate snaps. He must also be able to block an oncoming linebacker. Played without face masks or kidney pads, football required stamina and versatility as much as size and speed. There were no special teams reporting to Coach Gettings. If the ball was downed near the sidelines, that is where the next play unfolded. Field goals were rare, perhaps because they required drop-kicking the ball.

For most of his high school years, Ford and his teammates struggled to keep their footing at Island Park, a groundskeeper's nightmare located on the banks of the Grand River. Lacking proper drainage, the field was often waterlogged, with yard lines and player jersey numbers obliterated by mud. Coach Gettings, only a few years older than the student athletes he was guiding, is remembered for his punishing work ethic. "He couldn't stand to lose," recalls Jim Trimpe, another South High player. One of Trimpe's teammates, responding to the coach's demand for more assertive play, broke his collarbone on an ill-timed tackle. Anyone late for practice—Gettings included—was condemned to run laps around a nearby track, one for every minute he kept his teammates waiting. The use of profanity exacted the same penalty.

Though Gettings frowned on curse words, he wasn't above motivating players through unabashed ethnic slurs ("I would give three of you Hollanders for one good fighting Polack"). Gettings recognized a kindred spirit in Jerry Ford. "He doesn't like to think of himself as a loser anytime," explained his coach. "He's got that inner drive that if you're going to do it let's do it right." The *Grand Rapids Herald* described Ford as "a field marshal, always thinking for himself and for his teammates. He was a step ahead of most of his opponents all the time." His approach to the game reflected Gettings's philosophy of a team without stars. "It made no difference" who scored the winning play, says teammate Art Brown, "the thing was to get it done." The same might be said of Ford's collegial, grind-it-out approach to politics in later years.

After playing a full schedule on the freshman squad, Junie replaced an injured center in his sophomore year. Named an All-City player at sixteen, he suffered through a junior year during which he missed three games on account of a torn meniscus, forerunner of the trick knee that gave rise to his later reputation as something of a presidential bumbler. After a dropped pass cost South High the first game of the 1929 season, and with it the chance to repeat as city champions, Ford organized a weeklong training camp in advance of his senior year. Participants assembled at his father's remote fishing cabin, a necessary subterfuge since the entire exercise skirted league rules banning practice sessions more than a week before the start of school.

B Y THE FALL of 1930 an even more bruising challenge tested popular resilience during the worst economic downturn in Grand Rapids history. As recession spiraled into depression, several banks crashed and half the city's furniture factories closed their doors. Unemployment spiked above 25 percent. City Manager George Welsh launched an ambitious public works program that earned Grand Rapids a national reputation as "the city where everyone has a job." For fifty cents an hour, the unemployed painted city hall, built a municipal swimming pool, laid sewer pipes and water mains. Fire stations became distribution centers for milk and bread.

Women canned fruits and vegetables for marketing in city-run stores where families using scrip in place of currency could redeem their pride along with their next meal.*

On airless summer nights, those seeking relief from the heat climbed aboard the Cherry Street trolley line for the round trip to John Ball Park. For a dime you could stick your head outside the streetcar window and savor the artificial breeze stirred by the motion of the car. Now even that simple pleasure became for many an unaffordable luxury. As home foreclosures mounted, social status fell faster than the Dow Jones. After Richard DeVos's parents failed to meet the $25 demanded each month by their landlord, the future cofounder of Amway moved with the rest of his family into his grandparents' West Side attic.

The Fords were hardly immune to want. Lacking the ten-cent admission fee, Tom Ford missed out on a school play. Meat and potatoes, former staples of the family diet, made way for monotonous mounds of spaghetti and noodles. No one was heard to complain, however, especially in light of the ongoing struggle being waged by the man at the head of the table. Tiring of his position as office manager with the Grand Rapids Wood Finishing Company, Gerald Ford Sr. had convinced his boss to diversify the business by creating a division to market paint to builders and institutional clients. Within a year this was spun off by Ford as a separate enterprise. Ford Paint and Varnish opened its doors at the start of October 1929, three weeks before the Wall Street crash that ushered in a decade of economic holocaust.

Market advocates like to extol the heroic qualities of the lonely entrepreneur. As far as Gerald Ford—Sr. *and* Jr.—was concerned, this was no idealization. At Ford Paint and Varnish, sweat equity substituted for the real thing. Short on funds with which to buy paste for labels, "we'd go down and drain the shellac drum," says Tom Ford, "because it stuck just

---

* Welsh's response to the emergency was, if anything, too energetic for some in the business community, who criticized the city for entering into competition with hard-pressed store owners and building contractors. Led by the Reverend Alfred Wishert of Fountain Street Baptist Church, a harshly critical Committee of 100, drawn from the commercial and spiritual elite of Grand Rapids, denounced work relief aimed at "the usual indigent class" with whom the slightest contact might "contaminate" the better sort of workman unaccustomed to receiving public assistance. The same group assailed the use of scrip as discouraging patronage of local merchants and attendance at Sunday church services. It bears noting that Gerald Ford Sr., though a fervent advocate of free enterprise, did not align himself with the Committee of 100.

like glue." Gerald Sr. showed equal initiative in keeping his little workforce intact. He planned to give each man $5 a week with which to buy groceries and other essentials, the same amount he set aside for himself. As soon as conditions improved, "however long it takes," Ford vowed to reimburse his workers the difference between this meager sum and their regular income. Taking him at his word, the men of Ford Paint and Varnish avoided the swelling unemployment lines.

While still in high school, young Jerry found work at Ramona Park, a lakeside amusement complex boasting the obligatory roller coaster, midway rides, and multiple concessions, which the husky football player kept supplied with Coca-Cola and Cracker Jack. His wages may have been modest, but there were unforeseen dividends. One Fourth of July, owner Alex DeMar discovered Jerry shouldering the entire workload while his coworkers availed themselves of the nearby Marcus Girl Revue. DeMar, a respected figure in the local Greek community, was to provide critical support to his onetime summer laborer as the adult Ford waged an underdog congressional campaign against the GOP establishment.

Ramona Park was summer work. During the school year Ford waited on customers and washed dishes in a burger joint across the street from South High whose owner, Bill Skougis, made a practice of hiring football players. Working the lunch shift, Ford could help himself to all he could eat, as long as it didn't exceed sixty cents. One day early in 1930 Jerry noticed a stranger lingering near the candy counter, his eyes fixed on the younger man. Eventually he approached the counter.

"Are you Leslie King?"

Ford barely had time to say no before the visitor answered his own question. "You're Leslie King. I'm your father. You don't know me," he declared. "I'd like to take you to lunch."

"I'm working."

"Ask your boss if you can get off." Dazed by what he was hearing, though unwilling to dismiss it altogether, Ford obtained permission to accompany King to a nearby eatery. On the way he was introduced to his reputed stepmother, and to a four-year-old girl who made less of an impression than the glossy new Lincoln whose purchase explained the Kings' annual journey from their Wyoming ranch to the car-making capital of Detroit. Over a strained lunch, King boasted of his extensive landholdings. His continuing failure to pay court-ordered child support to Jerry's mother went unmentioned. Instead the visitor pumped Jerry for details of his athletic exploits. The boy feigned interest when invited to visit the Kings in Wyoming. He flatly rejected an invitation to relocate there permanently. Lunch concluded, and Junie was driven back to South High, where King handed him $25 and told him to buy something he couldn't otherwise afford. "Terrified" by the prospect, Ford nevertheless confronted his mother and stepfather with the story that same evening. In his memoirs he devoted a single sentence to their discussion, contrasting the love and support shown by Dorothy and Jerry Sr. with the callous neglect of Leslie King. Afterward he cried himself to sleep.

The next day, flush with King's parting gift, Jerry paid $13 for a stylish pair of golf knickers. Appearances mattered to Ford; years later an observant journalist noticed that he had a weakness for fine Swiss watches: "Apparently he didn't have a good watch as a kid." In fact, material deprivation was less of a trial to the youthful Ford than the emotional poverty he associated with the name on his birth certificate. His biological identity might be as fixed as his blood type, but that didn't prevent Jerry from showing the world another side to his character. Dressing for success communicated the upward mobility pursued by Dorothy and Gerald Ford Sr. It helped to distinguish the former Leslie King Junior from his raffish namesake, much as his stepfather had spent a lifetime distancing himself from the scandalous George Ford.

Years passed before any of Jerry's siblings were let in on the secret. Dick Ford was a twenty-five-year-old veteran of World War II by the time he discovered that his big brother was in fact his half brother. Incredibly, Dick's only knowledge of his mother's unhappy first marriage came from what he read, much later, in books about his famous sibling—an information deficit repeated in the next generation of the Ford family, as Susan Ford was sixteen before she learned the true relationship between her father and her uncles. She, too, would rely on published accounts for much of her family's complex history. "I don't want to say Dad hid it, it was just not talked about."

F ORD KNEW THE glare of celebrity from an early age. Whether the venue was a Grand Rapids backlot, Ann Arbor's fabled Big House, or the 1935 College All-Stars game whose roster was voted on by seven hundred thousand fans nationwide, the gifted center from Michigan was accustomed to being noticed. This had its drawbacks, as Jerry discovered in the spring of his junior year at South High. Elected handily to the student council, whose organizing principles defined partiality to one's friends as an offense "subject to impeachment," in 1929 Ford ran for senior class president about the same time he and younger brother Tom stood outside a Grand Rapids polling place and distributed campaign literature promoting successful mayoral candidate John D. Karel. Jerry drew inspiration from Percy "Pop" Churm, his modern history teacher and track coach, who also supervised student electioneering. Churm insisted on a strict code of ethics, with candidates limited to $10 in campaign expenditures, closely supervised by a watchdog committee. To ensure the integrity of the balloting, he borrowed voting machines from city hall.

Jerry's chief opponent in the South High contest, future banking executive Bill Schuiling, laid claim to the Republican designation, a definite advantage in the years before Depression-era hardship scrambled party loyalties in Michigan as elsewhere. Running on the Progressive Party ticket, Ford argued that South High students should be permitted two dances a year, double the number currently authorized. (It didn't hurt Jerry's ballroom performance that his Aunt Marjorie ran a local dance studio.) Ironically, it was the candidate's athletic achievements that doomed his candidacy. Considered indispensable to the school's football program, Ford also excelled at basketball and track and field. In his senior year he somehow found time to swim competitively for the YMCA. Winning was his habit, and his Achilles' heel. "Here he was," said Coach Cliff Gettings, "out for three sports, and the coaches couldn't see him sitting in there directing a Student Council meeting when the teams he were [sic] on were out practicing . . . So we started campaigning. I had Physical Ed classes and I told the boys in the class to vote for Bill Schuiling because we needed Jerry out for football and basketball and track." Gettings made sure the other coaches conveyed the same message to their charges. It proved an irresistible argument.

Unsuccessful in his bid for class president, Ford got consolation of a sort by being elected captain of a senior-year football squad that was undefeated going into the climactic Thanksgiving Day game against archrival Union, a West Side powerhouse. At stake was nothing less than the state championship. A heavy snowfall the previous day led South High players to trade in their cleats for rubber-soled gym shoes supplied by a local sporting goods store. Dorothy Ford was among the twelve thousand spectators who paid $1 apiece—a local record—to watch both teams battle the elements on the coldest Thanksgiving ever recorded. A raw thirty-mile-an-hour wind blew across the field, adding to the misery as frozen hands struggled to pass or hold on to the football. The contest ended in a scoreless tie. Yet Ford's luck hadn't deserted him. Union forfeited the game following the revelation that one of its players had signed to play pro ball. With that Jerry Ford and his South High Trojans were crowned the best of Michigan football. That evening Jerry and his parents hosted over eighty school officials, players and their dates at Marjorie Ford's dance studio.*

When the music stopped, Ford had his future to consider. After seeing how he handled the discus and shot put at a statewide tournament, Michigan State athletic director Ralph Young hoped to recruit Jerry for State's track-and-field team. Harvard and Northwestern also expressed interest in the fiercely competitive athlete-scholar, who had by now reached his adult playing weight of 198 pounds, proportioned on his six-foot-tall frame in ways sure to attract female admirers. By his own admission a romantic late bloomer, Ford was fully appreciative of the opposite sex, yet inarticulate in conveying his emotions. For a while he dated Mary Hondorp of the Class of 1930, who lived in a Hollander neighborhood not far from Ottawa Hills High School. Deprived of his automobile, most evenings Jerry escorted Mary home on a streetcar. Weekends they joined friends in a warm-weather exodus to Ottawa Beach, a popular retreat on Lake Michigan. Under the strict code of conduct imposed by the Christian Reformed Church, swimming was banned on the Sabbath, as was reading a newspaper or tossing a ball. (One persistent wheeze insists that basketball was permissible "as long as you didn't score.") Religious liberals claimed conditional leave to enter the chilly lake water. "You can go up to your knees," said one otherwise devout dad, "but don't splash." Relations between the sexes were as tightly regulated, though here, too, there were backsliders like the Christian Reformed acolyte who allowed, "I don't dance—but I can neck."

Around South High 1931 was the Year of the YoYo. Two events in the run-up to graduation traced the evolution of Jerry's political sympathies, and a robust ambition lurking behind his congenial surface. Local Communist sympathizers, their ranks swollen by Depression-era hardship, chose May 1, 1931, to shake their fists at the capitalist power structure of Grand Rapids. In solidarity with the world's workers, a small band of Young Pioneers descended on South High demanding free books and student lockers. When the would-be revolutionaries inscribed these sentiments on the school steps—other versions of the story have them painting a hammer and sickle on the sidewalk—a contingent of varsity athletes answering to Jerry

---

* One legacy of the 1930 championship season was the 30–30 Club, composed of everyone who played or coached for South High at the time of the big game. For nearly seventy years club members met on Thanksgiving Day to relive their youth and renew old bonds of friendship. In 1974 they convened at the White House per invitation of their captain. That day they came from as far away as South Africa. Only one Old Boy deliberately absented himself, in protest of Gerald Ford's pardon of Richard Nixon.

Ford appeared on the scene to repel the radicals. "Just like that Ford had a bunch of us out there and made them scrub it off with a brick," recalls Art Brown.

Nothing in the May Day confrontation hurt Ford's chances in a second campaign he waged that spring. As part of a national movie chain promotion, a contest to pick the Most Popular Boy in Grand Rapids promised its winner a six-day, all-expenses-paid trip to Washington, DC. Having been robbed of the class presidency by the scheming of nominal friends, Ford wasn't about to take any chances. "It's the only time he ever stuffed the ballot box," claimed Tom Ford, who exploited his small stature to gain repeated admission to the Majestic Theater on his big brother's dime. More precisely, "he'd give me the ten cents to go in and pick up the ballots" discarded by patrons more interested in the action on the screen. "But my problem," Tom added with a chuckle, "was the other guys had their people in there, too." In the final tally Ford narrowly defeated his chief rival, the pride of Central High.

The Washington excursion marked his first travel outside of Michigan. Welcoming a student delegation to the White House Rose Garden almost fifty years later, Ford evoked the feeling of awe he had experienced on first entering the visitors' gallery of the House of Representatives. Nothing else on the trip compared with it, not Mount Vernon nor the state rooms of Herbert Hoover's White House. That he might one day call the latter place home would have struck Jerry and his South High classmates as the stuff of fantasy. The school yearbook, perhaps reflecting the May Day clash in which Ford had captained the forces of order, guessed that he would become a policeman. Fewer members of the Class of 1931 followed Ford to college than emulated his teammate Art Brown, whose employment at a nearby General Motors plant was to have unexpected benefits for his old friend. As a one-man campaign committee, Brown could be counted on to neutralize opposition to Congressman Ford from the local chapter of the United Auto Workers. Thus Ralph Conger's advice regarding Jerry's choice of a high school was validated every election day.

GERALD FORD NEVER joined the rugged individualist school of conservatism. For all his personal frugality, he conceded a role for government in assisting the needy, demolishing racial barriers, and advancing the national interest through programs like the GI Bill, which enabled millions of his World War II contemporaries to attend college. Such views were deeply grounded in his own struggle to attain an education from the University of Michigan. Getting into the university entailed more than self-reliance. He never pretended otherwise. By his senior year of high school Ford's exploits on and off the athletic field had won him a cadre of admirers. Michigan alumnus Paul Goebel was an All-American football player who had enjoyed a brief professional career in the game before returning to Grand Rapids to manage a sporting goods store. In May 1931 Goebel invited himself to Ann Arbor for a golf game with the school's dynamic young football coach, Harry Kipke. He planned on bringing along several Grand Rapids prospects, Goebel told Kipke, pointedly observing, "Everybody is after this boy Ford."

Kipke took the hint. At their first meeting, Ford more than validated his advance notices. A reciprocal visit by Kipke to Grand Rapids sealed the deal. Ford could scarcely believe his luck. "I've always wanted to be a student at the University since I was able to read about the prowess of their athletic teams," he wrote, "and now it is almost a certainty that it will come true. . . ." His confidence belied the dire state of his finances. The continuing struggle of Gerald Ford Sr. to keep Ford Paint and Varnish afloat precluded any help from that quarter. In the shriveled economy of 1931, the university's cupboard was just as bare. No athletic scholarships eased the burden for even the most promising of recruits. Football supplied Ford with the ticket to a college education. Now the game would, indirectly, provide him the chance to work his way through school. Through Kipke, Ford secured jobs waiting on tables in the intern dining room of the university hospital, and washing dishes in the nurses cafeteria. Together they generated $1.50 a day, enough to feed an athlete in training, but not to pay his $100 tuition bill.

That responsibility fell to Arthur Krause, a school administrator who channeled profits from the South High bookstore into a first-of-its-kind scholarship.* For spending money, Jerry relied on a weekly check for $2, the gift of a childless aunt and uncle. During the football season he made a few dollars scalping game tickets supplied to the players. He supplemented this three or four times a year by selling his blood to the university hospital at the going rate of $25 a transaction. In the fall of 1931, Jerry moved to Ann Arbor, sharing a tiny fourth-floor walk-up at 908 Monroe Street with a Grand Rapids basketball player named Don Nichols. Into these cramped

---

* In 1965, Congressman Ford calculated the current value of his $100 bookstore scholarship, employing an annual interest rate of 6 percent, and donated $1,000 to help underwrite the college education of an athlete in Grand Rapids' public schools. "It was an obligation I felt I owed, because it had been given to me at a time when I probably couldn't have gone to the University." TAFI, 710.

quarters, for which they were assessed $4 a week, the roommates somehow squeezed two army cots and as many desks and chairs. The bathroom was on another floor. Any privations were far outweighed by the novelties of the environment and newly made friends.

Standing in the registration line at Waterman Gymnasium, Ford spotted a face familiar through newspaper exposure. He went over to introduce himself. "I'm Jerry Ford. Aren't you Willis Ward from Detroit?" Thus the white football player from West Michigan first encountered the Black track-and-field star from the other side of the state. It was the start of a lifelong friendship, one whose historical impact would outstrip their athletic achievements. Before leaving the gym that day, Ford enrolled in four classes. By far the toughest was English Composition, whose students were expected to produce a weekly essay of a thousand words. Nothing in Ford's prior schooling had prepared him for such demands. Under the circumstances he considered himself lucky to walk away from the course with a C grade and a marked improvement in his previously atrocious spelling. He did considerably better in history, government, and economics. His overall B grade average landed Ford in the top quarter of his class, an accomplishment that becomes more impressive when measured against the requirements of his outside employment and, especially, the football program.

First-year students were banned from Michigan's varsity squad, but this didn't prevent a hundred of Jerry's classmates from trying out for its freshman counterpart. Within two weeks their number had dwindled to sixty, no surprise given the three hours spent each day in physical conditioning and learning the intricacies of Coach Kipke's "punt, pass and a prayer" system. "We *always* kicked off," Ford explained. "We *always* punted on third down inside our own 25, unless we had about a yard to go. We played tough defense, a straight 6–2–2–1, with none of the sliding and stunting you see today. We ran the short punt to death. We were dull." They were also successful, remarkably so during Kipke's early years, winning four consecutive conference titles and two national championships between 1930 and 1933. Calling him "the ideal player for any coach to work with," Kipke singled Ford out for his hustle. In May 1932 he became the first center to receive the Meyer Morton Award, presented annually to the most promising player on Michigan's freshman squad.

Admittedly "a bit overwhelmed" in his first months as a Wolverine, Ford soon adapted to the rhythms of college life. Writing Dorothy Ford on Mother's Day 1932, he apologized for not sending her flowers or candy, gestures ruled out by his "dreadfully insecure" financial condition. "Have a fine time and maybe next year I'll be able to do something more for you."

Empty pockets didn't prevent him that fall from joining Delta Kappa Epsilon (DKE), an academically impoverished fraternity that admitted Ford, so it was claimed, in hopes he might raise their grade average. Another dishwashing job enabled the newest Deke to earn his keep (he was the only member of his class, mused Ford as president, to suffer both dishpan hands and a football knee). In time he was made house treasurer, prelude to becoming a salaried house manager in his senior year. In the latter capacity he succeeded in retiring the chapter's extensive debt, even as he ran up one of his own for unpaid tuition, books, clothes and other expenses.*

Founded at Yale in 1855, the local chapter of Delta Kappa Epsilon ("Friends from the Heart, Forever") introduced Ann Arbor to the club system in which members lived and dined together. Ford first crossed the threshold of their sprawling mock-Tudor-style house in the fall of 1932. He submitted without complaint to the petty humiliations with which upperclassmen avenged their own abuse as neophytes. Deke rituals included midnight visits to the Chapter Temple, or Shant, on William Street. Here, in a bizarre structure its architect described as thirteenth-century French Gothic, Gerald Ford was initiated before the shrine of the Holy Goddess of Delta Kappa Epsilon. The sacred quickly yielded to the profane, as personified by a drunken Deke member atop the roof of the Shant, from which lofty perch he dropped glasses onto a pillow fifty feet below—an experiment devised, with apologies to Galileo, to test their survival rate at Deke parties.

These were staged with numbing frequency and often riotous result (after forty years Ford could still recall the recipe for a milk punch whose lip-smacking appeal was unrelated to lactose tolerance). Mindful of the restraints imposed by his athletic training, Jerry contented himself, for the most part, with nonalcoholic refreshment. A notable exception came in the spring of his sophomore year. The night before he was scheduled to undergo knee surgery, his friend Jack Beckwith insisted that Ford accompany him to the Spanish Room for some fortifying tequila and cigars. The resulting hangover was severe enough for Jerry's doctors to postpone the operation. Such behavior hardly squared with the "serious-minded fellow" who brought his books along to study on long train rides to away football games.

---

* By his senior year Ford's indebtedness was severe enough to prompt a written appeal to his biological father. Getting no response, Jerry approached family friends Ralph Conger and his wife. The Congers had recently come into an inheritance sufficient to loan Ford the $600 he requested, and later repaid with interest.

"You should have seen the campus this week-end," Ford notified a friend in October 1933, "more drunks than ever and they all claim that the year is just beginning—I'm afraid to imagine what it will be by the Ohio State game." Yet he was no prig. Indeed, if the testimony of another frat member is accepted, on at least one occasion Ford acted the part of a Prohibition-era Paul Revere. Given the ease with which banned alcohol found its way to thirsty undergraduates, fraternity raids were a regular feature of campus life in the dry decade. The Dekes had themselves been evicted from their house for an extended period after just such a sortie revealed large stocks of hooch secreted within its walls. Needless to say, the success of these incursions depended on the element of surprise, which sharp-eyed lookouts in turn did their best to thwart. One Saturday night a pledge at a nearby frat house answering the phone was greeted by a breathless warning.

"The Feds are making the rounds, dump your booze."

Fast as the alarm spread through the house, the offending libation vanished down sinks and other plumbing fixtures. Empty bottles were neatly stacked in the trash barrels of an adjoining sorority house. Moments later a knock on the door announced a trio of government agents, who proceeded to ransack the place seeking evidence more tangible than the alcoholic breath of its residents. Eventually the gumshoes left, though not before expressing frustration over their failure to apprehend the informant who had disrupted their carefully orchestrated plans. Seventy-five years later, the recipient of the telephone warning identified the anonymous caller as a student waiter from the Deke house named Jerry Ford.

I N TRUTH FORD had little time for frat house revelry, though he seems to have derived a vicarious enjoyment from the party animals around him for whom every week was Hell Week. Among his closest friends was Jack Stiles, a carousing dissenter from the prevailing sobriety of his native Grand Rapids, to whom Ford would entrust the management of his early political campaigns. At the other extreme of the behavioral code was a high school athlete from Sheboygan, Wisconsin, named Philip Buchen. Having contracted polio at the age of sixteen, Buchen received treatment for his illness at Warm Springs, Georgia. There he met Franklin D. Roosevelt, who was to be a lifelong source of inspiration. A softspoken intellectual, drawn to philosophy and religion, Buchen wrote poetry on the side. Like Jack Stiles, though for very different reasons, he seemed an unlikely confidant to the hyperactive jock, who, in rare moments of repose, was more likely to pick up a tennis racket than a textbook. Yet the two men hit it off from their first meeting during the 1938 Michigan summer school.

Like schoolboys everywhere, Ford dreaded the label of grind. So he downplayed his classroom exertions, even while rationalizing a late-semester burst of scholarship preceding final exams. "From now until two weeks from now all I'll be doing is studying," he confided in January 1934 to Frederica Pantlind, the daughter of a Grand Rapids hotel-keeping family. "Sounds awfully funny but I'm really serious. Now can't you imagine me slaving over my books." A trip to Chicago's Century of Progress Exposition provided a rare distraction, although Ford passed up its more lurid attractions. "Saw all of the fair in 25 minutes, at least all I wanted to see," he told "Freddie." "Didn't see Sally Rand either." Averting his eyes from the notorious burlesque queen and her trademark fan dance, Ford opted instead for the antics of vaudeville headliners Olson and Johnson (though he professed shock at the exorbitant ticket prices).

Notably absent from his correspondence with Miss Pantlind is any complaint about a football career put at risk by an excess of athletic talent among his teammates. The 1932 Wolverines went undefeated, outscoring opponents 123–12 en route to the national championship. The spectacular performance of All-American center Chuck Bernard limited Ford's play even before a knee injury sidelined the promising sophomore for much of the season. "I wanted to play," Ford later acknowledged. "But you learn sitting on the bench. You learn patience, loyalty and dedication. Those things came in handy later." His junior year brought more of the same as Jerry alternated between left guard and occasional substitute for Bernard. In the classroom, meanwhile, Ford's academic exploits earned him admission to both the junior honor society Sphinx and its senior counterpart Michigauma.

In his senior year Ford came into his own. "He was one guy who would stay and fight for a losing cause," recalls one Old Boy in justifying Ford's selection that year as the team's Most Valuable Player. As events would demonstrate, this was akin to winning the dance contest on the *Hindenburg*. Before the season opener, quarterback Bill Renner shattered an ankle. Then kicker John Regeczi—"The best college punter I ever saw" in Ford's estimation—hurt his knee. So much for punt, pass and a prayer. Michigan went scoreless in five of its eight games. Yet it was their lone victory that redeemed an otherwise wretched season. On October 20, 1934, the Wolverines took the field against Georgia Tech amid an escalating controversy, with national repercussions, over race. Faithful to their segregationist traditions, the Yellow Jackets had refused to play unless Michigan benched its sole Black player, the versatile track-and-field star Willis Ward. This was perfectly acceptable to athletic director Fielding Yost, himself the son of a

Confederate soldier, who in a storied coaching career stretching over forty years had never sent an African-American player onto the field wearing the blue and maize of Michigan.

Harry Kipke disagreed. He had personally recruited Ward away from Dartmouth with a promise that his playing time would be determined by his performance on the field, not the color of his skin. On the road, Kipke scorned the custom whereby teams separately housed any Black players with a sympathetic local family. The Michigan coach insisted that his players— *all* his players—stay under the same roof. When the team visited Chicago and the venerable Palmer House refused to register Ward, Kipke reminded management of his team's longstanding patronage. Failure to accommodate one player would prompt the rest to go elsewhere. Moreover, said Kipke, "I will tell all of the conference teams and urge them to follow suit." The hotel relented. From a housekeeper's whispered confidences, Ward learned that he was only the second African-American to be accepted there, after Marian Anderson, the celebrated contralto.

Ward shared a room with Jerry Ford, his best friend on the team, and the only person he allowed to address him as Willie. If anything, Ward was the superior athlete. A ten-second sprinter and Big Ten long jump champion, he had been christened Michigan's "one-man track team" by reporters rubbing their hands in anticipation of a 1936 Olympics—in Hitler's Berlin—featuring Ward and his equally gifted rival from Ohio State, Jesse Owens. *Time* magazine credited Ward and halfback Herman Everhardus for Michigan's undefeated season in 1933. Now, however, against Georgia Tech he found himself sidelined on account of his race. Left to his own devices, Harry Kipke would no doubt have honored his commitment to Ward with the same resolution he'd displayed in the Chicago hotel lobby. But the issue was taken out of Kipke's hands. "The administration wasn't eager to have Blacks on the team anyway," Ward recalled. Coach Kipke had, in effect, been ordered by Fielding Yost to implement his gentlemen's agreement with Jim Crow.

Fifteen hundred students, and a handful of faculty members, signed a petition calling for Ward's reinstatement. The NAACP weighed in, as did the National Student League. Demonstrators threatened a sit-down on the fifty-yard line. "I never dreamed there would be so much agitation about the matter," Yost wrote his counterpart at Georgia Tech prior to the October 1934 game. This he blamed on "a committee of five Jewish sophomore students, four of them from New York City and vicinity." Next to Ward himself, no one was more troubled by the treatment accorded his friend than Jerry Ford. His views on race were neatly summarized by Silas McGee, an African-American teammate on the South High School Trojans. Asked in later years what, if anything, had separated young Jerry Ford from his contemporaries, McGee replied, "He was color blind," adding, "Gerald Ford was my brother." The Georgia Tech game tested Ford's convictions. On learning of Ward's exclusion, he thought about quitting the team in protest. He consulted with his stepfather, who urged him to reconsider.

Ward echoed this advice. Ford should take the field, said Ward, if only to avert a humiliating loss to a despised opponent. Listening to a radio broadcast of the game at his frat house, Ward was treated to two mediocre teams on a muddy field struggling to avoid defeat. It took three quarters for the Wolverines to score their first points of the season.* For Michigan fans the game's most satisfying play originated with a trash-talking Georgia Tech linebacker who taunted Ford and his teammates by demanding, "Where's your ni**er player?" His answer came in the form of a vicious double hit delivered by Ford and guard Bill Borgmann. The offending Yellow Jacket was carried off the field on a stretcher. "That one was for you," Ford informed Ward after the game, won by the Wolverines 9–2.

The glow was short-lived. "That Georgia Tech game killed me," admitted Ward, for whom the university's very public betrayal inflicted permanent scars. Fearing a repeat of his ostracism, this time at Nazi hands, Ward yielded the spotlight to Jesse Owens, whose harvest of Olympic medals in Berlin won him a place among the immortals. After graduation, Ward was employed by the Ford Motor Company, where he promoted thousands of African-American workers with Henry Ford's blessing. Following service in World War II, he earned a law degree and fought racial discrimination from the Wayne County Prosecutor's Office. A Republican aligned closely with the administration of Governor George Romney, Ward served seven years on the state's Public Service Commission before capping his career as a Detroit-area probate judge.

His friendship with Ford ended only with Ward's death in 1983. Both men were reluctant to discuss the Georgia Tech game, a low point in Michigan football that the university, for reasons of its own, showed no greater desire to publicize. Ford did mention the incident in his memoir, *A Time to Heal*, but only in passing. Not until 1994, while attending a ceremony at Michigan Stadium retiring his father's number 48 jersey, did Steve Ford hear a fuller version of events as related by another player on the 1934 team.

---

* At the last minute, Georgia Tech benched starting end Emmett "Hoot" Gibson to compensate for Ward's sitting out the game.

"Son, do you know what respect is? Respect is what you do when no-body's watching. And nobody was watching your dad as a young kid at the University of Michigan."

The audience was substantially larger in 1999, as university president Lee Bollinger confronted multiple lawsuits challenging Michigan's admissions policies. These treated race as one of several factors (including ethnicity, athletic performance and economic status) that might be considered in reviewing an applicant's candidacy. Bollinger hoped to enlist figures of national stature to bolster the case for affirmative action. But with the Clinton White House clinging to its mantra of "mend it, don't end it," even liberal stalwarts on Capitol Hill were loath to be associated with an unpopular program. Then someone raised Gerald Ford's name. To Bollinger's surprise, the former president was more than willing to lend his support. Presented an op-ed piece drafted by the university, he set it aside and wrote his own, highly personal account of the injustice inflicted on Willis Ward sixty-five years earlier.

The result appeared in the *New York Times* on August 7, 1999. Recognizing the university's dual roles as preserver of tradition and hotbed of innovation, Ford reminded readers that so long as books are open, minds can never be closed. "But doors, too, must be kept open. Tolerance, breadth of mind and appreciation for the world beyond our neighborhoods; these can be learned on the football field and in the science lab as well as in the lecture hall. But only if students are exposed to America in all her variety." Rejecting quotas, Ford argued that demographic changes and a global economy required more, not less, diversity in the student body. By way of illustration he cited Willis Ward, a gifted athlete and above-average student, whose prodigious talents had not protected him from officially sanctioned bigotry. This victimized not only his friend and teammate, said Ford, but anyone deprived of knowing Ward as the man he was. "Do we really want to risk turning back the clock to an era when the Willis Wards were isolated and penalized for the color of their skin, their economic standing or national ancestry?"

Ford's intervention angered some, but he wasn't through. He encouraged the nation's military academies to lodge an amicus curiae brief with the Supreme Court, a document to which Justice Sandra Day O'Connor alluded in her June 2003 opinion upholding the admissions process at the University of Michigan law school. In a parallel case, the same court struck down Michigan's undergraduate formula with its points system explicitly weighted in favor of Blacks, Hispanics, and Native Americans. For affirmative action it was at best a reprieve; even Justice O'Connor expressed hope that such remedies would no longer be needed in twenty-five years.* Still, Gerald Ford had delivered the ultimate hit in honor of the teammate from whom he had learned more than any classroom instructor.

Even now their shared story was incomplete, pending a final act of reciprocal appreciation. Following his death in 2006, Michigan admirers of Ford began a campaign to honor the state's only president with a statue in the United States Capitol. Given the limit of two such commemorative figures per state, this meant evicting from Statuary Hall the marbled likeness of Zachariah Chandler, a Gilded Age lawmaker and former Detroit mayor. Not every GOP member of the state legislature was enthusiastic about the proposed switch. Their misgivings had less to do with any sentimental attachment to Chandler than to lingering resentments from the Ford-Reagan slugfest of 1976, compounded by unhappiness over Ford's support of affirmative action and his pro-choice views on abortion. Counting on legislative inertia and a ticking clock to carry the day, opponents of the Ford statue did not reckon with Buzz Thomas, the Democratic floor leader in Michigan's Senate. As the great-grandnephew of William Webb Ferguson, the first African-American elected to the state legislature, Thomas commanded more than the usual attention accorded speakers by his legislative brethren.

The reason why quickly became apparent, as Thomas spoke of Gerald Ford and Willis Ward, and of a friendship tested on the darkest day in Michigan football. There was fire in his voice, and sparks of recognition in the words that came tumbling out of his mouth. For Buzz Thomas, it turned out, was Willis Ward's grandson. Stunned into silence, opponents watched helplessly as the resolution authorizing the Ford statue was gaveled through the Senate on the last day of the legislative session. In May 2011 an eight-foot bronze likeness of the former president joined those of eight other American chief executives in the Capitol Rotunda. More reflective of the country's evolution, and of the lessons imparted by Willis Ward, a nearby bust of the Reverend Dr. Martin Luther King Jr. reminds four million visitors who pass through the Rotunda each year that America was founded on promises, the delayed fulfillment of which makes her at best a work in progress.

---

* In fact, less than a decade elapsed before the issue was again before the high court, the result of a 2006 referendum in which Michigan voters terminated even the limited consideration accorded race under the same court's 2003 rulings. With the retirement of O'Connor, David Souter, and John Paul Stevens, affirmative action was doomed. One consequence: Black enrollment at the U of M fell from 8.9 percent in 1995 to 4.5 percent in 2014.

FORD NEVER DOUBTED the contribution of college sports to his character or career. Teamwork and grueling preparation were educational values in themselves. The discipline to absorb defeat without yielding to defeatism, to distinguish between constructive criticism and the shouted abuse of armchair quarterbacks—these and other lessons served Ford as guideposts in his political travels. Beyond this, sports provided a lingua franca in place of conventional eloquence, not just the small talk of innumerable grip-and-grin encounters, but a secret handshake of sorts, admitting one to a fraternity of self-improving competitors for whom ideology mattered less than speed, endurance and the will to win. Nothing was more purely democratic, or so quick to demolish artificial distinctions of race or class or regional identity.*

If Ford displayed unusual civility toward adversaries on Capitol Hill and in the White House, it wasn't hard to find a precedent. Amid the wreckage of the Wolverines' 1934 season, he gave Michigan fans some of their few reasons to cheer. Against Northwestern, for example, he turned in "probably . . . the best game I ever played in college." Facing an All-American guard named Rip Whalen, an energized Ford "just blocked the shit out of him." It was hardly the sort of encounter that instills friendly feelings on either side of the line of scrimmage. Early in December of that year, Ford received an invitation to participate in the upcoming East-West Shrine Game in San Francisco. In common with other participants, he must supply his own pads and footwear. In return game organizers promised him a commemorative sweater and wristwatch to supplement the travel allotment of $7 a day.

Ford seized the opportunity to make his first visit to the West Coast. En route he was approached on the train by none other than Rip Whalen.

"You know why you're here?" said Whalen. "I told our coach after the game you were the best damn center I ever played against."

The seemingly endless rail trip, interrupted by two practice sessions, gave Ford plenty of time to observe Curly Lambeau of the Green Bay Packers and other coaches for whom the San Francisco excursion was a recruitment tour. His own presence among so many prospects was barely acknowledged

---

* As a boy, Ford knew the batting average of every player in the big leagues. His adult hero was Al Kaline—"Mr. Tiger"—who overcame youthful poverty and osteomyelitis to play twenty-two years for the same team, once rejecting a pay raise from $95,000 to $100,000 because his most recent batting average didn't justify it.

by these talent scouts. All that changed, however, two minutes into the game, when the East team's designated center broke his leg. The next fifty-eight minutes belonged to Ford, whose impressive performance sparked interest from several scouts, Lambeau among them. A formal offer from the Packers promised $110 per game for a fourteen game schedule. "We pay in full after each contest," Lambeau added, "and all players are paid whether they play or not." Following Ford's appearance in that summer's College All-Star game against the Chicago Bears, Bears coach George Halas joined the bidding for Ford's services, as did the Detroit Lions.

By then, however, his thoughts had turned to a less glamorous pursuit than professional sports. Phil Buchen, destined to be his first law partner, believed that in pondering a legal career Ford had more in mind than job security. "He was definitely interested in politics from the beginning," recalled Buchen. "That's why he chose law—it was an 'entree' to politics." First, however, there was the long grind through law school, with no obvious means of financial support. Harry Kipke had come to his aid before; might an assistant coaching job under Kipke subsidize Ford's legal education at the University of Michigan? Kipke said his budget ruled this out. Then he did the next best thing, recommending Jerry to Yale's newly hired football coach Raymond "Ducky" Pond.

As it happened, Pond's staff already included a recent Wolverine team captain, named Ivy Williamson. Williamson required no introduction to Ford, whose stoic acceptance of his position as Chuck Bernard's backup had greatly impressed the older player. His strong endorsement preceded an Ann Arbor lunch, and an invitation for Ford to visit New Haven. There Pond offered him multiple responsibilities as an assistant line coach, junior varsity coach and coach of the boxing team. "Of boxing I knew next to nothing," Ford later conceded. "No, that's not right, I knew absolutely nothing." Still, he wasn't about to kill his chances for the Yale job by acknowledging this deficiency. That summer Ford slipped away from his father's paint factory three times a week for lessons from a former amateur champion at the Grand Rapids YMCA. By fall his pugilistic skills were sufficient "to fool the Yale freshmen."*

Ford's final year at Michigan solidified his B grade average. Editors of *Michiganensian*, the student yearbook, nominated him for their university hall of fame with a wry testimonial. Jerry Ford deserved such recognition,

---

* But not all of the Yale freshmen, all the time. Yale athletic director Malcolm Fraser told naval investigators early in 1942 that Ford received, without complaint, "a fairly good beating every so often." Third Naval District Investigation Report, February 3, 1942, GRFL.

they concluded, "because the football team chose him as their most valuable player—because he was a good student and got better grades than anyone else on the squad; because (as house manager) he put the DKE house back on a paying basis; because he never smokes, drinks, swears or tells dirty stories—qualities quite novel among the rest of his fraternity brothers; because he's exceedingly bashful but broke forth in the middle of his senior year with a date; because he has decided to coach football at Yale, and incidentally study law; and because he's not a bit fraudulent and we can't really find anything nasty to say about him."

Passing up the elaborate rituals of commencement, Jerry spent the day close to home, toiling at Ford Paint and Varnish. Henceforth, he would miss few opportunities to celebrate his ties to the U of M. During and after his presidency Ford returned frequently to an institution that had long since outgrown the cloistered, car-free college of the 1930s. Over the years he raised millions of dollars for his alma mater, coming to the aid of successive presidents and relishing extensive student contacts. He frequently met with Michigan football players, each of whom he knew by name and number. As long as his health permitted, he delivered a pep talk to the team each year as it prepared to face archrival Ohio.

Ford's last public remarks came at the October 2006 dedication of Weill Hall, a $35 million facility housing the university's newly rechristened Ford School of Public Policy. His doctors had forbidden him from traveling to Ann Arbor, so Steve Ford read from a letter in which his father waxed nostalgic about the campus and the community where he had come of age. There could be no greater honor than to have a school bear one's name, said Ford. "Such recognition means all the more when it comes from an institution that you love." Disappointed as he was to miss the ceremony, the former president welcomed the opportunity, two months before his death, to acknowledge "the luckiest break I ever made."

3

YALE

———◆·◆·◆———

I knew I could do both—coach and go to law school at the
same time. But I must say I worked my ass off.

—GERALD R. FORD

GERALD FORD SPENT more time at Yale than he did at the University of Michigan. The second American president to earn a graduate degree from an Ivy League school—Rutherford B. Hayes completed three semesters at Harvard Law School—Ford did it the hard way, convincing skeptical administrators that an assistant football and boxing coach could hold his own against the intellectually formidable Class of 1940, in whose ranks were to be found future Supreme Court Justices Potter Stewart and Byron White; Secretary of State Cyrus Vance; Pennsylvania governors William Scranton and Raymond Shafer; Colorado senator Peter Dominick; Peace Corps director and vice presidential candidate Sargent Shriver; best-selling *Titanic* author Walter Lord; and Najeeb Halaby, director of the Federal Aviation Agency (and father of Jordan's Queen Noor).[*]

Ford's five and a half years in New Haven did much to broaden the perspective of one who before August 1935—save for his post–high school trip to Washington, DC—had never been east of Columbus, Ohio. With its leafy quadrangles and Gothic towers, Yale in the 1930s resembled a medieval cloister as re-created on a Hollywood backlot. Appearances could be deceiving, for while Harvard might be more venerable, no university more successfully blended the trappings of tradition with a lusty urge to experiment. Its ecclesiastical origins hadn't prevented Yale from hiring the nation's

---

[*] Ford traced his interest in the law to a high school course in Latin, a language dead to all but priests, pedants and attorneys. A more personal factor was the continuing fallout from his parents' acrimonious divorce, and a desire to avenge Dorothy Ford's mistreatment by a judicial system impotent to enforce its own rulings. "I never thought of myself as a great orator in the tradition of William Jennings Bryan or Clarence Darrow," wrote Ford. He much preferred Lincoln's view of the legal profession: "It is as a peacemaker that the lawyer has a superior opportunity."

first science professor, opening the first university art gallery, or awarding the first crop of American PhDs. This taste for innovation extended to college sports. Sons of Eli pioneered in the development of basketball, swimming and boxing. When completed in 1914, the Yale Bowl was the biggest stadium since the Roman Coliseum, while the cathedral-like Payne Whitney gymnasium could lay claim to being the world's largest indoor athletic facility.

Jerry Ford joined an athletic department still basking in the legend of Walter Camp, the celebrated "Father of American Football," whose contributions to the game included the two-point safety and line of scrimmage. Ford's first two years in New Haven coincided with back-to-back Heisman Trophies for end Larry Kelly and halfback Clint Frank. Under head coach Ducky Pond, the Bulldogs won thirteen of sixteen games before Pond's luck ran out in 1938. By then Ford's success with junior varsity football and other programs had earned him fresh responsibilities as an athletic talent scout. With these came a corresponding increase in salary. At $3,600 a year, he took home more than a junior instructor with a PhD in his pocket. Besides coaching future congressional colleagues like Wisconsin's Democratic senator William Proxmire—"He helped me," Proxmire remembered. "He knew how to get the best out of people and be pleasant at the same time"— Ford represented the university on scouting trips that gained him wide exposure to other campuses and cultures. Over time he shed his youthful discomfort around more affluent or worldly associates.

Determined to quickly repay the money he had borrowed to get through Michigan, Ford lived modestly as ever. His first New Haven address was a single room, costing $30 a month. It was a far cry from the Sterling Law Building, whose lacy stone turrets and lead glass windows inspired comparisons with the English Inns of Court. Under Dean Charles Edward Clark, the Yale faculty supplied such pillars of the New Deal as Thurman Arnold, a trust-busting assistant attorney general, and William O. Douglas, who chaired the Securities and Exchange Commission before FDR appointed him to succeed Louis Brandeis on the Supreme Court. Clark himself was largely responsible for crafting the seminal *Federal Rules of Civic Procedure*. At the same time, he and his Yale colleagues nurtured the movement called legal realism, a bold departure from the rigidities of case law and abstract theory in favor of real-world empiricism and social context. This, too, would find reflection in Ford's pragmatism and his disdain for zealots of any ideological stripe.

In his initial job interview with Coach Pond Ford had maintained that he could carry a full load of law courses without neglecting his responsibilities to the athletic department. A skeptical Pond shared Jerry's unusual request with law school officials. Their negative response did nothing to weaken Ford's resolve; he had surmounted greater obstacles to get where he was. At the end of a successful first year of coaching in New Haven, his debts paid and a 25 percent raise in his pay envelope, Ford welcomed an invitation from basketball coach Ken Loeffler to spend the summer of 1936 working for the National Park Service. Reasoning that it beat filling paint cans in his father's business, Jerry contacted Senator Arthur Vandenberg, the family friend regarded by many that year as a dark horse Republican presidential prospect. With Vandenberg's help he landed a ranger's job at Yellowstone National Park as one of the agency's so-called ninety-day wonders assigned to Canyon Station, a cluster of tourist facilities overlooking the scenic Grand Canyon of the Yellowstone River.

Before reporting for duty in mid-June, Ford paid a visit to the Riverton, Wyoming, residence of Leslie King and his second family. What he saw there—a comfortable home, extensive irrigated ranch lands and a prominent position with the local bank—reinforced the impression of his birth father as a man of hoarded wealth and dubious character. Ford's brief stopover was to have consequences scarcely imaginable at the start of his Yellowstone summer. Awed by the rugged majesty of the landscape around him, he developed a lifelong appreciation for the National Park system, whose budget he proposed to double had he won the 1976 election. Recalled by superiors as "a darned good ranger," Ford was as adept at battling forest fires as he was feeding grizzly bears before appreciative tourists. Other responsibilities included noting each day the make, model, state and license number of 150 or more automobiles on park property. Typically this entailed a predawn run of two hours or more. Ford welcomed the opportunity to stay fit.

Only once that summer did he register a protest. Assigned to greet arriving dignitaries at the posh Canyon Hotel and Lodge, Ford complained that it was both "undemocratic and un-American" to give special attention to such VIPs. At twenty-four, the handsome athlete in his flattering uniform attracted more than his share of female company. Frequently he was chosen for dance duty. "This meant spending the evening at the lodge or the hotel in uniform all cleaned and pressed with riding boots polished with a clean white dress shirt and green tie," according to one of Ford's coworkers. "Visitors really flocked around and would often ask pointless questions just as an excuse to say that they had talked to a ranger." During rare hours reserved to themselves, rangers became better acquainted with the park's wonders. There were picnics by day and wrestling matches to combat boredom after dark.

One evening Jerry and his roommate returned to their quarters only to encounter a black bear rummaging through the cupboard in search of food. Startled, the creature lunged at the intruders, who managed—just—to leap out of his path. Ford gave the bear a swift kick in the backside before chasing it out of the cabin. He showed equal resourcefulness when a park visitor slipped on some loose gravel and tumbled down a steep cliff. Clinging precariously to a narrow ledge five hundred feet above the Yellowstone River, the man appeared to be losing his grip by the time Ford arrived on the scene. Sizing up the danger, Ford slipped a rope around a tree, bound himself to one end and tied the other to fellow ranger Wayne Replogle. As he fed out the lifeline, Replogle lowered himself to a point where he could reach the visitor before Ford pulled the two men back up and over the precipice.

Afterward Replogle remembered his buddy's lack of emotion during this rescue and in other times of stress. "He would always say, 'Calm down everybody. It'll turn out all right.' He was always so reliable that we looked to him to get difficult assignments done right, even though he was one of our youngest rangers that summer."

NOTHING ABOUT HIS summer in the great outdoors caused Ford's interest in the law to slacken. He remained convinced that he could do justice to his employers in the athletic department while simultaneously shouldering a full complement of legal studies. To prove this to himself as much as to the skeptics in New Haven, Coach Ford returned to Ann Arbor in the summer of 1937. There he enrolled in two introductory courses—Common Law Pleading and Criminal Law and Procedure—scoring a B in each and gaining a toehold in his ongoing struggle for acceptance by the mandarins of Yale Law School. Or so he hoped. Of some 125 students who comprised the incoming class, Ford was told, 98 belonged to Phi Beta Kappa. Undaunted, he secured an interview with Myres McDougal, then a junior faculty member teaching property law. McDougal's evaluation was a small gem of legal realism. He judged Ford to be a "very mature, wise person of good judgment, good-looking, well-dressed, plenty of poise, personality excellent . . . informational background not the best, but interested, mature and serious of purpose . . . I see no reason for not admitting him."

McDougal's endorsement carried the day. Admitted on a trial basis to the Class of 1940, Ford's performance in two courses during the 1938 spring term confirmed Professor McDougal's assessment. Because he started well behind the rest of the class, Ford spent the summer at the University of North Carolina at Chapel Hill, where he studied federal jurisdiction and contracts. An occasional tennis game or fishing trip on the nearby Pamlico River supplied the only distractions from his coursework. That Ford had begun to think like a lawyer was demonstrated on his return to New Haven. He neglected to inform Coach Pond of his intention to carry a full complement of law classes, just as he withheld from the law school his continuing full-time responsibilities with the athletic department.

Such a demanding schedule precluded socializing among the future stars of the Class of 1940. A friend from Corby Court, a private eating club, recalls touch football games involving Ford and Byron "Whizzer" White, whom John F. Kennedy would appoint to the United States Supreme Court. To his classmates Ford was the personable jock in the front-row seat—a commanding presence because of his size, stingy with words though worth listening to when he did break his silence. They knew nothing of his personal life, or of any outside rivals for his time and energy. Saturday-night card games over beans and franks at Ducky Pond's home were pleasant enough, but no substitute for the sort of female companionship that caused Yalies each weekend to pile into jalopies headed for Smith or Vassar.

Sometime in the fall of 1937, a casual girlfriend Ford had known at Ann Arbor tipped him off to a prep-school classmate enrolled at the Connecticut College for Women in nearby New London. "The most beautiful girl I ever knew," she called her. Ford tested this assessment on a pair of Coast Guard cadets he encountered the next time he took Yale's freshman boxing team to New London.

"Do you know a girl named Phyllis Brown?" he asked them.

"We sure do. She's the best-looking gal at Connecticut College."

Given the paucity of female companionship around New Haven, it would have been surprising had Ford resisted such incentives. Still, he waited a couple of weeks before calling the mystery woman. After mentioning their mutual friend, he asked if he could take Miss Brown out to dinner. "Which I did," said Ford, stirring the ashes of memory a half century after the fact. "She was really a classy girl." Long pause. "Thank God I didn't marry her." This was said with the benefit of hindsight. The younger Jerry Ford had been more than willing to pursue the tall, lissome blonde, part Lana Turner and part Holly Golightly, a self-described "Eastern preppie" five years his junior, who exuded more charisma than was legal for a department store owner's daughter from Lewiston, Maine.

The words "Gerald Ford" and "supermodel" hardly seem to belong in the same sentence. Yet nothing else in his New Haven experience shaped him as much as this restless, intellectually curious future cover girl familiar to readers of *Cosmopolitan* and other popular magazines, a gifted athlete and vivacious flirt known to her numerous male admirers as PB—for Perfect Body. Phyllis Brown would play *Pygmalion* in reverse, initiating Ford in the polyglot vitality of New York City, testing his knowledge of the latest dance steps at the Rainbow Room and imparting her downhill skills on New England's icy slopes. During a volatile romance spanning nearly four years, Jerry came to rely on Phyllis as tastemaker and tour guide; bridge instructor; tennis, golf and sometime business partner; lover and, very nearly, spouse. Phyllis introduced Jerry to serious theater—one night early in 1941 they saw Orson Welles's Mercury Theater production of Richard Wright's *Native Son*, a shocking-for-its-time depiction of racial oppression and violence on Chicago's South Side. Together they discovered an affinity for the clotted prose of Thomas Wolfe. More than any professor or athletic mentor, Phyllis fueled Ford's transformation from Midwest hayseed—her word—to the junior establishment enrolled in the Class of 1940.

Of course, there was more to their relationship than progressive theater and *Look Homeward, Angel*. It was, in Ford's words, "a pretty torrid romance." Phyllis seconded his assessment, identifying Jerry as the first man she had ever gone to bed with. Meanwhile illustrator Bradshaw Crandell induced Phyllis to leave college halfway through her junior year to pursue a modeling career under Crandell's aegis. Two summers she tore herself away from Manhattan long enough to sample the charms of West Michigan. She quickly endeared herself to Jerry's family and friends, notwithstanding her failure to wash a dish or make her own bed. Jerry reciprocated by spending a Christmas holiday with Phyllis's family in Maine. Back in New York he agreed to accompany her to "some deb thing" for which he donned the only formal jacket he owned. Exhausted from his legal studies and coaching regimen and paying scant attention to where he was going, Ford fell down a manhole and ripped his coat.

"Oh great," Phyllis told him jokingly. "You can sue the city and use the money to rent tails to go to this party."

"Oh God, no," replied Jerry, credulous as ever. "The jacket was only worth about ten bucks."

Much would be made of a ski weekend the couple spent in Stowe, Vermont—complete with backgammon games, horse-drawn sleigh rides, and heaping plates of lumberjack fare—captured in nearly two dozen photographs for readers of *Look Magazine* in March 1940. The spread gave rise to persistent assertions, even now repeated on Internet sites and the *Jeopardy!* Daily Double, that Gerald Ford is America's only president-cum–professional model. Claimants cite a *Cosmopolitan* cover published four months after the Japanese attack on Pearl Harbor thrust the United States into World War II. Ford in his naval uniform can be seen kissing Phyllis Brown goodbye, a stand-in for millions of citizen warriors off to avenge the Pacific fleet or liberate Nazified Europe. Not so, Brown insisted. The man in the picture bears not the slightest resemblance to Jerry Ford. Actually, he was the first of her three husbands.

Long afterward, reflecting on her marital history and the one who got away, she lamented, "I don't think I have ever thought beyond the end of my nose my whole life." Jerry, by contrast, read the *New York Times* every morning and never within Phyllis's earshot said an unkind word about anyone. His silences inspired some counterintuitive theories about her onetime suitor. "He is a highly emotional man," she maintained. "And he never lets any of it out." Ford granted her point. In letters to Phyllis he occasionally expressed hurt over her seeming faithlessness. Otherwise, she claimed, "I never knew what was going on in his mind. Never knew what he was thinking." Ultimately Jerry Ford was too conventional for the free-spirited Phyllis. "I needed more of a challenge," she explained.

In later years she rationalized the course her life had taken by telling herself, "I wouldn't have liked to be the little congressman's wife." Yet she never outgrew feelings of remorse. She cherished the crested silver ring Jerry gave her one Christmas, and the bracelets and the locket with their pictures (while admitting that she couldn't remember ever giving him a present in return). Phyllis's modeling career lasted five years, about the duration of her first marriage. Her sole encounter with the then Betty Bloomer Warren of Grand Rapids—"A near disaster" in Jerry's pained recollection—inspired an unflattering portrait in Mrs. Ford's White House memoir in which she could not bring herself to identify Phyllis by name. Still, her description was longer than her husband's portrayal in *his* book of the woman who had completed his Yale education.

Far more revealing was an interview with Phyllis conducted late in life by James Cannon, purposefully left out of his Ford biography, though shared with the current author. Nearing the end of their conversation, Phyllis asked if Jim would take a message back to her former beau. "I still think about him. I still dream about him. I still love him." Cannon dutifully repeated her words to Ford when they next met. The reserve with which they were greeted made Cannon momentarily wonder if he had overstepped the bounds of propriety. He needn't have worried; Ford was saying nothing without weighing its implications. Finally he broke his silence. "She was quite a gal," he conceded. "But Betty and I have had such a wonderful life together . . ." Two compliments bestowed, he didn't need to finish the sentence.

I F FORD WAS sparing with details of his relationship with Phyllis Brown, he was even less forthcoming about the grudge match that engulfed his biological parents beginning in the spring of 1937. Characterized by the *Omaha World-Herald* as "one of the most flagrant cases of alimony dodging in the history of the country," their legal row, and the savage recriminations it spawned, feels grossly disproportionate to the monetary sums involved. In truth, Ford vs. King was less about financial redress than a belated quest for moral retribution by a profoundly wronged woman. We can only speculate as to why Gerald Ford made no reference to the court case in his memoirs. Undoubtedly it was a painful subject, like anything related to his mother's abusive first marriage. Moreover, the former president may well have blamed himself, however unintentionally, for reigniting a blood feud that showed none of its participants to advantage.

But there is another, equally credible explanation for his reticence. This was no ordinary family squabble. It suggests a pattern of behavior on Ford's part that would arguably repeat itself forty years later, with historic consequences. Long before a self-seeking Richard Nixon reprised the part of Leslie King Sr. by attempting to evade his Watergate responsibilities, his vice president revealed a susceptibility to emotional manipulation. Ford's early attempts to mediate his parents' differences outside the courtroom were as honorable as they were thankless. They might also be said to have straddled the line between trust and gullibility.

One thing seems indisputable. It was Jerry's description of his 1935 visit to Riverton, and the general aura of prosperity he observed there, that stoked the long smoldering embers of Dorothy Ford's resentment. Early in 1937 she asked her son to recommend a lawyer who might obtain justice from the man who had defrauded them both financially and emotionally. Jerry recruited a former frat house buddy named Bethel B. Kelley, who was to prove an aggressive and resourceful advocate. By July 1937 Kelly had secured a court order in Omaha, scene of the original King divorce settlement, directing Leslie King to pay his former wife $5,600 in long withheld child support. So began a maddening game of cat and mouse, with King and his lawyers pleading poverty even as their delaying tactics enabled the defendant to conceal much of his wealth. A second suit, brought in the United States District Court for Wyoming, produced another victory for Mrs. Ford.

By the summer of 1938 she was owed more than $6,000. Facing a possible writ of attachment authorizing US Marshals to seize his assets, King turned to his estranged son for help. "Arrange with your mother for her to make some fair settlement with me. This is to your benefit in the future." At this point outside factors intervened to blur Jerry's loyalties. A few days before Christmas 1938, Bethel Kelley asked Ford to confirm rumors of his impending marriage to "the blonde Venus," Phyllis Brown. Ford responded with an urgent request for $1,200 by March 15: "You see I have invested in a business enterprise here in the east." Phyllis had convinced him to provide seed money so that Harry Conover, a prominent New York model who often worked with Phyllis at the John Robert Powers agency, could open an agency under his own name. This explains Ford's sudden appeal to Bethel Kelley, and the rumblings of compromise that now found their way into his correspondence with Leslie King about the ongoing suit.

Quick to exploit the situation, King persuaded Jerry to endorse a $4,000 settlement to be paid in six-month installments over two years. When Dorothy's lawyers reported insufficient Wyoming property in King's name to validate any unsecured agreement, Jerry echoed their insistence that the slippery King post collateral. Once that was resolved, he told his birth father, "then the whole mess will be cleared up and I can feel that more amicable relations can exist on all sides." Having taken it upon himself to get his mother's consent to the settlement, "consequently I am believing in you as for your own honesty . . . Please Dad do as you agree or else I'm afraid our now splendid relationship will have a hard time. I know you will for I'm counting on it."

Characteristically, Jerry emphasized the positive. A more detached observer might accuse him of wishful thinking. Unreliable as ever, King initially promised compliance with the court order, then reverted to a hardline position of "no collateral." For good measure he rebuked his son for doubting his "good and honorable father . . . Nobody is trying to cheat you Gerald as I have told you several times." Again King hinted at nebulous legacies to come, this time linked to the anticipated marriage of Jerry and Phyllis Brown. The latter was "certainly a fine looking . . . girl and looks as if she had a lot of sense. Gerry, I am very much interested in you and Phil and if times would get better and when they do [sic] I am going to help you. Remember what I am telling you old boy."

King's decision to enroll his son Bud in a posh New England prep school did nothing to bolster his claims of financial hardship. Professing interest in Jerry's opinion, King made plans to visit him at Yale. "Swell," replied Ford. "I will let you know the first time I get some news from Grand Rapids." While King no doubt congratulated himself on co-opting his firstborn, Dorothy Ford's lawyers prepared a counterstroke. On the morning of May 4, 1939, the day before King's planned departure for New Haven, they had Leslie King arrested in Lincoln and taken to Omaha, where he spent seven hours in a jail cell as ordered by a district court judge. Released that night after posting $2,000 bail, his humiliating stint behind bars only hardened King's resolve to defeat Dorothy Ford, whatever the cost.

"I am going to fight this to a finish," King warned his son, "and I am sure it won't be very pleasant for all concerned as the publicity in all newspapers won't do anybody any good." He arranged for his lawyers to take Jerry's deposition. "I want you to show your colors for me," King told him. Referring to the proposed settlement as reimbursement for Dorothy's earlier expenditures, King wrote, "You know down in the bottom of your heart your mother never spent this money on you . . . and I am sure you being a *King* would not be a party of getting something for nothing." Fearing the negative publicity and its impact on Gerald Ford Sr., Dorothy was now inclined to agree with Jerry—the case should be settled for a onetime payment of $4,000. On June 12, 1939, Leslie King forwarded a check in this amount, which Dorothy sent, minus 40 percent in legal fees, to her son. "Dear Jerry," she wrote. "Have a good time! Now you know what side of the fence your bread is buttered on." Jerry just as promptly returned the check to his mother.

It was the last contact either of them had with Leslie King. Plagued by asthma, King spent his later years in Ohio before moving to Arizona. On February 19, 1941, he died in Tucson, aged fifty-six, reportedly a victim of allergic shock to aspirin. On learning of his death, Jerry wrote King's lawyers to inquire about a possible inheritance. He needn't have bothered. Nor should it have come as a surprise to learn that he had been excluded from any share of Leslie King's estate.

FOR JERRY FORD, 1940 was a year of decisions, the most important taken one hot summer evening over a sixty-cent bottle of beer at the Rainbow Room in Rockefeller Center. "We found a way we could enjoy the nightlife of New York City without spending much money," recalled Phil Buchen, then clerking for $25 a week in a downtown law firm whose lack of air-conditioning was the least of its deficiencies. Sitting at the bar, nursing their drinks and listening to a nearby dance band, Buchen and Ford contemplated their fast-approaching graduation from law school and what lay beyond. Having already declined offers from law firms in New York and Philadelphia, Ford saw his future solidly grounded in his past. He proposed that Buchen, head of his class at Michigan and already well disposed to

Grand Rapids because of his frequent jaunts there with other Dekes, join him in a two-man firm commencing in the spring of 1941.

Other resolutions, hardly less momentous, flowed from this one. Phyllis Brown had sensed in Ford "an innate awareness, keenness. When you really think he is not even paying attention, he is cunning . . . until people get to know him, they think big, dumb Jerry." Phyllis knew better. "Not bright, but smart . . . not quick but he mulls things over." After considerable such reflection Ford concluded that his relationship with Phyllis was hexed by differences in temperament no less than by conflicting career goals. She could never be happy trading the polish and pulse of Manhattan for the quiet domesticity of West Michigan. Nor could she accept a husband in politics, especially one "who was so preoccupied," Ford acknowledged, "spent the long hours at it, traveled so much. She just wouldn't sacrifice the nice things that she was able to get through marrying other people." Their separation was formalized after Ford received his Yale Law degree in January 1941.

Without telling anyone else, almost hiding it from himself, Ford was drifting toward a life in politics. Increasingly his attention was drawn to the European war thus far dominated by Hitler and the Nazi war machine. Not content to be a mere spectator of distant events, Ford hoped to influence them through grassroots activism. In June 1940 he was part of a quartet of Yale Law students who formed America First, a national campus-based movement giving voice to the antiwar sentiments of those who, believing American participation in World War I to have been a colossal hoax, were resolved to keep Uncle Sam from repeating the mistake.

In later years Ford regarded his involvement with America First as youthful folly. "You know, when I was a youngster I was an isolationist," he told one friend in the White House. "I thought Lindbergh knew what he was talking about. I figured the last [thing] they wanted to do was go back to Europe and fight another war. Of course, I was as wrong as everybody else who perceived that we didn't belong over there." In this he had plenty of like-minded company, not just among student activists like John F. Kennedy and Gore Vidal, but political figures as diverse as Herbert Hoover and Norman Thomas; Hollywood icons Walt Disney and Lillian Gish; architect Frank Lloyd Wright and novelist Sinclair Lewis. As many as 850,000 supporters took up the cause in some 450 chapters. These were disproportionately centered in the Midwest, where Wall Street and Downing Street were often conflated, and flag-waving journals like the *Chicago Tribune* claimed a monopoly on patriotism for the Real Americans who would have their country shelter between two great oceans, divinely placed like moats with the drawbridge pulled up.

An understandable sense of urgency pervaded the organizers of America First. In May 1940 Hitler's armies had lunged westward to occupy most of France, from whose Normandy coastline a Nazi invasion force might yet succeed where Philip II, Bonnie Prince Charlie, and Napoleon had failed. In London the discredited Neville Chamberlain was replaced as prime minister by Winston Churchill, who attempted with fighting words to make up for his country's lack of more conventional weapons. In his first speech to Parliament, a defiant Churchill told his countrymen that he had nothing to offer them but "blood, toil, tears, and sweat." The fiery address was broadcast on both sides of the Atlantic. Gerald Ford heard it on his car radio. "I damn near cried. At the same time I was exhilarated."

However powerful the emotions generated by Churchillian rhetoric, trans-Atlantic solidarity was not among them. Besides strict enforcement of the 1939 Neutrality Act prohibiting American ships from transporting goods to belligerent ports, the instigators of America First looked askance at US aid to the embattled British. Grounding their opposition to foreign wars in fears of domestic oppression, they held that the authority of wartime command, on top of the powers already centralized in Washington to combat economic depression, posed a greater threat to American democracy than any European dictator. It was a thesis articulated most forcefully by Gerald Ford's first political hero, Wendell Willkie. The unlikeliest of presidential candidates, Willkie was a forty-eight-year-old interloper in the party he had only recently joined in protest, not of FDR's undisguised sympathy for Britain—which he shared—but of the president's seeming hostility to private entrepreneurship in his own land.

As president of the Commonwealth & Southern Corporation, the nation's largest utility, Willkie objected to the Tennessee Valley Authority and other government initiatives designed to generate cheap power in competition with existing providers. By the time a 1939 Supreme Court decision forced him to sell out to the TVA, Willkie had established himself as one of Roosevelt's more thoughtful critics. Stories by or about this rumpled, bear-like Hoosier, so unlike the stereotyped business tycoon with his batwing collars and Corona cigars, began to appear in many of the nation's most influential publications. Readers of the *Atlantic Monthly*, the *Saturday Evening Post*, and especially Henry Luce's media empire anchored by *Time*, *Life*, and *Fortune*, were introduced to a Wilsonian liberal whose heart beat to the competitive ethos of the New Freedom, rather than the improvised collectivism of Roosevelt's New Deal.

It surprised no one when Willkie took his act to radio, where his quick wit and unpretentious manner captivated listeners like Gerald Ford. His support of Willkie as a long-shot candidate for the presidency in 1940 was all the more noteworthy given the competing ambitions of Grand Rapids' own Senator Arthur Vandenberg. Certainly Vandenberg's isolationist views were more consistent with Ford's allegiance to America First. Alone among GOP contenders, Willkie endorsed FDR's efforts to relax the 1939 Neutrality Act in order to aid the Allies. Later he would back the president's controversial call for a peacetime draft, as well as a Rooseveltian scheme to swap fifty aging US destroyers for long-term leases on British military installations in the New World.

Yet none of this diminished Gerald Ford's enthusiasm for the brash outsider, whose standing in public opinion polls skyrocketed with the deteriorating military situation overseas. Registering just 3 percent support on May 8, 1940, Willkie soared to 44 percent by the time Republican delegates convened in Philadelphia on June 24. Their opening session coincided with issuance of a written appeal to students throughout the land. "We are not pacifists," insisted the founders of America First. Their platform was simple: "hemisphere defense rather than European intervention." Ford wasn't in New Haven for the launch of this manifesto. He was two hundred miles away, one of thousands of youthful enthusiasts packing Philadelphia's convention hall and shouting, "We want Willkie," with a passion that, refracted through a million telegrams and an unrelenting media barrage, made for the most exciting political convention in memory.

On the sixth ballot Willkie edged out Ohio's isolationist senator Robert Taft to become the GOP's standard-bearer. It wasn't logical. It was scarcely rational. Antiwar, anti-Roosevelt delegates were seized with enthusiasm for a convert to their ranks, one who insisted in his acceptance speech on addressing them as "you Republicans." Ford, delighted by the result, returned to Grand Rapids eager to join the Willkie campaign on his own turf. But where to begin? His stepfather had an idea. If Jerry wanted to get involved in Michigan politics, he had best go to the source of Republican power in the state. That meant a visit to Frank McKay.

A STOCKY, UNPREPOSSESSING clerk turned mutant entrepreneur ("I'm a businessman first, politician second"), with his trademark pince-nez and pearl stickpins, Frank McKay was the prototypical party boss who dominated Michigan government for twenty years, and Grand Rapids for another decade after that. McKay did not invent municipal corruption in Grand Rapids. In truth he was part of an ignoble tradition of civic mismanagement almost as old as the city itself. At the turn of the twentieth century the mayor, city attorney, city clerk and fourteen aldermen and representatives of the city's three leading newspapers were criminally implicated in the Great Water Scandal, a bipartisan exercise in bribery growing out of a proposed pipeline connecting thirsty Grand Rapidians to the limitless supply of pure Lake Michigan water. Thirty years later, Frank McKay would succeed where they failed, and at considerably greater benefit to himself.

In business as in politics, McKay was a monopolist. "He sold surety bonds, and all the banks that wanted city and state deposits knew where to buy their bonds, as did road contractors and a host of others," recalled Willard "Doc" VerMeulen, a Grand Rapids dentist and active member of the Morals Committee at Bethany Reformed Church. It was VerMeulen who first recruited Jerry Ford for a civic crusade against the boss whose far-flung organization more nearly resembled an organized crime family, down to the trio of McKay bodyguards supplied by Detroit's Sugar House Gang. McKay had his roots in Grand Rapids' West Side, where it was variously claimed he was of Lithuanian or Polish ancestry (he was in fact the son of a Scottish father and German mother). A sometime factory worker and iron molder, young Frank ingratiated himself with his neighbors by operating a foreign money exchange and making loans from a private bank. At the age of twenty-three he secured dual appointments as deputy county clerk and assignment clerk for the circuit court. Neither position made onerous demands on his time.

From his office in the old County Building, McKay sold insurance and real estate, laying the foundations of a personal fortune estimated at $15 million by the time of his death in 1965. "Frank made a lot of friends from local attorneys, big attorneys," says one steeped in Grand Rapids lore. "In the course of time these people owed Frank favors." Always willing to slip a $10 bill to someone collecting signatures for a favored candidate, McKay paid all expenses for delegates attending state or local GOP conventions. By 1920 he was Republican county chairman, promoting his own slate of candidates for city hall. Two years later he was elected state treasurer.

Once established in Lansing, McKay joined forces with grafting allies in Detroit and Flint to establish a shadow government for the state of Michigan. Under his protégé, Governor Frank Fitzgerald (in office 1935–36; 1939), McKay's fortune expanded with his proximity to the throne. Besides the McKay Lumber Company and the McKay Building Company, he owned the Akron Tire Corporation, its product required on all state vehicles (and for anyone interested in a state job). McKay reputedly got a kickback on every sack of cement poured into the great Blue Water Bridge linking Michigan to Canada at Port Huron. He did a lucrative trade selling road graders and other equipment to the state. His biggest windfall, however, came with the end of Prohibition. Control of the state liquor authority enabled McKay to squeeze a $1-per-case commission out of distributors who unavoidably bought their supply from him. Those who refused found the market for their product literally dried up.

At the 1940 Republican convention in Philadelphia it was McKay who swung the decisive votes behind Wendell Willkie's nomination. His price was characteristically steep—personal control over the appointment of all federal judgeships in Michigan. Willkie didn't hesitate: "Tell McKay he can have them if we get the votes." Thus did the worst elements in his party facilitate the triumph of young Jerry Ford's would be Galahad. Returning to Grand Rapids after the convention, Ford made his way to the elaborate office suite and apartment that McKay maintained in the former Michigan National Bank Building. (The tallest structure between Detroit and Chicago was about to gain two stories and a new name—McKay Tower—following its purchase by the Rotarian caudillo.) As Phil Buchen told the story, Jerry Sr. made the appointment so that his son could tender his services in what was sure to be a hard-fought campaign.

The meeting was not a success. The younger Ford never forgot his humiliation as McKay kept him waiting four hours in a reception room, before giving him the brush-off in a cursory session lasting scarcely five minutes. Nothing better symbolized McKay's hubris, or the insignificance he attached to Jerry Ford and his loyalties, unless it was the follow-up visit to Ford Paint and Varnish by a salesman peddling ad space in McKay's kept newspaper, the *Michigan Times*, whose primary function was to heap abuse on troublemaking reformers like Doc VerMeulen. His encounter with McKay left the junior Ford permanently soured on the most influential Republican in Michigan. Meanwhile, back in New York the Willkie campaign gladly availed itself of his volunteered services. Phyllis Brown's presence in the city provided additional incentive for the political novice.

That summer was one of shifting ground and unexplained reversals. On July 1, Ford received a progress report from Robert D. Stuart, his fellow Yale Law student and the primary organizer of America First. Recruitment, publicity and community outreach all showed promise, according to Stuart, as did efforts to enlist a "national committee of prominent citizens" to lend prestige to the cause. Less than two weeks later, Stuart again wrote Ford, this time to accept his resignation from the group's executive committee. "We took your name off the letter and understand perfectly," Stuart's wife assured him at the end of a friendly note acknowledging Ford's recent reference of potential supporters in Michigan and Indiana.

What could account for his abrupt departure from the organization he had so recently helped to establish? Included in the group's archives housed at the Hoover Institution in Palo Alto are undated letters (easily traced to the first half of July) in which Ford, citing unnamed members of the Yale administration, related their warnings of potential embarrassment to the university stemming from his high-profile activism. Concerned for his job with the athletic department, Ford said nothing to indicate any change of heart regarding the war; indeed, he left open the possibility of rejoining America First once he secured his law degree the following year.

On closer examination, Ford's action raises more questions than it answers. There is scant evidence that Ford was dependent on his coaching paycheck in the summer of 1940. He had acknowledged as much two years earlier, when informing his employers in the athletic department that he had saved enough to pay his way through law school should they deny his request to combine legal instruction with his existing coaching responsibilities. Moreover, not only had he been able to capitalize Harry Conover's modeling studio; he had also returned to his mother the court settlement pried out of Leslie King. Whatever his rationale for quitting America First, it seems unlikely that it was financial.

Nor, in all likelihood, was it grounded in fear of professional repercussions. If anything, Yale and its law school were hotbeds of antiwar feeling as personified by Bob Stuart and classmates like Sargent Shriver and future college president Kingman Brewster. Moreover, Ford had already secured Phil Buchen's commitment to join him in a Grand Rapids law partnership. Had Jerry's embrace of the Willkie campaign superseded his loyalties to Stuart and the other architects of America First? Had Churchill's speechmaking, and the increasingly desperate plight of his countrymen under attack from German bombers, led Ford to see aid to the British in another light? Most of that summer he was in New York City. Mingling with Phyllis and her set, volunteering at Willkie's campaign headquarters, Ford encountered attitudes radically unlike the views he was accustomed to hearing in New Haven.

Yet another factor, hidden until now, may help to explain Ford's ambivalence toward the isolationist movement. Six weeks before election day, with a Gallup poll showing for the first time a narrow majority of Americans favoring aid to Britain—even at the risk of being dragged into the broader European conflict—Gerald Ford Sr. appealed to his friend and fellow isolationist Arthur Vandenberg for help in securing young Jerry an appointment as midshipman in the US Naval Reserve, and a berth on the training cruises then being filled by seniority of enlistment. In Washington the chief of the Navigation Bureau, one Chester W. Nimitz, acknowledged Vandenberg's inquiry, but could offer little tangible assistance. He urged the younger Ford to keep in touch with the commandant of the Great Lakes Training Station near Chicago.

On November 5, Franklin Roosevelt cruised to an unprecedented third term, though Willkie managed to reduce the president's margin to less than half his 1936 victory. The campaign behind them, the former adversaries could revert to their true positions, with Willkie lending crucial support to FDR's Lend-Lease infusion of military aid to the British. In New Haven, Jerry Ford completed his final semester at law school. He ranked 56th in his class of 127, with an overall B average. Early in 1941 he returned to Michigan, living with his parents and two brothers in their home on Santa Cruz Drive in East Grand Rapids. After passing their bar exams in March, Jerry and Phil Buchen established the firm of Ford and Buchen as a shoestring operation. Their sixth-floor office suite in the Michigan Trust Building was decked out with new furniture, bought on discount and carefully deployed to convey an air of established prosperity. The partners chose this address because of the free access it afforded them to the local bar association's extensive law library housed in the same building. A life insurance agent working next door tipped Buchen off to the cheap meals and discounted war bonds offered new members of the nearby Peninsular Club, an upscale watering hole that doubled as an excellent place to make contacts.

Exploiting his athletic celebrity, Ford organized Curbstone Quarterbacks, a weekly luncheon series that attracted hundreds of die-hard football fans. Around Grand Rapids he would never lack for visibility; clients were another matter. A friend of his parents wanted a title search conducted. Rejecting the $15 fee as exorbitant, the petitioner eventually settled his bill for $10. Dorothy Ford's part-time maid enlisted the help of Ford and Buchen in her divorce. That fall the trickle of business freshened, as each partner grew into his respective role—Ford the classic joiner and outside man, employing his people skills in ways that foreshadowed a congressman's attention to constituent needs; and Buchen, turning his limited mobility to advantage, minding the store while also handling the bulk of research and drafting work. They made a good combination, if not an especially profitable one. Buchen estimated their first-year income at $1,200 apiece.

Ford, instinctively drawn to trial work, welcomed the opportunity to interpret and apply the law, one client at a time. "I enjoyed all aspects of the law," he would recall. "I enjoyed working in adoption problems. I didn't necessarily like divorces, but that was part of the spectrum. Wills. I enjoyed the opportunity to go over to the courthouse, to negotiate or try cases." One Sunday morning he was called to the county jail by a man who— "Obviously drunk"—had injured several other motorists in a multicar accident. Clearly at fault, the man volunteered to enter the army if doing so would spare him from criminal prosecution. "I talked to the judge," Ford remembered. "I talked to the prosecutor and I got them to suspend the sentence." In return for this strenuous advocacy, "we never got paid." He wrote it off as a learning experience, giving rise to Ford's First Rule of Legal Representation: "In criminal cases . . . get the cash up front."

I N THE SUMMER of 1941, Ford maintained a close watch on Congress, where the military draft enacted a year earlier was up for renewal. Its passage by a single vote left him convinced that his future belonged to Uncle Sam. Believing that his legal training would be best put to his country's service by the Office of Naval Intelligence or the FBI, he applied to both. On June 14, precisely one week after his admission to the Michigan bar, Ford submitted his application to be a special agent in the nation's premier crime-fighting organization. Immediately he became the subject of an exhaustive FBI investigation. Legal records were sifted in vain for evidence of possible character flaws—unless you count two arrests for speeding as evidence of moral degeneration. Other than a gimpy pair of knees, recently reinjured on Yankee ski slopes, Ford's physical condition was superb.

Accompanying his application was a sheaf of personal recommendations far exceeding the formulaic compliments usually found in such testimonials. His old high school principal Arthur Krause noted in Ford a rare ability to "keep his mouth shut." To Special Agent in Charge Wilson McFarlin, the applicant was "a superior type of young man" with "an analytical and keen mind." When several months had passed without any official feedback, McFarlin took it upon himself to remind Washington of Ford's interest, and to reiterate his own view of the Yale-trained lawyer as "one of the best applicants who has ever appeared at the Grand Rapids office." Not until February 1942 did J. Edgar Hoover, replying to a direct inquiry from Ford, send him the bureaucratic equivalent of a form letter rejection. Ford's application would be retained in the bureau's file for future consideration should his services be deemed necessary.

No further explanation was provided, then or later. White House documents compiled during Ford's presidency attributed his failure to the applicant's oft-injured knees. Yet neither the Grand Rapids field office of the FBI nor, subsequently, the navy ever mentioned this as a disqualifying condition. Politics was something else. On January 17, 1942, the FBI's New Haven field office received information relating to Ford's organizational work with America First. Informants hastened to vouch for his loyalty. They insisted that any isolationist sentiments he may have harbored at Yale were extinguished long before Pearl Harbor. This intelligence was duly forwarded to Washington, where, three days later, Director Hoover personally terminated Gerald Ford's candidacy for the FBI. Ironically, the Grand Rapids draft board *did* take notice of Ford's bad knees. In classifying him 1B (defined as "unfit for active military service but available for limited duties"), the board significantly narrowed his options.

Even as Ford kept a wary eye fixed on foreign battlefields, closer to home he was drawn into another kind of combat, a decidedly uphill fight to purge Grand Rapids of Frank McKay's baleful influence. "Grand Rapids," muttered one visiting physician. "Little Jerusalem. Everywhere you look in Grand Rapids you see a church. They fill them up Sunday morning, Sunday evening, prayer meeting in the middle of the week. You're always praying over there in Grand Rapids and you haven't got the guts to stand up for what's right."

Doc VerMeulen was tempted to prove him wrong. "If I could get ten guys I think I could take them," he vowed in opposition to the McKay machine. His recruitment process began one snowy morning in 1941. With several patients having canceled their bookings, the dentist invited himself to Paul Goebel's nearby sporting goods store. There he described in detail the sordid political situation arising from McKay's stranglehold over the Michigan Republican Party. Goebel was appalled. If VerMeulen was even 10 percent correct in his claims, Goebel said, he could count on his support. "Who else you got lined up?"

"Nobody, just me."

Just then the door to the shop opened and Goebel pointed to a strapping young blond customer brushing the snow off his coat. "That's Jerry Ford. He's come back to Grand Rapids to open a law office here. He'd be a 'natural.'" The two men approached Ford, who agreed on the spot to join their insurgency. Goebel urged him to think it over. "This may not be the place for you."

"Why not?"

Anyone bucking the machine could expect to come under intense pressure, political and economic. Could Ford afford to lose clients?

"I haven't got any clients. All I've got is an office." Ford grinned. "I'm with you. Let's form the group and get underway."

True to his word, Ford became a mainstay of the organization styling itself the Republican Home Front. He was a regular attendee at Friday night meetings in the basement of Goebel's store, where the insurgents interviewed newspaper reporters, prosecutors and a former state insurance commissioner before going public with their indictment of municipal corruption. They sought out like-minded allies who could "get up and make a talk," and who enjoyed economic immunity from the long arm of the McKay machine. "They had to own their own business," according to Doc VerMeulen. Salaried employees were considered too vulnerable to pressure from the head office. Coming on the heels of his volunteer experience in the Willkie campaign, the Home Front provided Ford invaluable lessons in grassroots organizing and the art of public persuasion. It showed him how to convert lofty ideals to self-interest of a kind average voters could embrace. In the process he made far more friends than enemies, while his prominence in the reform movement advanced his reputation for probity and applied effort.

"He's such a hard worker," marveled Phil Buchen, himself no slacker. "I don't know if he ever left the office before I did." Consequently there was nothing unusual about Ford's being at his desk on Sunday, December 7, 1941, or remaining in the office long after the rest of Grand Rapids learned of Japan's surprise attack on the US naval base at Pearl Harbor. Included in the collateral damage was the budding partnership of Ford and Buchen. Phil's youthful polio guaranteed his exemption from the war Congress made official on December 8. For Ford the conflict promised yet another chapter in the education of a young man called on not only to defend his country, but to determine its place in a world his generation would redefine.

4

THE VIEW FROM THE BRIDGE

If you are going to go to war, I was in a good
place to go to war. I learned a lot.

—GERALD R. FORD

SMOKE WAS STILL billowing from the twisted wreckage of America's
Pacific fleet, the pride of which lay submerged in shallow Hawaiian
waters, when Gerald Ford volunteered to join the navy. With his FBI application unresolved, he cheerfully submitted to a fresh round of interviews
and examinations as Naval Intelligence agents probed his character and
credentials. At the end of March 1942 he learned of his acceptance by V-5
Preflight Training, popularly known as the Tunney Fish Program in deference to former heavyweight boxing champion Gene Tunney, since December 1940 the nominal head of navy fitness. Improvised like so much of the
US war effort in the aftermath of Pearl Harbor, V-5 was intended to create
a nucleus of highly qualified physical training instructors, star athletes who
would in turn condition naval aviation cadets assembled by the thousands
at select college campuses and navy facilities.

Given the predominance of college football coaches and athletic directors among the program's overseers it was hardly surprising that the regimen
they devised resembled that of a preseason training camp. This hard-nosed
approach, with its emphasis on athletic competition as a spur to the warrior
spirit, was warmly embraced by Gerald Ford. Commissioned an ensign in
the Naval Reserve on April 13, 1942, he reported for active duty a week
later. Greener than grass—"I didn't know a gun from a bow and arrow," he
later acknowledged—Ford spent a month at the Naval Academy in Annapolis, followed by a year at the University of North Carolina at Chapel Hill
flight school.*

He welcomed the chance to return to the verdant campus where he had

---

* UNC was one of four college campuses chosen by the navy as a preflight training school.
Almost nineteen thousand men went through the program at Chapel Hill.

passed the summer of 1938 preparing for Yale Law School. Campus housing being in short supply, Ford shared a one-room cabin on Hidden Hills Drive with three roommates, one of whom, Earl Ruth, later served in Congress with him. Another roomie, a UNC basketball letterman named Bill McCachren, pronounced his fellow jock "as easy to wear as an old shoe." Attending a Chi Omega tea dance one Saturday afternoon early in 1943, Ford was immediately attracted to the hostess. "The prettiest girl here," he told Emily Irby, "and you are running the whole show, too." Their subsequent dates included bridge games with Boston Red Sox slugger Ted Williams and his wife.

Others sensed in Ford an affable detachment, something first noted by his South High School classmates. John Foster West was a journalism major and recent transfer student to Chapel Hill, who also ran on the UNC cross-country team. It was West's habit to end his day with a twilight jog around campus. One evening he chanced upon a laconic stranger, a young soldier who failed to identify himself or ask West's name. None of which prevented the two men from becoming running partners and friendly, if not friends. In November 1942, West joined a capacity audience in Memorial Hall for a live radio broadcast featuring the popular "Songbird of the South" Kate Smith. At one point in the evening he was spotted by his mysterious jogging partner, who asked if West would like to go backstage after the show and meet the performers. Among those on the bill that night to whom Gerald Ford introduced West was movie star Ronald Reagan, fresh off his screen triumph in *King's Row*. It was the first known encounter between the future rivals, the memory of which eluded both men in later years.

Surrounded as he was by amateur Lindberghs, it was only a matter of time before Ford caught the aviation bug. Together with two of his roommates, he obtained a Piper Cub, which the trio then set about teaching themselves to fly. "Ford crashed the plane," according to Bill McCachren. "He flew it straight into the ground and didn't pull up. He tore the wing off." If he felt any embarrassment over the incident, Ford hid it from his buddies. After all, he hadn't joined the navy to be a pilot, or, for that matter, to put those who had through their athletic paces. Increasingly bored with his land-based assignment, Lieutenant (junior grade) Ford applied for sea duty. "In a war, single, no responsibility," he later explained, "you might as well be where all the action was." In 1943 that meant the vast arena of the Pacific, with its countless atolls and island chains arranged as so many stepping-stones to Formosa, the Philippines and, finally, the aggressors in Tokyo who had provoked the sleeping giant that was the United States.

Ford wrote letters to anyone who might pull strings on his behalf. His lobbying efforts paid off on May 2, 1943, with a berth on the USS *Monterey*, one of nine so-called Independence-class aircraft carriers devised almost overnight to help fill the gap left early in the war by the loss of such behemoths as the USS *Lexington*, *Hornet*, and *Wasp*. Six hundred twenty-two feet in length, housing a crew of fifteen hundred, the *Monterey* was an ungainly hybrid combining the narrow hull of a cruiser with the flight deck and island superstructure of a flattop. So unstable was the resulting craft that tons of cement had to be poured into the port side in an effort to achieve balance. "When we were fully loaded with all kinds of aviation gas and fuel oil," Ford remembered, "we had a three degree list to starboard." Other changes to the *Monterey* would have lethal consequences. With fresh air vents left covered, men in the boiler or engine rooms risked death from smoke inhalation in the event of a shipboard fire.

The payoff for such alterations was measured in speed. "We were fast," boasts Frank Weston, a mechanic and carburetor specialist aboard the *Monterey*. "We could go thirty-five, thirty-six miles per hour." This was to prove invaluable in a far-flung series of hit-and-run missions against Japanese outposts in the central and south Pacific, part of the island-hopping strategy adopted by Allied warlords at the 1943 Cairo Conference. But Ford, so eager to join the battle then raging on the other side of the globe, came perilously close to missing the boat. Following a six-week shakedown cruise to Trinidad in August 1943, the *Monterey* underwent final repairs and modifications in a Philadelphia shipyard. Leaving Philly for good on September 20, the ship paused in Norfolk to load a full complement of torpedoes. From there the navy's newest light carrier was bound for the Panama Canal, and a thirty-six-hour tie-up on the Pacific side of the country.

Ford wasn't on the *Monterey* that night. He had accepted an invitation from some old football buddies on the Atlantic side to sample Panamanian nightlife, a movable feast of nightclubs "where you all were drinking and having a good time and telling war stories." Sober enough to calculate the distance between Naval Air Station Coco Solo on the Atlantic coast, and the Pacific berth of the *Monterey* at Balboa, Ford slept little that night. Allaying his concerns was naval pilot Richard Brooks. Not to worry, Brooks told him; come Wednesday morning he would fly him across the Canal Zone, making sure to deliver him to his ship before it weighed anchor and set sail for San Diego the next morning. Nature had other plans, beginning with the "God damnest rainstorm you ever saw." Waiting helplessly for the tropical downpours to slacken, Ford and Brooks risked more than an on-time arrival. The ensuing flight, ordinarily a twenty-minute jaunt, became a windblown nail-biter flown too close to the ground for comfort. Adding to the danger, direct overflights of the canal were theoretically prohibited by the US military. A series of predawn phone calls eventually roused the commanding army general, who helpfully instructed gun crews along the route: "Don't shoot at this guy, he's trying to get across the isthmus!" As it was, Ford barely made it. Landing at Albrook Field just outside Panama City, the lieutenant bolted from the plane and into a waiting navy car at the end of the runway. He reached the *Monterey* just as the gangplank was being raised. Thus Gerald Ford began his naval combat career by narrowly avoiding court-martial.

In San Diego the *Monterey* boarded additional planes and crew urgently requisitioned by a navy eager to take the offensive. With so many first-time sailors making the trip, Ford said, "you never saw so many people ill. By the time the ship got to Pearl Harbor it really stank." Worse, in complying with orders to travel at flank, or maximum, speed all the way to Hawaii, the overcrowded *Monterey* blew its boilers. Soon after rounding Diamond Head on October 9, crew members fell silent at the sight of Battleship Row. Ten months after the Japanese raid, the decomposing remains of the *Arizona* and *Utah* were fully visible, the scorched turret of the former breaking the waterline like a skeletal arm thrusting out of the grave.

Spirits rose at the end of the month when the *Monterey* left Hawaii as part of a flotilla that included the mighty *North Carolina*, hero of Guadalcanal, and dozens of other battleships, destroyers and support vessels. No one doubted their mission, even if their destination remained a secret. Since Ford's designated role as athletic director was not considered a full-time job, he doubled as the officer in charge of gunnery Division Six (of eight). At the shrill summons of a general quarters call, he and his fifteen-man crew raced to a forty-millimeter battery on the fantail, or overhanging rear deck. From this remote position they scanned the horizon for twin-engine Mitsubishi bombers, torpedo planes the men called "Bettys" or kamikazes resolved to die gloriously for their emperor.

Ford's shipboard responsibilities did not end there. When not recruiting musical talent for fraternity-like smokers, or amateur boxers for exhibition matches, he led his fellow officers in games of volleyball, medicine ball, and calisthenics on the flight deck. Named ship's historian, he retained its log long after the *Monterey* put into port for the final time. Ford edited a newsletter, *Flat-Top Flashes*, embellishing it with cartoons featuring Salty Joe the Boot, and what passed for the latest war news—an RAF bombing raid on Berlin, a Soviet victory over Nazi invaders in the Ukraine. There were frequent admonitions to conserve water and enforce strict codes of censorship. Onboard diaries were prohibited, as were weather reports, personal recordings or pictures of ships, planes, guns or other military equipment.*

Ford enlivened the pages of his newsletter with wry observations and doggerel ribbing those who had yet to find their sea legs, and whose mangling of the naval vocabulary—mistaking "front porch" for forecastle, and "downstairs" for below—was always good for a chuckle. Sometimes military humor was tinged with pathos, as in the "Psalm of a Carrier Pilot."

> *The landing signal office is my shepherd*
> *I shall not crash . . .*
> *He maketh me to land on green runways*
> *He waveth me off the rough waters*
> *He restoreth my confidence*
> *Yea, though I come stalling into the groove,*
> *At sixty knots,*
> *I shall fear no evil*
> *For thou art with me*
> *Thy hands and thy flags, they comfort me*
> *In the presence of mine enemies*
> *He attacheth my hook into a wire*
> *My deck space runneth over*
> *Surely safety and caution should follow me*
> *All my days in the fleet*
>
> *And I shall dwell in a fool's paradise.*

ON NOVEMBER 11, 1943, Captain Lester Hundt let the cat out of the bag. The *Monterey* was destined for Makin Atoll, a strategically important dot of land in the Gilbert Islands chain, halfway between Hawaii and Australia. Eight days later the crew was served a battle breakfast of steak and eggs. By six that morning the ship's planes were speeding toward their

---

* Needless to say, my own account of Ford's naval career is greatly enriched by diarists James Hill and John P. Priolette. Their comprehensive journals are invaluable to any re-creation of life aboard the *Monterey*. So are the extensive personal recollections, some of them dipped in acid, of Captain John Cadwalader, as well as an official history of the ship produced by the navy. Copies of all these documents can be found in the Gerald R. Ford Presidential Library at Ann Arbor.

targets; all returned safely within ninety minutes of takeoff. The *Monterey's* sister ship *Independence* was less fortunate, taking a Japanese torpedo that left seventeen of her crew dead and forced the carrier to return to Pearl Harbor for repairs. On November 26, a torpedo missed the *Monterey's* stern, and Gerald Ford's gunnery post, by twenty yards. Many nights attacking dive-bombers guided by hissing flares flew as close as fifty feet above the ship's flight deck. Like everyone else manning the big guns, the crew of Battery Six underwent frequent identification classes to lessen the likelihood of friendly fire. "Get their engine or their wings," they were told. A plane with a dead pilot could still hit the ship, with potentially devastating consequences.

The *Monterey* was not constantly under fire. Amid the often ferocious resistance put up by the enemy, there were stretches of boredom. A beard-growing contest broke the tedium. The flight deck attracted sunbathers. When crew members hung a couple of baskets over elevators designed to lift planes from the hangar deck, Ford officiated at the impromptu games that resulted. Crossing the Equator became routine, the *Monterey* managing the trick four times one night in January 1944. As supply lines lengthened, logistical miracles were daily occurrences. Some involved refueling missions, during which as much as four hundred thousand gallons of fuel were transferred between violently rolling vessels. "One time we didn't meet our tanker to refuel and get food," recalled Frank Weston. "We didn't get meat and all we had to eat for about a week was green beans and butter." Adding to the misery, erratic mail deliveries meant delays of up to six months between letters from home.

Ford witnessed firsthand the daily attrition of battle, as menacing planes swarmed overhead and tin cans (navy lingo for destroyers) dropped depth charges to minimize the submarine threat. He saw rescue crews pluck pilots from the drink after a tailhook failed to come down, and mechanics perform emergency surgery on planes like a bomber that made it back pitted with sixty-three newly acquired holes. Sailors watched in horror as an inexperienced deckhand walked into a furiously spinning propeller. Balancing the randomness of death were unexpected flashes of earthly paradise. Two weeks before Christmas 1943, the *Monterey* and the rest of the American fleet anchored in the broad harbor of Espiritu Santo in the New Hebrides. This idyllic setting, inspiration for James Michener's *Tales of the South Pacific*, had been cleared of its regular inhabitants. In their absence, Seabees had constructed baseball fields and volleyball courts. On the other side of the island a shark watch and raft were provided for swimmers. Before leaving the ship, each man received three chits, redeemable for as many bottles of warm beer. At night there were movies on the flight deck before sailors strapped themselves into bunks four layers high.

The celebratory mood changed abruptly on Christmas Eve, with the *Monterey*'s participation in a mass bombardment of Kavieng, a Japanese stronghold on Papua New Guinea. Retaliation was swift and brutal. A holiday meal of baloney sandwiches was interrupted by an aerial assault featuring twenty or more Japanese attackers. One Zero barely cleared the flight deck of the *Monterey* before crashing into the sea, eliciting cheers from the embattled bluejackets. More raids followed, "as much action as I'd ever hoped to see" in Ford's words. The *Monterey* acquitted itself admirably. By the first week of January 1944, the ship and its crew claimed credit for downing eighteen enemy aircraft, while also sending a Japanese cruiser and destroyer to the bottom.

Ford could take pride in his record as a gunnery officer, but he had no desire to spend the war on the ship's fantail. The action he longed to be part of was unfolding high above the flight deck, in the bridge that served as the ship's nerve center and strategic command post. His luck hadn't deserted him. The *Monterey* had barely put to sea before the former college football hero—in the words of one crewmate "an idol on the ship . . . a really macho guy"—came to the attention of Captain Hundt, a pioneering aviator in the First World War and a sports enthusiast who liked nothing better than talking football with his ship's athletic director. His disregard for the brass ensured that Hundt's command of the *Monterey* represented the apogee of his seagoing career. For Gerald Ford, he was a gift from the sea gods, the latest in a parade of older authority figures—from South High principal Arthur Krause and Coach Cliff Gettings to Paul Goebel, Harry Kipke, Ducky Pond and Doc VerMeulen—who, detecting in Ford the makings of a leader, had taken him under their wing.

Ford reciprocated in kind ("I cozied up to him") with results that fully validated Hundt's instincts. When the ship's assistant navigator got transferred off the *Monterey*, Hundt made sure that Ford replaced him, notwithstanding the younger man's inability to so much as take a sighting. With help from the current navigator, a red-bearded eccentric known as Pappy Atwood, Ford soon mastered the complexities of the job. This, too, was part of a recurring pattern, wherein an inexperienced student/athlete/officer, through his own dogged efforts, justified the faith of his more seasoned sponsors. Not content to rescue Ford from gunnery duty, Captain Hundt elevated him to a privileged position as officer of the deck during general quarters. "Which meant I was on the bridge with the captain, the navigator, the air officer during combat," Ford boasted. "Couldn't have had a better assignment. Hell, I was right there where everything was going on."*

Ford did not exaggerate. In the coming months the *Monterey* would earn a place in the record books, traveling more nautical miles in the Pacific theater than any other vessel flying the Stars and Stripes. Though overshadowed by the heroes of Omaha Beach, June 1944 would be remembered by its participants for the "Marianas Turkey Shoot," shorthand for the fiercely unequal Battle of the Philippine Sea, which in two days cost the Japanese over six hundred planes shot from the skies or sunk along with three aircraft carriers flying the flag of the Rising Sun. Ahead of the Third Fleet commanded by Admiral William F. "Bull" Halsey lay the hornet's nest of Formosa (Taiwan), a Japanese bastion with extensive air installations and land-based bombers. For MacArthur to fulfill his pledge to liberate the Philippines, Formosa must be neutralized and what remained of Japanese air power rendered impotent. In the vanguard of the American strike force, the *Monterey* paid a corresponding price. At one point successive waves of Japanese bombers, interspersed with suicidal kamikazes, forced the ship's crew to remain at their battle stations for fully twenty-four hours.

Ford and his crew took part in over a dozen significant campaigns aimed at the shrinking perimeter of Japanese sovereignty. Rabaul. Truk. Eniwetok. Saipan. Wake Island. Palau. Tinian. Guam. Formosa. The Philippines. Inevitably Ford's thoughts turned homeward. In September 1944 he sent congratulations to a former Grand Rapids girlfriend on her impending marriage. Imagining life after the "Mighty Monty," Ford expected Coach Potsy Clark, his prewar boss at Grand Rapids Community College, to buy him "a damn good drink. We haven't had much time out this way for the lighter things of life, and when we do return I hope there is a generous supply of wine, women and song."

He adopted a decidedly more businesslike tone in addressing two other issues, one the unwelcome legacy of his New York sojourn with Phyllis Brown and her circle, the other pointing toward an unknowable future in politics. Harry Conover's modeling agency, founded with Ford's capital, had begun to thrive just as Ford left Yale and returned to Grand Rapids to practice law. Then came the war. Geographical distance made it difficult to monitor Conover's management, yet Ford strongly suspected that the few hundred dollars he received each year fell far short of the 50 percent of

---

* In assessing Ford's overall performance, Captain Hundt awarded his protégé a 4.0, the highest possible score. Ford's lowest ranking, a 3.7, was for shore duty.

agency profits spelled out in their original agreement.* Even while island-hopping in the Pacific, Ford hired a law firm to demand an accounting. Eventually he settled for $5,000, significantly more than his original investment, but pitifully short of what he might have obtained had he not been ten thousand miles distant from the negotiation. In itself the incident may appear trifling, except as a harbinger of the obsessive attention to budgetary fine print that was to characterize Ford the congressional watchdog.

He hadn't forgotten the Home Front organization and its David-and-Goliath battle with the McKay machine. Doc VerMeulen had seen to that, passing on a steady stream of political intelligence at Ford's request. Seeking the strongest possible candidate for Republican county chairman following their victory over McKay in the September 1944 GOP primary, VerMeulen and his allies in the Home Front settled on Jerry Ford Sr. Spirits in the reform camp sagged when Ford declined the nomination. VerMeulen was contemplating the distasteful prospect of filling the chairmanship himself when his phone rang. On the other end was Mr. Ford, sounding not at all like yesterday's reluctant warrior. He had just received a special-delivery letter from Jerry in the South Pacific. "Dad," he read, "I can't get it out of my mind, how are the boys doing in the battle against McKay? With so many of us gone have they got the manpower to win? If they ever ask you to do anything, don't hesitate, just do it, and when I come back I'll take your place."

Under the circumstances, the senior Ford could only defer to his son's wishes. Thus it might fairly be said that Gerald Ford's political career was launched on a battle-scarred carrier, at a time when his attention was divided between MacArthur's coming invasion of the Philippines and another kind of liberation in a Grand Rapids struggling to free itself of boss rule.

A CHANGE OF command on board the *Monterey* saw Captain Hundt replaced in April 1944 by Stuart Ingersoll, another aviator who inspired derision as a flier, not a ship's man. Ford, for one, quickly changed his mind about Ingersoll: "He turned out to be a nice guy." More important, it was Ingersoll's stubborn refusal to abandon ship at a moment of maximum peril that prevented the *Monterey* and its weather-beaten crew from joining the casualty list attributed to Typhoon Cobra, a Category 3 tempest that severely tested "Bull" Halsey's seafaring skills in December 1944. Over the years the aggressive Halsey has been harshly criticized for blindly sailing

---

* At the time of Conover's death in 1965, it was said that the business at its height generated $2 million a year.

his fleet into the most devastating of storms. In fact, Halsey did the best he could with the meteorological information at hand. In that pre-satellite era, it was more crude guesswork than scientific surmise. Estimates of the storm's whereabouts varied, as did predictions of its strength and future course.

It was first detected by weathermen at Pearl Harbor on December 16, just as the *Monterey* and the rest of Halsey's fleet completed three days of intensive bombing of the Philippine island of Luzon. Within twenty-four hours barometers were falling rapidly in the central western Pacific. Yet even as conditions deteriorated, forecasters downplayed the threat. They were confident that any storm would recurve to the northeast, following the normal path taken by Pacific cyclones, leaving Halsey's fleet unscathed. It was said of Halsey that he understood only one thing—*attack*. His aggressive tactics caused him to pursue a Japanese decoy fleet, leaving MacArthur's beachhead at Leyte Gulf dangerously exposed to a much larger enemy armada. By mid-December much of Halsey's flotilla was running low on fuel. While some ships were refueled on the seventeenth, Halsey had particular reason to be concerned about the destroyers in his command. Top-heavy under the best of circumstances, when deprived of oil in their holds to serve as ballast they were especially vulnerable to high winds and water.

Whipsawed by constantly changing weather reports, Halsey repeatedly altered the proposed rendezvous site for his ships to refuel. The *Monterey* was more fortunate than most. Not only had it succeeded in taking on fuel from a storm-tossed tanker; it was able to help other vessels do the same before an angry sea ruled out such assistance. The clock struck twelve, ushering in the eighteenth of December, a new day that would be long remembered by everyone in the Third Fleet who survived until the next witching hour. On the bridge of the *Monterey*, Ford took the midnight to four a.m. watch. By now the ship had begun to roll from side to side, its lights flickering ominously. On the lower decks men in their bunks "couldn't help but wonder," said one, "just how far she could go over and still come back." The sounds of crashing dishes and utensils in the galley were suddenly displaced by an unearthly "wap, wap, wap" coming from a stern pitching so violently fore and aft that the ship's screws were frequently out of the water.

"That night was hell," Ford would recall. No Lake Michigan gale could have prepared him for winds that eventually exceeded ninety knots, or sixty-foot waves that put the flight deck under water. "If I was ever scared in my life I was scared then." Adding to the sense of helplessness were the sounds of twisting metal—of planes and the tractors and jeeps used to pull them around, self-powered engine cranes and bomb-hauling equipment. Bad as things were, they might have been even worse but for the foresight of Captain Ingersoll. During the gray and ominous seventeenth, he had ordered extraordinary precautions implemented on the flight and hangar decks. The ship's full complement of planes was secured with three-quarter-inch steel cables. Fuel tanks were emptied as a safeguard against fire. Now, as darkness fell, gale-force winds buffeted Halsey's flagship the USS *New Jersey*. The admiral, unable to sleep, resigned himself to a delay in the Luzon bombing campaign he had promised MacArthur. Of more immediate concern was the fate of the Third Fleet.

Elsewhere on the *New Jersey*, meteorologist George Kosco leaned into the winds raking the navigation deck. Having denied the possibility for two days, Kosco belatedly conceded the existence of a potentially lethal storm, though even now he was reluctant to call it a full-fledged typhoon. Orders for the fleet to rendezvous were canceled. In their place Halsey directed his ships to refuel by whatever method was feasible (for example, taking oil over the stern of the vessel rather than the conventional side-by-side arrangement). Unfortunately, contradictory instructions sent the fleet closer to the eye of the storm. On the bridge of the *Monterey* Gerald Ford peered through a driving rain, struggling to make out the profile of another ship. A frenzied ocean no longer supported bobbing hulls, but instead attacked them from crazy angles, isolating some, pushing others desperately close to sister ships, sweeping men and planes into the sea.

Gravity was repealed in those predawn hours. "We had a chief petty officer that washed overboard," recalls one member of the *Monterey*'s crew. "The next wave caught him and washed him back aboard." Leaving the bridge around four o'clock, Ford made for his bunk on the same level as the hangar deck. He wasn't there long. The ship's daily routine commenced at sunrise with a call to general quarters. No one was going to see the sun this morning. After half an hour or so on the bridge, Ford returned to his stateroom, hoping to get some much-needed rest. All his life he possessed the ability to close his eyes and instantly fall asleep. December 18, 1944, was no exception. One level down from where he dozed, crew members worked frantically to remove ammunition cans from the magazine, passing one hundred thousand rounds from one sailor to another, until at last they could be tossed overboard.

Shortly after eight o'clock, the general quarters call was again sounded. A bugler blew "Fire Call" over the public address system. "Fire on the hangar deck," an excited coxswain announced. "All damage control crews man your stations." Ford bounded out of his sack. "The minute I woke up

I could smell smoke." A plane on the hangar deck had snapped its steel cable, touching off a chain reaction that quickly engulfed its neighbors. Theoretically emptied of their fuel supplies the day before, most of the F-2 fighters and torpedo bombers contained enough gas—Ford guessed five gallons—to start their engines. As the storm worsened and the ship's roll became even more pronounced, one aircraft had crashed into another, sending sparks and igniting whatever gas or gas fumes remained in the area. Flaming gasoline in turn spread the fire belowdecks, accounting for the smoke that greeted Ford as he threaded the narrow passageway on the starboard side leading forward and outside to a catwalk that surrounded the flight deck.

He had just scrambled onto the rain-lashed surface of the deck when the ship lurched violently, causing him to lose his footing and skid a hundred feet or more toward the port side of the vessel. In that instant, Ford came closer to death than at any time when under fire by Japanese Zeros; closer, in fact, until the day in September 1975 when a disciple of Charles Manson dressed like Little Red Riding Hood, her girlish hand wrapped around a Colt M1911 pistol, approached the president of the United States outside the California state capital in Sacramento. Instinct took over. "I spread myself out to try and slow myself down or stop," Ford would recall. "On every aircraft carrier there's a ridge"—technically called a toe rail—"about two inches so that tools and people won't slip off. I hit that with my feet and thank goodness twisted and fell in this catwalk. Instead of going right over the side."

More than once that morning Ford had occasion to repeat the Bible verse he had learned as a child and that was to serve as a lifelong mantra in time of peril. *Trust in the Lord with all thine heart and lean not unto thine own understanding. In all thy ways acknowledge Him, and He shall direct thy paths.* Even as a shaken Ford retraced his steps and reported to the bridge, several planes on the flight deck broke their moorings in the hurricane-force winds and went overboard. On reaching the bridge, Ford sensed the most immediate threat was to those in the bowels of the ship who were trying to maintain the boilers that powered the *Monterey*'s engines and supplied pressure to the firehoses deployed against the flames above. As flying wreckage punctured ventilator tubes, toxic clouds of smoke were sucked into the boiler and engine rooms through vents designed to funnel fresh air. Blinding smoke overcame firefighters and stokers alike. Dozens suffered burns or smoke inhalation. Three died.

With three of its four boilers no longer functioning, the *Monterey* was nearly dead in the water. Critically low water pressure hampered the fire crews in their efforts to contain the blaze. Meanwhile, the hangar deck became a shooting range as aircraft in their fiery death throes sprayed 50-caliber bullets through the air. Men on the flight deck burned holes in its surface in order to train water hoses on the inferno below. By now it was a few minutes after nine o'clock. Up on the bridge Ford was part of the little group that heard Halsey authorize Captain Ingersoll to abandon ship. The cruiser *New Orleans* and several destroyers were directed to lend what assistance they could to the floundering *Monterey*. "We can fix this," insisted Ingersoll. With assistance from Pappy Atwood, he put the ship on a new course, one that allowed her to ride more easily in the heaving sea.

Ingersoll dispatched Ford below to assess the situation on the burning hangar deck. Against all odds, a single boiler continued to function, enabling fire crews to gradually get a handle on the flames. By nine forty-five, Ford was able to report the fire under control. Fifteen minutes later it was extinguished. As the plunging barometer bottomed out at 28.79 inches, shrieking winds forced those on the bridge to crane their necks and look *upward* to catch the next wave as it broke over the *Monterey*. Seemingly everything that was not lashed down, and much that was, had been consigned to the Pacific. More than half the ship's aircraft were destroyed by fire. The rest, twisted and pummeled beyond recognition, were either blown or pushed overboard. Still the "Mighty Monty" had defied expectations. "It was a lucky ship," concluded Ford.*

The same could not be said of the destroyers *Hull, Monaghan,* and *Spence.* They capsized in the turbulent seas, taking with them nearly eight hundred men, the loss of whom inspired a naval court of inquiry that largely exonerated Halsey for "errors of judgment." In addition to the three destroyers, seven other ships were extensively damaged, and 146 airplanes lost. At midday on December 19, a funeral service was conducted on the *Monterey* for the three sailors killed in the storm. Bruised and blackened, its hangar gutted and mechanical systems ruined by salt water, in January 1945 the *Monterey* limped into a dry dock at the navy yard in Bremerton, Washington.

---

* Some confusion exists regarding Ford's activities that frantic morning of December 18. In *Halsey's Typhoon*, their otherwise authoritative account of the storm, Bob Drury and Tom Clavin assert that he led a fire brigade that rescued crew members overcome by smoke or flames. Ford himself made no such claim, either in his memoirs or in a telephone interview conducted in 2004 by author Drury. Naval historians can find no evidence of such a foray. Instead, they buttress Ford's own account of being sent by Captain Ingersoll to evaluate the hangar deck fire—a dangerous enough assignment, without which Ingersoll could hardly have resisted Halsey's invitation to abandon ship.

FORD DID NOT accompany his battered ship to the West Coast. Leaving the crippled *Monterey* at Saipan, he took a fourteen-hour flight to Pearl Harbor, with a Christmas Day stop for refueling on Johnson Island. An even longer journey transported him by way of San Francisco to Grand Rapids, where he enjoyed a brief reunion with his parents. No doubt there was as much talk of local politics as of global strategizing. Jerry's return came hard on the heels of Frank McKay's defeat in the most recent GOP primary. It coincided with fresh disclosures of corruption in Lansing, where public outrage was sparked by the discovery of $1,350 slipped into a legislator's suit coat pocket, along with a note instructing him to vote against pending legislation opposed by chain banks.

In the ensuing uproar Judge Leland Carr, a letter-of-the-law jurist with aspirations to sit on the state supreme court, was impaneled as a one-man grand jury to investigate official wrongdoing. True to form, Michigan attorney general Herbert Rushton tried to sabotage the probe by naming as grand jury prosecutor a Grand Rapids lawyer close to McKay. Judge Carr would have none of it. Appointing his own prosecutor, he chose Kim Sigler, a flamboyant Battle Creek attorney said to possess forty-seven suits, a profile reminiscent of the great Barrymore, and presidential ambitions rivaling those of that other Michigan-born gangbuster Thomas E. Dewey. Not content to shoot fish in the legislative barrel, Sigler had bigger targets in mind. In December 1944, McKay was indicted on bribery charges relating to legislation that would have increased the state's take from horse-racing and pari-mutuel betting at the expense of his mob allies.

The investigation quickly widened to include the banking industry. Here Sigler's progress was slowed by spotty memories and a well-founded fear of reprisals. A vice president of Michigan National Bank, long suspect because of his lobbying activities, drew a blank on the witness stand. Next up was Earl Munshaw, a Grand Rapids state senator, who returned home after testifying and left his car in the garage, with the motor running, long enough to prevent his being called back. The president of the Star Transportation Company drove his car into an oncoming freight train rather than give evidence about the monopoly his firm enjoyed on all state liquor shipments. Finally Sigler caught a break in the form of Warren G. Hooper, a forty-year-old Republican, freshly elected to the state Senate after three terms in the Michigan House.

At first glance Hooper, the latest in a bipartisan parade of political mercenaries who acknowledged selling their votes in committee or on the floor, hardly seemed worth the effort. Under intense grand jury questioning, however, the senator from rural Albion admitted to having pocketed $500 for voting as Frank McKay demanded on a horse-racing bill. What's more, Hooper was prepared to implicate McKay in the attempted bribery of legislators by the state's chain banks. Speculation regarding his testimony ran rampant, nowhere more so than in Grand Rapids. "They said if he is called and opens up, you will see something hit the fan," according to McKay antagonist Doc VerMeulen. "Well, he was called up on a Wednesday." Whatever Hooper said, "it so upset Judge Carr that he ordered the state police to give him round-the-clock protection until Monday morning when he could appear before a court of record." Hooper brushed the offer aside. "Nobody is going to bother me," he told Carr and Sigler. "I don't want any police officers hanging around me with their guns."

The next day, January 11, 1945, Hooper was driving his green Mercury along a familiar stretch of M-99 pointing toward Albion, where he hoped to forget his troubles during a family weekend with his wife and two young sons. It was approximately four thirty in the afternoon as he approached the Springport grain elevator, a familiar landmark on the otherwise featureless route leading south from Lansing. Suddenly a maroon-colored car blocked the road in front of him. Forced off the highway by two other vehicles, Hooper was shot three times in the head with a .38-caliber pistol. His car was set ablaze, quite possibly by a cigarette dangling from his now-silenced lips. It was discovered, still burning, by the feed elevator operator. In response the state posted a $25,000 reward—the largest in Michigan history—supplemented by $5,000 from the *Detroit News* for information leading to the arrest of Hooper's assassins.

The late senator's death car was not the only thing going up in smoke. Having lost his chief witness against Frank McKay, Kim Sigler scrambled to redirect public attention toward his broader probe of statehouse corruption. On January 27, days after Jerry Ford left Grand Rapids to take up a post-*Monterey* training assignment in California, Judge Carr unsealed indictments on twenty state legislators and half a dozen finance company officials. Before they were through, the Carr-Sigler team combined for sixty-seven convictions, among them a former lieutenant governor, twelve state senators, eleven representatives, and dozens of lobbyists, police officials and prosecuting attorneys. Notably absent from the list was Frank McKay, whose attorneys persuaded a Jackson County judge in February 1946 that their client's manipulation of the state Liquor Control Commission was not criminal but entrepreneurial. McKay just wanted to sell more liquor, ruled the judge, and to claim a higher share of the proceeds. Pure Michigan.

His latest Houdini-like escape did nothing to allay suspicions that McKay had a hand in the senator's violent death. In Grand Rapids, Gerald Ford Sr., newly installed chairman of the Kent County GOP, purchased a revolver for protection. He stashed the gun under his pillow, at least until Dorothy Ford insisted he hide it in a safer place beneath a floorboard in the attic. In May 1945 four Detroit youths, all associated with that city's Purple Gang, were charged with conspiracy to commit murder (no one, then or since, has been accused of actually pulling the trigger that ended Warren Hooper's life). Hastily convicted, each of the men was sentenced to five years or less in Jackson County's Southern Michigan Prison. Subsequent investigations, the most thorough of which produced the 1987 volume *Three Bullets Sealed His Lips*, pointed the finger of blame at the Grand Rapids power broker whose reign, reputation and freedom were all at risk should Senator Hooper live to testify in open court.*

Though McKay would steadfastly deny any involvement in the crime, he could not prevent Grand Rapidians from speculating about his possible complicity in Warren Hooper's death. If nothing else the timing was awkward, given McKay's ongoing efforts to avenge his recent primary loss to Home Front reformers. The stage was set for a bitter rematch, pitting the erstwhile party boss against the Gerald Fords—father and son.

AFTER TWO MONTHS in California, the younger Ford was reassigned to the Naval Reserve Training Command in Glenview, Illinois. It was there, in the Chicago suburbs on April 12, 1945, that he learned of the death of his commander in chief. "I'm no Roosevelt man in domestic politics," Ford told a friend on hearing the news, "but it didn't help us any internationally at this time to have the old boy pass on." The fragility of life, even in peacetime, was brought home to Lieutenant Ford six months later when he joined his commanding admiral, like Ford an ardent football fan, in flying to Chapel Hill and a game pitting Navy against UNC. The naval pilot at the controls fought his way through a driving rainstorm, only

---

* In the scenario of authors Bruce Rubenstein and Lawrence Ziewacz, the murder plot may have been hatched in McKay's office, but it was implemented by Jackson Prison administrators, who agreed to release designated convicts in civilian garb long enough for them to eliminate the threat posed by Senator Hooper (after the fact, it was asserted that the maroon vehicle spotted blocking Hooper's car on M-99 belonged to a deputy warden). Their mission completed, the killers returned to the prison, where they disposed of the incriminating murder weapon and license plate before resuming their customary places in time for the next roll call.

to land on the wrong runway. The plane plunged down an embankment before coming to rest in a grove of trees. Along with everyone else on board, Ford scrambled to escape the wreckage, seconds before it burst into flames. For the second time in less than a year, he had defied deadly odds—first at sea, and then in the air. He was about to discover if his luck held on land, in love and in politics.

5

BETTY

Dancing is discovery, discovery, discovery.

—MARTHA GRAHAM

I N FEBRUARY 1946, Ford was discharged from the navy with eight battle stars to his credit and the rank of lieutenant commander. Within a
month of his return to Grand Rapids, honoring the pledge he had made
to his stepfather while still on the *Monterey*, he became president of the
Republican Home Front organization. As a faithful reader of a new weekly
journal, *World Report*—forerunner of *U.S. News & World Report*—Ford
had come to believe that the United States could not repeat its post–World
War I withdrawal from global responsibility. He said as much to Senator
Arthur Vandenberg, who had preceded Ford in shedding his isolationist
skin. Behind the scenes, the powerful lawmaker and the ambitious young
reformer were to forge a working alliance based on shared enmities no less
than mutual interests.

Meanwhile, Ford had to make a living. In contrast with many returning
GIs, his job prospects could scarcely have been more promising. Indeed, he
was still in uniform when a January 15, 1946, notice announced his association with Butterfield, Kenney & Amberg, the town's preeminent law firm.
For this he could thank his former law partner, Phil Buchen. As the war
drained the legal system of its most promising young talent, it left Buchen,
permanently exempted on account of his childhood polio, struggling to
meet local demand. Approached by the Butterfield firm about a permanent
position, Buchen conditioned his acceptance on assurances that a similar
invitation would be extended to Jerry Ford at the end of hostilities. Alternatively, Buchen told Ford, the old team could be reassembled practically
overnight should he prefer to resume their private practice.

What followed might be considered Ford's greatest stroke of luck yet,
surpassed only by the two-inch toe rail encircling the flight deck of the
Monterey, the existence of which made possible his survival to practice law

according to the demanding standards of Julius Amberg. Few who crossed paths, or swords, with Amberg would dispute Ford's characterization of his latest mentor as "a brilliant man, and a great public citizen." The only surviving son of a well-connected German Jewish family, Amberg had been groomed for success from the cradle. In high school and again in college he finished first in his class. He repeated the feat at Harvard Law School, where he also edited the *Law Review* and served as valedictorian to the Class of 1915. During the First World War, his Harvard classmate Felix Frankfurter enlisted Amberg's help in resolving labor disputes for the War Department. Returning to West Michigan and marrying into a locally prominent family, Amberg became the proverbial useful citizen who never let private interests blind him to public obligations.

A classic type A personality, in constant motion dictating to a secretary while twirling his Phi Beta Kappa key on a gold chain, Amberg scarcely hid his impatience with those who fell short of his intellectual criterion. According to his own law firm's history, this was almost everyone. Widely admired for his skills as a litigator, in 1939 Amberg was elected president of the Michigan Bar Association. Two years later he was persuaded by another Harvard classmate, Robert Patterson, to return to Washington as a special assistant to Secretary of War Henry Stimson. For this second round of wartime service, Amberg received the presidential Medal for Merit, at the time the nation's highest civilian award. Still in Washington when his law firm designated Phil Buchen a partner in 1943, Amberg welcomed the possible addition of Buchen's friend and associate Jerry Ford to the Butterfield roster of lawyers. While they couldn't make him a partner immediately, he told Ford, such a promotion would follow in short order.

But it was another Amberg pledge, one having little to do with money or professional status, that sealed the deal. "I want to train you to be a good lawyer," Amberg assured the younger man. That was all Ford needed to hear. Amberg's command of juries, his personal integrity, and his near-obsessive approach to the unglamorous tasks of research and preparation could not fail to make an impression on an aspiring lawyer. Although a lifelong Democrat, Amberg had much in common with Arthur Vandenberg. Patriots before they were partisans, both men demonstrated a capacity for creative statesmanship unconfined by party dogma. Each in his own way was to serve Ford as a role model. That Ford never became a partner in Amberg's firm had less to do with his legal skills than with the growing conviction held by all three men that he could make a larger contribution outside the courtroom.

COMMENCING IN FEBRUARY 1946, Ford worked overtime to validate Amberg's judgment of his abilities. "That guy would get down here at seven in the morning or earlier," recalled Niel Weathers, a Ford contemporary who also noted how quick his colleague was to assist newcomers. "I've never seen a fellow more at ease with all kinds of people." Weathers soon found himself, much to his surprise, pounding the streets and ringing doorbells for Republican candidates. "Jerry was a very enthusiastic, warm, magnetic guy," he explained. "And if he said, 'Let's do this' or 'Would you like to help me?' your immediate tendency was to say yes until you got in so deep you had to call a halt to it."

Only in retrospect did Weathers conclude that Ford's geniality masked a fierce ambition. "His law practice was a means to an end." Competitive as ever, Ford attached himself to a group of young lawyers who regularly met for a brisk game of squash or handball, followed by lunch at the University Club. If he couldn't beat them, he joined them—literally. By one count Ford served on the board or in a leadership position of thirteen civic organizations, from the Boy Scouts to the Humane Society. He joined the Kent County Farm Bureau and the NAACP; lent his name to the United Way and Family Services; and took an active part in the Junior Chamber of Commerce, which reciprocated by naming him its Man of the Year in 1948. He also lobbied city officials (unsuccessfully) for a new public museum, headed the drive to build a suitable war memorial, and volunteered his time to the Legal Aid Bureau.

As chairman of disaster preparedness for the local Red Cross, Ford dealt with a major hotel fire and multiple floods by a rampaging Grand River. Not content to run a small fleet of rescue boats on the city's waterlogged West Side, he recruited former army cooks to prepare meals for hundreds of residents plucked from their homes. At the behest of his close friend and fraternity brother Jack Stiles, Ford turned the newly formed Independent Veterans Association into a potent advocacy group and service center to help returning warriors adjust to civilian life. He organized the All-Veterans Housing Committee to press elected officials, Realtors, bankers and builders for affordable housing and the revision of local zoning laws stacked against first-time homeowners. Thanks to the committee's efforts, construction codes were modified to reflect new techniques and materials and thereby hasten the building process. Grappling with civilian bureaucracy for the first time, Ford undertook constituent services of the sort that would later be his congressional bread and butter. He signed up for the

American Legion, drafted a constitution and bylaws for the rival organization known as Amvets, and became a familiar face around VFW Post 850, located in a Polish neighborhood known for its loyalty to Frank McKay.*

Between work at the law firm and his multiplying commitments outside the office, Ford had little time for a social life. By 1947 all three of his stepbrothers were married and raising families, while Jerry, aged thirty-four, was living with his parents in a tidy brick residence in East Grand Rapids. "When are you going to start dating again?" his mother asked him. Repeatedly. "When are you going to settle down?" Ford did not lack for female company. Arguably the town's most eligible bachelor, "there were a lot of girls that were chasing him," according to Betty Ford. By his own estimate, he averaged a date a week. If he pursued prospective brides less fervently than legal or political soul mates, some assumed it was because he was still in love with Phyllis Brown.

Evidence of this came shortly after his return to Grand Rapids from the war. "I don't think I had even met Betty yet. I was sort of playing the field." One day he was surprised to receive a letter from his old flame— she remembered it as a phone call—containing news of her pending divorce. Ford made plans for the pair to get together. He had his plane tickets bought and paid for when he got a last-minute call from Lewiston, Maine. "You're doing what you want to do and I'm doing my life and shouldn't start this over again," Phyllis told him—though not, it might be added, on her own initiative. Phyllis blamed her change of heart on her mother, who had chastised her for yanking Ford's emotional chain. Jerry dryly observed that it was the first time Phyllis had ever listened to her mother.

By the fall of 1947, Ford was ready to settle down. When the publisher of the *Grand Rapids Herald* asked if he would help lead a Cancer Society fundraising drive, Jerry agreed after receiving assurances that he wouldn't have to do all the soliciting himself. For organizational assistance he could count on his friend Frank Neuman, an insurance salesman happily married to a vivacious girl named Peg. Ford arranged to meet the Neumans at their home after work. By nine p.m., their philanthropic labors concluded, he raised a subject of more personal interest.

"Who's around that a bachelor my age can date?"

As it happened, Peg Neuman had a candidate in mind.

---

* What most impressed Ford about the group was not its politics but its thirst. "Boy, those guys knew how to drink," he remembered. Their favorite concoction, a boilermaker and a helper, consisted of "a shot of whiskey and a glass of beer."

Much of the city was preparing to turn in for the night, but that didn't prevent Ford from enlisting Peg Neuman to run interference for him with Betty Bloomer Warren. A phone call was duly placed, Peg handed the phone to Jerry and the forceful young advocate made his pitch. Wouldn't Betty like to join him for a drink?

"Oh I can't do that. I've got a style show tomorrow."

She would get her work done a lot better, Ford insisted, if she went out for "just one drink." Betty reminded him that she had not yet finalized her divorce. As a lawyer "you ought to know better than to do that." Not to worry, Ford persisted, they could escape detection at "some little place around the corner." Betty let herself be persuaded to take a short break. Twenty minutes, she told herself. No more. Half an hour later they were chatting animatedly in a booth at Scottie's, a popular watering hole on Division Avenue. Though the encounter may have been unplanned, it wasn't quite the bolt out of the blue described in later accounts. According to her lifelong friend Lillian Fisher, Betty had first noticed Jerry Ford one summer day at Ottawa Beach ("I won't tell you what she said"), the lakeside colony favored by Ford and his classmates during the short Michigan summer. As an adolescent Betty made trips to Ann Arbor in order to cheer on the All-City athlete, five years her senior.

Their unscheduled date at Scottie's had its antecedents in a cocktail party hosted by one of Jerry's U of M classmates in the spring of 1947. Casting a competitive eye on Jerry and the woman with whom he came to the party, Betty couldn't help asking herself "how *she* ever got a date with him." Date or no date, Jerry had taken the occasion to engage Betty in a lengthy conversation. Believing a spark had been struck, she was disappointed by his failure to follow up. Indeed, several months were to pass before Ford intruded on her preparations for a Herpolsheimer's department store show benefiting the local Babies' Welfare organization. A few days after their impromptu meeting at Scottie's he invited her to dinner and bridge at the lakeside cottage of Phil Buchen and his wife, Bunny. It was the first of several semiclandestine visits to the Buchens, whose beach place held far fewer prying eyes than downtown Grand Rapids.

Not until Betty's divorce came through in September were they free to go public with their budding courtship, if that is what it was. "It was a very peculiar romance," Betty recalled. "Because he was always busy with meetings" he rarely crossed her threshold before ten in the evening. Keenly aware of the potential for gossip that neither wished to encourage, Jerry was careful to leave Betty's apartment on Washington Street by twelve. The first time he visited her there he knocked over a coffee table with flowers on it. "Oh great," thought Betty, "this football player is *sure* dumb." She soon changed her mind. For one thing, there was Jerry's persistence. One evening he planted himself on her living room sofa and read the newspaper, oblivious to the presence of several other candidates for Betty's affections. Only one of these remained by the time she announced her intention, the next day being a workday, to turn in for the night.

Once outside, Jerry demanded to know his rival's intentions, even as Betty eavesdropped on the full exchange through an open window. Secretly delighted, she was unprepared for Jerry's ardor. Much as they enjoyed each other's company, they told themselves, neither was looking to get serious. As for marriage, vowed Betty, "I certainly had had my fill of it." That fall the couple enjoyed several football weekends in Ann Arbor. Winter sports brought them even closer. Together with some friends, Jerry had bought a small parcel of land near Cadillac, 120 miles north of Grand Rapids. Everyone joined forces to build a rustic cabin-cum–ski lodge, with eight army cots arranged around a potbellied stove, and the nearest john a fifty-yard dash. "That didn't seem like a long way in the summer," Ford remembered, "but when it was ten below it was a hell of a long ways away." The skiing, with old-fashioned rope tows, was hardly worth the effort. But as a sign of her growing investment in their relationship, Betty got all the necessary equipment and taught herself the sport.

That Christmas she presented Jerry an overstuffed gift stocking. Inside was a cigarette lighter inscribed, TO THE LIGHT OF MY LIFE. At the start of 1948 a trial separation tested their feelings. Betty was in New York, representing Herpolsheimer's at a buyers' market. From Sun Valley, where he was enjoying a prolonged ski break, Jerry wrote her every day. Holed up in a Junior League Room at the Waldorf, Betty pondered her options. "This is either going to have to lead somewhere," she thought, "or I'm going to have to put an end to it." That she found Ford physically attractive went without saying. Beyond this, she was drawn to Jerry for his honesty and work ethic, maturity and sense of purpose. She admired him as an achiever whose burgeoning civic activities overlapped many of her own interests (on discovering they both belonged to the Urban League, they attended the group's meetings together).

At the same time, he was fun to be with. Much as she looked forward to raising a family, Betty welcomed a partner with whom she could go places and enjoy an active social life. It hadn't taken long for her to conclude that Jerry Ford was "the strongest man I have ever known." Unfortunately his strength was matched by his silence. "I was very immature in male-female relations," Ford acknowledged late in life. "I learned later that in this situation you must say you love her." Betty attributed his reticence to having grown up in a family of boys, with a "business-like" mother made uncomfortable by open displays of affection. He might be undemonstrative, but Jerry's cool self-possession neatly complemented Betty's more volatile temperament.

On at least one occasion she nearly put a premature end to their relationship. Still in New York during the first week of 1948, she made plans at Jerry's behest to meet the artist Brad Crandell at his studio overlooking the East River. Perversely, Crandell arranged for their visit to be interrupted by none other than Jerry's old flame Phyllis Brown, in Betty's indignant description "a gorgeous blonde, skinny as a rail," who peeled off a mink coat to reveal "a black satin dress cut down to her bottom rib."

"Oh, you're Jerry's new girl!"

At that moment the visitor from Grand Rapids in her maroon-colored suit and striped taffeta petticoat felt like country come to town. It was an impression Phyllis did nothing to alleviate as she fired a volley of questions confirming her prior claims to Jerry, as well as her intimate knowledge of Grand Rapids and the Ford family. That evening Betty vented her feelings. In a scorching letter to Jerry she denounced Crandell, Phyllis, and their entire circle. If they were the sort of people he wished to associate with, so be it, she fumed; he could count her out. Regretting her outburst almost as soon as she posted the letter, fearful that Jerry might be scared off by her ultimatum, Betty spent a sleepless night.

She needn't have worried; Ford was the last person to act in haste. Once reunited in Grand Rapids, he presented Betty a hand-tooled leather belt with silver buckle as a souvenir of Sun Valley. But he wouldn't give her the confrontational scene she desired, if only to clear the air. Instead their conversation was exactly what one might expect from a pair of recently separated lovers. It was what went unsaid that finally drove Betty to mention her letter with its criticism of Jerry's New York friends. Had he even received it?

"Yes I did. I gather you didn't care much for them."

Nothing more was said, though years later Jerry claimed to have the letter still, carefully preserved in his safe-deposit box.*

---

* Betty Ford got comedic revenge of sorts at a post–White House golf tournament named for her husband. Conversation after dinner inevitably turned nostalgic. At one point the former president ruminated on a handful of youthful dates he'd had with Mary Idema, whose father helped to found the office furniture giant Steelcase, and whose husband, Robert Pew, had since led the company to a position of global leadership.

Betty was unimpressed.

"Just think, Jerry," she remarked. "If you'd married Mary Pew, you could have been president of Steelcase instead of president of the United States." AI, Marty Allen.

BETTY FORD NEVER hid her admiration for Eleanor Roosevelt, the White House activist whose willingness to break with demure tradition in the 1930s had made her an iconic, if polarizing, figure among modern First Ladies. Indeed, she listed FDR's wife, along with her own mother and modern dance legend Martha Graham, as the most influential figures in shaping her identity. On first blush, comparisons may seem strained between the emotionally bruised ER and the outwardly sunny Betty Ford, whose mother liked to say that she had popped out of a bottle of champagne. On closer examination, however, one finds significant parallels between these reluctant First Ladies, who only belatedly came to appreciate the unstated power of their position. Each had a complicated childhood, with an alcoholic father loved from a distance before his premature death and a perfectionist mother who sent conflicting messages while implanting lifelong feelings of inadequacy. As an adolescent, the orphaned Eleanor Roosevelt fled to Europe, where she gained confidence from a free-thinking French headmistress, Marie Souvestre. In exchanging Grand Rapids for New York, Betty Bloomer found a similarly unconventional teacher and role model in the avant-garde dancer Martha Graham.

The parallels are inexact; for one thing, Betty's mother did not, like Anna Roosevelt, die before her daughter's ninth birthday. And Betty's elevation to feminist heroine came relatively late in life, just as her most significant impact in dispelling the stigma attached to alcohol and drug dependency occurred after she left the White House. In her own words "an ordinary woman who was called onstage at an extraordinary time," little about her early years hinted at the drama to come. Elizabeth Ann Bloomer was born in a Chicago hospital on April 8, 1918, the third and final child of William and Hortense Bloomer. Her Hoosier father, forty-three at the time of her birth, was a traveling salesman hawking conveyor belts for the Royal Rubber Company.

Bill Bloomer's family emulated his nomadic professional life. Besides the infant Elizabeth—the formal name vanished as quickly as the ink dried on her birth certificate—there was Bill's wife, Hortense, and their two sons, William Jr. and Robert. Together they lived briefly in the Chicago suburb of East Rogers Park, before moving to Denver. Betty's earliest memories were of a large frame house on Fountain Street in Grand Rapids, to which her family relocated when she was two. Bill Bloomer was a shadowy presence in his daughter's life, his frequent absences spotlighted by the nightly letters Betty's mother wrote to her distant spouse. Occasionally, with no more explanation than her missing him, Hortense—"Horty" to friends—would pack her bags and go to Bill, wherever he might be. It never occurred to the little girl with the Dutch-boy haircut that these impromptu journeys were in response to his emotional needs. Only after her father's death did Betty learn of his chronic alcoholism, a condition that would be replicated in Betty and her brother Bob.

Bloomer tried to compensate for his vagabond existence with gifts purchased on the road. Betty's favorite was a brown teddy bear to which she clung in place of the parent who gave it to her. Even when he was home, Bill Bloomer traveled in his imagination, as he sat hunched over a crystal radio set, adjusting dials until he picked up signals originating in Chicago—a discovery excitedly shared with the rest of his family. Summers were spent in a cottage at nearby Whitefish Lake, where Betty and her siblings perfected their swimming skills and Bill fished for walleye, bass and northern pike to the dismay of his youngest child, for whom a monotonous diet of lake fish bred a lifelong aversion to almost anything with fins or scales.

Far more vivid was the figure Hortense Neahr Bloomer cut in her daughter's memories. Decades after her mother's death, Betty still dreamed about the woman of whom she claimed that her own greatest desire was to be half as perfect. Later still, having reexamined her own upbringing and priorities as part of her treatment for alcohol and substance abuse, Betty acknowledged the price of meeting maternal expectations. Only Hortense Bloomer herself had paid a higher price. "There were so many things missing in her life." In her White House memoir *The Times of My Life*, Betty quoted Tennessee Williams's poignant characterization of a deserting husband as "a telephone man who fell in love with long distances." She had known men like that, the former First Lady acknowledged, "and they didn't all work for the telephone company."

Foreshadowing Betty's experience as a virtual single parent while her congressman husband took to the road, Mrs. Bloomer shared her gritty formula for raising the children of a traveling salesman. "If you have to do it," she said of her role, "you have to do it." No less than her friend Dorothy Ford, Hortense Bloomer kept up appearances. "A very dignified lady," recalls one of Betty's closest friends. "Regal. I mean, you stood up when she walked into the room." A music lover, Mrs. Bloomer enriched the cultural life of West Michigan by recruiting opera singers to perform on local stages. She frequently reminded her family of larger cities in which she had lived before coming to Grand Rapids, cosmopolitan centers like Chicago and Seattle. Presumably it was in these breeding grounds of refinement that Horty developed her exacting standards. At her dinner table, it was considered gauche to butter an entire slice of bread at one time. Apples were looked on as horse feed. The chewing of gum was likewise banned in her presence.

A vocal advocate of women's rights, Hortense nevertheless insisted that her tomboy daughter don a hat and white gloves when she went shopping. Other constraints would have consequences that can only be imagined. As a child Betty sometimes wandered away from the family's summer cottage to a nearby picnic ground, where she indulged her sweet tooth by accepting cookies and cake offered by strangers. To stem the flow of calories, Hortense resorted to humiliation, finally hanging a sign around her daughter's neck reading, PLEASE DO NOT FEED THIS CHILD. The incident carries faint echoes of the lifelong stigma imparted by Eleanor Roosevelt's mother, herself a celebrated beauty, who shamed her plain daughter by calling her "Granny."

None of this prevented Betty Ford, in *The Times of My Life*, from depicting hers as a happy childhood, free of complication. From an early age Betty identified with the underdog, invariably throwing herself into the fray on the side of whichever of her brothers was losing a wrestling match to the other. She often accompanied her mother, president of the Crippled Children Association of Grand Rapids, on visits to the Mary Free Bed Hospital. As a teenager entertaining youngsters in braces, her sympathy for their plight was equaled by the joy she felt in her own unrestrained movement. "Dance was my happiness!" she exclaimed. It would also be her escape to places more worldly than Grand Rapids. At crucial times it afforded her a means of support and self-expression. Most important, it introduced her to a surrogate mother, an austere genius even stricter in her demands than Hortense Bloomer.

Dance taught Betty the value of a disciplined body and a mind to match. It made her every bit the athlete that Jerry Ford was. Simultaneously it forced her to confront painful limitations, and the inescapable difference between good and good enough. Her passionate interest in the subject was kindled when, a child of eight, she enrolled in the Calla Travis Studio of Dance on Fulton Street. Her eponymous instructor was an elderly woman with dubious red hair and flashing castanets, the latter used to impose order on unruly children in ankle socks and black patent Mary Janes. Betty made an equally vivid impression on her peers. "Oh my, she's pretty and she's not too tall and she can kick," gushed one of her classmates after their first encounter in Calla Travis's ballroom. One of those rare youngsters who feel comfortable executing a foxtrot with a third-grade contemporary, Betty graduated from social dancing to study ballet, tap, acrobatics and Spanish dance. (She would have liked to take piano lessons as well, she acknowledged, but the onset of the Great Depression left her family's finances sadly depleted.) Scoring a ninety-seven in technique, Betty also impressed everyone with her costume sketches. On the other hand, she would never make the ballet corps ("I couldn't get my knees straight enough"). For the obligatory solo ballet number required of all Travis dance students, Betty designed an outfit that covered her knees as she whirled around the stage to a Tchaikovsky waltz.

Flowers didn't dance at Central High School. Like every girl in her class Betty was required to enroll in sewing and cooking classes. Neither brought out her best. She claimed to enjoy school, especially languages and math, though her fluctuating grades offer little evidence of sustained scholarship. Extracurricular activities engaged her more fully than classwork. She joined a sorority—informally dubbed the Good Cheers—that undertook a fundraising campaign for a French orphanage. Notwithstanding her evident curiosity about the world beyond Grand Rapids, Betty was excluded on account of her sex from the Central High debating society known as the House of Representatives. A contemporary edition of the *Helios*, the school yearbook, did little to encourage professional, let alone political, aspirations in Betty Bloomer or her female classmates. "In pioneer days, the man put powder in his gun and went out to look after deer," the editors observed. "Nowadays the deer puts powder on her face and goes out after a man."

An early notice in the *Grand Rapids Press* testified to Betty's popularity among her peers. She accepted her first ring from a boy—a ten-cent faux-diamond number from Woolworth's—at the age of ten. Thereafter she collected and returned a veritable hoard of fraternity pins. "We were always telling the girls to wait and see what boys were going to ask Betty for dates on Friday and Saturday to see who's going to be left over for the rest of us," Lillian Fisher recalls. Inevitably much of this adolescent flirtation transpired on the dance floor. Here Betty was in her element. With Miss Travis as impresario, she performed solo in a Mardi Gras–themed number, and danced the part of Salome in a homegrown Passion Play. One less-than-stellar turn elicited a maternal rebuke that might well have served as Hortense Bloomer's credo: "If you don't do it well," she admonished her daughter, "don't do it at all."

She did it well enough so that by the age of fourteen Betty was teaching the youngsters of Grand Rapids how to waltz and do the Big Apple. On Saturdays she earned $3 an hour as a teenaged model at Herpolsheimer's, eventually graduating to the rank of personal shopper. The extra money came in handy, given the onset of hard times after 1929. Summers at Whitefish Lake became a thing of the past, as did winter tobogganing parties at the local country club. Conditions in the Bloomer household worsened dramatically after July 18, 1934, a sizzling summer's day in West Michigan. Too far from Lake Michigan to derive any relief from its cooling breezes, Grand Rapids suffered through a humidity that wilted work clothes and the workers inside them.

Bill Bloomer was at home that Wednesday, the day before his sixtieth birthday. Excusing himself before the midday sun ruled out such exertions, he went to the garage to tinker on his car. Only the unexpected appearance of some out-of-town friends prompted Hortense to summon her husband from his labors. Entering the garage, she found its doors open to the outside, and the car's ignition turned on. Bill's lifeless body was under the chassis, a victim of carbon monoxide poisoning. Rumors of suicide bred like mushrooms in the overheated climate. Family members denied them all, citing insurance payments received by the dead man's widow. Still, it was a long while before the speculation subsided. Aside from a casual acknowledgment in her memoirs that "it was rougher for everybody after that," Betty avoided the subject of her father's death. In this as in so much else, she was her mother's daughter. Their shared characteristics almost guaranteed a clash of wills in the first meaningful test of Betty's independence.

Betty attended two graduation exercises in the spring of 1936. A traditional program at Central High School marked the end of her formal classroom education. That same month she performed for the last time as a student of Calla Travis. Dazzling the audience with her versatility, she danced the part of a clown in *Scaramouche*, and excelled in both classical routines and a modern dance ensemble. Far more than her Central High diploma, her Travis send-off pointed Betty in the direction she wanted her life to take. Other youngsters her age kept scrapbooks filled with pictures of Clark Gable or Joan Crawford, Lou Gehrig or Joe Louis. Betty Bloomer collected every story she could about the woman whose influence on twentieth-century culture would earn her comparison to Picasso, Stravinsky and Joyce.

"I worshipped her as a goddess," Betty wrote of Martha Graham. Forty years later, she still felt a shiver of awe on the night in October 1976 when Gerald Ford halted his reelection campaign long enough to present the nation's highest civilian award to the woman he credited with "using the human form to express human feelings." Graham came by her artistry as the daughter of a Pennsylvania doctor who taught her to look into microscopes and develop her lifelong credo: "Movement never lies." Early in the twentieth century, dance in America was defined by the synchronized high kicks of the Broadway stage, by vaudeville hoofers and ballroom exhibitions and Isadora Duncan and her Russian work songs. Graham charted a radically different path, starting with her seven-year apprenticeship as student and performer at the Denishawn School in Los Angeles. Her first New York recital in 1926 established Graham as an early exponent of what later generations would call "body language."

In later years she recalled visits to the lion's cage at the Central Park Zoo. For hours she studied the lion's movements, "those great padding steps" as he paced his narrow cell. "I learned from the lion the shifting of one's body. The shift of the weight is one key aspect of that technique, that manner of movement." Stripping dance down to bare essentials, Graham gave it a new vocabulary, infused it with political and social relevance and incorporated in her growing repertoire the cultural rhythms of African-Americans, American-Indians and others traditionally excluded from the recital hall. "I wanted to begin not with characters or ideas, but with movements," she would explain of her distinctive style. "I did not want it to be beautiful or fluid. I wanted it to be fraught with inner meaning, with excitement and surge."

Graham cut a striking figure in her tubelike jerseys and heavy gold kimono, black hair pulled back tight against her skull, her masklike face painted in the style of some ancient Egyptian priestess. One early reviewer likened her work to "motion pictures for the sophisticated." Broader audiences struggled to interpret the spastic motions resulting from her "contraction and release" system of muscle control. Nothing so invites caricature as originality. "Ugly girl makes ugly movements onstage," sneered one detractor, "while ugly mother tells ugly brother to make ugly sounds on drum." Other critics, scandalized by the pelvic thrusts of her overtly eroticized choreography, mocked Graham herself for being as sexless as Maude Adams in *Peter Pan*. Rejecting traditional composers, most of them French or Russian, Graham favored contemporaries like Aaron Copland (with whom she created her most popular work, *Appalachian Spring*), William Schuman and Samuel Barber.*

Her choice of composers reflected Graham's vision of American exceptionalism. "An American dance is not a series of steps," she contended. "It

---

* One reason for Graham's subsequent broad appeal was her eagerness to collaborate with such pop-culture icons as Halston, Donna Karan and Calvin Klein. Her influence extended to classicists Margot Fonteyn and Mikhail Baryshnikov, as well as the jazz-loving Twyla Tharp and iconoclastic Paul Taylor. Bette Davis credited Graham with helping her express character through movement; so did Kirk Douglas, Liza Minnelli, Gregory Peck and Madonna.

is infinitely more. It is a characteristic time beat, a different speed, an accent sharp and staccato."

Beginning in July 1934, Bennington College in southern Vermont offered her the ideal venue in which to expound her ideas. The Bennington School of the Dance pooled the creative talents of Graham and such leading exponents of modern dance as Hanya Holm, Doris Humphrey and Charles Weidman. Betty Bloomer studied under them all. Her original plan, broached to her mother when Betty was just sixteen, was to move to New York City, where she might pursue a career in dance under the tutelage of her heroine. Out of the question, Hortense Bloomer decreed, at least until her daughter's twentieth birthday. The Bennington School was a compromise acceptable to both women. Nestled between the Green Mountains of Vermont and New York's Taconic range, dominated by the soaring limestone shaft built to commemorate a Revolutionary War skirmish, Bennington was both history-drenched and a modernist magnet.

For six weeks each summer, 160 students from more than thirty states, Canada and Hawaii filled college dorms and tried, with varying degrees of success, not to offend Yankee sensibilities with their outdoor gambols and questionable wardrobe. After some locals confused flesh-colored leotards with the makings of a nudist colony, strict orders were issued—"Tunics are worn over leotards at all times outside of studies, and conventional dress is worn in public places off the campus." No one was happier to be at Bennington than Betty Bloomer, "Skipper" to newfound friends on account of the yachting cap she wore wherever she went. Nothing in her slightly priggish upbringing—at the time of her Bennington enrollment she neither smoked nor drank—dented her popularity. From her quarters in Woolley House, Betty enrolled in dance history and criticism, music composition and stagecraft.

Days were long, stretching from eight in the morning until dinner at six thirty, followed by an evening's worth of rehearsal and homework. "The dance is no easy solution to light entertainment," Martha Graham reminded her students. "It is not an effeminate art form." Betty could attest to this from personal, painful experience. Of her esteemed instructor she wrote, "If you got her knee in your back when you weren't sitting up straight enough, you never forgot it." Some notes she scribbled at the time convey the Graham regimen in its most basic form. "Sit on floor up straight. Place bottom of feet together and in front of you. Hang onto ankles and for 16 counts bend forward and touch head to feet with a bouncing movement."

Far from complaining, Betty said it was "ecstasy" to be able to dance eight hours a day, a view she clung to even after her muscles knotted up and she found herself unable to walk downstairs after dinner. Each year, one of the school's prominent artists conducted a workshop for approximately two dozen students, in the course of which a major new work was created, rehearsed and presented as part of a public recital in the local National Guard Armory building. During Betty's second year at Bennington she went through this cycle of composition and production under the guidance of Anna Sokolow, a Graham protégée who would go on to win acclaim for choreographing Broadway shows *Candide* and *Hair*.

Insisting that modern dance engage with offstage concerns, Graham rarely shied away from political controversy. In 1936 she rejected an invitation from Hitler's regime to perform in conjunction with that summer's Berlin Olympics. To do otherwise, Graham maintained, was to lend tacit support to Nazi persecution of German artists, many of them Jewish. A year later Betty's classmates raised $200 to assist the embattled forces of republican Spain. In dances like *Slaughter of the Innocents*, Anna Sokolow championed the workers' movement and antiwar themes. Her student Betty Bloomer, by contrast, minimized the role of politics in her upbringing. With a Democratic father and a Republican mother, any partisan loyalties in her household were bound to be muted.* Indeed, she wouldn't cast a vote until she was thirty and her husband was a candidate for Congress.

But that didn't absolve her from taking an active part in such organizations as the Junior League, which she joined after completing her first summer at Bennington. "I wanted to learn something about the welfare work of Grand Rapids," she explained at the time, "to help those needy who have problems affecting the community and cannot solve or adjust themselves." Among these challenges she mentioned unemployment, broken homes, child labor and racial discrimination. Her involvement with the League exposed Betty to the seamiest corners of urban existence. She encountered starving families in dwellings "shabby but neat," many threatened with eviction from homes already deprived of gas and light. The plight of a ten-year-old girl suffering parental abuse found its way into the same notebook in whose pages Betty defined "dance" as "movement in form with meaning."

Her return to Grand Rapids was short-lived. As a member of the Travis school faculty Betty led forty girls in a modern dance program that included *Three Shades of Blue*, part of a dance trilogy she choreographed to evoke Manhattan morning, noon and night. In March 1938, she was in the

---

* It bears noting that where religion was concerned, Betty opted to join the Episcopalian faith espoused by her mother in place of her father's Christian Science.

audience when Graham and her troupe performed in Ann Arbor. Going backstage after the show, Betty asked Graham to accept her as a student in her New York school. Her petition endorsed by Hortense Bloomer, Betty launched her assault on the Big Apple just as Jerry Ford was discovering its charms with the help of Phyllis Brown. That is not all she had in common with Phyllis. Since Graham couldn't afford to pay dancers a living wage, Betty supplemented her mother's financial aid by working as a model for John Robert Powers. In the summer she taught dance at a children's camp in Southwest Michigan.

Sharing a Chelsea apartment with a Bennington roommate, Betty ate frugal meals at the Automat. For more substantial fare she let college boys from Harvard or Princeton take her to dinner, with dancing to follow. One evening a self-professed talent scout gave her his card. He told Betty that he wanted her to take a screen test. Betty agreed to meet him after work one day, before second thoughts set in and she recruited a friend to go in her stead. In her estimation, Hollywood couldn't hold a candle to New York, a town that would always occupy a special place in her heart. If anything, she enjoyed it too much for her ascetic teacher. "You can't carouse and be a dancer, too," Graham reminded her. So while Betty's roommate was invited to join the main Graham company, the one that crisscrossed the nation in response to the growing popular interest in modern dance, Betty contented herself with a spot in the auxiliary troupe. She danced often before New York audiences, on one unforgettable occasion joining in a performance of Graham's signature *American Document* at Carnegie Hall.

All this won Betty recognition as "the Martha Graham of Grand Rapids," a title of little consequence beyond Kent County. Hortense Bloomer decided to pay her daughter a visit. Implicit in their original agreement was Betty's determination to become a great dancer, not one more gainfully employed New Yorker. For two weeks, Hortense regaled Betty with the latest weddings and showers among friends back home. She made sure to include her own remarriage in September 1940 to Arthur M. Goodwin, a Chicago banker and longtime family friend. By all accounts it was a successful match, with Betty genuinely happy for her mother, and equally grateful for the presence of a surrogate parent who returned her affection and respect.

Now Hortense offered Betty a fresh deal. Come home for six months, at the end of which, if Betty still preferred life in Manhattan, no one would stand in her way. Ultimately it was not a protective mother that put an end to her New York experiment, but Betty's own doubts regarding her commitment to the dancer's life. Dreams of a dance career were not entirely abandoned. Returning to Grand Rapids, Betty rejoined the Calla Travis faculty while teaching students of her own at $5 per ten-week course. She stepped up her work with disabled youngsters, learning sign language in the bargain. On Tuesday nights Betty braved her stepfather's disapproval to conduct classes in a predominantly Black neighborhood of Grand Rapids. For Travis she choreographed *Two Cigarettes in the Dark*—challenging stuff for audiences unexposed to modern abstraction. More shocking still was *Three Parables*, performed before the altar of the Fountain Street Baptist Church, and a congregation ill-prepared for its musical gyrations and scanty costumes.

Betty showed a rebellious streak in her romantic life as well. In the face of parental objections, she became seriously involved with William Warren, an insurance salesman who missed World War II on account of diabetes. "Bill Warren was one of the best looking men around town—like someone in an Arrow collar ad," a friend recalls. "And Betty was one of the most beautiful girls in town, but the marriage was a total personality mismatch." Even before their union was solemnized in April 1942, Betty entertained hopes of changing Warren's imperfections. In many ways these were the flip side of what drew her to him in the first place—his sense of fun and curly-haired blond good looks. Athletic and personable, Warren took to the road much as Bill Bloomer had before him. Consequently the newlyweds never really put down roots. In Toledo Betty taught dance and shot backyard rabbits for the dinner table. In upstate New York, she sorted spinach and peas in a frozen-food factory.

At length abandoning their nomadic existence, the couple had scarcely resumed living in Grand Rapids than Warren embarked on a career as a traveling furniture salesman. Betty went back to Herpolsheimer's, where she organized fashion shows with titles like "Poise Through Exercise." Over time her portfolio broadened to include the training of models, coordination of advertising—even the look of the store's display windows. Sadly, her professional success was not duplicated on the home front. Having forsaken the excitement of New York, Betty hoped to settle down and start a family, only to find no corresponding commitment in her playboy spouse. Three years into their marriage she concluded it was a mistake. "I was young and stupid," she later conceded. But if her judgment was flawed, her character was such as to make even Hortense proud. In the summer of 1945, Betty was writing her absent husband a "Dear John" letter terminating their relationship when she received a phone call from Boston. Bill was in a hospital there, in a diabetic coma and close to death. Betty raced to his bedside.

For the next two years, she put her own life on the back burner, nursing the man she had planned to divorce, visiting him when he was hospitalized, administering his insulin shots and supporting them both by adding YWCA dance classes to her responsibilities at Herpolsheimer's. Eventually, contrary to his doctors' predictions, Bill Warren recovered. Betty resumed the divorce proceedings so brutally interrupted by his illness. The grounds cited were the routine ones of "extensive repeated cruelty." She took away from the experience a feeling of failure. "For a long time afterwards, I felt very negative about men. I didn't want to have anything to do with them." Her reluctance could not have been too prolonged, for she knew, early in 1948, that she loved Jerry Ford. She didn't have to argue herself into accepting his proposal of marriage, made that February.

Yet one thing did strike her as odd. While there was nothing equivocal about Jerry's feelings for her, she was naturally curious when he suggested that they delay any wedding until sometime that fall. He could not say precisely when, Jerry told her, nor could he explain the reason for his vagueness. In fact, Jerry "What you see is what you get" Ford was about to spring a surprise on the local political establishment, and he wasn't letting anyone in on the secret until absolutely necessary.

6

TIME FOR A CHANGE

I like your style, young man, but Barney Jonkman is going to stomp all over you.

—Frank McKay to Gerald R. Ford, Summer 1948

Gerald Ford might never have run for Congress had he realized his original ambition. In December 1947 the death of Judge Joseph Gillard created a vacancy on the Kent County Probate Court. Ford hoped to fill it. He vigorously lobbied Republican governor Kim Sigler for the appointment, only to have a Grand Rapids rival carry off the prize. In retrospect, Ford considered himself fortunate. Had the job been offered to him, surely he would have accepted, with profound consequences for his place in history. "Because, you know," the former president wryly admitted, "once a probate judge always a probate judge."

Any regrets Ford might have had were quickly forgotten as he pondered a second campaign, more improbable than his chase after a seat on the bench. Throughout the winter of 1947–48, Ford weighed his prospects against Republican congressman Bartel J. "Barney" Jonkman, then in his fourth term preaching Dutch thrift and isolationist dogma of sufficient purity to satisfy the xenophobic *Chicago Tribune*. First elected in 1940, Jonkman had become a darling of the America First crowd by opposing that year's Selective Training and Service Act, a military draft designed to replenish the nation's depleted armed forces at a time when Hitler's armies strode the European continent and South America appeared vulnerable to Nazi subversion. In August 1941, Jonkman spearheaded congressional resistance to any extension of the mandated one year of military service. He and his allies came within a single vote of derailing American preparedness for a conflict increasingly viewed as unavoidable.

By 1947 both Ford and Senator Arthur Vandenberg had long since shed their earlier isolationism. Jonkman alone was unrepentant, clashing openly with Vandenberg, now chairman of the Senate Foreign Relations Committee, over the senator's support for the new United Nations. He denounced the European Recovery Program (ERP), more commonly styled the Marshall Plan, designed to contain the spread of Soviet-style Communism. The congressman had his own strategy for combating the red menace. As a self-professed "committee of one to investigate Communism in the State Department," Jonkman anticipated the more highly publicized anti-Communist crusading of Wisconsin senator Joseph McCarthy.

At sixty-three, Jonkman was yesterday's man to World War II veterans like Ralph Hauenstein, a former chief of intelligence on Dwight Eisenhower's staff during the run-up to D-Day. Returning home to Grand Rapids, Hauenstein shared his concerns with a group of his fellow officers. "I remember, it was right after the war . . . one of the first big meetings and we asked Barney to speak to us. He told us about how he flew over Alaska," then a territory as irrelevant to West Michigan as Tahiti. The discussion that followed reinforced Hauenstein's doubts. Having preserved Western civilization while letting the atomic genie out of its bottle, Hauenstein and his contemporaries were too focused on the developing Cold War to give much thought to Alaskan statehood.

Though few lawmakers were so confident of their hold on power, Jonkman was less invincible than he appeared. His close association with Frank McKay, once a source of political strength, was becoming a liability. The McKay organization had not had itself a good war. In 1945 the rival Home Front elected Gerald Ford Sr. as GOP county chairman. The following year McKay failed to prevent his courtroom nemesis Kim Sigler from winning the governor's office in Lansing. Still the machine retained vestiges of its former dominance. In George Welsh it had a pliant mayor in Grand Rapids City Hall, while Barney Jonkman's vote in Congress offset the bipartisan heresy of Senator Vandenberg.

Moreover, GOP voters in the Fifth District were not in the habit of turfing out entrenched incumbents, especially those with names suggestive of windmills and *Oliebollen* (Dutch donuts). As a prominent layman in the Christian Reformed Church, Jonkman appealed to his co-religionists inhabiting the fruit and dairy farms of rural Ottawa County, where the younger Ford enjoyed far less name recognition than in Grand Rapids. Even Phil Buchen offered only tepid encouragement to his onetime law partner. "You probably can't win," Buchen told Ford, "but lay the groundwork and see what happens." To manage his campaign, Buchen recommended Jack Stiles, their longtime friend and frat house brother from Ann Arbor. It was sound advice. Stiles, a frustrated author and prolific idea man, could be relied on for an unconventional outlook and boundless energy. His strategic gifts outweighed a mercurial temperament and fondness for alcohol.

Determined to avoid a split in the anti-Jonkman vote, Ford approached several potential rivals, deferring to each should he harbor ambitions of his own. One by one, mindful of the conventional wisdom that held Jonkman to be unbeatable, they demurred. Guaranteed a clean shot at the incumbent congressman, Ford and Stiles devised a game plan exploiting generational differences and encouraging Democratic voters to take advantage of the state's open primary law. Above all, they hoped to expose Ford to as many voters as possible, contrasting his youthful vitality and responsiveness with the chilly detachment of his opponent. Unable to outspend Jonkman— Ford's total campaign expenditures came to $7,000—the challenger would outhustle him instead. Exploiting the element of surprise, he was careful to delay any announcement of candidacy until just before the June 18 filing deadline.*

Ford's discretion concerning his marital plans was a matter of naked calculation. "There were those who said, 'Jerry, getting married to a divorced woman is not going to be the best thing for your campaign,'" he acknowledged. "Especially in that district." Much has been written, not a little of it tinged with condescension, regarding the Dutch influence on West Michigan and its politics. "If you were divorced, you were almost without a name," recounts one journalist familiar with the local culture, "particularly in the Dutch Reformed churches. And Catholics weren't fond of divorce, either." Hollanders frowned on Sunday politicking, and on seemingly frivolous entertainments seven days a week. "They never went to the movies or danced," claims another observer. As for commercial relations, "they didn't trade with you if you didn't have a Dutch name." (As late as 1965, lifelong friendships were severely strained, and charges of spiritual betrayal ruined many a conversation, when the locally dominant Meijer food and discount chain of stores became the first major business in the area to open its doors to Sunday customers.)

Sunday newspapers purchased on Saturday night were not read un-

---

* Over the years the secrecy surrounding Ford's intentions has been greatly exaggerated. For several months prior to his announcement he had been speaking before civic groups like the Grand Haven Rotary Club. In February 1948, its members heard the Grand Rapids lawyer strongly endorse the Marshall Plan committing the United States "to spend for peace about 5 percent of what we spent in an equivalent time for war." The same speech contained criticism of Congressman Jonkman for dubbing any European aid program "Operation Rathole." Political pundits were not slow to pick up the scent of a primary contest in the making. One scribe flatly predicted that incumbent Jonkman was in for "some tough going." "Ford Backs Europe Aid," *Grand Rapids Press*, February 10, 1948; Guy H. Jenkins, "Ford Eyes Congressional Seat of Jonkman," *Detroit News*, February 7, 1948.

til Monday, if they were purchased at all. After the *Grand Rapids Herald* folded, managing editor Werner Veit launched a Sunday version of the rival *Grand Rapids Press*. Soon he was visited by a delegation of eight Christian Reformed ministers. "They knelt before me and prayed for my soul because I was the worst kind of sinner," says Veit. "I caused other people to sin." That the Dutch might, in fact, be less dogmatic than their popular image was impressed on Veit after a rural paperboy took longer than his fellow carriers to complete his weekend delivery route. The fault lay not with the youth, but with customers who had identified for his benefit specific bushes and barrels in which to conceal the Sunday *Press* from the prying eyes of more observant neighbors.

A NOTHER SECRET, HARDLY better kept, involved Arthur Vandenberg. On the eve of the Republican national convention in June 1948, the senator met privately with Ford. He told the younger man that, although he could do nothing publicly for him in the run-up to the primary, he was nevertheless "very, very pleased" by Ford's decision to enter the race. Ford set about building a campaign organization. Art Brown, his former teammate on the South High Trojans, collected signatures on Ford petitions. At the Amberg law firm Niel Weathers became a one-man research bureau. Doc VerMeulen enlisted several noteworthy laymen in the Reformed Church. Eventually each of the city's precincts had at least two Ford workers distributing literature, with additional volunteers assigned to every veterans' post.

Then there was Ford's secret weapon. "You wouldn't believe the number of women who were out there working for Jerry," Jack Stiles boasted. "They really made the difference." Heading the list was Betty Warren. Initially skeptical of her fiancé's political aspirations—"I thought, oh well, he'll never beat Jonkman, who had been there for several years, and it's a Dutch town"—Betty changed her mind as she saw the impact of Jerry's tireless campaigning. She recruited models from Herpolsheimer's, and dance instructors like Kay Clark (then DeFreest), who attended morning meetings at which they offered frank assessments of Ford's performance as a candidate. His lack of polish was seen as evidence of sincerity. "There wasn't any way that you could look at that young man and not know that every word he said came right from his heart," Kay concluded. "There was no pretense there." If he didn't know the answer to a question, Ford was quick to say so, before educating himself on the subject thus raised, so as not to make the same mistake twice.

One Ford mannerism did come in for censure. "We kept praying he wouldn't use his hands," Clark said. "He was a great one for emphasizing everything with his hand motions." It was politely suggested that he might want to clasp them behind his back, or forcefully seize hold of the lectern. It was good advice, but Ford was loath to take it. "What are hands for?" he exclaimed in frustration. He never wholly conquered the habit.

Fittingly for a first-time candidate, many of Ford's supporters were political outsiders who saw themselves as bucking the establishment. Typical of these novices was Seymour Padnos, the son of an Orthodox Jew and junk peddler who, fleeing Czarist Russia at thirteen, had gone on to found a successful scrap metal company in the West Michigan town of Holland. The horrors of the Holocaust only strengthened Seymour and his brother, Louis, in their sense of public obligation. A youthful interest in politics had led to Seymour's election as a Republican state convention delegate while he was still in college. After the war he found himself drawn to the clean-cut lieutenant from Grand Rapids, whose upstart candidacy for Congress was an affront to the status quo.

"You've got to remember that I was part of a group of returning veterans . . . when I went to the American Legion I talked about Ford. My brother was a member of the VFW. These guys were activists—they were going to make some changes." Viewing Ford as a kindred spirit, the Padnos brothers slapped Ford signage on their company trucks, prompting a warning from Holland's mayor to their father.

"If these kids lose," Louis Padnos was told, "you may have some consequences to pay."

The signs stayed on the trucks.

UNFORCED ERRORS PLAGUED the Jonkman campaign. First the congressman got into a needless fight with Gerald Ford Sr., claiming he had opposed his son's decision to run. Not so, countered the elder Ford. "I told him in my opinion Jerry Ford, Jr. was old enough to make up his own mind." The challenger got another break when President Harry Truman called a special session of the Republican 80th Congress, all but daring it to make good on liberal promises contained in that year's GOP platform. With Jonkman tethered to Washington at a critical time in the campaign, Ford filled the resulting vacuum. In formal presentations before service clubs, farm groups, church meetings and labor organizations, he lashed out at Jonkman's isolationist voting record and prewar failure to embrace military preparedness, conveniently overlooking his own early flirtation with America First. At the dawn of the Cold War, Ford targeted a new enemy. "We cannot convincingly carry the torch for freedom against the Russian bully," he insisted, "unless we are adequately prepared from a military point of view." This meant a larger air force, higher pay for army reservists and a general reversal of the hasty demobilization ushered in with the surrender of Japan.

Yet for all his talk of reciprocal trade barriers and reforming the UN charter with its paralyzing veto accorded permanent members of the Security Council, Ford was careful to keep local concerns front and center. He blamed Jonkman for failing to include Grand Rapids in a major flood control survey by the Army Corps of Engineers. After attending a Chicago meeting at which Michigan farmers and resort owners protested the diversion of Lake Michigan waters to Illinois, he all but blamed his opponent for the lake's eroding shoreline. Jonkman, waking to the more immediate threat of political erosion, sought to moderate his image by endorsing a foreign aid bill he had only recently tried to eviscerate. He also supported a $65 million loan with which to construct a permanent home for the United Nations in New York. Beyond this, however, the incumbent stuck mostly to platitudes: economy in government, reducing high prices through "all institutional means," a "sound agricultural policy," etc.

Both sides played hardball. Anticipating the much later vogue for so-called fake news, McKay supporters accused Gerald Ford Sr. of profiting from reform politics through contracts for his paint company with the city of Grand Rapids. RETURN AN EXPERIENCED FAMILY MAN IN CONGRESS, Jonkman's newspaper ads proclaimed in a none-too-subtle reference to the challenger's unmarried status. Ford countered with billboards featuring his picture and the exhortation ELECT JERRY FORD TO WORK FOR YOU IN CONGRESS. These were intended to raise his visibility while contrasting the Dutch work ethic with Jonkman's reputed sloth. Evidence that the Ford campaign was getting under the congressman's skin came at an early candidates forum, where even mild criticism caused him to snap. "Jonkman lost his temper and just flew into a rage right on the stage," according to one of the meeting organizers. Then he stalked out of the room. His subsequent refusal to debate Ford—"There are no issues in this campaign," he maintained—invited fresh censure.

Jonkman could dodge his challenger's request for public debates, but he could scarcely hope to match him in vigor or personal rapport with the voters. Ford scored points with his relentless canvassing at factory gates, Kiwanis lunches and church socials. At least three times a week he visited rural parts of the district to chat up farmers on their tractors and dairymen milking their herds. "Got seven farms to do before eight o'clock" that morning, Ford informed Jerald terHorst, a reporter for the *Grand Rapids*

Press who tagged along with him one rainy day in June. At the first of these outposts, Ford whispered the owner's name to terHorst, adding "four votes in the family" before asking the journalist, sotto voce, "Now what kinds of cows are these?" The farmer, surprised to encounter a candidate for Congress in his dairy barn, especially in a rainstorm, listened as Ford acknowledged his agricultural shortcomings. Then, deftly changing the subject from himself to the potential constituent in his soggy boots and coveralls, Ford made his pitch.

"I'm sure you've got problems that Washington doesn't know about or isn't paying attention to," he said. "Now if I were your congressman—instead of Mister Jonkman—what would you like me to be doing for you?"

A litany of grievances ensued, the farmer expounding on irritants from equipment shortages to the price of milk. When he was finished, Ford slipped his host a card with his picture on it. "Remember the name—Ford. If I get to Congress I'll remember what you've told me." And so it went at the next farm, and the next. By nine thirty that morning Ford was back at his office, having introduced himself to all of seventeen voters. A skeptical terHorst wondered how such a listening campaign could possibly overcome Jonkman's advantages. Maybe he couldn't win, Ford conceded as he dropped the reporter off at the *Press* building. "But you heard them say they never had a candidate come to their farm before. Doing it on a rainy day was better than doing it on a nice one. They'll spread the word, wait and see."

Ford was able to rise at five in the morning to cultivate the farm vote because his employer, Julius Amberg, shared his antipathy to Barney Jonkman. Early in the campaign, Amberg told Ford he need only come into the office one hour a day to satisfy his obligations to the firm. Meanwhile he drew his full salary. Amberg's gesture wasn't purely altruistic. Recognizing that no Democrat stood a chance against Jonkman in the fall, Amberg went out of his way to assist Ford. Their alliance was sealed by an incident symbolic of Ford's insurgency. Underfunded and dependent on free press coverage to communicate his message, Ford and his supporters installed a surplus navy Quonset hut, bedecked in patriotic red, white and blue colors, in a parking lot next to Wurzburg's department store in downtown Grand Rapids. The eye-catching structure had Ford's likeness painted on the front, while its sides were emblazoned with the message:

JERRY FORD, JR.

FOR

CONGRESS

Designed to call attention to itself, and to Ford's efforts to ease the post-war housing shortage for veterans, the Quonset hut was just as likely to ruffle feathers. After hearing from an angry Jonkman, McKay telephoned Wurzburg's store president, Fred Schoeck, to demand the hut be removed. Well aware that Wurzburg's ranked high on Julius Amberg's list of clients, McKay expected the eminent lawyer to put the squeeze on Ford.

Summoning Ford to his office for a fatherly chat, Amberg said, "This is very embarrassing to the firm."

Ford replied that they had two options. "I either remove the Quonset hut voluntarily or we force Wurzburg's to go to court to throw us off. Because we have a contract." Much as he regretted any distress he may have caused the firm, Ford continued, "I'm telling you right now we are not going to move that Quonset hut." His unyielding stance made people wonder if Ford didn't have some Dutch blood after all. As for Amberg, he was only too happy to communicate Ford's ultimatum to store management. As news of the incident spread, even McKay must have realized that the game was up. "Hell . . . we couldn't have gotten better publicity," Ford chortled. As an attention getter, the stunt certainly beat allegations that Jonkman was abusing his franking privileges.

Editors of the morning *Grand Rapids Herald* and of its afternoon rival, the *Grand Rapids Press*, took note. Reluctant at first to offend a sitting congressman heavily favored to win reelection, both men kept their preference for Ford to themselves. This changed in the campaign's closing days. Provoked by Jonkman's continuing abuse of Senator Vandenberg, *Press* editor Lee Woodruff tossed caution aside and endorsed the challenger. Jonkman responded by attacking, not his ballot rival, but the newspaper that had for so long clung to neutrality in the race. "And that teed off Lee Woodruff," Ford recalled with a grin, ensuring that "for the last five days I got all the good publicity." More endorsements followed, including one from Leonard Woodcock, a future president of the United Auto Workers, then a regional organizer for the union and its allies in the Congress of Industrial Organizations (CIO). Again Jonkman lashed out, this time denouncing his challenger as an agent of organized labor. Voters should be on their guard, said Jonkman, if they didn't "want the CIO to run the Fifth Congressional District."

His warning backfired, spurring Democrats in unprecedented numbers to vote in the September 14 Republican primary. Uncertainty over the outcome drove nearly forty thousand voters to participate in the GOP slugfest. Such a heavy turnout was thought to favor the challenger. Still, no one was prepared for the landslide that gave Ford 62 percent of the vote. On the morning after the primary he performed the signature act of the campaign, a shrewdly timed gesture that would be talked about for decades. Rather than sleep in following his unexpected triumph, Ford honored his pledge to a local farmer—win or lose, he had vowed to return after the primary and help the man milk his cows and scour his barns. That Ford kept his word was confirmed by Ellen Grooters, the farmer's daughter, at the time a high school student of sixteen. "I had breakfast for two weeks with Jerry Ford," she revealed half a century later. "He was very nice and very quiet and he liked to eat. I sat next to him."

T HIRTY YEARS AFTER his stepfather wed his divorced mother in a ceremony banned from the Episcopal Church they both attended, Jerry Ford had no wish to see history repeat itself. From the Episcopal bishop of Western Michigan, a fellow member of the local Yale Club, Ford learned how the stigma of Betty's divorce might be removed in time for an October wedding in the church. Her lawyer duly notified the Reverend Donald Carey of Grace Episcopal Church that Betty was the innocent party in her divorce from Bill Warren. Five weeks later, on October 15, 1948, Jerry and Betty stood before the altar at Grace Episcopal and said their vows. Contrary to persistent rumor, the groom in his gray pin-striped suit arrived only a few minutes late for the four o'clock ceremony.*

He had spent the day on the campaign trail, Ford explained, and his shoes were dusty. Worse, one was brown and the other black. "We had a small wedding and a big reception," Betty remembered. From Kent Country Club the newlyweds piled into the front seat of best man Jack Beckwith's car for the drive to Ann Arbor. Jerry and Betty spent that Friday night at the Allenel Hotel, a four-story brick structure dating to the 1870s. "You never stayed above the second floor at the Allenel Hotel," claimed Ford's bride, "because it was such a firetrap you wanted to be sure you could jump." In the morning the couple were treated to a brunch by Jerry's U of M friends, in town for the Michigan-Northwestern football game. Betty, exhausted by the strenuous celebrating, skipped the game's first half, though she would attribute the Wolverines' come-from-behind victory to her belated appearance at the start of the third quarter.†

---

* Though Betty, herself no stickler for punctuality, gibed that Ford was so tardy for his wedding she almost wed the best man.

† "I thought I was giving her a great treat," Ford said long afterward. "I paid for that a thousand times." Betty remembered things differently. "We politicked on our wedding weekend," she said. "We did all the things he wanted to do."

That evening she and Jerry joined a large, thoroughly chilled crowd in Owosso to hear Republican presidential candidate Thomas E. Dewey address an outdoor rally in the town of his birth. As his party's newly minted nominee in the Fifth District, Ford earned a brief introduction from the stage. It was nearly midnight when the rally concluded; the newlyweds had a ninety-mile drive ahead of them. Stopping en route for donuts and coffee, they arrived at Detroit's Book-Cadillac Hotel around three in the morning. On Sunday they took in a movie. First, however, Jerry prowled around newsstands, hoping to find his name in freshly printed newspaper accounts of the Owosso rally. On Monday the couple left Detroit, pausing briefly in Ann Arbor so that Jerry could meet with supportive members of the Michigan faculty. Nearing Grand Rapids, Betty looked forward to "a nice quiet evening at home." Her reverie was interrupted by a request from her new husband; he had a political meeting to attend at seven thirty. "Do you suppose you could fix me a bowl of soup and sandwich before I leave?"

It's not as if she wasn't warned. Shortly before the wedding, her prospective sister-in-law Janet had taken Betty aside for a discussion of potential rivals. "Jerry's mistress will not be a woman," Janet assured her. "It will be his work." Ford had his own version of this conversation, in which his friend Jack Stiles predicted for him a long, mutually satisfactory marriage, assuming Betty was willing to accept his "damned work ethic." Throughout their courtship she had grown accustomed to the last-minute substitutions and unexpected absences imposed by a political campaign. These included a wedding rehearsal dinner from which the groom departed after cocktails and did not return until dessert. "But I really didn't expect it to be that way when we got married. Like every woman I thought that when you sign that certificate and walk down the aisle, all of a sudden everything changes, and you have all his attention and regular hours. Well, that wasn't to be."

At once "shocked" and "exhilarated" by Jerry's victory in the primary, Betty had a head start on other congressional wives, given the overwhelming advantage enjoyed by Fifth District Republicans. Yet Ford had no intention of slacking off before the last precinct was tallied on election night. His Democratic opponent was Fred J. Barr Jr., a Grand Rapids theater owner who also happened to be friends with both Jerry and Betty. Barr blamed the high cost of living on the Republican 80th Congress. Beyond this, his command of the issues was rudimentary. Asked to outline his campaign platform, Barr replied, "Well, I got nine kids and I need the job." On November 2, 74,191 West Michigan voters checked Ford's name on their ballots, giving him 60.5 percent of all votes cast. His showing was all the more impressive given substantial Republican losses elsewhere. With a powerful assist from organized labor, Harry Truman upset the heavily favored Thomas Dewey. At the state level, Governor Kim Sigler lost to G. Mennen "Soapy" Williams, a liberal Democrat who was to occupy the governor's mansion for the next twelve years.

Nationally Democrats picked up seventy-five House seats, including two in Michigan snatched from GOP hands. As a consequence, little attention was paid to the freshman Republican from the Fifth District, one of just fifteen members of his party who were new to Capitol Hill (compared to a freshman class of 103 Democrats). At the moment, visibility mattered less to Ford than finding a place to live in a city where affordable housing was scarce. Fortunately, Senator and Mrs. Homer Ferguson invited the Fords to stay in their Washington apartment while house hunting. There, on the evening of November 19, Betty learned that her mother was in a Florida hospital with a condition initially diagnosed as Ptomaine poisoning. Betty booked the first available flight out of Washington. Delayed by mechanical problems, she arrived at the Fort Lauderdale airport three hours after Hortense Bloomer succumbed to a massive cerebral hemorrhage. A sustaining mix of faith and fatalism reconciled Betty to losing her lifelong role model and best friend. But it could hardly dull the ache of imagined future joys— grandchildren being only the most obvious—denied to both women.

In any event, Betty had little time to grieve. Married less than three weeks, she and Jerry must exchange familiar places, people and professional careers for a city whose customs and neighborhoods were equally foreign to them. Suitable accommodations had to be found in Washington, and a second residence established in Grand Rapids. Office staff was required to assist Jerry in the business of representation. There were issues to master, colleagues to cultivate and committee assignments to pursue or avoid (Ford had no intention of repeating Jonkman's politically harmful membership on the House Foreign Affairs Committee). Betty was in the House gallery on January 3, 1949, watching as her husband was sworn into office along with the rest of the 81st Congress.* Not everyone in the legislative family got the message; that same week a Capitol guard shooed the Fords off the outdoor platform built for Harry Truman's upcoming inauguration. Only members of Congress, he barked at them, were permitted such access.

By then Jerry and his designated administrative assistant, a thirty-two-year-old Grand Rapids lawyer and former speech instructor named John

---

* The House Class of 1948 was distinguished by more than its unusual size. Besides Ford, the newcomers included Eugene McCarthy of Minnesota, Abraham Ribicoff of Connecticut, Texan Lloyd Bentsen and New Jersey's Peter Rodino.

Milanowski, had personally scoured Room 321 of the Old House Office Building (after 1962 renamed in honor of Speaker Joseph G. Cannon). Until recently occupied by the departing Bartel Jonkman and his staff, the space was cluttered with stacks of agricultural yearbooks and planting manuals, the distribution of which it was supposed would guarantee Jonkman's continued tenancy. The voters of the Fifth District had decided otherwise. Without neglecting the self-promotional possibilities presented by the Government Printing Office, Ford had his own ideas on how to avoid Bartel Jonkman's fate. Milanowski, a Catholic of Polish descent, would be his unofficial emissary to Grand Rapids' West Side, more broadly to the non-Dutch working class and other voters loyal either to the Democratic party or to Frank McKay.

"He was sort of a conscience," Ford said of Milanowski, a tough marine who didn't hesitate to object when Jerry suggested Betty might join his congressional payroll. Doing so would run counter to Ford's essential beliefs about public service, Milanowski insisted. Ford let the matter drop. During his first week in office, he attended President Truman's State of the Union address, conferred with Senator Vandenberg, and treated Michigan newspaper correspondents to lunch in the House Dining Room. Ford's maiden speech was a low-key affair, opposing repeal of an amusement tax on tickets to the forthcoming inaugural ball. The time would come when taxes of any kind outraged Republican doctrine. In 1949, however, fiscal responsibility was defined as raising sufficient revenue to pay for services demanded by the electorate. In any case, balanced budgets were holy writ in West Michigan, and no one expounded the gospel more fervently than the new congressman from the Fifth District.

"I cannot forget that 33 cents out of every dollar goes to the government in one form or another," Ford reminded his neighbors. "Anything that can be done to whittle down that expense without jeopardizing essential government functions should be done." The qualifying adjective "essential" hints at a pragmatism uncomfortable with labels. His early voting record reflected a middle-of-the-road outlook that, under Dwight Eisenhower, would be dubbed Modern Republicanism. Regarding public housing as a costly backup, justifiable only if private builders failed to supply market needs, Ford introduced legislation encouraging tenants and owners to set their rents instead of federal bureaucrats. Truman's cancellation of a Veterans Administration hospital planned for Grand Rapids came in for predictable criticism, as did the president's alleged indifference to the cost-cutting agenda of the Hoover Commission on government reorganization. Ford would have liked to reorganize Secretary of State Dean Acheson out of his job, if for no other reason than Acheson's stubborn reluctance to disown the alleged Soviet spy Alger Hiss.

Meanwhile, there was plenty in his early performance to justify muckraking columnist Drew Pearson's assessment of Ford as "a healthy replacement" for Bartel Jonkman. In his first significant test, Ford defied GOP leaders by voting to clip the wings of the powerful House Rules Committee, whose elderly mandarins had used their leverage to put a stranglehold on Truman's Fair Deal program. In an even greater break from orthodoxy, Ford added his name to a round robin, signed by eighty-three other House members, declaring support for "a world federation" realized through a strengthened United Nations. He might as well have proposed Esperanto as the official language of the United States.

A vocal proponent of the North American Treaty Organization, formed in 1949 to implement Truman's policies of collective security and Communist containment, Ford backed the president's Point Four initiative, under which the United States would share its technological and agricultural expertise with the underdeveloped world. On the home front, Ford cast repeated votes for civil rights legislation, against the poll tax and in favor of making permanent the wartime Fair Employment Practices Commission established under duress by FDR. He supported measures to prohibit segregation in veterans hospitals and the women's reserve of the Coast Guard. At the same time, he introduced legislation to raise the salary of FBI director J. Edgar Hoover, for whom he reserved some of his most lavish praise. Clearly the would-be G-Man of 1941 held no grievance against the agency that had turned down his application.

In his most personal gesture, Ford urged stiffer civil and criminal penalties for deadbeat fathers who crossed state lines to avoid court-ordered child support payments. Long relegated to the states, whose enforcement methods varied widely, the issue of child support did not become a federal responsibility until January 4, 1975. On that date, President Gerald Ford signed into law the Child Support Enforcement and Paternity Establishment Program. Of his own experience with parental neglect he said not a word.*

---

* On the same day, citing fears for the precedent it would set, Ford vetoed a bill that would affix his name to the Federal Building in Grand Rapids. "I would hope our Presidents will be remembered for their labors in building better Government," wrote Ford in his veto message, "rather than for their efforts in constructing public works projects to themselves." The renaming would take place, but only after Ford was out of office.

FORD'S DAYS BEGAN and ended at 2500 Q Street, a large brick apartment complex guarding the entrance to Georgetown. Thanks to the intervention of a Grand Rapids friend who worked for the insurance company that built the place, he and Betty were able to secure a one-bedroom apartment costing $80 a month, welcome news to a couple anxious to start a family on a congressman's salary of $12,500 a year. Their new home was a three-mile drive from Capitol Hill. Twice a week Ford covered the route on foot in fifty minutes. Approaching middle age, he maintained his college football-playing weight through frequent visits to the House gymnasium and a frenetic schedule governed by the buzzer that alerted members fifteen minutes ahead of a floor vote.

In his office most mornings by eight—"He was there early and stayed late," recalls secretary Anna Holkeboer—Ford could often be found at his desk on weekends as well. Frequently Betty joined him there, filing papers or just enjoying his company. Prospective Ford employees were informed that clock watchers need not apply. The congressman's work schedule was defined by his mail, with letters to be answered the day of their arrival, if at all possible, but in no case more than forty-eight hours after being received. Ford personally signed all outgoing correspondence. He also wrote his own speeches, as well as a weekly newsletter that emphasized his appreciation of constructive criticism: "One type of letter, however, burns me up . . . last week a letter came to the office which started out like this: 'Show your backbone if you still have one.' There was no signature at the end. If nothing else I would like to write this anonymous person that his Congressman recently had a physical examination and my doctor informs me my backbone is okay!"

No Ford newsletter was complete without a list of district residents who had recently dropped by Room 321. Few left without a souvenir of their visit. Ford kept a Polaroid camera mounted on a tripod in his inner office so that callers could have their picture taken with their representative. Women invariably received copies of the *Republican Congressional Cook Book*, while the entire family took away bumper stickers proclaiming, WE VISITED CONGRESSMAN JERRY FORD. Some guests were treated to lunch with Ford in the House restaurant. If he was unavailable, they might tour the FBI or be shown around town by Betty Ford, who made so many trips to Mount Vernon that she finally parked her charges at the main gate before retreating to the parking lot with a book in hand.

Backed by his staff of three, Ford established a reputation for constituent service second to none. Together they chased down missing Social Security checks; sliced red tape for small businesses; and promoted would-be government contractors based in the Fifth District. A Pentagon job was found for a Grand Rapids man recently graduated from Gallaudet University. A West Michigan couple was able to adopt a three-year-old French boy because Ford legislated the child's designation as a displaced person. Another couple, a polio-stricken veteran and his Irish war bride, were reunited after Congressman Ford personally accompanied the petitioning husband on his rounds of the Washington bureaucracy.

Recognizing that only a handful of voters ever made it to Washington, Ford hit on the idea of a mobile office, a thirty-five-foot-long trailer outfitted to accommodate anyone in the Fifth District with a case to plead or a bone to pick. Its coming promoted in advance, the Ford caravan became a popular attraction parked on a village main street or agricultural fairground. Legal pad in hand, Ford took notes as local officials lobbied him for federal dollars to build a water pipeline. A visiting group of eighth graders got help for their government class. Postal employees demanded justification for his vote against their 11 percent pay raise. The traveling office came to symbolize Ford's attentiveness to those who had sent him to Washington.

Anyone scratching his head in puzzlement over President Ford's premature liberation of Poland from Soviet domination in a critical 1976 election debate against Jimmy Carter need only scan Congressman Ford's remarks delivered on the floor of the House on September 1, 1950. He had chosen the eleventh anniversary of Hitler's invasion to contrast Poland's vanished democracy with the tyranny imposed on her and other Eastern Bloc nations by their Soviet overlords. "Is there any hope for a free Poland?" Ford asked his House colleagues. "Certainly no one would be sufficiently naïve to assume that the Polish people have embraced the tenets of Communism, nor would any of us further assume that the country has resigned itself to domination by Soviet Russia."

Even as he worked overtime to bond with the Fifth District, Ford played the inside game with surprising skill for a Washington rookie. As a rule, young men in a hurry found the House an uncongenial environment. "We are just worms here," complained Ford's House Office Building neighbor John F. Kennedy. The dashing representative from Boston was a frequent visitor to Room 321, where he chatted up John Milanowski, his fellow Catholic and bachelor. Then in his second term, Kennedy made no attempt to conceal his boredom with the arthritic procedures of Capitol Hill. He took particular exception to the seniority system that made longevity the chief criterion for advancement. "You can't get anywhere," said Kennedy. "You have to be here twenty years."

This was fine by Ford, who had every intention of making the House his career. Indeed, while still in his first term he settled on his ultimate goal, an ambition shared exclusively with Milanowski.

"We talked about how nice it would be for Jerry to be Speaker someday," Milanowski recalled. "That's where all the patronage was, the power."

Ford's gift for attaching himself to older, potentially helpful patrons paid rich dividends with Earl Michener, a veteran legislator entering his final term representing Southeast Michigan. Distilling the experience of thirty years, Michener told Ford he could become the proverbial workhorse respected by colleagues for his prowess and institutional allegiance. Or he could raise his external profile by pursuing an outsider's track, passing up the drudgery of committee work for more glamorous speechmaking and legislative dramatics.

For Michener the choice was obvious. "Pick something on which you want to be expert, so that when you speak on it, people listen."

Ford's initial assignment, as a junior member of the Committee on Public Works, carried little prestige and less clout. To be sure, his subcommittee on flood control and beach erosion enabled him to make good on campaign promises relating to the Grand River and Lake Michigan. And he never forgot his presidential tour of a crumbling White House in April 1949, capped off by Harry Truman's personal pitch for $5 million with which to rebuild the old house from the inside out. Exposure to sagging walls and a piano threatening to crash through the ceiling of the East Room convinced Ford that the residence of Jefferson, Jackson, Lincoln and both Roosevelts should not be replaced but restored.

Meanwhile he seized on the earliest opportunity to disprove John Kennedy's lament about the powerlessness of novice lawmakers like himself. His immediate target was John Rankin of Mississippi, a notorious bigot and anti-Semite in office since the Harding Administration. Not content to label Albert Einstein a "foreign-born agitator," Rankin blamed US battle losses in World War II on the alleged cowardice of Black soldiers. The most flagrant of congressional race baiters, as chairman of the House Veterans Committee Rankin was also the most tireless promoter of bonuses for veterans of both world wars. Specifically Rankin wanted to send every veteran over sixty-five a monthly check for $90. First-year costs of the program were estimated at $2 billion, a figure sure to rise with the aging of the warrior generation.

Horrified at the fiscal implications, junior Republican congressmen Glenn

Davis and Don Jackson, themselves veterans of the recent war, organized a campaign of opposition. One Wednesday afternoon at five o'clock, Jackson welcomed to his office thirteen of the most promising younger members on the GOP side, Jerry Ford and California's Richard Nixon among them. This was the nucleus of the Chowder and Marching Club, an old-boys' network for not-so-old boys. With their garish chefs' hats and slapdash organization, Chowder and Marching seemed an unlikely axis of power. Even its name invited ridicule ("We never ate chowder and we never marched," Ford pointed out). Yet in March 1949, its members succeeded, by a single vote, in defeating Rankin's pension plan.

Emboldened by their triumph, they evolved into the most desirable of House fraternities, the primary training ground for Republican leaders in Congress and beyond. Each year by secret ballot a handful of talented newcomers were invited to join their ranks. Geographically and philosophically diverse, the Chowder and Marching Club would eventually nurture the careers of three presidents, five vice presidents, plus dozens of senators, governors, Cabinet officers and judges.

WHILE JERRY ACCLIMATED himself to the ways of Capitol Hill, Betty Ford embarked on her own crash course in government. "I had to grow with him," she later said of her husband, "otherwise I would have been left behind . . . I have a lot of steel in my spine, too." Besides attending congressional hearings and sessions of the Supreme Court, protocol required that she pay a formal call, between the hours of four and six on prescribed days of the week, on the families of all twenty-seven members who served with her husband on the Public Works Committee. These demands left little time for homemaking. "My duties as a housewife," Betty told a Grand Rapids audience in the autumn of 1949, "are restricted mainly to keeping my husband's suit pressed." An African-American cleaning lady ("Miss Ida") helped out one day a week. She was succeeded by her daughter-in-law, Clara Powell, who quickly made herself indispensable, remaining with the Fords for the next twenty years.

Not all their time was taken up by official responsibilities. A typical night out featured dinner at the Great Falls Hot Shoppe, home of the eighty-five-cent Mighty Mo hamburger. Betty joined a dance class led by the wife of Connecticut congressman John Davis Lodge and became program director of the Congressional Wives Prayer Group and secretary of the Congressional Club, a bipartisan organization of spouses whose activities ranged from bridge games to Red Cross bandage making. Here she befriended contemporaries like Pat Nixon, Lady Bird Johnson and Muriel Humphrey. Called on to present a book review before the group, Betty suffered through her presentation, after which she enrolled in a public speaking class. On an inclement day in January 1950, she joined other congressional wives for tea at Blair House, the government guesthouse occupied by the Trumans during the White House renovations. Outside the rain came down in buckets. More than a bit nervous over her first encounter with a president's wife, a heavily pregnant Betty was instantly put at ease by her hostess. To Betty's mumbled words of appreciation, Bess Truman replied, "How nice of *you* to come out on such a rainy day." The exchange stayed with her, an object lesson in how to remain grounded amid the pomp of high office. Neither woman had ever expected to inhabit the White House. Both succeeded by simply being themselves.

Concerned about a possible miscarriage, Jerry insisted that his wife escape the debilitating heat of un-air-conditioned Washington by spending part of the summer of 1949 at her stepfather's Lake Michigan cottage. He, in turn, flew home most weekends to be with her. Her due date came and went. Two more weeks passed. Then a third. Physicians at Washington's Doctor's Hospital finally took matters into their own hands. Jerry was sent home, to be notified the moment the tardy infant made his long-delayed appearance in this ultramodern "hotel for the sick." On March 14, 1950, Betty got her wish with the birth of Michael Gerald Ford. Not only was her first child healthy, but any concerns she may have felt regarding Jerry's paternal instincts were quickly proved groundless. The new father lavished on his son all the attention and loving pride so conspicuously lacking in Leslie King Sr. "He'll be a center in the first University of Michigan football game in 1970," Ford predicted.

When a drivers' strike curtailed milk supplies in the nation's capital, mother and son were beneficiaries of some constituent service in reverse. A Grand Rapids dairy dispatched twenty quarts of refrigerated milk to their Washington apartment so Mike Ford wouldn't go thirsty. The baby's father did his share of diaper changing. Beyond that, his domestic talents were, by his own admission, severely limited. He demonstrated as much in hanging a screen door upside down. He wasn't entirely helpless. Once there were lawns to mow, Jerry mowed them without complaint. He could always be counted on to wash and dry dishes, skills he first mastered as an undergraduate working his way through the University of Michigan.

Betty had moved to Washington having no idea how long they might remain. Early signs pointed to a lengthy stay. More than a face in the crowd, Jerry stood out as the only member of Congress included in the 1949 roster of the nation's ten outstanding young leaders as compiled by the Junior

Chamber of Commerce (Jaycees). His 98.4 percent attendance record placed Ford among the most dependable of lawmakers. His campaign for a second term coincided with the outbreak of the Korean War in June 1950. From the outset, Ford backed the Truman White House in resisting Communist aggression on the Korean peninsula. Unopposed in the Republican primary, he came out swinging in the fall contest. Challenging his Democratic opponent, Grand Rapids attorney James H. McLaughlin, to debate the issues, he found McLaughlin no more eager than Bartel Jonkman had been to confront him one on one. Ford's total campaign expenditures amounted to $3,092. In November he defeated McLaughlin by a two-to-one margin.

Republicans across Michigan took note of his performance; privately, some speculated about his possible strengths as a statewide candidate. Ford had other ideas. He enjoyed his work in the House. Betty, too, had taken to her new life as a congressional spouse and mother. The couple looked forward to enlarging their family, just as Jerry anticipated assignments more gratifying than the lowly Public Works Committee. He couldn't imagine how soon his ambition would be realized.

Part II

EVERYBODY'S FRIEND

1950–1972

7

A MODERN REPUBLICAN

Do you need Papa's parakeet? He can say Hurray for Eisenhower.

—Dorothy Ford, to her son Gerald, July 10, 1952

GERALD FORD HAD every reason to smile as he entered the ballroom of Grand Rapids' Rowe Hotel on the evening of February 10, 1951. Fresh off his decisive reelection, Ford was already being touted as a potential candidate for the Senate in 1952. With increased visibility, he discovered, came demands to match. These included finding a Lincoln Day speaker of national repute, a celebrity statesman who could serve up sufficient oratorical red meat to energize the most sluggish of partisans. Ordinarily such a task might have been entrusted to Arthur Vandenberg. But the senator would not be attending this evening's festivities. At the moment he was confined to an upstairs bedroom of his house on Morris Avenue, his robust frame wasted by cancer. Originally detected two years earlier in his lung, the disease had more recently attacked his spinal column. Doctors hoping to dull the searing pain had unwittingly addicted their distinguished patient to morphine.

Mellowed by suffering, Vandenberg had set aside his senatorial dignity and shown Ford a warmth and vulnerability long hidden from the world. Indeed, to some observers his relationship appeared more paternal than professional. Vandenberg's wife, Hazel, had been equally helpful to Betty, notwithstanding her own failing health. In March 1950, the ailing lawmaker ensured that Jerry and Betty Ford were the only nonfamily guests invited to his birthday dinner. The knowledge that in Ford he had someone to whom his political legacy could be safely entrusted may have helped ease Vandenberg's intense suffering during his final weeks. For now his absence threatened to cast a pall over the approaching Lincoln Day dinner, increasing the pressure on Ford to secure a speaker of national stature.

The decision was made for him by the voters of California in November 1950. On the same day Ford's Fifth District neighbors awarded him a second term in the House, his ambitious colleague Richard Nixon breezed to victory in his race for a United States Senate seat against Hollywood actress turned congresswoman Helen Gahagan Douglas. Nationally recognized for his tenacious pursuit of former State Department official Alger Hiss—in Ford's words, "the leader in the crimson clique in the Department of State"—Nixon drew a capacity crowd to the Rowe ballroom, where he decried the Truman Administration for its seeming passivity toward Communist China, and mocked its handling of domestic subversion. Adding to the dramatics, halfway through Nixon's speech a blown fuse plunged the hall into darkness. Many a speaker would have been unnerved by such an interruption; Nixon passed it off with a joke. He displayed equal sangfroid in a lively encounter with two dozen Republican activists after the dinner. Amid a fusillade of questions about his role in the Hiss affair, the thirty-eight-year-old freshman senator made his case and kept his cool.

Ford was impressed. Reluctant for the evening to end, he mentioned a party at the nearby National Guard Armory. Would his guest care to put in an appearance? At the armory Ford witnessed a side of Nixon's personality, relaxed and sociable in this admittedly friendly setting, that would seldom be repeated in the years to come. Afterward the two men repaired to the home of Ford's parents (Dorothy Ford, then in Florida with her husband, later hung a sign over her four-poster bed reading, THE VICE PRESIDENT SLEPT HERE). Drinks in hand, their conversation turned to the 1952 election. Ford made no secret of his support, as yet unpublicized, for NATO Supreme Commander Dwight Eisenhower over conservative favorite Robert Taft. Nixon, as eager as Ford to reverse Truman's spending and economic policies, shared his host's skepticism about Taft's vote-getting ability.

The next morning Ford rose early and drove Nixon to the airport. By any standard it had been a highly successful visit for all involved. The party faithful of West Michigan were still buzzing about their dinner speaker, who clearly was headed for bigger things. Nixon left knowing that he had made a favorable impression, banking credits for later use. Ford solidified his relationship with a colleague he had known only superficially, mostly from meetings of the Chowder and Marching Club. What he had seen there suggested a man of mercurial temperament, one minute "joyous and very happy," banging out songs at the piano, and the next handicapping a forthcoming election or berating liberal opponents with grim intensity. At once gregarious and withdrawn, brash and reflective, in conversation "you couldn't tell whether he was listening to you or thinking about something else." Ultimately, Ford concluded, "his moodiness drained a lot from him," an estimate more revealing of the man who made it than the friend who inspired it. "Anybody who can get real down or real high," said Ford, "it takes something out of them physically or mentally."

T HE LINCOLN DAY dinner highlighted for Ford a season of professional recognition offset by personal loss. On January 23 his legal mentor Julius Amberg died. News of Arthur Vandenberg's death reached Capitol Hill on the morning of April 19, hours before General Douglas MacArthur, his legendary military career abruptly terminated by Harry Truman, was to address a joint session of Congress. Ford was in the audience that listened spellbound to MacArthur's melodramatic assertion that "old soldiers never die." His own tribute to Vandenberg was slightly stilted by comparison, perhaps because he found it harder to verbalize the loss of the closest thing in his life to a political father figure.

Refuting the cynical view of politicians as puppets on a string, dancing to the tune of some machine or corporate master, Vandenberg had gone his own way, said Ford, "intelligently independent and rightfully successful." On his office desk the senator displayed a sign declaring, AND THIS, TOO, SHALL PASS. The message was not lost on his Grand Rapids protégé. "Forget the glory of politics," Ford told his audience, "there's none in it. Just take it as you would any other difficult job, and if you like to work you'll be happy." Only the satisfactions of service could justify the insane hours, the mounds of correspondence and unending assaults on bureaucratic indifference. Whether in the office or on the campaign trail, it all came down to the human element. "An honest politician will rise or fall on the number of personal friends he can rally to his fold."

Embarking on his second term in the House, Ford had gained an unlikely mentor, as different from Arthur Vandenberg as chalk from cheese. John "Cash and Carry" Taber, senior Republican member of the powerful Appropriations Committee, was a celebrated skinflint. His crusty manner and tightfisted attitude toward federal spending clearly pleased Taber's upstate New York constituents, who had returned him to Congress fourteen times since 1922. Over the years he had achieved a place in the political lexicon; to "taberize" meant to drastically reduce what a president asked for. Harry Truman called it the Taber Dance, and with good reason. Taber's proudest boast was of slashing Truman's request for Marshall Plan funding nearly in half, compelling Arthur Vandenberg to battle in a House-Senate conference committee to restore most of Taber's cuts.

Appropriately, it was Taber's sleuthing after government waste that brought Ford to his attention. Tipped off to an Army Corps of Engineers officer in Rapid City, South Dakota, who was said to be building himself a $50,000 home at taxpayer expense, Taber smelled a scandal in the making. Lacking investigative staff of his own, he asked Ford if he could borrow John Milanowski, with his legal training, to probe the allegations and report back to the full Appropriations Committee. The resulting inquiry produced a series of recommendations designed to improve congressional oversight of the Corps of Engineers. It also reaffirmed Taber's hunch about Congressman Ford and his trustworthy deputy. Ford got a second break when Congressman Albert Engel cashed in sixteen years of seniority, and with it his seat on Appropriations, to make an unsuccessful run at the governorship of Michigan in 1950. Ford wasted no time in alerting Minority Leader Joe Martin to his interest in replacing Engel.

Around Capitol Hill it was said that Congress had three branches: the Senate, the House, and the Appropriations Committee. In the past the House committee had been chaired by the likes of Thaddeus Stevens, James Garfield and Joseph "Uncle Joe" Cannon. Occupying their place in 1951 was an irascible Missouri Democrat named Clarence Cannon. "He reminds you a lot of Scrooge in Dickens' play," Ford observed of Cannon. "Shrewd and cunning. Always manipulating things." For the most part Taber and Cannon worked in harness, notwithstanding a legendary argument in 1945 that had ended with the chairman bloodying Taber's lip. Almost as skeptical of government spending as his Republican counterpart, Cannon wielded a heavy gavel in opposing pork for everyone but the farmers in his rural district who produced the real thing.

Stringent criteria governed committee applicants. "They have to be willing to work," said GOP whip Charles Halleck. "They can't be lazy. They have to attend the meetings . . . And they have to be men of integrity." A bipartisan streak of fiscal conservatism bonded the group's membership. According to Halleck, "If we knew a fellow was openhanded, we wouldn't put him on the committee." That Ford could inspire equal confidence in globalist Arthur Vandenberg *and* penny-pinching John Taber suggested conciliatory talents of a rare order. Not every issue, however, lent itself to compromise. With the start of the 82nd Congress, Ford confronted an inescapable choice between conviction and ambition. At stake was a career-making promotion to the Appropriations Committee. Two years earlier, in his first recorded vote, Ford had joined a bipartisan majority of Northern Democrats and moderate Republicans to enact the twenty-one-day rule. Nothing less than a frontal assault on the omnipotent Rules Committee, often described as "the traffic cop" of the House, the new rule had made it easier for the full House to consider legislation not reported out by the Rules Committee within twenty-one days of its introduction.

Supporters of the rule change hoped to expedite the flow of legislative business. Southern Democrats and most Republicans had opposed it for the same reason. Charlie Halleck spoke for conservatives in both parties when he praised the Rules Committee for suppressing "unwise, unsound, ill-timed, spendthrift and socialistic measures." John Taber could not have put it better. Now a resurgent coalition of conservative Republicans and Southern Democrats moved to repeal the twenty-one-day rule. Facing the toughest vote of his fledgling career, Ford shared his dilemma with one person. "We discussed that for a good many nights," Betty would recall. "And I finally said, well frankly, Jerry, if you're not going to vote your conscience, you're no good as a congressman. And you might as well quit." Even if doing so risked his promotion to the Appropriations Committee, she argued, "you've got to vote for what you think is right."

On January 3, 1951, Ford did just that, as 1 of 180 lawmakers who voted, unsuccessfully, to retain the rule. Two days later he was summoned to a meeting with Taber. Once again Ford's luck held, his show of independence having only strengthened the older man's hunch about his personal and political character. On January 12, the Republican Conference ratified Taber's choice for the watchdog committee overseeing virtually every dollar of federal expenditure. So began an association, and an education, that would color the rest of Ford's career.

Taber's instincts were sound. For all his commitment to the internationalist agenda, and to foreign aid in particular, Ford fitted naturally the role of budgetary hawk. After a year on the job, Ford told the *Grand Rapids Press*, he blamed Washington's red ink on an ever-expanding federal bureaucracy, "the greatest lobbyists in the world." With a national debt exceeding $250 billion, the onset of fighting in Korea spotlighted Pentagon waste and pork barrel spending in general. Not content to reduce the budget for West Point's 150th anniversary to $5,000, Ford took to the House floor to fume over the car—"nothing less than a 210 horsepower sedan with white sidewall tires"—and driver provided to the architect of the Capitol. Henceforth that worthy must supply his own transportation.

Lawmakers habitually vow to eliminate waste, fraud and abuse from a bloated budget. Ford, applying lessons learned on the Public Works Committee, made good on the promise, much to the consternation of his log-rolling brethren. At issue was a $624 million budget request from the Army Corps of Engineers, the collective price tag for 130 individual river, harbor and flood control projects. Ford brought his concerns to Louis Rabaut, a Michigan Democrat who chaired the subcommittee on civil functions.

With Rabaut's blessing, Ford launched an exhaustive review of each project, assisted by Michigan State University economist Harry Brainard. Together they beat back every attempt to restore projects deemed wasteful, saving taxpayers $126,000,000 in the process. His dissection of the Army Corps budget further established Ford's reputation as someone who thought for himself, did his homework and yet didn't hesitate to seek counsel from more experienced lawmakers or outside experts.

Broadly supportive of Truman's war policies, Ford voted to extend the draft and to authorize wage and price controls in a belated attempt to combat wartime inflation. On questions of internal security, however, he parted company with the White House, voting to override Truman's veto of the McCarran Act, which required the registration of Communist and Communist front organizations. Hoping for a new occupant of the White House after the 1952 elections, Ford saw his own living arrangements upended in the summer of 1951, when Betty learned that she was pregnant and the cramped quarters of the Carlyn Apartments proved inadequate to the needs of a growing family. That summer the Fords moved to Parkfairfax, a large-scale housing development built to accommodate the influx of government workers during World War II, many of them employed in the nearby Pentagon. Occupying two hundred acres in suburban Alexandria, Virginia, this instant neighborhood of apartments and two-level townhouses included multiple swimming pools, tennis courts, a library and extensive landscaped grounds. It did not include any African-Americans or Jews, groups prohibited from renting there until 1963, when owner Metropolitan Life adopted less restrictive occupancy policies.

For the next four years the Fords called 1521 Mount Eagle Place home. Their move was perfectly timed. On March 15, 1952, Ford treated two-year-old Mike to a birthday lunch in the House restaurant and a haircut in the nearby barbershop. That same day Betty checked into Columbia Hospital for the birth of John Gardner Ford, known from infancy as Jack. (In his diary, the baby's father noted the arrival of his second son, shortly after midnight on a Sunday, adding, "worked in office 3 hours.") Money remained tight in the Ford household, as the congressman himself acknowledged in recommending a Department of Agriculture cookbook to constituents. "I took a copy home to Betty a few months ago and she uses it extensively in our home. I can testify that the meals are good, nourishing and of the thrifty type."

In addition to the needs of his growing family, Ford's thoughts turned to the coming presidential contest. With Republicans needing just nineteen additional seats to regain control of the House in 1952, the search was on for a candidate with coattails sufficient for the party to occupy both ends of Pennsylvania Avenue. Conservatives in his district wanted Ford to support Ohio senator Robert Taft. Much as he admired Taft, Ford doubted that his third attempt to become president would be any more successful than the previous two. Moreover, Taft remained at heart an isolationist, suspicious of foreign alliances like NATO, opposed to the Marshall Plan as a costly provocation to the Soviet bear, fearful above all that the national security state would end in bankruptcy and the curtailment of civil liberties. Subsequent events would lend much of Taft's critique a credibility missing at the time. To Ford's generation, however, the lessons of Munich took precedence over a fear of "entangling alliances" as old as the republic.

On February 22, 1952, Ford joined eighteen other Republican congressmen in a public appeal to Eisenhower in Paris. They wrote on behalf of their constituents, they told him, from all parts of the nation. "They want you to come home; they want you to declare yourself on the pressing issues of the day . . . we beg you to listen to them because we agree with them." The letter's authors, proud of their historical role, would gift Eisenhower as president a commemorative silver blotter containing their signatures and the presidential seal. Ike kept it on his Oval Office desk for eight years, during which he hosted an annual reunion of his original Capitol Hill supporters. But all that was in the future. More immediately, Ford's endorsement of Eisenhower stirred talk of a primary challenge to the congressman from the right. Although this failed to materialize, it may have cost him a place on the Michigan delegation to the party's July convention.

Long before then Ford had crossed swords with Republican national committeeman Arthur Summerfield, an auto dealer who fancied himself a GOP kingmaker. "Art always was suspicious of me," Ford recalled. "Among other things, he thought I was too liberal." That summer, Summerfield and Ford waged a proxy contest over Arthur Vandenberg's Senate seat. Urged to run himself, Ford deferred to state senator John Martin, a Grand Rapids ally who had stepped aside in 1948 so that Ford could have a clear field against Bartel Jonkman. In opposition to Martin, Summerfield backed Representative Charles E. Potter, a disabled war veteran and fierce proponent of "Americanism," who won a hard-fought primary and was swept into office on the Eisenhower wave that crested in November.

But Summerfield couldn't keep Ford out of the Chicago Coliseum, where Eisenhower won a first ballot victory on July 11. Delighted by the result, Ford lingered just long enough to congratulate Richard Nixon on his vice presidential nomination. From Chicago the congressman flew to a remote army base in Montana, first stop on a six-day, four-thousand-mile inspection tour of flood control projects on the turbulent Missouri River. Not satisfied with an aerial view, Ford traversed a hundred-mile stretch of water by tugboat. Congressional junkets being a sensitive subject, Ford took pains to account for the $134.04 his travels cost the federal Treasury. He often used his newsletter—"Your Washington Review"—to publicize examples of bureaucratic waste. The Office of Price Stabilization was a favorite target. Located at the foot of Capitol Hill, the OPS didn't think twice before communicating with members of Congress by airmail.

Though Ford generally downplayed partisan politics in addressing his constituents, his self-restraint was tested the nearer he got to election day. In particular his relations with organized labor had cooled considerably since 1948. In some ways this was unavoidable, as Ford's ambitions within the Republican caucus brought him into conflict with the liberal agenda of organizations like the United Auto Workers (UAW). Excluded from union halls, Ford redoubled his courtship of working families, "whether it was at a PTA, a church supper, a Farm Bureau meeting, any place where I could leapfrog some of the so-called 'bosses' in the political arena." Wherever he went, Ford stressed Ike's commitment to maintaining the social gains of the New Deal. As he explained it in his newsletter, "Eisenhower believes that Uncle Sam should provide a floor so that all our citizens would have protections against disaster. He wants the individual citizen in our economic system to have the incentive and the opportunity to provide for himself over and above the basic protection which the government should provide."

A pithier formula was needed to convey Ford's concern about the role of the modern state, and he wasn't long in devising one. "A government big enough to give you everything you want," he told voters repeatedly, "is big enough to take everything you have." Campaign audiences in 1976 who heard the then–president of the United States recite this mantra against the mildly progressive Jimmy Carter had no idea he was recycling language first employed before the Rotary Club of Coopersville, Michigan, a quarter century earlier.

O N SEPTEMBER 18 the Republican campaign was jolted by a sensational headline in the *New York Post*, then a mainstay of liberal journalism. "Secret Rich Men's Trust Fund Keeps Nixon in Style Far Beyond His Salary." The accompanying article, though considerably closer to the truth, still exaggerated the so-called slush fund of approximately $18,000 contributed by Nixon's original backers to offset political travel, postage and other incidentals not covered by his congressional allowance. From the outset Ford thought the controversy reeked of partisan hypocrisy. Politicians of both parties, himself included, relied on leftover campaign funds to underwrite nonofficial expenses. His conviction was strengthened when it was revealed that Illinois governor Adlai Stevenson, Eisenhower's Democratic opponent, maintained his own fund with which to pay for Christmas hams and supplement the salaries of state workers.

But Stevenson wasn't running an outsider's crusade to cleanse Washington of corruption. Nor did his survival rest on the whim of a running mate's insistence that he be "clean as a hound's tooth." With his career hanging in the balance, Nixon took to the airwaves on the evening of September 23. Appropriating for his own purposes FDR's Forgotten Man, by the time he was done Nixon had forged an emotional bond with millions of voters, the heart of what he would subsequently label the Silent Majority. In an act of political hardball he shifted responsibility for his continued presence on the ticket from Eisenhower to the viewing audience itself, whom he invited to register their views with the Republican National Committee. They responded in droves, swamping Congressman Ford's office and eliciting from him a strong endorsement of Nixon's "honest, clear cut presentation."

The crisis eased as Eisenhower embraced the youthful senator he had so recently held at arm's length. During the first week in October, Ford led the welcoming throng when Ike visited Grand Rapids. A crowd almost as large turned out to greet Nixon on a follow-up swing through West Michigan ten days later. In the campaign's final days, his partisan instincts aroused, Ford played the "soft on Communism" card, with an unhealthy dose of guilt by association, as he linked Democratic senator Blair Moody to the "ultra left wing" group Americans for Democratic Action. On Friday night, October 31, Ford staged his own televised encounter with the electorate, a three-hour telethon in which he and state auditor candidate John Martin were bombarded by more than six hundred questions. "I felt as though we were up against a firing line," Ford said afterward.

His reward came four days later, when Eisenhower scored a landslide win and the GOP narrowly regained both houses of Congress. Ford notched a personal triumph, running eight thousand votes ahead of Ike in the Fifth District. Victory brought not only spoils, but a resumption of the long-running conflict within the GOP between hard-shell conservatives eager to repeal as much as possible of the New Deal and the more pragmatic Eastern

Establishment, whose acceptance of Social Security, Wall Street regulation and a vigorously assertive foreign policy earned right-wing scorn for "me too Republicans." At the time Eisenhower appeared to have won a decisive mandate for moderation. In fact, Modern Republicanism proved to have a shorter shelf life than the Hula-Hoop. Less than twenty-five years would separate Ike's 1952 triumph from the skin-of-his-teeth nomination of Gerald Ford over Ronald Reagan in America's bicentennial year.

Ford's immediate response to the Eisenhower sweep included praise for the man he beat. "There's no question about it, Adlai Stevenson was a remarkable candidate," the congressman acknowledged in his first postelection newsletter. "A man of nobility and character . . . as admirable in defeat as he was in the throes of the campaign." Ford's magnanimity extended to the striped-pants State Department crowd, whose alleged inadequacies had fueled much of the Republican attack on the outgoing administration. "Try, sometime, to put yourself in the position of a high government official," Ford urged his constituents. "Place yourself at a long table with officials of foreign-speaking nations, each with many difficult problems to solve. For atmosphere, add bright lights, cameras, newsmen, secretaries, clerks and a general buzz of activity . . ." Recommending "a certain amount of mercy and a great deal of prayer" for American diplomats burdened with such responsibilities, Ford conceded the obvious—"They are all human, prone to the same weaknesses even as you and I."

He would not always be so forgiving, especially as the issue of domestic subversion gripped Washington, and he was urged to board Joe McCarthy's lurching bandwagon. Disapproving of McCarthy's tactics, Ford in later years expressed regret for not being more outspoken in his criticism, even as he praised Eisenhower for his hands-off approach to the Wisconsin senator and his anti-Communist crusade. For Ike to have gone head-to-head with his senatorial nemesis, claimed Ford, "would have . . . taken McCarthy out of the sewer and elevated him to a much higher public platform." Steering clear of McCarthy did not keep Ford from supporting legislation to outlaw the Communist Party; authorize government wiretapping in criminal cases involving national security; or endorse "anti-subversive" campaigns to root out potential security risks (many of them homosexuals) in government departments.

In the fall of 1953, Ford rallied to the defense of J. Edgar Hoover and the FBI after reports surfaced that Hoover had warned Truman against appointing Harry Dexter White, a former assistant Treasury secretary suspected of Communist sympathies, to head the newly formed International Monetary Fund. Less debatable than White's loyalty was Ford's reflexive support for the FBI under Hoover, a factor worth remembering in view of later controversies surrounding his activities as a member of the Warren Commission investigating the assassination of President John F. Kennedy.

W ITH THE FORMAL start of the 83rd Congress in January 1953, Ford came into his legislative inheritance. That is when Appropriations Committee chairman John Taber assigned him to the Subcommittee on Defense Spending, a budget- and policymaking elite said to handle more business than all twelve of the other Appropriations subcommittees combined. The navy man Ford was designated chairman of a three-member panel entrusted with oversight of the army. "All these damn Admirals will be after you, and you won't resist them," Taber explained to Ford. "But if you're with the Army, you will tell the generals to go to hell." Taber's job instructions were similarly blunt: "Give the Army only what it needs for national security, and no more."

It was pure myth to describe the Senate as the world's most exclusive club, contended John W. McCormack, a future Speaker of the House. "Why, that's minor league compared with a subcommittee on Appropriations. Not the full Committee, but a subcommittee. What a member wants there he gets, and there's no partisanship whatsoever." To McCormack they were "the hardest workers in Congress . . . You walk over in the morning to the House just before 10 o'clock, and you'll see them all pounding through the tunnel on their way to subcommittee meetings. Down that long tunnel they come, hurrying along and afraid they are going to be late. They shouldn't be going so fast—some of them are too old—but they do it just the same . . . working in secrecy and working hard. They're like missionaries."

Ford and his fellow panelists exhaustively questioned quartermasters and four-star generals. They slashed $33 million earmarked for horses and mules better fitted to George Custer's cavalry than a modern nuclear power, and scaled back daily ration costs to reflect a decline in consumer food prices. Before they were through, Ford and his colleagues hacked nearly $700 million from an army budget already reduced by the new president, a professional soldier who feared red ink more than red armies. The death of Stalin in March 1953 and a Korean armistice signed four months later paved the way for Eisenhower's New Look strategy emphasizing nuclear deterrence at the expense of more costly conventional land and naval forces. Again Ford threw in his hand with the reformers. If anything, he was more tightfisted than Eisenhower, who succeeded, with the aid of cheeseparing allies on Capitol Hill, in cutting defense outlays by nearly one-third.*

For Ford the defense subcommittee had additional significance, for it was there he made two of the defining friendships of his life. Appropriately bipartisan, they were grounded in shared priorities and unabashed patriotism. Democrat George Mahon, a tall, lanky Texan from Lubbock, was already something of a legend on Capitol Hill, where he served forty years on the House Appropriations Committee, most of them chairing the subcommittee on defense. As old as the century, as a youth Mahon had helped his father scratch a living from the parched soil around tiny Loraine, a former cotton and cattle shipping point on the Texas and Pacific Railway. From an early age he was taught to regard as sinful the purchase of something—anything—without the funds on hand to pay for it. In Congress, Representative Mahon emulated his father's self-denying habits, as well as his Methodist faith. Teaching a Sunday-school class during much of his congressional tenure earned him the derisive nickname "The Deacon" from Washington sophisticates.

His friendship with Ford grew out of common values. Both men gave priority to balancing the federal budget over cutting taxes. Both impressed elders with their ability to keep secrets. (Mahon was one of a handful of congressmen during World War II who were aware of the Manhattan Project to develop an atomic bomb.) Finally, nothing so bonded Mahon and Ford as the Texan's singular boast, "I have the most nonpartisan committee in Congress."

Even as Ford studied Mahon for his emollient approach to leadership, he also befriended a House newcomer whose rough edges failed to obscure his quicksilver intelligence and political savvy. Thirty-year-old Melvin Laird, like Ford born in Omaha, was the son of a Presbyterian clergyman who had enjoyed a second career in politics. The recipient of a Purple Heart for his World War II naval service—he and Ford had actually met on a Pacific Island, where the future statesmen enjoyed a couple of beers together—Laird succeeded to his father's place in the Wisconsin state senate at age twenty-three. In 1952 he rode the Eisenhower landslide to a seat in the House representing central Wisconsin's Seventh District.

Laird was still in his first term when his Republican brethren recruited him for the Chowder and Marching Club. Democrats were just as quick to recognize his talents. Indeed, Laird's work on a bill funding the new

---

* Eisenhower was the first American president since Calvin Coolidge to leave office having shrunk the federal budget in real dollar terms. No one since has repeated the feat.

Department of Health, Education, and Welfare so impressed the Democratic leadership that he was rewarded with a private dining room and office in a prime location of the Capitol. Named to the powerful Appropriations Committee as a rookie member of Congress, Laird renewed his wartime acquaintance with Ford. Sharing what Laird's biographer calls "an almost fanatical interest in golf," the two men were often partnered in tournaments at Burning Tree Country Club in suburban Maryland. As strong advocates of Eisenhower's New Look defense policies, both lawmakers displayed courage in voting to close or consolidate surplus military facilities.

Boasting children of similar age, the Ford and Laird families frequently vacationed together. Something else bonded Ford and Laird. They were part of a weekly prayer group, the brainchild of Minnesota congressman Al Quie, that met in a small room off the Capitol Rotunda set aside for the purpose. Charter members Laird and Quie each invited one of his colleagues to join them. Thus Ford of Michigan and Charles Goodell from a rural district in Western New York filled out the quartet. Laird's deeply felt faith might have surprised some House members who knew him chiefly as a born intriguer. "He doesn't scheme for any sinister reason," said Ford, "he just likes to keep the pot boiling." Bob Dole offered a somewhat darker view. Laird, he concluded, "is the kind of guy who would put poison in the river and then run down to the town and promise to save everyone." Ford was to be the beneficiary of Laird's plotting more than once. Trusting by nature, he took no offense when the outspoken Laird complained of his friend's relative passivity: "You had to kick him in the ass usually to get him to do something."

"He's a pusher," Ford said of Laird, "and I respond to that kind of challenge."

Where the two men agreed to disagree was on foreign aid. Impressed by his performance on the defense subcommittee, Chairman Taber gave Ford an even more thankless assignment—the subcommittee on foreign operations. Ford's latest exercise in forensic budgeting consumed most of July 1953, with committee members burning the midnight oil to dissect a $7.3 billion aid request submitted by the new president. One morning Ike invited the group over to the White House for breakfast and a little friendly lobbying on behalf of a friendless program. Declaring himself unalterably opposed to any foreign giveaways, Eisenhower reframed the issue as a matter of American self-interest. Foreign aid was inseparable from the nation's larger defense program. Ford was persuaded. "If Uncle Sam doesn't spend the money to help our allies," he recounted after the meeting, "our nation will be forced to spend many times more money and lives" to keep aggressors at bay.

Ford returned to the White House in the last week of July to witness Eisenhower sign legislation he had authored granting US citizenship to a twenty-two-year-old Polish flier, Lieutenant Franciszek Jarecki. The product of an elite jet training school, hailed as a model young Communist, Jarecki had risked his life to defect to the West and deliver a Russian MIG in the bargain. Ford made Jarecki's cause his own. He introduced the pilot to television viewers and to the president of the United States. And he took the occasion to reiterate yet again his belief in Polish exceptionalism. As it was, said Ford, Poles submitted to their Russian occupiers "only because there's not much else to do when someone issues orders at gunpoint and one has no weapons to defend oneself."

T HEIR STRATEGIES FOR combating the spread of Communism might differ, but the Truman and Eisenhower Administrations were in sync regarding the monolithic nature of the international conspiracy whose far-flung agents answered to Moscow. Four years after Communist forces loyal to Mao Tse-tung completed their occupation of mainland China, sending Generalissimo Chiang Kai-shek into exile on Taiwan, American policymakers treated Mao as a Soviet puppet. Even less was Korean nationalism credited as a factor in the bloody war only recently suspended by means of a negotiated armistice.

A similar disregard for history and indigenous culture applied to Indochina, where a century of French colonial rule was dwarfed by a thousand years or more of Vietnamese—and Cambodian and Laotian—identity. For three weeks in August 1953, Ford was able to form his own impressions on a Far Eastern inspection tour of army installations. His travels coincided with a regional visit by Secretary of State John Foster Dulles and Army Secretary Robert Stevens, imparting to Ford's mission an element of official recognition rare for a lone congressman with less than three years as an appropriator under his belt. The trip's emotional highlight came early. Wednesday, August 5, was reserved for Operation Big Switch, involving the exchange of POWs from both sides of the Korean conflict. From Seoul Ford and the military brass flew to Freedom Village, a United Nations outpost close to the demilitarized zone that separated the opposing armies.

The congressman blinked back tears as the first stretcher cases arrived by helicopter and in field ambulances, their disabilities in no way diminishing their joy at being set free. Together with Secretary Stevens, Ford toured the former battle lines by helicopter. On the ground he promised Michigan boys that he would call their wives and parents as soon as he returned home. He interviewed British, Canadian, and Australian soldiers before visiting the Seventh American Division, ably supported by battalions of Colombians and Ethiopians. Heaping praise on South Korean generals and trainees— "Not only top-notch soldiers, they are solidly anti-Communist"—Ford struck an optimistic note concerning US training programs and the progress shown by ROK (Republic of Korea) forces: "It is no easy task to take a man from a rice paddy and make him an efficient soldier in a relatively short period of time." Ford's words would return to haunt him and a generation of American presidents, diplomats, military planners and soldiers deployed because of their questionable reading of Southeast Asia.

It was five o'clock on a sultry Thursday, August 6, when Congressman Ford, accompanied by Secretary Stevens, first set foot on Vietnamese soil ostensibly claimed by France, the colonial power whose fading grip had been vigorously contested since 1946 by Communist insurgents calling themselves the Viet Minh. These rebels were in turn led by Ho Chi Minh, the son of a Confucian scholar who had refused employment in the French colonial bureaucracy. Deprived of a government scholarship, in 1911 Ho had made his way to Saigon, Marseille and, eventually, the United States, where he reportedly worked as a short-order cook in Boston's famed Parker House Hotel. While still in the US, Ho contacted Korean nationalists. Still later he was part of a group of Vietnamese patriots who wrote an appeal to Woodrow Wilson and other Allied leaders negotiating the post–World War I map of the world at Versailles.

Ignored by Western statesmen, Ho enrolled in a Communist university in Moscow, relocated to Canton, China, and traveled extensively through Europe before reaching Hong Kong. There he presided over the 1930 birth of the Communist Party of Vietnam. A decade later he secretly returned to his homeland, then occupied by Japanese troops, and organized the Vietnam Independence League, or Viet Minh. Ho the nationalist quoted Thomas Jefferson and the Atlantic Charter in justifying his war on French colonialism. The Truman Administration, burned by the "loss" of China and pledged to contain further Communist expansion, turned a deaf ear to his request for recognition. Instead, in February 1950 the United States and Britain simultaneously recognized the Saigon-based regime of Bao Dai, the puppet emperor originally installed by the Japanese and more recently reinstated on his throne by the French. Later that year Truman authorized $10 million in military aid to French forces battling the Viet Minh. By the end of his presidency, the United States was footing nearly half the cost of France's war effort.

Dwight Eisenhower inherited from his predecessor Indochina's "leaky dike." With French defeat a real possibility, and Western prestige on the line, the new president confronted a series of risky choices. What if, even with increased American support, the French were driven from the region? Should the United States pick up the standard dropped by colonial powers and commit troops to sustain an unpopular regime in Saigon? With hostilities in Korea officially suspended, Mao's government in Beijing increased military assistance to the Viet Minh. After July 1953 artillery and antiaircraft weaponry poured into Vietnam, where much of the countryside was already controlled by forces loyal to Ho Chi Minh. In response the US aided the beleaguered French to the tune of a billion dollars.

Ford arrived in Saigon at a critical moment in the escalating battle for Indochina. The climactic siege and ultimate French defeat in the valley called Dien Bien Phu was still several months in the future. As far as the congressman was concerned, he was in Vietnam to observe the deployment of American resources. Future shipments of military equipment would depend on French prospects for breaking the bloody stalemate. Shown a housing project built with American funds, he noted of the ceremony inaugurating the facility, "Speeches, pictures, bally-hoo."* A night on the town, notable for "good French food" and a "gambling joint," was followed by military and political briefings, capped by afternoon tea with General Henri Navarre, the recently appointed commander in chief of French forces throughout the region. The next day Ford toured supply depots and related facilities. "Neat and substantial in appearance," he jotted in his travel diary, "but somehow there is lack of 'esprit de corps' that is so vital to defeat the reds."

The rest of Ford's journey was, by comparison, anticlimactic. In Taipei a waiting limousine whisked the American visitor to the mountaintop residence of General and Madame Chiang Kai-shek, where he spent "an exceedingly interesting hour" sitting between his hosts while drinking minted iced tea and eating American ice cream. Two months had passed since the Generalissimo, determined to reclaim China from Mao's Communists, petitioned the Eisenhower Administration for help in training and equipping

---

* In some subsequent accounts of his visit, Ford's use of the term "ballyhoo" was mistakenly applied to his separate discussions with French military planners, leaving an impression of profound skepticism toward colonial rule. In fact, he welcomed a French declaration of independence for its Indochinese colonies—in return for which the United States announced an additional $385 million to help prosecute the war against the Viet Minh.

a Nationalist invasion force of six hundred thousand troops. From US diplomats, Ford learned of the "complicated" problem of two Chinas and of American efforts to rehabilitate Chiang's army, navy and air force. Clearly there was no invasion in the cards anytime soon.

During a brief stop on Okinawa, and a longer one in Japan, Ford marveled at the reconstruction work that had taken place since the war. Tokyo served as headquarters for the US Far East Command, which meant more briefings, with special emphasis on the efficacy of air power in Korea (this, too, was to have major ramifications in Vietnam). A concluding visit to Hawaii invited reflections on wars past and present. To Ford, the parallel was self-evident. "If the Japanese had known America would recover and fight with all its resources, the attack at Pearl Harbor undoubtedly would not have been launched," he wrote. Likewise, if the "Red Bosses" in the Kremlin understood American determination to resist Communist expansion, they would in all likelihood hold their fire.

On May 7, 1954, the French garrison at Dien Bien Phu surrendered to the Viet Minh. In Geneva, diplomats agreed to partition Vietnam, and to hold elections aimed at reunification within two years. In truth, neither Secretary Dulles nor the South Vietnamese government under Emperor Bao Dai and Prime Minister Ngo Dinh Diem welcomed the agreement. Diem staged a bogus referendum to justify his autocratic rule and reject the elections called for at Geneva. In February 1955 Ike dispatched a handful of US military advisers to help Diem build an army capable of resisting the Communist North; by the time Eisenhower left office in 1961 their number had swelled to nine hundred. Eisenhower's Indochina legacy, however unintended, may be summed up in a few unguarded sentences spoken at a presidential press conference in April 1954. After explaining the political and economic significance of Vietnam, a land most Americans would be hard-pressed to find on a map, the president spoke of "broader considerations." Imagine, he said, "a row of dominoes set up, you knock over the first one, and what will happen to the last one is a certainty that it will go over very quickly."

Gerald Ford completed Eisenhower's thought in a newsletter dated April 29. "The loss of Indo-China to the Communists would doom Burma and Thailand, making the conquest of the rest of Southeast Asia inevitable. This would threaten the Philippines and Japan. Our positions in Korea, Okinawa, Formosa and Hawaii"—all places visited by the congressman on his recent Asian swing—"would be endangered." Equally disastrous to the free world would be the loss of rubber, cotton, tea, sugar, coffee and pepper, not to mention tin, tungsten, zinc, lead, coal, iron and phosphates. Defeat in the Pacific would tip the scales decisively in favor of international Communism.

Twenty-one years later to the day, Gerald Ford would preside over Operation Frequent Wind, the largest helicopter evacuation on record, enabling thousands of Americans and their South Vietnamese allies to flee Saigon before it fell to the Vietcong.

8

THE MIDDLE OF THE ROAD

———◆·◆·◆———

*I always wake up in the morning with the thought that I've got some
problems to solve, and I'd better get up early and working at them.*

—GERALD R. FORD

MARCH 1954 CAME in like a lamb in the nation's capital, a premature
taste of spring that tempted the political class to an extended lunch
hour. Consequently the House chamber was half-empty at two o'clock when
Speaker Joe Martin gaveled into session a discussion of legislation admitting
Mexican farm laborers to work in fields disdained by most Americans. Up
one flight of stairs from Martin's rostrum, just down the hall from the Ladies'
Gallery, where out-of-town visitors lined up to observe their elected representatives in action, Jerry Ford was presiding over a routine army budget hearing. Entering his sixth year on the Hill, Ford spent more time in committee
rooms than he did on the House floor. This fact alone spared him from the
murderous protest of four members of the Puerto Rican Nationalist Party
who opened fire on the massed House membership at 2:32 that afternoon.

"*Viva Puerto Rico libre.*" The voice in the gallery belonged to the group's
ringleader, a thirty-four-year-old former beauty queen named Lolita Lebrón. Later it would be determined that Lebrón fired her .38 German
Luger in the direction of the ceiling. Her male accomplices practiced no
such restraint. As they sprayed the assembly below with some thirty shots,
Speaker Martin took refuge behind a marble pillar, and Republican floor
leader Charlie Halleck absorbed wood splinters from a desk scarred by flying bullets. Another lawmaker applied his necktie as a tourniquet to stem
the blood flowing from a colleague's leg. Oblivious to the scramble for
safety a floor below them, Ford and his fellow appropriators attributed the
popping sounds outside their hearing room to out-of-season firecrackers.
Pushing back the heavy committee room door, they gazed, slack-jawed, at
the pandemonium unfolding around them. In the nearby gallery a ragtag
posse of local and Capitol police, House staff, congressional pages and at
least one member gifted with lightning-quick reflexes seized three of the

four would-be assassins (the fourth was apprehended a short time later at nearby Union Station).

Ford hastened to the House floor. There he made a beeline for his Michigan colleague Alvin Bentley, the most seriously hurt of five lawmakers shot by the gallery marksmen. Bentley lay where he fell, his gaping chest wound concealed by a bloodstained blanket. Within minutes several young pages helped evacuate the congressman to a waiting ambulance. At the hospital doctors gave him a 50–50 chance of survival. In the event Bentley beat the odds; eventually all the shooting victims would recover.* In the wake of the attack, the worst since the British torched the Capitol in 1814, a proposal to install bulletproof glass in the visitor galleries was rejected lest it isolate the people's representatives from their constituents.

Certainly Ford remained accessible as ever to petitioners whose swollen ranks testified to his increased visibility. Some were corporate lobbyists or White House emissaries in search of a vote. Many more were ordinary citizens, remote from their government until Jerry Ford personalized it for them. On his frequent visits to the Fifth District he could be found in the Knife and Fork restaurant in the Pantlind Hotel. "I called it the Knife and Ford," says one friend who met him there. "He didn't really make the rounds as much as he did sit in the Knife and Fork and drink coffee and talk to people. He enjoyed it."

An Eisenhower loyalist, during the 83rd Congress, Ford voted with the new administration 93 percent of the time. The two men were natural allies. From his first day in office Eisenhower governed from the center right, with an emphasis on the center. Making good on his promises of retrenchment and reform, he set about pruning the overgrown garden of bureaucratic Washington while retaining and, in some cases, expanding the social safety net fashioned by his Democratic predecessors. With a thriving economy to pay the bills, it became possible for Washington to add ten million workers to the Social Security rolls, and to commit $1 billion a year toward construction of the Interstate Highway System. Another vast building project, the long-delayed St. Lawrence Seaway, promised an economic shot in the arm to the American Midwest, and to Michigan in particular. Modest tax cuts benefited parents with children, retirees, small businesses and farmers.

---

* The terrorists were tried, found guilty, and given lengthy sentences, commuted after a quarter century by President Jimmy Carter—in exchange, it was said, for the release of CIA agents held in Fidel Castro's Cuban jail cells. Today a bronze statue of Gerald Ford, flanked by a similar likeness of Barack Obama, guards the Walkway of the Presidents leading to the Capitol in San Juan. It commemorates Ford's December 1976 call for Puerto Rican statehood, and Obama's visit to the island three decades later.

Ford backed them all, reserving his highest praise for Eisenhower's first-year reduction of the federal workforce by over two hundred thousand workers. On education he threaded the needle, opposing federal funding for teachers' salaries while embracing a complex scheme using federal dollars to spur construction of secondary schools to accommodate the first wave of baby boomers. "The Republican Party is a humane party," Ford told GOP Lincoln Day audiences in 1955. "It is concerned with human beings and their welfare. But the Republican Party does not sponsor a wasteful, inefficient, socialistic scheme of things." He much preferred the Eisenhower formula of governance—"Liberal in matters of human rights and conservative in the sphere of economics."[*]

Ford praised the May 1954 Supreme Court decision in *Brown v. Topeka Board of Education* that outlawed segregation in public schools nationwide. Declaring the unanimous ruling "morally and constitutionally right," he predicted a peaceful transition to integrated schools "if the extremists on both sides will 'hold their tongues' and let the moderates implement the Court's decision." He supported an amendment introduced by Harlem Democrat Adam Clayton Powell that, by placing federal education dollars in escrow for up to three years, would penalize school districts that refused to comply with court-ordered integration. To Ford the civil rights issue had implications far beyond American soil. After *Brown*, he said, "No longer can Communists holler and scream that the United States legalizes racial discrimination in our public school system."

In the run-up to the 1954 congressional elections, Ford was urged to take on incumbent Republican senator Homer Ferguson. It did not escape his notice that some of those now imploring him to challenge Ferguson for being too liberal had voiced similar objections to Ford himself at the time Arthur Vandenberg's Senate seat fell vacant. Nor had he forgotten Ferguson's hospitality at a time when he and Betty were struggling to adapt as Washington newcomers. "I liked Homer . . . a very decent guy," he recalled. "And I thought he did a good job as a Senator." He would have no part in the right-wing "vendetta" aimed at Ferguson.

Ferguson's enemies may have misjudged Ford's priorities. But they were not altogether mistaken about the drift of his thinking. His youthful energy and willingness to uphold the Vandenberg tradition of bipartisan

---

[*] While Congressman Ford enjoyed a cordial relationship with the president, Betty Ford became closely acquainted with Mamie Eisenhower through Dottie Schulz, the wife of Ike's military aide, who lived near the Fords in Parkfairfax. The First Lady and her daughter-in-law Barbara were frequent visitors to the Schulz residence. On more than one occasion the president's grandson, David, joined Mike and Jack Ford to douse imaginary fires with toy fire trucks.

foreign policy masked a growing suspicion of overreaching government. An economics major in college, his work on the Appropriations Committee supplied Ford another kind of education, in a classroom where success was measured in unspent dollars. "The longer you stay on the committee the more conservative you get," claimed Ben Jensen, an Iowa Republican and self-proclaimed progressive, "till I saw how things were done down here. I even had to pull harder on the conservative side than my heart dictated to keep these liberals from pulling us over the brink."

Ford underwent his own evolution from insurgent to insider. By casting himself as a Treasury watchdog, he practiced shrewd politics while remaining faithful to his personal convictions. Blessed with a safe seat, he could afford to wait for the seniority system to work its will as he rode the congressional escalator upward toward his ultimate goal of the Speakership. Until then Ford and GOP colleagues hoping to institutionalize a Republican majority in Congress put their faith in a popular president for whom duty took precedence over party building.

FORD KNEW HE had arrived on Capitol Hill the first time Speaker Sam Rayburn nodded wordlessly in his direction. Such acknowledgment of a younger member, let alone a Republican, by the legendary Mr. Sam was rare. The House was Rayburn's life. Decades earlier he had entered into a disastrous marriage that lasted barely three weeks. The painful subject was buried fathoms deep. Only a handful of intimates sensed the profound loneliness that engulfed Mr. Sam out of session. "God, what I would give for a tow-headed boy to take fishing," he lamented. Gerald Ford shared Rayburn's devotion to the House. Unlike the older man, however, he had family responsibilities competing for his time and attention. In a Christmas season newsletter he described the challenges of living in more than one place. "We're getting to feel almost like gypsies having to pack up every few months to go back and forth from Washington to Grand Rapids," Ford confessed. During his early years in Congress, he often made the trip by car rather than pay for an airline ticket.*

An equally frugal Betty Ford put her Herpolsheimer's training to work

---

* Bob Michel, elected in 1956 to represent the central Illinois district once home to Everett Dirksen, described what it was like when the government reimbursed members of Congress for a single round-trip airfare between their district and the nation's capital. "Dan Rostenkowski, the Democrat from Chicago, and Phil Crane and Harold Collier and I, we were in a station wagon, we'd leave on a Thursday night and that's before the interstate, and we'd drive all night. I'd drop them off in Chicago and I would go 160 miles down to Peoria afterwards and then go back the following Monday night or Tuesday." Thus all four men became members in good standing of the T and T Club—"In Tuesday and out Thursday." AI, Robert Michel.

in assembling a wardrobe whose versatility exceeded its cost. Coming to her rescue in the spring of 1954, her mother-in-law Dorothy Ford described some discounted evening dresses—"Way way below even half price"—she had glimpsed on a recent visit to "Herp's." A store saleswoman had asked her to imagine the congressman's wife in a stylish pink beaded outfit. Dorothy didn't need convincing. "So I stretched my budget a little and I'm paying for all but $10 of it and charging that to you and I'll make it up to you," she told Betty. "If I charge the balance to Dad now he'll see how much I paid for it and I don't want him to know!! Can't be married *40 years* and not learn something." Around this time Dorothy's son gave at least passing consideration to abandoning politics for a more lucrative legal career. "There was some talk about people who were in law practice back home becoming very successful financially, and we had two boys then," Betty Ford confided to her husband's biographer in 1990. "And yet I knew that it was in his blood, that he loved every minute of it."

By 1954 increasing travel added to the demands of the job. "He was away—even then," recalled Betty, citing Jerry's three-week Asian trip the previous year during which she was left alone to care for two small children. By way of compensation Ford's decision to seek a fourth term coincided with his family's move out of Parkfairfax and into a home newly built to their specifications on an undeveloped tract in the Clover neighborhood of Alexandria. The modest brick-and-clapboard structure contained four bedrooms, two and a half baths, a finished basement/rec room, wood-paneled den and small enclosed porch overlooking the backyard. The master bedroom was perched above a two-car garage. Dissatisfied with the original landscaping, Betty employed her green thumb to good effect. When they got old enough, Mike and Jack scavenged sufficient scrap lumber from neighboring building sites to construct a shack for which their father supplied the roof and their mother furnished curtains. In 1961 a twenty-by-forty-foot swimming pool was installed.

Here Ford swam half a mile in daily laps, and hosted parties under the stars for the Chowder and Marching Club. Betty used the same setting to entertain friends in the Congressional Wives Club. The combined cost of house and lot was approximately $34,000, on which the Fords made a 20 percent down payment. Besides borrowing on his and Betty's life insurance, Jerry secured a loan through the House Sergeant at Arms and a Washington bank. When Betty's stepfather, Arthur Goodwin, died in December 1955, he left her an inheritance of $16,000. Much of the windfall was applied to outstanding debts. The rest funded a European trip in 1956, the first of several deferred honeymoons to compensate for their spartan weekend in Ann Arbor and Detroit. Cold War tensions shadowed the Fords in Vienna, where they met refugees escaping from Hungary's brutally thwarted uprising against its Soviet masters. On their return home Betty considered adding a fallout shelter to 514 Crown Point Drive, only to reject the idea, having determined that a postnuclear existence would hardly be living at all.

She got no argument from her husband. At the start of 1955, overwhelmingly reelected by the voters of West Michigan even as Democrats reclaimed majorities in both houses of Congress, Ford had moved to a larger suite of offices in the Cannon Building. On one wall of his private workspace he hung copies of the Declaration of Independence and Constitution. Next to them he placed a photograph of a recent hydrogen bomb blast on Bikini Atoll in the Pacific, used by the United States as a nuclear testing ground. Asked why he exhibited the fearsome image, Ford replied, "to remind me of atomic energy's menace."

O NE DAY EARLY in 1956 Clarence Cannon, the crusty chairman of the House Appropriations Committee (HAC), approached Ford on the floor of the House. Their exchange was brief, a few terse sentences instructing the younger man to appear the next morning at a place and time stipulated by his gruff senior colleague. There armed guards carefully examined Ford's credentials before admitting him to a nondescript hearing room where he encountered four other members, all of them familiar from their prior labors on Appropriations. Besides Chairman Cannon and his Republican counterpart John Taber, they included Ford's close friend George Mahon and Harry Sheppard, a California Democrat who hadn't let party loyalty stand in the way of his endorsing Richard Nixon over Helen Gahagan Douglas in their bitter 1950 Senate contest. Together they comprised the Intelligence Subcommittee of the full HAC, one of four such groups entrusted with oversight of the Central Intelligence Agency. (Besides a Senate counterpart to the HAC, the Armed Services Committee in each house had its own watchdog panel.)

There was nothing casual about Ford's designation. According to Walter Pforzheimer, the CIA's legislative counsel under Director Allen Dulles and his predecessor, General Walter Bedell Smith, he had been under surveillance, so to speak, for some time. Early in 1953, Pforzheimer assured Dulles that on the Hill, Ford, his contemporary at Yale Law School, was considered "extremely reliable" and "a real comer." His name surfaced again as Senator Mike Mansfield floated the idea of a joint congressional intelligence committee that would scrutinize not only the CIA budget but also the clandestine activities it funded. In April 1956, Mansfield's resolution was defeated by a vote of 59 to 27. It is against this backdrop of official deference that Ford's elevation should be understood. For nine years beginning in 1956 he belonged to Washington's most exclusive club, his membership prima facie evidence of the credibility he enjoyed on both sides of the aisle. Meetings of the subcommittee were rare. To guard against leaks, staff were excluded, and no transcripts were made (rules slightly relaxed in the 1960s).

His new assignment gave Ford a front-row seat on the Manichean struggle pitting the capitalist West against what he called "the worldwide atheistic communist conspiracy." The same outlook may have dulled his critical faculties when confronting CIA exaggerations of Soviet military strength or the agency's consistent misreading of US prospects for victory in Vietnam. Ford appeared in thrall to Allen Dulles, the urbane spymaster who reserved to himself what information to reveal, and to whom. "Dulles is a legendary figure," John F. Kennedy remarked, "and it's hard to operate with legendary figures." When not spinning tales of wartime derring-do, the Director of Central Intelligence (DCI) cloaked his objectives in a fog of erudition. Emerging from one oversight hearing, Ford turned to Dulles and inquired half-jokingly, "What was that meeting about?" Assured by a CIA functionary that it was a budget review, Ford replied, "That's what I thought, but the word 'dollar' wasn't even mentioned."

At Chairman Cannon's insistence, Dulles briefed Ford and other committee members about the U-2, a high-altitude reconnaissance plane created on Eisenhower's orders to photograph Soviet military installations otherwise shielded from Western eyes. In May 1960 the Soviets shot an American U-2 out of the sky, then paraded its captured pilot before the cameras to disprove a feeble US cover story and inflict maximum humiliation on the American government. Ford's response was to throw a lifeline to the Eisenhower White House: "Because this business inevitably involves deceit, misrepresentation, falsehood, intrigue, and every devious avenue of approach, public officials may not jeopardize the national security by publicizing the true facts," he asserted, "but when a given situation (no matter how embarrassing) becomes public knowledge, we commend a frank and honest disclosure."

In that single, convoluted sentence, Ford rationalized Eisenhower's original lie about the missing plane, even as he praised the president for his belated acknowledgment of the truth. His deference to the executive would be demonstrated a second time, following the ill-fated 1961 Bay of Pigs invasion organized by the CIA and approved by President Kennedy. He cut JFK some slack for the most honorable of reasons: members of all four CIA oversight committees had been told of invasion planning as it unfolded. No objections had been raised then. To do so after the fact would exploit national embarrassment for partisan advantage.

Kennedy himself was less forbearing. The bungled operation cost Allen Dulles his job. On Capitol Hill, angry lawmakers renewed demands for a joint intelligence committee. Henceforth Ford's questioning of CIA officials took on a harder edge. Mindful of the nonexistent "missile gap" used to considerable effect by Richard Nixon's opponents in the 1960 presidential election, Ford voiced suspicion when, early in the Kennedy presidency, the CIA reported a sharp decrease in the number of Soviet Intercontinental Ballistic Missiles (ICBMs) on launchers. "I must admit that my attitude toward the reliability of our intelligence is changing considerably," he told Secretary of Defense Robert McNamara in a February 1962 hearing. Ford asked Joint Chiefs of Staff chairman Lyman Lemnitzer if ending U-2 flights over the Soviet Union had weakened American intelligence. Lemnitzer readily conceded his point.

"It was a mistake to stop them," Ford grumbled.

The general's response is unrecorded, but it was presumably sympathetic, for Lemnitzer was among the first to be invited, thirteen years later, to join the presidential commission established by Ford to examine CIA abuses and forestall more sweeping actions by congressional investigators.

I T WAS NO ordinary bread-and-butter note Ford received from Richard Nixon in the second week of January 1956. Of some five hundred supporters gathered to celebrate his forty-third birthday, an appreciative Nixon singled out members of the Chowder and Marching Club, "my closest political and personal friends in this world." Among these, Ford was prominent. That Nixon should require his help flew in the face of political logic. By all accounts the vice president had acquitted himself admirably in the uncertain period following Eisenhower's heart attack in September 1955. Keeping a low profile, avoiding any actions suggestive of a power grab, Nixon displayed a maturity and judgment for which he had not always been credited. Still the president held back from any formal endorsement of his once and (presumably) future running mate.

Ike's silence contrasted with the very public encouragement offered by Ford and other Nixon partisans. They wrote the president extolling Nixon's virtues. They expressed delight when a write-in campaign in the pivotal New Hampshire primary generated nearly twenty-three thousand votes for the vice president. And when Harold Stassen, at the time Eisenhower's foreign aid administrator, launched a quixotic effort to dump Nixon and replace him with Massachusetts governor Christian Herter, Ford added his own name to a list of twenty GOP House members calling on Stassen to resign. None of this disturbed the equilibrium of Republican delegates assembled in San Francisco that August for a convention notably lacking in drama. In the end, Herter placed Nixon's name in nomination, and Stassen was prevailed on to deliver a seconding speech for the man he had tried to oust from the ticket.

So cut-and-dried were the proceedings that Ford skipped them altogether, the only such GOP conclave he missed between 1952 and 2000. Family obligations took precedence. With the birth that May of their third son, Steven, Betty had all she could handle, even with the vital assistance of Clara Powell. Betty and the boys spent much of the summer at a newly built cottage at Ottawa Beach in Holland, Michigan. Jointly owned by Ford and his three brothers, the single story, log-sided structure overlooked Lake Michigan, in whose chilly waters the Ford children swam and sailboated. In August 1956 Jerry joined them there in preference to the droning, boastful oratory of San Francisco.

After less than a decade in office, Ford was someone worth knowing, on national security and military affairs someone worth listening to. Official Washington took note. At the invitation of Admiral Hyman Rickover, Ford accompanied him and the crew of the USS *Nautilus* on an overnight trip aboard the world's first nuclear-powered submarine. He socialized with General William Westmoreland, the future commander of US forces in Vietnam, whom he had first befriended on his Korean inspection tour. Ford's position at the nexus of Appropriations and military oversight paid rich dividends for Michigan residents. With his assistance, $100 million worth of defense contracts were reallocated to help cushion the loss of jobs in the state's auto industry. Ford sided with Chrysler and General Motors, prime contractors for the army's Jupiter ballistic missile, over the rival Thor program entrusted to McDonnell Douglas and favored by the air force. He even leaned on the army to rescind its halt on the purchase of Michigan cherries.

In what was becoming a biennial ritual, Michigan Republicans who tried to draft Ford to run for governor found him no more seriously tempted in 1956 than in previous years. His reluctance to abandon the Fifth District for a costly campaign and a job he didn't particularly want says much about Ford's political judgment, and even more about his indifference toward the executive role. For most of his quarter century on Capitol Hill he administered nothing larger than a personal staff of four or five. Even after he became minority leader in 1965, his office payroll didn't exceed two dozen—and, strictly speaking, they didn't report to him at all, but to his administrative assistant. After eight years in the job, John Milanowski wanted to return to Grand Rapids to practice law and raise a family.

In his place Ford turned to Frank Meyer, a onetime city councilor and civics teacher from the lakeside community of Grand Haven, Michigan.* Meyer rapidly made himself indispensable to his new boss, handling personnel matters, supplying Ford with drafts of his biweekly newsletter and radio broadcasts, and allocating the congressman's modest campaign budget among print, radio and television advertising. A classic workaholic, Meyer thought nothing of calling farmers with legislative updates at five in the morning. He maintained his obsessive habits up to the day in August 1972 that he collapsed and died at his desk, the victim of a massive heart attack.

By the late '50s, members of Congress received an annual salary of $22,500, plus $3,000 in tax write-offs to help offset the cost of a second residence. Office budgets were even leaner: $35,000 a year to support the entire staff, plus a $1,200 allowance for stationery, and $1,800 to rent and maintain a district office. Until he became minority leader, Ford hired staff exclusively from Michigan, with a single exception. Mildred Leonard was a Californian drawn to Washington in search of work. Never married, the product of a large, fervently Catholic family, she had grown disenchanted with the pressure-cooker atmosphere surrounding the Republican leader, and sometime Speaker, Joe Martin. Seeking a less stressful alternative, she was directed to Ford's office in the Cannon Building. "A very wonderful lady" in the estimation of her new employer, Leonard was "loyal, efficient, tireless." Conservative in dress and habitually pleasant, she was shrewd enough to spot the self-seekers who trail after important men, and tactful enough to divert them from their objective. It was no wonder that Ford viewed her as the perfect personal secretary.

J ANUARY 20, 1957, falling on a Sunday, Inauguration Day was postponed twenty-four hours. Sitting with other members of Congress fifty feet from the podium where Dwight Eisenhower recited his presidential oath for the second time, Ford may well have reflected on how far he had come in the eight years since he and Betty were threatened with eviction from Harry Truman's inaugural platform. Even now, however, Republican feelings of triumph were muted by the party's failure to regain control of

---

* Tall and spare, a devout Christian Reformed churchgoer who didn't smoke, drink or swear, Meyer liked to quote an elderly parishioner on his decision to work for Congressman Ford in sinful Washington, DC: "I've known Frank a long time," she remarked censoriously. "He was such a fine Christian gentleman." AI, Gordon VanderTill.

Congress on Eisenhower's coattails. At the start of the 85th Congress the GOP counted twenty fewer House members than in 1953, and one less senator.

Ten days after the Eisenhower inaugural, Ford entered Bethesda Naval Hospital for treatment of a slipped disc. The effects were short-lived; within days he was back on the House floor, decrying the cost of a new legislative office building and the proliferation of overpaid government consultants. In case anyone should doubt his full recovery, in June 1957 Ford hit a grand-slam homer in the annual congressional baseball game. Symptomatic of his party's fortunes in this blighted year, it wasn't enough to prevent the Democrats from winning 10–9. One month later, America's pastime was to provide a semicomical backdrop to the final expansion of the Ford family. Like her mother before her, Betty had reached middle age without a daughter. Nearing forty, she considered her childbearing days behind her. More than satisfied with their three sons, husband and wife lost little time regretting what they did not have. (Betty did say that had she known there would be four Ford children, she would have designed a home with more bedrooms.) Much to her surprise, while accompanying her husband on his European tour in the last weeks of 1956 Betty had been plagued with recurring bouts of nausea. Could history possibly be repeating itself? she wondered. As soon as she returned to the States, her pregnancy was confirmed by her doctor.

On Saturday, July 6, 1957, Betty woke at dawn, "swollen and sweaty," and began to cry. Eventually she cried herself into labor, whereupon Jerry packed her into the family station wagon for a headlong ride to the nearest hospital. His haste had less to do with Betty's impending delivery than the one o'clock start of a baseball game between the sad-sack Washington Senators ("First in war, first in peace, and last in the American League") and the world champion New York Yankees. Ford had promised to take Mike and Jack to Mickey Mantle Day at Griffith Stadium. At the hospital Betty's attending physician watched the game on television, hoping to catch a glimpse of the Ford men in the stands. Unexpected as her fourth child may have been, the real shock came, conveniently enough, in the seventh-inning stretch. Then, or so she claimed, is when Betty Ford gave birth to a seven-pound, five-ounce girl she named Susan Elizabeth.

"There goes the guest room," remarked the child's father on learning the news. His flippant response in no way foreshadowed the special relationship he would enjoy with the little girl he first held in his arms that Saturday night. Back on Crown View Drive, waiting with open arms of her own, was Clara Powell. The Ford family housekeeper was much more than that title suggests. Jerry was only half joking when he said that he couldn't stay in Congress were it not for Clara. A childless woman who became a surrogate parent, to Susan Ford she was "the glue that kept us all together." More than that, "Clara kept Mother calm when she was out of control and frustrated when Dad was traveling and not available and Mother was trying to raise four kids. We all used to fake being sick so we could stay home with Clara instead of going to school."

Clara read to the Ford children, and indulged them in their passion for animals. In addition to dogs, ducks, chickens, rabbits, birds and tropical fish, there was a black snake that Susan kept under her bed until the day it escaped and terrified a cleaning girl who encountered it in a bathroom. Clara's attendance at Susan's christening in an Alexandria church led at least one family to withdraw from the parish in protest. Inevitably the subject of race came up in the Ford household. "There was just no discussion," Susan remembers. "I mean, Clara was family." When her employers traveled abroad, sometimes for a month or more, Clara moved into their bedroom on Crown View Drive. She taught the Ford kids about boxing and her hero, heavyweight champion Cassius Clay (the "slave name" by which he was known until he changed it in 1964 to Muhammad Ali). Clara Powell could laugh as hard as she worked, never more so than when Betty spun a favorite record of the Fred Waring Singers performing "Get Down on Your Knees and Pray" as she and her housekeeper scrubbed the floors and sang along until both women, covered in soap suds, exploded in laughter.

That Betty Ford's children meant everything to her was no secret, especially as Jerry's career demanded more and more of his energies. With a candor that would do his parents proud, Mike Ford describes a consuming drive in his father that "created great achievements, but it was also his blind side. And it was a weakness that I think contributed to some of the pain we experienced as a family." Susan agrees, but with a twist. "He liked people . . . he became the Energizer Bunny when he was around people. And I'm not going to say he didn't like being home and he didn't like reading his newspaper. He loved watching his boys play football; he loved being with the family. But the man liked people and it was like food to him, I guess is the best way to put it."

Ford himself conceded her point. "An old timer in Congress once told me a Congressman either ought to be a bachelor or wait until his kids are all grown up before running for election," he remarked after a decade on Capitol Hill. His congressional calendar juggled meetings of the Michigan Milk Producers with evening sessions at the local PTA. Most Saturdays Mike and Jack Ford accompanied their dad to the Capitol as he got his hair cut and caught up on office mail. The boys had license to explore the back rooms and alleys of the vast building, at least until they were apprehended by security guards, who escorted them back to the congressman's office.

At which point "he would always stick us in front of a typewriter and let us practice our typing and we'd type a letter to Mother," Mike says. "We'd hunt and peck and tell her what a great mother she is and how hard Dad is working." Ford proofed what they had written and made sure the letters were properly signed before they were delivered to their intended recipient. In time Steve and Susan would engage in the same ritual, with the added incentive of playing hide-and-seek in Statuary Hall. Newspapers figured prominently in Steve's memories of 514 Crown View Drive. After an early morning swim, coffee and the papers, his father left for the Capitol. In his wake, the Ford children would grab or wrestle for a favorite section of the morning paper. "And you didn't really talk much at breakfast . . . you had to sort of suck in some facts because you knew there would be a discussion that night at dinner and you had to defend your position."

Another Ford family tradition began in the winter of 1957. Both parents enjoyed skiing, and they wanted their children to share in the fun. That Christmas, and for many more to come, the family traveled to Boyne Mountain in Northern Michigan, where the five-hundred-foot vertical drop provided sufficient challenge for beginners and experts alike. Susan was five years old when she graduated from the skating rink to her first downhill run. Her brothers all played football, bonding them to their dad in ways that she resented. At the start of each season Ford circled all the home games in which his sons might play, and had Mildred Leonard mark those dates off-limits in his calendar. As his children grew into adolescence and Ford spent much of his time on the road pursuing his dream of the Speakership by electing more Republicans to the House, he made it a habit to call home every evening to talk with the youngsters lined up beside a phone in Betty's room. After telling the children how much he missed them, Mike Ford recalls, "then he'd say, 'Take care of Mom. Do what Mom says. Be good for Mom.'" With the kids in their beds, Ford would call back to catch up on the day's news with Betty.

The parallel with her parents' long-distance marriage could not have escaped her. Once again, it seemed, Betty had married a traveling salesman. "What are you doing here?" she allegedly asked her husband, after rolling over in the middle of the night when she thought he was on the road, only to discover his bulky form stretched out beside her. One week a month, Betty carpooled half a dozen neighborhood kids to school. Lunch might be with the International Wives Group, comprised of ten women from the foreign diplomatic corps and as many congressional spouses. Wednesdays were reserved for the 81st Congress Wives Club, Thursdays set aside for marketing after Betty found that by taking advantage of cut-rate pricing available exclusively on that day of the week she could trim twenty cents per pound from her weekly meat bill. On Friday, coffee with the Republican Congressional Wives Forum preceded afternoon tea at the Congressional Club. Once a week she attended a modern dance class, a reminder of the life she had once fantasized about.

For three years Betty worked with the Westinghouse Broadcasting Company on "The Government Story," a series of forty half-hour programs explaining their democratic institutions to millions of television viewers. Her own knowledge of the subject had nothing to do with textbooks. Jerry might be on the road, but that didn't stop phone calls from favor seekers trying to enlist her help. "Some of the constituents are convinced that the way to influence a congressman is through his wife," Betty confided to a reporter for the *Detroit News* in May 1961. On another occasion she observed, "Sometimes I think my life is one long succession of picking up and putting away. When I feel tense I just change jobs . . . and believe me, there is always a new job waiting." Amid the marketing and budgeting, the rides to football practice and the emergency room; between den mothering Cub Scouts and preparing Sunday-school lessons, seeds of suburban disenchantment were taking root.

F OR ALL HIS personal popularity, Eisenhower would not escape the second-term curse afflicting presidents since George Washington. White House chief of staff Sherman Adams, a famously frosty Yankee, made an exception for Ford, telling him, "You are one of the guys I happen to like to see around." He wouldn't see him after September 1958, when Adams was forced to resign following revelations that he had accepted gifts from a shady Boston businessman whose case he had raised with the Securities and Exchange Commission. Adams's departure capped a season of disappointments for the aging administration. Overseas, Ike's hopes for a suspension of nuclear testing were stillborn. At home the economy flashed early warning signs of recession, which in turn spurred the Democratic Congress to bust the Eisenhower budget. A White House–sponsored civil rights bill, the first since Reconstruction, was watered down by Senate Majority Leader Lyndon Johnson of Texas.

On September 11, 1957, Eisenhower took a moment from his escalating confrontation with Arkansas governor Orval Faubus over the court-ordered integration of Little Rock's Central High School to acknowledge Ford's stalwart defense of the president's foreign aid and military requests. "From all sides I had reports commending your effective efforts," Eisenhower wrote, "and for such responsible service to our country and loyal support of our Party's efforts, you have my heartfelt thanks." The crisis in Little Rock was quickly followed by something close to hysteria when the Soviet Union hurled *Sputnik*, the first spacecraft to achieve Earth orbit, into the heavens. Eisenhower refused to join the panic. His personal enthusiasm for space exploration was limited. "Look," he told his Cabinet, "I'd like to know what's on the other side of the moon, but I won't pay to find out this year."

He agreed instead to the establishment of a new federal space agency, the National Aeronautics and Space Administration (NASA). Speaker Rayburn appointed Ford to the ad hoc committee that would draft the necessary legislation. It was his first opportunity to work closely with Lyndon Johnson, whose controlling tendencies were evident from the outset. As Ford remembered it, "Johnson elected himself chairman, and boy, did he operate." With passage of the National Aeronautics and Space Act in June 1958, Ford joined a new, thirteen-member House committee overseeing the American space program. At a time when the US was playing catch-up to the Soviets, he urged creation of "savings bonds for science," enabling American citizens to help finance their country's participation in the developing space race.

His proposal went nowhere, but at least it was a constructive contribution to the national debate, unlike his scapegoating of former president Truman and his administration for blocking ICBM development by canceling a $75 million R&D contract with the Conair Corporation. Pressed by reporters for a reaction, the former president laughed off Ford's allegation. Collegiality still ruled Capitol Hill in the 1950s, even in an election year. John McCormack was a New Deal Democrat, closely associated with the pre–Vatican II Catholic Church that retained its spiritual grip on McCormack's South Boston and Dorchester constituency. Famously devoted to his wife, it was said that the couple had not missed a dinner together since coming to Washington in 1932. McCormack's allegiance to his party was just as intense.

Yet McCormack was also a gentleman of the old school. He demonstrated as much at the end of May 1958, following an hour-long meeting with President Eisenhower at the White House. The president was highly pleased with their joint efforts on the infant US space program, McCormack reported to Ford, singling out for praise the bipartisan decision to entrust NASA to civilian control. Beyond this, added McCormack, "I have always admired you as not only a person, but an able and courageous legislator." A very different message was being delivered to Ford's constituents by McCormack's Bay State colleague and sometime rival Jack Kennedy. Six years had passed since Ford's neighbor in the Old House Office Building made his escape to the Senate. By 1958, Kennedy's popularity in Massachusetts was such that he could take reelection for granted. This freed him to campaign for Democrats around the country, some of them in politically hopeless districts like Michigan's Fifth.

Ford's increasing prominence made him an attractive target in what was shaping up to be a Democratic year. Nineteen fifty-eight saw unemployment hit a postwar high of 7.7 percent. A decade after he was first sent to Congress with union support, Ford confronted his most serious opponent to date: thirty-six-year-old Richard Vander Veen, a Harvard-trained lawyer with an impressive war record and the not-inconsiderable resources of the United Auto Workers and other unions to draw on. Railing against the incumbent as an "Eisenhower echo," Vander Veen called for the resignation of Secretary of State John Foster Dulles and invited Senator Kennedy, already market testing themes for a 1960 presidential run, into the district to accuse Republicans of neglecting US military strength. Foreign policy dominated a series of debates between the rival candidates, with Vander Veen on the attack and Ford vigorously disputing Kennedy's claims that Eisenhower had ceded military superiority to the Russian bear. Beyond this, Ford supporters touted his near-perfect attendance record, his widely praised constituent service and his selection, along with Missouri's Richard Bolling, as subjects of a *Newsweek* cover story on the most promising members of Congress.

In November the GOP suffered its greatest thrashing since FDR's 1936 landslide. Ford, however, emerged from the wreckage with over 63 percent of the vote, a margin only slightly reduced from his 1956 record. Their rout at the polls bred dissatisfaction in the greatly reduced ranks of Capitol Hill Republicans. A scapegoat being required, much of the blame for the party's dismal showing was directed at Joe Martin, the seventy-four-year-old minority leader. In his blue serge suit and box-toe shoes, Martin hardly offered an image of fighting vigor. It didn't help that Martin and Sam Rayburn both lived at the Occidental Hotel, and were often seen dining together. "You could always tell what they had for dinner because Joe had it on his tie," claims one Capitol Hill staffer of the period. Under his somnolent leadership, the House Republican Policy Committee had grown rusty with disuse.

Demands for change began to percolate up through the ranks. No one wanted to humiliate Martin. "You can keep your office, keep the limousine . . . and the rest of it," he was told. The old man dug in his heels. Leon Parma, then chief of staff to Representative Bob Wilson of California, recalls a postelection request from his boss to round up Mel Laird and other members of the SOS group, Republican members first elected in the Eisenhower sweep of 1952. A consensus quickly formed regarding the need for fresh leadership. When the rebels approached Ford, they were told that, while the congressman would take the job if it was offered to him, he had no intention of campaigning for it, or of joining a cabal "to throw Joe Martin out."

Charles ("I am a gut fighter") Halleck entertained no such reservations. Sixteen years Martin's junior, Halleck was the son of an Indiana state senator whose name he amended to Abraham Lincoln Halleck because it sounded better in Lincoln Day speeches. At Indiana University, Halleck earned a Phi Beta Kappa key and served as student body president. He was still in law school when he ran successfully for prosecuting attorney of his native Jasper County. Following the death of a Republican congressman days after the 1934 elections, Halleck won the first of his seventeen terms on Capitol Hill.

His rise thereafter was rapid. At the 1940 GOP convention in Philadelphia, Halleck's rousing nominating speech set Wendell Willkie on the road to nomination. Eight years later, as majority leader in the Republican 80th Congress, "Available Charlie" delivered the Indiana delegation to New York governor Thomas E. Dewey in advance of the first convention roll call. Expecting to be Dewey's running mate, Halleck cried double cross when California governor Earl Warren got the nod instead. In 1952, Halleck was again passed over, though his name appeared on Eisenhower's short list of acceptable vice presidential candidates. With his party's return to minority status in 1955, Halleck lined up the necessary votes to replace Martin as Republican leader, only to be stymied by Ike's reluctance to sanction mutiny in the ranks. Three years later he notified the president of his intention to retire, then let Eisenhower talk him into running again. Having barely survived that year's Democratic tsunami, Halleck returned to Washington the presumptive favorite to replace Martin *if* he could rally to his side younger members moved to rebel by their all-too-fleeting taste of power during Ike's first years in office.

"I'm going to an execution," Martin remarked outside the GOP caucus the morning of January 7, 1959. Demands were made for a secret ballot, which Halleck won by a single vote—one vote more, as it turned out, than the number of accredited GOP lawmakers present. A second ballot produced a 74–70 victory for the challenger. The winning margin was supplied by Ford's Michigan. Yet Halleck made no effort to reward Ford or include him in the reconstituted leadership. In April 1959, Republican National Chairman Meade Alcorn quit after just twenty-six months on the job. Ford's name surfaced as a possible replacement, until the White House expressed its preference that he remain where he was. He didn't require much persuading; among his Republican colleagues a sufficient number had retired or been ousted in the November debacle so that Ford, overnight, climbed four places on the Appropriations Committee roster.

His working partnership with George Mahon on the defense subcommittee was closer than ever, too close for some who faulted the bipartisan rituals with which each year's military budget was rubber-stamped by the full committee. In June 1959, Ford joined President Eisenhower, Queen Elizabeth II and Canadian prime minister John Diefenbaker for the formal opening of the St. Lawrence Seaway. Afterward he journeyed to Ottawa and Montreal for joint discussions with members of the Canadian Parliament. Ford became active in the Inter-Parliamentary Union, the world's oldest organization promoting cooperation among nations while upholding principles of representative democracy.

The IPU had fifty-seven members in August 1959 when Ford arrived in Warsaw for the group's first meeting behind the Iron Curtain. Away from the conference hall, Ford spent an evening talking informally with a group of Polish students who were thoroughly disillusioned with Communism. The experience strengthened his belief in Polish resistance to the Soviet bear. From Warsaw, Ford traveled to Moscow for three days. Together with Betty he attended a performance of the Bolshoi Ballet and toured the special exhibition on American life in whose mock-middle-class home Vice President Nixon had conducted his famous "kitchen debate" with Soviet premier Nikita Khrushchev.

Ford's stock rose as his horizons broadened. Before a national gathering of Young Republicans earlier in 1959, Nixon had dropped his name as a prime example of the "new leadership" on which a forward-looking GOP could depend. Ford reciprocated, calling Nixon "the most able, competent individual in public life today." By then his domestic travel schedule was crowded with events far removed from West Michigan, in content as well as geography. Marveling at his stamina, less energetic lawmakers scratched their heads over the reluctant candidate who had spurned offers to run for the United States Senate, the governor's office in Lansing, and the House Republican leadership, an obvious stepping-stone to the Speakership he plainly coveted. As it turned out, that was not the only office to which Jerry Ford aspired. After a decade in Congress, he wanted to be vice president of the United States.

9

THE CONGRESSMAN'S CONGRESSMAN

---

*I have been pushing you as one of my two favorite candidates for
the presidency; the other one being Thruston Morton. Any man
who can take a thirteen on Pine Valley's number seventeen and
come home laughing, to me, is the kind of fellow I will back.*

—Barry Goldwater to Gerald R. Ford, July 1963

Too many people try to move too fast in politics." At the time he
made this observation, in December 1959, Ford hoped to be the
beneficiary of what the *Omaha World-Herald* called "a mild but interesting
boomlet" for his party's vice presidential nomination. First advanced by
Grand Rapids supporters four months earlier, the idea was taken seriously
enough by the presumptive Republican nominee for president. According
to Ford biographer Jim Cannon, Richard Nixon was behind a Ford trial
balloon raised in *Newsweek*'s January 10, 1960, issue by venerable political columnist Raymond Moley. His blind item described the ideal Nixon
running mate as "a young man from the House of Representatives who
has shown a capacity to learn fast." Moley was more explicit in a follow-up
column dated February 13, stressing Ford's potential value in promoting a
Nixon legislative agenda on Capitol Hill.

Bolstering the case for Nixon's involvement is a simultaneous endorsement of Ford by Earl Mazo, chief political correspondent of the *New York
Herald Tribune* and an early Nixon biographer. But then other publications,
well outside the Nixon orbit, weighed in with equally positive appraisals.
National Review labeled the congressman from Grand Rapids "young,
attractive and conservative." The *Wall Street Journal* pronounced him "a
smashing candidate . . . if only he were better known." Hoping to rectify
this deficiency, Ford took to the road. In the second half of 1959 he appeared before audiences in Boston, New York and Williamsburg, Virginia.
He addressed the AFL-CIO Legislative Conference, and spent a week at
the Aspen Institute in Colorado. In suburban Chicago he shared a debating
platform with former Democratic national chairman Paul Butler. A tour of

Southern California defense installations concluded with a speech before six hundred Republican faithful at a Los Angeles hotel.

"Some of my friends are talking me up as a favorite son candidate for vice president," Ford acknowledged. Whether he became one in fact, he continued coyly, depended on "how eager they are." Personally Ford thought it would be "a very healthy thing" for the GOP to have an open contest for the second spot on the ticket, much as Adlai Stevenson had generated excitement by throwing open the vice presidential balloting at the 1956 Democratic convention. Mindful of the attack-dog role assigned the bottom half of the ticket, Ford ratcheted up the partisan content of his speeches. Yet his voting record remained centrist, on issues of human rights slightly left of center. Praising Nixon as more supportive of civil rights than Ike, in February 1960 Ford signed a discharge petition releasing the administration's latest civil rights bill from the hostile grip of the House Rules Committee. He also endorsed a constitutional amendment extending voting rights to eighteen-year-olds.

In the same month Ford got his biggest break to date when a revitalized House Republican Policy Committee named him to chair a prestigious task force on American defense needs. Both the Eisenhower White House and the Nixon campaign hoped to counter criticism of US preparedness leveled not only by Democrats like Jack Kennedy, but also by Nelson Rockefeller, the first-term Republican governor of New York whose family-funded Rockefeller Reports did much to instill fears of a yawning missile gap between the United States and the Soviet Union. It was to refute such alarmist thinking that the Ford Committee was established. Consisting of fifteen House members and nearly two dozen academic experts, economists, military men and independent scholars, the group spent four months pondering the strategic challenges likely to confront the United States during the 1960s.

Its work took on added significance after the May 1 downing of an American U-2 spy plane by a Soviet surface to air missile. Two weeks later Soviet premier Nikita Khrushchev stormed out of a major power summit conference in Paris. Against this backdrop of escalating Cold War tensions, *American Strategy and Strength*—more commonly dubbed the Ford Report—was published in June to largely positive reviews. Its authors began by stating the obvious: with land- and sea-based nuclear missiles backed by 2,000 long-range strategic bombers, 16 wings of tactical aircraft, 14 aircraft carriers encircling the Soviet Union and 250 strategically located bases in allied nations, the American arsenal was unmatched in power and versatility. Describing such weaponry and its retaliatory/deterrent force as "largely indestructible," Ford and his experts sought to reassure a jittery populace with a blueprint for national defense that involved economic, scientific and diplomatic—as well as military—components.

They began by questioning the Truman-Acheson policy of containment, which risked attritional warfare (as in Korea) waged on multiple fronts and on the enemy's terms. By contrast, the report's authors claimed, the US under Eisenhower had resolved to punish Communist aggression "at the time, place, and with weapons of our choice." CIA-backed coups in Iran and Guatemala went unmentioned for obvious reasons, but the point was made. So, more explicitly, was the importance of psychological warfare. The Soviet Union's recent leap into space, no less than exaggerated Soviet claims of economic progress, was part of a larger plan to undermine Western confidence.

The Ford Report offered unqualified support of Eisenhower's Cold War stewardship. The same week it was published, Ford led a successful effort to restore $500 million in Mutual Security funding opposed by many within the president's party. Some of the money was earmarked for development of India's Indus River basin. The lion's share, however, took the form of military assistance for US allies in Europe and Asia. From Taiwan, a globe-trotting Eisenhower sent profuse thanks for Ford's help. The foreign policy establishment was equally generous with its praise. For the third time in six months, Ray Moley extolled the congressman's virtues in his *Newsweek* column. An interview on television's *Meet the Press* fell through, but the show's host hoped to have Ford as a guest on some future broadcast. "I have heard many nice things about him," Lawrence Spivak told Ford's schedulers. "I would certainly want to interview him if he got the Vice Presidential nomination."

It was to advance this long-shot possibility that Grand Rapids mayor Paul Goebel, whose association with Ford went back to their days in the Republican Home Front, announced formation of a campaign committee to promote his friend's candidacy nationally. Ford himself sent mixed signals. Discouraging any "big loud celebration" on his behalf at the Michigan GOP state convention in May, he voiced public support for Thruston Morton, the Kentucky senator who enjoyed greater recognition by reason of his stint as Republican National Committee chairman. Lacking funds to hire a staff or pay for advertising, Ford relied on Michigan's forty-six delegates to promote his cause. Goebel's Grand Rapids committee sent out two mailings to supposed power brokers. Responses were friendly if noncommittal. Among neighboring states, Illinois was already pledged to Senator Everett Dirksen. An Indiana businessman, though intrigued, remained wary:

"Ford's main difficulty is that no one knows him," he explained. The same complaint was voiced by Ohio party chairman Ray Bliss: Rockefeller, Morton and UN ambassador Henry Cabot Lodge were the names most often discussed in the Buckeye State.

No Ford endorsement was more surprising than that of Whittaker Chambers, the former *Time* correspondent whose sensational accusations directed at Alger Hiss had fueled the early stages of Richard Nixon's national political career. "Caliber-ability-fresh approach"—so Chambers summarized Ford's appeal in recommending him to his friend the vice president. Reinforcing his argument, a *Newsweek* poll of Capitol Hill correspondents ranked Ford the second-ablest member of Congress, behind Charlie Halleck but ahead of both Sam Rayburn and Jack Kennedy. Keenly aware that the ultimate choice of a running mate lay with the presidential nominee, Ford spurned a national headquarters or elaborate fundraising operation. "I am completely realistic," he told Henry Ford II, "but sometimes the 'underdog does win.'"

Betty Ford shared her husband's fatalistic outlook. Though "terribly exciting and wonderful to think about," the prospect of a national campaign raised fears about the demands on her young children and, implicitly, herself. At the end of June, Betty welcomed members of the Chowder and Marching Club to a backyard party at 514 Crown View Drive. Among those in attendance were Dick and Pat Nixon. Afterward Mrs. Nixon expressed pleasure at meeting the Ford children. "You must be mighty proud of them!" she told Betty. A week passed before Nixon subtly tipped his hand. Thanking Ford for his recent insertion of a friendly column in the *Congressional Record,* the vice president wrote, "As always, I am deeply grateful for an ally in the House whose friendship and support is so articulate and effective."

The message was implied, but unmistakable to anyone fluent in the language of political nuance: Ford in his current position was too valuable to spare, a condition that victory in November would only magnify. Unless he took such flattery at face value, Ford should have recognized the Judas kiss accompanying the demise of his vice presidential hopes. "If I feel I have a chance, and that the nod has not already gone to somebody else," he remarked on July 9, squeezing two qualifiers into a single sentence, "then my backers will do their utmost." In the convention town of Chicago, Ford accepted invitations to appear before half a dozen delegations. The exposure couldn't hurt, and it was sure to gratify his supporters back home. Ironically, his largest audience came on the last day of the convention, when he delivered a seconding speech for Henry Cabot Lodge, chosen as Nixon's running mate to emphasize foreign policy and appeal to the Rockefeller wing of the party.

Ford had reason to feel good about his debut on the national stage. Performing well as a Nixon surrogate and in his dealings with the press, he emerged from the experience with plenty of new friends, and no enemies. As the fall campaign progressed, he could take perverse satisfaction in the lackluster performance of Cabot Lodge, the Boston Brahmin whose credentials exceeded his energy level. Long afterward, confident that his comments would go no further, Ford betrayed a resentment nurtured for thirty years. "Forget whether I was qualified or not," he told a friend. "I would have been a hell of a lot better candidate than Lodge."

IN SEPTEMBER 1961, Gerald Ford was hailed as the "Congressman's Congressman" by the American Political Science Association. "A moderate conservative who is highly respected by his colleagues of both parties," the group's citation read, "he symbolizes the hard working, competent legislator who eschews the more colorful, publicity-seeking roles in favor of a solid record of accomplishment in the real work of the House: committee work." Grateful as Ford was for such recognition, the praise feels tepid in the context of a hyperbolic new era in American government. Call it the Age of Political Celebrity. Certainly no one did more than Dwight Eisenhower's successor and his glamorous First Lady to introduce star quality to the White House.

Long before his administration was immortalized through comparison to a Broadway musical about King Arthur and Guinevere, John Kennedy displayed a mastery of political theater at carefully staged press conferences and in Oval Office addresses to the nation. Hollywood handsome, as effective on TV as FDR had been on radio, JFK brought to the presidency a skill set ideally suited to the confrontational dramas of racial injustice at home and nuclear brinkmanship on the global stage. Along with youthful vigor he made personal charisma a prerequisite of the nation's highest office, rewriting the job description while raising popular expectations to a level few of his successors could hope to match. That included Gerald Ford. As it happened, the 87th Congress marked the halfway point in Ford's career on Capitol Hill. In January 1961 he stood seventeenth in seniority among House Republicans. Henceforth the lure of institutional advancement was to exercise a gravitational pull stronger than any lingering trace of youthful insurgency.

Ford demonstrated as much in his attitude toward the House Rules Committee, long considered a dead end for civil rights legislation and other items on the liberal agenda. As a newcomer to Washington he had upset his elders by supporting the twenty-one-day rule, effectively breaking the committee's stranglehold. A dozen years later, following the breathtakingly close election of a Democratic president *and* a Republican gain of twenty-one seats in the House, where the GOP remained very much in the minority, Ford announced a change of heart. "In light of additional experience and observation, I believe the Committee is generally responsive to majority wishes," he informed constituents. In the event, Republican votes supplied the margin of victory whereby the new administration succeeded in expanding the Rules Committee. This time, however, Ford was not among them.

A larger Rules Committee, theoretically friendlier to JFK's New Frontier, was part of the legacy bequeathed to the Kennedy White House by Speaker Rayburn. At the end of August 1961, racked by thyroid cancer, Rayburn left Washington for his Texas home. His death three months later deprived the administration of a hugely experienced strategist and advocate, one who enjoyed the universal respect of his colleagues. Ford liked to quote Mr. Sam's tribute to constructive statesmanship: "Any donkey can kick a barn door, but it takes an awfully good carpenter to build one up." For the first year of the Kennedy presidency, Ford balanced his carpentry skills against the growing demands of party loyalty and personal ambition. Believing that Kennedy should have the same authority to trade with Communist countries as his Republican predecessor, Ford was equally supportive of JFK's Alliance for Progress designed to improve living standards and prevent the spread of Communism in the Southern Hemisphere. Weeks after the Bay of Pigs fiasco, Ford stood in the Oval Office, the lone Republican in a high-profile group of New Frontiersmen that included Vice President Lyndon Johnson and former president Truman, as Kennedy affixed his signature to a $600 million aid package for South America.

After hearing a breakfast presentation from his Yale Law School classmate Sargent Shriver, Ford voted $40 million to launch the Peace Corps. He marked the tenth anniversary of Arthur Vandenberg's death by praising Kennedy's vow to resist Communist aggression in the remote Southeast Asian nation of Laos. Following a disastrous Vienna summit meeting between Kennedy and Khrushchev in June, the Soviets moved to expel US, British and French forces from the divided city of Berlin. With Cold War tensions escalating, Kennedy went before television cameras on July 25 to offer a sobering assessment of the US-Soviet standoff and to seek $3.25 billion to enlarge and modernize US forces. Speaking for the GOP, Ford broadly endorsed Kennedy's Berlin policy, with one significant caveat.

Accepting the need for additional spending on military readiness, he differed from the president in his call for offsetting cuts to domestic programs.

He was not always so economy-minded. Foreign aid dominated an hour-long meeting between Ford and Kennedy on March 23, 1961, part of a sustained courtship by the White House that included invitations to a white-tie dinner for the Shah of Iran as well as the legendary Mount Vernon soiree for the visiting president of Pakistan on a magical summer's night that July. Meticulously planned by Jacqueline Kennedy, the unprecedented event was staged under the stars in a candlelit pavilion overlooking the broad Potomac. An honor guard attired in Continental Army dress greeted guests on George Washington's bowling green, scene of an after-dinner concert by the National Symphony. When the last notes of Gershwin's *An American in Paris* died away, JFK beckoned to Gerald and Betty Ford to return to Washington on the presidential yacht *Honey Fitz*. Betty danced all the way back to the dock.

That the Fords should find their names on the most exclusive guest list of the season was no accident. JFK had long since taken the measure of his onetime House colleague. He knew that Ford's opinion carried weight among lawmakers of both parties, especially on issues of military spending and foreign aid. The latter, friendless as usual, was at the mercy of Representative Otto Passman, a Louisiana Dixiecrat who once declared, "If I had three minutes to live, I'd kill the Peace Corps." Passman chaired the House subcommittee dealing with foreign aid, the same committee on which Ford sat as ranking Republican. Faced with Passman's unrelenting hostility, JFK turned to the loyal opposition for help in passing a military–economic aid package worth $4.8 billion.

Summoned to the White House for a Saturday-morning strategy session, Ford told the president, "The only way I can square this with my kids is to get your autograph." Sitting at his desk, JFK wrote out, "Dear Jack: Wish you were here. Jack Kennedy." Then he repeated the gesture for Jack's brother Mike. Ultimately Ford supported the administration against his own party leadership. Kennedy got $1.6 billion for military assistance, only slightly less than his original request. Not long afterward the president asked Ford to join him for an overnight cruise on the nation's newest aircraft carrier, the USS *Enterprise*. The two veterans of the naval war in the Pacific watched as a contingent of marines staged an amphibious landing for their benefit. In September 1961, Ford visited Berlin, front line of a global conflict whose flash points he had previously observed in Korea, Vietnam and Poland.

Increasingly his thoughts turned to Asia. The United States must be prepared to send troops to the region, Ford asserted, to protect South Vietnam and the Philippines from Communist encroachment. Thoughts of his own wartime service prompted a plaintive inquiry on the floor of the House: "What right does this generation of Americans have not to expect to have to make sacrifices?" Behind the awkward syntax hovered the ghosts of Munich and the battle-tested conviction that appeasement only invited fresh aggression. Harshly critical of the Kennedy Administration's domestic spending, Ford opposed most forms of federal aid to education, directed assistance to economically depressed areas and creation of a new department of urban affairs. On foreign policy, however, he gave JFK the same level of support as he had extended to Ike. The Cold War consensus held both ends of Pennsylvania Avenue in its grip.

NINETEEN SIXTY-TWO WAS an election year, a chance to test Kennedy's soaring approval ratings against the traditional losses inflicted on the party in power during nonpresidential years. The new year was less than a month old when Ford took a phone call from his sister-in-law Janet in Grand Rapids. Five days after slipping on the ice outside Grace Episcopal Church, Gerald Ford Sr. had succumbed to a heart attack at the age of seventy-two. "His was a good life, despite the hardships which did befall him," Phil Buchen consoled his friend Jerry. "And I believe that his greatest satisfaction of all was in your accomplishments, of which he was always so proud." On the day of the funeral Ford was stunned to find Grace Church crowded with friends and seeming strangers, all come to pay their respects to a man whose modest success in business hardly explained such an outpouring of public sentiment. Dorothy Ford had her own ways of memorializing her husband of forty-five years. Recalling the elaborate celebrations with which they had marked Christmas, she made sure each December to decorate a small evergreen and place it on his grave.

The loss of his stepfather produced a final change of identity for Jerry Ford. Dropping the "Jr." from his name in no way lessened his connection to the man who had been so instrumental in shaping his character. In criticizing a 1962 Supreme Court ruling that effectively outlawed prayer in public schools, Ford might have been channeling the traditionalist views of Jerry Sr. In a similar vein he predicted that Governor Nelson Rockefeller's pending divorce would hurt his chances for the 1964 Republican presidential nomination. The governor's subsequent remarriage to a woman eighteen years his junior did nothing to alter this view. "My wife disagrees with me," Ford noted. Perhaps Betty Ford's personal history in divorce court colored her thinking.

Reflecting popular interest in fashion stoked by the elegant Jacqueline Kennedy, in April 1961 the *Ladies' Home Journal* introduced readers to Betty's stylish yet frugal wardrobe in a multipage spread featuring "the thinking man's Rita Hayworth." The congressman's wife looked smashing in navy wool skirts and evening dresses bought on sale for $40. Partial to simple hats and $8.95 shoes dyed to match her various outfits, Betty was not above wearing slacks around the house, though she invariably donned a dress before taking the children to the dentist. She again made news in the summer of 1962. According to the *Washington Star*, her buttermilk pancakes caused a minor sensation at a Republican women's breakfast attended by General Eisenhower. Breaking with precedent, Ike ate four of Betty's flapjacks and asked for the recipe.*

The former president figured prominently in a controversy involving Representative Ford and the military-industrial complex denounced by Ike in his January 1961 farewell address to the nation. At stake was the XB-70 Valkyrie bomber, a technological marvel capable of flying at three times the speed of sound. To advocates like air force chief of staff General Curtis LeMay, the XB-70 (renamed the RS-70 in 1961) wouldn't simply replace the aging B-52 in the American arsenal—it would supplement, and perhaps supplant, the ballistic missiles then establishing themselves as the most fearsome weapons on the planet. Skeptics questioned the plane's staggering cost as well as its strategic necessity. One month into his presidency JFK informed LeMay that, aside from two prototype aircraft reluctantly authorized by his predecessor, the RS-70 would not be taking to the skies.

LeMay had no intention of letting the matter rest there. A cigar-chomping caricature of the power-hungry generals planning a coup against their own government in the classic film *Seven Days in May*, LeMay had powerful allies on Capitol Hill, none more potent than Carl Vinson, the seventy-eight-year-old "Swamp Fox" from Milledgeville, Georgia. As chairman of the House Armed Services Committee, Vinson's allegiance to the military-industrial complex was surpassed only by his faith in Congress as the most reliable arbiter of national defense. On March 1, 1962, Vinson's committee colleagues voted unanimously to proceed with development of the RS-70. A vote by the full House was scheduled for later in the month. At the White House, Kennedy sensed a constitutional crisis in the mak-

---

* The following year Eisenhower welcomed Ford's invitation to meet his eldest sons during a Ford family trip to Gettysburg. As Jerry and Betty looked on, the old soldier shared stories of his youth and military career with Mike and Jack Ford. What had been planned as a ten-minute visit turned into two hours that none of the Fords would ever forget.

ing. Adopting his most deferential manner, the president invited the crusty Vinson to the Oval Office, followed by a private stroll in the adjoining Rose Garden. Whatever he said there convinced the chairman to accept a face-saving trade-off. In return for an administration promise to study the RS-70, Congress would vote to "authorize" rather than appropriate funds sought by the unwanted aircraft's sponsors.

If Vinson felt humbled, he certainly didn't show it. His resentments were directed not toward the young president who had called his bluff but at the handful of legislators who had dared challenge his views in the first place. Prominent among those questioning the Valkyrie, and its venerable patron, was Gerald Ford. "Dear Jerry," he wrote on the day his compromise amendment was to be voted on, having reviewed the latest issue of the *Congressional Record*, "I am again reminded of your expressed conviction that we should use proper words in our legislation and avoid awkward sentences." After quoting an inelegantly phrased passage from the current debate in which Ford had used the word "envious" in place of "anxious," Vinson adopted his most condescending tone. "I do not know how you would interpret this sentence as to being either awkward or precisely correct," he sneered. "In any event, I don't understand it. So I want to make a bargain with you. If you will tell me what your sentence means, I will tell you what my amendment meant."

Other men, more sensitive to ridicule, might have replied in kind. Ford laughed it off. After all, he had what he—and JFK—wanted. By a vote of 403 to 0, the House effectively relegated the Valkyrie to the status of museum exhibit.

That summer Ford's alliance with his fellow Cold Warrior in the Oval Office was tested amid rumors of unusual military activity on Fidel Castro's Cuba. On the last day of August, Republican senator Kenneth Keating of New York, citing unnamed intelligence sources, raised the possibility that Soviet-built missile bases were under construction on the island, and that offensive nuclear weaponry might soon be introduced barely ninety miles from American shores.

Keating's concerns were shared by CIA director John McCone. On August 10, 1962, McCone had attended a meeting of senior Kennedy Administration officials, the self-styled Special Group (Augmented), entrusted with the planning of operations directed against the Castro regime in Cuba. The president's brother Robert was present (thus the Augmented label) when someone raised the possibility of liquidating the Cuban dictator. McCone, red-faced, protested any such discussion. Of more immediate concern to the DCI were the same reports that so disturbed Ken Keating.

He had raised the issue with Secretary of State Dean Rusk, Defense Secretary McNamara, and McGeorge Bundy, director of the National Security Council. They all insisted the missiles were defensive in nature.

McCone thought otherwise. He said as much to Congressman Ford, who was well known to McCone through his oversight work and consistent support of the intelligence community. Ford gathered from McCone that "hard evidence" of Soviet duplicity was lacking. What about high-altitude surveillance flights over Cuba by U-2 spy planes? Ford asked the DCI. These were few and far between, said McCone. As a result the administration was being deprived of critical intelligence. The two men agreed that additional sorties were needed. Ford said as much to Secretary McNamara at the Pentagon. How much the combined efforts of Ford and McCone may have influenced subsequent events is impossible to know, yet McCone personally credited Ford with increasing aerial reconnaissance of Cuba. Out of these additional flights came the visual proof that enabled the Kennedy Administration to expose Soviet recklessness in the court of world opinion.

On October 20, claiming a cold and a touch of fever, Kennedy cut short a campaign trip to Illinois and returned to Washington. There plans were finalized for a naval blockade, or quarantine, of Castro's Cuba. The secret couldn't hold much longer. It became even more tenuous after Vice President Johnson scrubbed a visit to Grand Rapids originally scheduled for October 21. By then Ford knew far more than the average American about events transpiring in Cuba, if somewhat less than Kennedy and his inner circle scanning the latest aerial photographs. Over a nonsecure phone line at his Grand Rapids office Ford learned from Pentagon comptroller Charles Hitch of the developing crisis and the options under consideration for bringing it to a peaceful conclusion. On Monday, the twenty-second, Kennedy briefed the three living former presidents before appearing on live television to reveal the presence of offensive Soviet missiles on Cuba.* At the same time he described the naval quarantine intended to force their removal while avoiding nuclear confrontation.

Echoing the political establishment in both parties, Ford immediately pledged his support to the administration. It was not unequivocal; within days he was objecting to the proposed swap of aging US missiles based in Turkey in exchange for a nuclear-free Cuba. "They're Jupiter missiles, made in Detroit," Ford sputtered. "They pack quite a punch. And they're

---

* The presidents in question were Herbert Hoover, Harry Truman and Dwight Eisenhower.

strategically located." His complaints were brushed aside, and the Jupiters went deliberately unmentioned in the public settlement by which Khrushchev sacrificed his nuclear toehold in the Western Hemisphere and the US pledged not to invade Cuba. As Americans breathed a collective sigh of relief, attention turned to the midterm elections less than ten days away. Even opponents of the administration were quick to congratulate the president on his sure-footed performance during the most fearsome chapter of the Cold War. With partisan campaigning largely preempted by the missile crisis, Republicans could only warn voters against letting their guard down.

The results were predictable. On election day Ford ran up his customary two-to-one victory margin over a Democratic sacrificial lamb. His party enjoyed much less success, gaining just two seats in the House of Representatives, and losing four in the Senate. The day's biggest shock came in California, where former vice president Richard Nixon lost his comeback race for governor to incumbent Pat Brown. Ford had been among those urging Nixon to enter the fray, reasoning that Sacramento would provide the necessary political base from which Nixon might remain in the public eye. Nixon magnified the loss at his famed "last press conference" the morning after. "Just think how much you're going to be missing," he upbraided reporters who could scarcely believe their ears. "You don't have Nixon to kick around anymore."

Inspired by his petulant display, the ABC television network scheduled *The Political Obituary of Richard M. Nixon* for Sunday night, November 11. The half-hour program, hosted by newsman Howard K. Smith, featured comments from Nixon's original political guru, Murray Chotiner, as well as his 1946 congressional rival Jerry Voorhis. It gave pride of place to Alger Hiss, who portrayed his nemesis as an opportunist exploiting "an ugly time" in American history. When the network had difficulty in finding a Republican officeholder to defend Nixon, it turned to Ford, who knew nothing of Hiss's participation or the provocative title chosen for the broadcast. Besides recounting his long friendship with the former vice president, Ford disputed the idea that Nixon's political career was truly ended.

Afterward ABC was deluged with eighty thousand letters and telegrams, the vast majority harshly critical of the network. While a few letters to Ford criticized the congressman for associating himself with Nixon's enemies, many more lauded him for coming to the defense of a man wronged by a convicted perjurer and a biased press. Ford's act of loyalty said volumes about his relationship with the defeated candidate. "As I leave the political arena," a grateful Nixon informed his friend and defender, "I am greatly heartened by the fact that you will be in there fighting for our cause."

E LECTIONS HAVE CONSEQUENCES, especially for the losers. Their disappointing showing in November 1962 prompted murmurs of discontent among Republican lawmakers too young to remember the last time
their party had controlled Congress. Much of the cloakroom grumbling
concerned Charlie Halleck, the minority leader, whose obstructionist tactics made the GOP caucus appear lacking in positive alternatives. "This
party isn't divided between liberals and conservatives," said one Halleck
detractor. "It's divided between people who want to sit on their bottom and
not do anything and those who want to do something." Among the latter
group were Michigan's Robert Griffin, first elected in 1956, and Charles
Goodell, chosen in 1960 to represent a district in Western New York State.
Over a postelection lunch they hatched a plan to unseat two-thirds of the
existing GOP leadership—specifically, Les Arends of Illinois, the longest-
serving whip in House history, and Iowa's Charles Hoeven, then embarking
on his eleventh term, the last three of which had seen him double as chairman of the House Republican Conference.

That Hoeven had made no discernible mark in the position was hardly
his fault. Long before Charlie Halleck, the conference had atrophied, becoming "the toothless heir of the once great juggernaut known as the Republican caucus." In the nineteenth and early twentieth centuries the caucus
had been governed by the unit rule, as enforced by gavel-wielding tyrants
like Blaine, Reed and Cannon. Eventually their heavy-handed methods
provoked rebellion in the ranks. The stifling discipline of the caucus gave
way to the mild camaraderie of the conference. Lacking the votes to replace
Halleck, Young Turks Griffin and Goodell hoped to send their leadership
an unmistakable message about the need for greater consultation and a
more constructive public image for the party. Ford learned of the planned
uprising during the first week of January 1963. He advised the plotters to
forgo any move aimed at toppling the popular Arends.*

Hoeven was a different story. Lacking a substantial following of his own,
Hoeven had become synonymous with Charlie Halleck's status quo. Ford
waited until the last minute before consenting to run. Even then he insisted
that his supporters refrain from criticizing Halleck. In return Halleck limited himself to a pro forma statement of support for his embattled deputy.

---

* It should be noted that in his memoirs, Ford also listed Representatives Mel Laird and
Donald Rumsfeld among those who first approached him about running for the conference
chairmanship.

Perhaps he thought it unnecessary to do more; the usually reliable Les Arends predicted no more than twenty Ford votes when the caucus met on January 8. Hoeven, less confident of the outcome, tried to make the contest about ideology. Behind closed doors he portrayed Ford as a dangerous liberal, "practically . . . a member of ADA," said one participant in the party conclave.

The strategy failed. By a margin of eight votes Ford secured the third-ranking position in the GOP hierarchy. A chastened Hoeven took the occasion to warn other members of the existing leadership. "I happened to be in the line of fire when the shooting started," he told them. "They're really gunning for others, laying the groundwork for another showdown two years hence." Ford's increased visibility was soon reflected in the volume of speaking invitations crossing his desk. To his staff, no less than his family, it must have seemed as if he accepted all of them, from the Republican Women's Federation of New Hampshire to a $10-a-plate affair benefiting MacMurray College in tiny Jacksonville, Illinois. Scheduling demands aside, Ford's admission to the GOP leadership had the paradoxical effect of raising his profile while narrowing his options by binding him to an increasingly conservative caucus. Henceforth his words, no matter how casually delivered, would be parsed as an expression of party sentiment.

GOP successes in the 1962 elections were rare. One notable exception was the South, where a modest beachhead of House members grew overnight from nine to fifteen. For Republicans, 1963 represented what Ronald Reagan would later call a time for choosing. That spring, civil rights demonstrators led by the Reverend Dr. Martin Luther King Jr. focused the nation's attention on Birmingham, Alabama, a fiercely resistant stronghold of economic and political segregation. With Southern politics seemingly ripe for realignment, the question naturally arose—was the traditionally Democratic Solid South most vulnerable to the Eisenhower brand of Modern Republicanism, grounded in the region's fast-growing suburbs and moderate on issues of race, or to the more doctrinaire school personified by Arizona's fast-rising senator Barry Goldwater, with its appeal to former Democrats—and Dixiecrats—opposed to civil rights for Blacks and increasingly hostile to the federal government?

On June 11, a few hours after Alabama Democratic governor George Wallace tried to block the admission of two African-American students to the University of Alabama, the president of the United States addressed the country on an issue "as old as the Scriptures and . . . as clear as the American Constitution." Within a week, vowed Kennedy, he would introduce a comprehensive civil rights bill for congressional consideration. Ford's response was conciliatory. Republicans would "not try to make political capital out of this very tense and crucial issue," he told reporters. Instead they would look for potential areas of legislative agreement—voting rights, for example, and ending segregation in public places like airports and rail stations, department stores and restaurants.

Meanwhile the best way to paper over GOP divisions was to exploit fissures in the opposition. So Ford assailed Kennedy-sponsored tax cuts adding $5 billion to the federal deficit as "immoral" without compensatory cuts in spending. Raking over the Cuban embers, he charged the administration had been slow to act on CIA reports of Soviet activity ahead of the president's October 1962 quarantine. He also demanded that the White House release an internal report, said to be highly critical of events leading up to and including the aborted 1961 Bay of Pigs operation. In August 1963 Dwight Eisenhower included Ford's name on a list of those he considered qualified for the presidency. "Congratulations on making General Eisenhower's Hit Parade," jibed columnist Robert Novak. Rumors circulated that Ford might be Nelson Rockefeller's running mate in 1964, or a potential attorney general in a Goldwater Administration.

His newfound status as GOP insider did not prevent Ford from registering an occasional dissent to Republican orthodoxy. On Capitol Hill he took issue with Charlie Halleck over proposed cuts to the Kennedy defense budget. Addressing a convention of California Young Republicans, he warned delegates that the election of a YR chairman supported by the John Birch Society would diminish GOP prospects in the Golden State. Ford opposed the sale of American wheat to the Soviet Union, but not to Yugoslavia, the wayward satellite that under its president for life Josip Broz Tito enjoyed a degree of autonomy unparalleled within the Soviet Bloc. In September 1963, Ford was in Tito's capital of Belgrade, attending the latest gathering of the Inter-Parliamentary Union in the company of newly elected Massachusetts senator Ted Kennedy and Madame Ngo Dinh Nhu, the sister-in-law of South Vietnam's increasingly unpopular President Diem.

Madame Nhu's travels did not noticeably broaden her perspective, but they did extend her life. Having survived two earlier coup attempts, her husband and brother-in-law fell victim to a third uprising in November 1963. At the time the notorious Dragon Lady was in Beverly Hills, an unlikely refuge from which she accused the Kennedy Administration of complicity in the violent overthrow of the Diem regime. Kennedy himself was said to feel guilt over US involvement in the coup and subsequent murder of the Diem brothers. Deteriorating events in Saigon competed for his attention with domestic politics, as the president's belated embrace of civil rights cost him support, especially in the South. Polls showed Barry Goldwater leading JFK in much of the region, and competitive in Vice President Johnson's home state of Texas. A fence-mending trip to the Lone Star State was arranged for later in November. On their return to Washington, the president and First Lady were to host German chancellor Ludwig Erhard for a state visit. Jerry and Betty Ford were invited to the state dinner for Erhard, scheduled for the evening of November 25.

FORD'S CALENDAR FOR the weekend of November 23–24 was characteristically crowded. On Saturday, the twenty-third, he was to participate in dedication ceremonies for Grand Rapids' new airport. That evening's plans called for him to introduce Barry Goldwater before a thousand dinner guests, the largest number ever to attend a West Michigan political fundraiser. First, however, the congressman and his wife had more personal business to attend to: a midday conference scheduled with their son Jack's school counselors on November 22. Leaving the meeting, they heard via their car radio initial reports of shots being fired at the motorcade carrying President and Mrs. Kennedy through downtown Dallas. A disbelieving Ford hastened to his Capitol Hill office to learn more. By the time he got there the president was dead.

Ford's sorrow was genuine. Never claiming any special intimacy with Kennedy, the fact remained that they had enjoyed a friendly relationship going back fifteen years. More recently the two men had found ways to surmount partisan differences, particularly on issues of foreign policy and national defense. Sharing the popular horror that a lone gunman could rob the nation of its dynamic leader, Ford grieved for the widow and her young children. On Saturday, the twenty-third, he and Betty knelt before the late president's casket in the East Room of the White House. Two days later Jerry attended the funeral mass at St. Matthew's Cathedral. Betty meanwhile boarded a bus transporting members of Congress and their spouses to Arlington National Cemetery for the interment. There she and some friends took up a position sufficiently removed from the burial site that they wouldn't be intruding.

At least that is what they thought as they watched the funeral cortege, hundreds of cars long, crawling across Memorial Bridge toward the cemetery. Eventually Mrs. Kennedy's car stopped beside the grave site. Traffic backed up. Suddenly Betty found herself in "a mob of world leaders" striding over the hill where she had taken refuge. Charles de Gaulle, Ethiopia's Haile Selassie and several dozen royals and national representatives swept by her on their way to the final rites. In his sandpaper voice, Boston's

Richard Cardinal Cushing prayed for the president's soul. Mrs. Kennedy lit the eternal flame, and the Kennedy family departed, to be followed by the international congregation of mourners. Betty waited until they had left. Abandoning her hillside perch, she walked down to the now-deserted grave site. "There wasn't another soul around except for one man," she recalled. "I was standing looking at the casket and the little basket of flowers that the Kennedy children had left beside it, and this man pressed a button, and with that, the casket was automatically lowered into the ground . . . Up until that moment, it had been a nightmare; now it was real, and I started to shake." She turned around and made her way unsteadily to the waiting buses.

Six days passed. Grieving Americans heard their new president appeal to Congress to enact a civil rights bill in honor of his slain predecessor. Still numb with shock, they went through the rituals of Thanksgiving. On the day after the holiday, a few minutes before seven p.m., Mike Ford was on the phone at 514 Crown View Drive when a White House operator interrupted his call. Then the booming voice of Lyndon Johnson came on the line. He expressed surprise to find Congressman Ford in Alexandria and not Michigan. "Thank God I have somebody in town," he said. Ford complimented the president on his recent speech to lawmakers.

"Jerry, I've got something I want you to do for me." Johnson described a "top, blue-ribbon, presidential commission" he was creating to investigate John Kennedy's death. He identified Chief Justice Earl Warren as the group's likely chairman, with a membership including diplomat turned banker John J. McCloy and former CIA director Allen Dulles. "And I want it non-partisan," Johnson added, before mentioning in his next breath that the group would number five Republicans to just two members of his own party. From the Senate he was recruiting his old friend Richard Russell and the universally respected Kentuckian John Sherman Cooper. "I'm going to ask Hale Boggs and you to serve from the House . . ."

Ford didn't hesitate. "You know very well I would be honored to do it, and I'll do the very best I can, Sir."

"You do that," said LBJ before hanging up. "I'll be seeing you."

It seems a curious appointment. Certainly Charlie Halleck, livid at being passed over, thought so. Hale Boggs, Ford's Democratic counterpart on the commission, enjoyed the prestige and influence that accompanied his position as majority whip in the House. On closer look, Johnson's action was history repeating itself. The congressman from Grand Rapids, a junior member of the Republican hierarchy, had been recommended to the president by Cabinet secretaries Robert McNamara and Dean Rusk. Both men vouched for his nonpartisan approach to issues of national security. Less than a decade had passed since Ford's promotion to the House committee charged with CIA oversight. His performance since had fully justified the confidence of his original sponsors on both sides of the aisle. In choosing Ford to round out his commission of insiders, the new president recognized a rising star from Capitol Hill, a workhorse with a reputation for discretion and the imprimatur of the national security establishment.

THE WARREN COMMISSION

Gentlemen, our only client is the truth.

—LEE RANKIN TO WARREN COMMISSION STAFF MEMBERS, JANUARY 1964

NEXT TO HIS pardon of Richard Nixon, nothing in Gerald Ford's public life spawned more lasting controversy than his membership on the Warren Commission, the all-too-official investigation of John Kennedy's assassination convened by his successor to refute rumors of a plot hatched in Moscow or Havana. Over time the inquiry designed to quell popular doubts would have the opposite effect, giving rise to innumerable conspiracy theories and as many books, documentaries and websites disputing its conclusions. As with the Nixon pardon, Ford could expect wherever he went to be quizzed about the assassination and his role in its investigation. His repertoire of war stories included an off-the-rails interview with Lee Harvey Oswald's killer, Jack Ruby, and confrontations with Oswald's elusive wife, Marina, and her grotesque mother-in-law. Late in life he would theorize that a sexually impotent Oswald, a twenty-four-year-old former marine who defected to the Soviet Union in 1959 before returning to the US after less than three years, gunned down the president to prove his manhood to an estranged spouse.

In his memoirs Ford could spare only two and a half pages for the commission and the public skepticism its findings had generated. Minimizing disagreements within the group, he said nothing about his relationship with the FBI, or his initial suspicions of foreign involvement in Kennedy's murder. With the death of former senator John Sherman Cooper in 1991, Ford became the sole surviving member of the commission, and the most visible proponent of its conclusion that Oswald acted alone in killing the president. Repeated often enough, Ford's unvarying defense of the Warren Commission and its findings came to sound like boilerplate, to be called into question with every belated document release or critical volume to hit the bookstores.

His version of events was contradicted by the revisionist fireworks of filmmaker Oliver Stone, whose 1991 motion picture *JFK* implicated more suspects—among them Lyndon Johnson, the Mafia, the American military, the CIA, FBI, and Secret Service—than an Agatha Christie whodunit. The only member of the Warren Commission to see Stone's "counter-myth"— involuntarily, to be sure, as part of a captive audience exposed to *JFK* on an in-flight showing—Ford acknowledged the director's cinematic skills even as he disputed his fidelity to historical fact.

Needless to say, Oliver Stone had no monopoly on reinterpreting the past. During the last years of his life, Ford was targeted by commission critics who accused him of lending his name to a blue-ribbon cover-up. He was portrayed as a part-time commissioner, distracted by his day job and recently assumed chairmanship of the House Republican Conference.* More serious claims concerning Ford's ties to the FBI cast doubt on his integrity no less than his objectivity. It has even been asserted that Ford was blackmailed by the bureau into serving as an informant on Warren Commission activities for J. Edgar Hoover. Conspiracy theorists holding to this view cite as their source Bobby Baker, the Capitol Hill fixer and LBJ protégé whose dealmaking eventually landed him in federal prison for fifteen months on charges of tax evasion, theft and fraud.

In a sensational oral history done for the Senate Historical Office in 2009 and 2010, later excerpted in *Vanity Fair*, Baker claimed a sexual liaison between Congressman Ford and Ellen Rometsch, a purported East German spy pimped by Baker to the late President Kennedy. According to Baker, the FBI taped Ford's indiscretions, making him a useful fifth column on the commission Hoover viewed as a threat to his cherished bureau. It is a difficult story to accept for multiple reasons. It bears noting that Ford had been among the first to cry whitewash about the arrangement under which Baker, the so-called 101st senator, was allowed to resign his position as secretary of the Senate in October 1963. A year later, following the arrest of Johnson aide Walter Jenkins on a morals charge in the closing days of the 1964 presidential campaign, Republican candidate Barry Goldwater proposed an examination of FBI findings on both Jenkins and Baker. The review panel would consist of six members, with each presidential contender to choose three. Goldwater made it clear that his first selection would be Gerald Ford.

---

* A simple perusal of Ford's diaries, supplemented by the testimony of commission staff who tracked the attendance records of each member, refutes this accusation. According to David Belin, an assistant counsel to the commission who would go on to write two books upholding its findings, Ford was surpassed only by Chief Justice Earl Warren in his commitment of time and energy.

Setting aside questions of Baker's motive or credibility, his assertion that a sexually compromised Ford was coerced into serving as an FBI mole overlooks one salient fact. No member of the House was a more sincere admirer of, or apologist for, the agency and its seemingly eternal director. As a freshman congressman, Ford had lavished praise on Hoover, whose salary he proposed to increase by special legislation. "It is always extremely gratifying to learn of such favorable observations," an unctuous Hoover had responded. In time Ford was placed on the "Special Correspondents List" reserved for the director's favorites. He could count on congratulatory notes from Hoover after each successful reelection campaign, and an auto-graphed copy of the latest ghostwritten book published in Hoover's name. FBI agents performed a background check on a prospective Ford family maid. On another occasion they swept the phones in Ford's Alexandria home for possible bugs (none were found).

Ford's original contact at the agency was Louis B. Nichols, a fellow Wol-verine and assistant FBI director entrusted with the bureau's congressional relations. Nichols had been alerted to Ford's potential even before his 1948 primary challenge to Bartel Jonkman. Their ensuing friendship was no doubt enhanced by Nichols's connection to the Booth Newspapers chain, based in Grand Rapids. Throughout this period the Ford and Nichols fam-ilies socialized with each other. From such exposure Nichols formed an opinion of the congressman as "a religious man who is very happily married and is very conservative in his personal habits." This estimate would be echoed by dozens of colleagues at the time of Ford's nomination to replace Spiro Agnew as vice president. Many of these witnesses differed with Ford politically. Some disputed his qualifications to hold the nation's second-highest office. None questioned his conduct as a husband and family man.

Not that he was blind to the casual relationships that flourished on the campaign trail and in many congressional offices. "I was never intrigued with any of the females on Capitol Hill," Ford told Trevor Armbrister, with whom he collaborated on *A Time to Heal.* "Number one, because of our marital circumstances, and two, I knew that it was dynamite." He was even more explicit when journalist Tom DeFrank raised the subject following the exposure of Ford friend Frank Gifford's affair with a former airline flight attendant. "You have to think of the ten bad things that could happen to you from something like that and the one good thing," Ford ruminated, "and tell yourself the one good thing will get taken care of some other way."

In his Senate oral history Bobby Baker was notably stingy with dates. He provided a single clue to the timing of Ford's alleged liaison with Ellen Rometsch. "You know, his wife had a serious drug problem back then," said Baker, implying a rough patch in the Ford marriage during which Jerry momentarily forgot his wedding vows. In fact, Betty Ford's reliance on pain pills can be dated with considerable precision. On August 15, 1964, a day after she incurred a pinched nerve attempting, unsuccessfully, to raise a stubborn window in her Alexandria kitchen, Betty entered a nearby hospital for a week's stay. Unable to relieve her pain through traction and therapy, doctors eventually prescribed medication to ease her suffering. This in turn formed the genesis of a dangerous dependency on prescription drugs, the effects of which were magnified by alcohol.

Betty's hospitalization came one week after Republicans assembled in San Francisco to nominate Barry Goldwater for president. More important, it occurred exactly one year after Ellen Rometsch and her husband were deported to East Germany at the behest of then–attorney general Robert Kennedy. In other words, during the specified period in which Baker claimed an illicit relationship between Rometsch and Jerry Ford, the East German femme fatale was an ocean away from Washington. Murky as it may have been, Rometsch's dalliance with JFK did make it into the notorious "Official and Confidential Files" maintained by J. Edgar Hoover. The first person outside of Hoover's orbit to systematically examine this fabled collection of political sludge was Laurence Silberman, a deputy attorney general in the Ford Administration, later US ambassador to Yugoslavia and a Reagan appointee to the United States District Court for the District of Columbia. While at the Justice Department in 1974, Silberman was directed to review the late FBI director's documentary legacy. This meant three weekends wallowing in "all the dirt collected by agents under instructions to supply dirt directly to Hoover," according to Silberman.

In a 2017 interview with the author, Silberman was adamant—nothing in Hoover's secret files compromised Ford. Ironically, it was Ford's own withholding of information, specifically, his original doubts about Oswald and possible foreign complicity in the assassination, that led to a distorted perception of his role on the Warren Commission. Far from passively accepting the rush to judgment implicit in the initial FBI report prepared at Lyndon Johnson's request, Ford stood out among the commission's members for his suspicions of foreign involvement. In December 1963 Gerald Ford was among the original conspiracy theorists trying to make sense of the crime of the century.

THE WARREN COMMISSION scheduled its first meeting for December 5. By ten o'clock that morning, Room 105 of the National Archives building had been swept three times for possible explosives or hidden microphones. A slight awkwardness pervaded the large conference room, as commission members introduced themselves like college freshmen on the first day of the fall term. The white-haired chief justice claimed a brown leather chair at the head of the table. Inches away sat his legislative nemesis Senator Richard Brevard Russell. Bullied into serving alongside the man whose court was remaking his beloved South, Russell could barely bring himself to look at the chief justice. ("I don't like that man. I don't have any confidence in him at all.")

A jurist by profession, a former prosecutor and hugely popular governor of California, Warren was an executive by temperament. Moving with customary self-assurance to define the scope and duration of the inquiry, he told the men around the table that he wanted a final report no later than June 1964. After that date presidential politics would surely complicate their task. But it was Warren's attempt to install his own candidate as chief counsel that prompted Ford to complain of a "one man commission." Warren Olney, a former US assistant attorney general, more recently administrator of the federal court system, had an association with the chief justice dating to the 1930s, when Warren was district attorney in Oakland, California, and Olney his deputy.

It wasn't just Ford who found the appointment of Warren's protégé troubling. The introduction of Olney's name met with vehement opposition from Allen Dulles. Congressman Hale Boggs flatly refused to serve on the commission unless an alternative to Olney was found. In the face of such dissent Warren put up "a stiff argument," recalled Ford, but the group eventually compromised on J. Lee Rankin, a fifty-six-year-old Manhattan attorney who, as Eisenhower's solicitor general, had argued forcefully before the Warren Court on behalf of school desegregation. More threatening to the commission's fragile unity was the intervention of Deputy Attorney General Nicholas Katzenbach. Acting in place of a grief-stricken Robert Kennedy, within hours of Oswald's televised murder at the hands of Dallas nightclub owner Jack Ruby, Katzenbach wrote a memo to the new president's assistant Bill Moyers urging a quick endorsement of the preliminary FBI report declaring Oswald solely responsible for the deaths of President Kennedy and Dallas police officer J. D. Tippit. "Speculation about Oswald's motivation ought to be cut off," Katzenbach argued in making the case for a presidential commission, "and we should have some basis for rebutting thought that this was a Communist conspiracy. . . ."

With the Dallas police promoting just such a scenario, Katzenbach wrote directly to Earl Warren on December 9, all but pleading for the commission to place its stamp of approval on the hastily crafted FBI findings.

Professing "great concern" over whether Oswald had in fact acted alone, J. Edgar Hoover opposed the Katzenbach initiative. No doubt Hoover wanted to be sure of his facts before going public. But he also knew from the outset that his bureau was vulnerable to criticism for its seeming carelessness in monitoring Oswald's activities leading up to November 22. It was in this context—a developing tug of war with the FBI for control of the investigation—that Ford on December 12 reached out to Cartha "Deke" DeLoach, Lou Nichols's successor as the bureau's assistant director for congressional relations. Speaking in "the strictest of confidence," Ford vented his frustration over the chief justice and his high-handed methods. At the same time the congressman took strong exception to the Katzenbach letter. He might be "a minority of one," Ford told DeLoach, but he refused to rubber-stamp the FBI report that effectively preempted the work of the commission before it had begun.

Ford mentioned a recent visit from CIA director John McCone, who had appeared in the congressman's Capitol Hill office with "startling information" concerning Lee Harvey Oswald. According to McCone, an FBI tipster in Mexico City had witnessed the exchange of money—an amount later pegged at $6,500—between Oswald and "an unknown Cuban Negro." The revelation bolstered Ford in his suspicions of a foreign plot. His corresponding reluctance to accept the FBI report as an investigative road map survived a dousing of cold water by DeLoach, who told Ford that the CIA source in question had since recanted his story, only to reverse himself again and claim to be telling the truth.

But it was DeLoach's concluding words in the memo describing his encounter with Ford that touched off a minor tempest when they came to public light in 1978. "Ford indicated he would keep me thoroughly advised as to the activities of the Commission. He stated this would have to be on a confidential basis; however, he thought it should be done. *He also asked if he could call me from time to time and straighten out questions in his mind concerning our investigation*" (italics added). No one, then or later, had to ask why the FBI would welcome a back channel of communication into the commission. That such an arrangement might be as useful to Ford as to the bureau went largely unnoticed. Instead, the characterization of Ford as Hoover's informant was seized on by critics of the commission to validate their doubts about the group and its divided loyalties.

Disputing DeLoach's claims, Ford offered an alternative version of their relationship. By 1963 a fifteen-year veteran of Capitol Hill, he occupied a powerful position on the House Appropriations Committee. This alone invited cultivation by bureaucrats whose programs depended on his support at budget time. "Typical of Hoover, he had his representative up here dropping by people that at some point might be helpful to the FBI," Ford remarked to journalist Bill Kurtis in a 1992 interview. "So Deke De-Loach, before the assassination, would probably drop by my office once every month just to say hello; were there any problems that the FBI had that I could tell him about? Well, when I got on the commission, those visits to my office became once every couple of weeks." Long exposure to the Washington mindset convinced Ford that DeLoach was exaggerating their intimacy in the apple-polishing tradition of other Hoover sycophants who were "always anxious to go back to their boss and embellish their responsibilities and their actions. A little puffing. That's human nature. And I think that's what Deke DeLoach did."

It was hardly surprising that Ford should object to Warren's early assertiveness. To a friend he later confided that the chief justice had "treated us as though we were on the team, but he was the captain and the quarterback." Even before Dallas police on the afternoon of November 22 apprehended a Marxist-spouting misfit named Lee Harvey Oswald, Warren had publicly attributed Kennedy's murder to the "hatred and bitterness that has been injected into the life of our nation by bigots." According to Washington journalist and Warren Commission scholar Max Holland, this was a euphemism for "southerners opposed to civil rights." Under the circumstances, Holland added, Ford feared that Warren, "owing to his liberal bias, would attenuate the investigation into Oswald's political background and gloss over his motives." The same source alluded to a December 6 statement by the House Republican Policy Committee taking issue with Warren's thesis, and substituting as a potential Oswald motive the "teachings of communism."

Meeting on December 16 in their newly rented quarters in the VFW Memorial Building on Maryland Avenue, a stone's throw from the Supreme Court, commission members vied with one another in criticizing the FBI report. Before adjourning for the day the group made it official: they would undertake their own, separate inquiry. The next day Ford met again with Deke DeLoach. He mentioned in passing that at least two members of the group remained unconvinced that the fatal shots aimed at the president had come from a sixth-floor window of the Texas School Book Depository building. Wrenched out of context fifteen years after the fact, Ford's casual remark made *at the start* of the commission's investigation would be misinterpreted as a final, previously unreported dissent.

Nothing in the extensive documentary record unearthed since 1978 suggests additional information sharing with DeLoach. (Ford merits a single

passing mention in DeLoach's 1995 memoir, *Hoover's FBI*.) Talking with
FBI historians a decade later, DeLoach included Hale Boggs with Ford as
useful sources on the commission. But he also debunked the notion that
they were FBI plants. "They asked me questions from time to time and I'd
tell them the true facts and they would take it back to the Warren Commission," said DeLoach. "We weren't controlling the Warren Commission
but we were trying to protect our own operation . . . There were many
factions reporting to them and testifying before them." Just how far the
bureau would go in support of its Capitol Hill friends was strikingly illustrated when journalist Holmes Alexander falsely implicated Ford in leaking
Oswald's diary to the *Dallas Morning News*. Hurrying to Ford's defense,
Hoover proposed that the congressman be interviewed by a special agent—
the mere fact of which, coupled with an official denial of the allegations by
the agency, would establish Ford's innocence for all but the most jaundiced
of observers. Ford accepted the offer, and was subsequently questioned by
Cartha DeLoach himself.*

In common with his commission brethren Ford cited the constraints of
the calendar to justify their use of existing investigative resources—primarily
agents of the FBI—rather than hire an independent contingent of sleuths.
They may well have entertained second thoughts after being summoned to
an emergency meeting on January 27, 1964, at which Lee Rankin passed
on an unconfirmed report from Texas attorney general Waggoner Carr that
Oswald had been a paid FBI informant. A January 28 meeting between
Rankin and Hoover did not go well. The director subsequently provided
the commission with no fewer than eleven affidavits from himself and other
FBI personnel, all denying what commission staff member Howard P. Willens, writing long afterward, labeled "a third hand rumor."

Nor were relations between the bureau and the commission improved
by Hoover's withholding of facts concerning James Hosty, a Dallas FBI
agent whose name and telephone number had shown up in Oswald's address book. Hoover justified his action by declaring Hosty's information
irrelevant to the commission's work. His arbitrary move convinced Warren

---

* In April 1964, Jerry and Betty encountered Hoover at a party in DeLoach's home. In a
message hand-delivered to Ford the next day, Hoover expressed satisfaction over their informal discussion of "some vital issues of interest to you as well as the FBI. Let me say that
I found your observations both helpful and germane. It is always encouraging to know that
we have alert, vigorous Congressmen such as you, who are aware of the needs and problems
confronting our country, and I wish you every success in meeting your grave responsibilities." The old flatterer concluded by inviting the Fords for a special tour of FBI facilities,
which Betty Ford had expressed interest in seeing.

and his associates that any intelligence obtained by the FBI must be carefully cross-checked. Even someone as favorably disposed to Hoover's men as Ford harbored doubts about Jim Hosty. Writing in his diary after interviewing Hosty, Ford concluded, "Knew his answers but very selective about why Oswald not regarded as a person capable of violence."

FORD WOULD ALWAYS remember 1964 as one of the busiest years of his life. It was still December 1963 when Jerry and Betty were the only Republicans invited to an intimate dinner coinciding with the Johnsons' first full weekend in the White House. The Fords did their best to look inconspicuous as a half-dozen other couples present toasted LBJ's electoral prospects a year hence. The piper was not long in demanding payment, with Johnson appealing for Ford's help in spiking Republican amendments that would effectively scuttle a taxpayer-subsidized deal to sell American wheat to the Soviet Union. LBJ eventually got what he wanted, although he failed to persuade Ford, who also voted against an $11.5 billion tax cut and a substantial pay raise for government workers (members of Congress included).

In addition to the Warren Commission, whose investigative functions consumed three or four days of his week, Ford confronted an unusually crowded legislative calendar. For nine days in February 1964 he shuttled between the VFW building, where he joined in the questioning of Oswald's widow, Marina, and the nearby Capitol, scene of an intense debate swirling around landmark legislation that would outlaw racial discrimination in public accommodations, classrooms and any program receiving federal funds. Passage of HR 7152, better known as the Civil Rights Act of 1964, would also overhaul the voter registration process and establish a new Equal Employment Opportunity Commission. Reflecting its controversial provisions, the draft legislation inspired 155 amendments, most of them rejected by House members, though the belated addition of gender among the categories receiving legal protection dramatically expanded the bill's reach.

On days when commissioners did not meet, Ford attended lengthy sessions of his defense subcommittee, where discussions of aircraft procurement alternated with Secretary McNamara's latest assessment of the situation in Vietnam. Ford argued that the time had come for the United States to assume a more decisive role. South Vietnam's borders should be sealed off. "If that doesn't work then we should engage in hot pursuit and bomb the supply lines and bases in North Vietnam." At the moment there were fifteen thousand American military personnel in Vietnam. "With just a few more," Ford said, victory was achievable. On February 10, Ford and his Warren Commission brethren had their first encounter with Oswald's "aggressive, dogmatic, difficult" mother, Marguerite, a thrice-married practical nurse who did not hesitate to remind Lyndon Johnson that "he is only President of the United States by the grace of my son's action." Marguerite Claverie Pic Oswald Ekdahl insisted that her son was a secret agent of the United States government.

Marguerite's appearance before the commission coincided with the civil rights debate nearing its climax. Ford was part of the 290–130 majority by which the House approved HR 7152 and sent it on to the Senate. He hardly had time to catch his breath. There were Lincoln Day speaking commitments in Galesburg, Illinois, and Fort Wayne, Indiana, to honor. One dismal night a snowstorm stranded him at a motel near the airport in Toledo, Ohio. On weekends he made time to watch Mike and Jack play basketball for the Alexandria Public Recreation League. But these were rare distractions from the Warren Commission and its demands. Lacking any personal investigative staff, Ford enlisted the services of his friend and former campaign manager Jack Stiles. For legal counsel he tapped John Ray, a retired congressman from Staten Island. Together they reviewed the latest interview transcripts and fashioned questions for upcoming witnesses, of which there would be 552 in all.

Nearly one hundred of these were personally interviewed by members of the commission.* Ford's personal diary/calendar contains his impressions, hastily jotted down, of the witnesses who came before him. Thus we know that his sympathy for Marina Oswald peaked on the fourth day of her testimony, when Lee's widow broke down in tears while identifying some clothing that had belonged to her husband. A second round of questioning raised doubts, as Ford inquired about Marina's proficiency in English prior to November 22, and asked how she had obtained permission to leave Russia with such apparent ease. He had Marina in mind when urging the administration of polygraph tests, voluntarily, "where there appears from the record certain inconsistencies or a failure to be completely frank."

Ford postponed a trip to Williamsburg, Virginia, and a conference of the Bilderberg Group convened by the Netherlands' Prince Bernhard, so that he could participate in the interrogation of Ruth Paine, Marina's close

---

* To assist them, a dozen outside lawyers were hired to investigate six areas of interest: planning for the presidential trip to Dallas; the actual assassination and any evidence pointing to an assassin or assassins; Oswald's background; his suspicious foreign travels; his murder by Dallas nightclub owner Jack Ruby; and what might be done to improve the protection accorded future presidents.

friend, with whom she had been staying on November 22. He empathized with a nervous Helen Louise Markham, a thirty-nine-year-old waitress and eyewitness to the slaying of Officer Tippit. Of a Dallas cabdriver seeking government reimbursement for lost work Ford concluded, "Not too bright or articulate." Dallas Police Chief Jesse Curry left him ambivalent. "Seems to be telling truth," he scribbled in his diary. "*But*, always qualified answers (I believe, etc.)."

A parade of FBI experts testified about ballistics, fingerprinting, handwriting analysis and photography. Ford sat through repeated viewings of the silent color film, twenty-six seconds of unspeakable brutality on Elm Street, shot by Dallas dressmaker Abraham Zapruder. He asked that a pair of Secret Service agents be called as witnesses, along with the wife of former Dallas mayor Earle Cabell, who from her place in the Kennedy motorcade had spotted a gun protruding from the sixth floor of the Texas School Book Depository. Ford sought interviews with "any civilians or police who might have been on the triple overpass at the time of the shooting." He could not know that his request anticipated future conspiracy theories involving a shadowy second gunman targeting the presidential motorcade from the Elm Street overpass or nearby shaded slope, immortalized as the Grassy Knoll.

Ford's written questions, distilled from the flood of depositions he was reading, enable one to trace the evolution of his thinking. Zeroing in on the number and direction of bullets fired at the president, he inquired, "Could shots have come from that"—a reference to the railroad overpass— "Or any direction other than the Book Depository Building?" Why was there so much confusion initially over the make of rifle—Mauser or Italian Mannlicher—found on the sixth floor? What was known regarding Oswald's practice "at the gun range or anywhere else"? Had the accused man's jailhouse conversations with his wife, mother or brother Robert been monitored or taped? Where had Oswald obtained his ammunition? What about an alleged pre-assassination meeting at Jack Ruby's Carousel Club involving Officer Tippit, among others? Why on a warm day would Oswald after shooting Kennedy go home to retrieve a jacket? Was there anything to rumors that he had routinely received small sums of money at Western Union?

Ford requested copies of any material received from state and city officials in Texas. He examined not only all correspondence between Oswald and members of his family while he was in Russia, but letters exchanged between Ruth Paine and Marina from the time Marina joined her husband in New Orleans in the spring of 1963 until her return to Fort Worth that fall. "Our record seems pretty sketchy as to what Oswald did during his stay in Mexico City late in September to early October 1963," Ford complained. Indeed, questions had been raised concerning a possible side trip by Oswald to Cuba. "It is my recollection that there is some FBI information concerning these matters and if so could you send me copies of the interviews or the reports by the FBI or any other federal agency such as the CIA?" Had Treasury agents carried out a promised investigation of Oswald's finances? "I sincerely hope so because the source of Oswald's funds and the amounts and how and where they were spent will be a vital part of our records."

A S THE COMMISSION and its staff moved from evidence gathering to drafting a report, Ford told the lawyers he wanted no "fait accompli," but reserved the right to review drafts so that he and other members "could pass on every word and sentence." At the beginning of May, Lee Rankin informed him that Warren, Dulles and McCloy opposed publication of testimony compiled in the course of their probe. Ford took strong exception to this view. So did Senator Russell, Congressman Boggs and the legal staff assembled by Rankin. In time the commission's main report would be accompanied by twenty-six volumes of supporting materials. Another controversy involved Rankin's special assistant, a New York University School of Law professor named Norman Redlich. A native of the Bronx, Redlich had served in the army during World War II before attending Williams College and Yale Law School, where he finished first in his class. As a lawyer Redlich donated his services to death-row inmates. As a community activist he had stood with opponents of a Robert Moses–planned highway that would have destroyed Greenwich Village.

But it was another kind of political activism that caused Warren Commission members, goaded by Ford, to divert their attention from Lee Harvey Oswald to Norman Redlich. In the 1950s Redlich had belonged to something called the National Emergency Civil Liberties Committee, on whose behalf he had addressed a public rally calling for abolition of the House Committee on Un-American Activities. He had signed petitions to this effect, and more petitions appealing for clemency for two friends imprisoned for contempt of Congress after they refused to answer questions at a HUAC hearing on "Communist Infiltration and Activities in the South." Gerald Ford knew of Redlich's activities because the then-chairman of HUAC, Louisiana Dixiecrat Edwin Willis, brought them to his attention in February 1964. Others joined Willis in making an issue of Redlich's past. "There were three or four people in the House of Representatives,"

Ford would recall in 1997, "extreme rightwing Republicans" led by Iowa's redoubtable H. R. Gross, "who used to give me hell because Redlich was alleged to be an extreme liberal."

Armed with an FBI report documenting Redlich's history of political activism, the full commission took up the matter of his continued employment on May 19. Lee Rankin confessed his ignorance of Redlich's anti-HUAC record at the time he hired him. What he *did* know was that his assistant's work ethic surpassed that of anyone else on the commission staff.* Warren argued that Redlich was at the very least owed a hearing in which to defend himself. Plainly uncomfortable, Ford professed admiration for Redlich's talents and his contributions to the commission. "I like him," Ford said. "I don't want to hurt him in the slightest." Yet, he went on, his employers faced a wrenching choice between "the image of the commission and the problem of an individual." Redlich's history of political activism could conceivably tarnish their final report. With speculation rampant about foreign plots, and Oswald tagged as a defector to the Soviet Union and an apologist for the Castro regime, Redlich was vulnerable to criticism from the right.

Senator Russell conceded Redlich's loyalty, but not his objectivity. "He is a born crusader, and I don't think he can help himself." Allen Dulles and Hale Boggs seconded the Georgia lawmaker. The chief justice suggested that Ford might repeat on the floor of the House what he had just told everyone sitting around the table. Ford was agreeable, so long as he was free to add that he would not have hired Redlich had he known of his past, and consequently he could not justify retaining his services now. On his motion, security clearances were approved for the entire commission staff, Redlich included. Ford then moved to terminate Redlich's employment as of June 1.

"Jerry," Warren interjected, "there are no charges against this man."

The point was moot, as Ford's motion attracted no second. Exercising his right to go off the record, Ford asked that the tape recorder be switched off. What he said must remain a matter of conjecture. He may have attempted as graceful a retreat as the situation allowed. He may also have described the intense pressure applied to him by a handful of conservative House colleagues. Two and a half hours after it began, the meeting sputtered to a close. The experience was sufficiently traumatic to make Ford

---

* "It was said that he could work from eight in the morning until three the next morning, seven days a week," according to Vincent Bugliosi, author of the definitive *Reclaiming History: The Assassination of President John F. Kennedy*.

consider resigning from the commission. The next day Allen Dulles joined members of the congressman's staff in arguing against such a move. Whatever they said must have been persuasive. After May 20 nothing more was said of Redlich or resignation.

I N THE FIRST week of June the *Christian Science Monitor*, hard on the heels of the *New York Times*, quoted a nameless insider predicting that the commission in its final report would reaffirm Oswald's guilt. Moreover, according to the *Monitor*'s source, "There is no evidence that he was working in any way as an agent of a foreign government." This infuriated Ford. "I don't like being quoted when I have not made any final judgment," he told his commission colleagues. Citing "newspaper friends" who assumed that same-day stories leaked to the Associated Press and UPI were almost certainly the work of government agents, Ford referenced the early attempt by Nicholas Katzenbach to solicit an endorsement of the FBI's report identifying Oswald as the assassin. A sympathetic Warren echoed Ford's complaint about press accounts concocted out of "thin air and speculation." In response the Chief fashioned a public statement, acceptable to all ("It couldn't be better," said Ford) that made it clear the commission had yet to reach any conclusions, and that press leaks to the contrary were unreliable.

Ford's show of temper indicated that, where possible involvement by foreign or domestic conspirators was concerned, his mind remained open. On Saturday, June 6, he flew to Dallas, where he joined Warren, Lee Rankin, and two of the commission's staff lawyers, Joseph Ball and Arlen Specter, a future Republican, and Democratic, US senator from Pennsylvania. Their visit was brief—"I'll give you Sunday," the chief justice had informed Rankin—but it was to prove critical in dampening various conspiracy scenarios and validating the most hotly debated of all the commission's findings, the single-bullet theory. Most of the morning was devoted to the Texas School Book Depository. Here Ford traced Oswald's movements, running down four flights of stairs, stopwatch in hand, to prove that the alleged assassin had sufficient time to get from the sixth to the second floor, where he had been observed moments after the shooting. At one point Ford planted himself directly beneath the sniper's perch occupied by Oswald, listening to the sounds of ejected rifle shells hitting the floor, just as workmen had during lunchtime on November 22.

The HERTZ RENT A CAR sign atop the red brick Depository read eleven o'clock as Ford looked down at the spot occupied on November 22 by H. L. Brennan, a Dallas steamfitter who had observed Oswald in the southeast window of the sixth floor. It was Brennan's description of the gunman, flashed to police throughout the city, that led Officer Tippit to halt his squad car at the sight of Oswald, who had responded by firing three shots at the young officer, killing him instantly. Now, from the same window where Oswald had crouched waiting for the midnight blue presidential car to come within range, Ford sighted the crime scene through a duplicate of the assassin's rifle. From his vantage point he concluded that it required no special marksmanship to hit a moving target less than three hundred feet away. Prior tests had already convinced him that Oswald had sufficient time to get off the requisite number of shots—by himself—before ditching the murder weapon and abandoning the deserted sixth floor for a street-level exit and a bus ride back to his boardinghouse in the Oak Cliff neighborhood of Dallas.

Scanning the dip of Elm Street as it skirted the notorious grassy knoll before vanishing into the blackness of the art deco triple overpass, Ford's thoughts went back fifteen years, to Room 321 of the Old House Office Building, and his charismatic neighbor across the hall, whose violent death he was now investigating. Images from the Zapruder film played in Ford's mind as Arlen Specter outlined for the chief justice his single-bullet theory of the crime. He began by noting the discovery on this same floor of Oswald's Mannlicher-Carcano rifle, identified through ballistics testing as the murder weapon. At Parkland Memorial Hospital a bullet from this gun had been recovered from Texas governor John Connally's stretcher. (Connally and his wife, Nellie, occupied jump seats in front of the president and Mrs. Kennedy.) As evidence that this same bullet had caused the president's initial neck wound before penetrating Connally's chest, wrist and thigh, Specter cited the presidential autopsy, velocity tests conducted on similar bullets and the Zapruder film.

Most revealing was a recently staged reenactment of the crime, in the course of which Specter had occupied this identical perch as he clutched Oswald's gun with its four-power scope, pointed at Kennedy and Connally stand-ins, while stop-action photography confirmed the placement and timing of three shots. Of crucial significance was the clothing worn by the president and the governor. Besides traces of copper in the presidential suit jacket, there were cloth fibers in the back and front of Kennedy's shirt, as well as holes that perfectly aligned, all pointing to the trajectory of a bullet fired from above and to the rear. Similar holes in Connally's suit and shirt were consistent with a 6.5-millimeter bullet of the type used in Oswald's weapon. For eight minutes the chief justice listened as Specter outlined his theory of the case. When he finished, Warren stepped back wordlessly from the window. It was, arguably, the turning point in the investigation.

At eleven forty-five the visitors were driven the short distance to Dallas Police headquarters. Here they were ushered into a small room ordinarily used as a kitchen by Sheriff Bill Decker and his staff. Ford and Warren sat across from one another, while Jack Ruby occupied the end of the eight-foot table closest to the kitchen sink. As unprepossessing as the surroundings, at fifty-two Ruby was a balding, potbellied figure of medium height, wearing sandals and a white jumper. In the notes he made at the time, Ford judged Oswald's killer as "surprisingly rational and quite composed," with an impressive memory for names. "Everything seemed to go well for about 45 minutes" until Ruby took violent exception to something his attorney had written on a notepad. For the next half hour or so Warren did his best to placate the volatile witness, who had already extracted a promise from the chief justice that he could have his story confirmed by a polygraph test.

The thought of killing Oswald hadn't occurred to him, Ruby told Ford, until Sunday morning, November 24. That is when he spotted in the local paper a letter from a grieving reader addressed to the late president's young daughter, Caroline, and a story asserting that Kennedy's widow would have to return to Dallas to testify at Oswald's trial. Up to this point, Arlen Specter had been excluded from the overcrowded interrogation room. Following the departure of Sheriff Decker and his people, Ford passed a note to Warren suggesting that Specter be sent for. The chief justice demurred, relenting only when Ruby loudly demanded that there be another Jew in the room. By the time Specter appeared, Ruby was babbling that his life was in danger from the John Birch Society. As the stenographer paused to reload her machine with paper, Ruby abruptly rose from his seat, pulled the chief justice into a corner and ordered Specter to join them.

"Chief, you've got to get me to Washington," Ruby pleaded. "They're cutting off the arms and legs of Jewish children in Albuquerque and El Paso."

When Warren hesitated, Ruby mentioned Abe Fortas, Lyndon Johnson's close adviser and future Supreme Court justice. "Get to Fortas. He'll get it worked out." From the corner of his eye Ruby observed his lawyer slipping a note to Ford. He demanded to see it. Warren lent his own eyeglasses to the agitated witness, who read out the words scribbled on the page.

"You see. I told you he was crazy."

FOR ALL THE hours it siphoned from his schedule, the Warren Commission provided Ford a timely diversion from the internecine warfare consuming Republicans in the run-up to their San Francisco convention in July 1964. A string of bitter primary contests had left the party more divided than at any time since the Great Schism of 1912, when Theodore Roosevelt and his hand-chosen successor, William Howard Taft, waged a historic fight for the party's nomination pitting GOP progressives against standpatters who despised TR more than they appreciated Taft. Half a century after TR ran unsuccessfully as a third-party candidate, the shock waves from 1912 reverberated still. In San Francisco the opposing camps were represented by Mr. Conservative, Senator Barry Goldwater of Arizona, and his liberal nemesis, New York governor Nelson Rockefeller. Only the favorite-son candidacy of Michigan governor George Romney afforded Ford the luxury of remaining neutral.

As it happened, Goldwater's nomination coincided with Ford's fifty-first birthday. He marked the occasion by arguing the party's platform before a national television audience, resisting efforts to portray Goldwater as a reckless Cold Warrior who would outsource the US nuclear arsenal to commanders in the field. Ford's support for the national ticket that fall set him apart from Governor Romney, whose name he had placed in nomination in San Francisco with some predictable banter about Ford yielding to Rambler (Romney had been an auto executive before entering politics). In the run-up to November, Goldwater appealed to Ford in the hope that Romney might come around, but the governor was immovable.

Prospects for a Republican victory against Lyndon Johnson were remote at best. Yet Ford refused to concede the inevitable. His hopes ticked up in the closing days of the campaign. "Reaction to Ronald Reagan television show excellent," he wired campaign manager Dean Burch after watching Reagan's masterful speech titled "A Time for Choosing," which made the case for Goldwater far better than the candidate himself could. "Strongly urge it or something comparable by him be repeated," Ford advised. "This is vitally important."

Ford spent less time than usual on the hustings, during a period when Warren Commission members and staff were working feverishly to meet a September publication deadline. Increasingly confident of the facts in the case, Ford knew their work would not be done "until each of the hundreds of rumors were run to ground." For example, concerned that the State Department "appears to get off scot-free," Ford asked in writing, "Are State Department procedures sufficient in the transmitting of information on defectors to FBI and CIA?" Elsewhere he corrected the record to better reflect Oswald's drinking habits. In a section of the draft report dealing with the Dallas police and press behavior he took exception to use of the word "apparently"—"This is not a satisfactory way to summarize evidence."

One Ford memo contained a paragraph that could as easily have come from the pen of Earl Warren. "A significant question that occurs to me is the extent to which Oswald's legal rights (to counsel, against self-incrimination, present arraignment and bail, etc.) were prejudiced by police conduct. This is an excellent opportunity," Ford concluded, "to force public attention on the problems of suspects being detained by the police." The Ford-Warren relationship had improved markedly since their trip to Dallas. Both men agreed that the Secret Service should remain part of the Treasury Department. Both were appalled by the antics of Mark Lane, a flamboyant New York lawyer who claimed to represent the interests of the accused assassin, and would go on to make a lucrative career out of maintaining Oswald's innocence. In no mood to play an evidentiary game of cat and mouse with the publicity-seeking Lane, Warren blew his stack. It was "the one time I saw him get irritated" at a witness, Ford recalled of the chief justice, "and I mean really irritated."

Another questionable actor was Yuri Ivanovich Nosenko, a Russian defector who insisted that Moscow had never employed Oswald, a pathetic figure of unstable temperament and quixotic intelligence. The FBI was inclined to believe Nosenko. Not so the CIA. Its doubts about the purported defector's motives were shared by Ford and Warren, who pronounced himself "allergic to defectors." Equally skeptical was Allen Dulles, arguably the closest thing to a double agent in the commission's ranks. The former DCI discouraged Lee Rankin from reviewing CIA records touching on Oswald. He kept his erstwhile colleagues in the spy agency informed about the commission's progress, even dictating a statement denying any CIA contact or communication with the alleged assassin.

In his memoirs Ford mentioned contemporaneous rumors vaguely linking the CIA to the assassination, though hastening to add that no factual connection was ever established. This was not for lack of trying on his part. No commission member was more persistent in examining DCI John McCone and his deputy, Richard Helms. But Ford's hands were effectively tied by national security concerns, or so he thought. He had learned of Oswald's travel to Mexico City and his contacts with Soviet and Cuban representatives there "because of the manner in which our intelligence agents had obtained the information," a thinly veiled reference to bugs secreted in the Soviet embassy in the Mexican capital. Accordingly "we decided not to publish the details." Of CIA plots to assassinate Castro following the Bay of Pigs fiasco, agency officials said nothing. More than a decade would pass before enterprising journalists and congressional inquisitors exposed Operation Mongoose, a shadowy alliance of CIA operatives and organized crime figures with a shared agenda of eliminating Fidel Castro by whatever means necessary.

The shock and dismay with which surviving staff members of the Warren Commission reacted to the revelations seemed authentic enough. At
the same time it is hard to believe that the Cold Warriors on the commission to whom they reported were wholly ignorant of Washington's involvement in what Lyndon Johnson would call "a damned Murder, Inc. in the
Caribbean." Ford acknowledged as much, obliquely, in a 1992 interview
with newsman Bill Kurtis. "Our committee was told that there were certain actions involving Castro . . . they didn't go into the details, they didn't
describe precisely what was planned or what was executed." Nevertheless,
as a member of committees overseeing US defense policy and the CIA,
Ford was privy to "certain background information" confirming "that our
government was doing whatever was needed in order to make a change in
Cuba."*

Perhaps this helps to explain Ford's reluctance to sign on to the commission's draft report with its unequivocal denial of a conspiracy to murder America's president. "Senator Russell and Representative Ford were not
completely convinced that Lee Harvey Oswald had acted alone," remembered commission lawyer Bill Coleman. It was at this point that Warren the
conciliator rewrote the categorical assertion to read, in Ford's words, "The
commission has found no evidence of a conspiracy, foreign or domestic."

A second issue engendering heated debate was the single-bullet theory.
Later critics would seize on some Ford editing of staff language identifying
the president's back "at a point slightly above the shoulder and to the right
of the spine" as the entry point of the first bullet. To any reasonable person, Ford maintained, this sounded "very high and way off the side." He
revised the sentence to read "the back of his neck, slightly to the right of the
spine." Commission detractors accused Ford of tampering with evidence
to buttress the case for a single bullet striking both President Kennedy and
Governor Connally. Nonsense, said Ford; he had made the change, one
of hundreds suggested by commission members, in the interest of accu-

---

* William Coleman, a Warren Commission staff member and future Cabinet member in
the Ford Administration, said much the same thing, in equally nebulous language. "In my
many meetings with the CIA, from which we developed an excellent rapport, I believe I was
made generally aware that such planning had been undertaken and debated." In his 2010
memoir, *Counsel for the Situation*, Coleman described a prolonged slog through thousands
of pages of intelligence reports concerning Oswald's time in the Soviet Union, and his more
recent visit to Mexico City, where he had tried unsuccessfully to arrange a trip to Cuba. A
number of leads were tracked, according to Coleman, though none provided credible evidence of Cuban involvement in Kennedy's death. Among his employers Coleman singled
out Ford, Warren and Allen Dulles for their active involvement in pursuing the conspiracy
angle.

racy. Such controversies might have been averted had the commission been given full access to the Kennedy autopsy materials, however gruesome their contents. Arlen Specter called this oversight "the biggest mistake" made by the commission, a viewpoint widely shared by commission staff then and historians since.*

A third dispute involved the FBI, and a mildly worded rebuke that Warren wished to include in the final report. Other government agencies, most notably the Secret Service, had come in for their share of complaint. To overlook the failure of Hoover's agents to share their knowledge of Oswald's movements with the Secret Service and local law enforcement smacked of favoritism if not worse. Ford disagreed, but Warren carried the day by a vote of 4–3. Even now the commission's labors were incomplete. The draft report described events leading up to and occurring on the twenty-second of November. "Speculations and Rumors," as Appendix Twelve was titled, addressed itself to what did *not* occur, to conspiracy theories in the making—the number and origin of shots fired at the presidential motorcade; Oswald's prior knowledge of the motorcade route, published in both Dallas newspapers on November 19; his whereabouts on the morning of the twenty-second; the provenance of a chicken lunch attributed to a possible accomplice; Oswald's exact status as a defector to the USSR; and his alleged training at a Minsk school for assassins.

By the first week of September, commissioners were reviewing galleys in anticipation of a formal White House presentation later in the month. Unknown to his colleagues, Ford was also working with editors at *Life* magazine, with whom he had contracted in June to produce an insider's account of the Warren Commission, notwithstanding a prohibition on such tattle adopted by its members. Embarrassment nearly turned to scandal when a postponement in the commission's publication schedule risked the appearance of *Life*'s scoop before the official report could be handed over to President Johnson on September 24. Even worse, from Earl Warren's viewpoint, was a second literary project on which Ford was engaged. At the instigation of his friend and collaborator Jack Stiles, himself a disappointed author, Ford was laboring on a book-length manuscript for Simon & Schuster.

Portrait of the Assassin, published early in 1965, was exactly what its title promised, an Oswald biography culled from the testimony of commission witnesses, and a cut-and-paste job that the *New Yorker* nevertheless called

---

* Specter was equally harsh on the commission's failure to interview President and Mrs. Johnson, and the deference shown to Mrs. Kennedy in an exchange with Chief Justice Warren that lasted less than ten minutes.

"fascinating . . . a narrative as well as a portrait." Writing in *Book Week*, journalist Douglas Kiker was less generous. "Since it adds nothing new and fails to improve on the old," wrote Kiker, "is it really needed?" For his efforts Ford was paid $10,000, which he split with his coauthor. To the chagrin of its publisher, *Portrait* failed to sell out its initial run of ten thousand copies. Earl Warren never forgave Ford for exploiting popular interest in an investigation the chief justice found too painful to discuss in public. Interviewed by the FBI at the time of Ford's nomination to be vice president in October 1973, Warren said he was "shocked" by the nominee's authorship of *Portrait of the Assassin*, a book the retired chief justice had not troubled himself to read. Warren declined to evaluate Ford's fitness for the nation's second-highest office. He did allow that the congressman had impressed him as "an extremely partisan politician."

UNLIKE THE NIXON pardon, which gained public acceptance over time, suspicions about the Warren Commission multiplied with the passing years. Notwithstanding Ford's objections to Oliver Stone's *JFK*, the former president welcomed passage in 1992 of the President John F. Kennedy Assassination Records Collection Act. A by-product of Stone's film, it mandated the public release of all government records relating to the assassination. The same legislation established an Assassination Records Review Board to collect, preserve and make available these materials.

In January 1992, Ford wrote to Chairman Louis Stokes, who had chaired the House Select Committee on Assassinations. By then the committee's headline-making examination into the violent deaths of JFK and Martin Luther King Jr. was more than a decade in the past. Stokes and his colleagues had taken issue with much of the Warren Commission report, without ever disputing Lee Harvey Oswald's identity as the president's assassin. Yet Ford added his voice to the chorus demanding the public release of all documents acquired by the committee, and anything of relevance retained by the FBI or CIA. He did so believing this would only validate the original findings to which he had signed his name.

Even now he had misgivings. No doubt within the Archives' holdings were "unsubstantiated allegations" held back out of concern that reputations might be unfairly tarnished. "But I think you have to weigh that," Ford told the press, "maybe hurting some limited number of people, against the overall public good." Choosing his words carefully, Ford reminded his countrymen "we didn't say there *was* no conspiracy. We said we *found no evidence*. So if new evidence comes to the forefront that's credible, of course people ought to review it. And if it's truly sound evidence and it undercuts what the commission said, of course the commission's report ought to be modified."

Nothing greased conspiratorial wheels more than the search for Oswald's motive. So profoundly traumatizing an act, rechanneling history's course while nullifying the democratic choice of seventy million American voters, must be an expression of political calculation rather than random revenge for a menial existence in which fantasies of personal omnipotence could no longer be sustained. So it was assumed. At the end of his book, which was largely ignored then and since, Ford uncovered something essential about Oswald and his striking ability "to fabricate grievances." As a child, Lee had never been called on to face reality. Throughout his short life, meaningful relationships had eluded him, even as his inflated self-image chafed against the bleak obscurity of work in a Russian radio factory or Dallas warehouse. "The man who sets out to become famous irrespective of a real love for his fellow human beings, irrespective of the consequences of his acts upon other human beings, is a special kind of criminal," wrote Ford. "Lee Oswald was that kind of a person."

In his craving for posthumous recognition, Oswald belonged to a dark tradition of American assassins whose chief complaint was their own insignificance. At the same time he was the prototype of countless criminals and mass gunmen to come, whose senseless violence made sense only when gauged against their need for notoriety. In the end, it was not a legal but a moral verdict Ford pronounced on John Kennedy's killer, and the culture that produced him.

11

HOLDING THE LINE

⟼⟻

Don't ever tell me again what a nice goddam guy Jerry Ford
is. He beat me, and I'm king of the gut-fighters. Or I was.

—CHARLES HALLECK

MEL LAIRD WAS adamant. As he remembered it, "I went to Charlie Halleck in the 89th Congress [*sic*] to tell him that we were looking at replacing him if he didn't shape up out on the floor." It was typical of Laird to insert himself into the cockpit of events; less so to recognize the assistance of a copilot, in this case California representative Glen Lipscomb, whose intuitive reading of discontent among younger members of the Republican caucus helped spark the uprising of January 1965 that put Ford in Halleck's place. Under the circumstances one might have expected the party to showcase its junior varsity, if only to dispel the aura of old fogeyism that clung to the GOP brand. In fact most newcomers felt excluded from power, their program ideas and committee preferences ignored by a leadership that had forgotten the art of listening.

Still reeling from Goldwater's defeat, a badly fragmented party experienced something akin to shell shock. Additionally there was the personal element to consider. As Laird delicately phrased it, "Charlie had not coped well with the stresses of being minority leader." A well-connected Capitol Hill staffer of the period is less nuanced in his assessment. "He was a heavy drinker," says this insider of Halleck. Scotch, bourbon and rye flowed abundantly in the windowless basement room in the Capitol familiar to Republican lawmakers as the Clinic ("Charlie's drinking room," snorted his abstemious predecessor, Joe Martin). Here the conversation was mostly shop talk. "Halleck doesn't read books," explained one of his flock. "He doesn't even read the bills."

Certainly it wasn't from classroom texts that Halleck derived the tactical skills and slashing oratory he used to compensate for Republican weakness in numbers. A strict disciplinarian, not above threatening to cut off campaign funds to recalcitrant members, Halleck enjoyed the chilly regard extended to a successful drill sergeant. Ordinarily the fiercest of partisans, he welcomed the chance to cross party lines as long as his partners were from south of the Mason-Dixon Line. "He was constantly trying to set up coalitions with the Southern Democrats and sometimes this came at the expense of our members," according to Pennsylvania's Hugh Scott. To be fair, Halleck did shepherd his caucus toward support of the 1964 Civil Rights Act, though even here it was hard to tell how much of his leadership was grounded in conviction and how much stemmed from a deal he cut with the Johnson White House to have NASA fund a research building at Purdue University in Halleck's Indiana district. "Charlie Halleck was like Everett Dirksen," explained New York congressman Barber Conable in drawing comparisons with the Illinois senator famed for his mercenary skills. "He would have sold his grandmother for two federal judgeships."

After six years in the job, Halleck's zest for the deal had waned. By the autumn of 1964 a belief had taken hold that he lacked what Ford called "the affirmative . . . aggressive approach that we felt was needed to turn things around." Part of the problem was cosmetic. With his double chin, heavily seamed face and whiskey-red nose, Halleck cut a poor figure on television. Playing straight man to the roguish Dirksen in the *Ev and Charlie Show*, as their joint press conferences were dubbed, Halleck offered feeble competition to the charismatic JFK or his commanding successor. Even so he might have survived the 89th Congress unscathed but for external events that left him vulnerable to scapegoating by a shrunken GOP.

On election day Goldwater's dismal showing in Michigan—he lost the state to LBJ by more than a million votes—undoubtedly shaved a few points off Ford's usual blowout at the polls. Still, a victory margin of 61 percent for a candidate seeking his ninth term in office contrasted painfully with the performance of Republicans almost everywhere else. In the House, forty GOP incumbents were swept out of office on the Democratic tide. Not only had Ford bucked the trend, but the defeat of two senior Republicans on the Appropriations Committee had unexpectedly raised him to the status of ranking minority member. Presumably he could enjoy this exalted position for as long as he remained in Congress. Alternatively there was Charlie Halleck's job, the penultimate rung on the ladder ending at the Speakership. Though very much occupied at the moment, the minority leadership might be winnable *if* Ford could rally to his side a critical mass of younger members disenchanted with obstructionism and partnering with unreconstructed Southern Democrats.

With tiresome frequency it is asserted that all politics is local (as if *all* politics is anything). From the first whispered speculation to the final gavel ushering in a new era, the politics of the Ford-Halleck contest were less geographical than generational. No one illustrates this better than Donald Rumsfeld, a mainstream conservative first elected in 1962 to represent Chicago's northern suburbs. Having withstood the Johnson flood on November 3, Rumsfeld chose the day after the election to contact Thomas Curtis, a Missouri Republican and fervent supporter of both civil rights *and* Barry Goldwater. Standing in the rubble of their party, Rumsfeld mentioned the next meeting of the House GOP caucus, scheduled for the week of January 4, 1965. Why not move it up to mid-December, said Rumsfeld, thereby affording rank-and-file members an early opportunity to discuss reforms in House rules and increased staffing for the Republican minority? Infused with Rumsfeld's sense of urgency, Curtis composed an eight-page letter to colleagues proposing "that major decisions on party organization and policy matters be made on as broad a base as possible."

The implicit rebuke of Charlie Halleck's arbitrary style was hard to miss. On the day before Thanksgiving, Rumsfeld and Curtis met in the office of Michigan's Robert Griffin, along with Griffin, Charles Goodell of New York, John Anderson of Illinois, and Minnesota's Al Quie. After the meeting, Griffin contacted Ford, who, as chairman of the House Republican Conference, had responsibility for calling the membership together. Ford conditioned his assent on a groundswell of member support for such a gathering. Within hours Rumsfeld, Griffin and Company generated sufficient demand to force his hand. Ford's reluctance to summon a December caucus involved more than consideration for his colleagues' holiday plans. "I thought I ought to be discreet, and not push myself," he later acknowledged. He had no desire to relive the awkwardness of his pre-election visit to Warsaw, Indiana, when a fulsome tribute composed for "Charlie Halleck Day" had been eclipsed by stories out of Washington identifying Ford as a possible replacement for the man he was there to honor.

Tactics aside, Ford was understandably chary of taking on Halleck, the self-proclaimed gut fighter. Becoming minority leader would mean relinquishing his top-ranking slot on the Appropriations Committee. To grow the bloodied GOP from its present 140 members to the 218 needed to elect a Speaker of the House would almost certainly require more than one or two election cycles. Besides patience on a heroic scale, the leader's job would entail constant travel and fundraising to elect new members and reelect incumbents. The ultimate sacrifice would be made by Ford's family. Yet with Betty in near-constant pain from a pinched nerve, Jerry had no desire to leave her alone, or forgo the most important years in the lives of his children, ages seven through fourteen.

Stacked against these negatives were some powerful incentives. Although no one blamed Halleck for the Johnson landslide, no one was inclined to give him a pass, either. Even Halleck's stash of political IOUs was likely to be devalued, as the average member's need to distance himself from a tarnished leadership exceeded his gratitude for past favors. Not all was bleak, however. Many of the newly elected Democrats in the House were byproducts of the Goldwater debacle, their hold on power tenuous at best. Consequently whoever was minority leader in 1966 would be able to claim credit for likely GOP gains in that year's midterm elections. By failing to contest Halleck now, Ford could expect to wait up to six years for conditions anywhere near as promising—more than enough time for fickle House Republicans to transfer their allegiance to younger, more aggressive candidates.

To Mel Laird perhaps. At forty-two, Laird was nine years Ford's junior. In the immediate aftermath of Goldwater's loss, Laird had stepped forward to play the part of party unifier. "This is no time to hunt for scapegoats," he announced on November 11, "no time to indulge in intra-party bickering, no time to pursue personal ambition." In private, Laird conveyed a different message, telling intimates, "If Jerry doesn't go against Halleck, I will." He was notably absent from any of the early strategy sessions at which the Rumsfelds and Griffins pondered their options. "Everyone seems to think that Mel Laird was one of the plotters," Charles Goodell reminisced long afterward. In fact, "Laird had absolutely nothing to do with the nuts and bolts of getting Jerry Ford to run for minority leader. I personally called him and he wanted nothing to do with it."

Actually, it wasn't that simple—few things were once Laird began spinning his webs. After the Old Guard and the Young Turks, a third group of Republican members, predominantly northeastern liberals and moderates in the mold of Manhattan congressman John Lindsay, styled itself the Wednesday Club. Their numbers were modest, twenty or so by common count. But in a closely fought struggle, they were strategically placed to break the deadlock. Ford needed the Wednesday Club if he were to reach the seventy-one votes required to win. But as he admitted at the time to journalist Neil MacNeil, "if he should appear as their candidate, it would be the kiss of death" among conservatives still on the fence.

Wednesday Club members appreciated Ford's pragmatism. They were less forgiving of Laird for his role in crafting the Goldwater platform so decisively rejected by voters in November. And they fretted that a Ford-Laird leadership team would be too accommodating to the right. To counter these fears, the resourceful Laird devised a scenario emphasizing Ford's independence while simultaneously sowing seeds of confusion in the Halleck camp. In the run-up to the caucus, Laird made no secret of his impatience with the slow-to-decide Ford. This, too, was part of the charade intended to establish Ford's autonomy and Laird's neutrality. Deceived by the act, Halleck supporters dropped hints of their own: Charlie might only serve one more term as leader, before stepping aside for Laird in 1967.

The 119 Republicans who filed into the House chamber on December 16 barely recognized the genial Hoosier chieftain who went out of his way to solicit their views. Milking his postelection visit to Ike in Gettysburg, Halleck stressed the former president's desire for party unity. Did younger members feel estranged from policy discussions? Halleck floated a committee of House and Senate Republicans to foster program ideas and coordinate the party's response to LBJ's legislative juggernaut. At one point he even borrowed a favorite term from the Young Turks in describing himself as "forward looking." By the end of the day reporters were writing of a Halleck comeback. Not so the professional politicians in attendance, who were sufficiently shrewd—or cynical—to interpret their leader's sweet reason as a sign of weakness. That evening a dozen or so rebels convened in Griffin's office to share the latest intelligence. Their head count was encouraging enough to overcome any lingering doubts Ford might entertain.

Yet even now he conditioned his running on a promise to his family. As Ford explained it, a long-planned skiing vacation would preclude his campaigning in Washington during the critical week between Christmas and New Year's. "That way I didn't have to get involved in commitments to people," he later confessed. Clearly he had more in mind than navigating the slopes of Northern Michigan. Ford's colleagues accepted his stipulation without dissent. The home front was a different matter. "Betty is not going to be very happy about this," he observed ruefully. But she had always been a good soldier. "It seems to me this is the time. Let's do it."

Out in the open at last, the Ford drive quickly shifted into high gear. Bob Griffin was designated campaign "coordinator." Don Rumsfeld drafted Ford's announcement of candidacy for delivery the morning of December 19. He also prepared a battery of likely press questions (and equally useful answers), as well as a telegram for Ford to send to all members of the Republican caucus.* On the twenty-first a full house assembled at the Na-

---

* The candidate had promised to call Halleck personally to notify him of his decision. Failing to reach Halleck in Florida, he wired him the news.

tional Press Club to watch as Ford, in blue suit and blue television-friendly shirt, staked his claim to "a new era in American politics." Avoiding mention of Halleck by name, he stressed participation over personalities. On his team, insisted Ford, there would be no benchwarmers. Everyone would be assured of playing time, regardless of seniority. Coming off "the first negative landslide in American history," he vowed to travel "the high middle road of moderation."

Drawing on the ideas of his colleagues, Ford promised "attractive workable alternatives" to Lyndon Johnson's Great Society, with its centralization of power and resources in Washington.

He also took aim at his party's traditional coalition with Southern Democrats. "Ford Urges Split with Dixie," headlined the *Detroit Free Press*. In truth, he was merely restating before a larger audience his long-held vision of a Southern GOP that was more Eisenhower than Goldwater, and more Lincoln than anyone else. His deft performance reassured members of the Wednesday Club even as it allowed him to make inroads among the incoming class of Southern Republicans.

WITH THE ADVENT season in full swing, Christmas decorations imparted a festive look to Capitol Hill. But the colored lights and jolly Santas belied an increasingly bitter contest that no amount of holiday trappings could obscure. Taking no chances, on December 22, Ford met with the five-member Kansas delegation in Bob Dole's office. Later in the day he confided to a friendly reporter, "We're over the 50–50 mark. We've got over 70 votes." The next day he left town to hit the ski slopes of Michigan's Boyne Mountain, ceding center stage to his opponents. In his absence Halleck marshaled his forces, enlisting a potent network of Washington lobbyists and far-flung bankers, Chamber of Commerce executives and other business interests who could exert economic pressure where conventional persuasion fell short.

Reporters cautioned their editors against predicting the outcome of a race further complicated by use of the secret ballot first employed six years earlier in the Halleck-Martin contest. No one appreciated this more than Don Rumsfeld. It had not escaped Rumsfeld's notice that if Ford won on January 4, his place as ranking member on the Appropriations Committee would be taken by Ohioan Frank Bow. Suddenly the Buckeye State appeared up for grabs. On December 28, yielding to an urgent request from Rumsfeld, Ford returned to the Capitol for a twelve-hour burst of campaigning by telephone. He then resumed his vacation, only to be called back a second time amid reports that his support was eroding.

The decisive moment came on Sunday afternoon, January 3. "A few minutes after three," recalled Charles Goodell, "Jerry got a call from Betty and we all walked out of the room. About ten minutes later, Jerry called us back in and said he was going home because Betty wanted him to grill some steaks. I took him aside and told him that was impossible." Reneging on his promise of a family dinner, Ford spent the next seven hours on the phone seeking commitments from wavering colleagues. By Monday morning a mood of uncertainty gripped his supporters. "The votes have been shifting all night," Goodell confided. On the other side of the Capitol, Halleck's partner in the *Ev and Charlie Show* entertained no such doubts. "That fellow from New York," as Everett Dirksen referred to John Lindsay, had cut a last-minute deal with Halleck that was expected to bring much of the Wednesday Club with him. Halleck had it in the bag, chortled Dirksen, who predicted victory with thirty votes to spare.

As the caucus got down to business on January 4, choosing Mel Laird for conference chairman before listening to brief nominating speeches for Ford and Halleck, an anxious Betty Ford paced the floor of her husband's office in the building next door. Following custom, the counting of votes in the caucus was entrusted to a trio of female members.* What followed was virtually a carbon copy of the 1958 cliffhanger, down to the embarrassing single vote over and above the 140 accredited members present. Once again, a second, confirming ballot was called for. Ford won it 73–67. Bob Dole had come through with a bloc of four Kansans, and Frank Bow's Ohioans were equally helpful. A dozen years later Ford would return Dole's favor, for considerably higher stakes.

It was 11:35 when the phone rang in Ford's third-floor office in the Cannon House Office Building. Elated by the news, Betty raced to the House gallery. From his seat near the front of the chamber, Jerry motioned for her to come downstairs. They nearly collided rounding a corner, each rushing to greet the other. Any formal celebration would have to wait, Betty telling reporters, "He promised to take me out to dinner before the night session." She smiled. For now, she returned to Jerry's office, where she ordered a hamburger and a glass of milk, to be eaten, alone, at his desk. A few hundred feet away Halleck retreated to the Clinic to nurse his wounds with Grant's Standfast Scotch. "I'm from Indiana," he mused. "I'm used to blood on the floor." It was painful enough to hand over the keys to the Re-

---

* Among them Charlotte Reid of Illinois, a former radio vocalist who made history in 1969 as the first woman to wear pants on the floor of the House.

publican leader's office, with its spectacular view down the National Mall to George Washington's obelisk and the Lincoln Memorial beyond. Also surrendered: the plush black Cadillac and uniformed driver that went with the job. Charlie drove himself home that night, in his '65 Oldsmobile. He did obtain one concession from his successor: Ford promised to give Halleck's chauffeur a month's trial.

T EN DAYS AFTER his ouster Halleck enjoyed a measure of revenge. On January 14 the former minority leader again addressed the Republican caucus. Summoning oratorical gifts of old, he delivered a passionate nominating speech for GOP whip Les Arends against Peter Frelinghuysen of New Jersey. Honoring his pledge to Frelinghuysen, Ford suffered an embarrassing defeat when his candidate lost to Arends 70–59. He had no greater luck in promoting Charles Goodell over Arizona's John Rhodes for the vacant chairmanship of the Republican Policy Committee.

Rather than risk another public rebuff, Ford split the policy job in half. A new Planning and Research Committee, chaired by Goodell, was to focus on "long term solutions to national problems." Rhodes would concentrate on more immediate concerns. Not content to revise the organizational chart, Ford wanted to rethink the essential mission of his badly outnumbered minority. "I don't think the party can sit back and just be continually in opposition and wait for some catastrophe to come along and then try to pick up the pieces." To oversee the first House Republican research unit Ford recruited Dr. William Prendergast, research director of the Republican National Committee until he was turfed out in the Goldwater purge. Ford committed himself to raising enough money from private donors to support a permanent policy research staff of eight. He organized a Washington conference at which 250 college students from around the nation were exposed to Republican thinking on issues foreign and domestic.

Praise for his initiatives came from the New York law office where Richard Nixon was already planning a 1968 comeback. "I don't need to tell you that a leader has to take some risks and you can't win every time . . . what counts in my book is that you had the guts to take a chance—both in trying for the Minority Leadership and on the question of the Whip," wrote Nixon. While he didn't advocate rashness, the party's former standard-bearer lamented that "the trouble with many of our Republicans is that they are conservatives not only in their economics but also much too conservative in their tactics." Having sold himself as an agent of change, Ford could not escape criticism for a sometimes chaotic transition. "Jerry's first •

three or four months as Minority Leader were extremely shaky," acknowledged Charles Goodell. "Of course he was a nice guy. Even his critics liked
him . . . His secret was that he was almost totally non-vindictive and he
made the effort to make people feel that they were part of the leadership
team."

As House minority leader, Ford's primary function was to do whatever
he needed to do to become majority leader or, better yet, Speaker. Inevitably
the position reflects the personality and priorities of its occupant. No leader
can hope to succeed if he lacks strategic acumen, or the skills to unify a
fractious caucus and project a coherent message before a national audience.
Leading by consensus, Ford even extended consideration to members of the
opposition. "As a very young member of Congress," recalls Lee Hamilton,
an Indiana Democrat elected in 1964 on Lyndon Johnson's coattails, "I
made a big parliamentary mistake on the floor. Didn't even know I'd made
it. I knew so little about the rules." At Ford's instigation, or at least with his
consent—"I never knew which"—Hamilton's Hoosier colleague Republican Bill Bray gently set him straight. "Lee, you just made a mistake," said
Bray. "Here's the way you correct it." The issue in question was not the stuff
of headlines, and Ford had the tools to block it long before it became law.
Still, Hamilton was stunned that the Republican leadership should exert
itself "to help this newcomer Democrat not make a fool of himself."

In fact Ford's treatment of Hamilton was smart politics. Multiplied
numberless times across his twenty-five years in Congress, it helps explain
the bipartisan consensus that rallied around Ford when Richard Nixon
needed a replacement for his disgraced vice president, Spiro Agnew. "Jerry
Ford had a deceptively low-key approach to things," Hamilton reflects. "He
was ambitious, no question about that, I think. But he cloaked it nicely."

Historian Eric Goldman described Ford's strategy for making his party
relevant in spite of its diminished numbers. "Republican members would
not simply be against a Democratic White House bill," said Goldman.
"They would turn to other techniques—voting to recommend it to committee for 'revision and improvements,' which would give them a chance
to smother it; or introducing a substitute of their own, with GOP representatives under party discipline to vote for the substitute measure." Should
these efforts fall short, as they usually did, members of the loyal opposition
could always support the administration on a final roll call.

More than once Ford availed himself of this option, in marked contrast
to some recent additions to the GOP caucus. Among these five Alabama
freshmen stood out for their hostility to civil rights legislation passed and
pending. Beginning in January 1965 a series of demonstrations organized

by the Southern Christian Leadership Conference and led by the Reverend Dr. Martin Luther King Jr. focused attention on Selma, Alabama, a sleepy river town of thirty thousand inhabitants located forty miles west of the state capital of Montgomery. A majority of Selma's residents were African-American, yet Blacks accounted for just 1 percent of the town's registered voters. While the protests were peaceful, the backlash they spawned among defenders of the old order was not. On the night of February 18 a state trooper in nearby Marion shot and killed Jimmy Lee Jackson, a twenty-six-year-old church deacon, as he tried to shield his mother from the trooper's nightstick. Jackson's death led organizers to announce plans for a public march from Selma to Montgomery, commencing on Sunday, March 7.

From the outset Ford expressed sympathy for the demonstrators. "The patience of fair-minded people is wearing thin," he declared a few days before the marchers assembled on the outskirts of Selma, "when, after decades of waiting and three civil rights acts, the basic right of citizenship is still denied to a substantial number of citizens in defiance of the Constitution." A radically different tone was set by first-term Alabama representative William Dickinson, a Goldwater Republican who likened the Selma protesters to "a swarm of rats leaving an overturned hayrack." Worse, insisted Dickinson, "drunkenness and sex orgies were the order of the day" among those agitating for enforcement of the Fifteenth Amendment a century after its adoption. Pointing out to Dickinson the negative repercussions of such oratory, Ford pleaded with the Alabama congressman to make no charges he couldn't prove.

Then came March 7, Bloody Sunday in movement lore, when millions of Americans turned on their television sets to see a police riot at Selma's Edmund Pettus Bridge. Blocked from continuing their march, six hundred demonstrators were assailed by local police and Alabama state troopers wielding bullwhips and billy clubs. Some protesters were beaten senseless. Others were savagely kicked by law enforcement, or stomped on by police horses. As feelings of revulsion swept the nation, President Johnson scheduled an appearance before a joint session of Congress. In the most eloquent speech of his presidency, he likened Selma to Lexington and Concord, and paid tribute to the latter-day patriots whose blood now stained Alabama Highway 80. "Their cause must be our cause too," Johnson told a vast television audience on March 15. "Because it is not just Negroes, but really it is all of us, who must overcome the crippling legacy of bigotry and injustice. And *we shall overcome.*"

The effect was galvanizing. Two days later the administration introduced the Voting Rights Act as a remedy for the systematic exclusion of Southern

Blacks from the voting booth. The proposed legislation would suspend blatantly discriminatory literacy tests and abolish poll taxes, another device traditionally employed to disenfranchise poor Blacks in the region. Most important, it authorized federal registrars for any jurisdiction in which less than 50 percent of voting-age citizens were duly registered or cast a ballot in the 1964 elections. Following Senate passage of the act on May 26, attention turned to the House and to William McCulloch, ranking Republican on the House Judiciary Committee. As his party's chief spokesman and strategist on civil rights, the courtly Ohioan had been instrumental in passing the 1964 act.

Now, less than a year later, McCulloch entertained doubts, not about the moral urgency of extending suffrage to Black Americans, but the methodology employed by the administration and the number of jurisdictions that fell outside its 50 percent formula. His alternative, cosponsored by Ford, was nationwide in scope. In place of the 50 percent rule and its emphasis on literacy tests, the GOP would dispatch federal voting registrars to any jurisdiction reporting twenty-five or more credible complaints of voter disenfranchisement. The *Washington Post*, harshly critical of Ford's first weeks as minority leader, said he had redeemed himself with the voting rights debate. From the right, syndicated columnist William S. White complained that Ford went too far in federalizing election law, traditionally left to the states.

In the end, however, not the bill's critics but its nominal supporters doomed any chance of passage. Virginia congressman William M. Tuck, an unabashed segregationist, maladroitly endorsed McCulloch's legislation over the proposed Voting Rights Act, "this unconstitutional monstrosity" of federal intervention that promised a massive increase in voting by African-Americans. As Tuck ranted on, a stricken Ford listened, said one observer, "with the look of a man betrayed." On July 9, 1965, the House rejected the GOP alternative by a vote of 248–171. The same night it overwhelmingly passed the administration's bill, the margin of victory swelled by Ford and 111 other Republicans who lent their support on the final roll call. This last-minute change of heart did little to advance their party's standing among Black Americans. Neither did the appointment of a Ford-endorsed youth from Lincoln's hometown of Springfield, Illinois, as the first black page in House history.

That fall another symbolic gesture dramatized Ford's personal convictions. Having accepted a speaking invitation from the University of Mississippi, he readily agreed to add to his schedule a GOP fundraising dinner in Natchez. The latter request came from Prentiss Walker, the state's first

Republican congressman since 1885. Shortly after his visit was announced, Ford received a telegram from Charles Evers, successor to his murdered brother Medgar as NAACP field director in Mississippi. According to Evers, the proposed dinner was to be segregated, with Blacks denied the chance to buy tickets. Ford shared this information with Walker, who insisted that no Negroes had expressed interest in attending. A skeptical Ford asked Charlie Goodell to quietly investigate. Then he told Walker that bogus claims of a sold-out event would not be accepted if it meant excluding Black dinner guests. As a matter of fact, said Ford, he expected to see Blacks and Whites sharing tables.

A few days later another Evers telegram asserted that none of the promises made to Ford were being kept. Hoping to defuse the situation, Mississippi state GOP chairman Wirt Yerger urged adoption of a "local arrangement" employed during earlier visits to the state by Richard Nixon and Barry Goldwater. With a wink and a nod, Ford could be unavoidably delayed in arriving at the dinner, causing him to reserve his main remarks for a later public rally, which would be integrated. The offer was declined, Ford reminding event organizers that times had changed. What might have been tolerated by Republican candidates even a few years earlier was out of the question in 1965. The dinner was canceled two days before it was scheduled to take place.*

The incident reinforced a belief in GOP ranks that Ford was growing into his new job. "He has conviction and courage and generally good instincts," said Don Rumsfeld in assessing his friend's first year as leader. Ford strengthened his position through a tireless pace of travel—in his first six months alone he visited thirty-two states on behalf of Republican candidates. His Congressional Boosters Club raised $2 million to support candidates and underwrite the party's new research and policy apparatus. Ford encouraged his colleague John Byrnes to craft a GOP medical insurance plan in place of the administration's federally subsidized Medicare program. Reflecting the traditional Republican dread of "socialized medicine," Byrnes envisioned a voluntary program ("Bettercare") that would address

---

* All three of Ford's sons attended T. C. Williams High School in Alexandria, whose football team was celebrated in the 2000 film *Remember the Titans*, starring Denzel Washington. Williams resulted from the consolidation of three existing schools, each one largely segregated. Thus Mike Ford's first experience with integrated education occurred in his tenth-grade class. "And I kind of said, wow. And my Dad set me down and said, 'Son, this is America and this is what we are proud to have—ethnicities, all races. You know Clara . . .'" In time Mike introduced his father to George Wilbur, a much-admired world history instructor who happened to be Black. Parent and teacher became fast friends. AI, Michael Ford.

both hospital and doctor bills, as well as some patient services. Two-thirds of the costs would be covered by Washington out of general revenue. Monthly premiums scaled to individual patient income would cover the rest.

House Ways and Means Committee chairman Wilbur Mills, a fiscal conservative whose command of numbers included the 1964 election results, was sufficiently impressed by the Republican proposal to incorporate it in the "three layer cake" that became Medicare, Parts A and B, plus a federal-state partnership called Medicaid, designed for the most economically needy of patients. Medicare provided a classic illustration of how a badly outnumbered minority might still leave fingerprints on landmark legislation (even if an early attempt to codify this approach invited ridicule over the unfortunate moniker of Constructive Republican Alternative Policies and its resulting acronym).

Progress did not come without a price, one paid chiefly by Betty Ford and her children. In the month of March 1965, by no means exceptional in its demands, Ford was away from his family all but four nights.

"Daddy," said seven-year-old Susan Ford as she twisted her arms around his neck, "when are you going to be home more?"

"Things are going to get better, honey."

Tell it to Susan's mother. "I realize Jerry has all these people wanting to hear him speak, but he should spend some time at home," Betty complained to a reporter that spring. "The kids don't get to see much of him, and this is bad." The passage of time only made his chronic absence worse. As a key member of the Nixon White House congressional relations team, Max Friedersdorf ended his workday by dropping in to the minority leader's office to review the day's happenings and coordinate future plans. "I remember one evening he got a call from Mrs. Ford and he had to go catch a damn plane and go somewhere that evening to make a speech. And she was not feeling well. She said, 'Why don't you cancel that and come home?' But he said, 'Betty, I've made a commitment.' You know, he was torn and it was a sad situation."

Ford's punctuality was the stuff of family legend, as was Betty's chronic tardiness. Only much later did it occur to the younger Fords that their mother had weaponized time. "It allowed her to have some control," concludes Mike Ford. "For all the times that he was gone, and she had to trail after him, for all the late meals . . . when she could say, 'Jerry, it's going to be another ten minutes,' that gave her some power." If there was acrimony in the Ford marriage, it was concealed from the children. "I didn't see the tension between Dad and Mom," Mike notes. "But I saw it in just

Mother—her health was failing. She was beginning to get more dependent on her medications, and the alcohol started flowing."*

Then came the summer day in 1965 when Jerry was sailing the Potomac as a guest on the presidential yacht *Potomac* and Susan Ford found her mother in the house on Crown View Drive crying hysterically for reasons she couldn't articulate. Terrified, the young girl called out to Clara Powell, who managed to calm Betty down and somehow get word to Jerry that he was needed at home. Her mother was "very sick," Clara told Susan and her brother Steve. "She had to go to a psychiatrist." In *The Times of My Life* Betty identified some factors leading to her collapse. "I'd felt as though I were doing everything for everyone else, and I was not getting any attention at all." She fantasized about running away, driving to the beach without telling anyone. "I wanted them to worry about me. I wanted them to recognize me."

Betty had forfeited her sense of self-worth—arguably her sense of self. It was a condition familiar to many women of her age and station. "Their husbands have fascinating jobs, their children start to turn into independent people, and the women begin to feel useless, empty." Twice-a-week psychiatric visits enabled Betty to carve out some space for herself and address issues that, for whatever reason, she felt unable to raise at home. "My back hurts, there's dope in the schools, Jerry's away." At one point her husband sat down with her psychiatrist to discuss Betty's condition (Betty herself ruling out the doctor's suggestion that he meet with the Fords as a couple). Even as her mental health improved, the pain from her pinched nerve was complicated by a diagnosis of spinal arthritis. Hot packs and traction devices provided temporary relief. Prescribed drugs and furtive drinks dulled her suffering and masked the sources of her despair.

A S THE BIBLE of Midwestern conservatism, the *Chicago Tribune* had not always looked with favor on Jerry Ford. His close ties to Senator

---

* In his 2018 memoir of the Ford presidency, *When the Center Held*, Don Rumsfeld cites an Oval Office meeting on March 3, 1975, in which Ford, insisting that an upcoming vacation in Palm Springs was vital to Betty Ford's postoperative health following her surgery for breast cancer, revealed that his wife had once "damn near died from alcoholism," a condition requiring her hospitalization for a month. Fortunately in the years since she had made "a beautiful recovery." Besides confirming that Mrs. Ford's illness was more severe than generally portrayed, Rumsfeld's disclosure also raises the possibility that her husband's long absences on the road represented, in part at least, an escape from unpleasant truths as much as they did the pursuit of a Republican majority in the House.

Vandenberg, long confused in the pages of the *Tribune* with Benedict Arnold, stamped Ford as a midcentury Republican in Name Only. By the summer of 1965, however, the paper's flamboyantly chauvinistic publisher, Colonel Robert R. McCormick, had been in his grave for a decade. Writing from Washington, *Tribune* correspondent Willard Edwards wrote of Ford that "he has developed into a tough, resourceful, and persuasive leader" able to achieve near unanimity within his party's diminished ranks. Edwards's fellow scribe, Walter Trohan, concluded that the new House minority leader was succeeding for reasons the Colonel of all men would appreciate: "No man in Congress can escalate the temper of President Johnson as high and as quick as Ford."

"Lyndon on a couple of occasions got quite irritated at my nitpicking," Ford recalled. "I've forgotten the quote, but I think it was 'There's nothing wrong with Jerry Ford except he played football too long without a helmet.' Well, you know, that was kind of a backroom comment, but it got good publicity. From that [point] on, we had less than the best of relations." Taking the Republicans' support of his Vietnam policies for granted, Johnson couldn't resist needling Ford and his easily coopted minority.

"Jerry, if you and Ev Dirksen don't stop supporting my foreign policy so strongly, I'm going to wind up with a Republican Congress next year."

"That isn't the reason we're supporting you, Mr. President," Ford responded, "but if it happens to be one of the fringe benefits involved, we won't object."

Presidential gratitude was short-lived. Within days of this ribbing exchange, Johnson accused the GOP of foot-dragging on voting rights. Ford hit back, hard, at "Lyndon come lately," dredging up Johnson's record of opposition to civil rights legislation for most of his years in Congress. As the 89th Congress became an assembly line for Johnson's Great Society legislation, Ford claimed that Democratic lawmakers didn't know whether to clap their hands or click their heels. A resentful Johnson shot back that Ford "couldn't fart and chew gum at the same time" (an assertion cleaned up for publication). Much of this political trash talk had a ritualistic feel to it, with LBJ winning the battle of insult comedy even as Ford exposed the president's hair-trigger sensitivity to criticism.*

Acknowledging his initial modest estimate of Ford's abilities, Barber Conable said he changed his mind once he realized how often Ford suc-

---

* "LBJ made cracks like that all the time," noted Cokie Roberts, the daughter of House Majority Leader Hale Boggs. "He said some pretty insulting things about my father, too, and they were very close friends." AI, Cokie Roberts.

ceeded in getting under Johnson's skin. Especially prickly was the topic of America's growing military involvement in Southeast Asia, where LBJ was inclined to read disloyalty into every difference of opinion over his conduct of the war. Citing Eisenhower's reluctance to commit substantial armies to an Asian land war, Ford stressed the value of American air and sea power instead (a position he shared with General Maxwell Taylor, the US ambassador in Saigon until he resigned his post in July 1965). Ford wanted North Vietnamese ports blockaded to prevent foreign arms from reaching Ho Chi Minh's forces. He urged the bombing of Hanoi a year before Johnson authorized such raids.

On July 27, 1965, Johnson met with leaders of both parties before announcing the deployment of an additional fifty thousand US troops to Vietnam. At the same time he doubled the monthly draft call to thirty-five thousand. The next day reporters asked Ford about the White House meeting, at which Senate Majority Leader Mike Mansfield had voiced strong objections to the president's planned call-up of reservists. Ford made a passing reference to Mansfield's concerns, but said nothing about LBJ's response or his subsequent decision to rely on draftees instead of reservists. Within hours stories appeared in print, attributed to unnamed "congressional sources," that described Mansfield's intervention and the president's change of heart. Enraged by the leak and the implication that he required guidance from his party's congressional leadership, Johnson allowed his gift for sarcasm to smother his strategic judgment. Unburdening himself to reporters at his Texas ranch on August 1, the president accused "a prominent Republican" of having "broken my confidence" by exposing classified information. "Once in a while an inexperienced man or a new one, or a bitter partisan, has to play a little politics," he added, twisting the knife.

Immediately Ford's office was swamped with inquiries. Unnamed, but widely assumed to be the culprit, Ford was "furious," as he later recalled. "It really was a challenge to my integrity." Fortunately for him, *Newsweek* correspondent Sam Shaffer, among the guests at Ford's most recent luncheon for the press, happened to be the reporter who had interrogated him about the White House meeting in question. Shaffer was vacationing in New Hampshire when he read about Johnson's tongue-lashing of the Republican leader. He immediately dashed off a note to Ford absolving him of any breach of confidence. "For three days, I wandered around in a daze, Sam," Ford confessed to the journalist when he returned to Washington. "My leadership position was wobbly enough. Now I felt it had been destroyed. The President of the United States had said in almost so many words that

I was undermining the national security. I had made up my mind to resign as minority leader. The morning I made that decision your letter arrived."

If Ford sometimes felt like he was battling the Johnson Administration with one hand tied behind his back, he had Ev Dirksen to thank. As with LBJ, Dirksen and Ford were opposites who didn't attract. "The flute and drum," said columnist Mary McGrory, "velvet and sandpaper." "That young man has a double whammy problem with me," Dirksen confided to a friendly journalist. "You see, I liked Charlie Halleck and I also happen to like President Johnson." His disagreements with Ford ranged from the trivial—over a White House request for more time to produce a federal budget document—to Ford's game-changing proposal for a televised Republican response to LBJ's 1966 State of the Union address. Dirksen thought the latter idea "juvenile." He got his comeuppance when a thoroughly prepared Ford was interrupted by applause twenty times in thirteen minutes.

Being upstaged was a rare occurrence for the Illinois senator. With his tousled hair and time-ravaged face, Dirksen looked and, even more, sounded like a throwback to the Senate of Webster, Clay and Calhoun. Whether trumpeting the merits of the marigold as America's national flower, or recording patriotic ballads in the manner of William Jennings Bryan, whom Dirksen as a boy had met in Pekin, Illinois, the senator had long since realized his early ambition for a life on the stage. Accustomed to dominating the weekly *Ev and Jerry Show*, as their joint appearances were retitled following Charlie Halleck's exit, Dirksen may have entertained second thoughts after a stunning confrontation in April 1966. Before a television audience of millions Ford launched a stinging attack on the so-called credibility gap eroding popular trust in Johnson's wartime leadership. Almost as an afterthought the House Republican leader cited bomb shortages and shipping tangles to support his claim that the American war effort suffered from "shocking mismanagement."*

Then it was Dirksen's turn. "He went pretty far, didn't he?" the plainly unhappy senator remarked. As Ford bit his tongue to keep from responding in kind, Dirksen declared, "You don't demean the chief magistrate of your country at a time when a war is on." Within months of its debut, the *Ev and Jerry Show* was in danger of cancellation, not because it failed to draw an audience—Dirksen's gravel-voiced Cicero imitation was made for the TV

---

* Needless to say, Dirksen the public performer tended to overshadow the legislative craftsmen with an uncanny sense of timing, and the vision to recognize that some issues—like civil rights—demanded more than partisan gamesmanship.

cameras—but because his support of Johnson's war policies put the senator at odds with his own party. As the number of American combat soldiers in Vietnam topped three hundred thousand, and the weekly death toll reached one hundred or more, a rare consensus of left and right was forming against more of the same. Dirksen insisted on hard facts to justify dissent from the president's approach. In 1965 and again in the fall of 1966 House Republicans produced white papers on Vietnam that challenged LBJ's measured escalation. The same broadsides questioned the administration's truthfulness with the American people. "That had a tremendous effect on Dirksen," according to Mel Laird, the driving force behind much of the GOP criticism. "Dirksen and Ford had a kind of falling out about that."

On the eve of the 1966 midterm elections Johnson flew to the Philippines for a hastily convened summit of US allies in the region. When Ford dismissed the trip as "a political gimmick," Dirksen upbraided him for excessive partisanship. House Republicans decided they needed outside help if they were to compete with Dirksen for public notice. Enter Robert Hartmann, a forty-eight-year-old former Washington bureau chief for the *Los Angeles Times*. Hired by the House Republican Conference at the recommendation of his close friend Glen Lipscomb, Hartmann was initially seen as a latter-day Charlie Michaelson, the PR genius whose unceasing vilification of Herbert Hoover had made that star-crossed president forever synonymous with the Great Depression.

It was a job tailor-made for Hartmann, described by Richard Reeves, formerly chief political correspondent for the *New York Times*, as "nasty, vindictive and loud—and that was when he was sober." The *Washington Post*'s Sally Quinn in a warts-and-all profile of Hartmann evoked comparisons to Captain Bligh, Kennedy wordsmith Ted Sorensen and Ernest Hemingway gone to seed. Liberal references to his personal and professional insecurities were balanced by Hartmann's unparalleled access to Ford, who insisted, notwithstanding considerable evidence to the contrary, "Bob is a hell of a nice guy. And I happen to think a brilliant person. Shrewd. But he's got this irascible personality, particularly when he drinks . . . he is hypersensitive—yet very insensitive to the feelings of others. A prodigious worker when he puts his mind to it, but the tone of his voice, the words he will say, are like daggers in people."

A throwback to the Hildy Johnson *Front Page* school of journalism, Hartmann's gruff manner did nothing to endear him to the Ford staff, its members steeped in Michigan Nice. "We all said, who is that?" recalls one Ford loyalist. "And we found out who is that. He seemed to take over." Reciprocating their disdain, Hartmann barely tolerated Frank Meyer ("something of a prig"), the workaholic administrative assistant who ate lunch at his desk and insisted that young people on the staff write or call home regularly. Hartmann also clashed with Paul Miltich, the Ford press secretary whose prior career as a correspondent for the *Grand Rapids Press* and other Michigan papers was deemed inadequate preparation for his boss's expanded responsibilities.

Miltich had his own theory to explain the seeming hold Hartmann exercised on Ford. "He can write a hell of a speech when he wants to," allowed Miltich. Equally important, "he is as devious as Ford is guileless." As a political animal, "he's Machiavellian and I think Ford felt that he needed this in an adviser." With Hartmann on the payroll, Ford could indulge his habit of seeing the good in people, knowing he had another set of eyes to detect and defend against their less admirable qualities. Gail Raiman was an office intern who drew the short stick of working with Hartmann. Before embarking on her new assignment she was assured by Congressman Ford that, while Hartmann "can be a little difficult at times," she would enjoy working with him.

By and large Ford spoke the truth. Gail was able to overlook Hartmann's flashes of temper and strangely inverted hours. Like many a writer, he drew inspiration from a deadline. And if he sometimes tested Ford's patience, he could be counted on to deliver texts whose punchy rhetoric and political savvy made up for their author's eccentricities. At once protective and possessive, Hartmann avoided terms like "academician" or "nuclear," which Ford routinely mangled. On the road he surreptitiously watered his boss's martinis before a dinner speech. "He had Ford's back," said Raiman, "and he was someone who would tell him if something was wrong—this looks horrible, this should not happen, you cannot say this, you cannot do that." It is a point emphasized by Bob Hynes, a minority counsel on the House Rules Committee before he worked for NBC News (and married Ms. Raiman). "If there was such a thing as Mr. Ford's enforcer . . . Bob was that person." With a grasp of future events bordering on clairvoyance, he would alert Ford, "This is coming down the track."

Hired as a one-man brain trust, Hartmann evolved into Ford's chief political adviser. He was kept busy as the would-be Speaker of the House hit the road in advance of the 1966 midterm elections. Before election day Ford delivered over 200 speeches in 37 states. Sensing victory, Republicans enjoyed a unity absent in 1964, although at one point in the campaign Ford returned as tainted a $1,000 check from a California member of the John Birch Society. Increasingly his talk of inflation and "Potomac paternalism" yielded to criticism of how the war in Vietnam was being conducted.

Ford got help from an unlikely source when Johnson at the eleventh hour lambasted Richard Nixon as "a chronic campaigner" placing his private ambition ahead of soldiers' lives.

Ford sprang to Nixon's defense. The House Republican Congressional Campaign Committee purchased half an hour of television time so that the former vice president could return Johnson's fire. The controversy helped reestablish Nixon as the party's titular leader and a front-running presidential contender for 1968. "We're going to win forty seats," a confident Ford insisted to a skeptical David Broder of the *Washington Post*. Ford offered to bet him on the election's outcome. "And of course he collected on the bet," said Broder. "But he *knew*. I mean, he could taste it."

Ford's prediction turned out to be conservative, as Republicans gained back everything lost in 1964 and more: forty-seven seats in the House and three in the Senate. Fifth District voters returned him to office with 68 percent of their ballots, the high-water mark of his career. Barely able to restrain his delight, Ford vowed to redress the balance between Congress and the executive. "I feel a lot more like a leader," he enthused, "and a lot less like a minority."

O N JANUARY 19, 1967, a subdued Lyndon Johnson delivered his fourth State of the Union address before a skeptical Congress. Dispensing with guns-and-butter prodigality, Johnson described a "time of testing" to be funded by a 10 percent surcharge on personal and corporate income taxes. Breaking with tradition, it was Republicans in their nationally televised response who offered the more ambitious agenda. Sharing the podium with Dirksen, Ford made no fewer than thirty-nine legislative proposals. Some of the concepts he introduced that evening, like revenue sharing and block grants to the states, would underpin the new federalism pursued by GOP presidents from Nixon to Bush.

This was not the only challenge Ford posed to the existing consensus. Long before the Nixon Administration gave the term its negative connotation, Jerry Ford had a Southern strategy of his own. After November 1966 the GOP controlled fully a quarter of the region's House seats. Only Louisiana in the former Confederacy was without at least one Republican congressman. Raw numbers aside, Ford had reason to celebrate the victories of such promising newcomers as George H. W. Bush, elected from a suburban Houston district as a strong supporter of Planned Parenthood, and Howard Baker, the first Republican to represent Tennessee in the Senate since Reconstruction. Sensitive to allegations that the post-Goldwater party was hostile to the cause of civil rights, Ford preferred to dwell on Republican governors such as Winthrop Rockefeller in Arkansas and Spiro Agnew in Maryland, racial moderates who had defeated old-line segregationists running on the Democratic line.

Emboldened by his party's success below the Mason-Dixon Line, Ford announced plans to challenge right-of-center Democrats valued as collaborators by earlier Republican leaders. His strategy was ruthlessly simple, in his own words: "to drive southern Democrats into the arms of the Administration where they belong on votes that will hurt them in their home congressional districts." For someone not often credited with vision, Ford's party-building—and -wrecking—activities were as farsighted as they were disruptive. Accepting fewer legislative victories in the short term, he would make the GOP a truly national party for the first time in its history, even as he demolished the Democratic Solid South in place since the Civil War. The reaction among Southern Democrats can be easily imagined. Said one grizzled veteran, unaccustomed to competition at the polls, "I'm not going to coalesce with some s.o.b. who's out to whip me every two years."

Events would show that Ford was right to anticipate a Republican South. What he misjudged, like Eisenhower before him, was the willingness of white Southerners to discard racial politics and rally behind a moderate conservative alternative to the Great Society. Whatever he might think of it, the Dixiecrat-Republican coalition had shared interests, and enmities, that were to prove surprisingly durable. As an illustration of this, one might consider the fate of Adam Clayton Powell, the flamboyant congressman from Harlem who had gotten himself in hot water over charges of payroll padding and chronic absenteeism. At the start of the 90th Congress the bipartisan House leadership agreed to censure Powell, whose acidic comments about such civil rights icons as "Martin Loser King" and Roy "Weak-Knees" Wilkins had caused as much offense to the Black political establishment as did his professed kinship to the crucified Jesus to nearly everyone outside New York's Eighteenth Congressional District.

Following a vote by the Democratic caucus on January 8 to strip Powell of his chairmanship of the House Education and Labor Committee, Ford voiced concerns that the veteran lawmaker had not been given a chance to present his case before the full Congress. That few shared his outlook became apparent during a contentious GOP caucus meeting that lasted four and a half hours. The message was not lost on Ford. "You can leave the troops just so often," he remarked, "before they start to leave you." So when Majority Leader Carl Albert moved to seat Powell pending an investigation of the charges against him, every Republican voted no. They were joined in their opposition by all but nine Southern Democrats. A similar alliance of convenience stalled the passage of open housing legislation. As a prophet of Southern behavior, Ford proved less reliable than LBJ, himself a convert to racial justice, who had grimly predicted that in signing the 1964 Civil Rights Act he was relinquishing Democratic claims to the South for a generation. If anything, his forecast proved an understatement.

On August 8, 1967, Ford shattered the uneasy consensus legitimizing Johnson's policies in Southeast Asia. The president had opened the door by renewing his call for an income tax surcharge, coupled with the dispatch of an additional 45,000 troops to Vietnam, raising the US total there to 525,000. "Why are we pulling our best punches in Vietnam?" Ford asked on the floor of the House. "Is there no end, no other answer except more men, more men, more men?" Accusing the president of "handcuffing" US pilots in the bombing campaign dubbed "Operation Rolling Thunder," he wanted the target list expanded to include North Vietnamese oil depots, rail and canal points. In place of Johnson's "gradualism," Ford proposed a Kennedy-style naval quarantine of the North. Unless the administration was willing to bring "meaningful and concerted military pressure" to bear on Ho Chi Minh, the prerequisite to serious peace talks, "I can see no justification for sending one more American over there, let alone 45,000."

By the time he sat down Ford had reduced his party's Vietnam policy to a brutally simple choice: Win or Get Out. The public response to his tough talk was overwhelmingly positive, ten to one as measured by the mailbags cluttering his office. That October one hundred thousand antiwar protesters filled the streets of Washington. Hundreds were arrested attempting to storm the Pentagon. Popular divisions, said Ford, had reached a point greater than at any time in his nineteen years on Capitol Hill. "I don't know whether Lyndon Johnson is the chief architect or the principal victim of the bitterness. Perhaps he is a little of both. But I'm certain that he does not have the ability to heal the wounds that afflict our country."

Looking ahead to 1968, Ford's efforts to position the GOP as the peace party offered a distraction from personal grief. Having withstood two earlier heart attacks, high blood pressure, diabetes and a double mastectomy, Dorothy Ford had come to appear indestructible. In her later years people noticed a striking resemblance between Dorothy and the onetime Leslie King Jr. A shared hyperactivity further bonded mother and son. It surprised no one when Dorothy expressed a hope to die with her boots on. In the event, she did better than that. On a Sunday morning in September 1967, Mrs. Ford's heart stopped beating as she occupied her pew at Grace

Episcopal Church. In cleaning out her apartment, Dorothy's family discovered her appointment book, crowded with commitments for a month to come. Much as they grieved over the loss of their grandmother, Gerald Ford's children would always remember her funeral as the first time they saw their father cry.

12

THE GOOD SOLDIER

*The press wants advocates and combatants in the legislative
branch. They give short shrift to the mediators.*

—GERALD R. FORD

I T WAS IN anticipation of an oratorical mismatch that an all-male, white-tie audience filled the ballroom of a Washington hotel the night of March 9, 1968. They had assembled for the annual Gridiron Dinner at which Cabinet members, Supreme Court justices, and the leaders of Congress subject themselves to song and satire as offered by the Washington press corps. Filling in for a conspicuously absent President Johnson, Vice President Hubert Humphrey represented his fellow Democrats, while the Republican opposition made Gerald Ford its designated speaker. Defying expectations, Ford contested LBJ's putdown of the House minority leader for having played football too long without a helmet.

Not so, said Ford. "On the gridiron I always wear my helmet." He demonstrated this by donning, not without some difficulty, the actual leather headgear he had worn while playing for the University of Michigan. Acknowledging that it had fit better in college, Ford told the well-lubricated crowd that "everything's getting a little tight tonight." He then repaid LBJ for his numerous wisecracks by citing Henry Clay's immortal observation that he would rather be right than be president.

"Now President Johnson has proved once and for all—it really is a choice."

Ford praised the assembled political writers as "the zipper on the Credibility Gap" before paying the obligatory tribute to bipartisan civility and the healing balm of laughter. He concluded by assuring Humphrey that he had no designs on his job. Ford said he much preferred the House of Representatives, notwithstanding the irregular hours imposed by life on Capitol Hill. "Sometimes, though, when it's late and I'm tired and hungry—on that long drive home to Alexandria—as I go past 1600 Pennsylvania Avenue, I do seem to hear a little voice saying:

"'If you lived here, you'd be home now.'"

Ford sat down to hearty applause after remarks interrupted by laughter fifty-one times.* Laughter was a rare commodity in 1968, and the buzz engendered by Ford's Gridiron performance was quickly muffled by events still less predictable. Three days after the dinner, Minnesota senator Eugene McCarthy, running as an antiwar Democrat, stunned Lyndon Johnson with his strong showing in New Hampshire's first-in-the-nation presidential primary. On March 16 Robert Kennedy entered the Democratic race. Five days after that, Nelson Rockefeller defied predictions by pulling out of the Republican contest. The New York governor would change his mind five weeks later, and by an informal alliance with his ideological opposite Ronald Reagan maintain a degree of suspense right up to the GOP convention in August. But from New Hampshire on, the Republican nomination was very much Richard Nixon's to lose.

On March 31, President Johnson coupled the announcement of a bombing halt in Vietnam with his own withdrawal from presidential politics. The immediate shock of LBJ's renunciation had only begun to subside when, on April 4, Martin Luther King Jr. was assassinated in Memphis. America was "shamed" by King's murder, said Ford, who praised the civil rights leader for preaching the truth—"That only by working together and striving together in an atmosphere of goodwill can Negro and white Americans alike move ahead, and only in that way can America move ahead." King's slaying touched off rioting in dozens of cities. Nowhere was the reaction more violent than in Washington, where hundreds of businesses went up in flames that crept within two blocks of the White House. Troops with machine guns occupied the steps leading to the Capitol. Inside the building Ford walked carefully so as not to disturb the sleep of GIs sprawled on the marble floors after standing guard duty all night.

Even before King's death Ford had softened his demand for a conference committee to iron out differences between conflicting House and Senate

---

* Few in the ballroom had any idea of the preparation that went into Ford's speech. Recognizing the opportunities as well as the risks presented by such an event, the Republican leader had approached George Murphy, silver screen hoofer turned senator from California. Murphy referred Ford to his friend Red Skelton, who in turn recommended Bob Orben, the author of Skelton's monologue on his weekly television series. A veteran of the Jack Paar *Tonight Show*, the prolific Orben was also a regular source of comic material for Barry Goldwater *and* comedian and civil rights activist Dick Gregory. Initially unhappy with his latest assignment "because it smelled of no money. As, indeed, it proved to be," Orben became a frequent contributor to Ford speeches and, eventually, director of the White House speechwriting shop. AI, Robert Orben.

versions of open housing legislation. Now, deferring to the more liberal provisions approved by the Senate, he voted for a civil rights package that outlawed discrimination in the sale or rental of most housing units, including individual homes if sold through a real estate agent. Ford took other steps to try to defuse the crisis. He supported the candidacy of Republican James Farmer, a giant of the civil rights movement, for a Brooklyn House seat eventually won by Shirley Chisholm. He endorsed the Poor People's March on Washington—"A proper way to call attention to grievances"— and together with his children he visited Resurrection City, the plywood-and-canvas shantytown on the National Mall that housed some 2,500 demonstrators there to demand an end to poverty in the United States. As incessant rains turned the camp into a sea of mud, a leadership demoralized by King's death was further depressed by the assassination of Robert Kennedy on June 5. Three weeks later a dwindling camp population was forcibly evicted by capital police. A protest caravan made its way to party conventions in Miami and Chicago, but few paid attention.

As permanent chairman of the Republican gathering Ford observed a strict neutrality among the rival candidates for president. His personal preference was for Nixon, a conservative pragmatist acceptable to all but the outer fringes of their party. In a bizarre sideshow, a Republican committee woman from Missouri was approached in the lobby of the Fontainebleau Hotel by a representative of H. L. Hunt, the right-wing Texas oilman who once spent $150,000 in a failed campaign to make General Douglas MacArthur president. Hunt's emissary pleaded with the woman to place the name of Gerald Ford in nomination on Wednesday night, August 7. Nothing came of the overture, and it was just after one o'clock in the morning when Nixon claimed his first-ballot victory with twenty-five votes to spare.

Ford was among those invited to the victor's suite at the Hilton Plaza to review possible running mates. "I know that in the past, Jerry, you have thought about being Vice President," Nixon observed. "Would you take it this year?" Ford declined with thanks, as Nixon fully intended him to—a gain of thirty-one seats and Ford would be Speaker of the House. (In fact he was closer to his goal than it appeared. Hidden from the public was a list of seventeen Democratic incumbents, primarily Southerners, who had signaled a willingness to switch parties should the GOP get within spitting distance of a majority.) If Nixon was looking to balance the ticket, said Ford, he might consider New York City mayor John Lindsay, a charismatic former congressman and ardent champion of urban America. It would be hard to imagine a more unlikely pairing, given Nixon's personal insecurities and his disdain for the Eastern Establishment. The suggestion is important chiefly for what it says about Ford's continuing belief in big-tent Republicanism, and the need to appeal to constituencies considered off-limits to the GOP.

A less antiseptic account of what transpired at the Hilton Plaza is provided by Bob Hartmann in an early draft of his White House memoir *Palace Politics*. Nixon said he had been weighing a long list of qualified prospects for the vice presidency. As the roundtable discussion continued, with no one mentioning his favorite, Nixon was finally driven to volunteer the name of Maryland governor Spiro Agnew. "Ford's raucous whinny shattered the silence," Hartmann recorded. "His nervous, high pitched laugh had been known to penetrate the sandstone walls of the Capitol . . . a sure sign of nervous tension." H. R. Haldeman and the rest of Nixon's "Praetorians" stared at Ford, according to Hartmann, "as if he had farted at the communion rail."

Nixon had his own ideas on expanding the party, and they didn't include John Lindsay. Tapping disgruntled blue-collar workers and Southern Democrats alienated from their national party, his strategy for victory presumed that the race would be won or lost south of the Mason-Dixon Line. Thus 1968 became the first of several presidential contests in which cultural issues supplanted the economy as a determining factor. This helps to explain Nixon's choice of Agnew for a running mate. Agnew fitted neatly into the law-and-order motif of the Nixon campaign. To the extent that he was known for anything outside his state, it was the tongue-lashing he had administered to a group of Black leaders from Baltimore accused of failing to prevent two nights of rioting in that city following the assassination of Martin Luther King Jr.

Nothing Agnew would say on the campaign trail suggested any second thoughts on his part. His dismissive comment, "If you've seen one city slum, you've seen them all," summarized his grasp of the urban crisis. Early in September, Agnew earned a rebuke from both Ford and Everett Dirksen after he labeled Democratic nominee Hubert Humphrey "soft on Communism." Later that month it was Ford who was in hot water with the Nixon high command. Following a meeting with several Republican candidates in Winston-Salem, North Carolina, he artlessly told reporters that "we want all of them to support the national ticket if they can, but if they can't because of local sentiment, we will understand." This was code talk in a region where some GOP candidates were reluctant to offend supporters of third-party candidate George Wallace. "That's news to our campaign," retorted Nixon press secretary Herb Klein. His boss was lending vigorous support to *every* contender running on the Republican label, said Klein. He did not have to say that he expected such loyalty to be reciprocated.

On November 5, 1968, Richard Nixon squeaked by Hubert Humphrey, his popular vote margin the second closest in the twentieth century. Down ballot the winner's coattails were practically invisible. Instead of the forty to fifty seats he had once hoped to pick up, Ford had to settle for a gain of just five. Richard Nixon would be the first victorious candidate since 1849 to enter the presidency with both houses of Congress arrayed against him. Asked what it would take to promote successful relations between the incoming Nixon Administration and Capitol Hill, Ford's reply was unequivocal: "An absolute trust that each is telling the other the total truth."

The president-elect and his representatives took a rather different view of things. "When we get to Washington, there's a lot of people who are going to have to learn what it means to have a Republican president in the White House." So observed Nixon's putative attorney general, John Mitchell, to a friendly journalist.

"Who do you mean?"

"Jerry Ford. Everett Dirksen."

On Sunday morning, January 12, Ford was unexpectedly summoned to the White House. There, waiting for him in the Lincoln Sitting Room, was the outgoing president. Nearing the end of his stormy tenure, LBJ was by turns ruminative, resentful and forgiving. He showed Ford a draft of his farewell address, with its tribute to the Republican minority in Congress and its leaders for having shown him "the most generous cooperation." He reviewed the recent campaign, complimenting Nixon on the race he had run and praising his choice of Agnew as a shrewd nod to Southern voters. Johnson expressed puzzlement over the continuing refusal of liberals to welcome him into their ranks, despite his groundbreaking record on civil rights and anti-poverty legislation.

"Jerry, you and I have had a lot of head to head confrontations, but I never doubted your integrity."

"I never doubted yours, either," said Ford. "I didn't like some of the things you said about me, but I never questioned your patriotism."

Neither man was being entirely truthful, but the occasion called for strategic amnesia. Forty minutes after it began, the former antagonists ended their conversation, with Johnson draping his long arm around Ford's shoulders. "When I leave here, I want you to know that we are friends and we always will be, and if I can ever help you I want you to let me know."

JANUARY 20, 1969. Temperatures hovered in the midthirties, and Washington shivered under a pewter gray sky as Richard Nixon prepared to assume the office whose pursuit had defined his adult life. Ford had a secondary role in the day's pageantry. Together with an aging Everett Dirksen he escorted the new vice president from his hotel in Northwest Washington to the inaugural platform astride the west front of the Capitol. There, twenty years to the day after he and Betty watched Harry Truman sworn into office, they heard Nixon appeal for Americans to lower their voices as the prerequisite to addressing "a crisis of the spirit" that plagued the nation. While Ford thought the speech struck "just the right tone for this moment in history," a vigorous dissent was registered by antiwar demonstrators, some of whom hurled rocks and beer cans at the presidential limousine as it traveled Pennsylvania Avenue.

Ford did not need John Mitchell to spell out the job changes that would come with a Republican White House. Henceforth the needs of colleagues and constituents must defer to the man in the Oval Office and his collaborators. Habitual criticism of presidential initiatives gave way to (mostly) unquestioning loyalty, just as the aggressive development and promotion of alternative programs yielded to the agenda put forward by the new administration. No longer de facto spokesmen for their party, Dirksen and Ford quietly pulled the plug on the *Ev and Jerry Show*. Reviving a custom established in the Eisenhower years, the president-elect set aside Tuesday mornings for weekly meetings with the GOP congressional leadership. These were spartan affairs, no breakfast being served except once a year, when participants' wives were ushered into the Cabinet Room, as in December 1969, to hear Ford offer "the official alibi as to why we have been so busy all through 1969 and have accomplished so little."

According to White House Counselor John Ehrlichman these gatherings were in fact little more than dog-and-pony shows at which "our dynamic legislative leaders drank coffee and occasionally took notes, but they rarely were given a chance to say anything." Ehrlichman's caustic appraisal was typical of the condescension shown legislators of both parties by members of Nixon's inner circle. On the rare occasion when administration spokesmen did venture as far afield as Capitol Hill, the results were often counterproductive, as when Ehrlichman reportedly fell asleep in the middle of a strategic review held in the minority leader's office.

In September 1969 an unexpected loss deprived Ford of a powerful ally in his dealings with the White House. He and Everett Dirksen had long since patched up their differences over Vietnam and Dirksen's special relationship with LBJ. Away from the cameras Ford affectionately addressed the senior lawmaker as "Coach." So when Dirksen died on September 7 following surgery for lung cancer, Ford's sense of loss was genuine. "He was the kind of man who not only filled the canvas but spilled over from it," he told reporters. "There was nobody like him before. There will be nobody like him after." On September 11 Ford joined Vice President Agnew and seventy members of Congress who journeyed to Dirksen's hometown of Pekin, to bury him in the fertile black soil of the Illinois prairie.

Ford's relationship with Dirksen's successor, Hugh Scott of Pennsylvania, was friendly but professional. Scott, a portly, pipe-smoking collector of Chinese art objects from the Tang dynasty, was an outspoken supporter of civil rights and an unabashed Rockefeller Republican who never won the trust of the Nixon White House. Indeed, at times it seemed as if Scott's urbane presence was barely tolerated at the weekly strategy sessions attended by the president and his party's Capitol Hill leadership. At one meeting, dominated by the latest congressional attempt to cut off Vietnam War funding, Scott engaged in some verbal handwringing about all the difficulties presented by such a resolution. "Hugh, I understand the problem," said Nixon in a dismissive tone. Swiveling his chair in the opposite direction from the Senate Republican leader, "he started talking to Ford," recalls one who was present. "Here's Hugh Scott sitting there and Nixon is turning around just sticking it up Hugh Scott's rear every time he had a chance of getting Leader Ford to help him with one of the 'end the war' issues."

By contrast with his Senate counterpart, Ford's loyalty to the new administration was never in doubt. It was first demonstrated through that perennial hot potato, raising the national debt ceiling. In 1967 House Republicans had voted unanimously to deny Lyndon Johnson's request to accommodate a rising tide of red ink. Now, two years later, the shoe was on the other foot, and much as it offended their conservative ideology, Ford convinced three-fourths of his caucus to accede to a similar appeal from one of their own. A still more odious test of responsibility confronted lawmakers when Nixon decided that his first budget could not be balanced unless Congress extended the 10 percent income surtax LBJ had sold its membership the previous year as a wartime imperative. Working in league with RNC chairman Rogers Morton, Ford provided arguments that skittish colleagues might cite to justify a vote for fiscal responsibility. Even so, the outcome was by no means certain when, on the last day of the fiscal year, the full House voted 210–205 to retain the surtax for another year. This, in turn, set the stage for a much more popular tax reform package that lowered rates for single persons while establishing the first alternative minimum tax to prevent high-income earners from avoiding taxes altogether.

In Nixon's chosen field of foreign affairs, Ford worked closely with the White House to rally popular support for the president's Vietnamization policy, under which US military responsibilities would be turned over to the Saigon government as it demonstrated an increased capacity for successful combat. Ford welcomed the initial reduction of 60,000 in US ground forces before the end of 1969, with an additional 140,000 to be withdrawn in 1970. At the same time, he encouraged his office interns to follow their conscience, even if it led them to join thousands of mostly youthful protesters in the streets of Washington demanding an immediate end to American military involvement in Southeast Asia.*

As Nixon settled into the job, his global perspective was matched by an increasing desire for solitude. "I must build a wall around me," he told chief of staff H. R. Haldeman. On their first full day in office the two men located an alternative work space in the neighboring Executive Office Building. There Nixon could escape with his yellow legal pads and wallow in what he called "brainwork." A self-described introvert in an extrovert's profession, Nixon struck speechwriter Aram Bakshian as a latter-day Willy Loman—"the door to door salesman with bad feet." Scheduling White House Christmas parties for when he would be out of town, Nixon reportedly watched a taped replay of a Washington Redskins game by himself in the third-floor solarium while a holiday reception for lawmakers unfolded on the ground floor. The president did not lack for self-awareness, lamenting to Haldeman that "it would be goddamned easy to run this office if you didn't have to deal with people."†

Ford complained that he had enjoyed greater access to LBJ than to Nixon. "What Nixon wanted from the House minority leader was to do little errands for him, most of them slightly dirty," recalled Bob Hartmann.

---

* Ford's sons were either too young to be drafted for military service in Vietnam or drew a lucky number in the draft lottery instituted in December 1969. That did not prevent them from questioning their father's support of the war, as the Ford family experienced the same dinner table arguments over Vietnam that riled millions of other American homes.

† "Tell me the purpose of that meeting. What did that accomplish?" a resentful Nixon demanded of Bill Timmons after an Oval Office photo op with a congressional petitioner, all the while jabbing a finger into his aide's chest. "Mr. President, that's money in the bank," Timmons insisted. "That's an investment that will come back and reward you." ("Which was partly true," he observed long afterward with a rueful smile, "but it also got me off the hook and saved my chest from more bruises.") AI, William Timmons.

"If he was attacked by Teddy Kennedy, we were supposed to get up and denounce Teddy on the House floor." White House hatchet man Chuck Colson directed a foul stream of defamatory speech texts to Ford's office, where most were tossed into the wastebasket. Equally heavy-handed was the treatment accorded Vice President Agnew. At an early session with GOP leaders in the Cabinet Room, the agenda included discussion of the president's desire to consolidate hundreds of federal grant programs, most encumbered by extensive red tape, into a no-strings package of revenue sharing for state and local governments. The vice president presumed on his status as a former governor to criticize the scheme. "That was the end," says Tom Korologos, who worked on the Hill as part of Bryce Harlow's legislative affairs shop. "They went to him and told him, 'Never say another word in a leadership meeting again.'"

Agnew enjoyed greater success on the hustings, his talent for polemical speechmaking winning him a legion of admirers among Republican partisans. Thousands of the faithful were only too willing to fork over $100 to dine on chicken cutlets and feast on the vice president's denunciation of Vietnam War protesters as "an effete corps of impudent snobs who characterize themselves as intellectuals." Warming to his task, in Des Moines on November 13, 1969, Agnew turned his rhetorical firehose on the three television networks and "this small and unelected elite" of print and electronic journalists reflecting the biases of New York and Washington. Overnight Agnew had a national constituency.

Gerald Ford was not part of it. The vice president's language was "not my style," he told reporters. Personally he saw no threat to the republic in the so-called instant analysis offered television viewers following a presidential speech from the Oval Office. Still, he was careful not to put too much distance between himself and the suddenly popular Agnew: "I don't believe in muzzling demonstrators and I don't believe in muzzling vice presidents." When Dwight Eisenhower died just ten weeks into the Nixon presidency, Ford emphasized the old soldier's interest in young Americans, many of them alienated from their government over the Vietnam War. He recalled Ike's dissent when his Republican brethren, "tempted by an excess of partisan zeal," leveled a particularly harsh criticism at a Democrat in the Oval Office. "Well, gentlemen," Eisenhower had reproached them, "if I were sitting in that chair I wouldn't like that one bit. Remember, he is the President of the United States."

THAT RICHARD NIXON intended to be a foreign policy president surprised no one (he once likened domestic social programs to "building outhouses in Peoria"). A notable exception to this rule was the federal judiciary. Nixon was the first in a string of GOP presidents to run against an activist Supreme Court for its alleged liberal bias. Almost half a century before Senate Republicans refused to hold hearings on Barack Obama's nomination of Judge Merrick Garland to fill the seat vacated by the late Antonin Scalia on the nation's highest tribunal, a coalition of conservative Republicans and Southern Democrats blocked Lyndon Johnson in the twilight of his presidency from elevating Supreme Court Justice Abe Fortas to succeed the retiring Earl Warren as chief justice.

Fortas's confirmation hearing in July 1968 featured tough questioning about the justice's continuing relationship with the man who appointed him. The most damaging opposition came from South Carolina Dixiecrat-turned-Republican Strom Thurmond. In league with a Catholic Church lay group styling itself Citizens for Decent Literature, Thurmond seized on First Amendment rulings by the Court, some predating Fortas's original appointment, which appeared to sanction the production and distribution of materials widely regarded as obscene. Fortas withdrew his name from consideration a month before election day. His ordeal was just beginning, however.

Three months after Nixon took office, *Life* magazine exposed the justice's acceptance of a $20,000-a-year retainer from a foundation bearing the name of financier Louis Wolfson, a convicted stock swindler who served nine months behind bars in 1969. Fortas also came under fire for accepting speaking fees from groups with interests before the Supreme Court.* On May 14 he resigned his seat, reportedly "to save Bill Douglas," his closest friend on the high court and an ideological ally since their first meeting as young New Dealers. Talk of impeaching Justice Douglas had first surfaced in 1953, when he briefly stayed the execution of convicted atomic spies Julius and Ethel Rosenberg. It had again bubbled to the surface as recently as 1966, when five members of the House asked the Judiciary Committee to investigate Douglas's "moral character" following his divorce from his twenty-six-year-old third wife and subsequent remarriage to an even younger woman.

Hard on Fortas's departure, Nixon moved to replace Earl Warren with Warren Burger, a Minnesotan appointed by Eisenhower to the US Court

---

* Generously reimbursed by American University for occasional classroom appearances, Fortas offered nothing to the judges who substituted for him when he missed sessions. Such behavior caused the mild-mannered majority leader Mike Mansfield to declare "that unprincipled SOB is finished." Lawrence G. Meyer to author, February 17, 2016.

of Appeals for the District of Columbia. With his snowy white mane and toplofty bearing, Burger looked every inch a chief justice. His appointment foreshadowed a dramatic recasting of the Court, as aging holdovers from Franklin Roosevelt's Washington made way for more conservative successors. The process was far from smooth. Twice Nixon struck out by nominating Southern jurists whose views on race and organized labor made them vulnerable in a Senate where liberalism remained dominant. In November 1969 lawmakers rejected Judge Clement Haynsworth of South Carolina, nominated by Nixon to replace the disgraced Fortas, on a 55 to 45 vote. Five months later a second Nixon nominee, Floridian G. Harrold Carswell, suffered an even more humiliating defeat.

Among those pressed into service by the White House to defend Haynsworth and Carswell was Clark Mollenhoff, a Pulitzer Prize–winning journalist for the *Des Moines Register and Tribune* before he accepted a position as presidential "ombudsman"—putting his investigative skills to work identifying wrongdoing or ethical lapses within the Nixon Administration before they became public embarrassments. Prior to joining the administration in August 1969, Mollenhoff had been toiling on a major exposé of Abe Fortas's judicial colleague William O. Douglas.

The exact sequence of events matters, since nothing in Ford's congressional career did more to tarnish his reputation than his apparent complicity in efforts by the Nixon White House to avenge Haynsworth and Carswell by driving Douglas from the bench. In truth Ford's interest in Douglas predated Nixon's election. During the fall campaign Ford had approached his friend Bob Griffin, then spearheading Republican efforts to deny Abe Fortas Senate confirmation as chief justice.* Had Griffin or his investigators turned up anything in their probe of Fortas that might reflect harshly on Bill Douglas? Griffin assured him they had not. A full year passed; senators were still debating the Haynsworth nomination when the Associated Press in October 1969 linked Albert Parvin, whose eponymous foundation was funded with proceeds from the sale of Las Vegas's Flamingo Hotel, to gangster Meyer Lansky. At the time Justice Douglas, financially drained by multiple alimony settlements, was being paid $12,000 annually to serve as president of the Parvin Foundation, whose mob connections invited comparisons with Abe Fortas's work for the Wolfson Foundation.

Additional stories from the wire service fueled random attacks on

---

* Griffin owed Ford his Senate seat, vacated by the death of Democrat Pat McNamara in April 1966. When Governor George Romney approached Ford about the appointment, he disclaimed any personal interest, while strongly advising the governor to name Griffin.

Douglas by House Republican gadfly H. R. Gross, who got his information from Clark Mollenhoff, the same White House ombudsman who had been investigating Douglas at the time he joined the Nixon staff. "He didn't have any place to peddle this story," Bob Hartmann said of Mollenhoff, "so he peddled it to H.R. Gross." Mollenhoff also spoke with Ford, already the recipient of numerous tips concerning Douglas in the wake of Abe Fortas's resignation. In a late-night phone call, a former Parvin employee passed along names of potential witnesses who could purportedly attest to "what a son of a bitch this Parvin really was, how bad the Douglas connection really was." A no-doubt-groggy Ford jotted down the information before handing it the next day to Hartmann with instructions to follow up "without any stir" on the leads thus provided.

Ironically, it was Ford himself who sparked a controversy when asked by reporter Haynes Johnson about rumors that he considered Justice Douglas a possible impeachment target. Candid to a fault, Ford confirmed that an investigation was underway. Never mind that his investigation, so-called, was until then limited to whatever time and effort Bob Hartmann could spare from his regular duties; coming on the eve of Haynsworth's confirmation vote in November 1969 the minority leader's remarks were inevitably construed as part of a larger attempt by the White House to pressure wavering moderate Republicans and some Democrats into casting a reluctant vote for the president's embattled nominee, if only to avoid a tit-for-tat campaign of revenge directed at Douglas.

As events would prove, this was no bluff. In the event of Haynsworth's rejection by the Senate, Nixon told aide John Ehrlichman, the president wanted Ford to remind everyone of more serious conflicts of interest involving the liberal lion of American jurisprudence, William O. Douglas. To make sure Ford got the message, Ehrlichman raised the subject of Douglas's impeachment at a supposedly off-the-record breakfast with reporters. At the urging of Representative Joe Waggoner, a Louisiana Democrat sympathetic to Nixon, Ford hired a young investigator named Benton Becker. Formerly a criminal attorney in the fraud section of the Justice Department, Becker had impressed the minority leader by his evenhanded approach to earlier charges leveled at Representative Adam Clayton Powell.

With his arrival the pace of events accelerated. The same Associated Press scribes who had revealed Parvin's underworld connections now approached Ford's office with a proposed information swap. A request from the minority leader for additional assistance went out to John Mitchell's Justice Department. In response Assistant Attorney General Will Wilson offered to share material in department files, some of it compiled for the purpose by J. Edgar Hoover's FBI. A meeting was arranged in Ford's Capitol Hill office for December 12, 1969. Afterward Ford, Hartmann and Becker agreed that the evidence provided by Wilson fell far short of what they had been led to expect. "Five or six sheets of no letterhead, just plain typewriting," said Hartmann. "Sort of a brief of the case against Douglas." Becker agreed: "It would be generous to call it gossip," he complained.

Wilson remembered things differently. "I gave him the files as I had been instructed to do," he claimed, "but I cautioned Congressman Ford that the data were raw and that the entire matter needed much more time and a more thorough investigation."

Meanwhile, Douglas's resignation from the Parvin Foundation had done little to mollify his critics. Anything but contrite, in February 1970 the justice published *Points of Rebellion*, a ninety-seven-page ode to what Ford would label "the militant hippie-yippie movement" and the need for modern revolutionaries to rise up against an oppressive establishment. The same text was excerpted in the April issue of *Evergreen Review*, juxtaposed between a full-page caricature of Richard Nixon as George III and a seven-page photo section of soft-core porn featuring nude models embracing.

On April 8, 1970, the Senate rejected the Carswell nomination 51–45. Immediately Ford came under renewed pressure from conservative firebrands in his caucus. Louis Wyman of New Hampshire publicly threatened to introduce a motion of impeachment, an action that would take precedence over all other House business. This was Ford's greatest fear—that a premature move on Douglas would divide the fractious Republican family while enabling the Democratic speaker to assign any such motion to the Judiciary Committee chaired by Emanuel Celler of Brooklyn, a Douglas partisan. Backed into a corner, Ford asked Hartmann what his investigative efforts had turned up. "I've got a little more than is on the public record, but nothing conclusive," Hartmann told him. "We're not ready to go."

On Monday, April 13, Ford met with several of the red hots in his ranks who were braying for Douglas's scalp. He outlined his preferred strategy: to empanel a bipartisan committee of six, repeating the precedent established by the full House in dealing with payroll padding allegations against Adam Clayton Powell. One hundred five members, equally apportioned between the parties, signed on to Ford's resolution. On April 15 he took to the floor for ninety minutes. Making the best of the hand he had been dealt, Ford reminded lawmakers that, contrary to popular opinion, judges were not appointed to life terms, but for a period of service "during good behavior." Over the years that Delphic phrase had been subject to differing interpretation. Nine times in the history of the republic the Senate had sat as a court of judicial impeachment. Four of the jurists so tried had been removed from office; one resigned before his trial began.

Careful review of the record, Ford claimed, showed that "an offense need not be indictable to be impeachable." While the Constitution stipulated "treason, bribery or other high crimes and misdemeanors" as grounds for removal from office, the standard applied to unelected judges was as broad as the term "good behavior" itself. "An impeachable [offense]," said Ford, in words that would come back to haunt him and successive presidents, "is whatever a majority of the House of Representatives considers [it] to be at a given moment in history; conviction results from whatever offense or offenses two-thirds of the other body considers to be sufficiently serious to require removal of the accused from office."

Ford's bill of indictment was a scattershot affair. The justice had failed to recuse himself from an obscenity case involving publisher Ralph Ginzburg, who had paid Douglas $350 for an article on folk singing. Ford portrayed *Points of Rebellion* as "an inflammatory volume," and *Evergreen Review* as "hardcore pornography." He reminded members of Douglas's earlier moonlighting on behalf of the Parvin Foundation and linked him to Bobby Baker, the convicted tax evader suspected of worse. Finally he assailed as a "leftish" organization and incubator for student militancy the Santa Barbara–based Center for the Study of Democratic Institutions, from which Douglas continued to receive consultant's fees as chairman of the board of directors.

Ford was still speaking when Democrat Andrew Jacobs of Indiana hastened to the well of the House. Seizing the initiative, he dropped in the hopper—a wooden storage bin attached to the clerk's desk—a resolution calling for the outright impeachment of Justice Douglas. His tactic ensured that Ford's proposed independent panel was strangled in the cradle. Jurisdiction in the matter would instead rest with the Judiciary Committee under Manny Celler. Badly outmaneuvered, the minority leader was still fuming two days later, when an uncharacteristically cheerful Nixon, buoyed by the successful return to Earth of the crippled *Apollo 13* lunar mission, called him to share the good news. Before hanging up the president told Ford to give his best to Justice Douglas.

The presidential witticism concealed a change of plan, as Nixon had concluded that it was better to outlast the septuagenarian Douglas than to attempt his forced removal from the bench. Rather than convey this message to Ford directly, however, Nixon outsourced the job to John Mitchell, who was no more inclined to speak frankly to the minority leader. Instead, he took the occasion of a Law Day address to call for an end to "irresponsible and malicious criticism" that could damage the Court and with it public respect for the legal system.

What followed was predictable. With Celler calling the shots, no public hearings were held, and no testimony taken under oath. The committee's 924-page final report was not even circulated to the full membership because, it was explained, the December 16 due date was too close to the end of the session to allow a thorough review.*

By then Celler's handpicked subcommittee had voted 3 to 1 that no grounds existed for impeachment. Ford was not alone in crying "whitewash," but when a handful of anti-Douglas diehards attempted early in 1971 to revive the impeachment drive, the minority leader's name was conspicuously missing from their ranks. His campaign against Douglas had not been a total failure, however, as the justice recused himself from participating in an obscenity case involving the film *I Am Curious (Yellow)* and a libel action against *Look* magazine. Moreover, posterity would be less generous than his contemporaries in assessing the legacy of the longest-serving justice in Supreme Court history. Douglas biographers duly noted his haphazard legal craftsmanship, compulsive womanizing and inventive approach to autobiography. With rare exceptions, it was agreed, Douglas the outdoorsman had displayed more concern for his beloved trees than for the human beings who sheltered under their branches.

As president, Ford went out of his way to demonstrate that there were no hard feelings, inviting the justice and his fourth wife, Cathy, to a White House state dinner, and dispatching a military jet to transport Douglas's personal physician to the Bahamas after the justice suffered a crippling stroke on the last night of 1974 (Ford telephoned Cathy Douglas from his holiday retreat in Vail, Colorado, to offer assistance). The ravaged warrior, pain-racked and increasingly paranoid, refused to consider retirement. "Ford will appoint some bastard," he muttered as justification for grimly hanging on. Not until November 12, 1975, did Douglas finally bow to the inevitable and submit his resignation to the man who had once tried to compel his departure under a cloud of scandal. In his response, Ford combined an expression of "profound personal sympathy" with the nation's gratitude for a lifetime of public service matched by few Americans. "It is my sincere hope that your health will soon be restored so that you can enjoy your well-deserved retirement among the natural beauties you love

---

* Asked by a former clerk how he'd survived the impeachment effort, Douglas replied, "Manny Celler is a very good friend of mine." Bruce Allen Murphy, *Wild Bill: The Legend and Life of William O. Douglas* (New York: Random House, 2003), 442.

and have helped to preserve. Future generations of citizens will continue to benefit from your firm devotion to the fundamental rights of individual freedom and privacy under the Constitution."

The week before Christmas, Douglas in his wheelchair looked on as Chicago lawyer and Court of Appeals judge John Paul Stevens was sworn in as his replacement. At the conclusion of the brief ceremony Ford approached his old antagonist.

"Good to see you, Mr. Justice."

"Yeah," said Douglas. "It's really nice seeing you. We've got to get together more often."

G RAND RAPIDS WAS changing. By 1970 the Furniture City boasted four college campuses with an estimated ten thousand students, many of them too young to recall Ford except as an establishment figure whose primary allegiance was to his party and the man in the White House. That summer brought Ford the stiffest electoral challenge of his career. His opponent was Jean McKee, a forty-seven-year-old attorney and Democratic activist running on a platform ("Ford Speaks for Nixon. Who Speaks for You?") extolling open housing, federal aid to education and cutbacks in military spending. Vietnam was the biggest issue separating the candidates, with Ford defending Nixon's policies, including the April 30 "incursion" by US and South Vietnamese forces into neighboring Cambodia. Intended to disrupt enemy forces using border sanctuaries as a vast staging ground from which to subvert their neighbor, the operation was seen by critics as an expansion of the war and a reversal of Nixon's exit strategy from the region.

Nixon insisted that the Cambodian incursion bought time for Vietnamization to work. Ford agreed, calling the operation a "tremendous success." This was not a viewpoint widely held on Capitol Hill. In Washington the Senate voted to cut off funding for the Cambodian operation after sixty days. The House might have gone along, had Ford not demanded at a critical juncture in the debate to be put through to the president at his Western White House in San Clemente, California. Over the phone Nixon offered assurances that all US forces would be out of Cambodia by June 30. With that, the minority leader was able to return to the House floor, where his colleagues voted 237–153 to take Nixon at his word.

The Cambodian incursion was the tipping point for Jean McKee. Harboring no illusions about her prospects for victory, she nevertheless managed to tap into changing demographics and a war weariness that was increasingly bipartisan. McKee's appeal to younger voters drove Ford to recruit his own cadre of collegiate supporters. For the first time in his career he hired a professional pollster. In sharp contrast with the law-and-order campaign waged by the national GOP, slick television commercials depicted Ford as a man of peace dedicated to meeting local needs. At the same time, much was made of his vote to override Nixon's veto of legislation that would limit campaign expenditures on TV advertising. In a climactic debate with McKee on October 14, Ford contrasted his opponent's demand for a reordering of priorities with a series of changes already being implemented, beginning with a nearly 50 percent reduction in US armed forces in Southeast Asia by May 1, 1971.

On election day, fears that the incumbent was losing touch with his home base proved groundless, as Ford amassed 61 percent of the vote. If some of his neighbors resented the demands made by his party leadership position, they appeared to be greatly outnumbered by those for whom his success was a source of local pride.

THE SAME CAMPAIGN bared Ford's unhappiness with White House hard-liners who equated dissent with disloyalty. In New York the administration went out of its way to promote the Conservative Party candidacy of James Buckley over GOP incumbent senator Charles Goodell, a staunch critic of the Vietnam War and no friend to the president's rightward-leaning judicial nominees. Deserted by the national party, Goodell turned to Ford, his longtime House ally, for political validation. The minority leader didn't hesitate to campaign for Goodell, whose subsequent loss to Buckley gave the Nixon White House one of its few reasons for election night gloating. Actually, the outcome of the midterms was perfectly creditable for a governing party, as the GOP gained two senators while their minority ranks in the House shrank by twelve seats.

Still, with unemployment creeping above 5 percent and polls showing him vulnerable in 1972, Nixon had powerful incentive to make changes in policy and personnel. When Ford made a vote on funding an American SST, or supersonic transport plane, a test of loyalty to the GOP leadership in the House the gesture backfired, as five members of his own team voted to kill the plane. "They weren't voting against the SST," concluded one Nixon adviser. "They were voting against John Ehrlichman and the German combine in the White House." They weren't alone. "I wonder if the President knows what some of his people are doing in his name?" Ford asked a reporter he trusted not to repeat their exchange. In public at least, his displays of independence were muted—meeting with the newly formed Congressional Black Caucus, for example, or opposing White House efforts to fund mass transit by raiding the highway trust fund.

Not political outlook but values and temperament separated Richard
Nixon from Gerald Ford. A Secret Service man assigned to then vice president Nixon recalled his Black Irish volatility. Nixon was the sort of traveling
companion who might, without warning, begin pounding his fist into the
armrest of an airplane seat while muttering to himself, "You know, I'm not
tough enough. I've just got to make myself tougher." This self-assessment
may help to account for the strange dichotomy of Nixon's personality, the
coarse language employed on his White House tapes, the Prussian guard
with which he isolated himself as president—even his addiction to the film
Patton, in which tough guy George C. Scott embodied the great warrior,
whose bloodlust was matched by his erudition.

On the evening of June 17, 1971, one year to the day before the Watergate break-in lit the fuse on the century's most explosive political scandal,
Nixon telephoned Ford to thank him for his part in defeating the latest resolution setting a cutoff date for US military aid to Vietnam. Ford returned
the compliment, praising Nixon for his leadership in the war on drugs.
Awkward pauses punctuated their recorded conversation, until Nixon hit
on a subject he assumed would spark consensus.

"What do you think of all this hullabaloo about the *New York Times*?" he
inquired of Ford. "Isn't that the god damnedest thing?"

The president was referring to the so-called Pentagon Papers, an officially commissioned history of American involvement in Vietnam from
1945 to 1967, replete with the deceptions and misjudgments that governments prefer to bury. The searing study, originally requested by LBJ's
defense secretary, Robert McNamara, was spared that fate by defense
analyst–turned–war critic Daniel Ellsberg. Having copied much of the report in secret, Ellsberg passed it on to *New York Times* reporter Neil Shee-
han, among others. The *Times* ran the first of nine installments on June
13. In the days since, Nixon's anger at the paper of record had been stoked
by John Mitchell and Henry Kissinger. Both men had reason to fear leaks
exposing their part in wiretapping journalists and members of Kissinger's
National Security Council staff.

Ford, his judgment unclouded by such illegalities, admitted to having
"mixed emotions. In some respects [I'd] just as soon they publish," he told
the president. After all, the events described in the *Times* occurred well before the Nixon presidency. While acknowledging that "it really exposes the
other side," Nixon pushed back against Ford's argument. A fundamental
principle was at issue. The *Times* was "trafficking in stolen goods."

Ford conceded his point. "Still, other than the legal issue," he mused,
"we should sit back and . . ."

"Let them fight," Nixon completed Ford's sentence for him.

"Let them explain," Ford corrected him.

Lacking the killer instinct so prized in the Nixon White House, Ford was nevertheless lauded by the president as "the only leader we've got on either side in either house." Nixon sent him a recent life of Thomas Jefferson, and a quotation from former Indiana senator Albert Beveridge ("Partisanship should only be a method of patriotism . . . he who is the partisan of principle is a prince of citizenship"). July 14, 1971, was Ford's fifty-eighth birthday; Nixon called from San Clemente to wish him many happy returns of the day. The president said nothing of his audacious plans, to be revealed in a national television broadcast the next day, to visit Communist China early in 1972 at the invitation of the Chinese government. By the time he finished his televised speech, Nixon had succeeded in overturning a quarter century of American policy in Asia.

Equally radical was the economic U-turn executed a month later by Nixon and his recently installed Treasury Secretary John Connally. "Nixon continually looked for people who were bigger than him in the areas they knew," notes White House speechwriter Pat Buchanan. "He once told me, 'Pat, I want a national security adviser who can teach me something.'" Connally, a three-term Democratic governor of Texas with a swaggering charisma and a penchant for bold thinking, was less teacher than role model. As charming as he was ruthless, Connally brought to his new responsibilities a brass-plated assurance that Nixon could only envy. Bill Timmons perfectly captured their relationship in a single encounter growing out of the president's wish to have his new secretary (and would-be successor) participate in a Republican leadership meeting. Twenty-four hours in advance Nixon requested that Timmons go to Treasury and convey his personal invitation for Connally to attend and speak to the assembled legislators.

Big John was having none of it.

"Bill, you know I like you, but tell the President that if he wants me to come over there, he should call me."

A chastised Timmons, fearing the verbal backlash awaiting him in the Oval Office, repeated Connally's message as delivered. "Don't worry about it, Bill." Nixon chuckled. "I understand. I'll call John."

Abandoning Republican austerity as political poison, on August 15, 1971, Nixon unveiled his New Economic Policy incorporating a ninety-day mandatory freeze on wages and prices. The same package included a 10 percent surcharge on imported goods, and the closing of the gold window, ending the convertibility of US dollars to gold and effectively scuttling the postwar system of international finance in place since the 1944

Bretton Woods Conference. Had it been the product of a Democratic White House, the NEP, with its full employment budget and unequivocal embrace of Keynesian economics, would have drawn withering fire from Ford and other traditionalists. As it was, the Republican leader permitted himself an initial gasp of surprise, quickly amended to acknowledge "the wisdom of these decisions."*

Presidential politics saturated the Washington atmosphere long before Ford joined a caravan of Nixon surrogates fanning out across New Hampshire in advance of that state's March 7, 1972, primary. The president faced token opposition from Congressman Pete McCloskey, a maverick Californian running on an antiwar platform, and Ohio representative John Ashbrook, a conservative critic unhappy with the opening to Communist China and Nixon's abandonment of balanced budget orthodoxy. Both men struggled for media attention as the president's groundbreaking China visit that February dominated news coverage in the run-up to New Hampshire. Even with the support of the influential *Manchester Union Leader*, Ashbrook polled just 10 percent of Granite State Republican voters. McCloskey dropped out of the race after falling short of the 20 percent threshold he had set for himself.

Their open challenge to the president didn't prevent Ford from stumping for both men as they sought reelection to the House. Nixon, meanwhile, turned his attention to other matters. On the cusp of his final campaign, he had solid grounds for optimism. Together with a headline-making summit in Moscow later in May, to formalize the SALT I arms control treaty, his dramatic overture to the People's Republic dramatically shortened the odds on a Republican landslide in November. Having opened their door to the American president, Chinese leaders were eager to broaden their acquaintance with their Washington counterparts. In May, Senate leaders Mike Mansfield and Hugh Scott made the trip. A month later it was the turn of Hale Boggs and Gerald Ford, accompanied by their wives, representing the House. Their ten-day journey was an exercise in time travel as much as public diplomacy. Landing in Shanghai on June 26, the visitors encountered a society whose long isolation from the West was reflected in the two-wheel carts and bicycles navigating urban streets without sidewalks.

---

* Nixon's flexibility in public policymaking horrified nominal allies more wedded to ideological conviction. Arthur Burns, then chairman of the Council of Economic Advisers, voiced alarm over the discrepancy between the administration's proposed Family Assistance Plan, with its federally guaranteed annual income, and the president's "philosophy." "Don't you realize," replied John Ehrlichman, "the President doesn't have a philosophy?" Reichberg interview, Bryce Harlow, GRFL.

Shapeless clothing and indistinguishable straw hats added to the drabness of the physical environment.

For variety, one turned to the cuisine, and enormous meals of duck wrapped in lotus leaves, bean curd with shrimp, chicken porridge and turtle soup with ham, all washed down with endless cups of strong green tea and stronger red wine. Entering into the spirit of the occasion, Betty Ford managed to down sea slugs, though only at dinner. At a Beijing medical school the Americans marveled at the use of acupuncture on a woman who remained fully conscious while having a large ovarian cyst removed. Another patient undergoing an appendectomy cheerfully answered their questions as the procedure unfolded before their eyes. On his second day in the Chinese capital a jeep factory tour enabled Ford to quiz workers about their housing and wages.

It was almost ten o'clock on the night of June 28 when the official visitors were ushered into the Great Hall of the People overlooking Tiananmen Square. There Premier Zhou Enlai hosted yet another sumptuous banquet, at the end of which Ford, in his obligatory toast to the health of Chairman Mao, remembered boyhood visits to the beaches of Lake Michigan, where he had been assured that if he dug into the sand with sufficient perseverance he would eventually reach China. "We were not industrious enough to dig that deeply, but we are the beneficiaries of great developments in the history of mankind," said Ford. He raised a glass "to the growing and hopefully broadening friendship between the Chinese people and the people of the United States."

By now it was after midnight. Acknowledging that Mrs. Boggs and Mrs. Ford must be tired, the premier excused them for the night. He then convened a three-hour discussion with their spouses. "Well, a small Ping-Pong ball has brought us together," said Zhou, in referencing the so-called ping-pong diplomacy that signaled a thaw in US-China relations commencing in the spring of 1971. What followed was the verbal equivalent of table tennis, with China's functional leader offering a conversational tour de force. Zhou shared painful memories of the snub administered by American secretary of state John Foster Dulles, who had refused to shake his hand at a 1954 Geneva conference. Surprising the Washington lawmakers by his knowledge of US domestic politics, Zhou expressed alarm over possible cuts to the American defense budget, as advocated by Senator George McGovern, the likely Democratic nominee for president, and his supporters. Zhou was adamant; the men in Moscow would never rein in their military spending. "Never, never, never," he reiterated in English. Ford was impressed. China's premier, he concluded, was a leader "of steel will, high intelligence, and super sophistication."

The Americans spent July 4 at the Canton Children's Palace, watching youngsters aged seven to fifteen sing and dance to the strains of "We Are Little Red Soldiers." The next day the group left for home. Ford wasted no time before going under the surgeon's knife, without benefit of acupuncture, to correct the effects of an old football injury to his knee. Six weeks later he was still using a red, white and blue cane to negotiate the Miami convention hall that served Republicans as a stage setting for a carefully scripted Nixon infomercial. Repeating his gig from four years earlier, Ford was invested as the convention's permanent chairman after receiving the gavel from temporary chairman Ronald Reagan. Leaving nothing to chance Nixon retainers, fearing that Vice President Agnew might draw louder cheers from the red-meat crowd, rearranged things so that Agnew, and not Chairman Ford, would introduce the president. That way both men could share the stage and bask in the crowd's approval.

An odd bit of business, the substitution of Agnew was one of the week's rare flirtations with spontaneity. To thirty million television viewers, however, it was a welcome contrast to the chaotic gathering of Democrats in the same hall a month earlier. If George McGovern could not control his own party's convention, so went the Republican argument, could he be trusted to govern a nation, or honorably end the Vietnam nightmare? Ford left Miami confident of Nixon's reelection and hopeful that his victory margin would be sufficient to install a Republican House of Representatives for the first time in a generation. The Speaker's gavel was almost in his grasp.

Part III

THE REPLACEMENT

June 1972–October 1974

———◆———

13

AHEAD OF THE CURVE

In politics when the train comes by you'd better get on it.
Because the odds are it does not come a second time.

—GERALD R. FORD

WATERGATE PLAYED BUT a minor role in the 1972 campaign. "Mc-Govern really thought the war would win the election," according to Joseph Califano, then counsel to the Democratic National Committee, which had wasted no time in filing a $1 million suit alleging White House involvement in the June 17 break-in. "I don't know why any Republican, official or unofficial, would want to bug Democratic headquarters," Ford told reporters pressing him about the so-called Watergate Caper. "So far as I know, and I've talked to some pretty high persons about it, nobody had anything to do with it." He revised this blanket denial only slightly after September 15, when a federal grand jury indicted the five Watergate burglars originally arrested at the DNC, along with their coconspirators E. Howard Hunt and G. Gordon Liddy, both of whom were easily tied to the Nixon White House.

In his memoir *Blind Ambition*, published on the eve of the 1976 election pitting Ford against Jimmy Carter, Nixon's former White House counsel John Dean accused Ford, acting on White House orders, of blocking an investigation of Watergate by the House Banking, Currency, and Housing Committee under its venerable chairman Wright Patman. A scourge of Wall Street since the Hoover Administration, Patman had his suspicions aroused when $114,000 drawn on a Mexican bank found its way into the Miami bank account of Watergate burglar Bernard Barker. Convinced that the money trail would lead to the Oval Office, Patman submitted a list of forty potential witnesses to be called before his committee in the closing weeks of the 1972 presidential campaign.

Republican members of the same committee were not alone in reading partisan political motives into Patman's actions, especially after the chairman on his own authority had the committee staff interview Maurice Stans, finance chairman of the Committee for the Re-election of the President. Ranking Republican committee member Garry Brown questioned the legal propriety of Stans testifying before the Banking Committee, on the grounds that anything he said might prejudice the rights of those indicted by the Watergate grand jury then in progress.

Disregarding Brown's complaint, Patman scheduled a full meeting of his committee for September 14, at which time he expected to receive Stans's testimony concerning the possible laundering of campaign money. When Stans declined the chairman's invitation to appear, Patman said he intended to subpoena the former commerce secretary. This would require authorization from a majority of his committee. John Dean first raised the subject of Patman's hearings with Nixon in an Oval Office conversation on September 15. Ironically in light of subsequent events, Dean expressed unhappiness over Ford's failure to take "an active interest" in the matter.

Now he had the president's attention. "Jerry's really got to lead on this. He's got to really lead," Nixon observed. "Tell Ehrlichman to get Brown and Ford in and then they can all work out something, but they ought to get off their asses and push it. No use to let Patman have a free ride here . . ." As it happened, Ford and Dean had a history, dating to the younger man's brief service (1966–67) as chief minority counsel to the House Judiciary Committee. Initially impressed by Dean's brisk competence, Ford entertained second thoughts as he watched the boyish lawyer with the ingratiating manner. "A real conniver . . . trying to cuddle up to the boss," concluded Ford, who was surprised to encounter Dean as White House counsel after July 1970, when he succeeded John Ehrlichman in the position.˙

When the subject of Patman's aborted hearings came up in his June 1973 testimony before Senator Sam Ervin's Watergate committee, Dean failed to mention Ford by name. The controversy resurfaced in the pages of Dean's 1976 book, with the author claiming that Ford had worked closely

---

* With John Mitchell as his sponsor, Dean had evaded all but "the most cursory background check," according to Ehrlichman, who soon harbored his own suspicions of the flashy dresser with his Alexandria townhouse and sporty Porsche in the driveway. Informed that Dean had been fired by the Washington law firm of Welch & Morgan over his handling of a broadcast license case, Ehrlichman asked to see his successor's personnel file. Dean's assistant duly appeared with a large envelope tightly wrapped in cellophane tape. The file was "sealed," he told Ehrlichman. Why did he want to see it? Ehrlichman mentioned Dean's prior legal employment and termination. The assistant clutched the envelope all the tighter. He promised that Dean would call him, which he did the next day, insisting that any rumors Ehrlichman may have heard were unfounded. He reiterated this claim a few days later, contending that his former partners had misunderstood his actions. John Ehrlichman, *Witness to Power* (New York: Simon & Schuster, 1982), 85.

with Nixon's congressional affairs staff, led by Bill Timmons and his assistant Richard Cook, to forestall Patman's Watergate fishing expedition. The allegation was denied by both Timmons and Cook. From Los Angeles, John Ehrlichman weighed in with his own denial that Nixon or anyone else had told him to raise the Patman hearings with Ford, as Dean implied.

Also refuting Dean's charges was J. William Stanton, an Ohio Republican who served on the Banking and Currency Committee chaired by Patman. According to Stanton, it had been committee Republicans who asked Ford to meet with them, and not the other way around. "At the meeting Ford mostly sat, smoked his pipe and listened. The discussion went along the following line: If Patman limited probe just to laundered money in South America, the Committee had no objection. The Committee was adamantly opposed to unlimited use of subpoena power which gave the impression that it was a political witch hunt two days before adjournment of Congress and one month before the election . . . Ford's only remarks were to concur with our decision that a month's delay made sense and that we were taking the right action."

Even had Ford been more aggressive in the matter, it is hard to see how he could have affected the outcome, since it was six of Patman's Democratic brethren who accounted for the 20–15 margin by which the committee denied its chairman the subpoena powers he sought. To be sure, Ford had signed a staff-generated letter to Republican members urging them to attend the meeting, held on October 3, at which the full committee was to consider Patman's request. But this was boilerplate, no different from hundreds of similar appeals he had approved since his election as House Republican leader. Dean was closest to the truth when lodging his original complaint about Ford's detachment from the Patman inquiry.

One reason Ford hadn't paid more attention to the matter was his virtually nonstop campaigning for a House Republican majority on the crest of that fall's anticipated GOP landslide. He did not overlook the voters of West Michigan, where he faced a rematch with his 1970 challenger, Jean McKee. This time around, the McKee campaign boasted star power in the form of celebrity endorsements from Candice Bergen and Gloria Steinem. At the other end of the cultural spectrum from Steinem's *Ms.* magazine, *Playboy* awarded McKee a straight-A grade compared to Ford's dismal C–. In the traditionalist Fifth District, home to not one but two Calvinist sects that traced their roots to nineteenth-century Holland, such an endorsement did more for Ford than for his Democratic adversary.

In mid-October a personal tragedy diverted Ford's attention from the campaign trail. House Majority Leader Hale Boggs, emulating his Republican counterpart, had made the long trip north to Alaska to campaign for first-term Democratic representative Nick Begich. On October 16, Boggs, Begich and two others on board their twin-engine Cessna 310 aircraft were reported missing in the wilderness somewhere between Anchorage and Juneau. The news devastated Jerry and Betty Ford. Jerry's relationship with Boggs went back to their days as junior members of the Warren Commission. More recently the two couples had grown close in the course of their two weeks in China. Boggs's daughter, the future broadcast journalist Cokie Roberts, would never forget Ford's frequent calls and visits to her family's home in the Maryland suburbs, "checking on everyone and seeing how everyone was doing." A massive search failed to locate the plane or its passengers. Not until the first week of January 1973, two months after both men were posthumously returned to office, did Congress officially recognize their deaths.

IRONICALLY, IT WAS a December 1972 holiday gathering at the suburban home of Hale Boggs's son, Tommy, that set Gerald Ford on the path to the nation's highest office. Ford made no mention of this in *A Time to Heal*, in whose pages he described, or more accurately implied, when he first learned that the vice president of the United States might be in legal trouble. The date was August 4, 1973, the place Groton, Connecticut, East Coast home port of the nation's nuclear submarine fleet. Ford had traveled to Groton for the launching of the USS *Glenard P. Lipscombe*, named for his longtime friend and legislative colleague. Flying back to Washington, he fell into conversation with former secretary of defense Mel Laird. The two men discussed the metastasizing scandal code named Watergate that was sparking talk of presidential impeachment for the first time in more than a century.

At one point in their dour dialogue, the name of Spiro Agnew surfaced. "You think things are bad now," said Laird. "Well, they're going to get worse."

"Tell me about it."

"I can't. I would if I could, but I can't."

Three days after this cryptic exchange, the *Wall Street Journal* told Ford more than he wanted to know. The vice president was under investigation by a Baltimore grand jury for extortion, bribe taking, tax evasion and related crimes dating back a full decade to his term as Baltimore County executive. In sharp contrast with Nixon, whose response to negative press coverage had been to withdraw even more than usual behind a shrinking cordon of White House loyalists, Agnew came out swinging. Denouncing the allegations as "damned lies," he claimed to have first picked up rumors early in 1973 about a corruption investigation being conducted by US Attorney George Beall, the younger brother of Maryland Republican senator J. Glenn Beall.

He was not alone in this. Newly uncovered evidence suggests that Ford became aware of Agnew's legal problems around the same time as the vice president. To learn how we must return to Tommy Boggs's holiday party across the street from the posh Chevy Chase Club a few days before Christmas 1972. Son of the late Democratic majority leader, the younger Boggs had been privy to official secrets ever since, as a teenager, he operated Speaker Sam Rayburn's private elevator in the Capitol building. Mentored by the likes of Tommy Corcoran and Clark Clifford, Boggs would go on to give Washington influence peddling a sleek coat of respectability, while billing clients $550 an hour. With Mardi Gras and the Louisiana bayou in his blood, Boggs found the party circuit a convivial place to do business.

This evening was no exception. Among his guests was Lawrence Meyer, a thirty-two-year-old newcomer to the law practice of Patton, Boggs and Blow (PBB). Fresh off managing the successful reelection campaign of Michigan senator Robert Griffin, Meyer was a natural addition to the firm, which, in its brief history, had established itself as a lobbying and legal powerhouse in the nation's capital. Bipartisanship guaranteed profitability, and Republican Meyer boasted impressive credentials—stints at the Antitrust Division of the Justice Department, as legislative assistant and legal counsel to Senator Griffin, at the Federal Trade Commission and in Griffin's uphill battle for reelection in 1972 against Michigan's immensely popular attorney general Frank Kelly. Useful as Meyer's political experience would prove to PBB, it was for his legal skills, specifically his background in antitrust law, that he was recruited in time to attend the firm's annual Christmas party.

Also sampling Tommy Boggs's hospitality that evening was John Childs of the Baltimore engineering firm Matz, Childs and Associates. In between seasonal carols and speculative buzz about personnel changes in the second Nixon Administration, Childs confided to Meyer and Boggs his anxiety over a federal corruption investigation launched by United States Attorney George Beall, for which a federal grand jury had been impaneled in the first week of December. Beall's apparent target was Baltimore County executive Dale Anderson. A Democrat of the old school, Anderson had succeeded Ted Agnew in 1967 when Agnew left county government for the governor's office in Annapolis.

Such probes have a way of spilling over their original channels as unexpected facts are uncovered. Although they had yet to be subpoenaed by the

Baltimore prosecutors, Childs and his partner Lester Matz were feeling the heat, at once victims and perpetrators of a bipartisan pay to play culture in which politically favored contractors and engineers reaped millions of dollars in exchange for kickbacks to Maryland officeholders. On January 4, 1973, a first wave of grand jury subpoenas went out, twenty-eight in all. A week later one of the dreaded documents was served on Matz and Childs. On January 15, Matz met with Joseph H. H. Kaplan, a former assistant United States Attorney, now a litigator with the Baltimore firm of Venable, Baetjer and Howard. Kaplan showed no surprise at the federal subpoena Matz presented for his inspection. By then "every lawyer in town" was aware of George Beall's investigation, according to Agnew biographers Richard M. Cohen and Jules Witcover.

Urged by his lawyers to tell all in the hope of receiving immunity from a US Attorney pursuing bigger fish, Matz said that was impossible for a simple reason: "because I have been paying off the vice president." Their association dated to the late 1950s, when Agnew the night-school lawyer and accountant was reviewing zoning decisions for a Republican-controlled county council. Agnew's 1962 election as county executive inaugurated a lucrative business relationship between the two men. As housing tracts replaced farmland in Baltimore's fast-growing suburbs, developers with the right connections could grow rich off public contracts. So could sticky-fingered public officials. For Matz and Childs every engineering contract with the county carried a 5 percent premium, discreetly paid to the county executive. On surveying jobs the cost of doing business was a more modest 2.5 percent.

Agnew's election as governor of Maryland in 1966 sweetened the pot. Indeed, Matz and Childs were soon victims of their own success. As their financial obligations to the governor involved sums too large to conveniently hide from their bookkeepers, they devised a complex scheme involving bonuses to key employees of the firm. Say an engineer received $5,000 in special recognition of his professional accomplishments. In reality, the honoree kept only enough to pay any additional income taxes on the windfall. The rest was recycled to management, which handed it off to the governor, initially via a trusted intermediary and later by Matz himself. As vice president, Agnew's influence over state contracts all but evaporated, yet Matz and Childs figured that the veep was still owed for contracts awarded during his governorship. This would explain Matz's delivery to Agnew in his vice presidential office of an envelope containing $10,000.

Such was the lurid tale related in outline by John Childs over eggnog and punch to Larry Meyer and Tommy Boggs. Childs, fearing imminent ensnarement in the Baltimore probe, wanted immunity from prosecution as the minimum price for cooperating with the US Attorney's Office. Meyer put Saul Friedman, an influential *Detroit Free Press* columnist, on the scent. His inquiries were bound to get back to the Baltimore prosecutors, bringing pressure to bear on them to immunize Childs *and* thoroughly investigate the vice president for corruption. Convinced that Agnew was in serious trouble, late in January 1973 Meyer scheduled a secret briefing for his former boss Bob Griffin, then number two in the Senate Republican leadership. Griffin voiced no objection to his doing the same for Ford, the ultimate party loyalist who Meyer feared might instinctively rally to Agnew's support if he chanced to hear the same cocktail party chatter to which the vice president alluded.

Taking pains to emphasize the secrecy of grand jury proceedings, Meyer urged Ford to exercise caution "because we do know for a fact that Agnew was taking cash from Matz" in his Capitol Hill office. Stunned by the news, Ford asked Meyer to repeat it for Les Arends and Illinois congressman John Anderson, the second- and third-ranking members of the House GOP hierarchy. As it happened, Arends had already been forewarned by Agnew the previous day, a Sunday. Arends had just returned from church when the vice president phoned him with assurances that he could "go to his grave and his God" knowing Agnew had done nothing wrong.*

Confirming Meyer's account is another key player in the developing drama, one who, like Meyer, has remained silent for over forty years. In February 1973, John W. Hushen was director of public affairs for the Justice Department, where he handled press relations for four attorneys general— John Mitchell, Richard Kleindienst, Elliot Richardson and William Saxbe. A former *Detroit News* reporter, Hushen's Michigan connections included a long-standing friendship with Larry Meyer, his former colleague on Bob Griffin's staff, who now invited him to a K Street lunch. The date was February 6, 1973.

"Jack, you're not going to believe what I'm about to tell you," said Meyer. "Spiro Agnew is taking cash payments in his vice presidential office up on Capitol Hill."

With a journalist's skepticism, Hushen instantly surmised a political

---

* "I heard you had a great meeting with Ford," Tommy Boggs told Meyer a few days later. With uncanny timing, Boggs had chanced to be in the office of Democratic whip John McFall when Ford called McFall to relate the visit of "a trusted young lawyer from Grand Rapids" who had convinced him that, in Ford's words, "we need to talk succession." AI, Lawrence Meyer.

motive behind this lurid report. "Smells to me like your Democratic law firm is trying to shoot down a grand jury investigation, pulling out all the stops to save your people in Baltimore" by invoking Agnew's name. His initial suspicions did not prevent Hushen, on returning to the Justice Department, from requesting an immediate audience with the attorney general. The two men met in a cramped space behind the department's main conference room. There Hushen repeated for Kleindienst what Larry Meyer had said to him at lunch. Appalled by what he was told, the attorney general minced no words in response to Hushen's inquiry about potential consequences.

"If he's done what you say he's done, he's going to get what he deserves."

The AG's first call was to George Beall in Baltimore. Kleindienst asked how the grand jury was proceeding. The US Attorney assured him that the only witnesses heard from to date were low-level contractors.

"Has Spiro Agnew's name ever come up?"

It had not, said Beall. "Not a bit."

W HILE THE BALTIMORE prosecutors methodically built their case out of the public eye, the Nixon White House was becoming mired in events bracketing the "third rate burglary" of the Democratic National Committee the previous June. Postelection demands for hush money took a tragic turn on December 8, 1972, when the wife of E. Howard Hunt, herself a CIA employee, died in a Chicago plane crash. Dorothy Hunt had been carrying $10,000, in hundred dollar bills, purportedly a payoff to one of the burglars' families. At the White House Nixon calculated that Mrs. Hunt's death could justify a presidential grant of clemency for her husband, an essential step in the cover-up that was beginning to consume Nixon's presidency. A grieving Hunt agreed to plead guilty to three counts of conspiracy, burglary and eavesdropping charges, only to have the deal rejected by DC Court of Appeals chief judge John J. Sirica, a tenacious Eisenhower appointee whose harsh sentencing practices had earned him the courthouse tag of Maximum John.

The trial of the burglars began the second week of January 1973. Faced with a conspiracy of silence, Sirica applied the screws to five defendants who pled guilty early in the proceeding, as well as to the two holdouts convicted on January 30. By issuing provisional sentences, punitive even by his standards, Sirica hoped to loosen tongues in exchange for lessened punishments. His hunch was confirmed when the judge was handed a letter from defendant James McCord, the date of which coincided with Sirica's March 21 birthday.

Appreciative of "the best damned birthday present I've ever gotten," Sirica told his law clerk, "This is going to break this case wide open."

And so it did, but not as Sirica anticipated or as most historians since have described. In his letter to the judge McCord confessed that he had committed perjury, that the break-in was not, in fact, a CIA operation, and that he had been pressured to remain silent as to the truth. Most important, McCord implicated others, in the campaign and at the White House. This was sensational stuff, but it was also mostly hearsay. The true value of McCord's letter was the publicity surrounding it, which helped persuade both campaign official Jeb Magruder and presidential counsel John Dean that they should cut an early deal with the prosecutor. "They were the ones who blew open the cover up by coming in to meet with us secretly," United States Attorney Earl Silbert acknowledged in a 1992 oral history.

Nixon's immediate response, later rescinded, was to invoke the doctrine of executive privilege to prevent his aides from testifying before the Senate Watergate Committee established in February 1973 and chaired by North Carolina Democrat Sam Ervin. Ford urged a different course on the president and his staff. "Go before the Senate committee, take an oath and deny it publicly," he advised them. At the same time Ford harbored a secret of his own. Denied the Speakership he coveted, in the wake of November's disappointing election results Ford told his wife, "Maybe it's time for us to get out of politics and have another life." Over the Easter holiday the couple flew to Palm Springs for a few days of relaxation in the desert sun. At a breakfast with John Connally, a recent Democratic convert to the GOP, Ford confided his intention to retire from Congress at the end of 1976. He also offered his support to Connally should the Texan seek the Republican presidential nomination that year. The two men agreed to stay in touch.

Ford returned to a capital city awash in rumors. At the end of April a despondent Nixon, likening the action to cutting off both his arms, asked for the resignations of Haldeman and Ehrlichman, along with Attorney General Kleindienst and John Dean, the unwilling scapegoat fired after Nixon belatedly realized the extent of his cooperation with Watergate prosecutors. Ford shed no tears for the departed presidential aides, though publicly he praised Nixon's housecleaning as "a most courageous act." He welcomed the choice of Elliot Richardson, Mel Laird's short-term successor at the Pentagon, to replace Kleindienst, not least of all because the new AG was authorized to appoint a special prosecutor to examine Watergate and related offenses. At last, it appeared, the White House was seizing the initiative.

Ford's optimism was fleeting. Thanks to Dean and other witnesses before the Ervin committee, a vast television audience was soon introduced to the Plumbers operation formed in the aftermath of the Pentagon Papers affair, as well as the September 1971 break-in at the Beverly Hills office of Daniel Ellsberg's psychiatrist. Disclosure of a White House "enemies list" left Ford incredulous. "If you have so many enemies you have to keep a list," he observed, "you have too many enemies." Worse, Dean raised the possibility that presidential conversations were being taped, a suspicion confirmed on July 16 by Haldeman assistant Alexander Butterfield. With Butterfield's reluctant admission, the Watergate investigation was transformed over-night into a Battle over the Tapes. Believing them to be exculpatory, Ford urged the White House to make available to Special Prosecutor Archibald Cox anything relevant to Watergate. Nixon refused, an action the minority leader called legally defensible, but poor politics.

Wanting to believe the president's claims of innocence, Ford took heart from the willingness of trusted hands like Mel Laird and Bryce Harlow to return to the White House in an effort to restore credibility. In the mean-time he hadn't forgotten what Larry Meyer had confided to him earlier, and Laird now confirmed in deepest secrecy: Agnew was in serious legal trouble. After months of sparring with the Baltimore prosecutors, both Lester Matz and John Childs had detailed for investigators their venal relationship with the vice president. Their story formed part of "an open and shut, cut and dried case," Attorney General Richardson related to Nixon's new chief of staff, General Alexander Haig. Faced with the possibility of a double im-peachment, Richardson quickly concluded that an Agnew resignation was in the national interest. He would get no argument from Nixon, much less from Haig.*

That summer Agnew was conspicuously friendly to Ford, inviting him to play golf at the Burning Tree Country Club in the Maryland suburbs and attending a meeting of the Chowder and Marching Club on July 18. It did not take long to ascertain his motive. Hung out to dry by the Nixon White House, the vice president would entrust his fate to Congress, more specifically to the House Judiciary Committee. An impeachment proceed-ing there would take precedence over the Baltimore grand jury. It would presumably be drawn out, affording Agnew time in which to influence public opinion. And it would be political, making his survival a test of

---

* At one point Pat Buchanan demanded to know why the administration was not rallying to Agnew's defense. Haig invited him to come by his office. There he crisply informed Buchanan, "We've got him taking envelopes in the basement." Patrick J Buchanan, *Nixon's White House Wars* (New York: Crown Forum, 2017), 352.

party loyalty. To execute this constitutional equivalent of a Hail Mary pass, Agnew required the cooperation of Speaker Carl Albert and the House leadership in both parties. He first approached Ford, who told Agnew to put his request in writing.

A meeting was scheduled for September 25 in the Speaker's office. "The vice president wanted to consult with you," Ford announced at the outset. "You have agreed and now I'm going to let him carry the ball." While Albert appeared sympathetic to Agnew's request, both Tip O'Neill and Judiciary Committee chairman Peter Rodino were reluctant to intervene. It took less than twenty-four hours for them to confirm Ford's intuition. Agnew should not expect Congress to come to his rescue.

WITH THE IMPEACHMENT route foreclosed, both sides entered into serious plea bargaining. Richardson scratched his head in bewilderment over the vice president, whose pride was offended by prosecutors who referred to him as "Agnew," leaving aside the title he had tarnished through his cupidity. Agnew's lawyers offered his resignation in return for no jail time and the retention of his government pension. Richardson insisted on a public statement acknowledging guilt. In a contentious Oval Office meeting, Agnew reportedly threatened to denounce Richardson deputy William Ruckelshaus over prejudicial leaks to the press. "Then I'm going after Richardson," he told Nixon. "Then I'm going after you." Alerting a friend that "I'm caught between number one and number two," Ford begged off a presidential request for help in eliciting Agnew's resignation.

This may account for his tongue-in-cheek appeal to the White House congressional affairs shop during the first week of October. Having received a press clipping consigning him to Nixon's blacklist, "along with Senators Percy and Baker," Ford inquired whether he might be transferred to the company of "a lesser light" such as Mel Laird. No trace of irony remained, however, when his Wisconsin colleague John Byrnes proposed that they promote Laird to fill any vice presidential vacancy. He couldn't agree to Byrnes's suggestion, Ford replied matter-of-factly. "I'm interested in the job for myself."

His comment might have caused a sharp intake of breath among members accustomed to his "Good Old Jerry" persona. Yet on reflection, it should not have come as a shock. In 1960 Ford had openly campaigned for a spot on that year's GOP ticket with Richard Nixon. Now, thirteen years later, with the Speakership clearly out of reach, it was perfectly logical that he should wish to round out his career as Nixon's vice president. Such a promotion would afford colleagues in both parties a means of acknowledging

Ford's services to the House, and to them individually. It would give Nixon a richly experienced advocate to promote his legislative agenda, then dead in the water. And it would supply Ford a prestigious platform from which to launch a postcongressional livelihood as lawyer-lobbyist.

After the fact, Ford would recall hints of presidential interest in his candidacy that eluded him at the time. There was a White House meeting the morning of October 10 at which Nixon briefed congressional leaders about Egypt's surprise attack on Israeli forces in the Sinai. At one point in the discussion he interrupted himself to observe, "I'd like to be in the shape with the American public that Jerry Ford is." Later that day Ford received a phone call from General Haig inviting him to join the president in his Executive Office Building hideaway. Shortly after noon Ford entered Room 180 of the EOB to find Nixon, casually dressed in sports jacket and slacks, his feet propped against an ottoman. He was surprised, said Ford, to see the president puffing on a pipe. He only did it when alone, replied Nixon, "or when I'm with an old friend like you." In the ensuing conversation, perhaps the most confiding the two men would ever have, Nixon outlined the charges against his vice president and referenced ongoing negotiations between Agnew's lawyers and the Justice Department.

Simultaneously, in a rain-lashed Baltimore courtroom, the vice president of the United States was pleading no contest to a single felony count of evading taxes on $29,500 of income. The presiding judge sentenced Agnew to three years' probation and fined him $10,000. Of these proceedings nothing penetrated Room 180. After an hour or so Ford told Nixon he had to return to the House, where important roll-call votes were pending. Hastening to the minority leader's seat near the front of the chamber, he hadn't been back ten minutes when an agitated Michigan colleague, Elford Cederberg, slapped him on the back.

"Jerry, Agnew just resigned."

Ford thought Cederberg was pulling his leg.*

"No, I'm telling you he resigned." The two men hastened to the nearby Republican cloakroom, where a blaring television set confirmed the news.

Within minutes a letter was hand delivered to Ford. "Dear Jerry," it read. "Today I have resigned as Vice President of the United States. After an extremely difficult weighing of all the facts, my deep concern for the country required this decision. You have been a staunch friend. I shall always count your friendship as a personal treasure. My gratitude and affection

---

* "What I did was act very surprised," Ford would later confess. "Because I didn't want to reveal that I knew a lot more than everybody else."

will always be yours, Sincerely, Ted." At the White House, calls went out to congressional leaders of both parties. Shortly after four o'clock, Ford, Hugh Scott, Les Arends and Bob Griffin assembled in the Oval Office to discuss the search process for Agnew's replacement. Under the Twenty-Fifth Amendment to the Constitution enacted in 1967, the order of presidential succession was rearranged to preclude a vice presidential vacancy and guard against the possibility that some superannuated lawmaker of the opposing party might claim the office for himself.

Nixon's visitors found him remarkably businesslike for a man who had just lost his handpicked deputy to scandal, and whose own prospects for survival were dimming by the day. Over coffee he briefed the leaders on the sequence of events leading to Agnew's resignation. The talk turned to finding a successor.

"What are the procedures?" asked Nixon.

"This is something that the Democrats will have to decide," said Ford.

It was a telling exchange. To a beleaguered president hoping to avoid a prolonged confirmation battle, Ford's greatest asset was his standing with the majority party. Nixon acknowledged as much in spelling out his criteria for a successful vice president. He—and it was all but certain to be a he, notwithstanding his request to presidential counselor Anne Armstrong to provide the names of qualified women—must be capable of doing the president's job should the responsibility become his; he should be broadly sympathetic to Nixon policies, especially where foreign affairs were concerned; and he must be confirmable by a Democratic Congress. Though he made no mention of it before his GOP brethren, Nixon was equally mindful of the need to avoid dividing his already weakened party by choosing a polarizing figure like New York governor Nelson Rockefeller, anathema to Republican conservatives, or his California counterpart Ronald Reagan, who was only slightly less objectionable to party liberals and some moderates.

Nixon asked Scott and Ford to canvass their colleagues, each of whom should submit up to three names in their preferred order. These recommendations were to be delivered within twenty-four hours to his personal secretary, Rose Mary Woods. A mile away, House Republicans were already mounting a campaign on Ford's behalf. Their enthusiasm was premature, warned Mel Laird, and quite possibly counterproductive. Laird preferred to play the insider's game, impressing on the president that his first choice, John Connally, the Democratic turncoat, would never pass muster with the majority party in Congress. By contrast, Laird assured Nixon, Jerry Ford could be confirmed "within two weeks."

This argument carried extra weight with Nixon, whose attention was increasingly focused on Watergate special prosecutor Archibald Cox and his aggressive pursuit of the president's White House tapes. "I'm going to clear the decks," Nixon alerted Alexander Haig. He was determined to eradicate "this snake who Elliot Richardson put in here."

T HIS PROLONGED WEDNESDAY, October 10, was far from over. A few minutes after six o'clock Nixon welcomed Speaker Carl Albert and Senate Majority Leader Mike Mansfield to the Oval Office.

"Have you got any recommendations?" he asked the opposition leaders. Mansfield mentioned former secretary of state William Rogers and Kentucky's respected ex-senator John Sherman Cooper, both retired from active politics.

"We have a man in the House who could be approved," said Albert.

"Who is it, Jerry Ford?"

"Yes, sir."*

At a caucus of GOP House members the next morning, Massachusetts's Silvio Conte sparked an ovation by reading aloud his fervent endorsement of Ford. Elsewhere on the Hill the leadership of both parties grappled with questions of jurisdiction and procedure. The idea of a joint House-Senate committee to assess the president's nominee was opposed by Ford, who presumably liked his chances better with the Judiciary Committee chaired by Peter Rodino. Shortly before six p.m., Ford delivered to the White House letters from 163 Republican colleagues. Nixon took them with him to Camp David, where he enjoyed a solitary dinner before winnowing the list of candidates to four (a fifth contender, Elliot Richardson, had preemptively removed his name from consideration on the grounds that Agnew's prosecutor should not benefit from his downfall).

That same evening his Michigan colleague Guy Vander Jagt encountered Ford at the National Press Club. "Jerry expected to be asked," Vander Jagt reported later. "He was like a bride who is about to be proposed to." Returning to 514 Crown View Drive in Alexandria, Ford had dinner with Betty. Around nine o'clock the phone rang. Ford picked it up to hear Mel Laird's familiar voice. Would his old friend accept the vice presidency if Nixon offered it? Ford asked for some time to talk it over with Betty. It was close to midnight when he called Laird back.

"I'll do whatever the president wants me to do . . . But I reiterate, Mel,

---

* Unknown to Nixon, both Mansfield and Albert had been prepped for the meeting by Mel Laird, who wasn't taking any chances in the developing contest between Connally and Ford. AI, Mel Laird.

I'm not promoting myself. I am not campaigning for it. We have our plans, Betty and I. We are happy with what we have decided."

Even this modestly worded disclaimer could be said to advance Ford's prospects, for by emphasizing his decision to retire at the end of 1976—a point Laird was sure to highlight in reporting back to the president—Ford removed any potential obstacle to Nixon's preferred successor, John Connally, getting the Republican nomination three years hence. Ford reiterated as much the next morning, October 12, when he and Hugh Scott separately visited Nixon in the Oval Office. He appreciated all the letters from Ford's House colleagues, the president told him. "I'm pleased that you had so much support." With Al Haig taking notes, Nixon got up from his desk, shook hands, and declared, "Jerry I want you to be my nominee for vice president, and I want you to know that I would be for John Connally . . . in 1976" (so Ford recalled the exchange; in his memoirs Nixon denied making any formal offer). Pressed as to his own plans, Ford repeated the assurances he had previously given Laird: Nixon's commitment to Connally "doesn't bother me at all." For proof he cited his recent visit with Connally in Palm Springs, and his offer of support should the Texan run three years from now.

The conversation shifted to the logistics of announcing Ford's selection. Nixon said that he would call the Ford residence at seven thirty that evening, less than two hours before a televised East Room ceremony introduced Agnew's successor to the nation. Until then Ford must not breathe a word of the news to anyone, Betty included. The president summoned a photographer to record the moment.

"This might be a historic picture," he said.

POLITICS HELD NO allure for sixteen-year-old Susan Ford. "I've lived with it all my life," she confided to one reporter that fall, "and I think I sort of want to get away from it." This did not prevent her from betting $5 against her mother's certainty that Nixon would choose someone other than her father to replace Agnew. For Betty Ford, promised at last a stay-at-home husband come January 1977, the wager with her daughter carried more than a little wishful thinking. She remained convinced throughout that Friday, October 15, notwithstanding a local news reporter who repeatedly called her home with the message, "I know it's Jerry." She wasn't alone. The *Detroit News* made much of an unnamed Ford staffer who said her boss returned from the White House looking "glum and downcast." By contrast Bob Hartmann described an employer "on cloud nine . . . trying to suppress his exhilaration."

At midafternoon a rumor swept the *Washington Post* newsroom that Virginia governor Linwood Holton was the surprise pick. Reporters scrambled to locate the governor, who knew as little about unfolding events as they did. Bob Clark of ABC News had a different hunch. Dismissing Hugh Scott's claim that Nixon had ruled out of consideration *anybody* from Capitol Hill, when in fact he had merely excluded Hugh Scott, Clark prevailed on Ford press secretary Paul Miltich to let him speak to the minority leader. Taking advantage of Ford's well-known aversion to deception, Clark backed him into a corner.

"Can you deny that Nixon has picked you?"

Lengthy pause.

"No, I can't, I can't deny it."

Clark had his scoop.

Arriving home around six fifteen, Ford repaired to the backyard pool while Susan cooked steaks for dinner. When she asked what was going on, Ford replied, "Nothing. You will know tonight at 9 pm." The family was at the dinner table when the downstairs phone rang, the one with five extensions, reflecting the talkative habits of adolescent children. On the other end was Mike Ford calling from the Massachusetts theological school in which he had enrolled following his recent graduation from Wake Forest. He, too, had heard reports that his father was about to be nominated by the president.

"It's not going to happen," said Betty. "We don't want it. Just forget it."

At 7:25 the upstairs, unlisted phone rang. It was immediately answered by Susan, who announced from the head of the stairs, "Dad, the President wants to talk to you."

Ford dashed upstairs to hear Nixon's unmistakable voice telling him, "General Haig has something to say to you." Haig got on the line. "Do you want Mrs. Ford to hear what I'm going to tell you?" Complex explanations ensued, clarifying nothing until Ford asked Nixon to call back on the other number.

"Betty, get off the line, the president wants to call."

A minute later Haig made it official: "The president wants you to be his nominee."[*] They were due at the White House in little over an hour. Betty, in slacks and no makeup, wept at the news, though it's impossible to know

---

[*] The entire sequence whereby Ford learned that he was the president's choice to succeed Agnew, not from Nixon directly but via Haig, seems odd even by the standards of the Nixon White House. Ford himself gave differing versions of the phone call that changed his life. Mine is based on a contemporary account he provided to Bonnie Angelo of *Time* magazine.

whether they were tears of joy, pride or frustration. Two of their sons later offered differing reactions to the news. Mike sounded a cautionary note, telling his father, "Mother's been so good to us through all the years. Are you sure you want to put her through this?" Steve Ford was more reassuring, quoting back to his mother a paternal assurance that "vice presidents don't really do anything."

The Marine Band was playing selections from *The Sound of Music* as two hundred hastily invited guests filed into the East Room for an oddly festive gathering—"Until you stop and think about what made it all necessary," wrote Martin Schram in *Newsday*. Nixon milked the suspense like a game-show host speculating over the contents behind Door Number Two. A gust of genuine delight swept the room as Ford's name was announced. In his brief acceptance remarks, Ford stressed his desire to help rebuild trust between Congress and the administration.

Outside he ran a gauntlet line of reporters and photographers. "What about ['76]?" one newsman asked.

"I have no intention of seeking the presidency," Ford replied. "I told the president that tonight."

"Why aren't you going to run?" pressed another reporter.

"I said I have no intention of running," said Ford. Even now, for anyone who cared to parse his words, he was keeping his options open.

Returning to Crown View Drive, he and Betty encountered a small crowd of neighbors on their sidewalk singing "For He's a Jolly Good Fellow." Inside Ford took a congratulatory call from the man he was replacing. From Bethesda Naval Hospital, Agnew's predecessor offered his own, characteristically upbeat, reaction.

"It's good for the country," Hubert Humphrey told Ford.

Humphrey's enthusiasm was echoed by those who knew Ford best, Democrats as well as Republicans. Carl Albert pronounced him "a splendid nominee." Dan Rostenkowski, the congressman from Mayor Richard J. Daley's Chicago, declared, "I'm as happy for him as I would be for my own brother." Don Riegle, the former Republican lawmaker who had switched parties despite Ford's best efforts, characterized his erstwhile leader as "Nixon's Better Idea." Anticipating a Ford presidency, Riegle predicted that the task at hand would involve recovery, not renaissance, "and for this Gerald Ford may well be the right man."

The press was less generous. Although *Time* labeled Ford "a good lineman for the quarterback," to the *Wall Street Journal* his selection represented a squandered opportunity. "Agnew without alliteration," sniffed *Washington Post* columnist Nicholas von Hoffman. England's *Guardian* bannered the news under a particularly brutal headline: "After a Crook, a Mediocrity." The initial public response appeared to validate Nixon's judgment. Sixty-six percent of those surveyed by George Gallup approved of Ford's selection. Just 7 percent voiced opposition. Snap polls showed a majority of both houses inclined to vote for confirmation.

The vice president designate reinforced his common-man image by flying to tiny Cedar Springs, Michigan, the day after his nomination and keeping a prior commitment to march in the town's annual Red Flannel Festival parade. Twenty-five thousand people lined the six-block parade route. Their cheers were still ringing in his ears as Ford acknowledged the irony of the situation. "Here I have been trying . . . for twenty-five years to become Speaker of the House," he told reporters. "Suddenly I am a candidate for President of the Senate, where I could hardly ever vote, and where I will never get a chance to speak."

Still more dramatic changes overtook the Ford family. Secret Service agents requisitioned the two-car garage on Crown View Drive as a protective command post. Ford was personally billed over $4,000 for the cost of realigning his driveway. Secure phone lines were installed, along with bulletproof glass windows throughout the house. A self-described "lone wolf" accustomed to traveling with no entourage, Ford was henceforth surrounded by four to eight agents wherever he went—in his car, outside his office, looking down from the House galleries.

With Senate hearings on his nomination scheduled for the first week of November, Ford was about to become the most thoroughly investigated public official on record. In Grand Rapids, newsman Maury DeJonge was kept busy entertaining representatives from the *New York Times*, the *Boston Globe* and other out-of-town media directed his way by the local Republican headquarters. "They were all looking for junk—for the worst thing they could find" on the vice president designate, he explains. After all, as one network correspondent said hopefully, "This guy couldn't have been an Eagle Scout all his life."

"If you find anything on Jerry Ford," DeJonge shot back, "you come back and tell me."

Three hundred fifty FBI agents from thirty-three field offices took up DeJonge's challenge. Forty years after the fact, two agents tracked down a onetime Ohio State halfback tackled by Michigan linebacker Gerald Ford, who had been penalized for unnecessary roughness. The Ohio player volunteered that it was a clean hit. Other investigators poured over Ford's finances and friendships. A dozen auditors conducted a forensic examination of the nominee's tax returns going back a decade. "Anything you want you'll find it there," Ford told agents as he pulled open a desk drawer containing a quarter century's worth of checks neatly filed. The nominee's net worth was pegged at $256,378, most of it tied up in his Alexandria home and a Vail ski condominium purchased for $50,000 in 1970. The latter transaction had been made possible by borrowing against the congressman's life insurance policies and tapping the savings accounts of the Ford children, each of whom looked forward to Christmas on the Colorado slopes.

After twenty-five years in public office, their father's assets included a personal bank account containing $1,282; stockholdings valued at $13,500—primarily shares in the family paint and varnish company—plus a modest life insurance policy and congressional retirement plan. In recent years Ford had earned up to $30,000 annually in honoraria from paid speeches before corporate and political audiences.* As federal investigators fanned out across West Michigan, their sheer number and aggressiveness prompted calls to the Justice Department from Ford's neighbors. Assistant Attorney General William Ruckelshaus decided to visit Grand Rapids and personally supervise the investigation. Approaching his boss to obtain permission, he found the AG preoccupied with other matters.

"We've got a worse problem than Agnew," said a stony-faced Richardson. Impossible, replied Ruckelshaus. "There's no way we could have a worse problem."

"Oh yes, the White House . . . wants to fire Archibald Cox."

THE FINAL CHAPTER of Spiro Agnew's public life had overlapped with another drama, involving Nixon's White House tape recordings and a budding confrontation with "that fucking Harvard professor," in Nixonian parlance, chosen by Elliot Richardson, his onetime student, to investigate the Watergate break-in and any related crimes. An initial request from Cox for nine Watergate-related tapes had been refused by the White House, upheld by Judge Sirica and reaffirmed by the US Court of Appeals in a ruling issued the same day as Ford's nomination to succeed Agnew was made public. Nixon then proposed a compromise under which White House lawyers would provide the special prosecutor written summaries of subpoenaed conversations, their accuracy vouchsafed by Mississippi senator John Stennis.

---

* In a transaction that reaffirmed his honesty even while raising questions of judgment, Ford had accepted a place on the board of Grand Rapids' Old Kent Bank and Trust without anticipating conflict of interest charges to which it might give rise. Stung by the resulting criticism, he terminated his board membership after just six weeks.

Seventy-two years old and hard of hearing, Stennis would appear less than ideally suited for the task. Yet his personal integrity was unquestioned. "Stennis would rather have had his right arm cut off than to have participated in any knowing misrepresentations," said Richardson, who agreed to present the scheme for Cox's consideration. The special prosecutor was highly skeptical. Only actual tapes would pass muster in any criminal trial, he pointed out. Beyond this, Cox refused to rule out future demands in addition to the handful of recordings presently being sought. Nixon hit the ceiling. "No more tapes, no more documents, nothing more," he told Al Haig and Fred Buzhardt, a Defense Department lawyer installed as Nixon's special counsel for Watergate matters in May 1973. "I want an order from me to Elliot to Cox to that effect now."

By Friday, October 19, Haig and, by extension, the president for whom he worked believed they had Richardson's assent to fire Cox if he continued in his quest for presidential tapes or other records of sensitive conversations.* Richardson denied the existence of any such deal, even as he endorsed the Stennis compromise about which Stennis himself was starting to have second thoughts. "Elliot Richardson was a consummate politician," according to Tom Kauper, then a deputy attorney general in charge of the Antitrust Division. "I've never known how much he actually knew in advance of what happened that Saturday night, but knowing him, it would have been clear that he was going to come out clean."

On Saturday morning Cox held a televised press conference at which he formally rejected third-person summaries in lieu of the actual tapes. Privately Richardson offered the special prosecutor assurances of his continued support, in effect acknowledging that both men would soon be out of a job. Nixon, preoccupied with the Yom Kippur War then raging between Israel and her Arab neighbors, felt betrayed by his patrician attorney general. Sustaining presidential credibility, in Moscow and in the Middle East, at a time when Henry Kissinger was trying to enlist the Soviet Union in promoting a regional cease-fire, took precedence over Richardson's Yankee conscience. Needless to say, Richardson had a different definition of the public interest. At the Justice Department, he discussed his impending

---

* Mel Laird insisted that, as late as two p.m. on Saturday afternoon, the twentieth, the White House had Richardson's consent to fire Cox. Characteristically, Laird had prepared for the ensuing vacancy by quietly recruiting Houston lawyer Leon Jaworski to replace Cox. The former defense secretary was getting ready to attend a charity event with his wife when he got a call from Richardson.

"Mel, I can't do it." The fuse was lit for what Laird called "a gol-dern explosion in the Justice Department and the whole deal was off." AI, Mel Laird.

departure with Bill Ruckelshaus and the chain-smoking Solicitor General Robert Bork, the number three man in the departmental hierarchy. When Ruckelshaus made it plain that he, too, would resign rather than carry out the president's order to fire Cox, it left Bork as the reluctant instrument by which the king would rid himself of a most turbulent priest.

"Well," said the former Yale Law professor often mentioned as a candidate for the national's highest tribunal, "there goes my Supreme Court nomination."*

On the Sunday morning after the so-called Saturday Night Massacre, Ford tried to minimize the political damage, insisting, "The President had no other choice. Mr. Cox refused to accept the compromise issue." Privately he harbored doubts. In any event nothing he or anyone else might say could avert the storm of popular anger that broke over Washington that week. Stoked by feverish television coverage of what one network anchor solemnly intoned was the gravest constitutional crisis in the country's history, Americans flooded their elected representatives with messages of protest. Two hundred seventy-five thousand telegrams overloaded Western Union circuits in the nation's capital. In Grand Rapids, a petition denouncing the Cox firing and bearing 750 signatures was delivered to Ford's congressional office.

By Tuesday, October 23, two dozen resolutions calling for Nixon's impeachment had been filed with the House Judiciary Committee. That same day the president reversed himself, agreeing to the appointment of a new special prosecutor and yielding up the tapes as demanded by Judge Sirica. In little more than a week Ford had come to be seen less as Agnew's replacement than as a president in waiting. Adding to the sense of urgency was Speaker Albert's personal desire to hand off the baton of presidential succession lest he and the Democratic majority in Congress be accused of scheming to seize through partisan maneuver what they could not win at the polls. Behind the scenes just such a scenario was being promoted by a handful of liberal firebrands on the Judiciary Committee led by Bella Abzug, the impassioned New Yorker who directed the speaker, "Get off your goddamned ass, and we can take this Presidency."

The Abzug strategy was simple enough: delay Ford's confirmation until

---

* "Elliot came back from the White House that night just before I went over and said the one thing he was sorry about was perhaps he hadn't made it plain to the president that he couldn't fire Cox. And, indeed, he couldn't," Bork explained in a 2010 oral history. But the strangest reaction to the events of that tumultuous night came from Fred Buzhardt. "What's all the fuss about?" asked the White House lawyer. "People leave the federal service every day." AI, Robert Bork.

Nixon was driven from office, leaving a vacancy to be filled by the Democratic Speaker of the House. The full Democratic caucus was less willing to engage in a constitutional coup. Going into the public hearings, Ford was operating at something of a disadvantage, since none of the FBI's investigative findings would be shared with him. By prearrangement these were entrusted exclusively to the chairman and ranking member of the responsible committees in each house of Congress. This did not prevent the White House from attempting to coordinate testimony with its candidate. To help him prepare for the grilling to come, Ford turned to Phil Buchen, his onetime law partner from Grand Rapids, and Benton Becker, the hard-nosed investigator with whom he had worked on the Douglas impeachment probe.

The two men enjoyed a classic good-cop-bad-cop relationship, as Becker demonstrated one day in seizing the phone from Ford's hands rather than let him talk to Al Haig. "As sure as I'm alive, General, one of these senators or . . . members of the House are going to ask him at some time if the White House has fed him any information about the FBI report and investigation," Becker reproached the imperious chief of staff. "And when they do, I want him to be able to say no."

Barely pausing to let the impact of his words sink in, Becker crisply notified Haig, "Mr. Ford's not going to take these calls anymore."

Senate hearings commenced on Thursday, November 1. "I feel that I am among friends," Ford told the Rules Committee, chaired by Nevada's Howard Cannon. Apart from demonstrating his personal competence and character, success in the nationally telecast proceedings depended on Ford's agility in balancing support for the president who nominated him with suitable deference toward the lawmakers who might soon constitute a court of impeachment. Over two days of testimony he disarmed senators with his modesty and candor. He endorsed détente with the Soviet Union and China, opposed court-ordered busing to achieve racial balance in the nation's classrooms and readily acknowledged the positive role of the media in exposing the Watergate scandal.

More surprising was his willingness to stake out positions at odds with the embattled Nixon White House. "I do not think a vice president should just rubber stamp what a president proposes or believes," Ford told the senators. Employing the inevitable football analogy to describe his new role, Ford said he expected to confine his differences to private counsels. After all, he explained, "If a play has been made, you don't go out and tackle your own quarterback." Pressed by committee members, Ford said that he would probably change his mind if the quarterback was running the wrong way. Questioning the use of executive privilege to withhold information from congressional investigators, he expressed sympathy for Elliot Richardson and Bill Ruckelshaus. Had he been in their positions on October 20, and ordered to fire Special Prosecutor Cox, he probably would have acted as they did.

"Can Richard Nixon save his presidency?" asked Oregon's Mark Hatfield.

"It's going to take a lot of help from a lot of people," Ford replied. "And I intend to devote myself to that." Offering himself as "a ready conciliator" between the executive and legislative branches, the nominee was careful to acknowledge that each had a mandate from the voters. The next day's Associated Press headline said it all: "Ford Shows He's His Own Man."

From the outset, his managers hoped to address any potential controversies in front of the senators, who were considered friendlier than the more ideologically driven membership of the House Judiciary Committee. This strategy was fully validated as Ford dealt with the criticisms of a shadowy lobbyist and self-styled influence peddler named Robert Winter-Berger, whose 1971 book *The Washington Pay-Off* had enjoyed a modest run on the *New York Times* bestseller list.*

In his book and again in a sworn affidavit Winter-Berger made a number of sensational claims. He said it had cost him $1,000, paid to a mutual friend, to secure an introduction to Congressman Ford. Between 1966 and 1969 Ford was alleged to have accepted $15,000 in Winter-Berger "loans" to cover, among other things, Betty Ford's medical expenses. None of the lobbyist's money had been repaid. Ford was accused of selling an ambassadorship for $125,000 deposited with the national GOP. The congressman's Capitol Hill staffers had reportedly accepted Christmas wallets containing $100 bills, courtesy of Winter-Berger. Finally, the would-be whistleblower asserted that Ford had, for a year or more, been under the treatment of New York psychotherapist Arnold Hutschnecker.

Taking the allegations seriously, the *Grand Rapids Press* had flown a reporter to Washington to interview Winter-Berger, whose purported documentation turned out to be little more than some form letters and a pile of old newsletters. Then muckraking columnist Jack Anderson got into

---

* Subtitled "An Insider's View of Corruption in Government," Winter-Berger's gossipy account identified an incumbent member of the United States Senate who was allegedly arrested in a Greenwich Village gay bar raided by police. In its most improbable scene, a distraught LBJ sought Speaker McCormack's help in channeling $1 million to Bobby Baker, lest the president take Baker's place behind bars—a frenzied appeal made in full view of author Winter-Berger.

the act, giving Winter-Berger's claims wide exposure on the eve of Ford's confirmation hearings. Had he dug a little deeper, Anderson would have presumably uncovered the 1960 New York court order banning Ford's accuser from the securities business after that state's attorney general found him guilty of peddling shares in the nonexistent Saniphor Sales Corporation. With Wall Street off-limits to him, Winter-Berger transferred his scam to Capitol Hill. Taking advantage of Ford's open-door policy, he had sought to ingratiate himself by handing out compacts to secretaries and cheap wallets—devoid of cash—to more senior staff members.

Several times he had managed to penetrate Ford's inner office, dubbed the Jungle for its profusion of plants and GOP elephants. On at least one occasion he enlisted the congressman's help on a Dutch immigration case. In the process he also wore out his welcome. Tipped off that Winter-Berger was under investigation by the Justice Department, Ford gave directions to keep the dapper con man away from him. Winter-Berger would not be put off so easily. Having familiarized himself with the minority leader's daily routine, he knew just when to be outside the House chamber with clients he hoped to impress.

It should not have surprised anyone that the self-promoting lobbyist resurfaced in time for Ford's confirmation hearings. Fully prepared to refute his latest charges, Ford offered to take a lie detector test. Short of that, his handlers sought permission to cross examine witnesses appearing before the Senate committee. Both requests were denied. So they did the next best thing, providing Bob Griffin, Ford's closest friend in the Senate, with a list of questions to raise during closed-door testimony.

For three hours on November 7, senators subjected Winter-Berger to a withering interrogation, beginning with the $15,000 "loan" supposedly made from the lobbyist's own pockets. Chairman Howard Cannon cited tax returns showing Winter-Berger's gross personal income to be $14,096 in 1966; $7,615 in 1967; $1,643 in 1968; and in 1969 $4,912. Confronted with such numbers, Winter-Berger changed his story—90 percent of the money in question had been loaned to him by fellow lobbyist Nathan Voloshen, convicted of perjury and conspiracy charges arising out of his misuse of Speaker John McCormack's office and telephone. Unfortunately Voloshen was dead, and Winter-Berger had nothing to back his charges. Ford, by contrast, was able to prove through meticulous record keeping that Blue Cross Blue Shield had covered nearly all of his wife's medical expenses.

Winter-Berger's claim to possess detailed diaries supporting his allegations was just as easily refuted, leading the author of *Washington Pay-Off* to take refuge in what he called "literary license." Presumably this included two lawmakers identified as recipients of his largesse, though both had died before the shadowy transactions could take place. The Ford friend named by Winter-Berger as his original conduit to the congressman forcefully denied selling access, and threatened her accuser with a lawsuit. Finally, there was Dr. Hutschnecker, who confirmed that his sole encounter with Ford was a fifteen-minute conversation about politics, not medicine. "I had a feeling he was not quite sure why he was there," the therapist told the senators.

Ford invited criticism for seeing Winter-Berger at all, let alone an out-of-town psychotherapist recommended by him. Yet that was not how members of the Rules Committee, many no doubt having been trailed by similar pests, chose to interpret the lobbyist's claims. They had his testimony transcribed and sent to the Justice Department for possible charges of perjury. "You have a reputation for openness and honesty in reply," Senator Claiborne Pell told Ford, "and it appears these hearings justify it." Pell's assessment was shared by millions of television viewers who concluded that, whatever Ford lacked in charisma, he appeared to compensate for in personal integrity and substantive knowledge of the governing process. *Time* magazine reported him as "growing in stature." Even some of Ford's harshest critics began to have second thoughts. Writing in the *New York Post*, the acerbic Harriet Van Horne wrote of the nominee, "He puts one in mind of a big sloppy dog, not sly or clever, but capable of diving through the ice to rescue a drowning child."

N O SOONER HAD the Senate committee finished its work on November 14 than a second round of hearings was gaveled into order in a far more contentious atmosphere. Representative John Conyers Jr. and two of his Democratic colleagues on the House Judiciary Committee protested that it was "totally inappropriate" for Congress to act on the nomination of a president whose credibility had been so irretrievably damaged. Chairman Peter Rodino ruled the motion out of order, along with a second Conyers motion to make the entire FBI file on Ford, all 1,700 pages of raw data, part of the public record. In his opening remarks Ford reprised his Senate testimony, endorsing "friendly compromise" and paying bipartisan tribute to former Speakers Sam Rayburn and Joe Martin. The only thing that caused him concern, said Ford, was that old friends on both sides of the aisle might stop calling him Jerry.

Reasserting his support for the president, Ford urged Nixon to hold more press conferences and show greater flexibility concerning access to the

Watergate tapes. He rebutted complaints about his own lack of foreign policy experience and alleged indifference to civil rights. "I'm actually proud of my civil rights voting record," Ford told the committee. To those who called him a "final passage man," one who had sought to weaken proposed legislation before supporting bills whose ultimate passage was assured, he offered a frank explanation. "The voting record reflects Grand Rapids," said Ford. "On final passage I did vote for all those bills. I think that's a better indication of my true feelings." Such arguments failed to pass muster with the NAACP and other liberal groups that testified in opposition to his nomination. Civil libertarians offered criticism of the Douglas impeachment effort.

In a special category was Elizabeth Holtzman, a combative New Yorker who dissented from the five-minute rule under which the hearings were being conducted—necessarily, according to Chairman Rodino, if all thirty-eight committee members were to have a chance to question the nominee. Holtzman moved to table the nomination. Her motion was rejected 33–5, clearing the way for the full committee to approve Ford by a vote of 29 to 8. On November 27 the Senate followed suit with only three dissenting votes cast. "I have never before voted for a Republican vice president," declared Hubert Humphrey. "I did so today."

The House scheduled six hours of debate and a final vote on December 6. Chairman Rodino, unable to forget Ford's aborted campaign to remove Justice Douglas, cast a last-minute vote against the nominee. On the other hand Andrew Young, a close associate of the late Reverend Dr. Martin Luther King Jr., was loudly applauded when he broke with his colleagues in the Congressional Black Caucus to announce his support. "Decent men, placed in positions of trust, will serve decently," said Young. "I believe that Mr. Ford is a decent man." Young's was the 387th vote to confirm the nation's first appointed vice president; thirty-five members voted no.

Strains between the vice president designate and his White House sponsors marred Ford's formal induction into office. Accustomed to calling the shots, Nixon and his image makers planned for an East Room ceremony. Ford, keenly aware of the symbolism involved, preferred an inauguration held on his own turf. Having been introduced to the nation as the president's man, he wished to be sworn into office before his Capitol Hill colleagues. It was an unprecedented scene that played out before a national television audience at the dinner hour on December 6. Richard Nixon and Ford entered the House chamber together, but Nixon was too experienced in Washington's ways to believe the roars of approval greeting them as they made their way to the rostrum were intended for anyone but the balding, familiar, seemingly unremarkable figure at his side.

For all the history being made that winter evening on Capitol Hill, the day's most revealing moment had come a few hours earlier when *Time* correspondent Neil MacNeil observed a dress rehearsal of Ford's brief valedictory to his colleagues. Hardly a speech at all in the formal sense, it expressed in nine minutes the congressman's gratitude to those whose sacrifices over the years had enabled him to stand before the American people in "this moment of visible and living unity," pledging his best efforts "to do what is right as God gives me to see the right." Biting his lip, Ford wept openly as he recognized Betty Ford "for standing by my side as she always has." Given the challenges confronting the new vice president, MacNeil reflected after the practice session, "it was perhaps fitting that he should weep."

That evening House Majority Leader Tip O'Neill offered a consensus view of Ford's oath taking and brief address. "It was a very impressive ceremony," said O'Neill. "We won't see anything like it for maybe six or eight months."

14

"THE WORST JOB I EVER HAD"

———◆•◆———

I know something about power. At least I did before I became Vice President.

—GERALD R. FORD, JUNE 1974

F ORD'S INAUGURATION AS vice president coincided with Wall Street's biggest two-day gain to date, as the Dow Jones Industrial Average soared nearly fifty points. Few observers interpreted the market rally as a vote of confidence in the beleaguered Nixon Administration. In Washington a lobbyist for the AFL-CIO greeted Ford's confirmation as "our go ahead" to accelerate labor's campaign for impeachment. For their part Republicans in Congress hoped the new vice president might help stave off electoral disaster in 1974. Private polling indicated a bloodbath in the making, with the GOP losing up to seventy-five House seats and half a dozen senators. The same polls made Ford an early favorite for his party's 1976 presidential nomination. A Lou Harris survey at the end of December had him leading both Ted Kennedy and Washington senator Henry "Scoop" Jackson among Democratic contenders.

Ford's position was in truth less enviable than it appeared. From his first day on the job the self-professed "Instant Vice President" struggled to balance his loyalty to the man who chose him against growing suspicion of presidential complicity in the Watergate cover-up. "If Gerald R. Ford had decided to become a professional poker player, he would have been a billionaire, because the guy knew how to play his cards. He held them so close, including from his family and the closest advisers," according to David Kennerly, a Pulitzer Prize–winning news photographer who was to become a virtual member of the Ford family. "He was almost pathological about not talking [about] or hinting at or suggesting that he might take over the presidency." True to form, on December 7, Ford told reporters that the president had no intention of resigning, this following a forty-five-minute Oval Office meeting in which Nixon had promised that "when all of the facts are out" he would be exonerated. Ford repeated these assurances in appearances on NBC's *Today* show and ABC's *Issues and Answers*.

Away from the cameras he was more equivocal, confiding to friends, "What I have to watch out for is not to become Nixon's apologist. That won't help either of us." Throughout the confirmation process Ford had deliberately avoided setting foot in the vice presidential suite on the second floor of the Executive Office Building, believing that to do so would be seen as presumptuous. Now he moved into a spacious office, with high ceiling and fireplace, that had once housed General John J. Pershing. Overawed by the surroundings, a visiting Grand Rapids reporter told him, "Jerry, this is scary."

"Why?"

"It's all so different. You're vice president, you're in this huge office."

A hint of irritation crept into Ford's voice. "Maury, nothing has changed—remember that."

This would come as news to Chief of Staff Bob Hartmann, who staked his claim to office space used by FDR as assistant secretary of the navy. Suspecting the presence of wiretaps, Hartmann adjourned sensitive conversations to a nearby corridor. Hartmann's territorial instincts clashed with White House retainers who presumed to instruct the new vice president in his responsibilities. Alexander Haig set the tone even before Ford's confirmation hearings. "You'll want an advance man," Haig briskly informed the nominee, "and we'll be glad to lend you one of our best, young fellow named Dewey Clower. He'll tie in closely with the Secret Service. And probably some help in the speechwriting department; we can do that, too . . . Let's see, have you a good scheduler? John, see what Dave Parker can do there. And, of course, you'll need a foreign policy guy and a political guy . . ."

"You know," Ford interjected, "what we really need is some more good secretarial help."

Uncharacteristically he took an instant dislike to would-be handler W. Dewey Clower, a thirty-two-year-old Haldeman acolyte who did not engender confidence by officiously guiding his charge through Capitol hallways Ford might have negotiated blindfolded.

"Thanks," said the former minority leader to the eager beaver steering him by the elbow, "but I know the way."

Clower notified Hartmann of four areas in which Ford could be helpful. Besides congressional relations, and a surplus of speaking engagements that didn't fit into the president's curtailed schedule, Spiro Agnew's replacement was expected to campaign extensively for Republican candidates in advance of the 1974 midterm elections, and to undertake such foreign travel as Nixon assigned him. Key policy advisers were available to brief

Ford as desired. These included Dr. Henry Kissinger, the era's preeminent diplomatist, fresh off his bureaucratic triumph in acquiring Secretary of State William Rogers's title to go with his existing one of national security advisor. Brilliant and brooding, with insecurities to match his formidable brainpower, Kissinger would appear to have little in common with the unassuming congressman from Grand Rapids. In fact the two men had enjoyed a cordial if distant relationship in the decade since Kissinger first invited Ford, then ranking member of the Defense Subcommittee of the House Committee on Appropriations, to address his Harvard seminar on US defense policy.

On that occasion the visiting lawmaker had impressed the professor and his students with his expertise and lack of pretense. Subsequent encounters only confirmed Kissinger in his positive opinion of Ford's judgment and work ethic.* Beginning in December 1973 either Kissinger or his National Security Council deputy, General Brent Scowcroft, briefed the new vice president on a weekly basis. Under Kissinger's tutelage Ford hosted several receptions at Blair House for members of the Washington diplomatic corps. Soviet ambassador Anatoly Dobrynin, as eager as anyone to size up the potential occupant of the Oval Office, reiterated to Ford his country's interest in securing favorable trade legislation and Middle East peace. The latter subject dominated a visit by Jordan's King Hussein. "What we need is three more Henry Kissingers to cover all the problems—Egypt, Syria and Jordan," Ford told the king, in whose honor the vice president hosted a state dinner.

Displaying a touch of ruthlessness notably lacking eight months later when he confronted Nixon staff members reluctant to vacate their White House jobs, Ford decided there was no place on his payroll for Agnew holdovers. "I want my own team," he explained. "I feel sorry for them, but we can't have anything to do with them." Roger Porter was one of several White House Fellows who expressed interest in working for the vice president. In his job interview Porter asked Ford what qualities he most valued in the workplace. He was surprised by Ford's response, a subtle analysis of how to balance individual strengths and weaknesses. Only later did Porter, a graduate of Brigham Young and Harvard, learn that he had been selected

---

* The latter was on occasion carried to extremes. Flying into Andrews Air Force Base late at night, Ford insisted on being met with enough paperwork so he wouldn't have time on his hands, but could sign letters or review mail during the twenty-five-minute drive to his home in Alexandria. TAI, Jack Marsh.

over the objections of Bob Hartmann, for whom his Ivy League connection was suspect. "No, he's from Utah and he can be trusted," said Ford.*

Given the friction between old and new regimes, conflict was unavoidable. Ford staffers were denied access to the White House mess. Parking spaces proved hard to come by. In marked contrast with Spiro Agnew, who had maintained an official air force JetStar and crew on permanent standby at nearby Andrews Air Force Base, his successor traveled in a variety of downscale aircraft. The original *Air Force Two* was a lumbering little Conair Prospect whose top speed of 330 miles per hour earned it ridicule as Slingshot Airlines. "It was a twin engine prop plane," says Phil Jones of CBS News, "that's all the Nixon White House would give him." Ford claimed it was the only plane in the air force that stopped for red lights.

There was more to the story, according to Tom DeFrank, then a twenty-eight-year-old *Newsweek* correspondent instructed by his New York editor "to live with him until he's president." Well aware that Ford's status entitled him to a four-engine Boeing 707, DeFrank asked the vice president, "Why do you fly this piddling little airplane?" To which Ford replied, deferring to the peripatetic secretary of state, "Well, Henry asked me if he could have that other airplane and I said sure, no problem." Sometimes the veep and his entourage would upgrade to a windowless C-135 tanker affectionately known as the Flying Sausage. Whatever their means of transport, veterans of the Ford road show retain fond memories of the experience. "We would get on first," recalls Jones, "and then the vice president and his staff would come on. Well, they had to pass us to get back to the vice president's cabin in the back. It was always banter back and forth, and we might have done something that he would needle us about, but it was fun." On one early trip, Ford ordered the plane held thirty minutes to accommodate a tardy reporter from *Time*.

That magazine's Bonnie Angelo and photographer David Kennerly were regulars on *Air Force Two*, along with DeFrank, representatives of the three television networks and Marjorie "Maggie" Hunter of the *New York Times*. Hunter, a particular Ford favorite, gave the vice president a copy of former LBJ press secretary George Reedy's *The Twilight of the Presidency*, which he read twice over. Taking to heart Reedy's warnings of presidential isolation amid an expanding and self-important White House staff, Ford

---

* Porter would go on to serve three Republican presidents in positions requiring discretion as well as judgment. He eventually returned to Harvard, where he taught a popular course on the American presidency.

recommended the book to Bob Hartmann and others around him. For now his schedulers, led by a former All-American basketball player from the University of Arizona named Warren Rustand, discouraged high school bands from greeting the vice president with ceremonial ruffles and flourishes. More to Ford's taste: the University of Michigan fight song "Hail to the Victors."

When in Washington, Ford's calendar reflected his vow to promote bipartisan consensus. He was conspicuous in presiding over the Senate, many of whose members he had previously worked with in the House. One night during his first week on the job he attended a dinner honoring Tip O'Neill; the next morning he breakfasted with banker David Rockefeller, synonymous with Manhattan's cultural and economic elite. Ford also met with the Commission on Critical Choices for Americans organized by David's brother Nelson, who chose that month to resign his governorship of New York to wage a final campaign for the presidency in 1976. Accepting Rocky's invitation to join the commission, Ford enjoyed listening to idea men like theoretical physicist Edward Teller, a latter-day Dr. Strangelove, as he expounded on the possibilities of controlling the weather for military advantage.

A few days before Christmas, Ford undertook a more traditional vice presidential function by flying to Madrid for the funeral of Spanish prime minister Luis Carrero Blanco, assassinated by Basque separatists in a brazen assault on the Franco dictatorship then entering its final chapter. Bowing to the energy crisis brought on by an Arab oil embargo, Ford abandoned his early morning swim in the heated pool at 514 Crown View Drive. But he had no intention of forgoing his family's annual Christmas ski trip to Vail. Before departing Washington for Colorado, Ford filled ten otherwise empty seats on his government plane with servicemen he transported home for the holidays at no charge.

In their Vail condo, the vice president and his loved ones observed the rituals of the season. Cameras followed them everywhere, from the ski slopes to their Christmas Eve opening of presents, the latter broadcast by Phil Jones and a CBS camera crew. The trip drew criticism from columnists Rowland Evans and Robert Novak, who wanted Ford to remain in the nation's capital through the holidays in order to project a "take charge" image. This was exactly the opposite of the impression he intended to create. Nearing the end of their Vail holiday, Ford accompanied a dozen family and friends to a local Chinese restaurant. When the inescapable round of fortune cookies was served, the vice president took one look at the message in his and slammed it down on the table. Then he hid it under his hand, refusing to reveal its contents until Dave Kennerly finally pried it out and read it to the entire party.

"You will undergo a change of residence in the near future."

"No, no, I hope not," said Ford, fearful even then, according to Kennerly, that people might think "he was trying to force Nixon out."

L OOKING BACK, FORD would describe his eight months in the vice presidency as "the worst job I ever had." It is not hard to see why. "Ford had to walk this terrible tightrope all the time," says Tom DeFrank. At the outset his defense of the president bordered on sycophancy. After all, the two men had a friendship dating back twenty-five years. "But as time went on, and as he would learn things . . . I think it slowly began to dawn on Ford that Nixon was lying to him." Friends on the Hill advised him to maintain his distance from the burgeoning scandal. As one trusted associate from the Warren Commission cautioned him, "The way to save the Presidency is not to jump in the middle of the quicksand." In practical terms this meant refusing to listen to the Nixon tapes or read transcripts, Ford concluded, "because it would inevitably suck me in."*

Adding to his discomfort was the faultless personal conduct of the president and First Lady. To Betty Ford, Pat Nixon wrote warmly of a friendship that went back twenty-five years: "You won the hearts of congressional wives then and will capture the hearts of the American people now." The president arranged for Betty's husband to participate in meetings of the Cabinet and National Security Council, to sit in on discussions with the congressional leadership and the economic policy group known as the Quadriad, led by Federal Reserve Board chairman Arthur Burns. As often as his onerous travel schedule permitted, Ford attended the weekly luncheon meeting of the Senate Republican Policy Committee.

Following Bob Hartmann's recommendation, Ford spent much of his time on the road. Among those who went along for the ride, no one made a greater impression on the vice president than David Kennerly. Brash, irreverent and extravagantly talented, the lensman who had anchored *Time's* Saigon Bureau at twenty-five became an instant favorite of Jerry and Betty

---

* Ford's disillusionment would have been far greater had assistant Watergate prosecutors Jill Wine-Banks and Richard Ben-Veniste acted on impulse at a black-tie gala also attended by Vice President Ford. They debated sharing with Ford some of the mounting evidence against Nixon. "And then we decided, no, because of Grand Jury secrecy it would be illegal for us and very inappropriate to share the information." They did, however, chat briefly with Ford, recalled Wine-Banks in 2014, "and I actually got to dance with the Vice President." Ken Gormley interview with Jill Wine-Banks, October 2, 2014, GRFL.

Ford, and consequently a target of envy from other journalists who resented his special status. Ford liked photographers ("They're the kind of people you can trust"). He indulged Kennerly's beard and blue jeans, his hyperactive social life and his public displays of flippancy after the vice president repeated once too often a feeble joke or stale anecdote from his limited repertoire. Mrs. Ford established a relationship with the photographer that was at once motherly and coconspiratorial, as demonstrated on the Sunday morning a plainly hungover Kennerly showed up at 514 Crown View Drive to do a photo layout of the Fords on their way to church.

Alerted to his coming, Betty Ford stood at the front door, a cold beer in her hand. "Drink this," she said, "you'll feel better."

Her own encounters with the fourth estate were a mixed bag. "I sympathize with the press," she confided. "After all, they're just trying to make a living." She held to this view even after a catty reporter quoted in *Women's Wear Daily* concluded, "you can tell she's on something. Her eyes are starry." As a matter of fact, the Second Lady of the land publicly acknowledged taking Valium "three times a day, or sometimes Equagesic. That way I'm more comfortable." She was decidedly uncomfortable when Barbara Walters, after agreeing to a nonpolitical interview, led off with an explosive question about the Supreme Court's decision legalizing abortion. Betty expressed frank admiration for the justices who had, in her words, taken abortion out of the backwoods and placed it in hospitals, where it belonged. A flood of angry letters anticipated controversies to come.

She was similarly candid when talk show host Dick Cavett and a television crew visited 514 Crown View Drive for an extended conversation with the vice presidential couple and their children who still lived at home (son Mike was enrolled at Gordon-Conwell Theological Seminary, north of Boston, while his brother Jack was studying forestry at Utah State University in Logan). Asked by Cavett how the family dealt with the prospect of moving into the White House, Betty said it was "inconceivable . . . none of us talk about it or think about it." The self-imposed ban carried over to the workplace. "We had an understanding, Ford, Hartmann and myself," says Jack Marsh, a former Democratic congressman from Virginia who served as Ford's assistant for Defense and National Security. "We would never discuss the status of the Watergate proceedings because all three of us felt that if we started doing it, invariably it would get out and be leaked and that would be the worst position Ford could be in. By us taking that position," Marsh continued, "every member of the vice president's staff was put on notice—'This is off limits. You're not to talk about it because Ford doesn't talk about it.'"

Nineteen seventy-five brought no relief from what *Time* called "the worst political scandal in U.S. history," notwithstanding President Nixon's plaintive assertion in his State of Union address that "one year of Watergate is enough." Appearing on NBC's *Meet the Press*, Ford said that he could envision a compromise between the White House and the Senate Watergate Committee over hundreds of presidential tapes being sought by the committee. His comments were immediately disowned by Nixon's spokesman, leaving the vice president to protest unconvincingly, "I am sure the President feels if I have a conviction, I should say it. He doesn't want me to be a rubber stamp." Whether or not Ford believed this, he chose to act as if it were true. On January 11 he convened a two-hour lunch meeting attended by leaders of the African-American community, including civil rights mainstays Roy Wilkins and the Reverend Jesse Jackson, plus Congresswoman Barbara Jordan. Separately he reached out to Black appointees within the Nixon Administration, assuring them, "My door will always be open." Ford's attitude contrasted with the seeming indifference of a president and staff too preoccupied with defecting friends to court traditional adversaries.

Viewing the vice president as a team player, Nixon's palace guard scheduled Ford for far-flung speeches, often without prior consultation. Then they put words in his mouth, attesting to the president's innocence and denying that his effectiveness had in any way been compromised. Burdened by administrative duties, Bob Hartmann could hardly give major speeches the attention they deserved. With so few staffers of his own to rely on, Ford naturally turned to White House wordsmiths for assistance. The drafts they prepared for him were frequently late. More than once he was already seated at the event's head table by the time a copy of his speech was placed in Ford's hands.

In a category by itself was his January 15, 1974, appearance before the American Farm Bureau Federation in Atlantic City. A draft of the vice president's remarks reached Hartmann barely twenty-four hours before its scheduled delivery. Hastily scanning the work of his counterparts on the other side of Executive Avenue, the short street that separates the EOB from the West Wing much as the East River divides Brooklyn from Manhattan, Hartmann was appalled. Only this time he had no one to blame but his boss. "I told 'em I wanted a real tough speech," acknowledged Ford, his ire roused by harsh anti-Nixon rhetoric employed by organized labor in its drive to impeach the president. With Ford's permission Hartmann watered down some of the more overheated language in the text. But enough remained to raise doubts about the vice president's vaunted independence.

Only days after insisting that he was perfectly free to speak his mind, Ford denounced Nixon's critics in slashing prose reminiscent of Charles Colson at his most strident. He accused a "small group of political activists" of exploiting Watergate to achieve "a total victory for themselves and the total defeat not only of President Nixon, but of the policies for which he stands." Their motive was transparent: "to cripple the President by dragging out the preliminaries to impeachment for as long as they can, and to use the whole affair for maximum political advantage."

The blowback was immediate and scorching. At a Grand Rapids reception on January 17, Ford shook hands with hundreds of admirers, many of whom stood for hours in a numbing cold to greet him. "It was amazing how many people he knew on a personal basis," recalls NBC correspondent and future White House press secretary Ron Nessen. More than faces in the crowd, they were friends to be addressed by their first names—like Ed, who worked in a local automobile plant, and Sally, mother of two, with a son enrolled in college. Many in the long line echoed the concerns of Cliff Gettings, Ford's high school football coach, who told reporters, "We don't think he should get involved in Watergate." This advice was reinforced by Phil Buchen. Ford had become "a most precious political commodity," his old friend insisted, and he must not diminish his value, or further divide the country, by approaching impeachment as he would a partisan political campaign. Buchen's judgment was sound, and Ford knew it.

A White House meeting on January 21 crystallized his doubts. Nixon took the occasion to deny any responsibility for a mysterious eighteen-and-a-half-minute gap in one of the tapes being sought by the special prosecutor. Pointedly, Ford declined an offer from the president to review selectively edited transcripts that were said to disprove John Dean's claims of presidential complicity. In a rambling monologue that consumed the better part of two hours, Nixon revisited past campaigns and controversies as Ford stole furtive glances at his watch. He couldn't just get up and plead business elsewhere. Only gradually did it register—he was being used as an escape valve by a president for whom even a captive audience was better than none at all. What might have been tiresome in another context was oddly humanizing.

In the last week of January, Ford hit the campaign trail in Western Pennsylvania, where the GOP was struggling to retain a House seat following the death of longtime incumbent Joseph Saylor. On election day Democrat John Murtha scored a major upset, snatching away a district that had been in Republican hands since the Truman Administration. It was the first of several special elections testing the appeal of a greatly diminished president and his party. On February 18, voters in West Michigan went to the polls to choose Ford's successor. "You can't mean that," he stammered on hearing the news that Richard Vander Veen, the same sacrificial lamb defeated by Ford in 1958, had carried the district, the first Democrat to do so since 1912. Hoping to minimize the impact, the White House attributed the result to local unhappiness over the economy. "No, Mr. President, it is Watergate that is responsible, at least in part," said Ford.

On March 1 a Washington grand jury indicted Haldeman, Ehrlichman, Colson, Mitchell and three lesser-known accomplices for their role in the Watergate cover-up. (That the same jurors also named Nixon as an unindicted coconspirator remained under wraps until the first week of June, a stunning development whose eventual impact would be all the greater for its delayed announcement.) Four days later Democrats prevailed in Ohio's First Congressional District, yet another traditional GOP stronghold. A Gallup poll that spring found just 24 percent of the electorate identified as Republicans, the lowest share since 1940 (paradoxically, the number of self-identified conservatives rose to 38 percent, an all-time high). Speaking to the Harvard Republican Club on March 11, Ford tempered his defense of the president. "I don't happen to believe on the basis of the evidence I am familiar with—and I think I'm familiar with most of it—that the President was involved in Watergate *per se* or involved in the cover up, but time will tell."[*]

Within days of his Harvard appearance, Ford received some contradictory counsel from the president. Releasing him from his vow not to seek the presidency in 1976, Nixon urged Ford to concentrate on his current job and not engage in any self-promotion or delegate harvesting. In response Ford assured the president he had every intention of retiring when he said he would, at the end of Nixon's second term. Until then, however, he would exercise caution in taking the administration line. With the Atlantic City debacle fresh in memory, Ford sought redemption before the Midwest Republican Leadership Conference at the end of March. This time the vice

---

* The author chanced to introduce Ford on this occasion as "the next president of the United States," a phrase he no doubt heard many times that spring and summer. I later observed that the vice president's visit was a learning experience for all involved. Republican Club members were frankly surprised that Richard Nixon's vice president would venture so deep into enemy territory, while Ford was surprised to find there were enough Republicans at Harvard to form a club.

president and Bob Hartmann wrote the speech themselves, with no outside involvement.

It is easy to see why neither man ran the text by the White House in advance of delivery. Addressing a room full of party activists, Ford said it would be "improper" to speculate "on the criminal and legal aspects of this sorry episode." And yet one conclusion was unavoidable. "Never again must Americans allow an arrogant, elite guard of political adolescents like CREEP to bypass the regular party organizations and dictate the terms of a national election." Long-dammed resentments were loosed as Ford denounced the grudging support extended to Republicans down ballot by a reelection committee as selfish as it was unethical. "The fatal defect of CREEP was that it made its own rules and thus made its own ruin," he said to a crescendo of applause. "It violated the historic concept of the two-party system in America . . . if there are any more cliques of ambitious amateurs who want to run political campaigns, let the Democrats have them next time."

For Ford it was a liberating moment. Henceforth he would rely on speechwriters of his choosing, beginning with Milton Friedman (no relation to the famed economist), a former Washington correspondent for the Jewish Telegraphic Agency and press secretary to New York senator Jacob Javits. As someone who had done occasional work for Ford before Zionist audiences, the self-effacing Friedman was a welcome addition to an over-matched vice presidential staff staggering under the weight of nine hundred speaking invitations a week. Simultaneously Bob Hartmann was relieved of his administrative duties. In his place Ford turned to William Seidman, a hugely successful Grand Rapids businessman and accountant with a taste for politics and close ties to George Romney. Acting on the advice of Republican national chairman George H. W. Bush, Ford employed Gwen Anderson, GOP committeewoman from Washington State, as his liaison to the party and other groups outside government.

With each additional recruit, Ford seemed to grow in confidence. On occasion he even joked about his admittedly awkward status. Together with Betty he attended a tango party at the Argentine embassy in Washington. "Will you dance in the White House?" a reporter asked him on impulse.

"If I'm invited," Ford shot back.

He was not always so verbally agile. His growing doubts about White House truth-telling fed tensions between his camp and those around Nixon who resented his zigzag defense of the president. "I think he went as far as he felt he could," says Roger Porter. "At no point did he signal that he was disloyal in any way. But he had to have at least in the back of his mind the idea that it could happen." Porter didn't have to spell out the meaning of "it." At the start of April, Ford's son Jack won headlines by expressing his personal disillusionment with Nixon, for whom he had vigorously campaigned in 1972. What's more, said Jack, "I'm not so sure my father disagrees with me."

His remarks could be explained away as youthful indiscretion. One could hardly make the same argument after the vice president a few days later gave a not-for-attribution interview to John Osborne of the *New Republic*, in which he shared thoughts of what a Ford presidency might look like. His unflattering assessments of White House press secretary Ron Ziegler and Secretary of Defense James Schlesinger were offset by speculation on those members of the Nixon Cabinet he would retain, a list topped by Henry Kissinger, George Shultz and Interior Secretary Rogers Morton. Hiding behind, he thought, a cloak of anonymity Ford voiced sympathy for Al Haig, a capable manager forced to fritter away time indulging the president's seemingly endless chatter.

Osborne's article appeared as the Fords were enjoying an Easter holiday at Sunnylands, the Palm Springs estate of billionaire publisher Walter Annenberg. No one was quicker to pounce on the vice president's misplaced candor than William Safire, the former Nixon speechwriter who had reinvented himself as a *New York Times* columnist. Safire got Ford on the phone and asked him if he was in fact Osborne's unidentified source. Acting on Hartmann's advice, the vice president confessed to his involvement. The title of Safire's published rejoinder—*Et tu, Gerry?*—said it all. In Palm Springs, Ford maintained an office in the International Hotel. That is where he greeted Tom DeFrank, one of the original travelers on Slingshot Airlines, who had rejoined the Ford road show following the birth of his son. A solicitous Ford expressed relief that all was well following a difficult pregnancy.

After a while, DeFrank got up to leave.

"Put your notebook away. I want to show you something."

Ford reached for his copy of the Safire column. "Have you seen this?"

"Yes. I just read it."

Reluctant to offer a substantive opinion on another scribe's work, DeFrank noted only that the *Times* had misspelled Ford's name. Ford would not be put off so easily.

"I want to know what you think about this. Why did Safire do this?"

DeFrank, reluctant to fault the vice president's poor judgment in saying the things he had said to Osborne, held back. "Why has he done this?" Ford repeated. "Bill Safire knows that I've been damned loyal to Dick

Nixon." Now it was DeFrank's turn to speak frankly, as an agitated vice president, for whom loyalty was a cardinal virtue, sought absolution from one of Safire's tribe.

"Well, Mr. Vice President, it's very simple. They know Nixon is finished. They know he's a goner. They know he can't survive and they know sooner or later—probably sooner—you're going to be president."

"You're right," Ford blurted in response. "But when the pages of history are written, no one will ever be able to say that Jerry Ford contributed to it."

Suddenly realizing the impact of his words, Ford came around the table separating the two men. "You didn't hear that," he barked.

"But I did."

Fearing that his budding journalistic career might be in jeopardy, De-Frank's nerves were not soothed as the looming vice president—"He had about eight inches on me"—reached out and grabbed the reporter by his tie.

"Damn it, Tom, you're not leaving this office until we have some under-standing." DeFrank froze, speechless with fright. "Write it when I'm dead," Ford told him. "You can write it when I'm dead."

"OK."

Ford extended his hand, and the two men made peace. Reverting to form, a genial vice president remarked, "Nice to see you, Tom. I'm glad your wife is doing okay . . . Coming with us to Monterey tomorrow?"

Not long after this brush with disaster the press contingent on *Air Force Two* presented Ford with the picture of a donkey caught trying to leap a fence, its ungainly posture perfectly captured in the accompanying caption: "You're damned if you do and damned if you don't."

O N **APRIL 29,** Richard Nixon released 1,254 pages of Oval Office transcripts edited by the White House in hopes of blunting the mo-mentum for impeachment. The ultimate document dump, the Nixon papers were notable for what they left out: eleven of the forty-two tapes requested by the House Judiciary Committee were reportedly missing or nonexistent. What remained was full of holes, nearly two thousand in all, monotonously plugged with the soon-to-be catchphrases "inaudible" and "expletive deleted."

"I realize these transcripts will provide grist for many sensational sto-ries," Nixon predicted in televised remarks accompanying their release. This was an understatement. Publication of even a bowdlerized Nixon sent shock waves through millions of homes unprepared for the crude ethnic slurs and scatological exchanges profaning the Oval Office. Indecisive and vengeful, this Nixon was nothing like the bold history maker who had single-handedly ended China's isolation while cobbling together a new political consensus at home. Decrying the president's "moral blind spot," William Randolph Hearst Jr. wrote in his weekly column, "God knows what the unexpurgated tapes would show." Calls for impeachment or resignation came from normally sympathetic journals like the *Chicago Tribune, Omaha World-Herald, Miami Herald* and *Kansas City Times*.

Ford, as surprised as anyone by the brooding misanthrope whose bitter tongue bore scant resemblance to his friend of a quarter century, initially tried to defend Nixon, stressing his offer to let Judiciary Committee chairman Peter Rodino and his Republican counterpart William Hutchinson listen to the tapes for themselves. As the full impact of the president's words made itself felt, however, even Ford changed his tune. Reading Nixon transcripts en route to a May 4 speech at the University of Michigan, he was reportedly "aghast" at the language they contained. Conceding that the published transcripts "didn't exactly convey sainthood on anybody," Ford told a student audience at Eastern Illinois University that Washington had been engulfed by a "crisis of confidence—a continuous series of revelations and reports of corruption, malfeasance and wrongdoing in the federal government—not the least of which is the sorry mess which carries the label of Watergate."

On May 9, 1974, the House Judiciary Committee began formal impeachment hearings. Awash in what one Ford aide called "unidentified flying rumors," members of his Washington staff denied that the vice president had asked for television airtime to denounce Nixon, or that he was joining his former House colleagues in calling on the president to resign. On the morning of May 10, Ford was abruptly summoned to join the president in his EOB hideaway. What passed between the two men during their fifty-six minutes together is unknown, although Ford recalled Nixon telling him, "Jerry, you're doing too much. You've got to slow down your pace."

He failed to take the hint. The very next day, in Dallas, reporters asked if there had been any discussion about a possible transfer of power. "None whatsoever," he replied. What about his staff? Was anyone on the vice president's payroll engaged in transition planning? Ford rejected the idea. Referencing an article in the Knight newspaper chain claiming "that somebody on my staff was working on something like that," Ford volunteered, "If they are, they are doing so without my knowledge and without my consent."

FORD'S NONDENIAL DENIAL was traceable to a seemingly unrelated presidential request. Early in 1974 Nixon asked his vice president to oversee a Domestic Policy Council examination of privacy rights and the threats posed by massive, impersonal computer data banks. As a White House priority it bordered on make-work, but Ford took the request seriously enough to recruit his onetime law partner and most trusted adviser to serve as the group's executive director. "Jerry Ford wanted Phil Buchen in this town because he did not know what was going to happen, but he wanted somebody to be the eyes and ears for Jerry Ford around town." So claims Brian P. Lamb, a familiar face to millions of cable TV viewers since 1979, when he founded C-SPAN to initiate a dialogue between Americans and their government. Prior to that, Lamb's résumé included service in the navy, disc jockeying in his native Indiana and stints as a radio news reporter, Capitol Hill press secretary and White House military aide during the Johnson years. More recently, he had handled congressional relations for Clay T. Whitehead—Tom to friends and coworkers—in the White House Office of Telecommunications Policy (OTP).

A media revolutionary, Whitehead was one of those inner directed people whose entire life seems foreordained in purpose. Growing up in Southeast Kansas, Tom saved $90 from his work as a bicycle delivery boy. With this he bought a ham radio kit, then strapped a fifteen-foot antenna to his parents' chimney. On long winter nights he listened spellbound to stations as distant as New York and Nashville. Not that he neglected his schoolwork. When his son scored the highest IQ ranking in the state, Tom's father mortgaged his farm so that the boy could attend MIT, where he earned two degrees in electrical engineering, and a PhD in management. "I love the act of creating things," Whitehead enthused. As Richard Nixon's telecommunications czar, he changed the landscape of American television by introducing competition to satellite service and blocking attempts by the Federal Communications Commission to promote COMSAT as a domestic monopoly.

In the first week of May 1974, Whitehead was ensconced in OTP's quarters at 1800 G Street, midway between the White House and the State Department. Three days a week Phil Buchen occupied an adjoining office as Ford's handpicked director of the Committee on the Right to Privacy. At fifty-eight, Buchen was a full generation older than Whitehead, on whom he relied for guidance in navigating the minefield of official Washington. Such was their mutual trust that Buchen confided to the younger man his fears about the escalating Watergate scandal and its implications for his friend the vice president. "We have to do some planning for Jerry," said Buchen. "We may have to face the fact the president may resign." Understandably reluctant to comply, Whitehead's caution was no greater than

Ford's own. Fearful of the consequences should millions of Richard Nixon's admirers believe their man the victim of a coup, the vice president steered clear of anything that smacked of disloyalty.

Ford always claimed ignorance of Buchen's initiative, a detachment that others found hard to swallow. "My understanding is that it was Buchen who went to the Vice President and suggested to him that he ought to do some planning," according to Roger Porter. It is a view seconded by the surviving members of the ad hoc committee Whitehead's wife, Margaret, wryly dubbed the Ford Foundation. "My clear impression is that Ford was aware of what Buchen was up to," says Laurence Lynn Jr., the group member with the greatest government experience. "I'm not sure I . . . would have signed on if we didn't think Ford knew."

More fluent than Buchen in the euphemistic patois of Washington, Whitehead put his own spin on the vice president's public disavowal of transition planning. "We decided the statement was an implicit one," he explained. In his translation, Ford meant to say, "I hope someone out there is doing some planning, but please don't tell me about it." This was sanction enough for Whitehead. As a result the first chapter of the Ford presidency was written over Tom and Margaret Whitehead's dinner table on the evening of May 7, 1974. Together with their guest Phil Buchen, they had a candid discussion of Nixon's fading prospects to beat impeachment, and the corresponding need to prepare Gerald Ford for whatever might be his fate.

To assist him in this clandestine enterprise Whitehead recruited three contemporaries, each well versed in the intricacies of government yet sufficiently obscure to fly under the radar. Besides Brian Lamb, then editing an industry newsletter called *The Media Report*, there was Jonathan Moore, long associated with Elliot Richardson, and before that a precocious veteran of the State Department, the Capitol Hill office of Massachusetts senator Leverett Saltonstall and presidential politics as practiced by George Romney and Nelson Rockefeller. In the spring of 1974 Moore occupied a comfortable niche as director of Harvard's Institute of Politics. Whitehead's shadowy transition team offered him a unique opportunity to test his theories about public policy and government organization.

Likewise with Laurence Lynn Jr., Moore's former colleague at the department of Health, Education, and Welfare, and a Whitehead friend since the two men worked differing sides of the transition from Johnson to Nixon. Not yet forty, this graduate of Berkeley and Yale had already compiled an impressive record at the National Security Council and as an assistant secretary in the Departments of Defense, the Interior and HEW.

He was the sole Democrat in the quartet, which met four times during June and July in the un-air-conditioned Whitehead townhouse at Twenty-Eighth and N Streets. Doffing their coats in the stifling heat, the ad hoc planning group gulped down soft drinks while dissecting the structure of Richard Nixon's government amid bluish clouds generated by two pipes and one cigar smoker.

Unable to communicate with the would-be president from whom they derived their tenuous claim to legitimacy, Whitehead and his brain trust could only speculate as to future events and potential replacements for the existing Cabinet and White House staff. A consensus formed rapidly that Henry Kissinger should surrender one of his titles so that the National Security Council might reassert its historic independence of the State Department. The suggestion first made by Brian Lamb that White House press secretary Ron Ziegler be replaced by Jerald terHorst, Washington bureau chief for the *Detroit News*, met with unanimous approval.

The group's members blocked out a First Week for Ford that was heavy on reassurance, not as an endorsement of the status quo, but as part of what Whitehead called "a longer range transition from something profoundly bad in government to something better." Here he posed the defining test of Gerald Ford's first months in the presidency—how to make the office credibly *his*, even while retaining much of "the philosophy, personnel, and staff procedures" of his predecessor. "There will be great pressure from without to do this too soon," wrote Whitehead. "There will be greater pressure from within not to do it at all."

As for Ford, talk of such contingencies was easily avoided thanks to a travel schedule bordering on the manic. During his first six months in office the vice president covered 80,000 miles, visiting 30 states and making 375 speeches and appearances. In the month of May alone he was on the road twenty-eight days, few of which lasted less than sixteen hours. At one point the accompanying press corps sent Ford a three-word memo. "Subject: Complete Exhaustion."

At midmonth they got something of a respite in the form of a three-day stopover in Hawaii. Besides the inevitable speeches and grip and grin receptions, there was time to enjoy the verdant surroundings. Guests of Honolulu's Mauna Kea Hotel were astonished to find the vice president in the Pacific surf, outdistancing Secret Service agents half his age and shaking hands with other swimmers. Back on land Ford relaxed security procedures, inviting startled civilians to squeeze into the same elevator as the vice presidential party and putting an end to the rough treatment administered by some holdovers from Spiro Agnew's protective team. "I had a chat with the

Secret Service," Ford told Bonnie Angelo of *Time*. "I told them I wanted them to be as considerate of other people as they are of me."

RICHARD NIXON ENJOYED a somewhat more traditional relationship with his protective agents. As spring overtook Washington, the beleaguered president increasingly found refuge in his retreats at San Clemente or Key Biscayne. Alone but for his friend Bebe Rebozo, Nixon would direct agents, "Bring the car around, Bebe and I are going for a ride." Did he have a designation in mind? agent Richard Keiser inquired.

"I just want to be back in an hour."

Keiser recognized the routine—"About thirty minutes in one direction, turn around and come back." Sometimes Nixon and Rebozo would carry on a conversation; just as frequently the hour would pass in silence. Or the president might call out from the back seat: "Keiser"—Nixon rarely employed first names—"Get the music on the radio. You know what I like." The agent readily complied. But Nixon wasn't through giving orders.

"That's good music . . . when the news comes on, you turn that radio off. Do you hear me?"

"Yes sir."

"I mean it, not one word."

Nixon had good reason to avoid the news. His vice president was publicly urging compromise with Special Prosecutor Leon Jaworski over Jaworski's demand for sixty-four additional White House tapes. Nixon had no intention of surrendering the tapes, and with them the principle of executive privilege—the hard-won, if ill-defined, right to confidentiality first asserted by George Washington. When Judge John Sirica on May 9 ruled in favor of Jaworski, there was never any doubt that the administration would appeal his decision to the Supreme Court, four of whose nine members were Nixon appointees. It was against this backdrop that simmering tensions at the staff level boiled over.

Of the vice president one White House loyalist complained, "He's got a congressional point of view about releasing materials."

"He's got to live with the president," retorted Bob Hartmann. "But he's also got to live with his own friends on the Hill. If he doesn't tell the president what they are thinking, then he's going to lose his clout with the Congress." At a White House meeting on May 23, Ford questioned the administration's refusal to comply with a congressional subpoena: "It's a bad call, Mr. President, and it complicates the whole situation" by undercutting efforts to defend Nixon on Capitol Hill, and sowing fresh doubts among the public and the press.

Standing in for the president at the annual White House Correspondents' Dinner, Ford issued a droll apology for his recent criticism of reporters who had the habit of "putting the worst foot forward." He asked the assembled journalists to replace this unfortunate phrase with "unintelligible, inaudible, or, if you must, (expletive deleted)." The dinner guests responded with a standing ovation. A vigorous dissent was registered by Richard Reeves in *New York Magazine*. Lamenting Washington's seeming preference for honesty and decency over IQ or "pedal dexterity," Reeves decried Ford as "small gauge, very partisan, obsessively loyal, reflexively conservative, but essentially good. Unfortunately the thrill of high goodness is beginning to wear off and the rest of Ford is beginning to rattle." His complaint was picked up by columnist Mary McGrory. "The vice president should go underground," she wrote after developing whiplash from Ford's on-again, off-again defense of his embattled chief.

Wherever he went, Ford insisted that nothing in the president's conduct warranted his removal from office. Yet his protestations had a grudging quality. Yielding to White House pressure to participate in a pro-Nixon shivaree organized by Rabbi Baruch Korff's National Committee for Fairness to the Presidency, Ford made sure to arrive late, depart early and say nothing to inflame passions or defame the other side.

His measured tone bred resentment among some of the president's defenders, though few were as harsh as New Hampshire's *Manchester Union Leader* in declaring, "Jerry is a jerk." Nixon himself was forced to deny to his vice president a *Newsweek* story that he had welcomed Nelson Rockefeller to the Oval Office by pointing to the chair behind his desk and asking, "Can you see Jerry Ford sitting in this chair?" Nor was this the only embarrassment inflicted on Ford. Though his official salary of $62,500 exceeded his pay as minority leader, a ban on paid speeches reduced his annual income by perhaps $20,000. When his friend Bill Whyte, a vice president of U.S. Steel, arranged the aforementioned Hawaiian resort stay, there was never any question about Ford paying his fair share of the tab, amounting to $500. "I hope you don't mind till the end of the month when I give you a check," an apologetic vice president told Whyte.

Where public expenditures were concerned he remained frugal as ever. Since early in the century sporadic efforts had been made to secure an official Washington residence for the nation's vice presidents. With Ford's ascension to the office, Senator Robert Griffin spearheaded legislation seizing for this purpose the recently vacated quarters of the chief of naval operations. Located on the grounds of the Naval Observatory in Northwest Washington, the turreted three-story white brick structure dated to 1893.

Though badly in need of renovation, "Ford said to me that we can't let this cost any money," according to Jack Marsh. "Well, it happens that once you make it the Office of the Vice President of the United States, you immediately have security concerns that you don't have with the Chief of Naval Operations."

Marsh outlined a three-stage plan, with an initial expenditure requested by the Secret Service totaling $33,000. Ford frowned but said nothing. Marsh moved on to issues of interior design—"Painting and drapes and stuff like that . . . By now he's really having gas pains." What about the third stage? Ford demanded. "How much is that?"

"That's $450,000," replied Marsh. Ford removed the briefing book from his friend's hands.

"No, we can't do this, Jack. You go tell Betty."

Eventually the money would be found, and Betty Ford was enlisted in finding china and wallpaper for a house she would never call home. From a personal standpoint, this was just as well. Betty would much rather stay on Crown View Lane, where two stewards from the navy mess relieved her of most kitchen chores and Susan babysat for the Abbruzzese children across the street. Yet for all her attachment to familiar places, friends and family detected in Mrs. Ford changes reflecting her newfound responsibilities. Having dieted off thirty pounds, she received the ultimate compliment from her son Jack.

"Gee Mom, you've gotten good-looking."

"There's no new Betty Ford," countered her close friend Abigail McCarthy. "It's the real Betty Ford standing up . . . When Jerry was Minority Leader, she was just extra baggage. Now she's finding she's a personage in her own right." Among other things, her new role allowed Betty to indulge her interest in the arts. In April 1974 she traveled to Atlanta at the invitation of Rosalynn Carter, the wife of Georgia governor Jimmy Carter. The two women launched Art Train, a museum on wheels transporting works of Picasso, Matisse, Remington and Warhol on a six-month tour of the Southeast. As they posed for photographers, an impatient Governor Carter muttered under his breath, "Do you ever become accustomed to this?" Betty replied, sotto voce, "Twenty-five years, twenty-five years, twenty-five years . . ."

She and the governor were to meet again ten weeks later, under more somber circumstances. On Sunday morning, June 30, a twenty-three-year-old college dropout with a professed hatred for Christians walked into Atlanta's Ebenezer Baptist Church. Intent on killing the Reverend Martin Luther King Sr.—Daddy King—the assassin changed his mind when a much closer target presented itself. At the church's newly acquired organ sat Alberta King, the pastor's wife and mother of the slain civil rights leader whose funeral had taken place in the same violated sanctuary. She had just completed playing the Lord's Prayer when the gunman opened fire, killing her and another parishioner before escaping through a side door of the church. He was quickly apprehended. In Washington, the news of Mrs. King's death shocked the Fords. With the vice president summoned to Maine to greet Nixon on his return from an arms control summit in Moscow, Betty volunteered to represent the administration in Atlanta. "Go on down," Ford told her. "Somebody ought to be there." On her arrival in the Georgia capital Mrs. Ford went to the yellow brick house where the Kings had lived since 1941. There an inconsolable Daddy King took her in his arms and wept.

Two days later Betty returned to a church in the Baltimore suburbs for the wedding of her eldest son, Mike, to Gayle Brumbaugh, a vivacious classmate at Wake Forest who shared Mike's spiritual interests. Determined to avoid the limelight that surrounded her new in-laws, Gayle Ford worked in a bank under an assumed name. Susan Ford had no such option. Much to her father's distress, and to her own exasperation, the Secret Service assigned guards to protect the sixteen-year-old after her name turned up on a list of potential targets compiled by the same Symbionese Liberation Army that had kidnapped newspaper heiress Patricia Hearst the previous February.

As Washington baked in the summer heat, the Fords may have been one of the few families in America whose casual conversation excluded discussion of Watergate. Appearing on William F. Buckley's *Firing Line* on June 28, Ford said there were not then sufficient votes in the House to impeach the president. But what if Nixon were to defy a Supreme Court decision requiring him to yield his tapes to the special prosecutor?

"Then it's a totally different ballgame," he told Buckley. "It then becomes an institutional conflict."

On July 8 Nixon and Ford met in the Oval Office for seventy-five minutes, their conversation revolving around domestic issues and the recently concluded Moscow summit with Soviet leader Leonid Brezhnev. The two men met again five days later at the president's San Clemente retreat. This time their talks were "99 and 9 tenths" about the economy, said Ford. Responding to reporters' inquiries, the vice president rejected press secretary Ron Ziegler's characterization of the House Judiciary Committee as a "kangaroo court." In mid-July Ford returned briefly to Grand Rapids, where he met privately with an unofficial kitchen cabinet that included Bill Gill, news director of WOOD-TV.

"I have a feeling that he's going to be impeached," Ford confided to the group. "I just don't know what to do. I'm supporting him, yet I wonder where we're going with this thing."

"You've got to back off," Gill responded. "You've got to maintain your credibility, or you won't be able to handle it once you become president."

"Nobody else will support him," Ford protested. "Nobody else is behind him."

"Doesn't that tell you something?"

Perhaps influenced by such counsel, Ford said he would not twist arms, even among his fellow Republicans, in an effort to sway Judiciary Committee members to the president's side. He made an exception when Representative Lawrence Hogan, locked in a tight race for governor of heavily Democratic Maryland, called a press conference on July 23 to announce his support for impeachment.

"Larry, I think you're making a mistake* but you're your own man and I just called you as a friend . . . trying to give you some advice that I think is sound."

Overnight the headlines generated by Hogan's press conference were superseded by rumors of an imminent Supreme Court decision on Nixon's tapes. Ford's schedule for Wednesday, July 24, was typically crowded. In an interview with the *Christian Science Monitor* Ford gave his by-now-standard response when asked if he was ready to be president. "I think I'm ready for whatever might happen. And I think the best way to be ready for anything is to do the best job I can as Vice President." Moments after the *Monitor* reporter was ushered out of Ford's office the justices were heard from. By unanimous vote they rejected the president's claims of executive privilege. Rubbing salt into the wound, the court's opinion was written by Chief Justice Warren Burger, handpicked by Nixon to steer his brethren away from judicial activism of the Earl Warren school.

"The decision was not what we expected it to be," recalls White House staff secretary Jerry Jones. "I think most of us thought we would win the case." The shock of defeat was still fresh when Jones received a phone call from White House communications director Ken Clawson, like Al Haig following events from Nixon's Western White House.

---

* Ford's advice was borne out ten weeks later, when Hogan unexpectedly lost the Republican primary to GOP national committeewoman Louise Gore, a Nixon loyalist.

"Jerry, you think it is possible that the president could stiff the Supreme Court on this decision?"

As it happened, Clawson's inquiry reached Jones in the office of Bill Timmons, the president's chief emissary to Capitol Hill. One of the best headcounters in the business, Timmons had just completed his latest poll of Nixon supporters in the Senate. On the basis of its depressing results, Jones told Clawson that Nixon had no choice but to obey the Supreme Court.

Simultaneously Haig reached out to Robert Bork at the Justice Department.

"We're thinking about not obeying the order," said Haig.

"That is instant impeachment if you do that."

While these and other soundings were being taken, Ford was in the Executive Office Building, trapped in a predicament of his own making. Outside were Phil Jones and a CBS camera crew granted permission to follow him through a typical day in the life of the vice president. The timing of the Court's ruling called for all the guile Ford was said to lack if the TV crew's attention was to be redirected away from camera-shy presidential counselor Dean Burch, who was impatiently waiting to discuss the latest developments with the vice president. Luckily for both men, Ford's deputy press secretary, Bill Roberts, was able to lure Jones and his crew a mile away to the Capitol by promising them images of Ford arriving at his *other* office (only with difficulty was Jones dissuaded from pursuing Ford into his weekly prayer meeting with Mel Laird, John Rhodes and Al Quie).

Adding to the surreal atmosphere, Ford had committed to an off-the-record lunch with Katharine Graham and her editors at the *Washington Post*. Here he succeeded by making no news, though his parting words signaled changes to come. "Before I leave, I want you to know one thing," Ford told the assembled journalists. "I'm a person who likes to have adversaries who are not enemies." Hours passed, and still there was no comment from the West Coast. Could Nixon really be entertaining thoughts of non-compliance?

Back at his EOB office, Ford stuck doggedly to routine business. He was being photographed with Miss Virgin Islands when Fred Buzhardt, acting on presidential orders, checked out a potentially troublesome tape recording made on June 23, 1972. Only twice before had the tape in question been removed from the EOB storage vault; both times it was played exclusively for Richard Nixon. After listening to its contents a second time on May 6, Nixon had ruled the tape off-limits to his lawyers or anyone else.

Now this same recording singled out by the president was hand-carried by Jerry Jones to Buzhardt's office in the EOB. "Fred usually kept the tapes to the end of the day and I'd gather them all up," Jones would recall long afterward. "That day he called me back in, oh, [thirty, forty] minutes . . . met me at the door of his office, handed me the tape and he said, 'Jerry, it's all over.'

"So he knew," Jones concluded. "He knew."

15

"MY GOD, THIS IS GOING TO CHANGE
OUR WHOLE LIFE"

———

So here you have an imminent resignation of the president of the United
States and . . . the vice president has to act like it's not happening.

—JACK MARSH

ON JULY 24, 1974, the Supreme Court upheld the principle of executive privilege while overruling its practice in matters of criminal investigation. That same day thirty-eight members of the House Judiciary Committee began six days of televised debate over possible articles of impeachment. On the twenty-fifth, President Nixon appeared before the California Chamber of Commerce in Los Angeles to discuss "the major problem confronting America—inflation." A continent away, Nixon's vice president attended a breakfast meeting at the State Department hosted by Secretary Kissinger. Presumably he did not overhear Kissinger telling another guest, Federal Reserve chairman Arthur Burns, that "we have no government at present." Either Congress got on with the business of impeaching a president he claimed to "despise," the secretary warned Burns, or his own resignation was inevitable.*

Ford observed a different code of loyalty. That afternoon he left for a three-day campaign swing on behalf of endangered Republicans in the Midwestern heartland. His first stop: Muncie, Indiana, home base of Representative David Dennis, an ardent Nixon supporter on the Judiciary Committee. Recycling earlier claims of presidential innocence, the vice

---

* At the same breakfast Kissinger, wearing his second hat as national security advisor, confided to Burns that Al Haig was spying on him through the National Security Council. This did not prevent the secretary from approaching Haig the next day, July 26, with an earnest appeal that they work together "to end the agony . . . and to bring about a smooth transition." How they might affect this result was left purposefully vague, since Kissinger regarded it as the responsibility of senior elected officials in the president's own party to convince Nixon that the national interest required his early departure.

president warned of dire economic consequences should Nixon be driven from office. Even now Ford had no trouble rationalizing—or localizing—his defense of colleagues tainted with the wrongdoing of others. Dennis was typical of beleaguered GOP incumbents in that Watergate summer. Embroiled in a scandal not of their making, buffeted by fresh disclosures and plummeting polls, they had no one to turn to but Good Old Jerry. Ford, more than obliging, was in Canton, Ohio, stumping for GOP congressman Ralph Regula when the House Judiciary Committee on Saturday, July 27, voted 27 to 11 to impeach the president for his alleged obstruction of justice.

Six of the minority party's seventeen members voted to make Richard Nixon only the second American president to face a Senate trial and possible removal from office. Expressing surprise at the margin of victory, Ford chose to read sectarian bias into Democratic unanimity while minimizing the break in Republican ranks. By the time the Judiciary Committee resumed its work the following Monday, July 29, Ford was preparing to take the stage at the San Francisco convention of the National Urban League. During his brief stay in the city, Committee members approved a catchall second article of impeachment alleging presidential abuse of power. The next day Chairman Rodino and his colleagues adopted a third article, charging the president with contempt of Congress over his refusal to comply with its subpoenas for White House tapes.*

With his prospects for survival looking increasingly grim, even some presidential allies were beginning to ask themselves whether Nixon might be persuaded to resign rather than suffer a humiliating loss in the Senate. This was the course favored by White House counsel Fred Buzhardt, now joined by James St. Clair, the president's Boston lawyer whose belated exposure to the smoking-gun conversations from June 23, 1972—contrary to the popular memory of a single, fatal exchange, Nixon spoke with Haldeman three times that day—had wrenched him into an awareness of his own potential vulnerability. After hearing Nixon's explicit instructions to Haldeman on how to thwart an FBI investigation by asserting spurious CIA claims of national security, St. Clair insisted that all subpoenaed tapes be turned over immediately to Judge Sirica. Beyond this, he proposed that he and Buzhardt make a joint appeal to the president, stressing the hopelessness of his case. Nixon, however, refused to see his lawyers.

---

* Two other articles—dealing with Nixon's taxes and publicly funded home improvements and the secret bombing of Cambodia—failed on identical votes of 26 to 12.

Ford had by now moved on to San Diego, where on July 30 he received a warm welcome from the National Junior Chamber of Commerce. When the phone rang in Ford's suite at the Town and Country Resort it was picked up by the vice president's close friend Leon Parma, a local business-man and onetime San Diego State quarterback. Parma recognized the voice of Al Haig on the other end of the line. Ford was in the shower, Parma explained, getting ready for a reception and dinner benefiting Republican congressman Clair Burgener. Did General Haig need to speak to him im-mediately?

"As soon as he gets out, have him call me."

Moments later, Ford called back. He told Haig that he planned on leav-ing immediately after the fundraiser. By flying through the night, he could be back in his Washington office at eight o'clock Wednesday morning. He wouldn't be there long, having accepted an invitation to play that afternoon in a Worcester, Massachusetts, pro-am golf tournament with House Major-ity Leader Tip O'Neill and GOP whip Les Arends. Whatever it was, Haig's business could wait until the vice president's return from Worcester.

Ford hung up the receiver. He turned to Parma.

"It's getting close. Something is happening."

I N THE PREDAWN hours of July 31, as Ford and his entourage dozed on their cramped Convair turboprop lumbering toward Washington, a sleepless Richard Nixon weighed the pros and cons of resignation on a bed-side notepad. Harboring no illusions about his prospects in the House, well aware that a Senate trial would convulse the country and in all likelihood devastate Republicans in the fall elections, the president chose to emphasize instead the damage his quitting would inflict on the office he held and the family he cherished. After three agonizing hours Nixon reached a predict-able conclusion.

"End career as a fighter," he wrote on the back of his lists.*

As chance would have it, Ford, too, had retirement on his mind. On the Massachusetts course, where he and Tip O'Neill were joined by pro-fessional golfer Dave Stockton, the vice president had waxed enthusiastic about "all the golf he wanted to play" once he was out of office. Deter-mined to extend the day, Ford stayed for dinner, postponing his return to

---

* A glance at this morning's *Washington Post* might have caused his resolve to waver. The paper pointed out that a vote to convict in the Senate would deprive the former president of his $60,000 annual pension, as well as $96,000 a year in staff salaries, free office space in a federal building and a widow's pension for Pat Nixon worth $20,000.

Washington until well after ten p.m. Even then he had nervous energy to burn. On reaching 514 Crown View Drive, Ford prepared to take a late-night swim before turning in. First, however, he returned a call from Bob Hartmann. General Haig hoped to see the vice president first thing in the morning, said Hartmann, who was just as determined that Ford have a silent witness to whatever the two men discussed.

And so it was, a few minutes after nine on Thursday morning, August 1, that Haig took a seat within spitting distance of a man he considered a drunken degenerate. Hartmann returned the favor, viewing the spit-and-polish general as a megalomaniac whose first allegiance was to his own career. Mindful of Hartmann's reputation as a leaker, Haig was less forthcoming than he might have been. Things were "deteriorating," he informed Ford. "There are some new developments that will dramatically change circumstances affecting the President." Among the tapes ordered turned over to Judge Sirica by the Supreme Court was one whose contents when revealed to the public would be "devastating" to the president's cause. Having studiously avoided listening to any tapes himself, Haig shied away from details. He did acknowledge contradictions between what St. Clair had been telling the Supreme Court and Judiciary Committee, and the newly uncovered proof of presidential complicity in the Watergate cover-up.

Ford expressed interest in talking with St. Clair, then asked, "How's the President holding up?" His mood was volatile, said Haig, veering between fierce resistance and a curious passivity. Hartmann sensed a Haig effort to enlist the vice president "in the army of those he thought the president would listen to who would counsel him to resign and stop fighting a hopeless fight." If so, the general would be disappointed. Ford told Haig he thought it would be "improper" for him to recommend any action to Nixon because "he would be involved in the result of that action."

Forty-five minutes after he arrived, Haig took his leave. Ford headed to the Capitol, and a meeting with Israel's foreign minister, Yigal Allon, to be followed by lunch with Republican National Committee chairman George H. W. Bush. A stalwart Nixon defender in public, privately Bush was deeply offended by the administration's moral tone. He was not blind to the prospects for personal advancement should Ford's sudden elevation to the Oval Office create a vice presidential vacancy to be filled. Ford had just walked into Senate Room 212 when he was called to the phone. General Haig wished to see him again that afternoon—alone. Hartmann's protective instincts went into overdrive.

"Never mind," Ford told him. "I'll fill you in."

Neither man could know what had transpired in the short time since

Haig's earlier visit to the EOB. Having at last read for himself transcripts from the June 23 tapes, the general grasped their lethal significance to the waning presidency of Richard Nixon. For three months, ever since donning headphones on May 6 to hear himself condone official lawbreaking in recorded conversation with Haldeman, the president had misrepresented the facts to a supportive family and a dwindling band of true believers in his administration and on Capitol Hill. These included his vice president and chief of staff, not to mention his own lawyers, who had been denied access to crucial evidence and encouraged to argue a false premise before the United States Supreme Court. Now, chafing under the disciplinary code instilled in him as a West Point cadet, Haig gave vent to his raw emotions.

"I've been lied to for the last time. This guy has got to go."

FORD THOUGHT HAIG more frazzled, if less guarded, than in the morning. Even now the chief of staff wondered if the detested Hartmann might be lurking behind a decorative screen in one corner of the room (he was not). Coming directly to the point, Haig told Ford that the newly available evidence was "ironclad . . . Nixon knew on the 23rd of June the whole story." Everything since was, to use a term made notorious by press secretary Ron Ziegler, inoperative. An unmistakable formality crept into the conversation. "Are you ready, Mr. Vice President, to assume the Presidency in a short period of time?"

"If it happens, Al, I am prepared."

Haig couldn't say with any degree of assurance what the next few days might bring. Only yesterday Nixon had dismissed the June 23 tape as "manageable." His subsequent change of heart might yet be reversed due to family pressure to stay or outside pressure to go. Clutching two pieces of paper, and disregarding Ford's earlier refusal to involve himself in Nixon's decision-making, Haig raised the treacherous subject of presidential options. Nixon could do nothing and let the judicial process take its course. He could wage a bitter, protracted campaign to stay in office. He could step aside temporarily, under the Twenty-Fifth Amendment, and reclaim power depending on the eventual outcome of his case. He could agitate for a less severe punishment, like formal censure. Alternatively he could issue pardons for himself and all those accused of Watergate crimes, and then quit. Finally Nixon could resign his office "in return"—Ford's words—"for an agreement that the new President—Gerald Ford—would pardon him."

There it was: the poisoned chalice that was Nixon's parting gift to his successor. Haig asked for Ford's assessment of the situation. "One thing is sure," he later quoted the vice president, "and that is if anyone tries to tell Dick Nixon to leave, he'll stay and fight it out. That's his nature." Ford might well have stopped there, but he didn't. Reflecting Haig's emphasis on the subject, he sought clarification of the president's pardoning power. Haig cited an unnamed White House lawyer who ascribed to the executive the authority to pardon someone even before "criminal action" had been taken against him. Did this extend to wrongdoers who had yet to be indicted, the vice president inquired. Haig pleaded ignorance.

"I'll have to think about all these things you've told me," said Ford. He needed time to talk things over with Betty and, eventually, with Jim St. Clair, whom he assumed, incorrectly, to be the White House lawyer referenced by Haig. On that note the two men shook hands then, spontaneously, embraced. They agreed to stay in touch. Haig left with Ford the two documents he had carried into their meeting. Jack Marsh, who saw them both, would never forget—each was written, seemingly in haste, on yellow legal paper. Marsh instantly recognized the handwriting of Fred Buzhardt, who had been a general counsel at the Pentagon when Marsh served as an assistant secretary of defense. One page included "a very, very accurate summary of the law of pardon," including pardons for crimes before prosecution. The second note was a scrawled form "where all you had to do was take the sheet of paper and type it up and you had a pardon."

Before Marsh could see the documents for himself, he was preceded to Ford's office by an agitated Hartmann. After pledging him to absolute secrecy, Ford described Haig's visit, not excluding the various options laid out for his consideration by the chief of staff. The mere mention of a pardon caused Hartmann to explode.

"Jesus!" he exclaimed. "What did you tell him?"

"I didn't tell him anything. I told him I needed time to think it over."

"You what?" With characteristic distrust, Hartmann guessed that this was exactly the desired response as far as Haig was concerned. One didn't have to share his cynicism to imagine the long arm of Richard Nixon manipulating the emotions of his loyal-to-a-fault vice president. Hartmann told Ford that he "should have taken Haig by the scruff of the neck and the seat of the pants and thrown him the hell out of your office." Ford, still processing the reality that he had steadfastly denied to himself for months, and well aware of the animosity between the two chiefs of staff, thought Hartmann was overreacting. In any event he couldn't discuss the matter now. It was almost four thirty; Betty was outside in an official car, waiting for him to join her for a tour of the putative vice presidential residence he strongly suspected would never be their home.

At the Naval Observatory, Ford feigned interest in paint samples and floor plans for a torturous three-quarters of an hour before fleeing the old house and returning to the EOB to review his upcoming schedule. He was due to leave Saturday morning on yet another campaign trip, this time to Mississippi and Louisiana. To cancel now would intensify the death watch enveloping the White House. Worse, it might upset the delicate balance of Richard Nixon's calculations. Jack Marsh was standing by to talk to him, Hartmann reminded his boss, "before this goes any further." It would have to wait, said Ford. He and Betty had a dinner date at the Georgetown home of *Washington Star* society reporter Betty Beale and her husband. The timing might be inconvenient, to put it mildly, but no more so than his earlier house tour. Ford promised to get with Marsh in the morning.

On arriving home Ford said nothing of the day's events to his wife. They would discuss things more fully after dinner. Whatever was said in Betty Beale's dining room on Garfield Street stayed in her dining room. At one point Beale's husband, George Graeber, remarked that the vice president seemed unusually quiet. Ford chalked it up to an oncoming cold.

I N *THE TIMES of My Life*, the famously outspoken Betty Ford is silent about the eye-opening conversation ("My God, this is going to change our whole life") she had with her husband when they were finally alone in 514 Crown View Drive. Her spouse was scarcely more forthcoming in his memoir, sparing less than two pages for their late-night dialogue and the intimate rituals of faith with which they girded themselves for whatever lay ahead. Although Ford was careful to minimize his wife's feelings of resentment, others detected flashes of anger. Where Richard Nixon was concerned, claimed Phil Buchen, "Mrs. Ford was far more eager than he to maintain a separate identity." Now, as Jerry related details of his long day, with emphasis on Al Haig's two visits, he rekindled Betty's fears that her husband might be tarnished through his fidelity to a president not noted for reciprocating in kind.

Recounting their discussion over a bourbon and water in the downstairs family room, Ford leaves the reader in no doubt as to Betty's preferences. She was, he tells us, "very firm in her view that because of the peculiar position I was in I shouldn't get involved in making any recommendations at all. Not to Haig, not to Nixon, not to anybody." Betty was more explicit in her comments to Trevor Armbrister, Ford's ghostwriter on his presidential memoirs. "He came home with the word from General Haig that Nixon would resign if he would pardon him. We discussed it, and I said, 'You can't do that' and he agreed. And he turned it down."

Her last sentence may well refer to an incident that has never received the attention it deserves, in part because of Gerald Ford's habitual understatement. As Jerry and Betty climbed the stairs to their bedroom the phone rang. It was Haig, informing Ford that nothing had changed since they had last talked. In *his* memoirs Haig treats the call as a routine courtesy toward someone who might sleep better knowing he was unlikely to be awakened in the middle of the night by White House operators.

Ford himself says as much, albeit with a twist. "I told Al I had talked to Betty and we were prepared; but we couldn't get involved in the decision making process."

The general related a very different story to Bob Woodward in a 1997 interview preserved as part of the Woodward-Bernstein Archive housed at the University of Texas. According to this version of events, it was *Ford*, influenced by his discussion with Betty, who placed the postmidnight call. And what he said was far more assertive than the mildly worded disavowal quoted in *A Time to Heal.* As recounted by Woodward in *Shadow: Five Presidents and the Legacy of Watergate*, the exchange was brief and to the point. "Al," Ford said, "our discussion this afternoon, I hope you understand there was no agreement, no decision and no deal."

Ford's comments rattled Haig sufficiently that around two in the morning he called Fred Buzhardt to chew him out.

"Goddamnit, what did you do to me," Haig chastised the president's counsel. By all appearances the normally trusting Ford smelled a rat. But they had done nothing wrong, Buzhardt protested.

"All we did was give the options."*

By now the Fords had retired for the night, though not before reaching out to clasp hands in their twin beds with their single headboard—Jerry on the right, lying on his stomach, and Betty on the left, as they extemporized a prayer. Jerry concluded, as was his nightly habit, by silently mouthing words from the third chapter of Proverbs, first imparted to him in a Grand Rapids Sunday school half a century earlier. "Trust in the Lord with all thine heart, and lean not unto thine own understanding. In all thy ways acknowledge Him, and He shall direct thy paths."

---

* The reader is perfectly free to draw his own conclusion as to Haig's motive for making such a call, establishing for the record that any initiative for a Nixon pardon originated with Buzhardt.

A FEW HOURS after praying with his wife for divine guidance in the uncertain life before them, Ford was back at the EOB meeting with Jim St. Clair. "Very deliberately, very lawyer-like, he said the new evidence significantly changed his impression as to the outcome," Ford would recall. The odds of Nixon's survival had changed "dramatically." The president's lawyer assured Ford that he was as surprised as anyone by the revelations in the June 23 tapes. St. Clair also denied being the source of any information relating to the presidential pardoning authority.

Next, honoring his promise to Hartmann, Ford called in Jack Marsh, an unflappable figure whose judgment he considered "incredibly good." There in the big office whose windows framed the White House across West Executive Avenue, Ford mentioned, almost in passing, his late-night call to Haig, interpreted by Hartmann as a message to "do whatever they decided to do; it was all right with me." In Hartmann's narrative, Ford is eager to wash his hands of the whole mess. "He just wanted to get it all over with."* This was dangerous turf. Hartmann and Marsh suspected that their boss was being played by Nixon's people, if not by Nixon himself. Believing that they needed a reinforcing voice, someone with the clout and credibility to break through Ford's residual faith in the president, they quickly settled on Bryce Harlow, the much consulted Wise Man who was to multiple Republican presidents what Clark Clifford and Averell Harriman were to their Democratic counterparts.

Before this could happen, however, Ford had a one o'clock meeting at the Capitol with Mike Mansfield and Hugh Scott. The respective party leaders in the Senate were anxious to establish procedures for the impeachment trial that appeared unavoidable. Ford made it clear that he had no intention of taking part in the proceedings—a role constitutionally assigned to the chief justice—though he did agree to Scott's request that he remain in town, easily accessible, during any trial. After a while Mansfield left the two Republican leaders to themselves.

"What are your plans?" asked Scott.

"I'm going to Louisiana and Mississippi."

"Are you going to say anything more?"

---

* The problem with this account is that it excludes Betty's explicit rejection of anything resembling a deal, and downplays the postmidnight rebuke by her husband that prompted Haig's angry query to Buzhardt.

As a matter of fact, Ford told Scott he was considering a brief statement pledging silence as far as the president's legal position was concerned.

Scott asked him to hold off until he returned to the capital on Monday. The senator then began to cry. "This is an awful thing to go through."

Ford put a consoling hand on Scott's shoulder. "I know, I understand. It's just hard on all of us."

"You're all we've got now," declared Scott, "and I mean the country, not the party."

A T THE WHITE House Al Haig and Jim St. Clair were alerting Charles Wiggins, the president's ablest defender on the House Judiciary Committee, to the coming bombshell. The gesture involved more than courtesy; Haig in particular hoped to manage the fallout from any explosion involving the silver-haired lawmaker. Wiggins scanned the transcript placed in his hands. He read it a second, and then a third time. "There is just no misunderstanding of what the President is talking about," he said. Haig and St. Clair agreed.

"Does he have another Checkers speech in him?" Wiggins inquired. Returning to his office, he canceled evidentiary briefings scheduled for himself and other Nixon diehards in anticipation of a final vote by the full House.

Meanwhile, at the vice president's office Bryce Harlow was adding his voice to the alarm sounded by Hartmann and Marsh. "I cannot for a moment believe that all this was Al Haig's own idea," Harlow told Ford. "It is inconceivable that he was not carrying out a mission for the President." The situation might yet be salvaged, but only if the vice president immediately reached out to Haig with an explicit disclaimer that anything they may have discussed was "purely hypothetical." Moreover Ford should reiterate that he had no intention of advising Nixon as to his future. Harlow wrote out a few sentences conveying these thoughts, which Ford then related in a phone call to Haig, witnessed by Harlow, Hartmann and Marsh.*

It was now a little after seven o'clock. For a second straight night the Fords were running late for a dinner date, this time with their photographer

---

* The question remains: Why, having implicitly reprimanded Haig during his late-night call to the general, would Ford reiterate his concerns a second time? The vice president may have concluded that his original message needed reinforcement. He might also have welcomed the opportunity to enlist three credible witnesses, led by Harlow, a figure of universally recognized integrity, in the little drama he let them script for him.

friend David Kennerly, who had assembled a dozen of his professional colleagues at the picturesque Old Angler's Inn in nearby Potomac, Maryland. During the short drive to Great Falls Park, nothing was said to indicate anything unusual was taking place. Therein lay the problem as far as Al Haig was concerned. Hoping to end the national trauma as painlessly as possible, Haig feared too much discretion now could only prolong the agony, as Nixon's family urged him to fight on.

Haig was at home that evening when the call came through from a defiant chief executive. "Let them impeach me," said Nixon. "We'll fight it out to the end."

Prepared for such a change of heart, Haig put in a call to Bob Griffin. As close a friend as Richard Nixon could claim in the Senate, Griffin was also a known intimate of Jerry Ford, with whom he had lunched earlier in the day. Revealing nothing of his inner turmoil, Ford had been as sphinxlike with Griffin as he was around Dave Kennerly's pals at the Old Angler Inn. Haig was more forthcoming. Putting a liberal interpretation on the president's request that he poll senatorial sentiment, the general supplied Griffin with sufficient hints about the still-unreleased tape to convey the depth of Nixon's legal predicament. Numbers aside, anything Griffin could do to help make the president see reality . . .

The rest was implied. Putting down the phone, Griffin left for National Airport and the first leg of the trip to Traverse City, his home on Michigan's Grand Traverse Bay. Following a sleepless night, he rose early Saturday morning and began outlining his thoughts on a legal pad. When he had refined them to his satisfaction, he tracked the vice president through the White House switchboard to Starkville, Mississippi, where Ford was campaigning for a trio of Republicans running for Congress. Ford was unable to return the call until his next stop, a Hilton hotel in the state capital of Jackson. "Dear Mr. President," Griffin read from his text. "There is no doubt in my mind that unless you choose to resign, the House of Representatives will adopt the Articles of Impeachment, making necessary a trial in the Senate." In the meantime senators would undoubtedly subpoena White House tapes as evidence. "If you defy such a subpoena, I shall regard that as an impeachable offense and shall vote accordingly."

Ford thanked Griffin for alerting him to his action. There wasn't much more that he could say.

"Bob, do what you think is right."

Griffin hung up, then called his office in Washington. Over the phone he dictated the letter to his secretary, with instructions that it be hand-delivered to the White House. He also ordered its release to the Michigan newspapers. The letter reached Nixon around one thirty p.m. Shortly after, Haig was summoned to the Oval Office by a furious president. "Pompous little jackass," Nixon raged. "Who the hell does he think he is?" Silent concerning his own part in triggering the presidential outburst, Haig subsequently told Fred Buzhardt that Griffin's offensive letter was probably inspired by the senator's longtime friend and political ally Jerry Ford.*

CAMP DAVID ACTED as a tonic on Richard Nixon. The presidential retreat was exactly that, both refuge and restorative for one who did his best thinking in its pine-scented isolation. True to form, on arrival Saturday afternoon Nixon secluded himself in Aspen Cabin, while Haig and Ron Ziegler shuttled between these quarters and the camp's nearby dining hall, where Pat Buchanan was debating the president's options with Jim St. Clair. In a separate cabin, speechwriter Ray Price toiled on a statement to accompany the public release of the June 23 conversations—the first of three involving Nixon and Haldeman on that fateful day, and the one that has entered history books as the smoking gun. It was not going well, for what was being asked of Price was akin to squaring the circle. Nixon wished to minimize his plainly illegal instructions to Haldeman, while stressing his later directive to FBI director Patrick Grey to proceed with the agency's Watergate investigation.

St. Clair, by contrast, demanded an explicit acknowledgment of Nixon's failure to reveal to his defenders what he had known since May 6. Anything less, and the president's lawyer feared that he, too, might be seen as complicit in the attempted cover-up. Price's draft statement pleased no one, Nixon least of all. "Damn it, Al, this is not what I asked for," he barked at his chief of staff. Haig pushed back. The lawyers would jump ship, he maintained, if he asked them to change it.

The exchange grew heated. Voices were raised, but no minds were changed.

"The hell with it," Nixon concluded. "It really doesn't matter. Let them put out anything they want."

Ironically it was Pat Buchanan, the most fervent of Nixon partisans, who now offered an exit strategy for the president and his supporters. Rejecting

---

* In *Inner Circles*, Haig makes no mention of his role in ginning up the Griffin letter. To the contrary—he speculates that "it could be a signal that Ford's friends were going public to force Nixon's resignation." If they were, it was because Haig had encouraged them to do so.

Haig's effort to hustle Nixon out of office as early as Monday, Buchanan argued, "we don't want the old man resigning right now. What you do is you just drop the tape—Boom. And the tape will blow a hole in the bottom of the ship and we'll be gone by the end of the week." The next day Buchanan was among those assigned by Haig to brief presidential counselor Dean Burch and other members of the inner circle. Holding nothing back, Buchanan detailed the discrepancies between Nixon's claims and the irrefutable evidence of the June 23 tape.

"Jesus Christ," said Burch. "Get the scotch."

Ford had had four days to brace himself for such news. Monday, August 5, began for the vice president in a New Orleans hotel ballroom. At the end of his speech to the Disabled American Veterans, Ford was visibly startled to hear the group's commander respond, "Thank you, Mr. President." He was still in the air, en route to Washington, when military aide Jack Walker handed him a message from General Haig. The president would be making a significant announcement later in the day. By three p.m., Ford was back in his EOB office. In an adjoining room Hartmann crafted a statement that would formally divorce Ford from Nixon's defense. As the afternoon progressed, and the White House deferred any public comments on the still-unreleased tape, the vice president cooled his heels.

It was after five when Nixon's carefully worded communiqué was finally made public, in which he conceded that the accompanying transcript was "at variance with certain of my previous statements." Acknowledging his failure to inform St. Clair and other attorneys about the true nature of his first June 23 conversation with Haldeman, Nixon maintained that the totality of evidence did not justify the "extreme step" of impeachment and removal from office. Ford's response was less muted. "The public interest is no longer served by repetition of my previously expressed belief that on the basis of all the evidence known to me and to the American people, the President is not guilty of impeachable offenses." The language may have been stilted, but its implications were unmistakable.

No one grasped this quicker than Tom DeFrank. Part of the expanded press contingent traveling with the vice president, DeFrank read portions of the June 23 transcript over the phone to *Newsweek* bureau chief Mel Elfin.

"It's over," said Elfin. "You're about to cover the biggest story of your life."

Elfin's instincts were sound. Confirming Pat Buchanan's forecast, release of the "smoking gun" tape caused all ten Republican holdouts on the Judi-

ciary Committee to recant their support of the president. House Minority
Leader John Rhodes called a press conference to announce that he, too,
would vote for impeachment. Haig relayed the news to the presidential
yacht *Sequoia* as it transported Nixon, his family and longtime personal
secretary Rose Mary Woods on a two-hour dinner cruise of the Potomac
River. Watching them board the vessel late on that sultry August afternoon,
one member of the First Lady's staff was reminded of Czar Nicholas II and
his doomed loved ones going into exile. On returning to shore Pat Nixon
spent the first of three sleepless nights packing their belongings. Characteristically she told no one, but continued to insist that her husband defy his
enemies by remaining in office.

I F HAIG'S OVERTURES about a pardon had done nothing else, they
helped sensitize Ford to the dangers of passivity. Recognizing a need to
be more proactive in distancing himself from the man who had appointed
him, Ford asked Jack Marsh to work up some talking points for use at a
Tuesday-morning Cabinet meeting. Under the circumstances Nixon's vow
to fight on was pure bravado. Some detected in his stance faint traces of
a bargaining position, one made all the more inflexible by Ford's belated
refusal to sanction a pardon. Let others twist themselves into knots trying
to fathom motives; to Ford it seemed plain enough: the president had lost
touch with reality, a view confirmed when he opened the meeting with
an extended discussion of inflation. George H. W. Bush passed a note to
White House counselor Dean Burch. "This is unreal," it said.

From the economy Nixon moved on to the scandal threatening to topple
his presidency. "If the man had made up his mind that he was resigning,"
recalled HUD Secretary James Lynn, "it was some of the best poker I've
seen." For an excruciating twenty-five minutes Nixon rationalized Watergate as the product of his preoccupation with the great issues of state, and
subsequent neglect of the details of his reelection campaign. In the resulting
vacuum, other men had abused their authority and dishonored the administration. His conscience was clear: "In my opinion and in the opinion of
my counsel, I have not committed any impeachable offense." To quit now
would permanently weaken the presidency. No—the constitutional process
should be allowed to play itself out, "whatever the end may be. Some of you
may disagree, and I respect that, but I have made the decision. I ask your
support of this decision."

The ensuing silence was abruptly broken by the reedy, distressed voice
of the vice president.

"Mr. President, with your indulgence, I have something to say."

"Well, Jerry, go ahead."

"Everyone here recognizes the difficult position I'm in," Ford began. "No one regrets more than I do this whole tragic episode . . . But I wish to emphasize that had I known what has been disclosed in reference to Watergate in the last twenty-four hours, I would not have made a number of the statements I made either as Minority Leader or as Vice President." As a party in interest, he would be making no further public comment. Unwilling to conclude on so damning a note, Ford offered praise for Nixon's foreign policy—"A super job, and the people appreciate it"—and a pledge of support in the fight against inflation.*

As he spoke, an expression of shock on Nixon's face softened into a sad, knowing smile.

"I think your position is exactly correct," he told Ford.

After the meeting, the vice president attended a luncheon gathering of the Senate Republican Policy Committee. "Is that *all* the President had to say?" demanded an incredulous J. Glenn Beall of Maryland. Suddenly Barry Goldwater, red-faced with anger, leapt to his feet. "I'm not yelling at you, Mr. Vice President, but I'm just getting something off my chest—*the President should resign!*" Other senators endorsed his stand. Norris Cotton of New Hampshire proposed that a group of lawmakers go to the White House and tell Nixon that the national interest demanded his resignation. Given the turn in the conversation, Ford said it would be inappropriate for him to remain. Senators stood and applauded as he left for his EOB office.

Determined to stem any gathering panic, Ford clung to his original appointment schedule. He was meeting with a visiting group of Japanese legislators when Goldwater was called out of the senatorial luncheon to take a call from Al Haig. Convinced that Nixon was listening in on an extension, the Arizonan delivered a pessimistic estimate of the president's evaporating support in the Senate. Nixon could count on fifteen votes at the most, he told Haig, a number that kept shrinking in response to the latest revelations.

Goldwater's message was reinforced by Henry Kissinger, who deployed all his diplomatic skills to hint at the toll the scandal was taking on US

---

* Here memories differ. Bob Hartmann and Woodward and Bernstein assert that Ford omitted the most accusatory sentence from Marsh's talking points—the one withdrawing his past support, based on the newly released evidence contained in the June 23 tape. Ford certainly *intended* to read the entire statement. The passage in question is underlined for emphasis in the original reading copy preserved in Ford's papers.

foreign policy. As the day wore on, and reports from the Hill grew more dire, Haig and Bill Timmons bluntly informed Nixon that his position had become untenable. For close to two hours, Haig listened as a war-weary executive gauged his prospects and cursed his fate. At the end of this cathartic monologue Nixon told Haig that he would announce his resignation in a Thursday-night television address, then leave for California Friday morning. *Air Force One* would be over Chicago, he calculated, as Jerry Ford was being sworn into office. Ray Price was directed to start work on a resignation speech. Unaware of Nixon's decision, Ford simultaneously passed word to Bob Hartmann to prepare a short speech of acceptance.

Promising a dignified exit, Nixon managed a bit of self-deprecatory humor, remarking to Haig, "I screwed it up good, real good, didn't I?"

Then a potential snag: an ABC News reporter declared on air that Nixon had decided to quit, a scoop he attributed, erroneously, to Barry Goldwater. An angry senator called his friend Ben Bradlee at the *Washington Post*. "Now for Christ's sake, don't go out on a limb and print rumors," he told the editor. "Just between you and me, I think he's going to leave, but if you say it, he won't and then we're stuck with six to eight months of impeachment procedures and this country is going to pot."

At the vice president's EOB office, Ford asked his personal secretary, Dorothy Downton, for the whereabouts of the Bible used in his December swearing-in. She retrieved it from the office safe.

"Do you want me to put it back?"

"No," Ford told her. "Why don't I hold on to it?"

E VEN IN TIMES of worst crisis he was the most serene man in the room." Never was Phil Buchen's assessment of his friend's temperament more fully vindicated than on that Tuesday night. Buchen was the Fords' dinner guest in their Alexandria home, a date arranged without regard to the crumbling Nixon presidency. With the vice president running late, Buchen nursed a martini while he and Betty watched the CBS evening news. A few minutes after seven Ford walked in. Saying nothing about the day's events, he dashed upstairs to change into swim trunks. After doing his backyard laps he dressed for dinner before accepting a drink of his own. Table conversation was almost surrealistically devoid of any reference to the topic on everyone's mind. Instead Ford talked about his son Steve's college plans (the boy had been accepted at Duke).

Only when they had finished eating did reality intrude, as Buchen settled into a corner of the living room sofa.

"Look," he said to Ford, "you'd better tell me what's going on."

"It will all be over in seventy-two hours," Ford predicted.

The tension broken, Buchen revealed the existence of Tom Whitehead's unofficial transition team. The group's accomplishments were modest, but it had created an agenda to guide the new president during his first days in office. Ford took the news in stride. He barely knew Whitehead, still less the other members of his ad hoc committee. When Buchen offered to assist in putting together a more senior group to organize the transfer of power, Ford proposed the names of Bob Griffin, Rogers Morton, and John Byrnes, former colleagues in Congress; plus Bryce Harlow and Bill Whyte of U.S. Steel, a trusted friend and adviser from the business world. Eager to include a non-Washington figure on the roster, Buchen suggested Bill Scranton, the former Pennsylvania governor who also happened to be a Yale Law School classmate of Ford's.

Betty retrieved her address book. She gave Buchen the relevant phone numbers to call as soon as he returned to his room at the University Club. As he got up to leave, Buchen kissed her and put his arm around her husband.

"It is happening; we'll rise to it," he told Ford. "I'm proud of you."

That night Richard Nixon made five separate phone calls to Henry Kissinger. Five blocks north of the White House, Phil Buchen completed his own round of calls from the University Club. To Tom Whitehead, preparing to leave town with his wife for a camping trip in Aspen, Colorado, Buchen relayed Ford's message about an imminent change at the top. This meant that Tom and the others on Buchen's call list would have to compress the work of a normal presidential campaign and transition into a few days at most. Wryly, Buchen said he could envision a future volume from that master chronicler of American elections, Theodore H. White: *The Making of the President in 72 Hours.*

T HE FIRST EVENT on Ford's schedule for Wednesday, August 7, was a Capitol Hill breakfast with old friends in the Chowder and Marching Society. He never made it. An urgent appeal from Al Haig prompted the vice president, halfway down Pennsylvania Avenue, to reverse course and return to the Executive Office Building. There, a little before eight a.m., Haig informed him "on the President's instructions" that he should prepare to assume the presidency by the end of the week. The general outlined a tentative schedule of events surrounding the transition. He urged Ford to retain most of the current White House staff for at least thirty days.

Haig paused. Was there anything in particular that he could do for Ford?

"Look after the President, Al."*

By midmorning that Wednesday the engines of a Ford presidency were sputtering to life. The vice president received the initial draft of "a little straight talk" on which Hartmann had labored since three a.m. Hartmann himself reached out to Ford's San Diego friends, Leon Parma and his wife. "Your friend wants you and Barbara to be back here. That's all I can say. Get here as fast as you can." Shortly before noon Ford left for the Capitol and his weekly prayer meeting with Mel Laird, John Rhodes and Al Quie. From the Prayer Room Ford walked the few steps to S-212. There he secured Bob Griffin's agreement to join the transition team, whose first meeting was set for five p.m. at Bill Whyte's townhouse in the Spring Valley neighborhood of Northwest Washington.

During the afternoon Ford was briefed on pending legislation by Bill Timmons, and on foreign policy issues by Brent Scowcroft. Navy Secretary J. William Middendorf and Admiral James Holloway dropped by to make a pitch for their service in the forthcoming budget. They presented Ford with a framed image of the USS *Monterey*, prompting reminiscences about his December 1944 brush with death in the storm-tossed Pacific. Typhoon Cobra seemed a fitting metaphor for the American ship of state, and trials old and ongoing. Outside the White House, undeterred by intermittent rain showers, crowds had begun to assemble along the Pennsylvania Avenue fence. Inside Al Haig greeted the delegation of Scott, Goldwater and Rhodes prior to their five p.m. appointment with the president.

"Please give him a straight story," Haig urged the lawmakers. "If his situation is hopeless, say so. I just hope you won't confront him with your own demands." He needn't have worried. The word "resignation" was never mentioned during their half-hour visit. Goldwater marveled at Nixon's self-possession. Sitting back with his feet on the desk, the beleaguered president might have been a golfer who had just made a hole in one. A single bitter-sweet comment betrayed his emotions. In recalling past campaigns, Nixon couldn't resist a jab at officeholders who now presumed to sit in judgment

---

* Of greatest concern to the general was a budding movement among Capitol Hill Republicans to force Nixon's hand. "We had inferences that there were Ford supporters at work trying to make the inevitable happen sooner," Haig told Ford's ghostwriter Trevor Armbrister in a March 1978 interview. No one around the president wished to be party to a coup. Any decision Nixon made to resign must be voluntary, and perceived as such, if it was to be accepted by the 25 percent or more of the populace that still believed in his innocence. Haig planned on delivering this message over lunch with Barry Goldwater, chosen by his Republican colleagues to convey their dissatisfaction to the man in the Oval Office.

on him. "Some were turkeys," he observed, "but I campaigned for all of them." After twenty minutes of this he could no longer defer the business at hand. "How do I stand in the Senate?" the president asked Goldwater. "I'm going to level with you," the Arizonan replied. "You don't have more than twelve or thirteen votes, and you don't have mine." Nixon thanked his congressional visitors for coming. He would not make them wait long for his decision.

As the legislative trio parried with reporters in the White House driveway, a small fleet of company cars, government vehicles and Washington taxis discharged their passengers at Bill Whyte's front door on Rockwood Parkway. It was agreed that participants in that night's meeting would arrive at their destination separately, the better to throw inquisitive journalists off the scent. Even now public exposure of their activities could undercut Al Haig's efforts to hasten a dignified exit for the president. "The bar is open," Whyte informed his guests assembled in the paneled family room, though he was quick to add, "but this is a working session." In between answering the phone and monitoring television for any late developments, Whyte's son Roger grilled seven steaks purchased at the last minute by the lady of the house.

Working off Tom Whitehead's checklist, participants stressed the logistics of transition over grand themes. A consensus quickly formed around a low-key inaugural ceremony, minus the celebratory features that had marred Ford's earlier introduction to the nation as Spiro Agnew's replacement. Responsibility for coordinating plans with the Congress fell on Bob Griffin, who, ironically, was the member of the group most skeptical of Nixon's departure. Bolstering his doubts was a call early in the evening from the president's son-in-law, Edward Finch Cox. Events surrounding Nixon were still "touch and go," Griffin reported back to his fellow planners. Was it possible that they were wasting their time? "We can sit here and speculate," replied Buchen with Midwestern practicality. Alternatively they could assume Nixon was history and plan accordingly.

Dinner was delayed until nine, enabling latecomer Bill Scranton to join the team, which assigned him oversight of personnel matters. Several potential candidates to replace Ron Ziegler as White House press secretary were discussed. Jerald terHorst, Washington bureau chief for the *Detroit News*, was the consensus favorite. Don Rumsfeld's name surfaced as a potential chief of staff. "We constantly had one question in mind," according to John Byrnes. "How long do we have?" Bob Griffin provided an answer of sorts, returning from a second phone conversation with Ed Cox to announce, "We have to hurry up[,] guys."

There is no indication that anyone at the Whyte residence spoke to Ford directly. Had they tried, they would have found the vice president at home in Alexandria with Betty, pondering the radical changes about to engulf him and his family. Abandoning their modest split-level home, built to the specifications of a young couple and hallowed by memories of a growing family, promised to be traumatic. At one point in the evening, Ford glanced wistfully over the backyard of 514 Crown View Drive.

"I really hate to leave this pool," he said to no one in particular.

E VEN BEFORE HIS confirmation as vice president, Ford had requested of the Secret Service that any security requirements occasioned by his new job not inconvenience his neighbors. But with much of the street running by his home now closed to traffic, area residents had been given passes admitting them to their own neighborhood. On Thursday, the eighth of August, a rain-soaked contingent of local inhabitants and media folks kept vigil outside the Ford residence. As their numbers spilled into the street, Louise Abbruzzese allowed reporters to make use of her telephone, television and bathroom. Her husband, Peter, extended his own brand of hospitality. Each morning he hauled open his garage door, behind which a fresh pot of coffee beckoned. In the afternoon, it was replaced by a pitcher of martinis.*

At the EOB, Ford staffers noticed dark circles under the vice president's eyes. An uncharacteristic irritability testified to the strain of recent days. As with Nixon, Ford's options were dwindling by the hour. For most of the past eight months, out-of-town travel had offered an escape from depressing headlines and unwanted questions. No more—a twelve-day western tour scheduled to begin that afternoon appeared in jeopardy. In Palo Alto, California, the Hoover Institution would have to find a replacement speaker to mark the weekend's centennial celebration of its namesake, another American president who left office scarred by tragedy. Just before ten o'clock, Ford crossed Pennsylvania Avenue to Blair House for a Medal of Honor ceremony recognizing seven servicemen killed in Vietnam. His tightly strung emotions nearly gave way as he conveyed the nation's gratitude to each of the families.

He had barely returned to his office when Al Haig called.

"The President wants to see you. I think you know what he's going to say."

---

* A grateful press corps subsequently presented the Abbruzzeses with a plaque for their garage, identifying it as "First press room of President Gerald Ford, August 8, 1974."

Two Secret Service agents accompanied Ford as he made the short trek across West Executive Avenue. Inside the West Wing he strode through corridors lined with large color photos recalling happier chapters of the Nixon presidency. A short flight of stairs led to Haig's office. From there the chief of staff escorted Ford to the nearby Oval Office. This time Haig stopped at the threshold; the president wished to see his successor alone. Nixon rose from his desk. Extending his hand in greeting, he motioned Ford to take a chair to his right. Ford was startled by the president's appearance. He seemed noticeably older than at Tuesday's Cabinet meeting.

"I have made the decision to resign," said Nixon. "It's in the best interest of the country." He would not go into details; his mind was made up.

"Mr. President, you know that I am saddened by this circumstance. You know I would have wanted it to be otherwise, but I am ready to do the job and I think I am fully qualified to do it."

"I know you are too," said Nixon.

The hardest part of their meeting concluded, Ford sensed a change in the atmosphere, as if a burden had been lifted from both men. The president pushed his chair back from the desk and rested his hands on his lap. For the next seventy minutes he did most of the talking, conducting a tutorial on foreign policy and the domestic economy. He stressed the importance of NATO, and the need to honor US obligations to South Vietnam and Cambodia. He described Soviet leader Leonid Brezhnev as "bright and tough" but also capable of flexibility where a strategic arms deal was concerned. In Nixon's estimation, Henry Kissinger was "a genius," but that didn't mean Ford had to accept his every recommendation. "He can be invaluable, and he'll be very loyal, but you can't let him have a totally free hand."

From the world stage Nixon turned to the less tractable challenges of the home front. He complimented Fed chairman Arthur Burns and his old-time religion of balanced budgets. An expression of regret over his own 1971 imposition of wage and price controls was met by Ford's assurance—the experiment would not be repeated. Nixon dropped Nelson Rockefeller's name as a potential vice president. He advised Ford to retain Al Haig in his present role. The noon hour came and went. The Nixonian seminar proceeded, the instructor businesslike and seemingly without rancor.

At length it was time for class to be dismissed. Rising from their seats, both men struggled to repress their emotions. "I will not call you unless you call *me*," Nixon told his successor. Guiding him to a door opening on to the South Lawn, he said, "This is the last time I'll call you Jerry, Mr. President."

"I'll see you tomorrow," Ford managed. "Give Pat and the family my best."

Like a Realtor closing the deal, Nixon felt certain the Fords would enjoy living in "the mansion." A parting handshake, and the vice president stepped out into the Washington humidity.*

It was 12:13 p.m. In less than twenty-four hours Gerald Ford, until recently a legislator representing the interests of 467,543 West Michigan constituents, would assume executive responsibilities affecting 212 million Americans and the peace of the world. Silently measuring every step to his waiting limousine, Ford did not look back. A minute later he was in the EOB, where he would remain for the next seven hours. There he repeated to Hartmann and Marsh what Nixon had told him. The western trip was formally canceled.

At twelve thirty Ron Ziegler appeared in the White House briefing room to announce, in a voice thickened by emotion, "Tonight at nine o'clock Eastern Daylight Time, the President of the United States will address the nation on radio and television from his Oval Office." Ziegler exited the room as fast as he had materialized. In the nearby Oval Office Al Haig updated Nixon on his latest meeting with special prosecutor Leon Jaworski. Besides notifying Jaworski of the imminent resignation, Haig had promised the prosecutor access to Nixon's presidential papers and tapes following their removal to San Clemente. In return Haig had requested a public affirmation that no deal had been reached, or offered, involving the White House and Jaworski's office.

Fred Buzhardt put his own spin on the clandestine encounter. "Al said that he came away from the meeting with the impression that Jaworski was not out for the President's blood . . . that Jaworski would not prosecute him," Buzhardt recalled. Nixon's reaction to the news was "sort of blank." Obscured from the president and his agents was Jaworski's receipt that same day of a strongly worded memo from his staff recommending prosecution of the president on obstruction of justice charges.

Across the street someone at the EOB asked Ford how Nixon was far-

---

* "Brought a tear or two to his eyes," Nixon remembered in 1983. "I think to mine, too." For whatever reason, Ford withheld from publication the most emotional lines from their leave-taking, as recounted on tape, to his wordsmith Trevor Armbrister.

ing. "Better than I am," the vice president replied. Truth be told, Ford's chief emotion was one of relief. Preferring activity to introspection, he was eager to get on with his new job.

His first call was to Kissinger, who required no arm twisting to remain in his current positions. He and Ford made plans to meet later in the day for a more extensive discussion. Still to be finalized was the site of Ford's inauguration. Neither the traditional steps of the Capitol nor the White House Rose Garden seemed appropriate under the circumstances. The Oval Office was far too cramped to accommodate the numbers wanting to attend. On learning that Nixon had recommended it anyway as the backdrop to a private ceremony, Hartmann lodged a vigorous dissent. "You're the President of the United States," he reminded his boss. "You are going to have it where you want to have it, and you ought to have it in the East Room."

With that Jack Marsh began compiling a guest list, reaching out to members of the 1948 congressional class, to Ford's distant relations on the King side of his family and to Clara Powell, the Ford family housekeeper who had retired from their service to take care of her elderly father. Ford himself requested an air force plane so that John McCormack, the eighty-two-year-old former Speaker, could be present. He phoned Chief Justice Warren Burger, then attending an international judicial conference in the Netherlands, with an invitation to preside over the ceremony. The vice president apologized for interrupting Burger's travels, but the chief dismissed his concerns.

"Oh, I've got to be there," said Burger. "I *want* to be there."

Another military aircraft was dispatched to retrieve the chief justice. Ford meanwhile turned his attention to a replacement for Ron Ziegler. Sharing the enthusiasm of his transition team for Jerry terHorst, he transmitted a formal job offer via Phil Buchen. By late afternoon terHorst had joined the embryonic administration at Tom Whitehead's office, part of a group that worked until three o'clock Friday morning drafting memos and other documents formalizing the new regime. Rumors of terHorst's selection began circulating, to the discomfort of Ford's current press secretary, Paul Miltich, and his assistant Bill Rogers. Around two p.m., the vice president assembled them and other senior staff members for a brief pep talk. He thanked everyone for their labors to date and begged their indulgence during the looming period of uncertainty. He telephoned Mike and Gayle Ford, who left immediately for the nine-hour drive from their Massachusetts home to the nation's capital.

Another call went to Tip O'Neill, who wondered if his wife, Millie, was included in the invitation to Friday's oath taking. "She is now," said Ford. The Democratic majority leader shared with his Republican friend some comments he intended to make drawing a distinction between their close personal relationship and their "diametrically opposed" political philosophies. "That's fine, Tip,'" Ford responded. He knew he could count on him for advice and assistance. Seizing the opportunity, O'Neill urged Ford not to consider a Democrat for vice president in his pursuit of national unity: "This country doesn't work that way." Ford thanked him for his candor.

"Christ, Jerry[,] isn't this a wonderful country?" chortled O'Neill. "Here we can talk like this and you and I can be friends, and eighteen months from now I'll be going around the country kicking your balls in."

Ford feigned outrage. Was that any way to address a putative president of the United States? Inwardly he chuckled at the rough banter of the congressional fraternity.

There remained the matter of his speech at Friday's swearing-in. Well aware of Bob Hartmann's last-minute work habits, Phil Buchen had tried his hand at preparing a draft. So had Tom Whitehead. Neither man saw his efforts reflected in the final text, an eight-minute homily coupling the new president's personal modesty with a consoling pledge to restore public confidence in government. Hartmann was justly proud of the result, which he fiercely defended against Ford's attempt to excise the sentence for which the speech became famous.

"Our long national nightmare is over." In his White House memoir Hartmann professed ignorance of the line's origin, saying only that "it didn't struggle to be born. It just flowed naturally." Actually it flowed from the typewriter of speechwriter Milton Friedman, if Friedman is to be believed. Although "very little" of his suggested material was used, Friedman recounted in a 1998 oral history, he could lay claim to this fragment, an American catchphrase that would come to be as widely quoted as FDR's admonition that "the only thing we have to fear is fear itself." Time would validate both Friedman's eloquence and Hartmann's judgment. In the feverish emotions of the moment, however, Ford fretted that the wording was needlessly harsh toward his predecessor. Hartmann disagreed, as was his habit, vehemently. Any offense to Nixon, he argued, was vastly outweighed by the emotional connection with Americans hungry for hope. Ford let himself be persuaded. The nightmare stayed in.*

---

* Friedman could also claim responsibility for Ford's best-remembered line from his vice presidential confirmation hearings—"I am a Ford, not a Lincoln."

With his inaugural address put to bed Ford devoted the next two hours of his suddenly precious time to a freewheeling consultation with Henry Kissinger. The secretary proposed a series of introductory meetings with foreign representatives. He raised the subject of stalled arms control talks with the Soviet Union. Hoping to break the impasse, Nixon had invited Kissinger to arrange a fall meeting between the American president and Leonid Brezhnev. Did Ford want him to proceed with such an initiative?

"Yes, of course."

As Ford worked through the dinner hour, members of his family were coming to terms with their newfound status and visibility. In Alexandria Betty Ford debated the purchase of a new dress for the East Room inaugural. In the end frugality won out, as she settled on a pale blue, open-necked suit with white piping already hanging in her closet. The same storehouse was ransacked to dress Susan Ford and Mike's wife, Gayle, for the big event. A couple thousand miles to the west, Secret Service agents in a helicopter plucked Jack Ford out of a Yellowstone fishing stream prior to transporting him back to Washington, which he barely reached, still wearing cowboy boots, in time to see his father take the presidential oath.

The skill with which these and other preparations were rushed to completion could not hide fissures already appearing in Ford's administrative team. Assistant Press Secretary Bill Rogers learned from a CBS News producer that Jerry terHorst was the man to see regarding pooled network coverage of the East Room inaugural. Unable to reach terHorst, Rogers called Bob Hartmann seeking guidance.

"The Michigan mafia is taking over," growled Hartmann. "Let them run it."

Ford left for home at five minutes after eight. By then the crowd outside the White House had swelled to several thousand. From the steps of the EOB the departing vice president could hear cries of "Jail to the Chief" competing with the strains of "God Bless America." Ford was reminded of just how divided a country he was inheriting. At the top of the hour he joined Betty in the family room on Crown View Drive. They occupied their customary chairs flanking the television. Steve and Susan shared a nearby sofa. No one else was present, Ford having ruled out any photographs that might intrude on Richard Nixon's valedictory moment.

They watched as Walter Cronkite offered a brief recap of the tumultuous events leading to that evening's broadcast. Cameras on the South Lawn of the White House cut to a brooding image of the floodlit residence, its rooftop flag barely stirring in a clammy summer breeze. Then the picture dissolved, and 110 million viewers were suddenly face-to-face with the president of the United States.

"Good evening. This is the thirty-seventh time that I have spoken to you from this office, where so many decisions have been made that shaped the history of this nation . . ."

16

CHANGE AND CONTINUITY

~•◦•~

*We didn't have an election and three months before
going into office. We had twenty-four hours.*

—BETTY FORD, AUGUST 1974

I N NEW YORK, Broadway performances were halted so that audiences
might follow the real-life drama reaching its climax in the nine o'clock
hour. Normally crowded jazz dens and tourist-friendly eateries in New Orleans's French Quarter were largely deserted on that steamy Thursday night,
but it would take more than a constitutional crisis to deter the regulars
assembled at Graffagnino's Tavern to watch the World Football League's
Jacksonville Sharks take on the visiting Hawaiians. Protests erupted when
the televised kickoff was delayed by the presidential address. Griped one
impatient fan, "We have a new coach and a new assistant coach, that's all."

The crowd at Graffagnino's would not have long to wait. Richard Nixon's Oval Office farewell lasted all of sixteen minutes. Blaming his resignation on the loss of his political base on Capitol Hill—a sugarcoated
acknowledgment that he lacked the votes to ward off impeachment and
conviction—Nixon showed little remorse about his role in the vaguely referenced "Watergate matter" that brought him down. "It would have had a
better ring and a better response from the American people," Ford reflected
after both men were out of office, "if he had been more forthcoming, more
contrite, and asked for forgiveness." As it was, four-fifths of those polled
by George Gallup endorsed Nixon's decision to quit. Fifty-five percent opposed any further criminal investigation of the departing president (interestingly a *pre-speech* survey conducted by NBC revealed almost the same
number who objected to legal immunity for Nixon).

At the late-starting WFL game in Jacksonville, cheerleaders made no
attempt to hide their tears. More common was the reaction of twelve thousand baseball fans in Houston's Astrodome, who paused briefly to absorb
the news of Nixon's departure before breaking out in scattered applause. In
Lawrence, Kansas, additional telephone operators recruited in anticipation

of heavier than normal volume found themselves twiddling their thumbs. At Washington's National Cathedral, a chapel left open all night to accommodate those seeking spiritual consolation received only two visitors. On the second floor of the White House, Nixon retreated into the protective embrace of his family. His loyalists drowned their sorrows at an all-night wake in the Military Affairs Office. Presidential speechwriter Aram Bakshian, out of the country on August 8, heard stories on his return of a night of drunken abandon—"Like VE Day if you were German."

At the Federal Prison Camp in Allenwood, Pennsylvania, where he was serving time on conspiracy charges relating to the Watergate break-in, Jeb Magruder issued a stiff "no comment." "A great weight has been lifted," said Minnesota Republican national committeeman Rudy Boschwitz. "Dick Nixon doesn't have us to kick around anymore." More generous appraisals came from surprising corners of the political spectrum. George McGovern expressed sympathy for all that the Nixon family had suffered, "and for the ordeal still ahead." CBS White House correspondent Dan Rather discerned "a kind of class" in Nixon's farewell address, "more than that, a kind of majesty." As for any future prosecution of the soon-to-be ex-president, Rather doubted the public appetite "to shoot at lifeboats." Georgia state representative Julian Bond disagreed. "The prisons of Georgia are full of people who stole five or ten dollars, and this man tried to steal the Constitution of the United States."

The man whose opinion counted most wasn't talking. In his suite at the Jefferson Hotel, Special Prosecutor Leon Jaworski viewed the speech with mounting resentment. Nixon's remarks might have been appropriate for a president leaving office on account of illness, thought Jaworski, or one whose *policy* differences had led to a fatal hemorrhaging of congressional support. That was not what Jaworski heard. As Nixon on his television screen pivoted from vague regret to historical boast, claiming credit for the China opening and an easing of Cold War tensions, the prosecutor recalled Al Haig's confident prediction of a congressional resolution to shield the unindicted coconspirator from any further legal action.

"Not after this speech, Al, I thought," Jaworski would write in his memoirs. "He hasn't given Congress even a crumb of remorse to chew on."

The spotlight turned to a suburban street on the outskirts of Washington. Shortly after nine thirty, Vice President Ford navigated his congested front lawn to a bank of microphones. In a light drizzle he spoke with evident emotion, praising Nixon for "one of the greatest personal sacrifices for the country" and promising a continuation of his foreign policy in tandem with Henry Kissinger. On hearing that Kissinger—"A very great man"—

would be staying on as secretary of state, onlookers burst into applause. Nothing in his calm demeanor betrayed insecurity, yet Ford's words left the unavoidable impression that he needed Kissinger more than the other way around.

Inside the house the Fords had a late dinner, after which Jerry practiced his coming speech before an audience of one. Like any spousal couple pondering a job transfer, he and Betty talked over their abrupt departure from Crown View Drive. They watched the late news and shared a bedside prayer before turning out the light. Ordinarily the soundest of sleepers, Ford was uncharacteristically restless, his thoughts racing toward the unknown.

I N THE MORNING Ford prepared a light breakfast for himself and Steve. When Phil Buchen and John Byrnes knocked on his door at seven fifteen they found the vice president nattily attired in sober blue suit and matching tie. His visitors brought with them a four-page checklist distilled from the much larger transition playbook bequeathed by Tom Whitehead's planning group. Shortly before eight a modest caravan of cars, preceded by an Alexandria police vehicle and trailed by a press van, left the Ford driveway. Sailing past rush-hour traffic on Interstate 395, it afforded less favored commuters a fleeting glimpse of Ford in the back seat as he pondered the first decisions of his presidency.

Questions of personnel took priority. Mindful of recent events, Ford wanted no Haldeman coming between himself and his official family. Presented with three names to choose from for chief of staff, Ford selected Don Rumsfeld, his former House colleague then serving as US ambassador to NATO.* At his desk in the EOB by eight fifteen, Ford reviewed his speech once more with Hartmann; consented to multiple meetings with foreign diplomats later in the day, as arranged by Henry Kissinger; and recorded a telephone greeting for the residents of a new housing project in the South Los Angeles neighborhood of Watts, an impoverished working-class community, predominantly African-American, whose inhabitants had yet to recover from devastating riots in the summer of 1965. Originally scheduled to participate in that day's dedication ceremonies, Ford apologized for his absence—"Fate had something else in store"—and cited August 9 as an occasion for renewing a sense of pride, "not only in our individual homes but in our National home . . . the United States of America."

---

* The other two names on Ford's list of potential chiefs of staff were Frank Carlucci, secretary of defense under Ronald Reagan, and Bill Clements, James Schlesinger's quarrelsome deputy at the Pentagon and a future governor of Texas.

For Betty Ford the same date would always be remembered as "the saddest day of my life." Asked how she got through the demands of that improvised transition, she pointed to her *Living Bible* (not lost on her was the irony of its inspirational verse for August 9—"I will keep a muzzle on my mouth"). Meanwhile the Ford children nearly missed their father's inaugural after being stopped at the gate by a White House guard. His demand for identification prompted an embarrassing standoff, until Richard Frazier, the vice president's chauffeur, explained, "These are the President's kids." Shortly before ten Jerry and Betty were in a nearly deserted Diplomatic Reception Room, waiting for Richard Nixon to complete his East Room valedictory to staff and supporters. At length the muffled sounds of applause signaled the president and First Lady were on their way.

Showing remarkable composure, the Nixons wore their bravest faces as they entered the ground-floor parlor from which FDR had delivered his famed Fireside Chats. "Good luck, Mr. President," Nixon greeted Ford in a firm if slightly hoarse voice. Ford kissed Pat Nixon on her tearstained cheek. Words came haltingly. "You'll find the Lincoln Sitting Room about the way it was except for the big chair and footstool," said the thirty-seventh president to the thirty-eighth. "That belongs to me." Nixon checked his watch. "Time to go," he announced with feigned heartiness. Hundreds of White House staff, many fighting back tears, watched the Nixons and Fords lock arms as they strode out onto the South Lawn, past a stately magnolia tree planted by Andrew Jackson.

Judging from the wounded expression on her face, one ignorant of events might assume that it was Betty Ford who was abdicating her place of honor, with Pat Nixon offering verbal consolation. Taking note of the red carpet stretched before them, Mrs. Nixon remarked, "Well, Betty, you'll see so many of these . . . you'll get so you hate 'em." After running a final gauntlet of twenty-one uniformed servicemen, the two couples approached the olive-drab *Army One* helicopter. Here Ford briefly stepped aside so that Julie Nixon Eisenhower could kiss her mother goodbye. Nixon reached out to his successor for a parting handshake, reinforced by a quick brush of Ford's elbow.

"Goodbye, Mr. President."

"Goodbye, Mr. President. Drop us a line if you get the chance. Let us know how you are doing."

As Nixon turned away, Betty patted him lightly on the back.

"Have a nice trip, Dick."

Moments later the engines roared to life. The rotors accelerated and the chopper slowly rose above the South Lawn. Their faces set, grim and grieving, the Fords waved goodbye to their friends of a quarter century. As *Army One* swept past the Washington Monument and banked southeast over the Jefferson Memorial, the strain proved too much for Rose Mary Woods. She fainted and had to be revived by on-scene medical personnel. Ford took his wife's hand, and together they retraced their steps into their new home.

"We can do it," he told her.

Moments later an emotionally spent Bill Timmons, accompanied by Bob Hartmann, repaired to his West Wing office. From a stationery closet he retrieved a bottle of Chivas Regal scotch whiskey. Dispensing with ice cubes, Timmons poured out two healthy portions in water glasses.

"To the President," said Hartmann (meaning Nixon).

"To the President," Timmons replied (meaning Ford).

I**T WAS ALMOST** noon, and the Nixons were halfway to their California exile. In Ann Arbor the strains of "Hail to the Chief" rang out from a University of Michigan carillon. For the second time that morning the East Room was packed to overflowing. Attendees still benumbed from Nixon's emotive farewell less than two hours earlier mingled with first-time visitors and others, long barred from the People's House on account of political differences. Former New York senator Charlie Goodell entered the room for the first time since the 1970 campaign in which Nixon discarded him for the more conservative James Buckley. Al Haig was there, and Rose Mary Woods, at what psychic cost one could only speculate. Ford's half brothers from Grand Rapids were shown to their seats, along with Phil Buchen and Leon Parma and Susan Ford's current boyfriend, Gardner Britt, the son of an Alexandria car dealer. Claiming the front row, the Nixon—soon to be Ford—Cabinet sat adjacent to the congressional leadership from both parties. The Joint Chiefs of Staff took their places near Lindy Boggs and Peter Rodino.

No strolling strings supplied background music. No trumpets blared in exultant greeting. A few minutes before the hour, Ford military aide Colonel Jack Walker approached the director of the Marine Band to review the musical program.

"We're going to play our four ['Ruffles and Flourishes'] as the Vice President comes in and ['Hail to the Chief'] as he comes out."

Walker was dubious. "I don't think that's on the schedule."

"Oh yes it is."

Walker left to consult his superiors. He returned shortly, his hunch confirmed: there would be no "Ruffles and Flourishes" and no "Hail to the Chief."

Unheralded by the President's Own, Gerald and Betty Ford entered the East Room just after noon to sustained applause. They took their positions on a raised platform flanked by heroic portraits of George and Martha Washington. "And then the whole mood changed," George H. W. Bush observed. "It was quiet, respectful, sorrowful in one sense, but upbeat." Sensitive to the power of first impressions, Ford knew what he had to do in introducing himself to a traumatized nation. "I wanted to appear strong, confident, assured," he would recall. With Betty holding the Bible used in his earlier inaugural, Ford repeated the words first validated by Washington, bobbing his head slightly and giving special emphasis to the concluding "So help me God." The chief justice offered congratulations, the first to address Ford by the title that would set him apart until his dying breath. Betty resumed her seat as her husband waited for the applause to subside.

Promising himself, "I am not going to be emotional," Ford opened his reading copy of—what? Neither inaugural address nor fireside chat, he characterized it as "just a little straight talk among friends." His resolution was tested early in the text as he acknowledged the "extraordinary circumstances" that gave rise to his presidency. Acutely aware that his listeners had not "elected me as President by your ballots," Ford asked them "to confirm me with your prayers." On the other hand, he continued, having waged no campaign for either the vice presidency or the presidency, "I am indebted to no man, and only to one woman—my dear wife." Ford paused a moment, his voice close to breaking as he recognized the Congress and members of both parties who had exercised their constitutional mandate in making him vice president. He expressed hope, "even though it is late in an election year," that there might be more such bipartisan cooperation to advance the people's business. Addressing friendly nations—"And I hope that could encompass the whole world"—Ford pledged an uninterrupted search for peace. On his watch, American strength would be dedicated to global security "as well as to our own precious freedom."

Five minutes into his speech, continuity yielded to change. Although couched in inspirational language, the contrast he drew with the recent past was unmistakable. "I believe that truth is the glue that holds government together. That bond, though strained, is unbroken at home and abroad." Speaking for himself, Ford said he expected to follow lifelong instincts of openness and candor, confident that "honesty is always the best policy in the end." He came to the lines most likely to outlive him. "My fellow Americans, our long national nightmare is over. Our Constitution works; our great Republic is a Government of laws and not of men. Here the people rule." Less well remembered is the qualifying plea that followed. Deferring to a higher power "who orders not only righteousness but love, not only justice but mercy," Ford asked his countrymen to restore the golden rule to their politics.

In the next breath he showed them how. "In the beginning I asked you to pray for me. Before closing I again ask your prayers for Richard Nixon and for his family. May our former President, who brought peace to millions, find it for himself." Eyes glistening, Ford reaffirmed the promise he made eight months earlier when sworn in as vice president: "to uphold the Constitution, to do what is right as God gives me to see the right, and to do the very best I can for America. God helping me, I will not let you down." The audience rose to its feet. Less than twenty minutes after it began, the formal transfer of power concluded to the strains of "America the Beautiful." "Magnificent," pronounced Hugh Scott. "Authentic Jerry Ford," said Mike Mansfield. "It was superb."

A reception in the State Dining Room posed a cruel test for Nixon partisans. Their wounds still raw, more than a few had trouble accepting this new leader, forced on them by a hostile Congress. "All the Nixon people left," recalls one participant, "and just the Ford people went through the receiving line." Among those who peeled off was assistant press secretary Gerald Warren. "I was standing there with Al [Haig] and I said I didn't want to go in . . . I was feeling so down and so beaten. I said to Al, 'You know, I've had my picture taken with President Ford.' And he said 'So have I.' So I left." Warren's mood improved perceptibly on receiving Ford's note asking him to remain in his job.*

Around one o'clock Ford walked over to the press briefing room in the West Wing. Only a few hours earlier these same hallways had been lined with photographic blowups depicting the Nixons in a kind of executive family album. Now they sported informal images of Ford and his family. Pausing briefly in the Cabinet Room, Ford pondered the presidential portraits with which his predecessor had defined himself. Then and there he decided to leave Dwight Eisenhower in his place of honor over the north fireplace. On the opposite wall, Nixon's Progressive-era heroes Theodore Roosevelt and Woodrow Wilson would make way for Abraham Lincoln and Harry Truman. The latter selection, a surprise to many, grew out of Ford's identification with the plainspoken Midwesterner who had supplanted a controversial vice president before tragic events elevated him to the Oval Office.

---

* Haig had a more astringent answer for the senior staffer who asked why he stayed away from the Ford reception. "I don't dance on dead men's graves," he replied. AI, Tom Korologos.

Moments later Ford introduced his new press secretary, Jerry terHorst, to a room full of applauding journalists. "A tiny little guy," remembers one who joined in the laughter when Pete Lisagor of the *Chicago Daily News* said, "Hold him up so we can all see him, Mr. President." To one observer it seemed like the first time in a year that anybody had laughed in that pressroom. "It was like after a hard rain and suddenly it had stopped raining. The air was clear and everything was different. It just smelled different in that pressroom after that." Contrasting his own interest in "aquatic activities" with Nixon's decision to cover over the White House swimming pool originally built for FDR, Ford told reporters they needn't worry. Their relocated quarters were safe. He reiterated his pledge of an open administration, saying he was too old to change his nature.

Just before entering a senior staff meeting in the Roosevelt Room, Ford was handed a set of talking points by General Haig. Under the Heading "DO NOT," the new president was instructed "do not commit yourself to dealing directly with anyone but Al Haig." Hartmann and Marsh of the Ford staff managed to crash Haig's meeting. Both cringed when the new president announced that the general had "unselfishly" agreed to remain in his current post. Afterward, seizing a rare moment when he and Ford were alone, Hartmann expressed his dissatisfaction over the latest turn of events. Ford responded by designating him and Marsh counselors to the president. Suitably credentialed, they were about to get reinforcements from a most unlikely corner.

EVEN AS FORD was presenting Jerry terHorst to the White House press corps, a military plane carrying Don Rumsfeld circled Dulles International Airport an hour's drive west of the nation's capital. Interrupting a French Riviera vacation with his wife, Joyce, Rumsfeld had already decided to return to Washington when Tom Whitehead reached him around three in the morning with a formal request from the vice president's office. As he deplaned, Rumsfeld was handed a sealed envelope by a White House courier. Inside was a letter, signed by Whitehead and Bill Scranton, inviting the former congressman and sub-Cabinet official–turned-diplomat to coordinate the transition now underway.

Ford's afternoon was consumed by economic and diplomatic meetings, the latter including nearly sixty representatives from the NATO countries, Latin America, the Middle East, Japan and the Soviet Union. Like a man feeling his way in the dark, the new president put Kissinger front and center, at one point telling fifteen Arab emissaries, "He has our blessing and his policy has my support." It was after five thirty when Ford finally sat down with Rumsfeld and the other transition planners. He briefly out-lined their responsibilities, quipping that "he wanted his good friends to give him hell." Shying away from conceptual road maps, Ford signaled his broad philosophical agreement with the Nixon policy agenda, resoundingly endorsed by American voters less than two years earlier. Declaring, "We don't want the White House emptied," he rejected any mass exodus of current personnel. "I already sent you my letter of resignation," Rumsfeld volunteered.

"I didn't get it," replied Ford. "And you're not going to resign."

Ford asked the group to review the existing White House administrative structure, relations with the Cabinet and the sometimes heavy-handed role played by Nixon's Office of Management and Budget (OMB) in dictat-ing government priorities. Equally important was what he declared out of bounds. "He said not to mess with the national security side," recalled Dick Cheney, a clear sign of his reliance on Kissinger. In describing his work habits, Ford said he listened better than he read. "I like to take papers home to read at night," he explained, "and I prefer brief papers." As for his general approach to decision-making, "I like to see the alternatives." This was something Al Haig had learned the hard way when raising the issue of a Nixon pardon. Unlike his predecessor, Ford welcomed spirited debate before making up his mind.

This may help to explain his preference for the so-called spokes of the wheel imported from Capitol Hill, wherein an all-powerful chief of staff was replaced by several "co-equal assistants" reporting on a daily basis to the president. In much the same spirit, Ford told Interior Secretary Rogers Morton, "I'll have an open door to the Cabinet, but when you come, talk about something. If you waste my time, it's going to be a long cold winter before you come back." For all his geniality, Ford had little patience for chitchat. To Bob Griffin's inquiry about the vice presidency, he said he hoped to select his replacement within ten days. He would be soliciting the views of key individuals as early as that Sunday, two days hence, seeing them one-on-one. That way, said Ford, "people could speak their minds honestly."

Periodically the discussion was interrupted as Ford was called away to the nearby Roosevelt Room to greet foreign diplomats arranged in regional groupings. With each departure and subsequent return the men around the table stood. "Sit down, sit down—SIT DOWN," Ford finally blurted out. "How many times do I have to tell you?"

"Mr. President," replied John Byrnes, "we're not standing up because you're Jerry Ford; we're standing up because you're the President of the

United States of America." At that moment, as described by another Ford intimate, "he knew things were going to be different."

It was almost seven p.m. by the time Ford reminded Leon Parma that Parma was expected at the house in Alexandria for a low-key celebration with family and a few close friends. "Just wear your golf shirt, it's going to be real informal." First, however, Ford had appointments to keep with Simcha Dinitz, Israel's ambassador to the United States, as well as South Vietnamese envoy Tran Kim Phuong. August 9 saw an upsurge in offensive activity by the Vietcong in blatant violation of the 1973 Paris Peace Accords. Ford promised his Vietnamese ally a personal lobbying effort to restore military aid slashed by a war-weary Congress. It was "disgraceful," said Kissinger, to let Vietnam go down the drain "for a lousy $200 million . . . after 50,000 Americans died there."

This deferred disaster, this peace in name only, was now Gerald Ford's responsibility. So was a dangerously overheating domestic economy, the latest government statistics pointing to an annual inflation rate exceeding 12 percent. One of Ford's first acts as president was to call General Motors chairman Richard Gerstenberg with an appeal to rethink a 9.5 percent price increase on its 1975 model cars (GM agreed to minor reductions, saving new car buyers an estimated $54 per vehicle). Severe drought conditions in the Midwest promised dismal crop yields and a corresponding reduction in the amount of corn and soybeans available for export. As Ford embarked on the presidency, twenty-three judicial vacancies on the federal bench required his attention. On Capitol Hill a potentially historic bill to establish national health insurance languished in committee, its prospects diminished by election-year partisanship. Trade legislation benefiting the Soviet Union was likewise stalled pending an increase in the quota of Soviet Jews allowed to emigrate to Israel and other safe havens. Elsewhere unity in the Western alliance long championed by Congressman Ford was threatened by a brewing conflict involving NATO members Greece and Turkey over the Mediterranean island of Cyprus.

In the morning Ford would learn of fresh obligations, unimagined flash points and decisions that could be neither postponed nor evaded. For now, though, his thoughts were centered on Monday's speech before the Congress. It should be kept simple, he told Hartmann before leaving for home. "My friends, we've got a lot of work to do," he proposed as an opening line. "Let's get on with it." It was nearly eight when Ford returned to 514 Crown View Drive. In her postage-stamp kitchen Betty Ford was dishing out lasagna for Mel Laird and other dignitaries sitting cross-legged on the living room floor. "Jerry, there's something wrong with this picture," she observed. "You're President of the United States, and I'm still cooking." In time the guests drifted away, all but David Kennerly, whom Ford asked to stay behind to discuss the job of White House photographer. True to form, Kennerly said that he had two requirements before signing on—"Total access all the time and to only work directly for you."

"What," Ford shot back, "no use of [*Air Force One*] on the weekends?"

Finding the television set in the den broken, the family, joined by Kennerly, adjourned to the master bedroom upstairs, where they watched the East Room inaugural replayed on a smaller set. As it ended Ford took Kennerly's hand in both of his. "Would working for me be a problem among your colleagues?" To the contrary, said Kennerly, no one among his buddies confused Ford for Nixon. He thought they would be happy to have one of their own in the White House.

The next day Kennerly was in *Time*'s mailroom when a startled receptionist alerted him to a call from the president.

"Tell him to call back, I'm busy."

"He's on the line."

Picking up the receiver, Kennerly heard the familiar voice inquire, "Would you like to do this thing?" To Kennerly's one-word assent, Ford replied, "Well, you better get over here then, you've already wasted half a day of the taxpayer's money."

SATURDAY, AUGUST 10, launched a week of activity "perhaps unmatched," said *Time* magazine, "since the most frenetic days of Lyndon Johnson." The first full day of the Ford presidency began in an Oval Office purged of Richard Nixon's pictures and personal effects. The presidential desk was bare except for a telephone console whose numerous buttons defied easy comprehension.* Ford's official day commenced with a briefing from the CIA's David A. Peterson, an experienced intelligence officer who had grown accustomed to delivering the President's Daily Briefing (PDB) around the kitchen table at 514 Crown View Drive and in the back seat of Ford's car as he was driven to work. This morning Peterson concluded his Oval Office business, then headed for the nearest exit, only to find himself trapped between the executive lavatory and a small kitchen. Forced to

---

* Ford was offered the use of the Resolute desk, a gift from Queen Victoria currently on loan to the Smithsonian; alternatively he could work at a 1903 desk made for Theodore Roosevelt and boasting the names or carved initials of Presidents Truman, Eisenhower and Johnson. He spurned both, content to use a plain Kittinger desk of no historical distinction.

retrace his steps, he apologized for interrupting Ford in midconversation with Al Haig.

The president laughed good-naturedly. "The first time I tried to leave this office," he told Peterson, "that's what *I* did."

"Continuity and stability," Ford told the Cabinet that morning, "that is what the people want and the country needs." He rejected the bluntly worded advice of Treasury Secretary William Simon—"I really think that you're better off getting rid of everybody—Henry, me, everybody—and starting with a clean sweep of new Secretaries, new everything."

"Just stop right there, please," said Ford. "You must promise me as we sit here that you will stay with me until I leave office."

Before this gathering of Nixon appointees Ford paid tribute to his predecessor: "[A] great president. Under no circumstances will I speak adversely of him." Questions about Nixon's legal fate dominated the group discussion. Ruling out any congressional grant of immunity, Attorney General William Saxbe guessed that Special Prosecutor Leon Jaworski was reluctant to prosecute the former president. Left unmentioned was Jerry terHorst's assertion, made the previous day to a clamoring press corps, that Ford opposed any short-circuiting of the judicial process. On hearing this, Jaworski drew his own conclusions. According to one member of the prosecutorial staff, "I don't think anybody wants to adopt the monkey that's on his back."

In dealing with reporters, Ford urged his department heads, "Please be affirmative with them. A little extra effort will make the reactions a little less critical and that will help." Though mildly worded, nothing he said that weekend signaled a more dramatic break with the recent past.

Just around the corner from the Cabinet Room, terHorst had already set about ridding the White House lexicon of such Zieglerisms as "photo opportunity" and "news availability." Henceforth the press would make do with photo sessions and news briefings.

Returning to the Cabinet Room in the afternoon Ford presided over his first meeting of the National Security Council. Topping the agenda, CIA director William Colby reported on a Cold War escapade worthy of James Bond. Six years after it sank sixteen thousand feet to the floor of the north Pacific, far from its home base of Vladivostok, the Soviet ballistic submarine *K-129* lay ripe for the taking. At least US intelligence agents thought so. In 1971 the CIA had contracted with Howard Hughes's Summa Corporation to build a floating drill ship platform capable of retrieving the lost sub intact. Hughes's well-known eccentricities supplied the perfect cover for a supposed deep-sea mining expedition to extract manganese nodules from the earth's crust. The venture's $350 million price tag would be more than justified by the recovery of Soviet nuclear secrets, including code books and missiles. Even as Ford met with the NSC, the *Glomar Explorer* was hovering three miles above its target. A Soviet trawler shadowed the six-hundred-foot mine ship from a close distance. Adding to the delicacy of the situation, in keeping with its elaborate cover story the *Glomar Explorer* was unarmed. Should the Soviet ship catch wind of what the Americans were really up to and attempt to board the vessel, its crew would be unable to provide more than token resistance.

Ford accepted the risk. The mission would proceed. Initially all went well, as the spy ship's crablike pincers closed around the hull of *K-129* and began raising the sub to the surface, where it could be secreted inside the hull of the *Glomar Explorer*. Throughout the long ascent the Soviet trawler kept its distance. Then, disaster struck when the Hughes-designed claws malfunctioned at nine thousand feet, causing the sub to break apart and two-thirds of it to fall back to the ocean floor. The undertaking was not a complete failure, as the CIA recovered a pair of nuclear-tipped torpedoes and other sensitive intelligence finds. Also retrieved were the remains of half a dozen Soviet sailors, who were accorded a military funeral and burial at sea. Most important, the secret of the *Glomar Explorer* held—for now.

*C*HANGE AND CONTINUITY. In his first week as president Ford notified George Bush at the RNC that "the whole gut-fighting attack group" led by Nixon communications director Ken Clawson would be disbanded. Soon it was noticed that Secret Service agents were sharing with the Press Office formerly confidential information, like which gate the president used to enter the White House complex. Guards manning the perimeter were reported to be acting "friendly and helpful instead of obstructive and nasty." One Ford associate offered a vivid analogy of his old friend's position: "He's on a bus full of people that's flying down the road pretty fast. It's got some mechanical problems, and all of a sudden the driver jumps out. Ford's got to get in the seat and keep driving."

His first road test came Monday night, August 12, with his nationally televised speech to Congress. Building on the goodwill generated by his oath taking, the new president hoped to establish a conciliatory atmosphere with his former colleagues on the Hill and begin a desperately needed healing process among his countrymen. This failed to reckon with the confrontational instincts of the Nixon speechwriting shop. "Milt Friedman . . . spent the whole day just shaking his head," noted one Ford staffer as the Nixon holdovers around him crafted verbal attacks on Congress for its resistance to executive dictation. By Monday morning Bob Hartmann had resolved to fire the entire crew of Nixon speechwriters.

Among them was David Gergen, not yet the Washington Wise Man, bipartisan counselor and cable TV analyst of later years, but a self-described young kid who feared that his association with the Nixon White House was akin to playing for the scandal-tinged Chicago Black Sox. "There continued to be an arrogance and smugness on the part of some of the Nixon people," he acknowledges. These were balanced by Ford people who insisted on change for its own sake. "If Nixon had walked on the left side of the road, they wanted to walk on the right side of the road." That Ford did not share this attitude is attested to by others besides Gergen. "Mr. Ford said we had to try and get along," according to one veteran of the minority leader's office. "*They* weren't excited about getting along," she adds, "because . . . they wanted to do it the Nixon way."

Case in point. Ford hadn't been president twenty-four hours when Al Haig, claiming that only one copy existed, deflected Ford's request for the White House personnel briefing book. To Terry O'Donnell, a young air force officer and Vietnam veteran, more recently gatekeeper and personal aide to the embattled Nixon, Haig made a request. "We've got people coming in who worked with the Vice President," the general told him, with no record of what they may have discussed or Ford decided. "We can't run the White House this way," said Haig. "You know how to do it. Would you go down there and enforce the schedule?" This was code for Doing It Haig's Way.

"Great, let's give it a try," said Ford on being introduced to O'Donnell, whose father was a sometime golf partner of the new president. In the event Ford and the younger O'Donnell hit it off, aided no doubt by O'Donnell's self-effacing refusal to take sides in the escalating staff dispute. Typical of those caught in the middle was Robert Barrett, hired as a military aide in Nixon's last days, and now keeping his head down as each camp sought to enlist his support. "Boy, I'll tell you, Bob, this is something else," observed one survivor from the ancien régime. "These damned Ford people are sure a bunch of bush leaguers." Barrett looked sympathetic. "Two minutes later some Ford people would come in and say to me, 'Once we get rid of those Nixon holdovers everything will be all right.'"

The climate of mutual suspicion was not helped by demands from San Clemente, as relayed by Haig. Holed up in the former Western White House overlooking the Pacific, Nixon expected to be treated the same as any other former president. It wasn't enough for his successor to return the Nixon family's personal effects left behind in the hasty retreat from

Washington. Nixon wanted his papers—all forty-two million pages—as well as 950 reels of tapes compiled during his presidency, a historical treasure trove subject to subpoenas from the special prosecutor and courts preparing to try Haldeman, Ehrlichman, John Mitchell and other Watergate defendants. The issue forced its way to the top of Ford's agenda that same Saturday, August 10. In discussions with Bob Hartmann and Benton Becker, the Washington lawyer-investigator originally employed by the minority leader during the abortive impeachment of Justice Douglas, Ford learned that Nixon staffers were being instructed to prepare administration records for shipment to San Clemente.

Earlier in the week Hartmann and Ford had invited Becker to check out reports that Nixon people were clogging "burn bags" with legally sensitive documents. Obscured from all three men were a dozen boxes already transported to San Clemente, courtesy of William Gulley, the hard-boiled director of the White House Military Affairs Office. During eleven years in service to four administrations, Gulley had tapped hidden funds to build an air-conditioned movie theater at the LBJ Ranch, and flown in Florida stone crabs at $500 a pound so that H. R. Haldeman and John Ehrlichman could enjoy a last supper at Camp David before they were axed by an emotionally distraught Nixon. With equal resourcefulness, Gulley now determined to get Nixon his papers before the Ford White House was sufficiently organized to stop him.

He had not reckoned with Benton Becker. Early that Saturday evening, alerted by sources in the Secret Service, Becker left his office in the EOB to watch as White House file cabinets and hefty crates were loaded into a convoy of military trucks lined up along West Executive Avenue. Approaching the air force colonel in charge of the operation, he determined that the truck's contents were, as suspected, Nixon records destined for Andrews Air Force Base, per orders of General Haig.

Becker ordered a halt to the proceedings and the return of everything removed from the executive premises.

"Sir, I don't mean to be disrespectful, but I work for General Haig."

"Well, you may do that, but I work for Gerald Ford."

Becker hailed a pair of Secret Service agents he knew. "That truck does not leave here. I don't care if you have to shoot the tires out," he told them.* With the floors of the Old Executive Office Building groaning un-

---

* It bears noting that Becker's account varied with the telling. In some versions he went to the Oval Office to register his complaint, while in others he confronted Haig in the latter's office.

der the weight of a thousand boxes of paper, Ford signaled his after-the-fact approval of Becker's bold intervention. For now at least, Richard Nixon's presidential records would remain where they were.

S UNDAY MORNING, AUGUST 11. The Fords occupied a pew at their usual place of worship, Immanuel Church-on-the-Hill in Alexandria. Following a sermon on the need to "pick up the broken pieces" of a discredited presidency, Ford repaired to the Oval Office. There Mel Laird headed a parade of afternoon visitors summoned to discuss the vice presidency. It was no secret around Washington that Nelson Rockefeller was Laird's candidate to fill the slot. Feelings toward Rockefeller had lost much of the hostility so memorably displayed at the 1964 GOP convention that nominated Barry Goldwater. A decade later, Goldwater's rebels had become the party establishment, even as Rockefeller's faith in government solutions was somewhat tempered by his state's fiscal reckoning. The governor's international profile was a plus. So was his seemingly limitless access to talent, and his demonstrated appeal to minorities and urban voters ordinarily off-limits to Republican candidates. His executive skills, honed during fifteen years in Albany, offered a useful contrast to Ford the legislator. Finally, in singling out the charismatic scion of America's wealthiest family to be his partner, Ford would demonstrate a quiet yet unmistakable self-confidence.

For half an hour Laird made his case. After he left, Ford tasked Bryce Harlow with polling Republican officeholders, national committee members and other party grandees. Each would be asked to rank prospective vice presidents according to their stature, experience and ability to broaden Ford's political base. Scanning Sunday's papers Ford could not fail to notice the assertion by his twenty-four-year-old son, Michael, that Richard Nixon owed his countrymen "a total confession" of his role in Watergate. "All my children have spoken for themselves since they first learned to speak," Michael's father explained to reporters, "and not always with my advance approval. I expect that to continue in the future." It was an early hint that the Fords of Grand Rapids might not conform to the Ozzie and Harriet stereotype found in their press clippings.

At a time when success in the presidency was defined by not being Richard Nixon, journalists put their usual skepticism on hold to celebrate "Grand Rapids homespun . . . a man who toasted his own English muffins for breakfast"—a custom, it must be said, born less of Trumanesque simplicity than of Betty Ford's lifelong aversion to rising early ("I can't imagine anything worse than starting off the day with conversation"). "An unabashed lowbrow," according to *Newsweek*, Ford read the sports page before the rest of the paper. His personal tastes ran to double-knit suits, the Dallas Cowboys, Edgeworth pipe tobacco and bourbon and water. He addressed visitors as "sir" and took copious notes while conversing with Oval Office visitors. A reporter trailing Ford watched as marines standing outside the West Wing snapped to attention at his approach. One of them opened the door and stood wordlessly by the threshold.

"Hi, I am Jerry Ford," said the president, extending his hand in friendly greeting. "I am going to be living here. What is your name?"

Under Ford, references to the "Executive Mansion" gave way to the less grandiose "residence." Nixon's East Room church services ("a social event, really," Ford thought, "not a religious experience") were halted. The silver-and-blue presidential aircraft awkwardly christened *The Spirit of Seventy-Six* in recognition of the forthcoming bicentennial reverted to the familiar *Air Force One*. A junior official from the Justice Department professed amazement over the treatment accorded him in White House discussions. "In Nixon's day you'd go over there and you'd sit in the anteroom of the Roosevelt Room, and when your item was on the agenda, you would be summoned in." Within days of the Nixon to Ford changeover, "the environment was 180 degrees different. You'd go over, you were entitled to speak on anything that was on the agenda, and you were in there for the whole meeting . . . it was the difference between being closed and open."

This was not an unmixed blessing. An internal debate over federally mandated no-fault auto insurance preceded an appearance on Capitol Hill by Secretary of Transportation William Coleman. "Secretary Coleman thought it was constitutional and there were a number of other people on the other side," recalls one participant. "I remember [Attorney General Edward] Levi being there, Don Baker from the anti-trust division; there were perhaps six or seven people there and it was a Saturday. The President, after half an hour or so of the discussion, was pretty clear that he didn't want to support it. But we stayed there for another two or three hours giving everybody a chance to come back at him as many times as they wanted to.

"Eventually, Secretary Coleman ran out of arguments to make. He had a chance to make every single one and he wasn't getting anyplace with the President. Finally, he said, 'Well, Mr. President, I realize that I've lost this and maybe if you buy me a beer, we'll call it quits.' But he ended it. President Ford was willing to sit and talk with him as long as he wanted to."

From the outset Ford grasped that his new responsibilities demanded a broader outlook than his Grand Rapids constituents were accustomed to. In his first days as president he signed legislation expanding federal aid to subsidized housing, while revamping the program to afford local officials greater latitude in the allocation of funds. Displaying a surprising knack for symbolism, Ford moved quickly to validate his claim to be a president for everyone. Amid rumblings of discontent from the right, he reversed his earlier opposition to a ban on the purchase of chrome from Rhodesia's Whites-only government. On his third full day in office he placed a call to Harlem's Charles Rangel, chairman of the Congressional Black Caucus. "Hi, Charlie!" he greeted Rangel, before inviting him and fifteen colleagues to the White House for what Rangel would later call a "fantastically good meeting." Another day Bella Abzug and Shirley Chisholm led a delegation of women lawmakers who looked on as Ford proclaimed August 27 Women's Equality Day. Their smiles broadened when the new president declared his personal support for the proposed Equal Rights Amendment to the Constitution.

In his August 12 address to Congress, Ford reminded millions of viewers that while he had once declared himself a Ford, not a Lincoln, "I am not a Model T." Recognizing Congress as his "working partner," Ford insisted, "I do not want a honeymoon with you. I want a good marriage." Departing from the usual formula, he acknowledged that "the state of the Union is excellent. But the state of our economy is not so good." Beyond adopting Mike Mansfield's proposal for a domestic summit conference to help coordinate a larger war on inflation ("public enemy No. 1"), Ford offered few specifics. To be sure, he rang the bell of fiscal restraint and hinted at vetoes to come, even while stating his preference for "reasonable compromise." His sharpest break with the past produced the evening's most fervent response. On his watch, Ford promised, there would be "no illegal tapings, eavesdropping, buggings or break-ins."

His call for mutual forbearance went unheard by Bob Hartmann, instigator of an embarrassing run-in with his fellow presidential counselors Anne Armstrong and Dean Burch. Joined by Jack Marsh, the group assembled in advance of the speech in a room off the House chamber. By right of seniority, Armstrong and Burch should have headed up the little procession escorted to their seats by House doorkeeper William "Fishbait" Miller.

Hartmann had other ideas. "As we walked out of the room," Burch recalled, "Hartmann pushed me and Anne back and stepped in front of us . . ."

"What in hell are you doing?"

"This is *our* night," said Hartmann.

O UT OF THE gate Ford maintained a pace that in a less deliberate man might be called frenzied. In one thirty-six-hour stretch following his speech to Congress, he scheduled fifty-one appointments, boxing the compass of American politics and diplomacy. Tuesday morning, August 13, began with an Oval Office visit from AFL-CIO chieftain George Meany. Organized labor subsequently dropped its opposition to a revitalized Cost of Living Council. "Today we have a new president," explained the union's chief lobbyist. "We have confidence in the integrity of the president." Wednesday the fourteenth brought with it Ford's first foreign policy crisis. It involved the Mediterranean island of Cyprus, which since gaining its independence from Britain in 1960 had been dominated by the charismatic Archbishop Makarios III. Under Makarios, a wily throwback to the prelate politicians of Renaissance Europe, fears were expressed that Cyprus could become a Mediterranean Cuba. In truth the island more nearly resembled a Mediterranean Ireland, racked by sectarian and ethnic violence between a Greek majority longing for *enosis*, or union with the mother country, and a long-oppressed Turkish minority.

On July 15, local forces loyal to the Greek military junta in Athens had staged a coup against Makarios and his regime. Their success was fleeting, as Makarios, against all odds, survived to rally world opinion against the Greek colonels. On July 20, a Turkish armada landed forty thousand troops on the northeast coast of the island. Three days later the military government in Athens was itself overthrown. In its place, former prime minister Konstantinos Karamanlis reclaimed power. Requiring a scapegoat to divert attention from his government's military weakness, Karamanlis withdrew Greek forces from NATO. He also refused an invitation to fly to Washington to meet with the new American president. While Kissinger coaxed each side to moderate its demands, privately he tilted toward Ankara, telling Ford, "Turkey is more important to us and they have a political structure which could produce a Gaddafi," the latter a reference to Libya's bombastic military dictator and Arab nationalist Muammar al-Qaddafi.

Multilateral negotiations in Geneva ended abruptly on August 13. Shortly after midnight Washington time, Turkish warplanes appeared in the skies over Nicosia, the Cypriot capital. At the same time Turkish armored columns knifed across the northern third of the island. Eventually an estimated two hundred thousand Greek Cypriots were driven from their homes in the north as the island was effectively partitioned. In Washington Ford was left to juggle the competing interests of two strategically valuable members of the NATO alliance. Beyond a feeble condemnation of the latest Turkish assault and occupation as "unjustified," the president could only hope that Kissinger might work his diplomatic magic on Premier Bülent Ecevit, formerly his student at Harvard.

Cyprus joined a rapidly growing list of priorities, foreign and domestic, competing for presidential attention. That same morning Ford welcomed to the Oval Office Senator Russell Long, a conservative Louisiana Democrat who wielded immense power as chairman of the Senate Finance Committee. Invited by White House lobbyist Tom Korologos to make a list of issues he might want to discuss before his meeting with the president, Long replied that he never made lists.

"Why not?" Korologos inquired.

"Lists have ends to them."

During Long's meeting with the president, Korologos observed their two chairs drawing closer and closer to each other. "Pretty soon Long's got his knee on Ford telling him what we're going to do . . . and he's banging on President Ford's knee. 'Here's what you need to do' . . . beating up on the President, telling the President what he was going to do," and what "we" were going to include in the next tax bill. Also bottled up in Long's committee: the administration's plan for national health insurance. With its employer mandate and subsidized coverage for low-income Americans, the package, originally introduced by Richard Nixon, anticipated major elements of the Affordable Care Act enacted under Barack Obama in 2010. Then as later, advocates of small government objected to federal mandates and new taxes.

As a congressman Ford had opposed Medicare because of its funding through payroll taxes. Now, however, he displayed surprising flexibility, hinting that he might accept a 1 percent increase in such taxes if that was the price of catastrophic illness coverage. By the time Long said his goodbyes, Ford had his commitment to report a health insurance bill out of committee, election year or no election year.

After Long came another mainstay of official Washington. As ambassador of the Soviet Union, Anatoly Dobrynin had probably lost count of his Oval Office visits since the day in March 1962 when he presented his credentials to President John F. Kennedy. "He just goes on and on and on," Ford marveled. Dobrynin's long tenure in the American capital had seen the superpowers flirt with nuclear war over Cuba, then negotiate a precarious truce and a nuclear test ban treaty. The latest thaw in their thirty-year Cold War, known in diplomatic shorthand as détente, was a jewel in the crown of Nixonian diplomacy.

The men in Moscow were far less confident that Ford, "a typical American congressman-patriot of the Cold War era," would emulate his predecessor's outlook. Ford was eager to dispel their fears. Even before sitting down with Dobrynin, he had offered general secretary of the Communist

Party Leonid Brezhnev written assurances of his commitment to peaceful coexistence. Not satisfied to renew Nixon's invitation for Brezhnev to visit Washington, Ford expressed his openness to an earlier meeting if it might expedite stalled arms control talks in Geneva. But it is what Ford did next that stunned Kissinger as much as Dobrynin, who pointedly excluded the incident from his otherwise exhaustive memoir of a diplomatic career spanning six US presidents. Sitting beneath a portrait of George Washington, the president brought up the case of Simas Kudirka, a Lithuanian seaman who, in November 1970, had boldly leapt from a Soviet trawler onto the deck of a US Coast Guard cutter off Martha's Vineyard.

His experience of freedom was short-lived. Instead of granting him asylum, the American captain, after consulting with his bureaucratically minded superiors, permitted Soviet crewmen to board his vessel and remove Kudirka by force. For his daring escape attempt, the would-be defector was sentenced to ten years hard labor in a Soviet prison camp. There the matter rested until Gerald Ford, faithful to his congressional training, did for Simas Kudirka what he had done for countless earlier applicants from the captive nations of Eastern Europe. Ford's intense interest in the case was shared by no one at the State Department or on the White House staff. His request was nevertheless forwarded to Brezhnev, who promptly granted this personal favor to America's new president. By the end of August events were set in motion leading to Kudirka's emigration and eventual US citizenship.

In his wish to flee Soviet domination, the Lithuanian seaman had plenty of company. Each year Soviet Jews by the thousands petitioned Moscow for the right to emigrate to Israel and the West. The subject was made to order for American politicians and human rights advocates like Washington's Democratic senator Henry "Scoop" Jackson. A defense hard-liner and likely candidate for Ford's job in 1976, Jackson had long campaigned on the need to link US trade with the Soviet Union with the emigration policies of that and other "non-market" (Communist) nations. This idea took legislative shape in the proposed Jackson-Vanik Amendment limiting presidential authority to confer most-favored-nation (MFN) trading status on countries that restricted the right to emigrate. The politically astute Dobrynin, mindful of the economic consequences should détente unravel, indicated to Ford his government's willingness to increase the flow of Jewish emigrants in return for MFN status as spelled out in pending trade legislation.

There was, of course, a catch: the agreement could not be formalized in writing, lest it publicly infringe on Soviet sovereignty and inflame other minorities suffering under the Communist yoke. Washington would have to accept Moscow's word. Ford thought he had the makings of a deal. The timing of Dobrynin's visit was no accident, any more than a breakfast meeting Ford scheduled for the following morning with Senators Jackson, Jacob Javits and Abraham Ribicoff. As Dobrynin exited through the West Wing lobby, he passed Mel Laird and other members of Ford's weekly Capitol Hill prayer group. For the next fifteen minutes politics and diplomacy were set aside as old friends said their devotions. It was a measure of the group's discretion, and the capital's political fixations, that reporters attributed their visit to vice presidential talk.

Back to work. A transatlantic phone call from Prime Minister Harold Wilson updated Ford on the Cyprus conflict from the British viewpoint. Lunch was with Egyptian foreign minister Ismail Fahmy, in town to share some peacemaking ideas from President Anwar Sadat. Since losing the Suez Canal in the Yom Kippur War ten months earlier, the Egyptian president had broken with his Soviet allies and turned to Washington to broker a disengagement of frontline forces. One form of encouragement involved $250 million of US weaponry to be disguised as Saudi Arabian purchases. The pursuit of peace did not come cheap. In his first hours as president, Ford learned of outstanding Israeli requests for military assistance totaling $7.5 billion over five years. He might be forgiven for thinking the White House in August 1974 resembled Rick's, the fabled *Casablanca* nightspot to which everybody came: Fahmy today, to reiterate the Egyptian position that the path to peace ran exclusively through Cairo; Jordan's King Hussein over the next two days, arguing the cause of stateless Palestinians; the foreign ministers of Saudi Arabia and Syria later in August. Only Israel's new prime minister, Yitzhak Rabin, resisted American hospitality, a reluctance born of the fragility of his recently installed governing coalition.

Ford's day ended with another lesson in presidential etiquette. Legislation had been passed allowing Americans for the first time in forty years to buy and sell gold as a hedge against inflation. It awaited only the president's signature to become law. As it happened, no two chief executives approached bill signing ceremonies in the same way. The ebullient LBJ had employed a fistful of pens, with the entire lot later distributed to onlookers as souvenirs. The introverted Nixon had been content to use a single writing utensil. In opting for the Johnsonian approach, Ford again betrayed his congressional roots—more pens translated into more shared recognition for headline-hungry lawmakers. It was a small but telling indication of Ford's adherence to the hoary assertion that in Washington there is no limit to one's potential for achievement, so long as you don't mind who gets the credit.

For the rest of his long life, Gerald Ford would justify his pardon of Richard Nixon by citing the all-consuming demands of the presidency. The needs of one individual were consuming 25 percent of his time, he explained, to the detriment of over two hundred million Americans who deserved his undiluted attention. Thursday, August 15, was a case in point. Ford's breakfast session with Senators Jackson, Javits and Ribicoff went well enough, as Jackson conceded "good progress" on the issue of Soviet emigration, even while upping the ante by calculating the number of prisoners of conscience seeking early release at three hundred thousand or more. "This is not just for the Jewish emigrants," insisted Jackson. "It's all of the ethnic groups. They are hot on this."

"I know," Ford replied. "I have five thousand Latvians in my old district."

The discussion ended with both sides professing a desire to work together.

In office less than a week, Ford was learning fast how unforeseen events could hijack a president's day. Unaware of the preceding weekend's confrontation involving Benton Becker and those hell-bent on removing Richard Nixon's White House papers and tapes, Jerry terHorst had stumbled through a volley of press inquiries concerning the records and their eventual disposition. Ford's press secretary was equally in the dark about a *second* attempt, two days later, originating in a Monday-morning phone call from Fred Buzhardt to staff secretary Jerry Jones.

Assuming that Buzhardt was acting in an official capacity, Jones toiled for hours in the oven-like heat of the EOB basement. He had the tapes all boxed and ready for shipment to San Clemente when, suddenly, he heard a knocking on the door of the vault where they were housed.

"Jerry, we can't do this," Buzhardt told him with no further explanation. "We'll all go to jail. Unpack them."

Inevitably rumors began to circulate. In addressing them terHorst relied on Buzhardt's assurance that the tapes "had been ruled to be the former president's personal property." In truth, there had been no "ruling" in any legal sense, merely the unwritten opinion of Nixon defenders Buzhardt and James St. Clair, based on the long-standing tradition, never tested in the courts, that presidents took their papers with them on leaving office. The inexperienced terHorst made things worse by declaring the Justice Department and the special prosecutor's office to be on board with the Buzhardt–St. Clair interpretation of Nixon's rights. This came as news to

Leon Jaworski, who had not hauled the Nixon White House before the Supreme Court so that a disgraced ex-president could make off with critical evidence under his nose.

Equally surprised was Gerald Ford. Because the attempted heist of Nixon's papers over the preceding weekend had remained uncharacteristically hidden from the press, so, too, had Ford's reluctant assumption of custody over his predecessor's documentary legacy. By asserting Ford's compliance with Buzhardt's theory of ownership, however, terHorst unwittingly linked Ford to a conspiracy he had actually nipped in the bud. Awakening to the danger, Ford had terHorst on August 16 announce Buzhardt's resignation, his duties as counsel to the president reassigned to Phil Buchen. The new White House lawyer's first assignment was to consult with Jaworski and Attorney General Saxbe, as well as Nixon's representatives and various constitutional scholars, on the legal status of the Nixon materials stored in the EOB and facilities off-site.

Haig, furious over Buzhardt's replacement, railed against "the little executioner" terHorst.

"Do you feel good, executing a sick man?" he shouted. Privately Haig acknowledged the changing of the guard. "I have lost the battle," he conceded. "But I will stay long enough to get Nixon the pardon."

FRIDAY, AUGUST 16. Ford marked one week in office by calling Mike Mansfield to congratulate him on becoming the longest-serving majority leader in Senate history. The president voiced no objections when Mansfield's Republican counterpart Hugh Scott urged him to reconsider the United States' long-standing boycott of Castro's Cuba. The morning's highlight was a visit from Hussein I, the Jordanian monarch whose nimble diplomacy had enabled him to maintain his footing in the world's most treacherous neighborhood since his coronation at the age of seventeen. His Royal Highness was in Washington seeking clarification as to Jordan's place in any Middle East peace process; that and American weaponry with which to ward off possible aggression from unneighborly Syria. "We need to be strong to negotiate withdrawal and to be moderate and reasonable," he explained to his American hosts. "Otherwise our options diminish." Yet a single US Hawk air defense battalion carried a price tag of over $100 million.

Most of all, Hussein wished to know Washington's attitude toward the Palestinians, stateless and combustible, over whose loyalties Hussein had done battle with Yasser Arafat and his Palestine Liberation Organization. The king went off to lunch with Kissinger. He and the president would meet again in a few hours, joined by their spouses and an improbable guest list as the Fords hosted their first White House state dinner. Only the day before Betty Ford had been introduced to the place by curator Clement Conger. In allocating space on the second-floor living quarters, Mrs. Ford made it clear that she and her husband would share a bedroom as they had throughout their marriage. The adjoining room, traditionally used as a separate bedroom, would be turned into a personal study and family room with Jerry's favorite leather chair and other homey touches. Susan Ford, already musing about White House partying with the Beach Boys or Bette Midler, settled on a third-floor suite after assurances it would be repainted yellow and furnished with a brass bed.

"This house has been like a grave," her mother confided to a reporter. "I want it to sing." And dance, too, she might have added, if the evening of August 16, 1974, was to count for anything. Dispensing with a formal entertainment program, the First Lady also did away with the old rule requiring dinner guests to remain on the premises until after the president and his spouse departed the public rooms. "I'm just going to tell the guests that they can leave whenever they want to," she revealed, "nobody's going to get us off the dance floor at ten o'clock." There ensued, in the words of Jordan's jitterbugging Queen Alia, "a swinging party," which grew less inhibited as the evening progressed. No one was more surprised to be there as an invited guest than Helen Thomas, a frequent thorn in the side of presidents, unless it was her journalistic brethren from the *New York Times* and *Washington Post*. Also spotted in the dinner crowd were Tom DeFrank and Ron Nessen, hardy veterans of Slingshot Airlines.

Mrs. Ogden Reid marveled that she had been invited to the dinner along with her husband, formerly a Republican congressman from New York, since "we are Democrats now." Pete McCloskey was there. So was Pennsylvania's senator Richard Schweiker, persona non grata since opposing a Nixon administration bill in 1970. Former secretary of defense Robert McNamara hadn't been in the house since 1968. "My God," Henry Kissinger joshed the former defense secretary, "you know things have changed when they let *you* in here."

It was after eleven when the Fords bid good night to their royal guests. Rather than retire upstairs, however, Jerry and Betty returned to the dance floor, which quickly cleared so that the new president and Cindy Nessen, wife of the NBC correspondent, could stomp their way through multiple choruses of "Bad, Bad Leroy Brown." With an equally spirited rendition of the Mexican Hat Dance as performed by CBS newsman Eric Sevareid and Oregon senator Mark Hatfield, official Washington marked the completion of Gerald Ford's first week as president.

Not long after the Hussein visit, Russell Long returned to the White House for a follow-up discussion of the fall legislative calendar. The body language was markedly different from Long's earlier session. "Jerry Ford had become President," asserts Tom Korologos. "He was telling Long what he, Jerry Ford, wanted to do. And Long was backing down. 'Now, Russell, here's what . . . I want to happen on this.'"

"Hey," thought Korologos. "This guy's going to be alright."

17

THE PARDON

⟞⟝

You can't pull a bandage off slowly.

—Gerald R. Ford

FORD HAD HOPED to reveal his vice presidential pick by the end of his first week in office. This schedule did not reckon with events in Cyprus, where demonstrators angered by Washington's professed neutrality between Greece and Turkey stormed the US embassy in Nicosia on the morning of August 19. In the ensuing melee, Ambassador Rodger Davies, a fifty-three-year-old State Department veteran, fell victim to sniper fire. An embassy receptionist who went to Davies's assistance was also killed. Ford learned of the deaths while flying to Chicago and the national convention of the Veterans of Foreign Wars. A verbal tribute to the slain diplomat was hastily inserted into the president's text. Otherwise the draft distributed to reporters on *Air Force One* was a fairly standard plea for military preparedness, coupled with the crowd-pleasing nomination of former Indiana congressman—and national VFW commander—Richard Roudebush to head the Veterans Administration.

Deliberately withheld from the advance copy was the first great surprise of the Ford presidency, the announcement of a conditional amnesty—Ford preferred the term "earned re-entry"—for an estimated fifty thousand to eighty thousand Vietnam War draft evaders and deserters, many of whom had fled to Canada rather than fight in a conflict they found morally repellent. The idea in various forms had been urged on Ford by Mel Laird and strongly endorsed by the Ford children. Advocates said it would contribute to the national healing process, even while dramatizing the contrast between Ford's administration and its hard-line predecessor.

"I really want to bind up the wounds," Ford observed early in his presidency. "If I'm going to do that, then I've got to reach those kids who dodged the draft or deserted." Avowedly opposed to unconditional amnesty—"Deserters can't go home scot-free when the kid next door might have been killed in Vietnam"—Ford assigned Jack Marsh to work with Defense Secretary James Schlesinger and Attorney General William Saxbe to develop a program "to let them earn their way back." On the cusp of his trip to Chicago, Ford's own household had become tangled in laws requiring young men on reaching the age of eighteen to register with their local draft boards as a contingency (the draft itself having ended in January 1973). Tiptoeing into the Oval Office, Marsh asked the president if he was prepared for some disturbing news. Ford nodded in wary acquiescence.

"Steve hasn't registered for the draft," said Marsh, who went on to explain how the president's youngest son had simply forgotten to comply with a statute many of his contemporaries assumed to have gone the way of conscription itself. Ford put his head on the desk, "gobsmacked," in Marsh's recollection. Luckily for the administration, Steve's oversight was quickly, quietly corrected. Somehow the incident remained hidden from the press. So did his father's putative VFW shocker. Al Haig first learned of the latter en route to Chicago, when Ford showed him the section of his speech held back until delivery lest it evoke organized protests. Leery of amnesty on any terms, Haig warned the president to expect booing from angry vets. Ford was prepared. In fact he had deliberately chosen the VFW as a tough audience, emblematic of others who would have to be won over if the country was to make peace with itself.

In the event he received a rousing welcome from several thousand old soldiers assembled inside the Conrad Hilton.* They raised the roof as the president decried "unconditional blanket amnesty" for those who broke the law to avoid military service in Vietnam. The applause quickly died away, to be replaced by stunned silence, as Ford reminded the vets of his inaugural petition to a higher power ordaining "not only justice, but mercy." Citing the examples of Harry Truman and Abraham Lincoln before him, Ford appealed to the heroes of Anzio and Guadalcanal to emulate their generosity in reconciling Americans ravaged by a decade of foreign conflict and domestic turmoil.

Recognizing raw courage when they saw it, his listeners gave Ford a standing ovation as he left the stage. That their applause was for the office he held, and not his willingness to entertain some form of amnesty, was demonstrated the next day when delegates formally went on record in

---

* Outside the hotel was a different story, as several hundred pro-Greek demonstrators protested Ford's Cyprus policy. Asked if their chanting disturbed him as he was revising his VFW speech, Ford got up and looked out the window. "No, they are good people. I can understand their feelings. They're friends." Memorandum for Ron Nessen from Bob Barrett, January 8, 1976, GRFL.

374 | THE REPLACEMENT

opposition to the idea. Other veterans' organizations seconded the VFW. A notable exception was John Kerry, formerly head of Vietnam Veterans Against the War. "The purpose of amnesty is to forget the war and heal the wounds," said Kerry, "and it may be that the way to do that is to demand a sacrifice on both sides." Most Americans shared his view. Of those polled by George Gallup, 59 percent expressed support for conditional amnesty, while 34 percent preferred a blanket amnesty. Only 5 percent thought draft evaders should serve time behind bars.

Gerald Ford, it appeared, had hit a sweet spot in tackling one of the most contentious issues dividing Americans. Could he find a vice president who would be equally well received?

ONE WEEK INTO Ford's presidency rival patricians waited anxiously for the phone to ring at waterfront compounds separated by two hundred miles of Maine's craggy coastline. At Walker's Point in touristy Kennebunkport, George H. W. Bush sliced up the choppy Atlantic in his sleek two-engine cigarette boat *Fidelity*. Far up the coast Nelson Rockefeller taught the rudiments of competitive sailing to his young sons in the waters off Northeast Harbor. In spare hours the art-loving Rockefeller busied himself arranging stone Buddhas and Japanese lanterns in the pine forest adjoining his cliffside retreat, with its cantilevered living room jutting out from granite walls.

Ford's delay in filling the vice presidential vacancy, publicly blamed on the Cyprus crisis, was due as much to unexpected snags in the vetting process. This was fine by long-shot aspirants who welcomed the additional time in which to promote their candidacies. From the island of Maui Elliot Richardson phoned Henry Kissinger to see if the secretary's long-standing relationship with Rockefeller precluded his putting in a good word for anyone else (it did). Then columnist Jack Anderson, relying on questionable sources, reported that Rockefeller had clandestinely bankrolled plans to disrupt the 1972 Democratic convention in Miami Beach. Evidence supporting this claim was purportedly contained in documents once belonging to Watergate conspirator E. Howard Hunt. But when members of Leon Jaworski's staff opened two safe-deposit boxes connected to Hunt, they proved as empty as the charges themselves—part of a clumsy smear by members of the right-wing Liberty Lobby. Rockefeller emerged from the bizarre episode untarnished.

Less fortunate was his chief rival for the vice presidential nod, RNC chairman George H. W. Bush. A favorite of grassroots Republicans, Bush's star was in the ascendant when he was hit with accusations regarding the funding of his unsuccessful 1970 Senate campaign. Details were sketchy, though he was said to have received $100,000 from a secret slush fund maintained by the Nixon White House to assist more than a dozen GOP candidates that year. Bush's reputation for personal probity survived intact. But the stories could not have come at a worse time for his vice presidential prospects. Bush reportedly suspected Mel Laird of orchestrating the hit job. Others bet on a dark horse, one whose name was belatedly included on the short list of contenders Ford submitted to FBI director Clarence Kelley on August 16 for a background review.

The last-minute addition of Don Rumsfeld had a sequel three days later. Prior to boarding *Air Force One* for the flight to Chicago, the supposedly guileless Ford gave Bob Hartmann a second list of five contenders—those already identified for the FBI plus Elliot Richardson and Ronald Reagan. He wanted Phil Buchen to run the names past Leon Jaworski, Ford said. Hartmann pointed out the absence of any sitting members of Congress on the list. The president, well aware that such omissions could make for hard feelings, immediately saw Hartmann's point.

"Let's say that I narrowed it down to six. We won't even tell who the sixth one was, and they'll all be happy."*

In the event, it was Rockefeller who received the coveted presidential phone call on Saturday, August 17. In the course of an hour-long conversation, Ford sought assurances that the governor would accept the job should it be offered him. Rockefeller replied that he would have to discuss it with his wife, Happy, and his family, but he said nothing suggestive of his previous disdain for the vice presidency as mere "standby equipment." In addition to attending all meetings of the Cabinet and National Security Council, Ford wanted his veep to oversee the White House Domestic Policy Council, symbolic of a domestic policymaking role akin to Henry Kissinger's management of foreign relations. This led Rockefeller, one of nature's original optimists, to infer that he would be a different kind of vice president, entrusted with program innovation on a scale even grander than he was accustomed to in New York. He rationalized his acceptance of Ford's offer as a self-sacrificing response to a constitutional crisis. That this was not his sole motivation became clear to a longtime associate who argued to his face that Rockefeller lacked the temperament to be anyone's understudy. Admitting as much, the three-time presidential candidate finally blurted out to his friend what he failed to understand—"This is the closest I'm ever going to get."

---

* Acknowledging "I've been a creature of Congress all my life," Ford told one staff member, "I don't think I'm going to pick another member of Congress."

Media reaction to Rockefeller's selection, announced on August 20, was overwhelmingly positive. NBC's David Brinkley interpreted the nomination as yet another Ford break with his predecessor, offering voters "an open, casual government looking for friends to work with rather than enemies to punish." On Capitol Hill the leadership of both parties welcomed the news. Conservative lawmakers by contrast gritted their teeth over the promotion of their battle-scarred antagonist. "Amnesty and Rockefeller in the same 24 hour period is a bit much," rasped Barry Goldwater. Blunter still was the response of Kentucky senator Marlow Cook, already staring at likely defeat in November's midterm election. "Any way we can get Nixon back?" he groused.

That same day the Rumsfeld-led transition team presented Ford a twenty-three-page blueprint for dismantling the centralized hierarchy that had isolated his predecessor and incubated Watergate. Specifically the group urged a reduction in the bloated White House staff; a return to Cabinet government at the expense of Nixon's imperious Office of Management and Budget; enhancement of the vice presidential function; and elimination of the chief of staff's position. In its place transition planners endorsed Ford's spokes-on-a-wheel concept of executive management, with half a dozen presidential aides enjoying access to the Oval Office (more, if one included Bob Hartmann and Jerry terHorst). In effect the president would be his own chief of staff.

Doubts about the new administrative structure were rampant even if it met with Ford's approval. "It was the impression and not the reality that counted," explained Mel Laird. The spokes of the wheel represented symbolic change—"An effort to portray the president and the White House as open and accessible." Laird spurned Ford's invitation to join his fledgling government in a more formal capacity "because, like Haig, I was associated with the Nixon administration." He wasn't alone in believing the general's days were numbered. "Rumsfeld stopped me in the hall on his way back to NATO," Jerry Jones recalled. "Don't let them change a thing until I get back," he remarked.

O N THAT SAME eventful Tuesday, the House Judiciary Committee issued its own conclusions, 528 pages documenting three dozen instances of alleged criminality by the former president. Even committee Republicans were unanimous in their findings: only his resignation had averted Nixon's impeachment and conviction by the full Senate. With the conspiracy trial of key Nixon aides including Haldeman, Ehrlichman and John Mitchell rescheduled for September 30, Special Prosecutor Leon Jaworski felt growing pressure to indict their boss as well. Hoping to take the decision out of Jaworski's hands, Nixon defenders looked to Ford, a reluctant conscript in the battle over his predecessor's papers and tapes.

Ford had "put a lock on his records," Nixon complained after being served with a subpoena by John Ehrlichman's lawyers. It wasn't because the new president relished the legal and political wrangling that went with their stewardship. Even had he wanted to hand them over to Nixon, he was prevented from doing so by at least two restraining orders aimed at the Ford White House. On August 22, Ford raised the issue with Phil Buchen and Attorney General Saxbe, from whom he requested a legal opinion regarding ownership of the Nixon papers and tapes. The task was farmed out to Antonin Scalia, a Justice Department lawyer destined for bigger things. In his memo Scalia recounted two centuries of tradition by which departing chief executives had carted off their presidential papers along with the family furniture and pets. To Saxbe and Haig, it was an open-and-shut case. They were not shy about arguing their position.*

Damn tradition, countered Bob Hartmann and Benton Becker. The circumstances of Nixon's departure, not to mention the still-unfolding criminal investigation and trials to come, made *this* presidential archive sui generis.

As each side disputed the other's legal reasoning, they challenged Ford's faith in teamwork. A product of his athletic, military and political training, Gerald Ford "had the feeling everybody was working for the cause. Well, they *weren't* working for the cause," contended Hartmann, whose worst suspicions were realized when Justice leaked Scalia's memo to the press *before* sharing it with the president. Other factors contributed to Hartmann's abiding distrust of the so-called Praetorian Guard. "They thought Gerald Ford wasn't really very smart. And so he needed them to protect him from his own mistakes . . . hell, these little shits, they talked down to him and around him . . . in their memoranda they would put things before the president in a way as if[, if] it wasn't reduced to one-syllable words he wouldn't be able to understand it. Well, I knew he was quite capable of grasping rather complex ideas in [three-syllable] words."

---

* Saxbe would be among the first Nixon Cabinet members to be replaced. He was no slave to desk work; his frequent absences from the department moved subordinates to develop an elaborate charade when approached about politically sensitive matters such as antitrust cases. Responding to a telephone inquiry, Deputy Attorney General Laurence Silberman would offer grave assurances to the caller—only the attorney general himself could resolve the question at hand. With that, according to Robert Bork, Silberman "would go out of the office and down the hall and into the bay for elevators and smoke a cigarette and come back and say, 'The attorney general says . . .' and he went with it." AI, Robert Bork.

The debate over Nixon's papers proved his point. As the attorney general's tone became more patronizing, Benton Becker felt the ground shift. Watching the Saxbe-Ford exchange, Becker said later, "I didn't see the Oval Office; I saw a senator talking down to a congressman." When his turn came, Becker forecast "a hell of a bonfire in San Clemente" should Nixon succeed in gaining title to the documentary record of his administration. But it was a second Becker prediction that riveted Ford's attention. Whatever he might accomplish during his remaining time in office, said Becker, it would be overshadowed by the damning conclusion of future historians that "Jerry Ford committed the final act of the Watergate cover up when he sent those records and papers and tapes to Nixon in California."

At this the president of the United States did a slow burn. According to Becker, "You could see from the neck up, literally see the color change." Finally Ford broke his silence. "Those damn things are not leaving here," he declared. "They belong to the American people." Behind the scenes he encouraged Tip O'Neill and other trusted lawmakers to address the issue through legislation. "We're going to get an act," Ford told Becker, "and it's going to be retroactive to Richard Nixon." Until then Becker proposed a radical solution of his own: Why not have Judge Sirica take possession of the vast archive? This would relieve Ford of his custodial responsibilities, while assuring Nixon of his rights and Jaworski that no one would tamper with potential evidence. Becker's idea fell afoul of Sirica's reluctance to further insinuate himself between two presidents, a special prosecutor, and untold lawyers and members of Congress.

A Gallup poll published on August 26 gave Ford a 71 percent approval rating. Gushing sentiments filled the presidential mailbag. "You have made a great beginning," wrote Ohio Republican congressman Wiley Mayne, for whom the new administration seemed "a fresh breeze blowing across this nation." Praise nearly as effusive came from prominent Democrats. "I think the thing is off to a helluva start," declared JFK's press secretary, Pierre Salinger. George McGovern promised his help to a president "who is doing all he can to find the best course for the country." Ted Kennedy offered assistance in working out details of the Vietnam amnesty program. Even *Pravda*, official newspaper of the Soviet Communist Party, pronounced itself pleasantly surprised by Ford's early performance.

The same could not be said of domestic publications like *Human Events*, the conservative Bible whose unhappiness with the new regime was expressed through headlines such as "Militant Feminists Find Friend at White House." In distant Sacramento, Ronald Reagan, nearing the end of his second term as governor, declared anew his opposition to amnesty, conditional or otherwise, for those who had evaded military service in Southeast Asia. Fanning disenchantment on the right, Reagan operatives urged GOP power brokers to avoid early commitments in the run-up to 1976.

A different kind of threat was posed by a shadowy adversary, his name unknown to the president's protectors. In the third week of August, representatives of the CIA and Secret Service joined other law enforcement agencies working out of a basement room in the White House. They pooled their skills in a feverish race to identify the Alphabet Bomber, so named for the different letters of the alphabet with which he had announced a series of bombings in and around Los Angeles. Ten days into the Ford presidency, this random killer declared his intention to come to Washington and hurl a nerve gas bomb at the new president. Such a device might kill hundreds of spectators in the bargain. With invaluable assistance from psychological profilers, authorities took just eighteen hours to narrow their search to a single individual. When arrested on August 20, Maharem Kerbequovic, a twenty-eight-year-old native of Yugoslavia, had in his possession nearly everything needed to manufacture the deadly weapon he had described. He lacked only an organic phosphate, which even then was waiting at LAX for the would-be assassin to pick up. Americans never knew how close they came to losing a second president in as many weeks.*

A YEAR BEFORE his death in 2011, ABC News political director Hal Bruno reflected on a conversation he had with John Mitchell in mid-July 1974. Promised anonymity, the former attorney general said of Nixon's palace guard, "Those characters have finally learned to count. All they've got to do is iron out the Jerry Ford deal and Nixon will resign." Based on this exchange, Bruno alerted his disbelieving colleagues at ABC, "This is the end. He's going to be out." Then there is Jerry Jones. Thirty-five years after the fact, White House staff secretary Jones described insistent phone calls from a "frantic and distressed" Richard Nixon in his San Clemente compound. Directed to Al Haig, they conveyed an undeviating message: "Where in the hell's my pardon?" Careful to label his comments as speculative, Jones guessed that Haig, resolved to bring down the curtain on Nixon's tarnished presidency without inflicting additional damage on a traumatized nation, had encouraged the former president to believe that Ford would save him from still-greater humiliations.

Ford's presidency was still in its infancy when the head of the Secret Ser-

---

* Kerbequovic was subsequently tried and convicted for the deaths of three people in the August 6, 1974, bombing of the overseas lobby of Los Angeles International Airport.

vice alerted Bob Hartmann to the continued existence of listening devices in the Oval Office. At least two were embedded in the president's desk, with two more hidden by brass fixtures in the wall above the fireplace. "Dammit!" Ford said on learning the news. "They told me definitely there weren't any—that they'd all been removed long ago . . . you remember, I asked Al Haig specifically." The general was full of apologies. The Secret Service hastily extracted the bugs and replastered the wall. Unfortunately the agents were incapable of protecting Ford from an escalating power struggle between Haig and Hartmann. "Nothing irritates me more than having people I trust argue with other people I trust," said Ford.

Perhaps he trusted too much. Intent on replacing Haig as chief of staff, Hartmann exploited his journalistic connections to plant unflattering stories about the general. Readers of Evans and Novak were treated to leaked allegations tying Haig to the bogus shipment of jeweled gifts to Pat Nixon. The same columnists accused Haig of trying to slip past an unsuspecting Ford the appointment of Pat Buchanan as US ambassador to South Africa.* Asked why he skipped the general's daily meetings of senior staff members, Hartmann snapped, "Fuck Haig. I work for the President." Haig retaliated by repeating salacious reports from Secret Service agents describing late-night sessions in which an inebriated Hartmann summoned women to his West Wing office for impromptu dictation. Ford issued verbal reprimands to the offenders, but took no more forceful action.

In preparation for his first White House press conference, set for the afternoon of August 28, Ford answered mock inquiries from staffers who pressed him about Cyprus, his still undefined Vietnam amnesty, Rockefeller's nomination and the fight against inflation. The same handlers warned Ford that none of these topics was likely to interest reporters as much as the fate of Richard Nixon. Ford was skeptical. As he said later, "I thought we'd gotten over the hump of Nixon." It was a view held by few around him. Nearing the end of Ford's first month in office, what Jerry Jones called "the whole Nixon thing" was occupying more, not less, of the president's time. "It's distracting," Jones told Haig. "We're not getting the country's business done. I don't see how, without a pardon getting this off the agenda, the president is going to get to be president."

Nixon's legal fate was the least of it. The disposition of his papers and

---

* Furious over an Evans and Novak column in the *Washington Post* headlined "Mr. Ford's Advisers: Gen. Haig Must Go," the general threatened Novak with a $5 million libel suit. "Al, you're out of luck," replied the columnist, clearly unrepentant. "I don't have five million dollars." Robert Novak, *The Prince of Darkness* (New York: Crown Forum, 2007), 257.

tapes was no closer to resolution than on the day of his resignation. There was growing opposition to an $850,000 appropriation earmarked for the former president's transition to private life. Some lawmakers wanted to eliminate Nixon's government pension. Len Garment, Nixon's onetime law partner, personified those on the White House staff whom Ford refused to sacrifice in what he called "a Stalin-like purge." On Tuesday, August 27, Garment lunched with CBS News commentator Eric Sevareid and John Osborne of the *New Republic*. Garment asked his lunch partners, neither regarded as a Nixon apologist, whether Ford should pardon his predecessor. Both men replied in the affirmative. Garment shared this fact with Haig, who told him to draft a memorandum incorporating their arguments and address it to Phil Buchen, with a copy to the general. Garment worked through the night outlining the case for an immediate pardon. Wednesday morning he was at his White House desk early to finalize and submit copies of his memo to Buchen and Haig.

Garment remembered a midmorning call from Haig confirming that Ford would use that afternoon's press conference to announce a pardon. Half an hour later the general called back with dispiriting news, informing Garment, "They're lawyering it now." Phil Buchen told a different story. "You don't want to read this thing," he assured Ford, who was busy preparing for his encounter with reporters. Instead Buchen relayed the contents of Garment's memo to the president verbally. The two men agreed that any intervention in the legal process on Nixon's behalf would be premature. Thirty minutes of intensive questioning before television cameras was to inspire second thoughts. At the initiative of Jerry terHorst, the East Room had been physically transformed from the bear pit in which Nixon and hostile journalists had done battle. Gone was the bulky lectern that had served as protective shield, and the cool blue draping that supplied a regal backdrop for TV viewers. In their place terHorst relocated Ford before an open doorway leading to the red-carpeted Grand Hallway and State Dining Room just out of camera range.

Some tribal customs remained inviolate, like the opening question reserved for Helen Thomas of UPI. Unsurprisingly it concerned his predecessor's legal standing. Did Ford agree with the American Bar Association "that the laws apply equally to all men," Thomas asked, or with Nelson Rockefeller's recently stated belief that his forced resignation and subsequent disgrace should earn Nixon immunity from prosecution? Acknowledging "I am the final authority," Ford said that "in the last ten days or two weeks" he had prayed for guidance on the subject. No charges had been filed, nor any action taken by the courts. Until such time as this changed, it would be "unwise and untimely" for him to make any commitment. So far, so good. Then a second questioner wondered if Ford fretted about a double standard of justice, one for the former president and another for his aides about to go on trial for alleged crimes committed in his name. Ford reiterated his legal right to pardon Nixon, while simultaneously stressing Leon Jaworski's obligation to take "whatever action he sees fit in conformity with his oath of office." The impression left, intended or accidental, was of a man disinclined to let his predecessor see the inside of a jail cell, although perfectly willing to have the judicial process run its course.

"Well, that's the first one," Ford said to terHorst as they left the East Room. Inwardly he seethed, his disappointment in the press exceeded only by the anger he felt over his own confusing and contradictory responses. "Yet for me to say that Jaworski shouldn't do his duty would be telling a law enforcement officer . . . that he shouldn't carry out his oath of office . . . the allegation could have been made that I was signaling the Special Prosecutor that he ought to treat one differently from another."

For the rest of his life Ford would misremember a near monopoly of questions about his predecessor (in fact, only eight of twenty-nine queries put to him involved Nixon). The harshest adjective the *Washington Post* could come up with for Ford's performance was "workmanlike." Even then, the paper credited him for restoring civility to the face-off between president and press. The *New York Times* went further, portraying a relaxed, confident executive batting away inside fastballs and making news by signaling his willingness to emulate the Organization of American States should that hemispheric body decide to relax its economic and political boycott of Castro's Cuba. As for Nixon's unresolved legal status, the *Times* concluded that Ford had struck "just the right balance" in assessing the national mood. "There is neither desire nor intention—on anyone's part—to see Mr. Nixon persecuted," the editors asserted.

That was not the message Ford took away from the East Room. "Is this the way it's going to be for the next two years?" he mused. "Am I going to be dogged at every press conference with questions about Nixon and a pardon, and should he go to jail, should he stand trial?" Personally Ford expected Nixon to be indicted, tried and convicted of obstructing justice, at the least. "And this would take three years at a minimum, and that would just take me through '76 and beyond."

"That's when I called Phil Buchen in."

FORD SUMMONED HIS old friend for a task he could no longer put off. In his memoirs he makes only passing reference to a late-afternoon meeting on August 28 attended by Buchen, Marsh and Haig. Diary notes kept by Buchen's wife, Bunny, provide a window into Ford's thinking. Reluctant to saddle Jaworski with "this terrible burden of prosecuting" Nixon, the president said he was strongly inclined to pardon his predecessor. "My conscience tells me I must do this. I have followed my conscience before—and look where it has brought me." Ford continued, "If what I am doing is judged wrong I am willing to accept the consequences." Turning to Buchen, the president sought legal advice "as to how, when, and under what circumstances I can grant a pardon."

Beyond questions of his legal authority, Ford asked Buchen to get from the special prosecutor a full accounting of any offenses for which Nixon might be prosecuted. He didn't want to issue a blanket pardon only to be surprised by fresh allegations. The same emissary was to obtain Jaworski's estimate of how long it might take before the former president could have his day in court. Sufficiently familiar with Ford's decision-making to realize he was dealing with a fait accompli, Buchen made no effort to dissuade him. In the meantime, he had a suggestion of his own. "Look, if you're going to do this—to sort of put Watergate behind you—I think you also ought to let me see how far we can go to getting an agreement on the papers, and have that in place at the same time."

"Well," said Ford, "I don't want you to condition the pardon on his making an agreement on the papers."

Buchen deferred to the president's wishes, without backing off his request. "Can I work it so by *his* agreement it will come at the same time?" To this formulation Ford readily agreed. Buchen said he couldn't do the necessary research alone. "How about my going to Benton Becker?" Again the president gave his assent. Convinced that Washington law firms were leaking sieves, Buchen reached out to the Grand Rapids firm of Warner Norcross + Judd. He needed some research done, he confided to senior partner Harold Sawyer. "I don't need anything in writing. In fact, I don't want anything in writing."* That evening the Fords hosted a hundred guests at a White House dinner honoring members of the Cabinet. Showered with flattering comments about his earlier press conference, Ford affected a jaunty, even boastful tone. After the first five minutes of journalistic interrogation, he told his guests, "it was just like any football game."

The analogy was more apt than he knew. As vice president Ford's on-

* So great was the concern over leaks that Benton Becker's findings were presented to Ford in longhand: "Wouldn't let anybody type it . . . Even my secretary who was super loyal. They wouldn't even let her type it." AI, Benton Becker.

again, off-again defense of Nixon had earned him scorn in some quarters as a "zigzagger." Now it appeared that history was repeating itself, with Ford sowing enough doubts about his intentions to prompt his predecessor to hire himself a new lawyer. Herbert "Jack" Miller, fifty years old, was an alumnus of Bobby Kennedy's Justice Department, where he had led prosecution efforts against Teamsters boss Jimmy Hoffa and LBJ protégé Bobby Baker. An unsuccessful GOP candidate for lieutenant governor of Maryland, Miller was an accomplished plea bargainer who had already convinced Leon Jaworski to go easy on former attorney general Richard Kleindienst. At Jaworski's request Miller submitted in writing his arguments against a Nixon indictment. These emphasized massive pretrial publicity, which Miller contended made it impossible for his client to be fairly tried anywhere in the country that had twice elected him to its highest office.

Ford used similar claims in rebutting objections to a Nixon pardon from the handful of advisers taken into his confidence. After the fact he would be criticized for his failure to consult more widely, or prepare public opinion by raising the possibility of a pardon more directly than in his press conference of August 28. In the politically supercharged climate then prevailing, Ford believed that any trial balloon would be shot down before it cleared tree level. A fear of leaks prevented him from discussing the matter with members of Congress or his own attorney general. Additionally there was the glacial pace of justice to consider. A Nixon trial and subsequent appeals could stretch on for years, mesmerizing the press and diverting public attention from Ford's domestic and foreign agendas.

"And Nixon would not spend time quietly in San Clemente," Ford wrote in his memoirs. "He would be fighting for his freedom, taking his cause to the people . . . No other issue could compete with the drama of a former President trying to stay out of jail."

He could not argue with Ford's analysis, Buchen told him, "but is this the right time?"

"Will there *ever* be a right time?" Ford grimaced.

Hartmann was next in line to question Ford's judgment. "You should be aware of the liabilities," he told the president. "There will be strong editorial condemnation. Your standing with the public will suffer, because they won't understand." He reminded Ford of the upcoming congressional elections and the potentially ruinous impact of a Nixon pardon on Republican prospects. Citing recent opinion polls to bolster his case, Hartmann touched a raw nerve. "I don't need the polls to tell me whether I'm right or wrong," Ford barked. In fact he had reason to believe that a bipartisan consensus existed on Capitol Hill in support of a Nixon pardon, one grounded in earlier efforts to goad Nixon into leaving office voluntarily. On this point Judiciary Committee chairman Peter Rodino had been explicit, telling his Republican colleague Robert McClory on August 6, "If he resigns, we can drop all this, the impeachment, the threat of criminal proceedings. If he quits, that's the end, that's it."

Rodino's offer had been seconded by fellow Democrat Don Edwards, one of Nixon's harshest critics in Congress. Moreover, Rodino assured McClory, Speaker Carl Albert felt the same way, a claim the Illinois Republican had taken pains to verify. Eager to convey this olive branch to the Nixon White House, McClory had contacted Bill Casselman, formerly his legislative assistant and more recently a counselor to Vice President Ford. Casselman urged McClory to reach out to Nixon's director of congressional relations, Bill Timmons. By the time McClory tracked Timmons to his West Wing office, the president was behind closed doors with the delegation led by Barry Goldwater and Hugh Scott.

Bill Casselman may have closed his ears to McClory's bombshell; it defies logic that this bipartisan game changer would have been kept a secret from Ford during his first month in the White House. Since August 9 it had become apparent that neither the Speaker nor Congress itself could unilaterally halt Nixon's prosecution. That said, Albert's show of magnanimity may well have strengthened Ford in what he wanted to believe—that any outrage occasioned by a Nixon pardon would be shallow and short-lived.

SATURDAY, AUGUST 31, marked the start of an extended Labor Day holiday. Ford looked forward to an afternoon's golf at Burning Tree, prelude to two days at Camp David, the presidential retreat nestled high in the Catoctin Mountains of Western Maryland. Before availing himself of the camp's heated swimming pool, however, Ford hosted Attorney General Saxbe and Defense Secretary Schlesinger for a discussion of his Vietnam amnesty in the making. "I don't like the idea of amnesty," Saxbe told a *Newsweek* reporter in advance of his meeting with Ford. "But when the President tells you to do something, you do it." Reflecting this attitude, the plan devised by the two Cabinet officers provided only limited options for alternative forms of service in return for an amnesty. Its authors frowned on Ford's suggestion that these be expanded to include private institutions. To objections that such an arrangement would hardly constitute punishment, Ford snapped, "Amnesty isn't supposed to be punishment."

Forgiveness was a recurring theme this holiday weekend. From the Oval Office, Ford telephoned evangelist Billy Graham, who strongly urged a presidential pardon for their mutual friend. "There are many angles to it," Ford told Graham. "I'm certainly giving it a lot of thought and prayer." Beyond this he was unwilling to tip his hand. More temporal considerations drove Ford's designated agents in their search for legal precedent. In his West Wing office, Phil Buchen spent the weekend tracing executive authority to pardon someone as yet unconvicted or even accused of a crime. Six hundred miles away, Hal Sawyer convened a quartet of Grand Rapids lawyers to trawl the writings of Hamilton and Madison.

But it was Benton Becker, working out of the law library of the Supreme Court, who struck pay dirt. Rummaging through the legal attic Becker came upon an obscure 1915 Supreme Court ruling that redefined the pardoning power. The case involved one George Burdick, city editor of the *New York Tribune*, who had refused to divulge his sources in a corruption investigation of the New York Customs office. The first time he appeared before a grand jury, Burdick pled the Fifth Amendment and was excused by the US Attorney from giving further testimony. A month later, recalled to the stand, Burdick was handed a piece of paper signed by President Woodrow Wilson, pardoning him in advance for any wrongdoing he might reveal.

To accept the pardon would forfeit Burdick's constitutional right against self-incrimination; it would also eliminate the legal shield protecting him and other journalists plying their trade. In the event, Burdick rejected presidential clemency and was jailed for contempt of court. In due time his case came before the United States Supreme Court, which ruled 8–0 that Burdick was well within his rights to refuse the proffered pardon. Writing for the majority, Justice Joseph McKenna declared that a pardon "carries an imputation of guilt; acceptance a confession of it." As interpreted by Becker, the Burdick ruling gave Ford the legal and political rationale he was seeking for an act designed to refocus attention on issues greater than the fate of Richard Nixon.

Late on the afternoon of September 3, his Camp David holiday a pleasant memory, Ford met with Buchen and Becker in the Oval Office. That evening Buchen and his wife paid a social call on Leon Jaworski, their neighbor at the Jefferson Hotel on Sixteenth Street. To Bunny Buchen, Jaworski appeared a lonely, even "tragic" figure prone to repeat himself, especially on the subject of his beloved Circle J Ranch between Austin and San Antonio. The next day Bunny's husband returned to the Jaworski suite by himself. He found the special prosecutor still trying to decode the pres-

THE PARDON | 387

ident's conflicting comments at his press conference. "It sounded like he was saying that any action I might take against Nixon would be futile," Jaworski observed.

Buchen replied with a query of his own: "Can't tell you why I'm asking it . . . do you have any view on how long it would take to bring Nixon to trial if he were indicted? Can you give me your view in writing?"

Buchen got his answer later in the day. Jaworski estimated that nine months to a year would elapse—"And perhaps even longer"—before he could begin jury selection in any trial of the former president. Jaworski provided Buchen a comprehensive assessment of areas under investigation, ten in all, exclusive of the Watergate cover-up. An accompanying note from Assistant Attorney General Henry Ruth indicated that to date none of the allegations being probed were even close to being provable criminal violations. "One can make a strong argument for leniency," Ruth observed, "and, if President Ford is so inclined, I think he ought to do it early rather than late."

Similar sentiments were voiced by Henry Kissinger, prompted by a lunchtime conversation he had with Bryce Harlow. As Kissinger recounted it, Harlow "told me and I agreed, that Nixon would never make it through a trial. That he would have some sort of a breakdown, that he could not stand that humiliation . . . afterwards I told Ford that I'd had that lunch, that Bryce was of that view and that I wanted him to know that I was of that view, too. And that I thought the spectacle of putting the president of the United States before a criminal court was something that the country should not have to face. He asked one or two questions, and then the next time I heard from him was the night before he announced the pardon."

CONSCIENTIOUS TO A fault, Phil Buchen had spent the past few days debating with himself the morality and timing of a Nixon pardon. Referring to the former president, he mused, "If you know a man is about to walk a plank, when do you decide to save him? When he first steps upon the plank or when he gets to the middle or when he gets near the end or when he plunges over and you dive after him to save him from drowning?" To help resolve his doubts Buchen sought spiritual guidance from his minister in Grand Rapids. Duncan Littlefair (to more orthodox believers "Duncan Littlefaith") was no ordinary churchman. A charismatic critic of organized religion, Littlefair presided over an oasis of non-creedal spirituality in a fundamentalist town.

Boasting "I have never, even slightly, encouraged any supernatural idea or belief," Littlefair said he used the term "God" "poetically." During his ministry the Fountain Street Church had severed ties to the Baptist Church and renamed its Sunday school "Character School." Guest speakers Eleanor Roosevelt, Margaret Mead and Malcolm X had been welcomed to the massive Romanesque sanctuary where Old Testament prophets in stained glass shared space with Plato, Darwin and Thomas Jefferson.

Next to Reverend Littlefair, no one knew the Fountain Street story better than Phil Buchen, the author of three volumes collectively titled *The Only Church with a History Like This*. Given his legal, ethical and political reservations about a Nixon pardon, it was only natural for Buchen to seek out the clergyman he counted among his closest friends, even if it meant breaching the vow of secrecy exacted by his friend the president. Littlefair arrived in Washington on September 4, the same day army chief of staff General Creighton Abrams succumbed to cancer at Walter Reed Army Hospital. Abrams's passing created a vacancy ideally timed to resolve the issue of Al Haig's future. So it seemed to Ford. Taking exception to the idea, Secretary of the Army Howard "Bo" Callaway warned Ford that Haig's appointment would be met with "strong, perhaps even bitter" resistance from traditionalists who questioned the general's meteoric rise from colonel to four-star general while serving Kissinger and Nixon.

Other personnel changes aroused less controversy. Even as Reverend Littlefair settled into the Buchens' suite at the Jefferson Hotel, it was announced that George H. W. Bush would be taking over the US Liaison Office in Beijing. Replacing Bush as RNC chairman was Iowa's Mary Louise Smith, a self-professed "professional feminist" and the first of her sex to chair a national political party. On the afternoon of September 4 Ford looked on as economist Alan Greenspan was sworn in as chairman of the Council of Economic Advisers. It was the start of a history-making career that would include five terms as chairman of the Federal Reserve under presidents of both parties.

At the time he joined the Ford Administration, however, Greenspan generated far less attention than Betty Ford's simultaneous press conference in the State Dining Room, the first by a president's wife since Eleanor Roosevelt. For thirty minutes Mrs. Ford answered questions about her support for the Supreme Court's 1972 *Roe v. Wade* decision legalizing abortion, and for the proposed Equal Rights Amendment to the Constitution. That evening she and the president attended a Kennedy Center benefit performance of the musical *Mack and Mabel*. During intermission Ford prevailed on Nancy Hanks, chairwoman of the National Endowment for the Arts, to introduce him to choreographer Agnes de Mille. At one point he was seen deep in conversation with Julie Nixon Eisenhower.*

Ten minutes away, in Suite 805 of the Jefferson Hotel, Duncan Littlefair tossed restlessly in his bed. Around midnight the sleepless visitor threw off his covers and repaired to a nearby desk to fashion a statement Ford might use in pardoning Nixon. Americans in their hearts accepted "that however necessary Justice may be, it must be tempered by mercy," wrote Littlefair. "Forgiveness is at the heart of every world religion. Christians daily proclaim the ideal in prayer." Quoting Lincoln on the imperative of binding up the nation's wounds, Littlefair would have Ford justify a pardon as an attempt "to make us whole again . . . to help wash out the bitterness and resentment . . . to rise above vindictiveness and transcend the need to punish."

Not a little of Reverend Littlefair's midnight theology would find its way into Ford's September 9 announcement, with unintended consequences for the man whose signature on the Nixon pardon left his own presidency hanging in the balance.

THURSDAY MORNING, SEPTEMBER 5, began with a telephone call from Phil Buchen's West Wing office to the Jefferson Hotel. Could Bunny and the Reverend Littlefair please vacate Suite 805 by nine thirty? Phil and Benton Becker needed the premises to continue their negotiations with Jack Miller. These had reached a critical point, spurred by Buchen's revelation that Ford was considering a pardon for Miller's client. The lawyers agreed that a statement of contrition from Nixon would be in order, something, as Buchen put it, "that will tell the world, yes, he did it and he's accepting the pardon because he is guilty . . ." As they inched toward a deal splitting ownership of the disputed records between the former president and the General Services Administration acting as Ford's agent, the White House tapes presented an even thornier challenge. According to Al Haig, Nixon's fierce possessiveness stemmed in part from privacy concerns over intimate conversations touching on members of his family.

When the former president requested a meeting with Miller to review his options, the lawyer suggested that a representative of the Ford White House accompany him to San Clemente. Buchen was ruled out, his

---

* According to one onlooker, "She pulled him aside for a private talk about her father's health." Perhaps. But it is impossible to imagine Ford failing to raise the subject on his own, particularly with the younger Nixon daughter, a Ford favorite.

disability making travel difficult, and his face too familiar to risk journalistic exposure. Meanwhile the clock was ticking. Ford hoped to announce the pardon as early as the next day, and not just because he feared press leaks. "Leon volunteered to Phil that the grand jury is getting antsy . . . they want to move," recalled Benton Becker. "Some of my staff people are too," Jaworski had observed. So Becker became Ford's representative. He got little encouragement from Haig, who was convinced that Nixon would never surrender control over his records, nor sign his name to a confession of guilt. Jack Miller shared his pessimism. "Ron seems to be unwilling to make any concessions," he told Becker, referring to former White House press secretary Ron Ziegler. Having followed Nixon into exile, Ziegler more than lived up to his pugnacious reputation when, shortly after midnight, the two lawyers finally appeared at the deserted presidential compound to review the next day's schedule.

"Hello, Mr. Becker," said Ziegler. "Let's get one thing straight immediately. President Nixon is not issuing any statement whatsoever regarding Watergate. Whether or not Jerry Ford pardons him . . ." Becker, weary from the long flight and feeling the three-hour time difference between coasts, was determined not to cede the upper hand to Nixon's adjutant. Pleading ignorance of protocol—"I don't know how to reach the pilot that flew me out here, or the driver that dropped me off here"—Becker asked if he could use Ziegler's phone. "Because I'm going home." At this point Jack Miller intervened, figuratively separating the combatants. It was two in the morning when Miller and Becker checked into the San Clemente Inn for a few hours' sleep. Becker rose at five thirty (eight thirty Washington time) to call Phil Buchen and relate his initial run-in with Mr. Ziegler. Buchen responded with a tentative schedule leading to a Sunday-afternoon pardon announcement.

Could the secret hold until then? At the hotel's breakfast bar Miller was approached by a *Los Angeles Times* reporter curious as to the reason for his visit. Becker excused himself while Miller explained to the inquisitive journalist that they were in town to negotiate the issue of Nixon's papers and tapes. It was this same white lie that Buchen was to repeat to Jerry terHorst that afternoon, after David Kraslow, Washington bureau chief for Cox Newspapers, asked Ford's press secretary about rumors of a pardon in the works. Based on Buchen's disclaimer, terHorst told both Kraslow and *Newsweek*'s Tom DeFrank that no pardon was being negotiated. With this terHorst unwittingly set in motion a train of events leading to his resignation and a crippling loss of credibility for the Ford White House.

Yet what was the alternative? Sensitive to the relationship of trust between terHorst and his journalistic colleagues, Ford thought he was protecting his press secretary by leaving him out of the loop until the last possible moment. "We didn't tell Jerry, not because we didn't trust him, but because we knew he couldn't tell a lie," Ford explained. In fact the last thirty days had exacted a heavy toll on the career reporter whose prior experience left him ill-prepared for the brutal schedule and administrative demands of his new job. Deputy press secretary Jack Hushen witnessed the transformation of a "clear eyed steady handed guy" into one with bloodshot eyes and tremors. "He was looking for a way out," says Hushen. "The job was overwhelming," concurs Ron Nessen, who inherited the position following terHorst's resignation in the wake of the pardon. "There was a fifty-five-person staff to manage; it was a job where you'd go in there in time for the seven o'clock staff meeting in the morning, and if you were like me and you tried to answer all reporters' calls before you went home, you left there at 9:30–10:00 or later at night."

For terHorst the option of resignation might not have presented itself had Nixon and the lawyers restricted their negotiations to White House papers and tapes. Under the tentative deal arranged by Becker and Miller, the records, though physically removed to the West Coast, would remain subject to subpoena for three to five years; in addition Nixon was enjoined from destroying any originals. The tapes would be preserved in their original form, accessible to courts and congressional committees, for five years at least—maybe ten. Only then could they be destroyed (earlier in the event of Nixon's death). Frequent interruptions slowed the legal dickering, as Ziegler or Miller disappeared into a nearby office where their client lurked, insistent on reviewing every word of the emerging settlement. Becker took advantage of these intervals to telephone Buchen at the White House. Discretion was the order of the day, since both men believed their conversations were being monitored.

By midday Friday the focus had abruptly shifted. Ziegler handed Becker a draft statement for issuance at the same hour the pardon was revealed in Washington.

"What do you think?"

Adopting his best poker face, Becker scanned Ziegler's brief ascribing Watergate to the president's preoccupation with more important matters, and a misplaced faith in staff who betrayed his trust. "No statement is better than that statement," Becker concluded. Not only did it challenge the unanimously held conclusions of the House Judiciary Committee, in which 67 percent of the American people concurred. Such a sweeping denial of responsibility invited state or local prosecutors to fill the void created by a presidential pardon.

"Ron, he's right," volunteered Jack Miller. "He's absolutely right."

Three drafts later, Becker held in his hands a grudging acknowledgment of poor judgment, which he chose to interpret as an implicit admission of guilt: "I was wrong in not dealing with Watergate more forthrightly and directly, *particularly when it reached a judicial stage*" (emphasis added). To Becker this was the critical sentence affirming Nixon's complicity in the obstruction of justice. The same language would strike nonlawyers as opaque and evasive. Ford himself, pressed in later years to identify anything about that weekend that he would do differently, said he should have insisted on a more forthright statement of contrition from his predecessor.

As the afternoon wore on, the two sides narrowed their differences. By four o'clock only one task remained. Mindful of Ford's instructions, Becker requested a personal meeting with the former president, both to drive home the full implications of his acceptance, and to assess for himself—and Ford—Nixon's physical and mental state. Overcoming Ziegler's resistance, he was ushered into a barren office, its walls stripped of any reminder that the tremulous, prematurely aged occupant at his desk was only recently the most powerful man on the planet. To Becker, Nixon's head seemed freakishly swollen on his shrunken frame, the famous jowls more pronounced than usual. Standing, his stooped posture was reminiscent of the nursing home.

"Mr. President, your aides and I have accomplished some very good objectives for the nation and for you with respect to your papers."

Nixon did not respond. He stared vacantly as Becker mentioned the pardon and the relevance of the 1915 *Burdick* ruling, reiterating what Nixon already knew—to accept mercy from his successor was to concede an element of guilt he did not feel. On learning that Becker lived in Washington Nixon asked the White House lawyer how the football Redskins were likely to fare this season. Refusing to be diverted from the task at hand, Becker spelled out the full ramifications of their agreement. After a few more monosyllabic exchanges, he got up to leave. Once outside he was called back by Ziegler.

Nixon had thawed sufficiently to compliment Ford's emissary as "a gentleman. We've had enough of bullies." Glancing forlornly around the room, he said he wanted to give Becker something as a souvenir, but "They took it all away from me. Everything I had is gone." Opening the desk drawer

Nixon removed two small containers "from my personal jewelry box." Inside were a tie pin and presidential seal cuff links, the last in his possession. Handing them over to Becker he seemed close to tears.*

Their business transacted, Becker and Miller flew back to Washington. They arrived at Andrews Air Force Base not long after Ford returned to the same field from Philadelphia, where he had previewed the nation's bicentennial celebration with a dinner and speech commemorating the First Continental Congress. Sharing a head table with House Minority Leader John Rhodes and Pennsylvania governor Milton Shapp, the president said he was tempted by Rhodes's invitation to join him for a golf game at Burning Tree.

Much as he would like to accept, Ford told his successor as House Republican leader, "I have to make an announcement tomorrow."

"Am I going to like it?" Rhodes asked with a nervous chuckle.

Ford looked at his old friend with grim sincerity. "I hope so."

"*Oh shit,*" said Rhodes, proficient in the nonverbal language with which politicians communicate bad news.

A T SEVEN A.M. Saturday, September 7, Becker was back at the White House to be debriefed on his mission. "I am not a doctor," he told Ford, "but I have really serious questions in my mind that President Nixon is going to be alive at the time of the election."

Ford pointed out that 1976 was a long way off.

"I don't mean '76," said Becker. "I mean '74."

The president scanned the draft agreement governing Nixon's records. A call had already gone out to Art Sampson, head of the General Services Administration, which oversaw the nation's presidential libraries. Told to be at the White House by one o'clock—and nothing else—Sampson wondered if he was about to be fired, or appointed US ambassador to Jamaica. Ford was no more revealing to his inner circle. Becker concluded that he was "really pleased" with the San Clemente negotiations. No doubt that was the impression Ford wished to leave. Only much later would he admit to surprise and disappointment over Nixon's "nit picking" response to the offer of a pardon.

It could have been worse. When Becker called San Clemente to give formal notification of Sunday morning's announcement, he learned that

---

* On his return to Washington Becker presented Nixon's gift to Ford. "You take these and give them to the kids," he told him. AI, Benton Becker.

Ziegler was rewriting the feeble communiqué on which they had agreed less than twenty-four hours ago. "We can't tolerate any weakened statement," Ford told Becker. "Call Ziegler back and tell him that."

At midmorning Ford joined Mel Laird for the first round of a two-day golf tournament at Burning Tree. Conversation was more strained than usual, with Laird declining his friend's invitation to join him that evening in entertaining members of the upcoming Apollo/Soyuz space mission. Concerned that Laird might try to talk him out of it, Ford steered clear of the Nixon pardon.

Laird remembered things differently. Never lacking for confidence, he felt certain he could get thirty of the top people in the Congress "to visit with Ford and urge him to pardon Nixon for the good of the country"— after the fall elections. If Laird's scheme had been implemented, history would reflect a Thanksgiving pardon or, better yet, an act of Yuletide mercy enjoying bipartisan support. He was not the only former Cabinet official with a plan to spread the political pain of a Nixon pardon around. Drawing upon his experience in the case of Spiro Agnew, Elliot Richardson stressed prosecutorial latitude, including the option not to prosecute if it was deemed to be in the public interest. Jaworski should complete his investigation, Richardson argued, before sharing his findings with the attorney general, the White House and Congress. Ensuing discussions among the president, party leaders on Capitol Hill and the chairman and ranking member of the Judiciary Committees could then produce a consensus in support of the Justice Department's recommendation against prosecution.

Richardson's proposal, first broached at the end of August over lunch with representatives of the *Wall Street Journal*, was sufficiently well received to warrant a direct approach to the White House. On second thought, Richardson decided that Ford might prefer to gauge public reaction from a distance. As it happened, the hero of the Saturday Night Massacre had the perfect forum in which to test-market his ideas—a previously scheduled breakfast meeting with Godfrey Sperling of the *Christian Science Monitor*, whose weekly on-the-record sessions with Washington newsmakers had supplied the launchpad for many a trial balloon. Convinced he had a politically painless solution to Ford's dilemma, Richardson mentally circled the date of his upcoming Sperling Breakfast, Tuesday, September 10.

N O ONE ATTENDING the annual crab pick of the Alexandria Police Association on Saturday evening, September 7, could have guessed from his public demeanor that the president of the United States was harboring the biggest secret of his career. Mallet in hand, a relaxed Ford showed a trio of Soviet cosmonauts how to deconstruct hard-shell crabs. He was plainly enjoying himself in these familiar surroundings, relishing even a brief respite from the issue that threatened to engulf his month-old presidency. A decidedly grimmer atmosphere prevailed at the White House, where Jerry terHorst had just learned of Ford's intent to pardon Nixon for any crimes he may have committed during five and a half years in office. Angered by his last-minute notification—there was barely time to prepare press materials and summon reporters on their day off—terHorst could scarcely believe his ears.

"You know what this means?" he demanded of Bob Hartmann. "Jerry Ford is throwing away his presidency to do a favor for Richard Nixon."

That night as Ford prayed over his decision, terHorst was challenged by his wife and Vietnam War–protesting children. How, they asked, could he defend a nonconditional pardon for a president who broke the law while denying equal justice to young men who were true prisoners of conscience? Swayed by their arguments, terHorst was up much of the night crafting a letter of resignation.

Ironically the Sunday morning papers for September 8 marked the apogee of Gerald Ford's brief idyll with the press. A Gallup poll pitting Ford against Senator Edward Kennedy in a mythical 1976 matchup gave the Republican appointed president a crushing 57–33 percent advantage. In the *Washington Post* a prominent Democratic economist, Walter Heller, opined on how refreshing it was to be in a White House "open to a little laughter, open to a little dissent, and willing to face the unvarnished facts of life." A more sober appraisal in the *New York Times* from author Joe McGinniss, whose 1969 bestseller *The Selling of the President* was considered a classic of campaign reportage, questioned the durability of the Ford honeymoon. "In our lust for decent leadership," McGinniss wrote, "we are creating an idol which, history suggests, we will eventually feel compelled to destroy. That this idol is named Gerald Ford is accidental."

A FEW MINUTES before eight, Ford left for St. John's Episcopal Church, its pale yellow bulk visible across Lafayette Square from the White House. Joining the small congregation drawn to early Communion, Ford occupied Pew 54, used by presidents beginning with James Madison. Outside the church he was confronted by reporters inquiring about his plans for the rest of the day. "You'll find out soon enough," he told them.

By eight thirty he was back at the White House. A final review of his remarks announcing the pardon produced only one change. Recalling the grim portrayal of Nixon's physical and mental state supplied by Benton Becker, Ford scratched in a passing reference to the former president's precarious health. A routine courtesy in any other context, Ford's expression of concern would only add to public confusion over his motivation for sparing Nixon from any legal reckoning.

With little more than an hour until his eleven o'clock announcement, Ford called congressional leaders to give them a heads-up. Of those he reached, only Barry Goldwater and Tip O'Neill registered a protest. "What are you pardoning him for?" asked Goldwater. "It doesn't make any sense." Ford said he wouldn't argue the question with Mr. Conservative. "It's done."

"All right. History will probably thank you for it."

O'Neill was more pointed in his comments, bluntly warning Ford that his action would likely cost him the 1976 election.

Also on the phone was Jerry terHorst. He apologized to David Kraslow, the enterprising reporter whose potential scoop had been scuttled when ter-Horst deflected him from the pardon negotiations. "If I had known on Friday what I know now I would not have guided you away from that story," terHorst confessed. On the floor below the Oval Office, Jack Hushen marshaled a handful of reporters who comprised the hastily assembled press pool. Mystified by their Sunday-morning assignment, some speculated about possible military action being taken by the president. "Oh no," said one member of the group, clearly in jest. "He's going to pardon Nixon." This sally was greeted with hearty laughter.

Ford was about to vacate the Oval Office so that the television crew could set up its equipment when he was approached by terHorst, manila envelope in hand. "Mr. President, I have something here that you need to see." He handed Ford a single sheet of White House stationery conveying what terHorst called "the most difficult decision I ever have had to make." He could not support a pardon for Nixon "even before he has been charged with the commission of any crime." Nor could he defend such an act while denying similar relief for those who had evaded service in Vietnam or for Nixon associates imprisoned for their role in the Watergate scandal. "Try as I can," terHorst wrote, "it is impossible to conclude that the former President is more deserving of mercy than persons of lesser station in life whose offenses have had far less effect on our national well-being."

As he finished reading the letter Ford spun around in his seat, fixing his gaze toward the east door and windows opening onto the Rose Garden. His face a mask of equanimity, betraying no trace that he had just been kicked in the stomach, Ford said he was sorry for any misunderstanding. "I hope you would reconsider."

"My decision is firm, Mr. President. I've expressed my views in the letter."

Walking around his desk, Ford put his arm around terHorst.

"I hope our friendship will continue."

TerHorst had barely left the room when Ford dispatched Jack Marsh in an attempt to change his press secretary's mind. "Don't hurt the President this way—not today," Marsh pleaded. Moved by his sincerity, terHorst reluctantly took back his letter. Whether Ford could say anything that might induce him to take back his resignation as well seemed unlikely.

As it was, the president's televised statement reawakened doubts put to rest by his East Room inaugural. Perhaps Bob Hartmann had been right in urging Ford to forgo an Oval Office address ("Put out a press release"). The last-minute text contained more of Duncan Littlefair's Sunday-morning mercy than the pragmatic calculation with which Ford justified the pardon to himself.

Repeatedly invoking the Almighty, with whose help "I am sworn to uphold our laws," Ford referenced massive pretrial publicity that could endanger Nixon's right to "enjoying equal treatment with any other citizen" accused of violating the law. Lengthy delays and protracted litigation would inevitably rouse ugly passions, further polarizing the American people while calling into question the credibility of their government. And after all the legal proceedings were concluded, the courts might well rule that Nixon had been denied due process. Ultimately, Ford claimed, the fate of his predecessor concerned him less than "the immediate future of this great country." Deprived of domestic tranquility, Americans were trapped in "bad dreams" that he alone had the constitutional power to terminate.

Insisting, "I cannot rely upon public opinion polls to tell me what is right," Ford restated his conviction "that I, not as President but as a humble servant of God, will receive justice without mercy if I fail to show mercy." He concluded by reading the paragraph formally absolving Nixon of legal culpability "for all offenses against the United States" committed during his term of office. As he affixed his signature to the document, in a nearby room terHorst dashed off a note reiterating his earlier decision to resign. "I can see nothing in the next 24 hours that would change my mind." To those around him who disparaged the press secretary's motives, Ford vouched for terHorst's sincerity. "You just don't understand these evangelical Michigan Dutchmen," he told them.*

One thing was beyond debate—in quitting when and how he did, ter-Horst compounded the damage done by Ford's pardon. Admiring journalists harkened back to the Saturday Night Massacre of October 1973, with the genial terHorst cast in the Elliot Richardson role of conscientious objector. Incredibly Ford welcomed terHorst back to the Oval Office less than eight weeks later to help promote his newly published biography of the president. Like the pardon itself, his treatment of terHorst affirmed Ford's generosity, even as it raised doubts about his political judgment.

In time many would come around to the view, sanctioned in 2001 by the JFK Library when it presented Ford its prestigious Profile in Courage Award, that Ford's pardon was in the national interest. Yet even then questions lingered about the timing of his decision, and the hurriedly arranged Sunday-morning announcement. Why not wait? It was a question put to Ford by Bob Hartmann as well as Tip O'Neill. "Why must it be tomorrow?" Hartmann asked the president on Saturday night, September 7. "Why not Christmas Eve or a year from now, when things quiet down?"

"Some of the news people might ask me about it again," Ford replied.

"But all you have to do is say you haven't decided," said Hartmann, on the edge of exasperation.

"But I *have* decided," said Ford.

Some armchair psychologists professed to see history repeating itself, with Ford reenacting the peacemaker's role he had assumed during the turbulent court battle pitting his victimized mother against a morally deficient Leslie King. Arguably more relevant is a long-forgotten incident, trifling by itself, that nevertheless sheds light on Ford's personal as well as political character. The year was 1950. Then in his first term representing Michigan's Fifth District, the youthful congressman had his car stolen from a downtown Grand Rapids parking lot. Local police quickly spotted the vehicle and arrested the thief, a youth from nearby Battle Creek offering friends a drunken joyride before enlisting in the armed forces.

---

* Apparently even he had second thoughts about terHorst's motives. Jack Hushen recalls an encounter outside the Oval Office about a month after the pardon was announced, when Ford turned to him and abruptly inquired, "Why did Jerry terHorst quit?" Hushen offered his own explanation in which the pardon played a smaller role than terHorst's frustration with the punishing schedule and unforeseen demands of the job. "Maybe you're right," Ford murmured in response. AI, Jack Hushen.

Notified that his car had been recovered, Congressman Ford declined to press charges. "No need to keep this youngster out of the Air Force," he remarked.

Ford's merciful response gave evidence of a faith more formalized than the code of ethics espoused by Duncan Littlefair. Steve Ford, himself dismayed by the pardon, later described his father sitting him down for a discussion in which the elder Ford likened the American president to a paterfamilias: "At times the father of a family has to give grace and mercy to his kids for the betterment of the whole family. To keep the family together."

Still, the loss of his press secretary affected Ford more than he let on. Press aide Bill Rogers accompanied the president to the Burning Tree golf course that Sunday afternoon. Afterward Rogers observed that Ford imbibed two or three drinks in the clubhouse, "far more than he usually does at the end of a round."

"Do you know that Jerry terHorst is thinking of resigning?" the president had asked Rogers. He returned to the subject after dark. "I wish that Jerry could appreciate how difficult this decision was for me," said Ford. "I don't think he did." Conceding that good people could disagree with his course of action, Ford remained convinced that he had done right by the country. Suddenly he felt the full weight of his unsought office.

"Well, it's a much bigger ball game, Bill."

"Yes, sir," said Rogers. "but there's a pretty good batter at the plate."

AMONG FORD'S PRIVATELY stated reasons for pardoning Nixon was his desire to relieve Special Prosecutor Leon Jaworski of responsibility for deciding the former president's legal fate. The irony of this position was revealed only after he left office. In June 1982, ABC News marked ten years since the Watergate break-in by assembling seven of the twenty-three grand jurors originally summoned to investigate the Nixon White House. As they told it, prosecutors answering to Jaworski had drawn up a four-count indictment charging Nixon with "bribery, conspiracy, obstruction of justice, and obstruction of a criminal investigation." All it required for implementation was a formal jury vote and two signatures—those of the jury foreman and the special prosecutor. The grand jurors were unanimously prepared to endorse such a course.

Jaworski was not. Arguing that no precedent existed for indicting a sitting president, he had persuaded the grand jury to instead forward its evidence to the House Judiciary Committee for possible impeachment proceedings.

With Nixon's August 1974 resignation, the issue of presidential guilt resurfaced. Jaworski's deputy Philip Lacovara argued for an indictment of the now-former president, a view widely held among the prosecutorial staff and seconded by members of the grand jury. Jaworski thought he had suffered enough.

In reality, Jaworski had never intended to put Nixon on trial if he could possibly avoid it. Ford's pardon saved him from publicly disclosing this reluctance, something he confided before his death in October 1982 to Mel Laird and Houston congressman William Archer, both instrumental in his original appointment. By then the Ford Administration was receding in memory, an accident of history bracketed by Duncan Littlefair's mercy and Leon Jaworski's scruples.

Leslie Lynch King, around the time of his 1912 wedding to Dorothy Gardner. Charm and family connections masked King's dark side, which brutally revealed itself on their West Coast honeymoon. Two weeks after the birth of his son, Leslie Jr., in July 1913, King threatened mother and infant with a butcher knife.

Escaping her abusive spouse, Dorothy filed for a divorce and moved to Grand Rapids, Michigan. There she met and, in February 1917, married businessman Gerald Ford Sr.

Below: Ford Senior and Junior on the doorstep of their rented home, circa 1917. Even as Dorothy cloaked her failed marriage in mystery, Ford had his own secrets to keep. These included a bigamist father and the parentage of "Junie" Ford—or was it King?

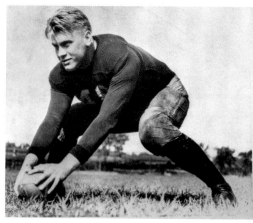

At the University of Michigan, Ford won national recognition for his football exploits. But his most enduring lesson involved an African-American teammate sidelined on account of his race, an incident that would have unexpected repercussions six decades later.

Four-year-old Junie Ford shows his support for US doughboys in World War I.

As a park ranger at Yellowstone in the summer of 1936, Ford rescued hikers and chased grizzly bears. As president he would propose a doubling of the National Park system budget.

Ford's involvement at Yale with the isolationist group America First led J. Edgar Hoover to blackball his application to join the FBI. After Pearl Harbor, Ford pulled strings to get on the USS *Monterey*, where he was the ship's assistant navigator.

An ungainly improvised aircraft carrier, the *Monterey* nearly capsized in a December 1944 typhoon.

Returning to Grand Rapids, Ford emulated hometown hero Senator Arthur Vandenburg in casting off his prewar isolationism. He also decided it was time to settle down and get married.

Tomboy Betty Bloomer, age nine, with her formidable mother, Hortense, outside the family's cottage in Pierson, Michigan. The cottage would become an unaffordable luxury after her father's death, a probable suicide, in July 1934. *Right:* Trained by Martha Graham, whose influence she ranked alongside that of her mother and Eleanor Roosevelt, Betty strikes a dance pose at Camp Bryn Afon in Rhinelander, Wisconsin, circa 1940.

Made skittish by the failure of her first marriage, Betty reconsidered when Jerry Ford entered her life. Jerry missed most of their wedding rehearsal dinner in October 1948—a preview of the all-consuming demands of a life in politics.

Campaigning against an isolationist GOP congressman in 1948, Ford vowed to return after the primary—win or lose— and help a farmer with his chores. He scored an upset win and kept his promise.

Ford told his congressional staff they would campaign 365 days a year. True to his word, when one elderly visitor to Washington fell and broke her ankle, no one knew how she would get back home to Michigan, until the congressman drove her there himself.

Jerry and Betty join Dwight and Mamie Eisenhower on the campaign trail in Grand Rapids in October 1952. Ford was the quintessential Eisenhower Republican—fiscally conservative and socially moderate to liberal. *Courtesy of Grand Rapids History and Special Collections, Archives, Grand Rapids Public Library*

Ford (*fourth from left*) at a formal presentation of the Warren Commission report to President Lyndon Johnson in September 1964. Initially inclined to believe theories of a foreign conspiracy in the assassination of John F. Kennedy, Ford went to his grave suspecting that a sexually impotent Lee Harvey Oswald killed JFK to impress his estranged Russian-born wife.

To Gerry Ford - with thanks for dedicated service, Lyndon Johnson

"Don't ever tell me again what a nice goddamn guy Jerry Ford is." All smiles in public only weeks after Ford ousted him as Republican Minority leader, Charlie Halleck (*left*), self-proclaimed "king of the gut-fighters," makes nice with Ford and Speaker John McCormack (*right*).

The *Ev and Jerry Show* was a weekly showcase for the loyal opposition to Lyndon B. Johnson's Great Society. Senator Everett Dirksen didn't always agree with his fellow Republican—especially on Vietnam.

"A very impressive ceremony," House Majority Leader Tip O'Neill said of Ford's swearing-in as vice president in December 1973, as Richard Nixon looked on. "We won't see anything like it for maybe six or eight months."
Courtesy of David Hume Kennerly

Bearing out O'Neill's prediction, the Fords, accompanied by David and Julie Nixon Eisenhower, watch the Nixons depart the White House for a California exile on August 9, 1974. Moments later Ford whispered into Betty's ear, "We can do it."
Courtesy of David Hume Kennerly

The Commuter President. Before moving into the White House, the new president and First Lady remained in their modest suburban residence. "There's something wrong with this picture," Betty told her husband one night as she prepared dinner. "You're president of the United States and I'm still cooking."

In the first surprise of his presidency, Ford announces a conditional amnesty for Vietnam draft evaders before a stunned Veterans of Foreign Wars convention in Chicago.

Ford chose Nelson Rockefeller to be vice president for his unparalleled experience and international credibility. It wasn't enough to prevent Rocky from clashing with Donald Rumsfeld and GOP conservatives who wanted him off the ticket in 1976.

The Ford family, enlarged by the addition of Mike's wife, Gayle (*far left*), celebrates Christmas in 1975 in a Vail, Colorado, ski resort that still bears their stamp.

Ford meets with Donald Rumsfeld, his reluctant, ambitious chief of staff in all but name, and Rumsfeld's deputy and eventual successor, Dick Cheney.

On September 8, 1974, a grim-faced president prepares to announce his pardon of Richard Nixon. Sharing the moment is Robert Hartmann, Ford's chief speechwriter and alter ego.

Caught up in controversy over the pardon, Ford became the first sitting president to testify before Congress. There the friendships and credibility earned over twenty-five years convinced most—though not all—that there was no deal between Nixon and Ford.

Days after undergoing surgery for breast cancer, Betty Ford shows she hasn't lost her sense of humor. By going public with her illness, the First Lady dispelled the stigma and secrecy that had traditionally surrounded the disease.

Ford welcomes ex-Beatle George Harrison and keyboardist Billy Preston to the Oval Office. Harrison got a WIN (Whip Inflation Now) button from the president. In return, Harrison gave Ford an OM button—a sacred incantation in Sanskrit. Afterward George said he got "good vibes" from the Ford White House.

In November 1974 at the Vladivostok summit, fearing Soviet-planted listening devices, the American negotiating team, led by Ford and Secretary of State Henry Kissinger, takes a strategy break outside the grim conference hall. Together with Soviet leader Leonid Brezhnev, they crafted a framework for limiting nuclear arsenals on both sides of the Iron Curtain.

On April 5, 1975, Ford holds one of the first Vietnamese orphans flown to the United States as part of Operation Babylift. Ultimately some 3,300 Vietnamese children found homes in the United States.

April 28. "Well Henry, we've done the best we could and you and I are taking it on the chin and just hope the Good Lord is with us." A grim-faced president authorizes a helicopter evacuation of Americans and as many South Vietnamese collaborators as can be rescued from Saigon before it falls to Communist forces.

May 15. One day after the rescue of the USS *Mayaguez* from Cambodian pirates, Ford explains the operation to the visiting Shah of Iran. "We perhaps overreacted," the president acknowledged, "to show the Koreans and others our resolve" following the North Vietnamese victory.

"The type of man I could establish a great rapport with." Ford's rapport with Egyptian president Anwar Sadat, seen here at their first meeting in Vienna in May 1975, would lead to Sinai II, a diplomatic breakthrough that paved the way for Jimmy Carter's more sweeping Camp David Accords, ending hostilities between Egypt and Israel.

At Helsinki in August 1975, Ford and Brezhnev failed to make progress on a SALT II deal, but they set in motion a series of events that would lead to the collapse of the Soviet Union and its Eastern European empire.

On September 22, 1975, on a San Francisco street, Ford hears the crack of a would-be assassin's gun, narrowly escaping the second attempt on his life that month.

"Jefferson did not define what the pursuit of happiness means for you or for me," Ford told a group of naturalized citizens at Monticello on July 5, 1976. "Our Constitution does not guarantee that any of us will find it. But we are free to try." It was arguably the finest speech of his presidency.

Afterward the president paid his respects at Jefferson's grave.

In Kansas City, Ford makes a last-minute appeal to a room full of Hawaii Republicans as he tries to nail down his party's nomination against formidable competitor Ronald Reagan.

Subdued in victory, Ford departs Reagan's hotel following an awkward postmidnight meeting at which the vice presidency was neither offered nor sought.

Right: The first televised debates since 1960 are remembered for Ford's denial of Soviet domination of Eastern Europe. Stubborn to a fault, he waited three days before acknowledging his error, halting the momentum that had brought him from over thirty points down to a virtual toss-up.

The Day After. On November 3, 1976, the First Lady tries to cheer up her son Jack before reading a concession statement from her voiceless husband. As for the man who defeated him, Ford croaked, "I've got to give him the White House in better shape than I got it."

Fully recovered from his post-election funk, Ford shares a belly laugh with Nelson Rockefeller, John Connally, and Ronald Reagan. Their mirth barely concealed each man's resentment over the recent election.

On January 20, 1977, the departing president bids farewell to members of the White House family.

Below: The torch is passed. Jimmy Carter's gracious words about his predecessor could not heal the raw emotions of the campaign. Few that morning could have predicted that Ford and Carter would become close friends, their autumnal reconciliation reminiscent of Adams and Jefferson. *Courtesy of David Hume Kennerly*

"That's my real home," Ford observes as he takes a final glance at the US Capitol en route to Andrews Air Force Base to begin life after the White House. *Courtesy of David Hume Kennerly*

On September 14, 1993, Ford joins Presidents Carter, Bush (41), and Clinton in a bipartisan push to enact the North American Free Trade Agreement. Five years later, Ford would anger Republicans by opposing a doomed impeachment drive based on Clinton's affair with a White House intern.

On May 21, 2001, after receiving the Profile in Courage Award from Ted and Caroline Kennedy, Ford said that people stopped asking him the question he had been asked for a quarter century. At last he could put the Nixon pardon behind him.

18

CLEARING THE DECKS

❧

It is tough to govern in the best of times. This is the worst of times.

—DONALD RUMSFELD, AUGUST 1974

EVEN THOSE SUPPORTIVE of Ford's action in pardoning his predecessor were stunned by the timing and manner of its announcement. "I
was driving a car up in New Haven going to visit a friend of mine," recalled
Robert Bork, "and when the news came over the radio, I damn near ran the
car into a ditch." No less surprised was Mel Laird. Tipped off by a reporter,
one of a group of female journalists he addressed that Sunday morning at
the Kennedy Center, the master plotter stammered out some blandly supportive words, then left for Burning Tree and a second round of weekend
golf with his friend the president. He left it to Ford to bring up the unpleasant subject over which much of the nation was rubbing its eyes in disbelief.
"Jerry, you know what I think about the way it was handled. If you don't
you're blind, deaf, and dumb, because I've talked about that." The pardon
could in any event wait. "We've got a chance to win this tournament, we're
only two shots behind." They lost the match by two strokes.

As the Ford-Laird duo struggled on the links, Washington switchboards
were being put to the test. Senator Howard Baker reported calls to his
office running nine to one against the pardon. Thirty thousand telegrams
flooded the White House, most unabashedly hostile. Recanting their earlier
support for the new president, actors Paul Newman and Joanne Woodward registered "a scream of protest and a plea to exercise wisdom." From
Nashville, Johnny and June Carter Cash offered friendlier words, reassuring Ford that "you can't do the will of God without getting a lot of people
against you." Presumably this included journalists whose sense of betrayal
was heightened by the uncritical coverage many had showered on the unelected president during Ford's first weeks in office.

"Two days ago he owned the country," wrote columnist Mary McGrory.
"Now he has to pacify it." In Grand Forks, North Dakota, county judge
Kirk Smith parodied the leniency extended to the former president by

freeing a pair of traffic violators as "an act of clemency." The California State Bar Association denounced Ford's action. So did Vladimir Pregelj, foreman of the Watergate grand jury, whose long-running efforts to hold Nixon accountable for his alleged crimes were effectively stymied by Ford's pen. Those around Ford whose names were publicly linked to the pardon did not escape abuse. "The phone calls," remembers Benton Becker, "the anti-Semitic remarks at three in the morning. To me, to my wife—'you kike Jew bastard, what did you get paid for this?'" The mild-mannered Phil Buchen was viciously rebuked, one correspondent writing the president's counsel, "Obviously your brain is as twisted as your legs," a crude reference to Buchen's youthful bout with polio.

On Capitol Hill, Republican senator Edward Brooke of Massachusetts spoke for many in decrying Ford's failure to obtain a full confession from his predecessor. The former president himself was out of sight, having taken refuge behind the high walls of Sunnylands, the sprawling Palm Springs estate of Walter Annenberg. Triple-digit temperatures and a persistent case of phlebitis kept Nixon off the eighteen-hole golf course on this, "the most humiliating day of my life." Adding to the indignity, a snap poll by the Gallup organization found the public opposed to the pardon by a two-to-one margin. More ominous was the overnight plunge in Ford's personal approval rating, from 71 percent to 49 percent, the largest such drop ever recorded. It was the kind of history no president wishes to make.

Ford had his defenders. Elliot Richardson went ahead with his news-makers breakfast, insisting that Americans would never stand for a Nixon trial. Taking the long view, the *Saint Petersburg (Florida) Times* praised Ford for "an act of courage that will ultimately benefit the country." It was a sentiment echoed by the Parisian daily *Le Figaro*. "We were perhaps too quick to affirm that President Ford was not a man of great decision," concluded the venerable journal, long regarded as the voice of center-right opinion in France. At the White House, the irrepressible Tom Korologos walked into a cheerless Monday-morning staff meeting. Exactly one month after drawing laughs by verbally disowning the old regime of which he had been an essential cog, the president's emissary to Capitol Hill addressed the same newcomers in his best voice of doom:

"Hey, you Ford people—welcome to the NFL!"

Ford himself was in Pittsburgh that morning, speaking before a conference on urban transportation. Outside the downtown Hilton hotel a small group of protesters waved signs reading NO AMNESTY FOR NIXON and DOES FORD KNOW THE MEANING OF JUSTICE? He tried to repair the damage at a September 16 press conference. Pressed to explain what had caused him to change his mind about intervening in the judicial process, Ford laid out a detailed sequence of events since his last such grilling by the press on August 28. More disciplined than in his earlier comments, he referenced input from Leon Jaworski and officials of the Justice Department among the factors contributing to his decision.* As for Nixon's guilt, Ford said it had been firmly established by a unanimous House Judiciary Committee, reaffirmed by the findings of the special prosecutor and conceded by Nixon himself through his acceptance of the pardon.

The next day Phil Buchen scribbled a note for the president's eyes only. "Leon Jaworski called me this a.m. Wanted you to know he thought you did superbly at your press conference and did much to lay low critics of your action." That night Ford took a call from San Clemente. "President Nixon called me. Said at outset he would reject Pardon if that would help: Sorry he caused me so much trouble . . . Seemed in reasonably good spirits but not as strong as usual in his conversation." Nixon's offer was rejected, as he knew it would be.

S EPTEMBER 8 MARKED the end of Gerald Ford's consensus presidency. Within days of the pardon, Congress rejected his inflation-fighting appeal to postpone a 5.5 percent raise for federal workers. The same lawmakers, responding to the influential pro-Greek lobby, brushed aside administration opposition and moved to cut off arms sales to Turkey. Ford's sudden vulnerability emboldened visiting statesmen like Israeli prime minister Yitzhak Rabin, who spurned the idea of simultaneous peace talks with Egypt and Jordan. Rabin showed no more enthusiasm when notified of an impending Kissinger mission to the troubled region. To Ford's appeal for "a good strong commitment" to work with the secretary in his peacemaking efforts, Rabin replied feebly, "We will do our best to move at that time."

Reporters dusted off their earlier reservations about the president's leadership abilities. "We were crushed. I mean, we were just crushed," says CBS News correspondent Bob Schieffer. "We fell headlong in love with Gerald Ford in the first weeks because he was so different." Acknowledging that "the press went too far, as we always do," Schieffer hinted that the journalistic reaction to the pardon was just as excessive: "The pendulum never stops in the middle."

Nothing had done more to bolster Ford's credentials as a change agent

---

* Ford could hardly reveal to the assembled reporters that it was their earlier barrage of questions concerning the fate of the former president that had acted as the catalyst for his subsequent pardon.

than his willingness to reconsider the status of Vietnam draft evaders and deserters. Unfortunately when the details of his "earned reentry" initiative were spelled out on September 13, almost no one was satisfied. To organized veterans any relaxation of military justice dishonored heroes dead and alive. Potential beneficiaries of the Ford program were just as unhappy over its required vow of loyalty to the United States, followed by a period of public service employment stretching as long as two years.

For those already convicted of draft evasion or desertion, Ford established a clemency review board, chaired by former New York senator (and outspoken Vietnam War critic) Charles Goodell. Liberals were further cheered by the inclusion on the board of Urban League president Vernon Jordan and Father Theodore Hesburgh, president of Notre Dame University and a former chairman of the United States Commission on Civil Rights until Richard Nixon summarily removed him from the position. For critics of Ford's middle-ground approach, Father Hesburgh had a bluntly worded response. "As long as Nixon was in," he reminded them, "these guys could rot as far as he was concerned. It's the difference between no chance and some chance."

According to Goodell, Ford was the only member of his administration sympathetic to the program and its objectives. At the urging of Goodell and Father Hesburgh he accepted hundreds of recommendations for individual pardons. In the end, of the 15,468 appeals lodged with the clemency board, 94 percent were approved. In most cases the period of alternative service required by the government was six months or less. Twice Ford extended the original application deadline. Nevertheless only about one-fifth of the eligible number actually submitted an application. Consequently it fell to Jimmy Carter on the first day of his presidency in January 1977 to issue a blanket pardon to Vietnam-era draft evaders, tens of thousands of whom had taken up permanent residence in Canada.

Ford's limited amnesty illustrated the perils of traveling the middle of the road. Speaking to the United Nations General Assembly on September 18, the president outlined "two basic, practical lessons" he had learned as a legislator. "First, men of differing political persuasions can find common ground for cooperation. We need not agree on all issues in order to agree on most . . . Second, a majority must take into account the proper interest of a minority if the decisions of the majority are to be accepted." Democracy, Ford continued, "thrives on the habits of accommodation, moderation, and consideration of the interests of others." Never did he more succinctly outline his Eisenhower-like approach to governing. As with Ike, Ford inherited a nation polarized by an unpopular war and inflamed by partisan division. Both men sought to lower the temperature and restore public confidence eroded by scandal.

There the parallels end. The 1970s were to the era of Elvis and Sam Rayburn what the legislator's function was to executive management—"Totally different job skills," contended William Brock, a Tennessee Republican who served in both houses of Congress, and whose name figured in vice presidential speculation in 1974 and again two years later. "You move from a very comfortable place where everybody around you are friends—in both parties . . . And you go to a place where you're leading a couple hundred million people." A place where success required "the ability to communicate, market, sell, as well as to guide the entire Congress." Don Rumsfeld offered a less flattering description of the legislative mindset: "Kind of holding your cards close to your vest, making your decisions at the last minute, going into the center—fashioning a coalition of some kind—as opposed to strategic leadership, directional leadership, bold leadership."

Lacking the mandate of popular election, deprived of a national campaign or transition period in which to determine priorities, formulate themes and recruit talent, Ford operated under yet another handicap. In the twentieth century a combination of foreign wars, crippling depressions and festering injustices—racial, economic and gender-based—had transformed American government. Not only had power been centralized in Washington; it had been personalized in a presidential office whose occupants were increasingly judged as public performers. Governing in the long shadow of Theodore and Franklin Roosevelt, haunted by the telegenic ghost of JFK, by 1974 success in the modern presidency required what political scientist Mark Rozell calls the capacity "to articulate a vision for the future characterized by far-reaching, innovative policies to alleviate public problems."

Call it the Arthur Schlesinger School of Leadership—a spellbinding, agenda-setting domination of Congress and the national conversation. Little in Ford's past experience prepared him to occupy TR's bully pulpit or undertake the kind of mass persuasion Harry Truman defined as the chief function of the modern presidency. Steeped in the virtues of legal pragmatism at Yale, on Capitol Hill Ford had learned to value conciliation and compromise. Shunning the inflated rhetoric employed by more colorful

---

* Congressional liaison Max Friedersdorf recalls the downside of having so many friends, and their inability to adapt to Ford's new station in life. "They were so familiar with him, they would call him Jerry," Friedersdorf remembered. "And Bob Griffin used to just go bananas. Bob would say, 'Damn it, that is Mr. President. That is not Jerry!'" AI, Max Friedersdorf.

leaders to raise expectations, only to suffer disillusionment when extravagant promises went unrealized, Ford instead would focus his presidency on redistributing power and resources through revenue sharing, tax cuts, block grants, welfare and regulatory reform.

Whatever its merits as public policy, this pursuit of smaller, more entrepreneurial government left Ford at a distinct disadvantage in the theater of politics. "Headlines can only be written about bold new initiatives," maintained White House Cabinet secretary Jim Connor. Revitalizing Cabinet government at the expense of a power-hungry White House staff was hardly the stuff of public drama. It was only a matter of time before journalists who had scorned Nixon as a threat to constitutional government were heard to grumble that his successor was dull, and his presidency lacking direction or theme. To this, Ford defenders had a ready retort. As one put it, "The guy was firefighting the whole first six months he was in office." As it turned out, this was a considerable understatement.

SECRET SERVICE AGENT Richard Keiser had some explaining to do. At least he thought so, after the *Washington Post*, noting his uncanny resemblance to the president he was sworn to protect, concluded that he was a decoy for Jerry Ford. The presidential party was at Camp David when the story appeared. Keiser immediately sought out Ford "because most famous people would not want someone around them who was getting some publicity as well . . . So I purposely went out to encounter him" on a morning walk in the woods.

"Good morning, sir."

"You seen the papers?"

"Yes sir, I have."

"Well, what did you think?"

"More importantly, sir, what did you think?"

"I thought it was funny."

Ford's sense of humor survived the presidency intact. An indifferent orator, he excused his delivery by telling audiences, "I told my wife Betty that I knew this speech backwards, and I'm proving it." When the *Grand Rapids Press* mistakenly ran a picture of Caroline Kennedy next to a short bit about Susan Ford's latest escapade, the paper received from the White House several pages stapled together. On top was a nine-by-eleven-inch photograph with the caption "This is Susan Ford." Next was a cutout of the offending article labeled "This is Caroline Kennedy." Underneath it was the president's official portrait. "This is Susan Ford's father," it read. "Caroline Kennedy and her mother do not read the *Grand Rapids Press*.

Susan Ford and her father do read the *Grand Rapids Press*." Finally, at the bottom of the paper pile, a second copy of the presidential portrait carried the warning, "Susan Ford's father is watching."

Ford had only been in the job a few days when he took a call from his close friend William Whyte, chief lobbyist for U.S. Steel.

"Mr. President?"

"Is this Mr. Whyte?"

"It's Bill Whyte."

"Well, it's Jerry Ford."

Breaking with Oval Office tradition, Ford passed up the high-back "Presidential" chair with arms that lent its occupant as much authority as the majestic portrait of George Washington that dominated the room. Displaying the politician's knack for establishing instant intimacy, Ford much preferred to sit beside visitors on one of the twin sofas flanking the fireplace. That is not all that set Ford apart from his predecessors. As Jack Marsh related it, "He always accepts responsibility when things go wrong—it's always 'I,' but whenever things succeed, it's always 'we.'" Not only was Ford underestimated, Marsh believed he "may work at getting you to underestimate him," reading people for character and motives "but not indicating that he's read them."

As minority leader, Ford had rarely left the White House without pocketing a handful of elegant matchbooks for deposit in a bowl on the receptionist's desk in his Capitol Hill office. Constituents scooped them up, the perfect souvenir of their brush with history. Ford continued the practice while acclimating himself to his new status, no longer a tourist but not yet at home within the walls that had sheltered Jefferson, Lincoln and the Roosevelts. Scarcely able to believe the latest turn of events, recalled newsman Bob Schieffer, "he would go around and every once in a while he would just pick up a handful of those White House matches because he'd just been doing it all his life."

To nineteen-year-old Steve Ford in his torn Levi's and yellow jeep, 1600 Pennsylvania Avenue was more museum than home. It was most definitely not a place to put your feet up, as the youngest Ford son discovered one night after plunking himself down beside a coffee table on which issues of *Sports Illustrated* were stacked next to an official-looking document stamped FOR THE PRESIDENT'S EYES ONLY. TOP SECRET. Soon after, Steve decamped for a Montana ranch, where, protected by a covey of Secret Service agents, he perfected his cowboy and rodeo skills.

His father, more adaptable, soon settled into a routine that was to govern his life until January 20, 1977. Ford's workday began at five fifteen a.m., when the shrill summons of an old alarm clock brought from Crown View Drive roused him from the bed formerly used at the same address. In the adjoining sitting room designed by the First Lady to reproduce, as nearly as possible, the casual comforts of their Alexandria home, Ford scanned the morning papers, any overnight intelligence briefings and a news summary prepared by the press office. Then he sweated his way through half an hour of calisthenics. During the week an exercise bike imperfectly substituted for his daily swim. Pedaling an uphill mile was not an activity Ford enjoyed, yet he submitted to the grueling regimen with an athlete's self-discipline. On weekends there was fiercely competitive tennis on the White House courts or a dip in Camp David's heated pool.

Unpleasant surprises were met with equanimity, as Ford demonstrated one Sunday morning in calling the Usher's Office, where former Secret Service agent Gary Walters happened to be on his first day substituting for the vastly experienced Chief Usher Rex Scouten.

"I didn't know whether to stand up or salute," says Walters.

"Yes, sir, how can I help you?"

"Well, I don't have any hot water in my shower."

Fearsome words these, conjuring up the ghost of Lyndon Johnson raging like Lear because the water pressure in his bathroom shower fell short of his sandblasting demands. On that occasion Scouten had personally seen to the installation of an additional water tank and other modifications, even testing the result for himself when the president was out of town, only to be pinned against the shower wall by scouring jets of water. Gerald Ford was more easily placated, as befitting a man who shined his own shoes until he became president. No plumbers were on duty Sunday morning, Walters explained apologetically. But he would gladly round up some engineers to undertake immediate repairs.

That wouldn't be necessary, said Ford. "I haven't had any hot water for two weeks." Rather than disturb anyone, he had simply availed himself of the First Lady's shower.

Walters's experience would come as no surprise to other members of the residence staff, unaccustomed as they were to a president who compared golf scores with butlers he invited to join him in watching the World Series on television. In response they made themselves indispensable, beginning with the stewards who toasted Ford's favorite English muffins for breakfast, served with orange juice and grapefruit or melon in Jacqueline Kennedy's Family Dining Room. A miniature television set beside his place setting distracted the president's attention from the nineteenth-century French wallpaper depicting grisly scenes from the Revolutionary War (Betty Ford would have it covered over because "I couldn't stand to look at soldiers fainting and dying while I was eating my soup").

Breakfast concluded, Ford left for the Oval Office and his daily intelligence update from David Peterson of the CIA and Brent Scowcroft, Kissinger's deputy at the National Security Council. During his abbreviated tenure as vice president, Ford had been impressed by Scowcroft's keen intellect and apparently limitless capacity for work. In the White House he would rely on the self-effacing soldier-scholar as the ideal staff man, one who didn't hesitate to speak truth to power. For his part Scowcroft appreciated Ford's integrity, no less than his transparent eagerness to master the foreign affairs portfolio. The nine o'clock hour was customarily claimed by Henry Kissinger. By then Al Haig or, later, Don Rumsfeld would have outlined for the president the day ahead and tagged any issues demanding immediate attention. In exchange Ford handed over his paperwork from the previous evening—"Ford's probably the only man in history that's ever read tab 'G' of a memo," recalls one aide—boxes checked and instructions furnished in a vigorous left-handed scrawl.

At midmorning Jack Marsh appeared in the doorway to the Oval Office with the latest gossip about feuding staffers and ideas on how to bridge the gap between Ford and Nixon factions. After Bill Timmons left in December to start his own lobbying outfit, the diplomatic Marsh assumed responsibility for congressional relations. It proved to be among Ford's best appointments. "Marsh was the kind of a guy," observes one coworker, "that you gave the worst conceivable problem to and he kind of wandered out of the room and you forgot about giving it to him, but you never heard about the problem again."

Jerry terHorst was less easily replaced than Timmons. The Ford White House flirted with history by sounding out *Time* correspondent Bonnie Angelo about becoming the first woman to fill the role of press secretary, but she begged off, pleading the greater demands of mothering a twelve-year-old boy. Other print journalists were considered for the post, under the assumption that their television counterparts would be reluctant to trade celebrity and a network salary for $42,500 a year. Thus it came as a distinct surprise when Ron Nessen expressed interest in the job. Forty years old, Nessen boasted a distinguished journalistic career, including six years with UPI and more than a decade covering Washington for NBC, with memorable detours to Vietnam and sleepy Southern towns where Black Americans demanding long-deferred rights confronted club-wielding sheriffs and uniformed authority on horseback. For all the history he had observed, by 1974 Nessen felt restless and underutilized; ninety seconds of exposure on

NBC's nightly newscast hardly satisfied his appetite for the limelight. He revealed as much, unwittingly perhaps, in his White House memoir, *It Sure Looks Different from the Inside*, when he complained of "always standing on the sidelines watching others perform."

Nessen had earned Ford's trust as part of the corporal's guard that regularly traveled with him on Slingshot Airways. He aced his Oval Office job interview, vowing in the event of a policy disagreement to defer to Ford's judgment, on the theory that it was the president's opinion that mattered, not that of his public spokesman. In return Ford promised Nessen the kind of access enjoyed by such storied predecessors as Pierre Salinger and Bill Moyers. He also spared him the task of partisan promotion. "If I can't sell the program," said the president, "then the press secretary can't sell the program." At his first White House briefing on September 20, Nessen identified himself as "a Ron, but not a Ziegler." This did not spare him from the pervasive distrust that was a Watergate legacy.

With the perspective of almost forty years Nessen attributed the hostile mood he encountered to envy of celebrity journalists Woodward and Bernstein. "You are a White House reporter. You're in the press room; twenty feet down the hall there is the President of the United States. Every day you come to the White House—you never get a whiff of that story. Two guys outside of the White House, who never went into the White House, broke the story. You are really frustrated, and some of that, I think, was taken out on me and on Ford." Readily accepting his share of blame for operational shortcomings, Nessen reflected, "I don't think I had exactly the right personality for that job. I was short-tempered, sort of full of myself. I didn't suffer fools easily, and it was a difficult time."

As the first television journalist to fill the position, Nessen was widely panned for hosting *Saturday Night Live* in April 1976. This may have satisfied Nessen's craving for airtime, but it obscured his considerable efforts to deliver on the promise of greater accessibility. These included the introduction of follow-up questions at presidential press conferences, a long-sought change that enabled reporters to probe beyond White House talking points. More one-on-one interviews with the president were arranged. At his urging, Ford provided author John Hersey near total entrée for a week in April 1975, the genesis of a lengthy *New York Times* magazine article and book. Likewise unprecedented was Ford's live appearance on NBC's venerable *Meet the Press*.

Full-scale presidential press conferences took place on average every three weeks. Ford enjoyed matching wits with reporters, going out of his way to recognize gadflies like the redoubtable Sarah McClendon, whose shouted inquiries had "made the veins stand out" on Dwight Eisenhower's forehead. Much to the delight of *Washington Post* editor Ben Bradlee, Ford autographed a mockup from the paper's latest ad campaign declaring, I GOT MY JOB THROUGH THE WASHINGTON POST. He reinstated the custom, abandoned during the Nixon years, of inviting journalists as guests to White House social events. True to his word, Ford included Nessen within his inner circle, reviewing sensitive topics with him prior to the daily press briefing.

This preceded a quick lunch, generally eaten off a tray at the presidential desk or in his adjacent private office. Accustomed to watching his weight, the former college jock offered no argument when White House physician Dr. William Lukash put him on an 1,800-calorie diet, or replaced his habitual scoop of butter pecan ice cream with ice milk. Also rationed was Ford's alcoholic intake. The issue arose after some noontime imbibing at an event in Denver caused Ford to skip over much of a lunchtime speech text.

"You're the president of the United States," Dr. Lukash reminded him sternly. "Stop drinking. Especially stop drinking martinis at lunch."

Afternoons when he was in town Ford made himself accessible to congressmen, diplomats, beauty queens, corporate executives, Republican office seekers and West Michigan cronies, not all of whom bothered to schedule an appointment before appearing in the West Wing lobby. His cluttered docket prompted complaints that Ford had yet to outgrow the congenial habits of Capitol Hill. "In a congressman's office, everything comes to him," noted Jack Marsh, elected four times to represent Virginia's Seventh District before joining the Nixon and Ford Administrations. In the White House this was a formula for chaos. "I was trying to do too many things," Ford conceded much later. "I was trying to see too many people. And too many people whom I did see were oblivious to time." Ten-hour workdays customarily stretched to twelve hours or longer. Only rarely before eight o'clock did he sit down to an evening meal of crab soup or liver with onions, a presidential favorite not inflicted on Mrs. Ford or any of the Ford children and their friends who might join them at the dinner table.

In the hours before bedtime Ford reviewed paperwork in the nearby hideaway where no Secret Service agents penetrated. Here the size-six Betty Ford exacted comical revenge after being told by her husband that she was too thin. For Halloween she dressed a skeleton borrowed from Dr. Lukash's infirmary and placed it in Jerry's pale blue leather chair ("a sad looking Barca Lounger," sneered White House curator Clem Conger, "the most god awful thing"). The First Lady's stunt elicited the desired chuckle from an overscheduled executive more accustomed to tricks than treats. On rare nights when Ford had difficulty sleeping, he sometimes wandered down the hall to the Lincoln Bedroom (in point of fact, the wartime leader's office and Cabinet Room). There, sitting alone where Lincoln had signed the Emancipation Proclamation, Ford meditated on the decisions made within its walls by the president against whom all others are measured. It helped to put his own tests in perspective.

A FTER A MONTH in office, Ford grudgingly acknowledged that his administrative wheel had too many spokes for comfort. He also thought he had the perfect solution. By making Al Haig Supreme Allied Commander of NATO, a position that did not require Senate confirmation, he could reward the general for his past services and, hopefully, halt the infighting that threatened to overwhelm his young presidency. In place of a Haldeman-like taskmaster, Ford envisioned a manager who would coordinate the flow of paper, control access to the Oval Office, and work comfortably in harness with a reinvigorated Cabinet. On the morning of September 19, Ford placed a call to Don Rumsfeld, who had returned to the Chicago suburbs for the funeral of his father. The president expressed his and Betty's sympathies on their friend's loss, and invited Rumsfeld to drop by the White House before resuming his post in Brussels as ambassador to NATO.

When the two men met in the Oval Office they shared paternal memories, before getting down to business. Someone was needed to bring order to the uneasy alliance of Nixon loyalists, Ford recruits, congressional staffers and friends from Grand Rapids that comprised Ford's White House staff. Unwilling to engage in a massive housecleaning, Ford nevertheless recognized that the status quo was unsustainable. Starting at the top. "I've got to change Haig," he told Rumsfeld. "You've got to do it." Plainly reluctant, Rumsfeld conditioned his acceptance on abandonment of Ford's spokes-of-the-wheel approach to White House administration. "A terrible concept," Rumsfeld later asserted. "It worked fine for a minority leader. It works fine running a congressional office." In the executive branch it bred endless confusion, with decisions inadequately staffed out and therefore subject to embarrassing reconsideration and reversal.

This was not the only point of dispute between the two men. By his own acknowledgment Rumsfeld's politics were to the right of Ford's. Believing the pardon to have been mishandled, Rumsfeld thought the choice of Nelson Rockefeller as vice president gave needless offense to GOP conservatives essential to Ford's chances in 1976. By publicly declaring his reliance on Henry Kissinger, in effect anointing him as prime minister for foreign policy, Ford had unwittingly yielded leadership "in the single area that is unique to a president." More damaging still was Ford's reluctance to expel Nixon holdovers en masse—cutting out the tumor, Rumsfeld said, "so that you save the whole body."

Ford, characteristically, didn't bridle at his friend's criticism. Just as he valued Bob Hartmann's wordsmithing and sensitivity to political booby traps, so he appreciated Rumsfeld's mix of tough-minded competence and charm that led *Newsweek* to designate him "a Haldeman who smiles." Even his adversaries respected Rumsfeld's political instincts. "One of the smartest political operators you could ever see," conceded one colleague whose loyalties lay elsewhere. "He can figure out moves and bank shots and lateral moves better by far than anybody else" in the Ford White House.

Like Eisenhower before him and Ronald Reagan after, Ford was most comfortable in the role of good guy, surrounded by accomplices with sharp elbows. Donald Rumsfeld—Princeton wrestler, navy aviator, crew-cut congressman and improbable friend of Elvis Presley and Sammy Davis Jr.— had more in common with Ike's Sherman Adams or Reagan's James Baker than he did with H. R. Haldeman. Indeed, Rumsfeld blamed the obsessive work habits of Haldeman for contributing to the so-called imperial presidency. He said as much to Ford: "If we keep getting consumed, and everyone here works fifteen hours a day, we're going to begin to think we are indispensable. And we're not. It's not healthy for you. It's not healthy for us and our families. So I am going to require that everyone have a deputy."

Anticipating Ford's job offer, Rumsfeld had made a pitch of his own. The night before he met with the president, Rumsfeld reached out to his most trusted subordinate, a laconic thirty-three-year-old named Richard Bruce Cheney. Nebraska-born and Wyoming-reared, Cheney had twice been evicted from Yale University ("The second time they said, 'Don't come back'"). An internship with the Wyoming State Senate had kindled Cheney's interest in politics. In 1964, having earned twin degrees in political science from the University of Wyoming, he married his high school sweetheart, Lynne Vincent. The following year Cheney won a fellowship and a year's employment with the governor of Wisconsin.

Yet another grant, from the American Political Science Association, enabled him to work on Capitol Hill, where his initial application to the office of Congressman Donald Rumsfeld was not well received. "It was the worst interview of my life, and he threw me out," Cheney would remember. "Thought I was some kind of fuzzy-headed academic. I thought he was about the most arrogant young man I had ever met. And we were both right to some extent." Cheney quickly found alternate employment with Wisconsin representative Bill Steiger, another policy wonk widely respected for his legislative skills and intimate knowledge of the tax code. In January 1969 a new president tapped Rumsfeld to oversee the Office of Economic Opportunity, linchpin of LBJ's war on poverty. Seizing the initiative, Cheney crafted a twelve-page memo offering Rumsfeld unsolicited advice on his upcoming confirmation hearings and the OEO agenda.

Much to his surprise he was invited to join a group assisting the new director in the transition process. So began a relationship that was to have historic implications. Cheney's experiences at OEO and especially at Nixon's Cost of Living Council were instrumental in shaping his conservative outlook. With Nixon's reelection in November 1972 came White House demands for the mass resignation of political appointees. Conspicuous for his resistance to a policy he thought callous and self-defeating, Rumsfeld was exiled to Brussels, an outpost too distant even for Cheney, who spent the next eighteen months in a DC-based investment advisory firm.

He had just returned to Washington from a Wyoming vacation with Lynne and their two daughters when, on August 8, 1974, he received a phone call from Rumsfeld's secretary in Brussels asking him to meet the NATO ambassador when he landed at Dulles the next day. Now, seven weeks later, Cheney was again accompanying Rumsfeld to the White House, this time to check out their West Wing office arrangements. That evening Rumsfeld introduced his protégé to Ford, whose immediate acceptance of the newcomer impressed Cheney deeply. The president's affable greeting masked the tensions of the moment, as Betty Ford's husband grappled with the gravest crisis of his married life.

S ETTING ASIDE HER original reluctance to live in the White House, Betty Ford had insisted on going as herself: "If they don't like it they can kick me out. But they can't make me somebody I'm not." Thrust center stage with no script, she replied with characteristic frankness when asked by a prospective press secretary to describe more fully the job for which she was applying.

"I don't know," the new First Lady responded. "What am *I* supposed to do?"

Her worst fears were confirmed early on, when guards and other household employees failed to return her cheerful morning greetings. "Nobody ever speaks to me," she told Rex Scouten. Scouten assured her that the silent treatment was nothing personal, merely a vestige of the more formal atmosphere surrounding the Nixons. The problem was easily resolved; in short order the Fords knew the names, and the names of the children, of maids and pastry chefs, florists and doormen, whose warmth and professionalism did much to ease the transition from suburban anonymity to president's wife. Of all the adjustments required by her husband's new job, none was more welcome to Betty than his schedule. "He was *never* home before, when the kids were growing up and now he's home all the time." From the second-floor West Hall with its great arched window overlooking the West Wing and Rose Garden, she could wave to Jerry as he took part in an outdoor ceremony or completed the three-minute walk from the residence to the Oval Office.

His self-confessed "best critic," Betty reviewed drafts of his more important speeches, not hesitating to fault a presidential statement or decision "unless he looks as if he's had a hard day. When he looks that way, I wait." On the occasions she didn't wait, her comments could be quite pointed, as when Ford in a televised conversation with Walter Cronkite expressed views on abortion that were notably more conservative than her own.

"Baloney," she wrote on the transcript of the interview, "this is not going to do you a damn bit of good."

Betty urged the president to appoint more women—especially women of color—to jobs in his administration. There was nothing subtle about her lobbying efforts: "If he doesn't get it in the office, he gets it at night in his ribs."

Even as the White House brought the Fords closer together, it also activated what one member of the First Lady's staff called her long-dormant "performance gene." Old *Washington Post* clippings attesting to her preference for "quiet" hats and "slightly more talkative" dresses didn't do justice to the stylish fashionista, whose eclectic tastes ran to silk scarves, tailored jackets, caftans and capes, floral prints and gauzy chiffon gowns from New York designer Albert Capraro. Betty Ford wore them all, with a flair that made her a favorite of the American fashion industry.

It was in her role as White House hostess that she appeared most self-assured. "Wow. The new Dolly Madison has arrived!" one congressional wife wrote Betty after a particularly lively party. "I've never seen the White House jump like that—even in the swinging Johnson era." Accustomed to official entertainments as stiff as the gilt chairs on which visitors endured them, Washington was ready to let down its hair. And the former Sunday-school teacher from Grand Rapids was more than accommodating. Dispensing with the rigid protocol of her predecessors, Mrs. Ford allowed unmarried, divorced or widowed invitees to bring their own escorts to

White House social events. She bent a long-standing rule by letting Senator Ted Kennedy's fourteen-year-old daughter, Kara, attend a state dinner in place of her mother.

Receiving lines were abandoned at these soirees so that the Fords could mingle comfortably with their guests. Smaller round tables replaced the U-shaped head table favored by previous administrations. Not only was the new arrangement more conducive to intimate conversation; henceforth, said Betty, status-conscious Washingtonians needn't worry "who sits above and below the salt." In place of tired floral centerpieces, she displayed Steuben glass, antique silver, Native American woven baskets and, for British prime minister Harold Wilson, a dedicated ornithologist, an array of hand-carved duck decoys. Besides showcasing the work of American craftsmen, these served as conversation pieces, part of Betty's determined effort to engage the most bashful or nervous of guests.

Guest lists mirrored the differing priorities of host and hostess. "The president always kept his little notebook with a list of people he would like included," recalls Maria Downs, the second of Mrs. Ford's social secretaries. After Ford suggested baseball players Joe Garagiola, Johnny Bench, Yogi Berra and Carlton Fisk as potential guests, Maria observed, "You must have a thing about catchers." His partiality for athletes extended to Olympians Bruce Jenner, Dorothy Hamill and speed skater Sheila Young. At one dinner, Spain's King Juan Carlos, an auto racing enthusiast, was spotted deep in conversation about pit crews and tires with racer Cale Yarborough. By contrast Mrs. Ford favored luminaries from the arts community (think sculptor Louise Nevelson and conductor Arthur Fiedler). Her affinity for such Hollywood legends as Cary Grant, Fred Astaire and Gregory Peck backfired after the president teased her about an invitation to actor and *Cosmopolitan* centerfold Burt Reynolds. She turned the tables one evening in November 1974, when singer Vikki Carr opened her set by performing "One Hell of a Woman" and concluded her visit to the president's house by asking her host to identify his favorite Mexican dish.

"You are," said Ford, within earshot of the First Lady, who hissed, "That woman will never get into the White House again!"*

As a rule entertainers were chosen to reflect the interests of a visiting head of state. Thus classical pianist Eugene Fodor performed for Israeli prime minister Yitzhak Rabin, and jazz musician Billy Taylor took the stage

---

* Only one movie star failed to pass muster with Mrs. Ford. To her social secretary's proposal that Hollywood bad boy Jack Nicholson be included on an upcoming guest list she replied, "I don't think so[,] Maria." AI, Maria Downs.

for Pakistan's Zulfikar Ali Bhutto. "The members of my trio are just like Congress," Taylor announced with a smile. "I tell them what I want them to do, and then they go about it in their own way." Ann-Margret's Las Vegas–style gyrations before the Shah of Iran raised questions of taste ("The Shah likes attractive young ladies," explained Mrs. Ford). Yet the Ford White House had its full share of more traditional artists, from Beverly Sills and Arthur Rubinstein to Carol Burnett and Tennessee Ernie Ford. On one memorable occasion Pearl Bailey coaxed a reluctant Anwar Sadat onto the same dance floor where she had enticed Gerald Ford into performing a spirited Charleston.

Mrs. Ford's support for the performing arts did not detract from her stewardship of what she called America's House. Dissatisfied with the eagle sconces and eye-popping blue-and-gold color scheme of the Nixon Oval Office, she raided the Yellow Oval Room for a temporary replacement rug until the entire space could be done over in more muted earth tones. Otherwise she regarded this second-floor parlor, where Warren Harding had played poker and FDR first learned of the Japanese attack on Pearl Harbor, as fussy and uninviting. Frequently used to receive visiting heads of state, the room's décor included a pair of Greek goddesses supporting a ceramic bowl. The First Lady was not above putting a cigarette in the outstretched hand of one deity each time she passed by.

Clem Conger would have been scandalized, though probably not surprised. He battled Mrs. Ford over a planned makeover of the elegant Queens' Bedroom, its centerpiece a grand four-poster from Charleston purchased by the curator while the Nixons were still in residence. Betty preferred to retain the historic bed used by the future Queen Elizabeth and other royal visitors. Perhaps her favorite spot in the house was the glass-walled solarium—Grace Coolidge's "sky parlor"—with its comfortable furnishings and fabulous rooftop views of the Washington Monument and Jefferson Memorial. Here the First Lady gossiped over tea with old friends, and handcrafted Christmas ornaments with Susan ("No tinsel, no sequins," she decreed for the White House Christmas tree). The same space provided a backdrop to Jack Ford's vegetarian lunch with George Harrison, an invitation extended to the former Beatle after they met backstage at a concert in Utah.

Betty had a staff of twenty-four to assist her in planning dinners, drafting speeches, coordinating travel and maintaining a vast and varied correspondence with the American people. Her de facto chief of staff was Nancy Howe, a bubbly military wife a decade younger than the First Lady. The two women had first met in the summer of 1973, when Susan Ford worked as a salesgirl for the White House Historical Association beside Nancy's daughter Lise. With the selection of Susan's father to be vice president, Betty hired Nancy to be her full-time personal assistant. Thereafter Mrs. Howe was never far from the vulnerable woman she cloyingly called Petunia (after the rare blossom able to withstand the heat of a Washington summer). No one was more delighted than Nancy when Mrs. Ford was cheered by six hundred Republican women on a rare solo visit to Chicago during the last week of September 1974. "This does a lot for the ego," the First Lady admitted. "It gives me an independence after years of just coping."

Betty's boisterous reception was still fresh in memory when Nancy mentioned a doctor's appointment she had scheduled for Thursday, September 26. Telling Betty, "You're due for a checkup," she invited her to come along. On the morning in question the women were driven nine miles to the Bethesda Naval Hospital in suburban Maryland. There Betty underwent a seemingly routine breast examination. Returning to the White House, her suspicions were aroused when Dr. Lukash asked her to meet him in his ground-floor office at seven o'clock. As it happened Lukash was not alone. Seeking a second opinion, he had invited J. Richard Thistlethwaite, a prominent surgeon at George Washington University School of Medicine, to conduct his own examination of the First Lady. This confirmed the presence of a lump in Mrs. Ford's right breast.

By now they had been joined by the president, summoned from the Oval Office by Dr. Lukash. The doctors wished to conduct a biopsy, the sooner the better. Betty insisted on going ahead with the next day's itinerary as planned. Together with Jerry, she was scheduled to participate in the groundbreaking ceremony for a memorial grove honoring Lyndon Johnson. She looked forward to hosting LBJ's widow, two daughters and son-in-law Chuck Robb for tea and an informal tour of the Ford family quarters. Determined to make their return to the White House a joyous occasion, Betty wanted nothing—her own health included—to distract attention from the Johnson family on their big day.

I'M NOT SO good on the small things," Betty professed, "but on the big ones, you can always rely on me." She demonstrated as much the morning of Friday, September 27. The woman who displayed the Prayer of St. Francis in her bathroom, where she was sure to see it first thing in the morning and last thing at night, kept by her bedside a booklet of spiritual meditations called *Forward Day by Day*. Turning to that morning's passage she read, "Thou has turned my laments into dancing." Certainly no one seeing Betty at the LBJ groundbreaking on the banks of the Potomac would imagine the secret she was carrying. She smiled broadly as her husband was introduced as the National Park Service's most illustrious alumnus ("Sometimes I wish I were back out in Yellowstone Park," he confessed to the crowd). Following the ceremony the president left for the Washington Hilton, scene of the White House economic summit. Farther up Connecticut Avenue, Betty addressed a Salvation Army luncheon at the Shoreham Hotel. She was back at the White House in time to greet Lady Bird Johnson and her family, none of whom noticed the small black suitcase parked at the foot of the bed in the Fords' bedroom.

Only after seeing her visitors off did Betty slip away, accompanied by daughter, Susan, and Nancy Howe. A few minutes before six she walked into Bethesda Naval Hospital, where the third-floor presidential suite had been readied for her use. There Jerry found her, dining on steak and fries, as soon as he was able to excuse himself from a White House reception for delegates to the economic summit. "I see you are having a party!" he exclaimed. By now Mike Ford and his wife, Gayle, had joined the group. Everyone wore their bravest face, in keeping with the positive outlook of the First Lady. By coincidence the Fords' Crown View Drive neighbor, Louise Abbruzzese, was on another floor of the hospital having given birth to her third child. Susan left her mother long enough to deliver a pair of satin pillows made by Betty as a gift for the new baby, named in her honor Elizabeth Ann. After her visitors departed Mrs. Ford was given a sedative. Thus she was fast asleep at ten p.m., when Richard Nixon phoned with words of encouragement from the Long Beach hospital where he was being treated for a blood clot in his right lung.

For Gerald Ford the overnight hours were, by his own admission, "the loneliest of my life." Before turning in he arranged to have three dozen red roses delivered to Betty's hospital room. They were at her bedside when she woke at six o'clock Saturday morning, along with a note in his handwriting that spoke for the entire Ford family. "No written words can adequately express our *deep, deep* love," it read. "We know how *great* you are and we, the children and Dad, will try to be as strong as you. Our Faith in you and God will sustain us." Waiting for her in the sitting room of her suite, a supportive Susan was joined by Nancy Howe and Billy Zeoli, a Grand Rapids gospel filmmaker and family friend who could be counted on for a weekly devotional message sent to the president's attention. Thinking he might help to raise the First Lady's spirits, Zeoli had chartered a plane and flown through the night, only to find the patient wisecracking about her toeless white hospital stockings ("This will be a great item for *Women's Wear Daily*") and telling her anesthetist, "Good night, Sweet Prince."

Betty's surgery was scheduled for eight o'clock. It took surgeons less than thirty minutes to cut open the cancerous nodule, not quite an inch in diameter, and examine it in a nearby lab. Their findings were transmitted to the president in the Oval Office, where he was halfheartedly reviewing his concluding remarks to the economic summit. Putting down the phone, a hoarse-voiced Ford told Bob Hartmann, "That was Dr. Lukash. It's malignant, and they're going ahead. Excuse me a minute."

He disappeared Into his private bathroom. Emerging a few minutes later, red-rimmed eyes betraying his emotions, Ford picked up the speech draft. But it was no use.

"Go ahead and cry," said Hartmann. "Only strong men aren't ashamed to cry."

Ford surrendered to his feelings, and his fears.

"Bob, I don't know what I'd do without her. I just don't know what I'd do . . ."

Tearing up his schedule, Ford said he would deliver his speech as planned, but only after skipping the morning session of the conference. He hoped the delegates would understand his need to be at the hospital with Betty when she regained consciousness. Outside the rain fell in buckets, and lightning bolts made the helicopter ride to Bethesda the second-most-frightening experience of the morning. Kneeling on the chopper floor, Mike Ford prayed with his father, no longer the most powerful man on earth but a frightened husband trying to prepare himself for any eventuality. On arriving at the hospital Ford was told that Betty was still in the operating room. Outside the weather worsened, grounding the presidential helicopter. A motorcade was hastily summoned. It appeared just as the First Lady was wheeled into the recovery room. The president was only able to spend a few minutes with his semiconscious wife before he had to depart for the Washington Hilton.

Initial reports from her doctors were upbeat. The operation to remove Mrs. Ford's right breast and underlying chest muscles had gone off without a hitch. (Subsequent tests would reveal traces of the disease in three of the thirty lymph nodes extracted as part of her radical mastectomy. In medical lingo, Betty Ford was a stage-two cancer patient, with a 50 percent chance of surviving five years.) Returning from the hospital on rain-slicked roads, Ford's party narrowly escaped injury when a tailing car rear-ended the presidential limo at a Wisconsin Avenue intersection. Unflustered, Ford entered the Hilton ballroom to a standing ovation. Struggling with his emotions, he told conference delegates he had just come from his wife's hospital room. "Dr. Lukash assured me that she came through the operation all right." He bit his lip before adding, "It has been a difficult thirty-six hours. Our faith will sustain us; Betty would want me to be here."

The Fords' next move was never in doubt. Keeping in mind the precedent set by Mamie Eisenhower in the wake of her 1957 hysterectomy—"Mother and Dad could have said she was having female problems," observed their daughter Susan, "and the press would have left it alone, because I don't think you could even say breast on TV then"—Mrs. Ford decided otherwise. "She sent word that she wanted this to be announced while she was still on the operating table," according to Ron Nessen. "And that's what we did." By refusing to hide behind euphemisms, Betty Ford introduced the topic of breast cancer to polite conversation. When postoperative depression briefly darkened her spirits, she was bolstered by media reports of women in large numbers heeding her example and scheduling their own examinations (among them Ron Nessen's mother and Happy Rockefeller, wife of the vice president designate, who underwent a double mastectomy that fall).

At the White House, the president invited a few former colleagues to allay his loneliness. "You know, I don't have any business to discuss with you," he told one guest. "I just wanted to get a couple of friends in here to have dinner with me. I'm a bachelor." Dessert was being served in the Family Dining Room when Ford voiced his apologies for getting up from the table. "I've got to leave. I've got to go to the hospital." Ahead lay strenuous rounds of physical therapy for Betty, part of the laborious process of regaining the use of her right hand and arm. She would be taking prescribed medication—bottled chemotherapy—for two years. The pills did not come without side effects, especially if mixed with alcohol. For now she concentrated on getting well. Jerry buoyed her spirits through his frequent visits to Bethesda. Her sons called often, and Susan came every day after classes let out at the nearby Holton-Arms School, where she was in her senior year.

Mother and daughter forged a special bond as Susan shared some of the thousands of letters addressed to Betty's hospital suite. "You have been through a rough ordeal," wrote Richard Nixon, "But I know you will be sustained by the realization that Jerry needs you, the children need you, and millions of Americans are pulling for you." Other correspondents shared homegrown remedies, one well-wisher promising Mrs. Ford she could prevent any recurrence of cancer by massaging the instep of her foot while drinking carrot and apple juice. The most touching messages came from women like Alice Roosevelt Longworth, Washington's self-proclaimed "topless octogenarian," who had triumphed over the same disease and disfigurement.

Joan Kennedy likened Betty's courage to the bravery displayed by her son Ted Jr. during his recent bout with cancer. *Washington Star* columnist Mary McGrory, a journalistic mainstay of Nixon's enemies list, wrote admiringly of a First Lady who had come into her own ("Everybody sat in the hospital waiting room last Saturday morning . . ."). A grateful president sent McGrory a handwritten note of appreciation.

From Southern California Walter Annenberg offered the use of his Sunnylands estate for Betty's recuperation. Lady Bird Johnson sent a pink robe that became a familiar sight in hospital corridors, where, four days after her surgery, Betty surprised Jerry by tossing him a football signed by the entire squad of the Washington Redskins. One presidential visit to the hospital coincided with the changing of Betty's postoperative dressings. As recounted by her biographer Lisa McCubbin, it was a pivotal moment for husband and wife. "With all the drains and the ugliness of surgery, it was not pretty," says their daughter, Susan. Jerry did his best to offer reassurance. When Betty worried about not being able to wear evening gowns tailored to her figure, he told her, "Don't be silly. If you can't wear 'em cut low in front, wear 'em cut low in back."

On the night of October 5, resplendent in a $50 red chiffon gown bought for the occasion, seventeen-year-old Susan Ford substituted for her mother at the annual White House reception for the diplomatic corps. Having reluctantly donned long white gloves loaned her by her mother— "They are miserable to get on and off"—Susan joined her father in a Blue Room receiving line. For the rest of the evening she gamely danced with a procession of bemedaled, mostly elderly diplomats. Declining her father's request for a jitterbug, Susan yielded gracefully when the Marine Band broke into "Thank Heaven for Little Girls." "Not bad," she concluded as guests departed into the night, "but I would rather have Mother do it."

T HE JOKE WAS attributed to Art Buchwald, resident humorist at the *Washington Post*. It involved a pair of Washington lawyers who, as amateur sailors, had crashed their yacht in remote waters before swimming to a deserted island. Contemplating their bleak prospects, the first lawyer complained about the intense heat and all the creature comforts denied him. "And I was just getting to be successful in Washington," he lamented. As proof, he cited a recent lunch with Minority Leader Gerald Ford. Forgetting his wallet, the attorney had been forced to borrow $30 from the congressman to pay their tab, with a promise to reimburse him the full amount "this week."

"Oh my God, we're saved," the other lawyer rejoiced. "If you owe Jerry Ford thirty dollars, Jerry Ford will find us."

No one laughed harder at Buchwald's jibe than its intended target. Depression-bred thrift heavily influenced Ford's approach to economic policy. (Late in life he could boast that, even during the worst of hard times in 1930s Grand Rapids, Gerald Ford Sr. had never availed himself of government assistance.) But there was more to it than that. According to economist Arthur Burns, adviser to presidents from Eisenhower to Reagan, Ford displayed the surest grasp of any where the dismal science was concerned.*

Ford didn't require two decades as a congressional appropriator to appreciate the fragility of his economic inheritance. Determined that his own campaign for a second term should unfold against a backdrop of robust growth, Richard Nixon had leaned heavily on Burns and the Fed to relax their grip on the nation's money supply, which expanded at an annual rate of 9 percent or more in the run-up to the 1972 election. The inflationary fires thus kindled were not doused by Nixon's subsequent wage and price controls. They were worsened by crop-killing droughts and the 1973 Arab oil boycott. In a matter of months the cost of imported oil tripled. As prices at the gas pump and in the supermarket soared, so did basic interest rates. By the summer of 1974 these were topping out at 10 percent or higher. In the resulting credit crunch, some banks tottered and the stock market lost almost half its value between January 1973 and December 1974—"Panic on the installment plan," according to one Wall Street wit.

All this made Ford the first White House occupant since Herbert Hoover to preside over a period of declining expectations. His two-day Summit on Inflation at the end of September 1974 was the culmination of a monthlong series of public meetings designed to elicit input from economists, lawmakers, businessmen, consumer advocates and representatives of organized labor. By casting a wide net and televising the resulting clash of ideas, the Ford White House drew yet another contrast with Nixon and his economic U-turns, hatched in secret and sprung on an unsuspecting nation. In the event viewers were treated to a litany of demands by groups whose naked self-interest was catalogued by journalist Richard Reeves.

---

* He was also the most scrupulous in respecting the central bank's independence. Ford once apologized to Burns for raising the issue of troubled banks, the subject of a recent *Washington Post* series. Should the chairman of the Fed consider the question improper, said Ford, "please attribute it to my ignorance . . . just ignore it." Arthur F. Burns, "Ford and the Federal Reserve," in *The Ford Presidency, Twenty-Two Intimate Perspectives of Gerald R. Ford,* ed. Kenneth W. Thompson (New York: University Press of America, 1988), 137.

"Labor leaders led by George Meany, president of the AFL-CIO, argued against wage controls and proposed money-moving anti-recession programs," Reeves observed. "Farm organization leaders argued against restrictive export regulations designed to keep more American food at home and reduce food prices. Business leaders called for greater depreciation allowances to increase industrial production. Oil executives wanted an end to price controls on domestic petroleum production. Airline executives wanted more fuel price controls . . .'Spokesmen' for the poor attacked defense spending, depreciation allowances and business tax credits, advocating more social programs."

If anything like a consensus emerged from this dissonant chorus, it echoed the 70 percent of Americans who told pollsters that inflation was their country's greatest challenge (far more than those who cited Watergate or other concerns). At one preliminary meeting no one took a harder line than the liberal icon John Kenneth Galbraith. "Let us not be beguiled by our fears of unemployment," he told the assembled experts. "Let us keep in mind the damage that is also done by inflation." Agreeing with Galbraith on the need to reimpose wage and price controls, Senate Majority Leader Mike Mansfield urged Ford to resurrect the Depression-era Reconstruction Finance Corporation as a vehicle for channeling credit to beleaguered airlines, railroads and defense contractors.

Other voices carried greater weight with Ford and his economic team. Among the most influential was economist Milton Friedman, the anti-Galbraith famous for his theories equating economic growth with limits on the growth of the money supply. Friedman urged Uncle Sam to spend, as well as print, less money. "We must stop kidding ourselves that we can all tap the public purse at the expense of someone else," he exhorted a hotel ballroom crowded with advocates for doing exactly that. Moreover, said Friedman, no one should expect victory over inflation to be quick or painless. His weapons of choice—spending cuts, coupled with a significant tightening of monetary policy—were likely to produce "a temporary period of low growth and relatively high unemployment." In short, a recession.

Far more politically palatable was the proposal from Sylvia Porter, the *New York Post*'s nationally syndicated personal finance columnist. First broached at a White House meeting on September 20, it was reiterated a week later before a television audience of millions. Porter wanted to enlist the American consumer, whose spending accounted for two-thirds of the nation's GNP, in a grassroots campaign to lasso inflation through organized boycotts of price-gouging merchants. The president could help the effort, she suggested, by promoting energy conservation and recycling initiatives.

Unknown to Porter, she was pushing against an open door. Before there was Whip Inflation Now (WIN), there was IF (for Inflation Fighter), the brainstorm of White House speechwriter Paul Theis and a Philadelphia business executive named William J. Meyer. Over lunch, Meyer, president of the Central Automatic Sprinkler Company, spun a scenario wherein millions of consumers used their purchasing power to restrain the demands of business and labor alike.

The scheme drew inspiration from FDR's National Recovery Administration, with its morale-boosting Blue Eagle emblem stamped on countless storefront windows, and the "E" awards handed out during World War II to help spur industrial production. Under Meyer's plan, merchants who signed up to participate in the voluntary campaign would be recognized with a trademark sticker marked IF. More tangible incentives included the patronage of consumers steered toward those cost-conscious establishments that bore the IF stamp of approval. That a war might be waged against inflation on the cheap, without government mandates or costly new bureaucracies, appealed to the conservative in Ford. "We discussed it in the Economic Policy Group," he remembered. "Nobody really spoke up against it. There were some who were less enthusiastic than others . . . Everybody at that point was grasping for some way, either by law or by any other means, to do something about 13% increases in the cost of living. So there was nobody inside who said, 'Oh, forget it.'"

Ford's interest was rekindled when Sylvia Porter urged adoption of something virtually identical to the scheme first proposed by Paul Theis. Seizing on the voluntary concept, Ford sought out Porter at the September 28 conference and asked if she would help lead a national citizens campaign. Feeling sympathetic toward a man who had just returned from visiting his cancer-stricken wife in the hospital, Porter agreed to help out. Her acceptance was hastily added to Ford's closing remarks, in which he encouraged members of the television audience to share with him their own lists of ten ways to save energy and fight inflation.

Dismissed at the time by critics as a media event, the summit changed the way Washington dealt with the economy. In what Ford called the most important institutional innovation of his presidency, he established the Economic Policy Board, chaired by Treasury Secretary William Simon, with William Seidman as executive director. Henceforth not only the fight against inflation, but all government efforts to combat unemployment, formulate a coherent energy strategy and promote global trade fell within the broad mandate of the EPB.

The summit also foreshadowed Ford's pursuit of economic deregulation, a policy shift with historic implications for democratic capitalism. Ford didn't shy away from government intervention when he thought the public interest required it. Less than a week after the summit adjourned, overriding objections from Agriculture Secretary Earl Butz, he canceled $500 million in corn and wheat sales to the Soviet Union. (Under Nixon, Moscow had been able to purchase up to one-quarter of all US grain production.) Such transactions became harder to justify as supplies dwindled and American consumers staggered under double-digit increases at the supermarket checkout counter. The nation's farmers, dependent on exports for much of their income, took a dimmer view of Ford's jawboning. Their anger was likely to make itself felt on election day, barely a month away.

Farmers were not the only voters being asked to put patriotism ahead of their pocketbook. Appearing before a joint session of Congress on October 8, Ford served up a mulligan stew of thirty-one separate proposals, many inspired by the recent summit—and all jealously guarded by Bob Hartmann, who had prevented officials at the Treasury Department and Office of Management and Budget from reviewing Ford's text before its delivery.*

Fiscal restraint was the predictable keynote of Ford's strategy to combat inflation and stimulate growth. Before his former colleagues he reiterated his call for modest cuts to federal spending, though he readily acknowledged the short-term impact of a $300 billion budget cap to be mostly psychological. Far more tangible, and politically risky, was Ford's proposed 5 percent surtax on corporations and higher-bracket taxpayers. The proceeds of this temporary levy would in turn fund tax cuts for lower- and middle-income Americans, along with extended unemployment benefits and a new Community Improvement Corps to provide work for the unemployed whenever the jobless rate exceeded 6 percent nationally. Hoping to placate farmers upset over the loss of Soviet trade, Ford promised the elimination of government rules curbing rice, peanut and cotton production. For business he offered investment tax credits; for first-time home buyers, an additional $3 billion to help relieve the credit crunch; for consumers, a renewed commitment by the Justice Department to prosecute price fixers, bid riggers and antitrust violations.

---

* "Bob was the kind of guy on writing speeches who would say, 'You can't see it because the president hasn't seen it yet!'" recalls Dick Cheney. "And after the president had seen it, he said, 'You can't see it because the president signed off on it, and we're not changing it.'" According to Cheney, the day before Ford addressed Congress, Alan Greenspan expressed "deep concern" over the speech and the proposed WIN campaign. Thus alerted, Don Rumsfeld urged postponement of the event, but Ford brushed off his objections. AI, Dick Cheney; Dick Cheney, *In My Time* (New York: Simon & Schuster, 2011), 76.

Ford's energy strategy was simply stated: produce more, consume less—enough to reduce imports of foreign oil by one million barrels a day. The president previewed legislation facilitating cleaner coal, safer nuclear power and the deregulation of natural gas supplies in order to stimulate domestic production. Looking further into the future he emphasized the potential of non-fossil fuels—"the power of the atom, the heat of the sun and steam stored deep in the earth, the force of the winds and water." Almost in passing Ford unveiled yet another new commission, this one charged with regulatory reform, beginning with "a long overdue total reexamination of the independent regulatory agencies."

None of this fit on a bumper sticker. So Ford opted instead for a red-and-white WIN button, prominently displayed on the presidential lapel. Whip Inflation Now had supplanted Inflation Fighters as Gerald Ford's Blue Eagle. The slogan originated in the New York advertising agency Benton & Bowles, which also crafted the WIN pledge appearing in newspapers the day after the president's speech. Ford concluded his remarks to Congress by hyping an upcoming address to the Future Farmers of America, in which he promised to elaborate on the promise of WIN. In the meantime he thanked Sylvia Porter and others in the newly formed Citizens Action Committee to Fight Inflation for their willingness to enlist under its somewhat mysterious banner.

Reaction to the speech was tepid. "Neither surprising nor bold," declared the *Wall Street Journal*. The *New York Times* took the president to task over his failure "to recommend the uncomfortable for fear it may prove to be unpopular." That was rich, thought Ford, fresh off the canceled grain deals and his call for higher taxes less than a month ahead of election day. As it happened, his introduction of the WIN button was eclipsed by the collapse that same day of the Manhattan-based Franklin National Bank. The largest bank failure in US history to that time did nothing to bolster consumer confidence.* With September's unemployment rate hitting 5.8 percent, up half a point in one month, Whip Inflation Now struck many economists as a misplaced priority.

Wherever he looked that first week of October, the president faced

---

* In sharp contrast to policies embraced by the Federal Reserve during the financial crisis of 2008, the Fed in 1974 loaned the troubled institution $1.75 billion. When this proved insufficient to halt its death spiral, the Federal Deposit Insurance Corporation consulted with sixteen healthy banks about a takeover. Eventually Franklin was purchased by a firm called European-American. Its managed liquidation would not be repeated thirty-four years later, when the giant Lehman Brothers investment banking firm fell victim to the largest bankruptcy in American history.

mounting troubles. Overseas Henry Kissinger's latest rounds of shuttle diplomacy had failed to sway the OPEC oil cartel or advance Middle East peace. Ford's deal to return Nixon's papers and tapes to the former president was coming unglued on Capitol Hill. Nelson Rockefeller's nomination was snagged in press leaks revealing substantial gifts from Rockefeller to Kissinger and other associates on the public payroll. The embers of Watergate had yet to cool, as the House Judiciary Committee, unhappy with the White House response to its questions about the Nixon pardon, demanded a fuller accounting of events preceding Ford's September 8 thunderbolt.

Not all the news was bad. On October 11, four days before their twenty-sixth wedding anniversary, Betty Ford left Bethesda Naval Hospital to return to the White House. Hundreds of staff members assembled on the South Lawn to welcome her home. There she was also introduced to the newest member of the Ford household, a six-month-old golden retriever pup, presented only the day before to the dog-loving president by David Kennerly and Susan Ford. Renamed Liberty by its new owners, the copper-colored canine was a living reminder of Harry Truman's wry admonition that a man in official Washington who wants a friend should get himself a dog. Ford would have ample opportunity, in the exigent months bracketing the end of 1974, to test Truman's adage for himself.

Part IV

RIDING THE TIGER

October 1974–December 1976

19

"RUN OVER BY HISTORY"

———◆———

*That big Michigan Christmas tree in the Blue Room and I have
a lot in common. Both the tree and I never expected to be in the
White House. We both came in green. We both were put on a
pedestal. And then, a little while later, we both got trimmed.*

—GERALD R. FORD, DECEMBER 1974

PHIL BUCHEN WAS worried. Already disturbed by Ford's seemingly abrupt decision-making and verbal missteps—like the October 9 press conference at which he criticized Boston's court-ordered busing at a time of violent resistance to the plan from white parents—the presidential counselor had begun to question his friend's capacity to outgrow his congressional past. Doubts gave way to feelings of betrayal when, in the course of drafting answers to pardon-related queries from the House Judiciary Committee, Buchen learned about Ford's conversations with Al Haig on August 1, and with Jim St. Clair the next morning. It was his first exposure to the intrigues surrounding Nixon's final days in office, and his anger at being kept in the dark prompted thoughts of resignation.

To lose his White House counsel so soon after Jerry terHorst quit in protest of the Nixon pardon would cause irreparable damage to Ford's presidency. Mindful of the consequences, Buchen was at length talked into staying by Jack Marsh, Ford's counselor-without-portfolio. Not even Marsh, however, could persuade Buchen to endorse Ford's latest jaw-dropper, a proposal that the president himself answer the committee's questions—in person. "We were talking about something they had written to get sent up to the Hill," Marsh recalls, "and I said, 'You can't sign that.' He said, 'I know I can't.' Because it stated that there had been no contacts or conversations in advance of Nixon's resignation."

Showing more acumen than he had in gauging popular reaction to the pardon, Ford dispatched Marsh to establish ground rules with the Democratic leadership. "Tell [Speaker Carl Albert] anything he wants to know and tell him I need his advice, or what he thinks would be best for me to do."

"He isn't going to get hurt," Albert promised Marsh, "because the most important thing for the United States is the success of Jerry Ford as president." No previous chief executive had so blurred constitutional distinctions, notwithstanding murky claims surrounding Abraham Lincoln's appearance before the same committee in February 1862. A reluctant witness, Lincoln denied that his wife had leaked, or sold, an advance copy of the president's annual message to Congress. But that encounter had occurred behind closed doors, long before modern communications made presidents ubiquitous in our living rooms. Ford *wanted* to appear before the Judiciary Committee's subcommittee on criminal justice chaired by Missouri Democrat William Hungate, and to do so in front of live television cameras. Counting on relationships developed over twenty-five years, and an element of trust that crossed party lines, Ford dismissed warnings from experienced pols like Representative Charles Sandman. "The whole world knows why you pardoned former President Nixon," the New Jersey Republican wrote. "No earthly good can come of the entire inquiry. They will badger you from pillar to post and make it necessary for you to come back again and again."

Heeding another lesson from the pardon debacle, Ford set aside two hours a day for preparation during the run-up to his October 17 appearance on the Hill. On the eve of the hearing he got an unexpected boost when the *Wall Street Journal* published an interview with Leon Jaworski in which the departing special prosecutor downplayed the amount of new information that would have surfaced in a Nixon trial.*

Shortly before ten on the morning of October 17, Ford entered Room 2141 of the Rayburn House Office Building and took his place at the long witness table. After pouring himself a glass of water, he began reading from his forty-three-minute statement. "I am here not to make history, but to report on history," he said.

For the most part that is what he did, describing for the first time his two meetings with Al Haig on August 1, and his introduction to the so-called

---

* That same day Ford swallowed his doubts and signed the Federal Election Campaign Act amendments designed to prevent future scandals by limiting individual contributions and campaign expenditures in presidential and congressional contests. The same legislation established a Federal Elections Commission to enforce the new rules.

In January 1976 the Supreme Court validated Ford's reservations by declaring limits on campaign spending an infringement of free speech. The justices also struck down the formula imposed by House Administration Committee chairman Wayne Hays, the most reluctant of reformers, who had conditioned the bill's passage on Congress appointing four of the six FEC commissioners, the remaining two to be named by the president.

smoking-gun tape that doomed Richard Nixon's presidency. As the sub-committee members—five Democrats and four Republicans—listened intently, Ford related his initial feelings of shock, tinged with embarrassment over his unwavering defense of a mendacious president. He recounted the various options, including a presidential pardon, outlined for his consideration by Haig. His response to this overture, Ford told the committee, had been to defer any response, at least until he could discuss the matter with his wife. In pardoning Nixon, Ford said, he had hoped "to change our national focus . . . to shift our attentions from the pursuit of a fallen President to the pursuit of the urgent needs of a rising nation."

Few members challenged this premise, a notable exception being Brooklyn's Elizabeth Holtzman. Already dismayed by the committee's lack of investigatory zeal ("They never asked for one document, they never asked for one witness"), Holtzman alluded darkly to "suspicions that have been created in the public's mind" regarding a possible deal between Nixon and Ford. "There was no deal, period, under no circumstances," Ford retorted. Unconvinced, Holtzman rattled off half a dozen queries without affording the witness time to answer. Why had Ford acted in such haste, without consulting the special prosecutor or his own attorney general? Why hadn't he obtained a full confession of Nixon's wrongdoing? And what, precisely, had transpired during his conversations with General Haig?

Unable to stem the flow of insinuation, Ford could only hope that Holtzman's confrontational stance turned off the viewing audience. His performance garnered mostly favorable reviews, with the *Washington Post* pronouncing the hearing "pretty anticlimactic" and Leon Jaworski making another of his congratulatory phone calls to Phil Buchen. Finding little to dispute in the president's testimony, Chairman Hungate voted with the majority of his colleagues to pull the plug on any further proceedings. This spared Al Haig or any other member of the White House staff from answering questions under oath about their role in the pardon affair. As Nixon faded from the nation's front pages, and polls showed an uptick in his own approval ratings, Ford turned his attention to a sagging economy and an electorate less persuadable than Bill Hungate.*

N O ONE CONTRIBUTED more to the education of Gerald Ford than his new chief of staff in all but name. "My purpose always is to go

---

* In retrospect Ford believed that his deficiencies as a public performer had actually enhanced his credibility before the committee. "I am not that good an actor," he maintained; he might as well have said, "I am not that good a liar."

in," Don Rumsfeld explained in a 1997 oral history, "immediately get the staff fixed the way you want it, get the organization arranged, set up priorities." Ford responded in kind. As soon as Rumsfeld came on board at the end of September, he recalled, "we sat down and laid out a schedule for the resignation or retirement of the people we felt ought to go." By the end of the year, it was agreed, they would have in place not only a new administrative team, but a legislative program and budget bearing Ford's stamp. None of this dulled Rumsfeld's belief that Ford was excessively soft-hearted, a distinct liability in a president, and one for which the newly fashioned staff coordinator compensated with gusto. Pat Buchanan, his ambassadorial career scuttled by a malicious press leak, was far from alone in believing it to be open season on Nixon holdovers. Resolved to make the best deal possible with the new regime, Buchanan assured Rumsfeld that he had no desire to stay on.

"So I'll tell you what. What about me going off the premises November 15th and off the payroll December 15th?"

"What about off the premises October 15th," Rumsfeld responded, "and off the payroll November 15th?"

Some retirements involved score settling by outgoing officials. "Claude Brinegar was Nixon's secretary of transportation and he didn't want to leave," according to Dick Cheney. At least not unless he was preceded through the exit door by Alexander Butterfield, the FAA administrator fiercely resented by Nixon loyalists for having revealed the existence of the White House taping system. In this instance it was Cheney who fired Butterfield. At other times Cheney went to the rescue of Nixon holdovers like David Gergen, a talented wordsmith caught up in Bob Hartmann's purge of White House speechwriters. "So I got Gergen a job over at Treasury, working on Bill Simon's staff," Cheney recalls. It was only a matter of time before Gergen was surreptitiously enlisted in a bootleg speechwriting operation answerable to Rumsfeld and Cheney.

Determined to shield the administration from scandal, Rumsfeld worked closely with Phil Buchen in devising a rigorous code of ethics. Ford's reputation was to be jealously guarded; nearly as important, so was his time. After three consecutive days on which Henry Kissinger showed up late for his regular morning session with the president, Ford suggested moving the starting time of their meeting to accommodate the secretary. Rumsfeld thought this would send the wrong signal. It was up to Kissinger to adapt to the presidential schedule, not the other way around. The message was not lost on Kissinger, who was transformed overnight into a model of punctuality.

The flow of paper was entrusted to Jim Connor, a Rumsfeld protégé since their days together at OEO. Several times each week Connor delivered to the Oval Office stacks of reading material outlining policy alternatives, the necessary prerequisite to spirited discussions involving the president, Cabinet officers, program advocates and tightfisted budget makers. Ford adapted quickly to the system, not hesitating to return a decision paper whose analytical content fell short of his standards. "Look, these guys aren't ready for the meeting yet. Scrub it," he instructed Dick Cheney. "Don't let them in until they get their act together." To open a dialogue with the academic community Rumsfeld recruited Robert Goldwin, until recently his assistant at NATO, and before that, dean of St. John's College in Annapolis. Goldwin protested his characterization as the administration's intellectual in residence. "There is something fishy about the word 'intellectual,'" he contended. Most of those so identified were more accurately defined by their distaste—"Sometimes even contempt"—for popular values.

Described by historian Tevi Troy as "the first conservative to serve as intellectual liaison in a modern White House," Goldwin exposed Ford to the ideas of Irving Kristol and Gertrude Himmelfarb, Thomas Sowell and futurist Herman Kahn. One evening historian Daniel Boorstin and Harvard government professor James Q. Wilson joined the president in the Red Room for a free-ranging dialogue about leadership in an era of pessimism. A Saturday-afternoon gathering in the White House solarium featured Daniel Patrick Moynihan talking about his book *Beyond the Melting Pot*, and the persistence of ethnicity in the United States. (After one such seminar, Alan Greenspan confessed to Goldwin "he didn't know that so many conservative Blacks existed.") According to one prominent academic, the level of discussion at these meetings was akin to what took place around the table of a faculty club at an elite university.*

"Ford always brought the talk down to practical examples," Goldwin said. "If a scholar would make a theoretical statement, Ford would relate that to some actual experience that he had in Congress," culminating in an unexpected "Yes, that's exactly what I mean" from the president's donnish company. Regrettably no one thought to consult Goldwin before Ford traveled to Kansas City on October 15 to promote his nascent WIN pro-

---

* Clearly impressed by Boorstin, a prolific historian, classroom instructor and museum administrator, in June 1975 Ford nominated him to be the twelfth Librarian of Congress. A transformational leader who more than doubled the institutional budget, Boorstin's accomplishments earned Ford's characterization of him as "one of the best appointments I ever made."

gram. Oversold to the broadcast networks as a significant policy address, the speech as delivered before the Future Farmers of America was a pastiche of household hints worthy of Heloise—"Stick to a shopping list." "Turn off lights when not needed." "Take all you want, but eat all you take." These were interspersed with homely counsel culled from the president's mail basket. "From Hillsboro, Oregon, the Stevens family writes they are fixing up their bikes to do errands." A fourteen-year-old in Pasadena, California, having unplugged his stereo to curtail the family electric bill, thought his classmates would benefit from classes in energy conservation.

Lost in the ridicule aimed at Ford's "Clean your plate" speech was the grassroots response it elicited. Over two hundred thousand Americans wrote the White House pledging their support for WIN. A thousand phone calls a day overwhelmed an executive director and four volunteers who comprised the staff of the Citizens Action Committee to Fight Inflation. The American Automobile Association announced a companion effort to reduce gasoline consumption. Budget Rent A Car put in an order for a million WIN buttons. McDonald's ran McWin commercials touting its cheaper burgers and shakes (though the hamburger chain failed to secure a presidential endorsement of the Egg McMuffin for its inflation-fighting properties). Composer Meredith Willson contributed a theme song to inspire Ford's "great citizen mobilization":

> *Win! Win! Win!*
> *We'll win together,*
> *Win together, that's the true American way, today.*
> *Who needs inflation?*
> *Not this nation.*
> *Who's going to pass it by?*
> *You are, and so am I . . .*

The Music Man's latest composition was unlikely to make audiences forget "Seventy-Six Trombones."

By November, with a spike in unemployment rivaling inflation for public attention, Ron Nessen took to wearing his WIN button upside down. Asked to explain the meaning of "NIM," Nessen replied, "No Immediate Miracles." Sylvia Porter, dismayed by partisan infighting and administration indifference, resigned from the citizens committee in March 1975. An uncredited memo in her personal papers summarized the ill-fated program characterized by Dick Cheney as Gerald Ford's Bay of Pigs. "You cannot talk your way out of inflation. You cannot talk your way into prosperity . . . You also dare not dismiss the fact that Americans no longer buy the quick scare."

F OR THE POLITICAL animal, nothing is more easily rationalized than a lost cause. During golf games that fall at Burning Tree Country Club, a tony all-male preserve with a $12,000 initiation fee, Tip O'Neill predicted disaster for Ford's party in the November elections.

"It's going to be an avalanche," chortled O'Neill.

Had the decision been left to Don Rumsfeld, Ford would have stayed off the campaign trail. By sticking to his official duties, it was argued, prioritizing economic management over partisan barnstorming, the unelected president might bolster his public standing while distancing himself from the impending wipeout. Truth be told, not many GOP candidates were clamoring for his help. Ford was a much less valuable political commodity after September 8 than before. In the end, however, the old warhorse could not resist the clang of the bell. So he visited nineteen states before election day, traveling twenty thousand miles and delivering at least eighty-five speeches before a numbing array of airport rallies and half-empty arenas. In Cleveland, former Ohio governor James Rhodes, trying for a comeback against Democratic incumbent James Gilligan, pleaded a scheduling conflict to avoid sharing the dais with Ford. A presidential fundraising breakfast in Oklahoma City drew an embarrassing fifty-nine patrons. Recycling lines he had used as vice president, Ford warned of "legislative dictatorship" and "one party government" should Democrats achieve their goal of a veto-proof Congress. Before one audience he even suggested that "peace could be in jeopardy" unless apathetic voters did their civic responsibility by voting Republican.*

Ford's campaign regimen made for a grueling schedule. "He didn't want to leave the White House until four or five o'clock," Special Assistant Terry O'Donnell remembers. "You know we'd fly out to Iowa to do a dinner for a congressman and a friend of his and he'd fly back because he wanted to be in the office in the morning because that's the people's business." He didn't want to spend the night "even though that would be the easy thing to do." The last weekend of the campaign Ford was in California, where polls showed state comptroller Houston Flournoy, the GOP's candidate for

---

* His fears on this score were less extreme than his rhetoric, since Ford personally authorized Speaker Carl Albert to use military aircraft that fall when campaigning for Democratic candidates.

governor, rapidly gaining on Jerry Brown, son of Ronald Reagan's two-term predecessor, Edmund "Pat" Brown. It was no accident that, before departing for the West Coast, Ford signed legislation making it illegal to deny credit to women on account of their sex. But if his gesture was intended to deflect talk of the pardon as a campaign issue, it was easily overshadowed by the simultaneous announcement of the latest monthly unemployment rate of 6 percent, a psychological tipping point whose timing could not have been worse for Republicans like Flournoy.

As for the man most responsible for his party's plight, Richard Nixon remained politically radioactive even while fighting for his life. Ford had earlier in October expressed a desire to call the former president, who was facing surgery to avert potentially lethal blood clots, only to be talked out of it by Rumsfeld. Hours after the operation, performed on the morning of October 30, Nixon went into shock due to massive internal bleeding. Frantic lifesaving efforts stabilized his condition, but he remained critically ill, his prognosis uncertain.

This time Ford angrily rejected advice to keep his distance. "If there's no place in politics for human compassion," he observed testily, "there's something wrong with politics." On Thursday evening, October 31, he telephoned Pat Nixon.

"I don't want to push it, but would it help if I came down there?"

"I can't think of anything that would do him more good."

The next morning Ford made the short journey from Los Angeles to Long Beach by military helicopter. In the hospital's cardiac care unit he was greeted by the former First Lady and her daughters, Tricia and Julie. Five minutes later, accompanied by Nixon's personal physician, Dr. John Lungren, Ford approached Room 704. Finding it locked from the inside, workmen were summoned to break the lock with a hacksaw.

Propped up in bed, surrounded by equipment monitoring his vital signs and with an air tube in his nose, Nixon managed a weak, "Hi, Jerry."

"Oh, Mr. President," Ford blurted out, unable to conceal his surprise at Nixon's ravaged appearance. He took a seat beside the bed.

"Did you have a good night?"

"None of the nights are too good."

Ford alluded to his forthcoming trip to Japan, South Korea and the Soviet Union. He promised to convey Nixon's greetings to Leonid Brezhnev. Of the midterm elections, inevitably a referendum on the man in the white hospital gown, he fudged his reputation for truth telling by claiming, "We're going to be fine." Nixon of all people knew better. He urged his successor "to hang in there with Rockefeller."

It was time to go. Rising from his chair, unsure whether he would see his friend again, Ford expressed appreciation to Nixon for "all you did for me."

Nixon in turn said he was "deeply grateful" for the presidential visit.

"Be well," Ford murmured as he took Nixon's right hand in his. Outside the hospital he told reporters that although Nixon "obviously" was a very sick man, "he's coming along very, very well."

Four days later Democrats scored their most lopsided victory since Lyndon Johnson's 1964 landslide. A gain of forty-three seats in the House gave them one more than the two-thirds required to override a presidential veto. Most of the House candidates for whom the president went to bat—many identified as staunch Nixon defenders—fell victim to the Democratic sweep. In the Senate, Democrats padded their existing majority by an additional three seats. They finished the night holding nearly three-quarters of the country's governorships. An embittered Houston Flournoy blamed his narrow loss to Jerry Brown on Ford's ill-timed visit to the former president at his Long Beach hospital. Ford watched the returns with Betty and a few invited friends on a trio of television sets brought into the Oval Office for the occasion. Before turning in at one thirty a.m. he placed calls to Bob Dole, narrowly victorious in Kansas, and to Governor William Milliken, whose Michigan was alone among the nation's largest states in voting Republican.

"The people have spoken," a subdued Ford acknowledged after rubbing a short night's sleep from his eyes, "and for twenty-six years I have accepted the verdict of the people, which is the essence of our system of free government." His successor as House Republican leader, John Rhodes, had a more visceral reaction, blaming the Nixon pardon for his party's electoral humiliation.

"Goddam it, Jerry, why couldn't you have at least waited until after the election?"

O N NOVEMBER 17, Ford boarded *Air Force One* for an eight-day Asian journey that marked his debut on the world stage.* After a fence-mending visit to Japan, its government still recovering from the shock of his predecessor's fickle economics and secret diplomacy involving

---

* Technically, Ford's first foreign trip was a hastily arranged cross-border meeting with Mexico's president, Luis Echeverría Álvarez, on October 21. In between ceremonial wreath laying and public remarks by both men, the eight-hour mini-summit revolved around issues of illegal immigration, the trade in illicit drugs and Mexico's recent oil discoveries and their implications for American energy policy.

Beijing, the schedule included a brief stop in South Korea to be followed by a two-day summit with Soviet leader Leonid Brezhnev in Vladivostok, the Far Eastern port city off-limits to foreigners since 1958. Before Watergate robbed him of political and diplomatic credibility, Nixon had negotiated with his Soviet peers a Strategic Arms Limitation Treaty (SALT) that modestly restrained development of antiballistic missile (ABM) systems as well as some strategic offensive weapons. SALT I didn't cancel any system then in development, much less reduce existing nuclear stockpiles. It did, however, suggest a workable formula by which to pursue genuine arms control—"First limitation," explained Soviet ambassador to the United States Anatoly Dobrynin, "then reductions of nuclear armaments under international control."

As détente took hold, the rival superpowers tested the limits of cooperation in science and technology, space exploration, and environmental protection. With Nixon in office they had agreed on a tentative settlement of outstanding World War II debts owed the United States, contingent on congressional approval of the most-favored-nation trading status eagerly sought by Moscow. One unintended consequence of this quid pro quo was to empower Senator Henry Jackson and other Capitol Hill skeptics who would condition any trade deal on the easing of restrictions that limited Jewish emigration from the Soviet Union to Israel and other countries. Hoping to finesse the issue, Ford had dispatched Kissinger to talk with American lawmakers and Soviet diplomats. Under a complex formula grudgingly accepted by the Soviets, up to forty-five thousand émigrés a year were to receive exit visas in return for which the USSR got US trading terms to its liking.

Meeting with Jackson on October 18, Ford thought they had a deal. In giving the senator license to speak to the White House press corps the president urged discretion; after all, the proposed settlement could only succeed as a tacit understanding. So when Jackson implicitly faulted the administration for not applying sufficient pressure to reach or surpass his personal "benchmark" of sixty thousand émigrés, he fanned suspicions on both sides of the Iron Curtain.

A month later Ford departed Washington for Asia. As if to confirm Kissinger's forecast of "a tough meeting" at Vladivostok, *Air Force One* encountered severe turbulence over Pacific waters once patrolled by gunnery officer Gerald Ford and his crewmates on the USS *Monterey*. The first American president to set foot on Japanese soil, Ford might be forgiven for believing he had never left Washington. Absent from the tarmac when the big silver-and-blue jet landed at Tokyo's Haneda International Airport was that country's embattled prime minister, Kakuei Tanaka. Mired in a personal Watergate of shady land dealings and tax evasion, Tanaka was in the last days of his premiership.

Demonstrations by leftist groups in advance of the president's arrival recalled Dwight Eisenhower's last-minute cancellation of his planned visit to Japan in June 1960. The Secret Service, citing security concerns, vetoed Ford's request to attend an exhibition baseball game between the New York Mets and a Japanese all-star team. Riot police were stationed at fifteen-foot intervals as Ford made his way to the Akasaka Palace, a three-hundred-room royal guesthouse modeled after Versailles. After a restorative night's sleep in his pink canopied bed, Ford was formally welcomed by Emperor Hirohito the next morning. One could only guess at the emotions these wartime enemies felt as their respective national anthems were followed by a sprightly rendition of the University of Michigan fight song, "Hail to the Victors." The powerful symbolism of the occasion failed to obscure the fact that the president's pants, part of a rarely worn suit of morning clothes stitched together by a Grand Rapids tailor in 1948, were two inches too short.*

Having done his homework, the former congressman found it easier than expected to make conversation with the constitutional monarch, who claimed descent from the sun goddess Amaterasu. Besides a shared interest in baseball, the two men talked about Hirohito's passion for marine biology. Ford invited the emperor to visit the United States (setting the stage for an unforgettable photo of the diminutive demigod touring Disneyland in the company of Mickey Mouse). On leaving the moated Imperial Palace, Ford ordered his driver to stop so he could work a crowd of fifteen hundred well-wishers on the other side of a restraining rope. His encounter with ordinary Japanese was more enjoyable than his talks with Prime Minister Tanaka ("just a hardline Japanese politician") in the ornate Rising Sun Room of the Akasaka Palace. The two leaders, one the beneficiary of scandal, the other its victim, discussed China, the Middle East, the pooling of scarce energy supplies and resumption of trade talks in Geneva. On the touchy subject of Japanese exports to the United States, Ford offered assurances of his commitment to free trade.

The rest of his visit showcased Japan's ancient culture and its modern democracy. Ford watched gymnastic exhibitions in a Tokyo sports arena, addressed Japanese legislators and removed his shoes before touring Kyoto's

---

* Journalists quick to report this sartorial faux pas failed to notice that the emperor's formal trousers obscured the tops of his shoes.

Nijo Castle, formerly home to Shogun rulers. That night he polished off a ten-course dinner with sake served by a pair of chalk-faced geishas. On November 22 a brisk, ninety-minute flight from Osaka to Seoul returned Ford to a land bearing scant resemblance to the war-ravaged country he had first visited as a junior member of Congress in 1953. A million flag-waving onlookers chanted "*Man-sei*" ("Long life!") as the American president traveled by motorcade from the airport to his hotel.

The boisterous welcome could not obscure popular resentment of President Park Chung Hee and his increasingly authoritarian government. After placing a wreath at South Korea's Tomb of the Unknown Soldier Ford paused at the black marble memorial to First Lady Yook Young See, killed three months earlier by an assassin's bullet intended for her husband. From there he rode a helicopter twenty-five miles to Camp Casey, near the demilitarized zone that separated the two Koreas. Lunching on burgers and potato chips with members of the army's Second Infantry Division, Ford received some unsolicited advice from a soldier sharing his table. "Tell those Russians who's number one," the enlisted man admonished his commander in chief. Ford responded with a noncommittal grin and a handshake. Back in Seoul, he conferred with South Korea's president at his official residence, Blue House. Two years after the withdrawal of some twenty thousand US troops from the Korean peninsula, Ford promised to keep the remaining thirty-eight thousand GIs in place as a bulwark against the Communist North. He also renewed an earlier pledge of US financial support to help Park modernize his six-hundred-thousand-man army.

At this point Ford asked everyone but Park to leave the room. Only when they were alone did he raise with the South Korean strongman the sensitive topic of Park's repressive rule. Careful to frame his remarks as an appeal to self-interest rather than outside dictation, Ford referenced two congressional committees that were already weighing substantial cuts in US military aid to Park's regime. The next Congress, elected earlier in the month on a platform of reform and retrenchment, was even less likely to provide the desired funding unless Park moved to reconvene the National Assembly, end press censorship and release hundreds of jailed political opponents.

His appeal fell on deaf ears. South Korea's president grew ever more dictatorial until the night in October 1979 when he was assassinated by the head of his security guard as part of a failed coup. Had he heeded the cautionary advice of Gerald Ford, Park Chung Hee might have lost nothing worse than an election.

WIDELY PORTRAYED AS mere layovers en route to Vladivostok, Ford's visits to Japan and South Korea were confidence builders for someone who had yet to establish himself in the field of international diplomacy. As it happened, the same people skills that had sustained him as a candidate and congressman were easily transferred to a Japanese palace or Korean army base. In both countries he navigated minefields of controversy with a common touch and uncommon sensitivity to the feelings of his hosts. Ford's emotional intelligence nicely complemented the academic virtuosity of Henry Kissinger, who in preparation for their upcoming meeting offered him a perceptive analysis of Leonid Brezhnev. The Soviet leader would probably remind his American counterpart of "a tough and shrewd union boss, conscious of his position and his interests, alert to slights."

Fond of earthy humor and luxury cars from abroad, the general secretary of the Communist Party preferred weekend hunting parties to performances at the Bolshoi. An avid soccer fan, at sixty-nine Brezhnev had exchanged ice skating, skiing, cycling and parachute jumping for more sedentary pursuits, reflected in a medical history believed to include multiple heart attacks and a possible stroke. None of which moderated his vodka consumption, nor his reliance on Nembutal to combat chronic insomnia. Kissinger's portrait of the aging leader reaffirmed the secretary of state's omniscience. Yet not even "SuperK" could factor in a Siberian snowstorm that prevented Brezhnev and his party from landing at the military airport closest to Vladivostok. Soviet officials considered moving the talks to Khabarovsk, whose location five hundred miles to the north resulted in freakishly dry winters. Unfortunately its proximity to the Chinese border—just nineteen miles distant—made the diplomatic climate unacceptable.*

A Soviet army division was deployed to the airport, ninety minutes by train from the original conference site, to clear a single runway for *Air Force One*. With virtually all military facilities around the airfield placed underground, it was a featureless vista that greeted Ford as his plane skidded to a halt among Siberian snowdrifts. Below-zero temperatures precluded any ceremonial welcome or soldierly parade. Instead Brezhnev, a bearlike figure as outwardly gray as his overcoat and astrakhan hat, led a small greeting

---

* Illustrating the complexity of US efforts at triangulation with rival Communist powers in Moscow and Beijing, American diplomats took care to run the proposed Vladivostok meeting site by the Chinese. According to Kissinger, "they preferred Vladivostok to Europe." Memcon, Ford-Helmut Schmidt, December 5, 1974, GRFL.

party that included Foreign Minister Andrei Gromyko and Ambassador Dobrynin. (Seeing Ford off at the White House five days earlier, Dobrynin had spontaneously removed his beaver-fur *shapka* and given it to the president. No doubt Ford appreciated Dobrynin's gift as he disembarked from *Air Force One* to be enveloped in the numbing cold.) After the obligatory hearty handshakes Brezhnev broke the conversational ice by commenting on the visitor's athletic reputation. "I understand you're quite an expert on soccer," Ford replied. The Russian leader said he hadn't played for a long time. Ford allowed that his own days on the gridiron were long past, adding, "I wasn't very fast, but I could hold the line."

A short ride in Brezhnev's ZIL limousine brought the two men to a waiting train, its green-and-gold, wood-trimmed cars heated by ancient potbellied stoves. Taking their places in the dining car, Ford and Brezhnev leaned across a table spread with fruit, soft drinks, red and black caviar and six brands of cigarettes. As the train sped across a frosted landscape dotted with scrub oak, pine and birch trees, Ford fell back on the oldest of conversation starters.

"How much snow do you get in Moscow?"

"It changes from year to year," Brezhnev replied. "Sometimes a half meter falls in one night."

In Washington, Ford confessed, even a modest snowstorm could produce chaos, since municipal authorities lacked equipment with which to clear the streets. Brezhnev returned his serve. "And that will be our first deal," he declared, wagging his cigarette for emphasis. "We send you Soviet snow plows."

"At a good low price," chimed in the famously dour Gromyko. *

The relative merits of tea versus coffee were discussed. "The Bulgarians like very dark and strong coffee," according to Brezhnev. Ford put in a plug for New Orleans chicory. He hoped that on his next visit to the United States the general secretary would be able to see something of the American heartland—"For example Chicago"—in order to get a broader perspective on the national character than that afforded by the two coasts. The passing scenery prompted Ford to inquire about Soviet agriculture, the length of the growing season and usage of hydroelectric power. A brief conversational

---

* Inevitably a snowplow dealer in northern New York, close by the Canadian border, saw in this exchange a threat to his livelihood and complained to the White House. He was duly assured by a presidential assistant that the Ford Administration had no intention of importing snow removal equipment from the USSR. Palmer Lockhart-GRF, December 3, 1974, GRFL.

detour to irrigation and the most effective use of fertilizers preceded an abrupt change of channels.

How, Ford inquired, did the general secretary wish to proceed? Certainly the two men might expand their personal acquaintance. "On the other hand, there are some subjects which we could usefully discuss in either a smaller or larger group." Brezhnev, accustomed to the more elliptical style of Ford's predecessor, was momentarily thrown off his game. Diplomacy was rarely so direct.

"Let us speak not as diplomats but as human beings," he told his American guest. "Both you and I fought in World War II. That war was child's play as compared to nuclear war."

Avoiding rhetorical flights, Ford complimented Brezhnev and "my predecessor" for all they had achieved in bettering relations between their countries. As for himself, he intended to continue on the path initiated by President Nixon. He conveyed the former president's best wishes to Brezhnev, who in turn revealed the existence of a Nixon letter, written at the time of his resignation, praising Ford and certifying his commitment to the relaxation of superpower tensions. Though fully prepared, he said, to discuss weapons numbers and classes—thereby accepting Ford's transactional approach to curtailing the arms race—Brezhnev stressed the latent dangers of armaments as yet unimagined. "I just don't know how much farther we can go in building up so-called security."

Ford thanked the general secretary for his "statesmanlike" views. "I think we could talk in this broader context at a later time. But I believe it important at this meeting to discuss these issues in specific terms and step by step." And so it went, with Brezhnev citing the existential threat of nuclear Armageddon, and Ford politely steering the conversation back to the immediate problem, broken down to its component parts as a legislator eschews soaring oratory for practical attempts to find common ground. By revealing his intention to run for a full term in 1976, something Ford had only hinted at publicly, the president scored points for frankness while making it plain that domestic political considerations would influence the ensuing negotiations.

It was around four in the afternoon when their train reached the Okeanskaya (Oceanside) Sanatorium, a hundred-acre compound encircled by a ten-foot fence and patrolled by guards whose fur-trimmed uniforms inspired comparisons to Czarist-era Cossacks. Dubbed the Comrade Hilton by a White House advance man, the place reminded Ford of "an abandoned YMCA camp in the Catskills." Brezhnev escorted the president to his temporary quarters, a stucco-and-frame dacha two hundred yards from the main conference center. Like any good guest, Ford availed himself of an indoor swimming pool refurbished in anticipation of his visit.

As he did his laps the Soviet delegation labored to conceal a health scare involving their leader. According to Dobrynin, Brezhnev suffered a seizure on board the train carrying him and Ford to Okeanskaya; other accounts say it occurred the day before in Khabarovsk. Rejecting suggestions that he postpone the summit, Brezhnev appeared healthy enough when Ford arrived at the conference hall shortly after six o'clock, fortified against the cold by a three-quarters-length wolf-skin parka gifted him during a refueling stop in Alaska. "I'm a sheep in wolf's clothing," he quipped. An envious Brezhnev admired the garment, leading Ford to insist he try it on. Only after he promised to get a second coat for the general secretary did the two men and their advisers take their places around a long table furnished with snacks and mineral water. The schedule called for a two-hour introductory session, followed by dinner. These plans were soon discarded, and the glass-sided winter garden overlooking Amur Bay became the proverbial smoke-filled room as Brezhnev chain-smoked white-filtered Novost cigarettes and Ford sucked on his pipe.

Lending urgency to their discussions was the blistering pace of technology, far outstripping the shuffling tempo of international diplomacy. Even before the ratification of SALT just two years earlier, its signatories had learned to equip their land- and sea-based missiles with multiple warheads known as MIRVs, each device targeting a separate objective. This multiplication of force threatened to destabilize an arms race characterized by radically different arsenals. The United States, relying on a blue-water navy and military bases around the globe, entrusted its security to a "strategic triad" of land, sea and air delivery systems. This was reflected in the Trident submarine, B-1 supersonic bomber, and pilotless cruise missile, systems well advanced in development.

The Soviet deterrent was less versatile, but just as deadly. Primarily land-based as befitting a continental power, it emphasized heavy long-range missiles (ICBMs) whose destructive force exceeded that of the smaller but more accurate missiles deployed by the US. The sheer variety of delivery systems made the concept of numerical equivalency elusive (imagine balancing the number of calories in rival store displays of apples and sausages). What Nixon had referred to as "offsetting asymmetries" was to be enshrined at Vladivostok as "equal aggregates." Further complicating matters were the independent nuclear forces commanded by US allies Great Britain and France. The Soviets insisted they be included in any overall accounting of Western military might. The same argument applied to some 468 forward-based US tactical bombers stationed close enough to the USSR to inflict a devastating counterpunch should a Soviet first strike cripple American missiles in their silos.

In his memoirs Ambassador Dobrynin portrays Kissinger as the de facto leader of the US delegation at Vladivostok, with Ford deferring to his expertise. Kissinger himself recalled things differently. Meeting transcripts corroborate his depiction of a confident American president fluent in the arcane language of Cold War armaments mastered during his years on the House Appropriations Committee. Employing a political bluff to seize the initiative, Ford announced at the outset that he was delaying final consideration of the Pentagon budget pending the outcome of the talks. By contrast Brezhnev required frequent promptings from his military and diplomatic subordinates. At one point Andrei Gromyko found it necessary to whisper in his leader's ear that an American president cannot serve three terms.

Ford let others debate the theoretical advantages to the US of widening existing missile silos by 15 percent. Negotiations stalled over a proposed ten-year ceiling of 2,400 launchers on each side, accompanied by an American promise not to exceed 2,200 before 1983. Objecting to such disparities, even in the short term, Ford cited millions of his countrymen for whom the very notion of equivalence was upsetting. This loosed a volley of complaints from Brezhnev about the imagined dimensions of US nuclear silos and the alleged threat posed to the USSR by American bases in Japan. Shortly after eight o'clock the general secretary proposed a ten-minute break so that each side could take stock of its position. To counter Soviet listening devices, Ford's advance team had installed "babblers," blocking devices that employed hundreds of recorded voices overlapping with, and presumably drowning out, any live conversation taking place.

If anything, they worked too well. "I can't talk or think with these things on," said Ford. He led his advisers outside for an impromptu parley in the frigid darkness, their words made tangible by the icy plume expelled with every breath. After fifteen minutes or so they came in from the cold. Resuming his place at the conference table, Ford offered a fresh twist on American domestic politics, and how the two adversaries might yet defy the odds to achieve "irreversible détente." Without mentioning any names— "And you can imagine whom I am talking about"—the president warned that a Soviet advantage in delivery systems, however temporary or qualified, "could be used against me and against détente and could bring in the elections an administration that would not be as committed to pursuing détente on a continuing basis."

Brezhnev lamented that the two sides seemed further apart than when Kissinger visited Moscow in October. At this point Ford played his trump card. "In the spirit of mutual understanding" and to ease Soviet concerns about US weaponry on European soil, he offered to abandon "most reluctantly" the US nuclear submarine base at Rota, Spain.* Instantly the clouds lifted. The agreeable Brezhnev reappeared. Bemoaning the voracious appetites of his nation's defense establishment ("I don't think there are any holy people in the military") the Soviet leader suggested they call it a night after fresh snags developed over how to count long-range bombers against the proposed ceiling of 2,400 launchers. He had to be in Mongolia in two days, said Brezhnev, his enthusiasm well contained. Ford said he had never been to Mongolia, "but I'm sure that in some ways it is more pleasant than Washington, D.C."

It was after midnight, and they had been talking for over six hours without dinner.

BACK AT HIS dacha, the mood was celebratory as Ford invited aides to join him for a hastily prepared meal of thick, salty *solyanka* soup, cold cuts and vanilla ice cream. As he climbed into bed at one thirty, Ford's thoughts were of another long-standing rivalry unfolding on a field five thousand miles away. He indicated as much on being awakened, a little after seven a.m., by military aide Bob Barrett.

"How did the game turn out?"

"Twelve to ten," said Barrett, no mean diplomat himself.

"Wait a minute. Who had the twelve, and who had the ten?"

Ford knew the answer even as he posed the question. While he slept, Woody Hayes's Ohio Buckeyes had defeated the Michigan Wolverines on a missed field goal with eighteen seconds to go. The news failed to dent the president's upbeat mood as he thanked the kitchen staff for their late-night cookery and donned his now-familiar wolf-skin parka and fur hat for the short walk to the conference center. There he opened the day's bidding by suggesting US flexibility on the issue of long-range bombers. Brezhnev barely acknowledged the offer. Instead he handed Ford a souvenir pipe, part of a gift set that would be prominently displayed in the Oval Of-

---

* In truth it was not much of a concession, owing to the development of longer-range nuclear missiles on Trident and Poseidon submarines that would render such installations superfluous. Moreover, American defense planners, anticipating an end to the Franco dictatorship, had concluded that a resurgent Spanish left would make any US nuclear presence at Rota untenable—but the Russians didn't know that.

fice. The subject of Brezhnev's prospective visit to the United States resurfaced. "I would like to take you to the Marjorie Merriweather Post estate in Florida," Ford ventured, a reference to the Palm Beach palazzo left to the United States by the late cereal heiress, and subsequently made famous as Donald Trump's Florida White House.

"Mr. President, where is your home?"

"I was born in Nebraska, but the home where I was born has been torn down."

Brezhnev was reminded of his birthplace, an apartment building in the Ukrainian city of Kamenskoye that was badly in need of repair by the time he joined the local city council. This was all very pleasant, yet Ford had not journeyed halfway around the globe to swap childhood recollections. About the long-range bombers . . .

"An interesting suggestion," Brezhnev mused. Anticipating such a gambit, he had an alternative plan in mind. Under its terms, bombers—"Specifically the [B-1s]"—would be counted as one launcher if carrying cruise missiles with a range of up to six hundred kilometers. Aircraft equipped with missiles in the range of 600 to 3,000 kilometers would have their individual missiles counted against the overall limit of 2,400. But when Brezhnev urged a ban on bombers with a missile range over three thousand kilometers it was a bridge too far for the American president. That is, unless the Soviets were willing to restrict their deployment of MIRVed heavy missiles to two hundred. A fifteen-minute break failed to break the impasse. "It seems that weapons designers have won," Kissinger observed.

At 1:40 the leaders went into executive session at Brezhnev's request. The general secretary wished to discuss his proposed "nuclear condominium," first raised with Nixon, whose response—"That's an interesting idea"—had been the diplomatic equivalent of clearing his throat. Now as then, no one wanted to give offense, not with so much riding on the fledgling relationship forged by two American presidents. Ford decided his best strategy was to run out the clock. Feigning ignorance, he asked Brezhnev to relate for him what he had said to his predecessor. The general secretary complied by reiterating his conviction that nuclear war was unthinkable. No one could object to that. Unfortunately, Brezhnev continued, not every nation had signed on to the Nuclear Non-Proliferation Treaty, placing both the US and USSR at risk of attack by a rogue power. His solution to this quandary was a separate treaty, exclusive to the world's two superpowers, in which each renounced the use of offensive nuclear weapons while pledging to go to the assistance of the other should it be targeted by third-party aggressors.

Only one member of the nuclear fraternity fit this description. That

Brezhnev viewed Mao's China, just emerging from its long isolation, as the common enemy came as no surprise. But did he really expect Ford to scuttle his country's historic opening to Beijing by joining Moscow in a pact quarantining the Maoist regime? Not to mention pulling the plug on NATO, for three decades the chief instrument of Communist (read: Soviet) containment? It defies logic, and yet this longed-for coupling of the eagle and the bear may help to explain Brezhnev's generally cooperative stance at Vladivostok. Kissinger offered his own interpretation of Russian motives. "With Watergate aside, the Soviets didn't know what would follow Nixon," he concluded. "Now they have a potential six-year President." Moreover, Ford's record in Congress made him a plausible Cold Warrior, theoretically willing to engage the Soviets in a renewed arms race. Consequently, said Kissinger, "They really have a stake in détente—and to defuse people like Jackson."

Whatever the Soviet leader's strategic objective, Ford had no wish to encourage his fantasy alliance. So he fell back on verbal rope-a-dope, asking Brezhnev to clarify his doomsday scenario. "Does it mean strategic nuclear attack, tactical nuclear attack, or any nuclear attack?" What was the difference? Brezhnev replied, not unreasonably. The result would be a nuclear war, "and we want to prevent that." Ford pressed on. "What about an attack by a nuclear power on a third party that is not an ally?" Brezhnev acknowledged it was hard to give "a precise answer. Perhaps we could agree to enter consultations as to the best course."

By all means, said Ford. Let's talk. In the meantime the US and USSR should make "a major effort" to prod nonsignatories into adding their names to the Non-Proliferation Treaty. This gave Brezhnev something tangible, if meaningless, to show for his efforts, while enabling Ford to steer a middle course between Moscow and Beijing. On this anodyne note, the sidebar summit concluded.*

In the final afternoon session, Gromyko played bad cop to Brezhnev's good on the Middle East, Cyprus and the much-debated Conference on Security and Co-operation in Europe (CSCE). The latter was important to the Soviets, who were counting on the CSCE to legitimize post–World War II boundaries originally imposed by the Red Army. Only passing mention was made of Jewish emigration and trade legislation. Neither side, it appeared, wanted to do anything that might put at risk the summit's main

---

* After his "clarifying questions," Kissinger would recall, Ford "referred the subject to Dobrynin and me for further exploration. Neither of us ever took it up again."

business of crafting a basic framework to guide Geneva negotiators in concluding a SALT II treaty ready for signing in the summer of 1975.

Their formal talks concluded, Brezhnev and Ford set off on an impromptu tour of Vladivostok, a city that reminded the American visitor of San Francisco, though it was questionable how much of the place he could see by the fading light of a short Siberian afternoon. On the drive back to Okeanskaya, Brezhnev seized Ford's hand. There was urgency in his voice as he recalled for his guest the brutalities suffered by Russians during World War II.

"I do not want to inflict that upon my people again."

Ford said he looked forward to building on the "very significant" progress made over the last two days.

Brezhnev's grip tightened.

"We have accomplished something very significant, and it's our responsibility, yours and mine, on behalf of our countries, to achieve the finalization of the document."

Ford nodded consent. "We have made so much headway. This is a big step forward to prevent a nuclear holocaust."

There was satisfaction on the faces of the two men as they walked to the ramp at the foot of *Air Force One*. Suddenly Ford peeled off his Alaskan wolf-skin parka and handed it to his Russian host, who seemed "overwhelmed" by the gift. The leaders parted, if not exactly as friends, then convinced they could do business together, to anticipate the measured compliment bestowed a few years later by Margaret Thatcher upon a very different Soviet leader. Ford left Vladivostok believing that seeds of trust had been sowed. By outlining a deal to cap nuclear inventories at 2,400 launchers on each side, the two sides foreshadowed additional talks aimed at finalizing a SALT II agreement. Thus Ford and Brezhnev left it to future negotiators to reduce, and not merely restrict, the world's nuclear stockpiles. "We have averted an arms race of unbelievable cost," Ford insisted at a nationally broadcast press conference on December 2.

But if he expected praise for fixing the parameters of peaceful competition he was in for disappointment. By the end of 1974, superpower summitry had lost its novelty, much as Project Apollo declined in public interest with every astronaut's footstep planted on the powdery lunar surface. Even staunch friends of the president like George Mahon, chairman of the House Appropriations Committee, gave only lukewarm reviews to his nuclear summitry. "The best that could be done at the time," Mahon said of the tentative agreement. Less sympathetic observers on the right accused

Ford of giving away to the Soviets more than his wolf-skin coat. Their ideological opposites in the pro-disarmament camp mocked the "conceptual breakthrough"—Henry Kissinger's expansive phrase—that on closer examination sanctioned a decade of new and costly weapons development.[*] Then there was Scoop Jackson, whose dissent was more predictable than the premise on which it was based. Hoping to appeal to the Democratic left, Jackson objected to the proposed ceiling of 2,400 launchers. Ford should have pressed for more substantial reductions, Jackson argued—say, to 1,700 or so.

That a Soviet leadership under pressure from its military establishment was in no position to make such concessions would become painfully obvious during the Carter Administration, which was compelled to accept terms negotiated by its predecessor in order to secure a SALT II treaty in 1979. When the Senate in turn failed to ratify SALT II, the more hawkish Reagan White House nevertheless observed the limits set by Ford and Brezhnev. Such restraint paved the way for the landmark INF (Intermediate-Range Nuclear Forces) agreement signed in 1987; four years later the first Bush Administration finalized the Strategic Arms Reduction Treaty (START). Anatoly Dobrynin did not exaggerate when he called the Vladivostok agreement the starting point for all subsequent attempts at nuclear disarmament.

V LADIVOSTOK AND ITS unrealized hopes illustrate a central truth about Ford's presidency. As a high-stakes diplomat he had gone toe-to-toe with far more experienced adversaries, winning their respect while protecting American interests. He was even more successful in the ceremonial aspects of his job, impressing nearly everyone he met with the genuine interest he showed in the people and cultures to which he was exposed. Yet the modern presidency unfolds before television cameras more than behind closed doors. "A leader has two important characteristics," said Harry Truman, whose bust, retrieved from a government warehouse by Betty Ford, appeared in the Oval Office while her husband was still overseas. "First, he is going somewhere. Second, he is able to persuade other people to go with him."

Here Ford operated at a disadvantage, his negotiating skills exceeding

---

[*] "For the first time the superpowers have stopped the quantitative race," Kissinger told Canada's visiting prime minister Pierre Trudeau on December 4. "The quality race continues, but no one argues that quality alone can give strategic superiority . . . Once an equal ceiling is agreed, it is easier to reduce, because the quest for quantitative superiority is over."

his political salesmanship. If the polls were to be believed, the president's travels inspired little confidence that he was making his countrymen safer, and more than a little resentment that he was neglecting a rapidly deteriorating American economy.* Contributing to the appearance of drift was a series of legislative rebukes that called into question Ford's special relationship with his onetime colleagues on Capitol Hill. To be sure, he negotiated a compromise mass transportation bill containing the first federal operating subsidies for urban rail, bus and subway systems. But on November 21 lawmakers overrode a presidential veto of the Freedom of Information Act, a post-Watergate attempt to demystify government by making public records more readily accessible. Ford, his attitude inflamed by a series of press leaks, thought the measure ceded too much authority to the courts. (Out of office he readily conceded government's tendency to classify information excessively.) A third piece of legislation would increase educational and training benefits for Vietnam-era veterans by 23 percent, $500 million more than the 18 percent increase advocated by Ford. Frank Carlucci, then an undersecretary at HEW, was invited to brief the president in the absence of Secretary Caspar Weinberger. "It's a bad bill," said Carlucci. "But I must tell you, if you veto it, you'll be overridden in a minute."

"Well that doesn't concern me," said Ford. "If it's a bad bill, I must veto it."

"And he vetoed it," Carlucci recalled, "and he was overridden in a minute."

The most ominous document awaiting Ford on his return from Vladivostok was a memo from Alan Greenspan, chairman of the Council of Economic Advisers. "The economy is now in the midst of a marked contraction in production, employment and incomes," wrote Greenspan. "The weakness that we anticipate in the economy during the first half of next year is consistent with rates of unemployment in the 7 to 7 ½ percent area." On December 6 the Dow Jones Industrial Average fell to levels unseen since October 1962. Especially hard hit was the auto industry. With the

---

* Ford's Whiggish belief in dispersed responsibility gave Cabinet members license during his absence to assert themselves as policymakers. On November 20, Attorney General William Saxbe notified AT&T (but not the White House) that the Justice Department was filing suit in federal court to break up the nation's telephone company. Five days later Treasury Secretary William Simon missed Ford's homecoming at Andrews Air Force Base, being enmeshed in talks aimed at settling a nationwide coal strike. The secretary's intervention was crucial in securing for striking miners a 53 percent increase in wages and benefits over the next three years, a result that did little to bolster the administration's credibility in the war on inflation.

new model year upon them, Chrysler dealers reported a 136-day backlog of unsold vehicles. Industry-wide, slowing sales caused the layoff of 285,000 autoworkers, nearly half the total workforce.*

Equally depressed was the housing market. Realtors in the Miami–Fort Lauderdale market alone counted eighteen thousand condominiums for which no buyer could be found. Nationwide housing starts were down one-third from the previous year. Pawnbrokers observed an influx of middle-class customers coming through their doors. Adding to the seasonal gloom, Montgomery Ward joined other department stores in scaling back Christmas lighting displays to save on bloated energy costs.

The need to coordinate energy policy, and perhaps rethink his anti-inflation agenda, dominated Ford's conversations with a series of Western leaders, beginning with Canadian prime minister Pierre Elliott Trudeau in the first week of December. An unabashed intellectual, trained in Keynesian economics by Harold Laski at the London School of Economics, the charismatic object of Trudeaumania would not, at first blush, appear a kindred spirit to Gerald Ford of Grand Rapids. In fact the two men got on like old friends, agreeing before they parted to a future meeting on the ski slopes. It helped that Ford, from a state bordering Trudeau's Canada, gave considerably more attention to his northern neighbor than did his immediate predecessors. (Informed that Nixon's White House tapes included references to him as an "asshole," Trudeau said, "I've been called worse things by better people.") By contrast Trudeau appreciated Ford's initiative in admitting Canada to the Group of Seven (originally six) leading industrial nations that first convened in Rambouillet, France, in November 1975. "For a middle power like Canada," wrote Trudeau in his memoirs, "membership on this powerful group is important, so we owe much to Gerald Ford."

The American president was to enjoy an even warmer relationship with West German chancellor Helmut Schmidt, who arrived in Washington just as Trudeau was leaving. Like Ford a political pragmatist, unexpectedly elevated to his current position after Willy Brandt resigned in the wake of an espionage scandal, Schmidt was a former defense minister firmly committed to NATO *and* to the easing of tensions with the Soviet Union. At the time of the Nixon pardon he had characterized Ford as "one hell of a strong man." That he found him personally trustworthy was demonstrated at their first White House meeting. In the middle of an extended riff de-

---

* In a sign of desperation carmakers early in 1975 instituted buyer rebates to help reduce the inventory glut. Nearly half a century later, the temporary fix is a habit few consumers are willing to break.

nying that European security interests had been neglected at Vladivostok, Ford was interrupted by his German visitor.

"You do not need to reassure us," said Schmidt. "We have no doubts."

Ford flattered Schmidt by inviting him to sit in on a meeting of the White House economic team. Like Trudeau before him, Schmidt welcomed Ford's collegial approach to the Western alliance. The president's detailed knowledge of German political history, especially Bismarck, was another revelation.* In his memoirs Schmidt devoted an entire chapter to their friendship, which grew even closer once both men were out of office. In 1982 Ford and Schmidt, joined by former French president Valéry Giscard d'Estaing and former British prime minister James Callaghan, assembled in the Colorado ski resort of Vail to inaugurate the annual World Forum sponsored by the American Enterprise Institute. Every summer thereafter, for as long as their health permitted, this politically diverse quartet would convene at the foot of Vail Mountain to expound, amid clouds of pipe smoke, on current events and relive old controversies over a glass of whiskey and a pinch of salt.

Few would have predicted such fraternal bonds in the autumn of 1974, when the German chancellor played diplomatic matchmaker to Ford and Giscard d'Estaing, the toplofty technocrat recently elected to succeed the late French president Georges Pompidou. Under Pompidou France, faithful to its Gaullist legacy, had boycotted the Washington Energy Conference held in February 1974 as well as the International Energy Agency (IEA) formed under American auspices to counter the oil monopolists of OPEC. More annoying to Washington was Giscard's competing call for a conference of oil producers, consumers and developing countries before the industrial West (plus Japan) could formulate a unified strategy.

Haughty though he might appear, the Frenchman was capable of surprises, as he demonstrated in the course of a two-day shirtsleeves-and-swimsuit summit on the French island of Martinique in December 1974. Dispensing with formality, Ford was housed at a resort hotel on the outskirts of Fort-de-France. There he was able to squeeze in several rounds of talks between dips in the hotel pool and a pair of state dinners at which the leaders staked out their positions regarding energy policy. "We don't plan to go to a producer meeting for a confrontation," Ford reassured his Gallic counterpart, "but we have to go to the meeting with a consumer position and an agenda." Otherwise, he argued, OPEC might pick off vulnerable nations one by one.

---

* Schmidt barely concealed his preference for Ford in the 1976 campaign, an indiscretion that permanently soured his relations with the Carter White House.

On their second day in paradise, Ford and Giscard, joined by their foreign secretaries, swam and dined at a restored sugar and tobacco plantation
in the shadow of a malevolent volcano named Pele. ("I have to admit,"
wrote Ford, "seeing Henry in blue bathing trunks was a new experience.")
A steel band serenaded the statesmen as they tucked into a lavish spread
of suckling pig and pineapple mousse, washed down with a trio of French
wines. That evening Ford returned the hospitality, relying on the musical
artistry of Sarah Vaughan to put the seal on a summit at which the Americans got the consumer solidarity they sought, while Giscard secured Ford's
commitment to dialogue with the oil producers. As a bonus, the two sides
settled a long-running dispute over the cost of the forced relocation of US
military facilities prompted by Charles de Gaulle's decision to withdraw
France from NATO's integrated military command. The $100 million
French payoff to Washington was less important than the symbolism of an
alliance revitalized.

THE OPTICS OF Ford's Caribbean summitry did little to convey presidential solidarity with furloughed autoworkers or would-be store
clerks excluded from the usual Christmas hiring rush. The slowing economy fostered a sense of urgency impervious to tropical breezes. Members of
the Economic Policy Board worked overtime in anticipation of Ford's first
State of the Union address and proposed budget. Their occasionally heated
discussions reflected the enormity of "doing something that never has been
done before," as *Businessweek* observed of the young administration. "It
must fight inflation and recession simultaneously." Either/or economics
followed a predictable course in which the remedy for inflation was a dose
of budgetary austerity and a tightening of the money supply. Conversely
in times of sluggish growth orthodox Keynesians relied on red ink and the
monetary printing press to spark recovery.

Now, a year after the 1973 Yom Kippur War and subsequent explosion
in oil prices, these rival dogmas appeared inadequate to the challenge of
stagflation. No one in the Ford White House was quicker to grasp this
fact, or to promote what amounted to an economic U-turn, than Bill Seidman, executive director of the newly established Economic Policy Board.
The politically sensitive Seidman made sure the EPB heard from outside
economists, not just reliable conservative voices like Herb Stein and Paul
McCracken, but Walter Heller and Arthur Okun, Keynesian policymakers
associated with the Kennedy and Johnson Administrations. One evening
early in December the search for new ideas led Rumsfeld and Cheney to a
rooftop restaurant on Fifteenth Street, overlooking the White House and

President's Park. There a little-known economist named Arthur Laffer drew on a cocktail napkin his soon-to-be-famous "curve" illustrating the relationship between lower taxes and increased economic activity. The seed planted that night at the Hotel Washington would blossom under Ronald Reagan into supply-side economics, and a federal deficit confounding Laffer's theory that tax cuts alone would beget sufficient economic activity to pay for themselves. More immediately, a memo outlining Laffer's unorthodox ideas was submitted for Ford's consideration; Laffer himself became an informal adviser to Treasury Secretary William Simon on Social Security reform.

Beyond this Ford was reluctant to venture. He held with the old school that tax cuts were rewards for fiscal discipline, and deficits were reluctantly accepted, if at all, as an emergency palliative. On December 13, Alan Greenspan informed the president of an 8 percent drop in fourth-quarter economic activity. Greenspan diagnosed a classic inventory recession, in which merchants hoping to beat the inflationary cycle amassed excessive stock, only to slam on the brakes as consumers grew skittish and faltering sales set off a ripple effect throughout the economy. Such contractions typically resembled the letter *V*, with plunging output and skyrocketing unemployment abruptly reversed as inventories were liquidated and demand reasserted itself.

"We're going to be in serious economic trouble for the next four to six months," Greenspan told Ford, during which "you're going to be under very heavy pressure to open up the taps and . . . spend our way out of the recession." As skeptical as Greenspan of government's power to deliver prosperity, Ford entertained slender hopes where the incoming 94th Congress was concerned. "The 75 freshmen Democrats are a different breed," warned his former House colleague Don Riegle. If Ford wished to retain any semblance of legislative authority, his administration "must move now to formulate plans that are bold enough to revitalize the economy and win genuine bipartisan support." The waning days of a lame-duck Congress offered few opportunities for cross-party collaboration. Nothing mattered more to Ford than confirming Nelson Rockefeller as vice president. Alarmed by reports that his nomination might be held over until the new Congress convened in January, Ford reached out to the Democratic leadership in both houses. "You just can't do that to the country," he reportedly told them. "You can't do it to Nelson Rockefeller, and you can't do it to me." His plea, unusually, had the desired effect. On December 10, senators voted to confirm Rockefeller 90 to 7. The House followed suit nine days later.

For Ford it was two steps forward, one step back as lawmakers enacted the most sweeping trade legislation in forty years, though not before adding the Jackson-Vanik Amendment explicitly tying most-favored-nation status for the Soviet Union to a dramatic relaxation of that country's emigration policies. Compounding the damage from the White House perspective were stringent new restrictions on Export-Import Bank loans to Moscow, with any loan exceeding $50 million requiring congressional approval. This had the effect of tying the president's hands by depriving him of diplomatic sticks and commercial carrots with which to incentivize Soviet good behavior. Unwilling to accept such infringements on their sovereignty, the Soviets disowned their 1972 trade agreement with the US. They also raised fresh obstacles to Jewish emigration, which slowed to a trickle. "The Soviets have to keep détente going for political reasons," a despondent Kissinger confided to the president, "but our hold on them is gone."

On December 16, Ford signed the Safe Drinking Water Act, establishing the first national standards to safeguard public water supplies. With considerably less fanfare he put his name to the Presidential Recordings and Materials Preservation Act, effectively nullifying the deal his representative Benton Becker had negotiated with Richard Nixon three months earlier. According to Rumsfeld, close to 150 additional bills required presidential action before the second of January. "It looks like we'll be pretty busy at Vail," Ford muttered, his Colorado ski holiday shriveling as each day brought fresh appeals from Cabinet officers objecting to their proposed budgets for fiscal year 1976. To replace the unpopular OMB director Roy Ash, Ford turned to feisty, competent James Lynn, a former secretary of Housing and Urban Development with whom the president had once engaged in a public shouting match at a Washington party over Lynn's refusal to name a Michigan candidate to a regional job at HUD. In time the two men laughed off the incident. Ford appreciated Lynn's willingness to accept a 30 percent pay cut and serve wherever he might be thought most useful.

By contrast no place at the table could be found for John Lindsay, the former mayor of New York (and former Republican), whom Ford had once promoted as vice presidential material to Richard Nixon. "That's all we need," the president said with a grimace, on being told that Lindsay had offered his services to the administration. It wasn't hard to imagine the roar of disapproval Lindsay would evoke from subscribers to *Human Events* and *National Review*. As for the most prominent reader of those increasingly critical conservative publications, Ford wondered if Ronald Reagan might be interested in Jim Lynn's old office at HUD. Not likely, said Mel Laird.

On the other hand, Laird advised, Reagan might just forgo a presidential run in 1976 if offered the department of transportation.*

Ford granted forty-seven pre-Christmas pardons, the most prominent recipient being James G. Synodinas, better known as "Jimmy the Greek" Snyder, convicted of gambling violations in the early 1960s. The holiday season made its usual social demands, in a White House decorated with homemade crafts reflecting a generalized austerity. The message was lost on AFL-CIO president George Meany. At a meeting of Ford's Labor-Management Committee, Meany argued for a $100 billion deficit as the minimum required to jump-start the sluggish economy. Nearly as contentious were sessions of the president's Economic Policy Board, their freewheeling discussions reminiscent of Capitol Hill hearings in which Congressman Ford had shown a knack for asking the right questions. "I think he is the kind of person who plays a good poker hand," said one participant. "It would offend him if he thought you were trying to guess what he wanted. He doesn't like yes men."

On December 21, less than twenty-four hours before he was to leave town for the Colorado ski slopes, Ford resigned himself to what the acerbic Ron Nessen called a 179-degree change in economic policy. Quietly buried was the 5 percent surtax proposed just ten weeks earlier to combat runaway inflation. Recession-fighting tax cuts were the new order of the day, pending discussions with his economic advisers in the shadow of Vail Mountain. Early the next morning Ford bounded up the stairs to *Air Force One*, his spirits buoyant despite the outsized headline on the front page of that morning's *New York Times*:

HUGE CIA OPERATION REPORTED IN U.S. AGAINST ANTIWAR FORCES, OTHER DISSIDENTS IN NIXON YEARS.

Contrary to later assertions, Ford was not blindsided by Seymour "Sy" Hersh's sensational article, having been forewarned to expect something of the kind by CIA director William Colby. The presidential plane had barely reached flight altitude when a White House operator patched through a phone call from Colby to Ford. Minimizing the likely impact of the *Times* exposé, the DCI claimed that Hersh had cobbled together a "few disconnected aspects" of past agency practices, all of them terminated by 1973.

---

* The White House acted on Laird's advice. And failed. Reagan was even more put out by a subsequent invitation, delivered in person by Don Rumsfeld, to become secretary of commerce. "They're working their way down the scale," one Reagan intimate remarked.

Ford prepared a brief statement for release as soon as *Air Force One* touched down in Colorado. In it he quoted Colby's assurance that "nothing comparable" to the domestic surveillance operation described by Hersh was currently in effect.

The *Times* story was enough to ruin any vacation. It was also part of a disturbing pattern. Four months into his presidency, just as he had begun to assert control of the national agenda, Ford was once again trapped in the gravitational pull of past presidential wrongdoing. He was "paying for the mistakes of the last four decades," observed George Reedy, onetime White House press secretary to Lyndon Johnson. "He's been run over by history."

20

A PRESIDENT IN THE MAKING

━━◆◆◆━━

Maybe President Ford played football with his helmet on after all.

—Tom Wicker, *New York Times*, February 21, 1975

Ford's Vail neighbors had no intention of letting past transgressions by the CIA eclipse their moment in the national spotlight. The town owed its existence to an enterprising pair of ski bums named Pete Seibert and Earl Eaton, veterans of the US Army's famed Tenth Mountain Division who glimpsed in some steeply graded sheep pastures in the Gore Range of the Rockies, one hundred miles west of Denver, the makings of Colorado's newest ski resort. On first scaling the future Vail Mountain (elevation 11,250 feet) in 1957, Seibert was dazzled by the alpine vista stretching out before him. "A virtually treeless universe, of boundless powder, open slopes, and open sky," he marveled.*

Within five years these unlikely visionaries had attracted sufficient outside investment to welcome a handful of ski enthusiasts to their fledgling resort. Wooden sidewalks hedged unpaved streets when, late in 1967, Gerald Ford paid his first visit to the faux–Bavarian village, with its half-timbered houses and white plaster clock tower; where snow bunnies and downhill racers routinely crossed paths with part-time residents Henry Fonda, Lowell Thomas and Leonard Bernstein. In 1970, Ford purchased a three-bedroom condominium in the freshly built Lodge at Vail. For his guide and ski instructor he turned to Pepi Gramshammer, a charismatic Austrian ski racer lured away from rival Sun Valley to be Vail's public face and tireless promoter. In the early '70s, Gramshammer attracted far more notice than Congressman Ford did standing in line to buy lift tickets or

---

* The denuded landscape was an ironic legacy of the nomadic Ute Indians, or Blue Sky People, who had been accustomed to retreating each summer into their beloved "Shining Mountains" of north-central Colorado. Angered over their eviction from the area by nineteenth-century fortune hunters, tribe members exacted revenge by setting fire to thousands of forested acres, an act of environmental vengeance that cleared the way for future developers like Seibert and Eaton.

sampling the bratwurst and Wiener schnitzel at Pepi's Restaurant and Bar. All that changed after 1974, when Ford's presence raised Vail's profile as a year-round destination and future home of the Jerry Ford Invitational Golf Tournament, the Ford Cup skiing competition, and the Ford-hosted World Forum of global leaders sponsored by the American Enterprise Institute.

For his first visit as president, security demands drove Ford and his family out of their condo and into the three-story A-frame dwelling of Dick Bass, one of the Texas oilmen who gave Vail its aura of political conservatism. Built around a soaring stone hearth and floor-to-ceiling windows framing the sawtooth Gore Range to the east, the Basshaus was located five hundred yards from one of the nation's earliest gondolas and adjacent chairlifts, to which Ford hastened on the afternoon of Sunday, December 22. Ruddy-faced in the six-degree cold, he shared a ride with a Secret Service agent handpicked for his skiing ability. Other agents commandeered chairs just ahead of and behind the president as he stabbed the air with his ski pole in acknowledgment of cheering onlookers. Ford's bright orange ski parka and red stocking cap made it easier for agents to follow his downhill progress through a telescope installed on the roof of a local eatery.

"He was fast, faster than the agents who were supposed to be protecting him," claims Billy Kidd, World Cup ski racer and Olympic medalist who skied with Ford on a number of occasions. Crouching in the shadows, cameramen hoped for a money shot of the president taking a spill on Vail's punishing slopes. Publicly Ford laughed off the taunts of rubber-legged journalists who trailed in his wake. "Every skier takes a fall once in a while," he remarked. "There are more skiers that fall down than stand up." Away from the cameras he vented to Press Secretary Ron Nessen: "Those reporters, the only exercise they get is sitting on a bar stool." A notable exception was NBC White House correspondent Tom Brokaw, who more than held his own when invited to race Ford to the bottom of the mountain.

Ford's days in Vail followed a set pattern, with paperwork and bill signings sandwiched between time on his K2 Winter Heat skis and holiday rituals shared with Betty and the children. One day the president attended an elk steak luncheon and awards ceremony hosted by the ski patrol. The closest he came to making news involved speculation about a proposed federal gasoline tax.

"Is that still a live option?" one correspondent pressed Ford.

"That's about as dead as any option I know."

When not on the slopes, Ford whittled down the stack of bills bequeathed him by the 93rd Congress. Mindful of the deepening recession that mocked holiday cheer in millions of American households, the president approved measures funding one hundred thousand new public-service jobs and extending unemployment benefits thirteen weeks into the new year. He drew the line, however, at cargo preference legislation requiring at least 20 percent of all imported oil to be transported in US ships (a figure that would rise to 30 percent in 1977). Politically he would have liked nothing better than to gratify the Seafarers International Union and its president, Paul Hall, a longtime friend whose past support made him and his union a rarity in the heavily Democratic ranks of organized labor. But a careful review had convinced Ford that cargo preference was both inflationary and protectionist. He called Hall to break the bad news personally. Their conversation, although civil, illustrated the limitations imposed by circumstance and conviction on "the most powerful man in the world."

On Christmas Eve the Fords attended services at Vail's interdenominational chapel. Refusing a front-row seat in the packed church, the president stood throughout the forty-five-minute program. The next morning saw the exchange of gifts and red felt Christmas stockings originally made for each Ford child by their grandmother Dorothy. In the afternoon, Ford telephoned holiday greetings to Richard Nixon in San Clemente. Christmas Day photographs show a beaming chief executive wearing a hand-knitted sweater garishly emblazoned with the letters WIN. A gag gift? Whatever its provenance, the same garment was nowhere to be seen three days later, when Ford assembled his Economic Policy Board in the two-story great room of his vacation residence to review policy options in advance of the State of the Union address, scheduled for mid-January.

Imagine being the first to ski, on a moonless night, the most treacherous of unchartered slopes. Ford and his economic brain trust confronted their own trackless landscape. Operating without precedent to serve as a guide, they somehow had to revitalize a free-falling economy without loosening an avalanche of renewed inflation. And make no mistake, inflation remained enemy number one, at least as far as Ford was concerned. "The more progress you make in winning the battle against inflation," he insisted, "the more confidence the private sector acquires, and the more it expands; and expansion means the creation of new jobs." All well and good in theory, but with his assembled experts forecasting unemployment to average 8 percent or more throughout 1975, Americans expected their government to prioritize job creation.

Similar divisions roiled the kitchen cabinet meeting under Dick Bass's Bavarian chandeliers. Bill Simon, exercising Adam Smith's proxy, opposed any action by Washington that might swell the federal deficit. Less doctrinaire friends of the market echoed Alan Greenspan's response to Ford's query about a $16 billion tax cut. "As long as it's a one-shot deal and doesn't become permanent," Greenspan said, "it's not going to do much harm." A motion to limit any tax reduction to calendar 1975 prompted a presidential objection—not to the one-year restriction but to the consequences of delay. "The people who need it the most are unemployed in 1975," Ford reminded the group, "and they wouldn't get anything." No: to be effective, tax cuts must be retroactive. Moreover, they should be delivered in tangible form rather than as credits on future tax obligations. "If you don't send a man a check, money that he can see and hold in his hand, you are going to lose some of the impact."

Gradually a consensus emerged: as a spur to economic recovery, tax cuts were preferable to increased government spending. Final decisions on their scope and means of distribution would await Ford's return to Washington and additional consultations with outside economists, representatives of business and labor—even the Democratic leadership in both houses of Congress.

NOT EVERYONE IN the administration shared the president's preoccupation with auto sales and housing starts. Seventeen hundred miles from Vail, Henry Kissinger spent his holiday trying to contain the brushfire ignited by Sy Hersh and his claims of massive domestic surveillance by the CIA. Kissinger's early morning call to Don Rumsfeld in Vail roused Ford's closest aide from his first night's sleep in the mountain resort. "I want to talk to you about the Helms matter," said Kissinger, neatly deflecting responsibility for any prospective scandal onto Richard Helms, DCI under two presidents until Nixon fired him in November 1972.

Claiming ignorance of any CIA wrongdoing, Kissinger said there was one thing he *did* know: "Sy Hersh is a son of a bitch." For a more politic response to the *Times* story, the secretary suggested that Colby write a report to the president addressing the issues raised by Hersh's exposé. Ford knew Capitol Hill too well to believe it would end there. A White House still smarting from its handling of the Nixon pardon braced itself for a probe rehashing CIA links to the original burglary, the Ellsberg break-in and other abuses charged to Richard Nixon and his agents. These fears were amplified on New Year's Day, when a Washington, DC, jury convicted John Mitchell, H. R. Haldeman and John Ehrlichman on charges of perjury and conspiracy to obstruct justice.

In his 1978 memoir, *Honorable Men*, Colby portrayed himself as an institutional reformer scapegoated by officials anxious to distance themselves from past sins. His self-portrait might have more credibility had Colby taken into his confidence the presidents he served as DCI. Instead he had kept them in the dark about the soon-to-be notorious "family jewels," 693 pages documenting illegal or embarrassing CIA actions going back over a quarter century, and commissioned by James Schlesinger during his brief stint atop the intelligence agency. Yet Colby willingly shared this same dirty laundry with his agency's congressional overseers. "It is curious that you don't think of such an obvious thing," he reflected long after the fact. "If you are going to brief the two chairmen (of the House and Senate Armed Services Committees), the least you ought to do is to brief somebody in the White House that you trust."

Colby was scarcely more forthcoming in the report requested by the president, a copy of which reached Ford at Vail the day after Christmas. In it Colby confirmed much of what Hersh had written in the *Times*, while pointing out errors of fact and emphasis. For example, by dwelling almost exclusively on the Nixon years in his initial reporting, Hersh minimized Lyndon Johnson's role in Operation Chaos, an illicit surveillance campaign grounded in Johnson's conviction that foreign support undergirded the domestic antiwar movement.

But it was Colby's passing reference in the same report to unspecified "activities which might be questionable" that fostered suspicion above his pay grade. Kissinger met with Colby on December 23. First exposure to what the secretary of state called "the horrors book" prompted an alert to Ford that some actions described in its pages "were clearly illegal." Others raised "profound moral questions." For the next week Colby waited for a summons to Vail that never came. Instead he spent the last day of 1974 at the Justice Department being debriefed on CIA misconduct by Acting Attorney General Laurence Silberman.

From their talks Silberman distilled eighteen areas of concern that merited criminal investigation. He shared these with the president in a January 3 letter. There was reason to believe Richard Helms guilty of perjury stemming from comments made during his confirmation hearings to be US ambassador to Iran. CIA agents had tailed columnist Jack Anderson after he exposed the Nixon Administration's "tilt to Pakistan" at the time of that country's 1971 war with India. The same agency had routinely opened mail coming from and destined for China and the Soviet Union; broken into private offices seeking evidence against former employees; and conducted drug tests on American citizens, one of whom committed suicide under the influence of LSD.

Almost as an afterthought, Silberman referenced agency plans to assassinate foreign leaders "which, to say the least, present unique questions."

F ORD RETURNED TO Washington on January 2. The next day he questioned defense secretary Schlesinger about the agency he had briefly led.
"Was the CIA involved in Watergate?"

"Not as an organization," said Schlesinger. "There may have been an old
boy network at work . . ."

Later that afternoon it was Bill Colby's turn. Fifty-five years old, trim
from morning calisthenics, in his gray pinstripes and horn-rimmed glasses,
the Princeton-educated lawyer-turned-spook effectively disguised the wartime commando who had parachuted into occupied France and outskied
Nazi pursuers to blow up Norwegian rail lines. "A character straight out of
a John le [Carré] spy novel," Henry Kissinger concluded, "the epitome of
discretion in thrall to anonymity." As station chief in Saigon beginning in
1959, Colby had been tasked with overseeing all CIA covert activities in
Asia. Later he supervised Operation Phoenix, a brutal counterinsurgency
program linked to the deaths of at least twenty thousand suspected Vietcong. Recalled to Langley in 1971, as deputy director for Plans—in effect
the CIA's number three man—Colby implemented sweeping changes resulting in the forced retirement of over a thousand employees (roughly 7
percent of the total agency workforce).*

Meeting with Ford in the Oval Office, the normally unflappable DCI
seemed off his game, at one point addressing the president as Mr. Chairman. "I think we have a 25-year-old institution which has done some things
it shouldn't have," Colby told Ford. Evading Ford's question about the reaction of lawmakers to the so-called family jewels, Colby instead quoted
himself telling his Capitol Hill overseers: "Here it is; we are not going to
do it again." As evidence, he cited the long-standing practice, recently terminated per his order, of opening first-class airmail letters from the USSR
("we have four to Jane Fonda"). He was "pretty sure" the surveillance of
Washington Post reporter Michael Getler had been authorized by Helms,
"but whether it was directed from higher up I don't know."

Ford responded with a brisk summary of his plans going forward, starting with a personal reminder to his intelligence chiefs of the oath they had

---

* In the process he made a bitter enemy of James Jesus Angleton, the legendary chief of
counterintelligence, whose Ahab-like pursuit of imagined moles within CIA ranks did not
cease when Colby in December 1974 showed him the door. In time Angleton was to accuse
Colby himself of being a Soviet agent, an absurd claim that was nevertheless accepted by
Vice President Nelson Rockefeller.

taken to obey the law. A blue-ribbon panel was in the works to examine past abuses and identify useful reforms. Left to his own devices, Ford would have preferred a latter-day Warren Commission, a bipartisan probe encompassing the executive, Congress and the judiciary. This being impossible in the current political climate, Ford could only hope that lawmakers would establish a single joint investigatory committee to complement the White House inquiry. However they decided to proceed, said Ford, they could count on his administration's cooperation.

"We don't want to destroy but to preserve the CIA," he told Colby. "But we want to make sure that illegal operations and those outside the charter don't happen."

With that, Colby unpacked the most unsavory of the family jewels. "We have run operations to assassinate foreign leaders," he revealed. "We have never succeeded." Fidel Castro's name was mentioned, as was the late Rafael Trujillo, caudillo of the Dominican Republic, and Chilean general René Schneider, killed in a bungled kidnapping attempt encouraged by the CIA. That night Phil Buchen informed his wife over dinner that he had been the recipient of shocking news. Beyond naming Colby as his source, he refused to divulge details. "Said he would not let me carry the burden of the knowledge which was secret," Bunny Buchen recorded.

T HE MAHOGANY SEYMOUR Tall Case Clock in the Oval Office chimed quarter to ten. It was Saturday morning, January 4, and Henry Kissinger was raging like Lear on the heath. "What is happening is worse than in the days of McCarthy," he fulminated. "You will end up with a CIA that does only reporting, and not operations." Ford should expect more damning revelations in the pipeline. "For example, Robert Kennedy personally managed the operation on the assassination of Castro." Still murky was the extent of CIA involvement in the 1973 military coup that overthrew Chile's Marxist president, Salvador Allende, resulting in his death and replacement by a right-wing military junta. "That is sort of blackmail on me," said Kissinger.

Shortly after noon Richard Helms arrived in the West Wing lobby. Before he could keep his appointment with Ford, however, he was shunted off to hear from Phil Buchen, acting on legal advice from Larry Silberman that anything the former DCI might say to the president about his potential legal exposure could be used against him in court. Ford's strained greeting did nothing to allay Helms's suspicion of a setup. Vowing to fight any attempt to blacken his reputation, Helms dropped hints that others might join him in the hot seat. "At the base is Congressional oversight," he told

Ford, who hardly needed reminding of the decade he had spent performing just that function in the House. Colby's tattling to Silberman elicited a more explicit threat. "If allegations have been made to Justice a lot of dead cats will come out," Helms snarled. "I intend to defend myself."

No one doubted his integrity, Ford assured him. "You have my pledge that everything I do will be straightforward. I plan no witch hunt. But in this environment I don't know if I can control it," said Ford, referring to his embryonic CIA commission and its deliberately limited mandate to investigate domestic wrongdoing. Unknown to Helms, Ford had spent much of the previous night on the phone, recruiting members for his blue-ribbon panel. Over one hundred names had been reviewed. Thus far six had accepted the president's invitation: LBJ's secretary of commerce, John T. Connor; former Treasury secretary C. Douglas Dillon, a veteran of the Eisenhower and Kennedy Administrations; former solicitor general Erwin Griswold; General Lyman Lemnitzer, former chairman of the Joint Chiefs of Staff; retired University of Virginia president Edgar F. Shannon; and—to the surprise of many—Ronald Reagan, days after completing two terms as governor of California. Don Rumsfeld thought the group leaned too far to the right. At his urging Ford appealed to George Meany, who in turn recommended his deputy at the AFL-CIO, Lane Kirkland.

To chair the commission the White House hoped to enlist Judge Henry Friendly, a much-admired jurist frequently mentioned in the same breath as Holmes and Cardozo. But Friendly declined Ford's request, citing the separation of powers and a potential conflict with his position on the United States Court of Appeals for the Second Circuit. Once again Rumsfeld had a replacement candidate in mind. In office barely two weeks, Vice President Rockefeller had yet to clash with the chief of staff over control of the White House Domestic Policy Council and its policymaking functions, their philosophical differences or their competing personal ambitions, which some thought put the two men on a collision course for 1976.

Suspicious by nature, at the mention of his name by Ford a wary Rockefeller asked to know exactly what the commission would investigate. If its members restricted their activities to the domestic side of the CIA, Kissinger told the vice president at an Oval Office meeting on January 4, "we are okay." On the other hand, should the Silberman inquiry into Helms's testimony before Congress generate press leaks, "we may have to have the panel expand its charter" and examine foreign covert operations.

"Helms thinks Colby shafted him," Ford interpreted. "Helms made it clear if there were any dead cats to be thrown out he would throw some of his own."

Mentally surveying the minefield before him, Rockefeller reminded the group of his own vulnerabilities. A decade earlier he had chaired Eisenhower's Covert Activities Advisory Committee. More recently he had filled a sensitive position on the President's Foreign Intelligence Advisory Board (PFIAB). "I am not trying to duck anything, just to point out the potential problems." Though he didn't say it, Rockefeller was far too close to Kissinger to inspire public confidence in his objectivity.

Kissinger made no pretense of objectivity. Colby must be brought under control, he declared. Otherwise "this stuff will be all over town soon."

Ford promised to talk with legislative leaders of both parties. Parrying Kissinger's call for an FBI investigation into leaks, an action vaguely reminiscent of the White House Plumbers, Ford said he had no intention of publishing Colby's Christmas-week report, at least not before the Rockefeller Commission completed its work.

Colby had other ideas. On January 15—without notifying anyone at the White House—he used an appearance before the Senate Subcommittee on Intelligence Appropriations to submit an opening statement that duplicated his report to Ford at Vail. For good measure he gave the committee, meeting in closed session, full license to share his findings with the public. By the end of January senators had voted overwhelmingly to establish a separate committee, chaired by Idaho's Frank Church, to look into CIA activities both domestic and foreign. The House followed suit on February 19.

T HE END OF 1974 found Bob Hartmann nursing more than his usual grievances against coworkers he suspected, not without reason, of trying to curb his influence. The president was still in Vail when his chief speechwriter was evicted from his modest work space chosen for its proximity to the Oval Office. It was Dick Cheney's idea to reclaim the area for an executive study, more conveniently located than Richard Nixon's hideaway across the street in the Executive Office Building.* With the dawn of a new year Hartmann's insecurities blossomed like an out-of-season poinsettia. Undercutting positive reviews for a recent Ford television broadcast, he described for the *Wall Street Journal* extensive practice sessions with a teleprompter, and a resulting improvement in the president's on-camera delivery of "about 2,000%."

At such times Hartmann's self-promotion took precedence over his lit-

---

* In its place Hartmann was given the West Wing quarters only recently vacated by congressional liaison Bill Timmons—ironically, the same space Cheney would occupy as George W. Bush's vice president.

erary output. "Hartmann is like a lot of newspaper guys," recalled one co-worker. "He writes only under pressure . . . he will stay up all night with a bottle of bourbon and bang something out and then come in with it at the last minute." What had been tolerated on Capitol Hill held the potential for disaster in the White House. In the event Hartmann's final draft of the State of the Union address—"Short on specifics and long on rhetoric," in Ford's characterization—did not reach the Oval Office until January 14, the day before its scheduled delivery to a joint session of Congress. Rumsfeld, Cheney, Alan Greenspan and others spent the afternoon in a frantic cut-and-paste operation. Confronted with rival texts, an exasperated Ford said, "Go back and give me one speech, not two speeches."

Bill Seidman recalls what happened next. "We sat down in a room with Rumsfeld and Hartmann and Zarb and Greenspan." Over the next three hours, ninety pages were pared to forty. At nine p.m. the two camps repaired to the Oval Office, where the composite draft was dissected line by line, a painful process further slowed as each side contended for its thematic vision or disparaged language in the text attributed to the other side. Six hours passed. If the end product, put to bed just before three in the morning, turned out to be "a damn good speech," as claimed by Seidman, "it was because he [Ford] did it. He sat there and he went over every word of it with this group."

The most memorable line harkened back to Ford's experience, as a rookie congressman from Grand Rapids, imbibing the official optimism voiced by Harry Truman on the same dais. A quarter century later, Ford as president offered a much bleaker assessment. "I must say to you that the state of the Union is not good," he blurted out. "Millions of Americans are out of work. Recession and inflation are eroding the money of millions more. Prices are too high, and sales are too slow." That year's federal deficit was on track to reach $30 billion. Its successor would be half again as large. Adding to the nation's woes, "we depend on others for essential energy."[*]

The prosaic language and flat Midwestern delivery muffled Ford's sense of urgency as he outlined his prescriptions for economic recovery. To combat the deepening recession he called for a onetime tax cut of $16 billion, three-quarters of it reserved for individual taxpayers, up to $1,000 per return. If Congress acted quickly, initial rebates could be in the mail as early as May, with a second installment coming in September. Looking beyond

---

[*] Dispensing with the usual laundry list of legislative proposals, Ford promised lawmakers he would send separate messages recommending extension of general revenue sharing and the 1965 Voting Rights Act.

this single shot in the arm for an ailing economy, Ford introduced elements of more lasting tax reform. He recommended a permanent reduction in corporate tax rates from 48 percent to 42 percent. And because inflation had pushed millions of taxpayers into higher brackets through no fault of their own, he proposed an additional $16.5 billion reduction for individuals, the bulk of it going to low- and middle-income taxpayers. Under Ford's plan everyone below the poverty line, then calculated as $5,600 for an urban family of four, would be removed from the tax rolls. Those too impoverished to pay taxes would receive an $80 check courtesy of Uncle Sam.

The president reserved his harshest words for congressional spendthrifts. "For decades we have been voting ever-increasing levels of government benefits," he asserted, "and now the bill has come due." Projecting the current rate of spending growth into the near future, Ford warned that by 1990 federal, state and local government outlays could easily consume half the gross national product. Only by controlling Washington's borrowing habits could interest rates begin to decline and the Federal Reserve halt its inflationary expansion of the money supply. Acknowledging the political risks, Ford urged his former colleagues to adopt a 5 percent limit on federal pay increases as well as entitlement programs including Social Security and food stamps. He promised to veto any new spending initiatives "except for energy."

Turning to energy policy, Ford outlined a complex program emphasizing market forces—i.e., higher prices—over government controls (rationing, gas taxes, quotas on foreign imports) to reduce energy consumption and incentivize domestic production of oil, coal, nuclear and synthetic fuels. To get Congress's attention, Ford said he would slap monthly import duties on foreign oil at the rate of $1 per barrel for three months beginning February 1. He asked lawmakers to tax domestic oil at $2 per barrel. With no more regard for public opinion Ford pitched the early decontrol of prices on all domestic oil and gas production. The resulting $30 billion windfall to producers would be returned to consumers through the tax cuts and rebates already outlined, funded by an excess profits tax on oil and gas companies. An additional $2 billion in revenue sharing would help states and cities cope with higher energy costs.

Ford did not ignore conservation. He supported fresh tax incentives for homeowners who insulated their dwellings, and legislation requiring all new buildings to comply with thermal efficiency standards. Any goodwill this bought him among environmentalists was instantly squandered through proposals to defer tough new standards governing auto pollution in return for a 40 percent increase in automobile gas mileage by 1980. Equally contentious was Ford's request to amend the landmark Clean Air Act to permit greater use of nuclear and coal-fired power plants, and to open naval reserves in California and Alaska to oil drillers. At the beginning of his speech Ford said he expected little applause for his proposed remedies. His audience did not disappoint him.

"Your conclusions were great, Mr. President," remarked Majority Leader Tip O'Neill as he escorted Ford from the chamber. "But we can't go down the same street together."

"Be charitable," the president replied with a grin. "See if you can give us a chance."

FORD'S STATE OF the Union performance was the inaugural address denied him in August 1974. The *Wall Street Journal* called it "his most presidential speech, and his most effective." Even critics conceded that he had offered a program at once comprehensive and credible. "He stepped forward, showed some initiative," allowed Senate Majority Leader Mike Mansfield. Granting that higher energy costs might rekindle inflationary fires, *Time* contrasted the specificity of Ford's policy blueprint with Speaker Albert's vague talk of "substantial" tax cuts. The majority's energy plans were equally nebulous, with consumers asked to pick from no fewer than seven different proposals.

If the opposition appeared at times unable to agree on anything but opposition, blame it at least in part on a party in transition. Regarding 1974 as a change election, freshman Democrats in the House were intent on overthrowing the seniority system and, with it, the unchallenged power exercised by entrenched committee chairmen (and Southern Democrats) like George Mahon and Wilbur Mills. A month before delivering his State of the Union address, Ford had returned to his old haunts on Capitol Hill to unveil a portrait of Mahon in the Appropriations Committee room long synonymous with the Deacon from Lubbock, Texas. On the same day a publicly shamed Mills surrendered the gavel he had wielded for sixteen years as chairman of the House Ways and Means Committee.

Drunk on more than power, Mills had embarrassed himself by carrying on a very public affair with a thirty-eight-year-old stripper calling herself Fanne Foxe, the Argentine Firecracker. One Saturday morning in October 1974 a befuddled Mills was pulled over by park police near Washington's Tidal Basin at two a.m. Suddenly the door of his limousine was flung open and there emerged an intoxicated woman, shrieking bilingual abuse before leaping, fully clothed, into the shallow waters lapping the Jefferson Memorial. The incident, though widely publicized at the time, did not prevent Mills's Arkansas constituents from reelecting him in November. Three weeks later Mills joined Ms. Fox onstage in a seedy Boston strip house, thereby sealing his fate. His resignation as chairman of Ways and Means was announced on December 10. Ford telephoned Mills at the hospital where he was receiving treatment for alcohol abuse. They may have been on opposite sides of the political fence, he reminded the onetime kingpin, but he had always acted with courage and strength—qualities that would see Mills through his current dilemma.

Thus did the curtain come down on Capitol Hill legends Mills and Mahon, one reduced to a national punch line, the other marooned in a party alienated from its crumbling base in the formerly Solid South. They were not alone. On January 22, emboldened Democrats in caucus deposed a trio of autocratic committee chairmen: Agriculture's W. R. Poage, seventy-five; F. Edward Hébert of Armed Services, seventy-three; and Wright Patman of Banking and Currency, at eighty-one the longest-serving Democrat in the House.

The drive for congressional reform created unforeseen opportunities for the Ford White House. Resurrecting a strategy first employed during his years as minority leader to rally bipartisan support for foreign aid, Ford would exploit regional, philosophical and, above all, generational differences within Democratic ranks. He had a name for what he was doing. He called it the floating coalition, and it would be tested often as he struggled to sustain a blizzard of vetoes—beginning with his rejection of gasoline rationing as authorized by the 94th Congress.

For a president holding such a weak hand, the veto was his ace in the hole. Skillfully deployed it allowed Ford, even while presiding over record deficits, to define himself in opposition to the free-spending ways of Capitol Hill. A constitutional roadblock was recast as the essence of constructive statesmanship. "The American people have a right to expect their president to protect their interests. That is one reason the veto power exists in the Constitution and why I will use it when necessary," Ford told a Cincinnati audience in July 1975. Far from the "negative, dead-end device" it might appear to be, "in most cases, it is a positive means of achieving legislative compromise and improvement—better legislation, in other words."

This was not mere rhetoric. Ford frequently coupled a message of rejection with more conciliatory language aimed at reaching accommodation with the Democratic majority. In the first week of March 1975, he simultaneously vetoed the latest congressional resolution repealing his $1-a-barrel tax on imported oil *and* voluntarily suspended the tariffs for sixty days, all but daring lawmakers to enact his energy program or produce a credible alternative. That summer Congress passed housing legislation subsidizing generous 6 percent mortgage interest rates and offering middle-class home buyers $1,000 toward their down payment. Representative Albert Ashley, an economy-minded Democrat from Ohio, complained to his caucus, "I do not understand why we insist on serving up these veto pitches that come over the plate the size of a pumpkin." Ford duly hit it out of the park. Within days Congress and the White House compromised on a housing bill that reserved mortgage assistance exclusively for unemployed Americans.

"Ford's greatest strength was the one that would make a great president," asserts Jack Marsh, "his knowledge of the process of governing, which arose largely out of his knowledge of the budget and appropriations process." Equally important were the relationships Ford had established on Capitol Hill. After Tip O'Neill lambasted him before a highly partisan audience, Ford got him on the phone. "Tip," he said, "I see all those things you said about me up there in Boston. You know what I'm going to do? I'm going to come up to Boston and tell them what a lousy congressman they got." Laughter to follow. O'Neill reciprocated, alerting Ford at the last minute not to attend an O'Neill birthday party being thrown by Tongsun Park, the shady Korean lobbyist subsequently indicted on bribery and conspiracy charges.

Ford's pragmatism enabled him to work with unlikely allies. Andy Biemiller was a case in point. A youthful campaign manager for Socialist presidential candidate Norman Thomas, and later a two-term Wisconsin congressman and cofounder of Americans for Democratic Action, Biemiller had been the AFL-CIO's chief lobbyist on Capitol Hill since 1956. "As liberal as you could get," said Marsh. But this did not prevent Ford from dispatching Marsh to cultivate Biemiller and enlist the giant labor federation's aid whenever possible.

"We had an agreement," Marsh explained. "They would help us on the Hill on issues that related to national security, but they had their own agenda." Where the minimum wage or right-to-work laws were concerned, "they knew we were going to oppose it." There were times when the White House and organized labor were at loggerheads in the Senate, and closely aligned on some national security issue in the House. "We'd be working both doors," is how Marsh put it, "but one side we'd be working against them, and on the other side we'd be working together." Ford had to pick his fights carefully. An ill-timed attempt to limit food stamps by requiring individuals to spend 16 percent of their income on food before they became eligible for assistance ran into a stone wall of resistance on Capitol Hill.

In response, Ford issued a nonveto message chastising the Congress for its unwillingness to make hard choices.

Working at the legislative equivalent of warp speed, by the end of February 1975, the House Ways and Means Committee under Wilbur Mills's successor, Oregonian Al Ullman, produced a $22 billion tax cut that directed the lion's share of benefits to families earning less than $20,000 a year. The result was more equitable than the administration's plan, according to Ford scholar Yanek Mieczkowski, and also quicker to take effect since low- and middle-income beneficiaries were more likely to spend than save, "thus stimulating the economy in a quick jolt." Ford's proposed budget for fiscal 1976 sparked fresh controversy. Bill Simon said the $349 billion spending plan, with its eye-watering $52 billion deficit, left him "horrified." The secretary's dismay was shared by conservative activists who gathered, five hundred strong, in a Washington hotel in mid-February to applaud Ronald Reagan's criticism of a Republican president whose tolerance of deficits raised "deeply disturbing questions about the quality and stability of his leadership."*

By the end of that month, the full House having adopted something close to the tax cut endorsed by Ullman's committee, attention turned to the Senate. There Ford's $16 billion proposal was transformed into the proverbial Christmas tree, garlanded with shiny new entitlements: a $100 windfall for each of the nation's thirty-four million Social Security recipients; fresh exemptions in the tax code worth $30 per personal deduction; an earned income credit enabling low-income taxpayers to deduct up to $400 of their earnings; additional tax credits benefiting the real estate industry; and partial repeal of the oil depletion allowance, reducing the amount of oil- and gas-generated income exempted from taxation to 22 percent. The authors of these promissory notes solemnly declared them to be temporary, although few students of the political process expected taxes to revert to pre-recession levels on the cusp of a presidential election. Around the White House fewer still believed Ford would accept a tax cut anywhere near the $29.8 billion figure adopted by the Senate.

Ultimately his threat of a veto, personally conveyed to the Democratic leadership of both houses, led members of a House-Senate conference committee to split the difference between Ford's original request and the senatorial giveaway. On March 29 the president went on television to announce

---

* Largely overlooked in the budget dispute was Ford's $15 billion increase in military spending—a marked reversal of recent trends and something many on the right had been promoting.

grudging approval of "a reasonable compromise" that was nevertheless marred by extraneous elements likely to swell the projected deficit to $60 billion. Worse, should Congress continue to spend at its present rate, Ford said the deficit could balloon to $100 billion, an unimaginable number in 1975, and a dagger aimed at the heart of any economic recovery. Declaring, "This is as far as we dare to go," he promised to veto any program that breached his $60 billion line in the sand.

A FTER SIX MONTHS in office Ford appeared to have caught a second wind. "A President in the Making," the *Christian Science Monitor* called him. Its venerable Washington observer, Joseph C. Harsch, glimpsed in Ford's dismantling of the imperial presidency the ghost of Dwight Eisenhower. The defining characteristic that set this president apart from his recent predecessors, wrote Meg Greenfield of the *Washington Post*, was a refusal to talk or think about himself in the third person. The public responded in kind. On February 11, ten thousand Kansans braved frigid temperatures to welcome Ford to Topeka. In his address to state legislators he announced the release of $2 billion in federal highway funds to expedite construction projects nationwide. Later in the week, appearing before an audience of Wall Street traders, Ford ruled out the use of inflationary policies to combat recessionary pressures. The markets liked what they heard, with the New York Stock Exchange registering its largest single-day volume on record.

On a two-day swing through South Florida, ostensibly planned around the latest "town hall" organized by the White House Office of Public Liaison,* Ford fit in an afternoon's play at the Jackie Gleason Golf Classic, where, to the delight of forty thousand spectators, he was joined by Gleason, Bob Hope and Jack Nicklaus. On the same trip, Ford used a televised press conference to rebuke members of the international banking community for conduct "repugnant to American principles." Language sanitized

---

* The brainchild of longtime Washington operative William J. Baroody Jr., the OPL fancied itself an instrument of public persuasion akin to the White House press and congressional liaison offices. Besides affording farmers and veterans, minorities and youth, Rotary Clubs and trade associations a direct pipeline to official policymakers, the OPL convened hundreds of public policy seminars in the White House and on the road. More than thinly veiled campaign rallies, at their best these gatherings produced tangible changes in how government operated. For example, discussions with advocates for the handicapped inspired an executive order modifying federally administered monuments—beginning with the Lincoln Memorial—to afford greater access to visitors with disabilities. An OPL-sponsored economic summit in Atlanta led the administration to quadruple, to $40,000, the amount of bank deposits guaranteed by the Federal Deposit Insurance Corporation.

for the broadcast audience understated his private anger over the boycott of Jewish investment houses by Arab-owned banks selling their bonds in the American marketplace. On learning of the Saudi Arabian blacklist, Ford dispatched Bill Simon and White House deputy counsel Rod Hills to Merrill Lynch and other financial institutions profiting from the exclusion of Jewish-owned competitors such as Goldman Sachs and Lehman Brothers. The message they were asked to deliver did not lack for point. As Hills recounted it, "The president of the United States has said that if it should happen again, he will do everything in his power to see that they are no longer in the business." ("Can we do that?" asked Simon.)

Meg Greenfield said of Ford that he was humanizing the presidency. Who, if anyone, was humanizing Ford? Volunteering for the assignment was John Hersey, Pulitzer Prize–winning novelist and author of *Hiroshima*, his landmark account of the first atomic bombing. Hersey, whose acquaintance with the president dated to their shared experiences with Yale football—where Assistant Coach Ford had judged Hersey the best punter on the team—proposed to the Ford White House that he undertake a sequel to his fly-on-the-wall portrayal of Harry Truman's presidency published a quarter century earlier. That memorable sample of instant history had drawn on several weeks of virtually unlimited exposure to an embattled president halfway through his second term. Now, confronting similar stresses, Ford was more than willing to grant Hersey privileged observer status.

For the week of March 10 it was agreed that the writer-journalist would shadow the president. Only one part of Ford's schedule was off-limits— "My daily meetings with Henry." The resulting account would be published in the *New York Times Magazine* (and eventually appear in book form). Hersey's title for his first day in Ford's company—"A Stubborn Calm at the Center"—might have summarized the entire series. To Hersey the president appeared businesslike and never less than fully prepared, whether discussing monetary policy with Arthur Burns or jousting with Chief Justice Warren Burger over proposed pay raises for the judiciary. In his role as greeter in chief, Ford showed unfailing courtesy accepting gifts from Shirley Cothran, Miss America 1975, and calming the frightened five-year-old girl in braces who was that year's model for the Easter Seals campaign.

Meeting with his speechwriters ahead of a visit to Notre Dame University, Ford wanted to counter the lure of isolationism that had so infected his student generation in the 1930s. As a rule Ford's command of history was limited to his own life span and experiences. Hersey detected in him a "glacial caution" that was the antithesis of Truman's snap judgments and reflections grounded in extensive reading about the past. Both men had been forced to learn the job, on the job. Truman's early mistakes had cost him control of Congress in 1946. Ford's party suffered losses nearly as grievous in November 1974.

But whereas Truman could draw on public sympathy and the immense reservoir of goodwill amassed by his predecessor, Ford was strictly on his own. Hersey had discerned in Truman a duality that baffled him, until he learned to distinguish between the poker-playing, bourbon-swilling partisan and his alter ego, only the thirty-second among his countrymen to occupy the position of leadership hallowed by Washington, Jackson, Lincoln and Roosevelt before him. This other Truman, with his semimystical belief in a strong presidency as the fulcrum of constitutional government, impressed Hersey as "a separation of powers within a single psyche."

Yet for all his personal modesty, Truman's lofty view of executive power foreshadowed the Imperial Presidency mythologized by court historians like Theodore H. White and exploited—and, finally, abused—by Ford's immediate predecessors. If Ford lacked Truman's sense of history, neither was he inclined to view a president as the personification of American power and moral authority. One could hardly imagine Ford seizing the nation's steel mills, as Truman did in reprisal for a wartime strike by steelworkers, only to be slapped down by the Supreme Court. Hersey frankly preferred Truman's cocky assertiveness to Ford's seeming passivity, although he modified his criticism in light of the disadvantages burdening an unelected president forced to deal with top-heavy Democratic majorities in Congress.

Where Hersey miscalculated was in applying to the Ford presidency rules that had governed New Deal and Fair Deal America. Passing an ambitious legislative program was not the only mark of a successful presidency. Under the circumstances, to restore public trust while drawing off the accumulated poisons of Vietnam and Watergate was no mean accomplishment. Of course there were important stylistic differences between Hersey's presidents. Whereas Truman had been "all nervous energy . . . emotion in harness," Ford was phlegmatic and slow to anger. The author had difficulty reconciling Ford's personal thoughtfulness with his apparent insensitivity to individual need as gauged through public expenditures.

Especially revealing were some Ford comments about tax-cut legislation slowly working its way through the Senate Finance Committee. "It's just a welfare thing, Mr. President," declared Bill Simon. In response Ford cited two of his sons, each of whom earned approximately $1,500 in summer jobs. Would they be entitled to a $100 rebate from the government?

Simon said they would.

"That's ridiculous."

"But if they go out and spend it," interjected Bob Hartmann.

"They'll spend it, all right!"

A Democratic jobs proposal clocking in at $6 billion elicited a similar rejoinder. Hersey was dismayed. "I have seen a first glimpse of another side of the man who has been so considerate, so open and so kind to me as an individual—what seems a deep, hard, rigid side." Hersey was perplexed. "What is it in him?" he asked himself in print. "Is it an inability to extend compassion far beyond the faces directly in view? Is it a failure of imagination?" Such comments revealed less about Ford than about Hersey, the New Deal liberal who admired antiwar students at Yale intent on purging "the whole station wagon load of junky white middle-class values the wagon carries on its chrome luggage rack." That so perceptive an observer should find conservative economics incompatible with social responsibility was to anticipate similar head-scratching as the old Roosevelt coalition yielded to Ronald Reagan's right-of-center consensus. It also placed Gerald Ford in an unfamiliar position: ahead of his time.

TIMED TO COINCIDE with St. Patrick's Day, Ford's visit to Notre Dame was something of a milestone, the first time in a decade, according to university president Father Theodore Hesburgh, that an American president had been able to set foot on "a first rate college campus." His enthusiastic reception by twenty-five thousand students and townspeople offered fresh evidence that, following a bumpy first few months, Ford had turned the corner of presidential credibility.

Strengthening this impression was his belated overhaul of the Cabinet he inherited from Richard Nixon. Topping Ford's list of priorities was a Justice Department severely tarnished by its Watergate associations and administratively hobbled by a succession of five attorneys general in six years. Bill Saxbe's replacement, Ford decided, must be someone "of towering intellect and spotless integrity. No campaign managers need apply, nor members of the family, official or political."

It was his fellow Chicagoan Don Rumsfeld who introduced the name of Edward Levi, the sixty-three-year-old president of the University of Chicago. A nationally recognized authority on antitrust law, Levi was a classroom legend, characterized by his former student Robert Bork as "the intellectual version of a Marine bootcamp drill inspector." The son and grandson of rabbis, Levi had abandoned his pursuit of a doctorate in literature after being told by a sympathetic professor not to expect a job offer from the humanities department at Chicago or any other university on account of his Jewishness. Switching his focus to the law, Levi had carved out a brilliant career as a wartime special assistant to Attorney General Francis Biddle; chief counsel to the House Judiciary Committee; dean of Chicago's law school; and university provost before he assumed the presidency in 1968.

While he may have practiced the Socratic method in the lecture hall, in the president's office Levi more nearly resembled Solomon—never more so than when student protesters staged a fifteen-day occupation of the university's administration building. On the one hand, Levi refused to negotiate with the trespassers until they vacated the premises. On the other, he said he would resign before summoning police to remove them by force. His evenhanded response called to mind a voyage he made with his wife, Kate, on a Caribbean cruise ship that happened to catch fire. As other passengers ran around in confused panic, Levi puffed away on his pipe, confident that crew members would extinguish the flames before they posed a serious danger.

In hopes of putting out an even bigger blaze, Rumsfeld had invited Levi to the White House early in December 1974 for what the visitor assumed would be a discussion of education policy. And so it was, for twenty minutes or so, until the conversation veered in the direction of the Justice Department and Levi was asked what qualities he looked for in an idealized attorney general.

The next thing Levi knew, he was being hustled off to the Oval Office for an unscheduled meeting with the president. The two men had never before crossed paths, but Ford had done his homework, conjuring up the names of faculty members familiar to them both as products of the Yale Law School. He further ingratiated himself by picking up a pipe. "If you're going to do that," said Levi, "then I'm going to get my own." Eventually, as with Rumsfeld, the talk turned to Justice, with Ford soliciting Levi's views on the troubled department's most pressing need. Levi replied without hesitation that it required nonpolitical leadership—nothing mattered more than restoring public confidence in the impartial administration of justice.

"Then I sprang my trap," Ford remembered. Would his visitor be willing to serve as attorney general? Taken aback, Levi sputtered something to the effect that he needed the job like a hole in the head. At this Ford wryly pointed out that he, too, had been thrust into a job he hadn't asked for.

But his university had only just embarked on a major fundraising campaign, Levi protested.

"I know about that," said Ford. "Henry Kissinger went out to give a talk opening the drive."

"Yes," Levi noted, "and he never mentioned the drive."

"You're right," Ford shot back. "I read the speech and you're right."

Outside the Oval Office a visibly shaken Levi turned to Rumsfeld. "Isn't that a little sudden?"

Strongly tempted as he was to say no, Levi kept replaying in his head something Ford had said to him: "With the administration of justice in such difficulty, how can you possibly turn your back on it?"

One subject Ford did not raise, then or later, was Levi's political affiliation. Promised a free hand in running and rebuilding the damaged department, Levi spent a weekend pondering his options. On Monday he informed the White House of his acceptance. Before his appointment could be formalized, however, Levi must pass muster with Senate Judiciary Committee chairman James Eastland of Mississippi, a fossilized champion of the Old South, and his Republican counterpart, Nebraskan Roman Hruska. The latter statesman was best recalled for his endorsement of Supreme Court nominee G. Harrold Carswell on the grounds that even mediocre people deserved representation on the high court. In a meeting with Ford on December 10, both men decried Levi's nomination as "terrible." Hruska went a step further. Conjuring up a liberal bogeyman, he asked Ford point-blank if Levi wasn't another Ramsey Clark.

In fact their objections had less to do with Levi's long-ago membership in the left-wing National Lawyers Guild than with the administration's failure to show proper deference by consulting them in advance. "They didn't tell me or I would have said, 'Let me tip off Hruska and Eastland,'" congressional liaison Tom Korologos said of his White House overseers. "Like you do anytime . . . tell them at ten o'clock in the morning and announce it at noon. Didn't happen.'" Ford apologized to the senators for the lapse in protocol. Turning on the charm, he described Levi's handling of student protests at Chicago, making sure to emphasize the nominee's unyielding stance on the expulsion of lawbreakers. By the time they exited the Oval Office, Ford had won over both men, although nothing he said could alter Eastland's bias against Levi's signature neckwear. Henceforth the chairman of the Judiciary Committee referred to the attorney general designate as "Bowtie." He did not mean it as a compliment.

A breakfast meeting was arranged so that Levi could personally debunk any lingering concerns. "Eastland ordered for all of us," according to Korologos. "So here came bacon, scrambled eggs and grits and toast . . . and Levi looked at this breakfast and I looked at it and I knew immediately."

His worst fears were confirmed when Eastland seized a bottle of Worcestershire sauce, which he proceeded to sprinkle over the white pasty mound Levi had been pushing around his plate in an unconvincing show of enthusiasm for regional cuisine. Taking advantage of Eastland's momentary absence, Levi mumbled to his White House handler, "What's that?"

"It's grits."

"Do I have to eat these things?"

"You want to be attorney general?"

Outside the Senate Dining Room the prevailing view was summed up by the *Washington Post*. Ford's choice of Levi was "a master stroke," said the paper. The full Senate agreed, confirming him by voice vote the first week of February.

Similar praise greeted Ford's candidate to head the transportation department, announced the same day as Levi's nomination. If Ed Levi made history as the nation's first Jewish attorney general, Bill Coleman was the first African-American appointed to the Cabinet by a Republican president. Coleman, a portly Philadelphian of dry wit and impeccable tailoring, was accustomed to making history ever since he finished first in his Harvard Law School class and Justice Felix Frankfurter made him the first of his race to clerk for a justice of the United States Supreme Court.

Together with his fellow clerk and best friend, Elliot Richardson, Coleman had difficulty finding a lunch counter in the nation's capital that would serve a Black man. That was in 1949. Two years later Thurgood Marshall, chief counsel for the NAACP, recruited Coleman to be part of the legal team preparing to argue *Brown v. Board of Education of Topeka* before the Supreme Court. Following their victory in that seminal case outlawing segregation in the nation's classrooms, Coleman went on to defend Freedom Riders in Mississippi and overturn a Florida law prohibiting interracial marriage. He and Ford became friendly during their work on the Warren Commission. At the time Ford asked him to succeed Claude Brinegar as secretary of transportation, Coleman was serving as president of the NAACP Legal Defense and Education Fund.

His confirmation by the Senate was a formality. Much the same could be said of Ford's next two Cabinet selections, their names made public only days apart in February 1975. John Dunlop, Harvard economist and former dean of the faculty, was widely considered the nation's foremost authority on industrial relations when Ford asked him to take the helm at the Labor Department. The president sweetened the offer by promising Dunlop a place on his Economic Policy Board. Carla Hills, a whip-smart forty-one-year-old Yale Law School graduate and mother of three, was initially reluctant to trade her position heading the Civil Division of the Justice Department for the bureaucratic quicksand that was the department of Housing and Urban Development (HUD).

"I'm not an urbanologist. Is this a good idea?" an anxious Hills questioned Don Rumsfeld when first approached about the HUD job.

"Well, the president thinks you're a good manager."

Housing advocacy groups protested her lack of experience, but Hills silenced her critics with a bravura performance at her confirmation hearings. Only five senators voted against her nomination when it reached the floor on March 5. No one took greater pleasure in Hills's appointment than Betty Ford. That she was the decisive factor in convincing her husband to make Hills only the third woman (after Frances Perkins at Labor and Oveta Culp Hobby at Health, Education, and Welfare) to administer a Cabinet department was their secret, not to be revealed until the president was out of office.

P RESIDENTS ARE JUDGED in part by the company they keep, the talent they attract and their ability to work with subordinates who may surpass them in IQ and charisma. In recruiting the likes of Levi, Coleman, Dunlop and Hills, Ford had steepened the grade curve of his entire administration. He didn't do it alone. Yet pillow talk, Betty Ford was to discover, had its limits. More public advocacy—of issues or individuals—entailed risks of the sort that had confined Mrs. Ford's predecessors to such anodyne causes as volunteerism and historic preservation. Betty Ford's schedule for February 14, 1975, reflected this fact. Her attendance at the Heart Association's annual Valentine luncheon and fashion show was as likely to generate news as her recipe for beef stew with walnuts. The first hint to the contrary came near the conclusion of the event, as reporters outside the hotel ballroom clamored to speak with Mrs. Ford. Their interest had been sparked by a thin line of sign-carrying protesters trudging the sidewalk in front of the White House. Dressed in funerary black, the little group was at that moment making history by picketing the First Lady of the United States.

Their placards spoke for them. BETTY FORD IS TRYING TO PRESS A SECOND-RATE MANHOOD ON AMERICAN WOMANHOOD, read one. Another demanded, BETTY FORD GET OFF THE PHONE. The demonstrators were upset with Mrs. Ford over her lobbying efforts for the proposed Equal Rights Amendment to the Constitution. Originally introduced in 1923, the twenty-three-word amendment would prohibit legal discrimination based on sex. Now, three years after securing passage in both houses of Congress, ERA supporters could check off thirty-four of the thirty-eight states needed to ratify. And with a flurry of state legislatures scheduled to vote on the issue in 1975, final victory appeared within reach.

This optimistic assessment did not reckon with Phyllis Schlafly, a fifty-ish St. Louis–area housewife, conservative activist and author (*Ambush at Vladivostok*, *Kissinger on the Couch*). In the 1950s Schlafly had sounded the alarm against international Communism and a nuclear test moratorium advocated by President Eisenhower. In the 1960s she was in the vanguard of political insurgents using the Goldwater campaign to conduct a hostile takeover of the Republican Party. More recently Schlafly had become convinced the ERA would deprive women of legal deference, while subjecting them to the military draft and unisex bathrooms. Her immediate complaint, and the motivation for three dozen of her acolytes to walk a picket line on a bitterly cold February day in Washington, was Betty Ford's use of a White House telephone to communicate her pro-ERA views to legislators in North Dakota, Illinois and Missouri. Nearly as offensive was the First Lady's reliance on taxpayer-funded staff to create a slideshow arguing the amendment's merits and shown to White House employees on their lunch hour.

The realization that she was the first president's wife to be picketed on her doorstep did nothing to weaken Mrs. Ford's zeal for the ERA. "I'm going to stick to my guns and will continue to do what I can as long as I feel I can be helpful," she told reporters at an impromptu press conference. Accused of appropriating government facilities to advance her lobbying efforts, she sounded dismissive. "I'm just using my own home," she insisted (in fact, her calls were made on a WATS line, a primitive flat-rate plan that allowed government and business to phone toll-free). The alternative, gibed one White House staffer, was for the First Lady to take a pocket full of nickels and a few Secret Service men "and go down to the pay phone at the corner drugstore."

What about her husband? someone asked. How did he react to her lobbying efforts?

"He just smiles," said Mrs. Ford.

No one in the counsel's office was smiling when Phil Buchen cautioned those working for the First Lady "about use of White House staff or facilities for activities not in the ordinary course of official operations." The same office volunteered to assist Mrs. Ford's team in answering letters or press inquiries touching on the propriety of "particular activities." More than a few letter writers took Mrs. Ford to task, one correspondent calling her "the only woman in the White House that we cannot refer to as 'Lady.'" Coming to her defense, the *Detroit Free Press* said it wasn't bothered by those who disagreed with Mrs. Ford's opinion on the ERA, unless they questioned her right to have an opinion at all. "Push on, Mrs. Ford," the paper advised. And so she did, after first installing a private outside telephone line to relieve any taxpayer concerns.

Of course, it was not the means by which she conducted her campaign of persuasion but the campaign itself that evoked heated discussion about the limits of free speech as applied to an unelected, unpaid spouse traditionally reluctant to lend her name to partisan controversy. It may have been sheer coincidence, but it bears noting that support for the ERA plateaued just as Mrs. Ford's lobbying efforts made her a foil for the media-savvy Phyllis Schlafly. After North Dakota in February, no additional states would ratify the would-be Twenty-Seventh Amendment in 1975.

For Betty Ford, housewife-turned-feminist, that spring witnessed a curious role reversal. As her husband became more comfortable in his job, Mrs. Ford experienced growing pains that all too often translated into the real thing. A flare-up of painful osteoarthritis required physiotherapy and hot packs. She wore a neck brace when telephoning wavering legislators on behalf of the ERA. Staff infighting added to her discomfort. It was bad enough that the possessive Nancy Howe, the First Lady's de facto chief of staff, had assumed the role of palace guard. Worse, Howe's habit of feeding inside stories to favored journalists nourished her ego at the expense of the woman to whom she professed devotion. "I know Nancy is a constant problem," Mrs. Ford consoled her press secretary, Sheila Weidenfeld. "Everyone tells me she is. My kids don't like her; they complain that she regulates my time so much she even keeps them from seeing me."

Annoyance turned to anger when *Family Weekly* early in March headlined an exclusive story: *Mrs. Ford's Best Friend: Nancy Howe.* "She is not my best friend!" Betty insisted to Weidenfeld, who needed little convincing. "I am going to fire Nancy!" Only the timely intervention of David Kennerly calmed the First Lady's frayed nerves. The storm passed. Yet it was instructive of the pitfalls that went with such prominence. The Ford children developed coping mechanisms of their own. Mike Ford and his wife, Gayle, kept the lowest profile, not surprising after a honeymoon encumbered by the presence of Secret Service agents. The Fords' eldest son concentrated on his divinity studies at Gordon-Conwell Theological Seminary in the Massachusetts town of Hamilton. Gayle Ford fended off periodic rumors, and a *People* magazine story, that she was pregnant with the Fords' first grandchild.

Nineteen-year-old Steve Ford used humor in dealing with the agents who shadowed him around the Montana ranch where he played cowboy

("It's bad enough on a first date trying not to say something dumb, without having three Secret Service guys with you"). One day Steve evaded his agent minder and, together with a friendly coconspirator, disappeared into the woods on horseback. After being gone long enough to plant seeds of concern over their whereabouts, the missing pair fired off four rifle shots. A few minutes later they wandered into camp with Steve melodramatically slumped over the saddle.

Horror-stricken, the agent assigned to protect the president's youngest son exclaimed, "My career! My career!"

Jack Ford, as ruggedly handsome in his Yellowstone Park ranger's uniform as his father had been forty years earlier, emulated his mother's candor when faced by inquisitive reporters. Protesting saccharine media accounts of his all-American family, Jack insisted that the Fords of Crown View Drive had experienced the same dinner table fights as any other clan. As for life since August 9, 1974, "I wish the whole thing with my father had never happened, but I'm resigned to making the best of it." That included pushing him "very, very hard" to support amnesty for Vietnam draft evaders. Jack's feelings about the Nixon pardon were at best ambivalent ("the timing was bad and maybe the whole approach was bad").

Somewhere along the way the forestry major from Utah State became sufficiently starstruck to hang out with Bianca Jagger and Andy Warhol, and to contemplate inviting Mick Jagger to join him for a postconcert cruise on the presidential yacht *Sequoia*. The ship remained in dock, but only because Sheila Weidenfeld discovered the passenger list was to include *Washington Post* rock critic Tom Zito, and she convinced Jack it might not be helpful to his father for *Post* readers to discover "that the President's son is smoking dope on the *Sequoia*." Typically outspoken on the need to relax marijuana penalties, Jack was reticent when pressed about his own use of pot. "Uhhh, you know, I would rather not say anything."

Of all the Ford children, seventeen-year-old Susan was most inclined to heed the advice of Alice Roosevelt Longworth to "have a hell of a good time" while in residence at 1600 Pennsylvania Avenue. To be sure, her mother insisted that she make her bed each morning so that this most basic of chores would not be left to a White House maid. That said, it wasn't hard to find some member of the household staff to help her bathe Liberty, the golden retriever that enjoyed the run of the South Lawn (while evading repeated requests from the *National Enquirer* to pose with its mascot, Lucky). A thoroughly normal teenager, Susan interrupted an Oval Office briefing by Henry Kissinger to obtain her weekly allowance from her father. Discovering the former cloakroom FDR had converted into a movie theater, she invited her friends to join her for private screenings of *The Sting* and *The Longest Yard*, the latter starring Burt Reynolds.

More public diversions included front-row seats at a Rod Stewart concert in suburban Maryland. This led to claims in London's tabloid press that the rock superstar had been invited back to the White House by a smitten Susan. The stories were untrue, but the president himself was hard-pressed to get his daughter off the dance floor, as illustrated by advance man Pete Sorum, whose beeper phone went off late one night at a dance Susan attended.

"This is the President. Tell Susan to come home. It's almost twelve o'clock." Sorum duly conveyed the paternal instructions.

"Well, I'm not ready," Susan replied. And she didn't leave until she was.

That spring Susan took up a new hobby, photographing her father on the job for a school project. After David Kennerly arranged for the Fords to meet the great landscape photographer Ansel Adams, Susan jumped at the chance to participate in a ten-day photography workshop taught by Adams.* Before joining Adams's master class, Susan achieved her own footnote in White House history as the East Room hosted its first and, thus far, only high school prom. On Saturday night, May 31, members of the Holton-Arms School Class of 1975, accompanied by their dates in rented tuxedos, speared Swedish meatballs and consumed nonalcoholic punch prepared by Chef Henry Haller's kitchen. (The evening's $1,300 tab was picked up by class members, who had raised the money at bake sales and school fairs.)

With Susan's parents out of the country, chaperoning duties fell to her aunt Janet and several favorite teachers designated by the class. Organizers had approached the Beach Boys about performing on the East Room stage. The band indicated a willingness to play for a reduced fee, but only if the event could be filmed for later broadcast. Out of the question. Taking their place was the distinctly countercultural Outer Space Band, one of whose members mooned the White House as he arrived in the battered Chevrolet Suburban that transported the musicians from Massachusetts. After a couple hours the scruffy ensemble gave way to Sandcastle, a Richmond,

---

* An outspoken supporter of the Sierra Club, Adams did not hesitate to challenge the president over his spotty environmental record. "Hey, wait a minute," Ford protested. "I'm not the enemy." The ensuing dialogue led to a personal friendship that lasted until Adams's death in 1984. *Snowstorm Clearing over Yosemite, 1944*, a gift from the celebrated artist, occupied a prominent place in Ford's hideaway office in the West Wing. At Ford's request, Adams submitted ideas for more than a dozen new national parks. He also inspired the president's bicentennial-year proposal to double the National Park Service budget. AI, David Kennerly.

Virginia, band with a Top 40 playlist. Before the music ended around one a.m., photographers captured Susan doing a spirited Bump with her date for the evening, a premed student from Washington and Lee University named Billy Pifer. The pair had first met at the recent Apple Blossom Festival in Winchester, Virginia, where a beaming Gerald Ford had crowned his daughter queen.

Pifer substituted for Gardner Britt, Susan's steady boyfriend until the Equal Rights Amendment came between them. "A job is all right if women can do an equal job," Britt informed readers of the *Ladies' Home Journal*, "but I don't think they can." Susan Ford might be Daddy's Little Girl, but as a budding feminist she was very much her mother's daughter. About the time she took off her crown, Susan declared her independence. Breaking up with Britt was her idea.

21

THE CRUELEST MONTH

❧

April is the cruelest month, breeding lilacs out of the dead land.

—T. S. Eliot

T HAT FORD'S GENIAL manner masked a formidable stubborn streak invariably surprised those encountering it for the first time. Dr. John Harper, the rector of St. John's Episcopal Church across Lafayette Park from the White House, once invited his most prominent congregant to take part in an evening service at the self-proclaimed Church of the Presidents. Ford accepted without reviewing the passage assigned him to read before the parishioners. Taken from the letter of St. Paul to the Ephesians, it included the admonition "Wives, submit yourselves unto your own husbands, as to the Lord."

Politely but firmly, Ford declined to read the offending words. Brushing aside the protest of Dr. Harper, he turned instead to the Beatitudes as recounted in the Book of Matthew.

Don Rumsfeld was more accustomed than Dr. Harper to bouts of presidential obstinacy. Keenly aware of Ronald Reagan's early success in fundraising and staff recruitment, Rumsfeld in the spring of 1975 pressed Ford to adopt a more thematic approach to governing, thereby giving restive conservatives an agenda to rally around in advance of the 1976 election. "You know as you look ahead, you have to have something that you are pushing for," he told the president. "Otherwise you are not leading." Ford said that for a quarter century, he had run for office on the basis of his job performance. He had no intention of changing strategies, even in the face of a prospective challenge from the Reagan right.

"Reagan doesn't feel that I am a candidate and he wants to position himself so that he can stop Rocky," Ford theorized. He would have an opportunity to test his thesis at the end of March, when the Fords flew to Palm Springs for their annual Easter holiday, and the Reagans joined them one evening for dinner.

Again Rumsfeld registered a polite dissent. Recognizing Ford's eagerness for a change of scene, his chief of staff questioned the optics of presidential golfing in the sun-splashed millionaires' colony while his countrymen remained in the grip of recession, and multiple foreign hot spots posed challenges reminiscent of October 1956, when Dwight Eisenhower simultaneously confronted the Suez Crisis and a brutally suppressed uprising against Hungary's Communist government. Ford's plate was hardly less crowded. In Europe a continuing standoff between Greece and Turkey over Cyprus threatened the NATO alliance, already strained by a left-wing revolution in Portugal and the recent inclusion of Communist militants in that country's government.

Relations between Washington and Moscow were visibly fraying, with post-Vladivostok arms control talks stalled and the Soviets nursing a grudge over the linkage of Western trade and congressionally mandated levels of emigration. Fresh embarrassments ensued when columnist Jack Anderson on March 18 broke the previously embargoed story of the *Glomar Explorer*, the phantom vessel commissioned by the CIA to retrieve a sunken Soviet sub in the first days of the Ford presidency. Keenly aware of the diplomatic ice age ushered in by Eisenhower's public acknowledgment of an American U-2 plane downed by Soviet missiles in May 1960, Ford would neither confirm nor deny the daring mission of the spy ship whose cover was blown by Anderson and an earlier, fragmentary report in the *Los Angeles Times*.

On March 23, Henry Kissinger returned to Washington with little to show for seventeen days of high-stakes diplomacy aimed at separating Israeli and Egyptian forces more than a year after the Yom Kippur War left Israel in control of the Sinai Desert and within shelling distance of the Syrian capital of Damascus. In return for withdrawing to a defensive line thirty-five miles east of the Suez Canal, and relinquishing control of the Abu Rudeis oil fields on the Gulf of Suez, the government of Prime Minister Yitzhak Rabin demanded from Anwar Sadat's Egypt a formal declaration of non-belligerency. For Sadat, successor to the pan-Arab torchbearer Gamal Abdel Nasser, to announce an end to hostilities while Israel continued to occupy large swaths of Egyptian territory was akin to putting his head on the chopping block.

According to Kissinger, this fact had been reiterated no fewer than twenty-four times to Rabin and his Cabinet. "So the negotiation was whether Sadat could give elements of non-belligerency without the declaration," the secretary explained to congressional leaders at the White House on the morning of March 24. Ford believed the Egyptian leader had more than met the test by promising during the life of any agreement to settle disputes through peaceful means; accepting the continued presence of UN peacekeepers; and signifying his willingness to ease economic boycotts and travel restrictions directed against Israel and the West. Plainly upset over Israeli foot-dragging, Ford wrote confidentially to Rabin informing him that failure to make progress in the negotiations then sputtering to a close would prompt a reassessment of US policy in the region, "including our policy towards Israel."

The Israelis promptly leaked Ford's letter to the press, in a transparent appeal over his head to influential Jewish Americans and sympathetic members of Congress. "This is no reflection on you," Kissinger informed the president. "But Israel doesn't think they have to be afraid of you." Ford reached out to Max Fisher, a trusted friend, Detroit real estate magnate and philanthropist whose work on behalf of the American Jewish Committee and the United Jewish Appeal had earned him global credibility. "Nothing has hit me so hard since I've been in this office," Ford told Fisher. Failure to break the current logjam would almost certainly lead to a reconvened Geneva conference involving the Soviets as well as Europeans espousing the Palestinian cause—perhaps even the terrorist Palestine Liberation Organization (PLO) itself. If the Israelis weren't afraid of him, Ford implied, they could always present their case before that hanging jury. Alarmed by the prospect, Fisher accepted the president's invitation to conduct his own brand of shuttle diplomacy, impressing on the Israelis that Ford meant business, and that his proposed policy rethink enjoyed bipartisan support on Capitol Hill.

Before Fisher could embark on his mission, however, fate intervened to further destabilize a region defined by its volatility. During a March 22 stopover in the Saudi Arabian capital of Riyadh, Kissinger had drawn parallels between the Middle East and Indochina, where the alleged failure of the United States to honor its commitments to vulnerable allies fed doubts about American trustworthiness—at the very moment the American secretary of state was asking rivals of biblical vintage to put their faith in his government's guarantees. The Saudi ruler, sixty-nine-year-old King Faisal, was a close ally of the United States, notwithstanding his endorsement of OPEC's oil boycott staged in protest of Western support for Israel. The desert monarch demonstrated as much with a stunning proposal to supply South Vietnam the financial lifeline Washington had promised but failed to deliver.

Three days later Kissinger and Ford were to experience a different kind of shock when Faisal was assassinated in his Riyadh palace by a disgruntled nephew. With him died any chance of an eleventh-hour miracle in Vietnam. At the White House the death of a king competed for presidential attention with the threatened extinction of whole countries in Indochina. Ford designated Vice President Rockefeller to represent the United States at Faisal's royal obsequies. He braced himself for heartache in Southeast Asia on a scale dwarfing the geopolitics of oil.

O N THE SAME day King Faisal met his violent end, Ford welcomed a group of South Vietnamese parliamentarians to the Oval Office. They brought greetings from their embattled president, Nguyen Van Thieu, along with rationalizations for Thieu's recent decision to abandon his country's strategic Central Highlands to rapidly advancing Communist armies, in effect conceding half of South Vietnam to invaders from the North. "I will do everything I can to get the aid that South Vietnam needs," the president blandly assured his visitors. They knew nothing of his meeting earlier that Tuesday with army chief of staff General Frederick Weyand, tagged as "the Savior of Saigon" for his successful defense of the city during the Communist Tet Offensive in February 1968.*

Accompanying Weyand was Ford's ambassador to South Vietnam, Graham Martin, a tall, courtly North Carolinian and Thieu apologist who once threatened to "cut the balls off" CIA station chief Tom Polgar for questioning his upbeat forecasts of military victory. The loss of his son, Glenn, in Vietnam combat had given Martin an intensely personal stake in the conflict. Some thought it distorted his judgment as well. "I've been in this business for forty years and I've never been wrong," Martin boasted. His presence in Washington—he had returned home for several days to keep medical appointments—testified to Martin's refusal to believe the more dire predictions of a Communist onslaught, or to prepare evacuation plans should these be validated. Even now the ambassador insisted the North Vietnamese had no designs on Thieu's capital.

Awash in conflicting reports from unstable battlefronts, Ford wanted a firsthand appraisal from someone he could trust. General Weyand was his chosen emissary. "You are not going over to lose," he instructed Weyand, "but to be tough and see what we can do." Ford conceded that any mili-

---

* At a Saigon cocktail party in 1967, Weyand, then deputy to General William Westmoreland, the commander of American forces in Vietnam, observed to a CBS news correspondent, "The war is unwinnable. We've reached a stalemate, and we should find a dignified way out." *New York Times*, February 13, 2010.

tary options were severely limited. "I regret I don't have the authority to do some of the things President Nixon could do," he remarked wistfully. As the Oval Office emptied, photographer David Kennerly stayed behind. "You know, I would really like to go with the General," he said. Ford needed no persuading. As a journalist, and a friend, with extensive knowledge of the region from his two-year stint as a combat photographer for UPI, *Life* and *Time*, Kennerly could be counted on for an honest assessment of events—more honest, perhaps, than that of diplomats and military men—and with it, the pictures to back him up. Kennerly returned to his first-floor office with a sign dangling from his neck. GONE TO VIETNAM. BACK IN TWO WEEKS.

That evening the irreverent photo hound dubbed Hot Shot by the Secret Service appeared in the upstairs family quarters to say goodbye. Ford threw a protective arm around the young man's shoulders.

"You be careful. You have everything you need?"

As a matter of fact, Kennerly's pockets were empty. Local banks were closed and he could use some cash. Ford opened his wallet and handed over its contents, $47, as Betty Ford gave Kennerly a hug. He was striding toward the door when the president called out his name. "Here," said Ford, tossing Kennerly a quarter. "You might as well clean me out."

B Y THE BEGINNING of April, Henry Kissinger was pondering the return of his Nobel Peace Prize awarded in recognition of the Paris Peace Accords, signed in January 1973 and flouted ever since by both Vietnams. Heavily resupplied by Moscow and Beijing, Hanoi reinforced its combat troops in the South, until their numbers far exceeded the 150,000 allowed under the Paris agreement. North Vietnamese engineers rebuilt the Ho Chi Minh Trail, trespassing on the neighboring kingdoms of Laos and Cambodia, formerly a crude logistical network of roads and bicycle paths, into a modern all-weather supply artery.

As part of his postelection campaign to bludgeon the Saigon government into accepting a negotiated settlement, Richard Nixon had promised President Thieu that Communist violations of the peace deal would trigger "severe retaliatory action" by the United States. Later it was claimed that, but for Watergate and a related surge of congressional assertiveness in foreign policy, the former president would have been quick to honor his pledge. Nixon's own words suggest otherwise. "Of course, we've told Thieu we'd do it and all that," he reminded Brent Scowcroft two months into his second term—"It" being a code word for retaliatory bombing by US B-52s.

"But we've also told the American people that we've gotten them ready to defend themselves, and they've got an air force and all the rest, and they say, why the hell don't they [the South Vietnamese] do it?"*

Moreover, Watergate was still a sideshow when Congress voted in the summer of 1973 to prohibit US military operations in Indochina as of August 15. In November it passed the War Powers Act requiring the president to consult with Capitol Hill before taking military action anywhere, and placing a sixty-day limit on the commitment of US combat forces unless Congress adopted a formal declaration of war or otherwise authorized an extension of the mission. By the time Ford replaced Nixon in August 1974, support had all but evaporated for any continued US military involvement in Southeast Asia. That fall's elections confirmed a popular desire to turn the page, regardless of past promises made by discredited presidents and unctuous diplomats in their gilded salons.

"You thought the last Congress was bad," Ford remarked to an Oval Office visitor early in 1975. "This one is beyond comprehension." By then the lawmakers of whom he complained had cut in half the new administration's budget request for $1.4 billion with which to honor US pledges to South Vietnam. In the first week of January 1975, Ford appealed to Congress for $522 million in emergency assistance to help stave off defeat in South Vietnam and Cambodia. His request never made it out of committee. That same week Hanoi's Fourth Army Corps overcame stiff resistance from ARVN (Army of the Republic of Vietnam) forces to capture Phuoc Long Province near the Cambodian border. No B-52s attempted to slow their advance. Convinced at last that any threat of American reprisal was an empty one, the North dramatically expanded its offensive (original plans envisioned a final assault on Saigon in 1976 or beyond). In neighboring Cambodia the situation was even more ominous, with Communist-backed Khmer Rouge forces tightening their noose around the capital of Phnom Penh, a city totally dependent on resupply by air.

On March 12 the House Democratic Caucus voted 189–49 against providing additional military aid to either nation. Five days later a flurry of contradictory orders from President Thieu led his forces to relinquish the Central Highlands. Poor generalship and sagging morale converted re-

---

* Nixon's most recent biographer, John Ferrell, quotes his comment to H. R. Haldeman about the need to enforce the recently signed peace agreement, at least "for a while. As far as a couple of years from now, nobody's going to give a goddamn what happens in Vietnam . . . Not one damn degree."

treat into a rout, with vast amounts of US-supplied military equipment discarded by fleeing ARVN soldiers. This was to have devastating repercussions when Ford floated the idea of a so-called terminal grant of $5.5 billion over three years to help redress the military imbalance reflected in the rationing of ammunition and fuel by Thieu's beleaguered armies. "The irrefutable minimum the United States owes any friend or client," declared the *Washington Post* on March 21, "is a reasonable opportunity to make serious plans for going it alone in the face of a flat out final timetable for terminating all aid."

On March 25, the same day Ford voiced his personal support to South Vietnam's ambassador and authorized General Weyand's Indochina reconnaissance, a distraught Thieu wrote the president asking for American B-52s to halt the Communist onslaught. According to Jerrold Schecter, chief diplomatic correspondent for *Time*, Thieu considered the port city of Da Nang a military asset second only to Saigon itself, "because the Americans might have a Normandy-type landing to come to rescue them." Clearly Weyand would have his hands full disabusing Thieu of his D-Day fantasies.

However grim the prospect, it wasn't enough to make Ford change his vacation plans. His stubborn streak overwhelming his political judgment, Ford expected to spend part of his time away from the capital working on a major foreign policy address to the Congress scheduled for April 10. His speech should "wake up the country" to its continuing global responsibilities, advised Kissinger. Hell-bent on disproving the axiom that history is written by the winners, Kissinger insisted to Ford, "ARVN didn't collapse because of an attack, but because of Thieu's fears about us" reneging on commitments of military assistance renewed at the time the Paris Peace Accords were signed. Less inclined to refight lost battles, Ford was more immediately concerned about an estimated seven thousand Americans still in Cambodia and South Vietnam. Their fate was in his hands, a fact he could not publicly acknowledge for fear of starting a panic endangering not only their safety, but that of thousands of South Vietnamese made vulnerable by their association with the departing Americans.

Ford's reticence extended to his approaching Easter retreat. The public knew nothing of his anxieties over Betty's health. Six months after her cancer surgery there was no sign of any recurrence. Yet recent events had exacted a visible toll on the First Lady's energy and outlook. On days when crippling arthritis kept her bedridden, Betty's mood reflected the grayness and chill of Washington in March. "It's important to get her out of here and into the sun," the president told Don Rumsfeld.

IT WAS AFTER dark on the evening of March 29 when *Air Force One* departed Andrews Air Force Base for California. The plane was still in the air when Ford learned that Da Nang had fallen to the Communists. Orders went out to US Navy transports off the Vietnamese coast to pick up escaping refugees. What ensued was neither D-Day nor Dunkirk, as desperate people swamped overloaded barges and South Vietnamese soldiers stormed planes intended for women and children. The horrific scenes fed anti-American feelings. In Saigon fearful residents railed at "those blue-eyed sorcerers," and President Thieu complained of Ford's seeming indifference: "He is going off on vacation to Palm Springs while we are dying. When is he going to respond?"*

On Sunday the thirtieth, the Fords attended Easter services at St. Margaret's Episcopal Church. In the afternoon the president hit the links with Hollywood director Frank Capra and legendary West Point football coach Earl "Red" Blaik. His outing enraged columnist William F. Buckley, who noted Ford's golfing "on the same day that young men and old women are falling from American airplanes because they can no longer cling to the undercarriage and would rather die than live under a regime four American presidents promised they would never have to live under." It got worse. Monday morning, Ford made the fifty-minute flight to Bakersfield, the jumping-off point for his visit to Naval Petroleum Reserve No. 1 at Elk Hills. Following an extensive inspection tour meant to promote his stalled energy program, he returned to Bakersfield by helicopter.

Waiting at the airport, wireless microphone in hand, was CBS reporter Phil Jones—"Because my bet is that he is going to come off of that [chopper] and walk the proverbial rope line," he explained. His hunch was quickly confirmed. *Marine One* had scarcely touched down when Ford veered toward the knot of friendly bystanders.

"And he finally gets to me and he says, 'Phil, what are you doing here?'"

"And I said, 'Well, Mr. President, I wanted to ask you about Vietnam.' And he says, 'Ohhh nooo you don't.' And he takes off running for [*Air Force One*], which is probably a hundred feet away. And as he takes off

---

* Thieu would have been even more anxious had he overheard Don Rumsfeld's counsel to Ford early in the president's Palm Beach furlough. "At some point you have to take your hand off the bicycle seat and see if they can ride," Rumsfeld said of the South Vietnamese. "And it turns out that they can't." Donald Rumsfeld, *When the Center Held: Gerald Ford and the Rescue of the American Presidency* (New York: Free Press, 2018), 141.

running, he looks at me and he says, 'Come on!'" Televised scenes of an American president fleeing the cameras, winded reporters in full pursuit, became, paradoxically, among the most widely broadcast images of the Ford presidency.

On the other side of the world another unelected president, Cambodia's Lon Nol, had already left his ravaged country on an American transport plane bound for Hawaii. With his departure, Ford quietly signaled Ambassador John Gunther Dean to activate Operation Eagle Pull, an evacuation of Americans and Cambodians marked for death by the Khmer Rouge (in fact the exodus fell far short of predictions, "because," said Ford, "the gutsy Cambodians chose to stay and die rather than leave"). Now, Kissinger advised the president, the time had come to develop similar plans for Saigon. Ford asked the secretary if he was hearing criticism of his California holiday. Not at all, Kissinger replied in his best courtier's voice. Presidents should be able to unwind once in a while. Besides, "There's not a damn thing you can do here. There's no decisions to be made while Vietnam is being dismantled."

Ford's long, dispiriting day was not over. Shortly after eight p.m. he and Betty welcomed Ronald and Nancy Reagan for cocktails and dinner. Afterward, Ford told a friend that much of the conversation revolved around football. Hadn't they discussed issues? the friend inquired. The '76 campaign? "He didn't bring anything up," said Ford. "We just had small talk all through dinner."

FREAKISH WEATHER PLAGUED Washington that first week of April. An out-of-season nor'easter threatened the city's famous cherry blossom trees, then at the peak of their short-lived splendor. A different kind of storm raged in the White House Situation Room, where a crisis management team calling itself the Washington Special Actions Group (WSAG), met on the morning of April 2. Based on the latest intelligence, Kissinger gave South Vietnam three months to survive. More like three weeks, Jim Schlesinger countered. As events would demonstrate, this was the least of the strategic and tactical differences separating Ford's secretaries of state and defense. Both summa cum laude graduates of Harvard's Class of 1950, neither man was inclined to hide his light under the bushel. But whereas Kissinger, a world-class flatterer, personified the kiss up, kick down school of advancement, his Cambridge classmate and future Cabinet adversary made no distinction between presidents, peers and the most inconsequential GS-4. Like Professor Henry Higgins, who prided himself on treating a duchess like a flower girl, Schlesinger didn't much care whom he offended.

During his short tenure at the CIA he had complained that the place had been run like a gentleman's club, "but I'm no gentleman."

With characteristic shrewdness, Kissinger expressed sympathy for Ford's plight. "It's a pity that you have to keep adjudicating disputes between Schlesinger and me."

"Schlesinger's fight is not with you, it's with me," Ford responded. "He thinks I'm a dummy, and he thinks I have to be run by somebody, and he thinks you're running me. And this won't stop until either I make him believe he's running me, or I fire him."

By the start of April 1975 his patience was running out. "Schlesinger was always pressing me and differing with Kissinger," Ford recalled after the fact. "He wanted to get our people out of Saigon at least two weeks earlier than I did." The Pentagon boss showed less concern for the fate of Vietnamese civilians. "Probably we will not get a significant amount," he forewarned the president. Kissinger, by contrast, stressed the moral obligation of the United States to extricate as many Vietnamese at risk of Communist reprisal as possible. A conservative estimate put their number at 174,000.

"In the last two months of Vietnam," Kissinger conceded long afterward, "the only debate was, do we pull the plug right away, or do we keep it going for as long as we possibly can to evacuate as many people as we possibly can[?]" Neither he nor Ford was in doubt as to the unequal battle now playing out. "We knew what the end would be," said Kissinger. "But it seemed important to Ford . . . that the United States not stab a dying ally in the back." Even if Congress turned down his appeal for last-minute military aid, "we would not end something in which thousands of Americans had been killed by turning on the people that we'd been supporting."

The Pentagon, in keeping with Secretary Schlesinger's priorities, scheduled flights out of Saigon earlier and with less discretion than either Kissinger or Graham Martin thought warranted. "We were sending planes, and they were coming back half and two-thirds empty," complained one Defense official. Ford suspected the empty planes were for show, enabling the Pentagon to disclaim responsibility should the Da Nang fiasco be repeated in Saigon. The imminent evacuation of Cambodia was another flash point. Schlesinger favored fixed-wing aircraft, while Kissinger sided with Ambassador Dean in Phnom Penh, who wanted to use helicopters should enemy artillery render the airport inoperative. Truth be told, Kissinger confided, he would hold off on the Cambodian operation altogether, "if we had a sane Secretary of Defense."

On one subject, at least, everyone in the Situation Room seemed in agreement—the performance of their rogue ambassador in Saigon. "A very difficult person," in Ford's estimate. Lauding Martin's "determination and guts," the president discerned in the ambassador a mulish insistence on his own version of reality. Martin particularly resented the latest directive from WSAG to produce within twenty-four hours a complete list of South Vietnamese nationals who should be included in any evacuation. Either his judgment prevailed over that of his alleged enemies in the State Department, Martin snorted, or the president should relieve him. In the latter case, "I can make an even greater contribution by speaking out with complete candor on how we got in this current situation." Thus did the American ambassador to South Vietnam threaten to blackmail his own government.

Members of Congress were only slightly more accommodating. Inspired by the recent debacle at the Bakersfield airport, House Majority Whip John J. McFall urged Ford to redouble Vietnam assistance "and not run from the press." Republican Frank Horton of New York wanted the president to immediately return to Washington and convene an emergency session of Congress to authorize humanitarian aid. Ford responded at a San Diego press conference on April 3. Rejecting calls for Thieu's removal, he decried "a great human tragedy" in the making and announced Operation Babylift, a series of mercy flights using C-5A and C-141 cargo aircraft to evacuate two thousand Vietnamese orphans, many of them fathered by American GIs. "This is the least we can do," Ford concluded, "and we will do much, much more."

Tragedy was compounded the next day when the first plane carrying the war's youngest victims crashed on takeoff from Saigon's Tan Son Nhut Airport, killing seventy-eight children and fifty adults. Ford rearranged his schedule so he could personally greet a second plane when it arrived in San Francisco on April 5 bearing 325 South Vietnamese orphans to new homes in the United States. First, however, he conferred with General Weyand, just back from his Vietnam fact-finding mission. Pulling no punches, the general said that South Vietnam stood on "the brink of a total military defeat." Only a resumption of US bombing could slow the Communist juggernaut until the rainy season provided the Saigon government time to regroup. Failing that, Weyand proposed a $722 million package of military aid, enough (in theory) to re-create four divisions and partially replace the ordnance lost when the Central Highlands were abandoned.

Anticipating the worst, Weyand told Ford there was no time to lose in evacuating US citizens and tens of thousands of South Vietnamese and third-country nationals owed protection. "The essential and immediate requirement is Vietnamese *perception* of U.S. support," he advised the president.

(Weyand's verbal observations, too sensitive to commit to print, blamed plummeting ARVN morale on President Thieu's erratic leadership and strategic misjudgments.) Ford thanked the general for his insights. Most of Weyand's suggestions would be incorporated into Ford's forthcoming speech to Congress.

Fulfilling the president's desire for an independent assessment of the situation, David Kennerly described in vivid terms his war zone experiences in Vietnam and Cambodia. These included being targeted by mutinous ARVN troops while riding in a chopper over Cam Ranh Bay. The most gut-wrenching part of Kennerly's journey involved Vietnamese friends and coworkers who had begged him to put their children on a return flight to the United States. "All my friends know they are going to get killed," he told reporters outside the rented house where the Fords were staying in Palm Beach. "It is really shitty, and you can quote me."

Inside, Kennerly let his grim portfolio of refugee youngsters and wounded evacuees speak for him—to a point. Cambodia, he insisted, was gone. Vietnam couldn't last more than a month, "and anyone who tells you different is bullshitting you . . . the party's over." Kennerly's images conveyed the indiscriminate horrors inflicted on the people of Indochina, their plight poignantly encapsulated in the photo of a little Cambodian girl with a dog tag around her neck and a vacant look in her eyes. Her visage stayed with Ford as he battled a complacent ambassador and Pentagon resistance to any mass evacuation of Vietnamese nationals.

After a half hour or so with Kennerly's pictures, it was time for the president and Mrs. Ford to leave for San Francisco. Ron Nessen accompanied them on the short flight north. At San Francisco International Airport he and Ford boarded the Pan Am jumbo jet chartered by Operation Babylift to transport "hundreds of babies . . . They'd been flying across the ocean for like seventeen hours," Nessen recalled. The cumulative effect of all those dirty diapers was overwhelming, "like a nerve gas attack." Pictures of Ford cradling in his arms one of the first infants to come off the plane evoked mixed reactions. Critics accused the president of staging a photo op to distract from the unfolding military disaster eight thousand miles away. The International Committee of the Red Cross said the entire operation violated a Geneva Convention rule that war orphans should be educated within their own culture. The flights continued; ultimately some 3,300 Vietnamese children found homes in the United States.

Returning to Washington on April 7, Ford had just seventy-two hours in which to devise strategic arguments capable of winning over a fiercely skeptical Congress. He didn't lack for advice in crafting what was billed as his

State of the World address. Richard Nixon, utilizing back channels, urged his successor to avoid the blame game. Ron Nessen submitted a speech draft ruling out further US military aid to the tottering regimes in Saigon and Phnom Penh. "Until now it is not 'Ford's war,'" cautioned Nessen. "But it will be if he requests more aid to keep the war going." Ford rejected his press secretary's counsel. "That is not the way I am," he told Kissinger. "My determination is to stay with it. Ask for the money and authority to evacuate. Then if everything happens as we foresee, I will have done my best."

As we foresee . . . Throughout the West Wing the usual hallway scenes of smiling first family members and well-attended presidential events gave way to Kennerly's stark blowups of the horrors he had witnessed in Vietnam and Cambodia. Some staff members objected to the images, which were admittedly disturbing. To Ford that was the point of displaying them in the first place. "Leave them up," he ordered. "Everyone should know what's going on over there."

M OST ACCOUNTS OF April 1975 cast Ford in the role of bystander to the chaos and recriminations attending the end of the Vietnam War. An improvised airlift of Americans and their South Vietnamese collaborators from the rooftops of Saigon to a protective flotilla of US Navy ships is remembered less for the thousands rescued than for the even greater numbers left behind. Likewise with Ford's campaign, in the face of congressional indifference and public hostility, to resettle over 130,000 South Vietnamese refugees on US soil. As reluctant as anyone to dwell on painful scenes, Ford devoted less than two pages of his White House memoir to the "exceedingly careful course" he charted during Vietnam's torturous final weeks in the hope of averting something even worse. Only much later would he acknowledge the struggles within his administration as he sought to mediate between "those who wanted an early exit and others who would go down with all flags flying."

This was code for Schlesinger and Kissinger. Asking Congress for money he knew he wouldn't get to facilitate survival of a rump South Vietnam incorporating Saigon and the Mekong River Delta prompted the verbal equivalent of a slap in the face from the secretary of defense. "It's all over, Mr. President. We lost." That, at least, is how Schlesinger recalled the conversation, his memories embellished with caustic references to Michigan fight song imagery and blood, sweat, toil and tears rhetoric. It is the exact opposite of what Schlesinger actually said to Ford at a meeting of the National Security Council on April 9. Then he urged the president "to give a call to the people. You could talk along the lines of Churchill's blood, sweat and tears." Exhibiting his trademark self-assurance Schlesinger insisted, "There are two kinds of people on the Hill. Those who want to get out of Vietnam and those who are just waiting for an inspirational message."[*]

Kissinger, naturally conspiratorial, was more adroit than his Cabinet rival in pursuing hidden agendas. "The president is aware it is in his power to trigger the collapse in Saigon," he briefed network correspondents a few hours before Ford's April 10 speech to Congress. Ford did his part, admonishing legislators with whom he met to avoid defeatist comments. "If they think we have given up, that will set them off," he confided. So closely held were American intentions that President Thieu was kept in the dark about the final evacuation until after Ford made his eleventh-hour appeal to Capitol Hill.

To many in his audience the speech was an exercise in denial, analogous to the first of Elisabeth Kübler-Ross's famed five stages of grief. Veering between the combative and the conciliatory, an unusually nervous Ford declared it was no time "to point the finger of blame." This did not prevent him from implicit criticism of lawmakers who had failed to provide South Vietnam "adequate material support" as promised by American officials in signing the Paris Peace Accords. His request for a last-minute infusion of military and humanitarian assistance to the beleaguered nation prompted a walkout by freshman Democrats Toby Moffett of Connecticut and California's George Brown.

Those who stayed heard Ford ask Congress to "clarify" existing restrictions on the deployment of US forces "for the limited purpose of protecting Americans," a request he now amended to include "those Vietnamese to whom we have a special obligation." Denial mingled with barely suppressed anger—the second of Kübler-Ross's way stations on the road to acceptance—as Ford reasserted executive claims of authority eroded by Vietnam and Watergate. He called for a resumption of military aid to Turkey; sought to waive restrictions in trade legislation that unfairly penalized US allies like Venezuela who also belonged to OPEC; proposed to revisit "self-defeating" limits on trade with the Soviet Union tied to that nation's

---

[*] Schlesinger was hardly alone in deviating from the written record. Historical accounts of this period routinely quote Al Haig's purported advice to Ford, delivered at a White House meeting in March 1975, that the president had to "roll up your sleeves like Harry Truman" and demand an emergency aid package for Saigon—polls be damned. Haig's actual words were far more equivocal. "I think we have to think whether additional money will solve Vietnam," he remarked. Ford should "explore" making the case to Congress and the American people, "so that the use of force is not discredited . . . right now it looks like we may be pouring money down a rathole." NSA Mem Com, March 27, 1975, 5:15 p.m., GRFL.

emigration policies; and, in a telling departure from his text, warned of "catastrophic" consequences should Congress in its investigatory zeal dismantle "our intelligence systems upon which we rest so heavily." Ford's rushed delivery was that of a man anxious to conclude a distasteful task. Perhaps the breach of etiquette by Moffett and Brown rattled the president more than he let on.

Ford's listeners knew nothing of a domestic drama unfolding a couple miles from the House chamber. Before leaving the White House for Capitol Hill he had been tipped off to the suicide of James Howe, a retired military officer married to Betty's personal assistant Nancy Howe. Howe had become despondent over questions raised by *Washington Post* columnist Maxine Cheshire regarding the Howes and their relationship with South Korean influence peddler Tongsun Park. Was it true they had visited the Dominican Republic as guests of Park? Had Nancy and her husband cashed in their plane tickets for a second Park-funded trip to Mexico when Jimmy became ill?* Their denials of wrongdoing had failed to dissuade Cheshire, and the *Post* went ahead with the story—or intended to, until Nancy's psychiatrist convinced editor Ben Bradlee that publication might cause his patient to do something drastic. Bradlee deferred to the doctor's warning, only to have Jim Howe, under the blowtorch scrutiny of the post-Watergate press, pick up a gun in his Northwest Washington home and point it at himself.

Ford waited until after his speech, but before the late local newscasts, to break the news to Betty. She immediately called Nancy to offer condolences. The next day her visit to a distraught Nancy and her daughter coincided with publication of Cheshire's original column in the same edition of the *Post* that carried James Howe's obituary.

It was too much for the First Lady. "Maxine Cheshire killed Jimmy Howe," she proclaimed, even as Phil Buchen acknowledged that his office had been looking into the allegations before Nancy's widowhood. Additional journalistic digging uncovered evidence of alcoholism and marital problems in the Howe household. Acting on the advice of her doctors, it was decided by the White House that Nancy would not return to work there. Such, at least, was the official explanation. "I hope everyone understands," Mrs. Ford remarked, "that I plan to continue my friendship."

---

* Park was to figure prominently in the so-called Koreagate scandal. Granted immunity from prosecution, the lobbyist closely linked to the dictatorial regime of President Park Chung Hee admitted making cash payments to thirty members of Congress, several of whom subsequently resigned or were censured or tried in courts of law.

Then another *Post* columnist, Myra MacPherson, asked a grieving Howe if it wasn't somewhat callous of the First Lady to discharge her "best friend" under such circumstances.

"You said it," Howe replied. "I didn't. I just don't want to talk about it."

In her White House memoir, Betty would relate how much she had wanted to be with Nancy during her ordeal. "It wasn't permitted," she concluded enigmatically.*

O N THE STREETS of Saigon Ford's speech appeared to have its intended narcotic effect. Uncle Sam might yet intervene to avert a bloodbath and, with it, the loss of American credibility in Asia and elsewhere. The alternative—to simply write off fifty-eight thousand lives and $150 billion sacrificed since 1959—seemed unthinkable. As it happened, Saigon was not the only capital gripped by wishful thinking. In Kübler-Ross's emotional road map for handling loss, denial and anger are followed by bargaining, depression and, finally, acceptance. On April 14, Ford sought accommodation with the Senate Foreign Relations Committee, whose members invited themselves to the White House for the first such confrontation since Henry Cabot Lodge and his colleagues grilled Woodrow Wilson on the Versailles Treaty and his proposed League of Nations in August 1919.

This encounter went no better than its predecessor. The issue of military assistance to the Thieu government was again raised. Failure to act on his request, said Ford, might put the remaining Americans at greater risk from their nominal allies than from sixteen enemy divisions massed at the gates of Saigon. Someone suggested a massive evacuation by sea. "Four thousand on one ship would pull the plug," Ford cautioned the senators—it would fuel hysteria and increase the likelihood of US troops being forced to shoot their way into Saigon to seize and hold possible evacuation sites. Mindful of his accountability to Congress under the War Powers Act, Ford insisted that any evacuation involving the American military would last at most a few days. "But," he emphasized, "we have to have enough funds to make it look like we plan to hold for some period."

Senator Jacob Javits had heard enough. "I will give you large sums for evacuation," Javits snapped, "but not one nickel for military aid for Thieu." Delaware's Joe Biden took it a step further. "I will vote for any amount for getting the Americans out," said Biden. "I don't want it mixed with getting the Vietnamese out." Even if they succeeded in rescuing thousands of South

---

* The First Lady did attend James Howe's funeral on April 14.

Vietnamese marked for death by the Communists, where would they go? One senator suggested Borneo in the Malay Archipelago—"It has the same latitude, the same climate, and would welcome some anti-Communists." Ford had a different destination in mind. "Our tradition is to welcome the oppressed. I don't think these people should be treated any differently from any other people—the Hungarians, Cubans, Jews from the Soviet Union." He thanked the senators for coming. "We've had a good discussion but the decision is my responsibility and I'll accept the consequences."

On April 17 the Cambodian capital of Phnom Penh surrendered to the Khmer Rouge, who immediately set about publicly executing Lon Nol's designated successors. In Washington the Senate Armed Services Committee formally rejected Ford's proposed military assistance for South Vietnam (although it did endorse $150 million for vaguely defined humanitarian purposes).* That evening Ford became the first American president since April 14, 1865, to see a play performed at nearby Ford's Theatre. The show, starring James Whitmore as a shrewd, salty, history-haunted Harry Truman, was called, fittingly, *Give 'em Hell, Harry!* Afterward Ford went up onstage to congratulate the actor and reminisce with Truman's daughter, Margaret. Often belittled while in office, by 1975 Truman was riding a wave of popular and scholarly reappraisal. Under the circumstances Ford could only hope that history repeated itself.

The next day he named L. Dean Brown, a diplomat richly experienced in crisis management (in Cyprus he had filled in for the murdered envoy Rodger Davies), to chair an interagency task force to coordinate Indochina refugee policy. Brown immediately set about marshaling the resources of thirteen separate government agencies. Confronting a bureaucratic brick wall in the form of the Immigration and Naturalization Service (INS), Brown and his allies enlisted Ford's help and that of Attorney General Levi in easing visa requirements. The Immigration and Nationality Act of 1965 capped the annual influx of refugees at 10,200. A loophole in the legislation empowered the attorney general to "parole" an alien into the United States if he judged it to be in the public interest. Citing the precedents of Hungarians and Cubans fleeing Communist regimes, the administration

---

* Kissinger dismissed as "absolute, total nonsense" retroactive claims by his former Cabinet colleague Mel Laird that Ford could have secured additional funding on Capitol Hill if only he had tried harder. "We couldn't get the other Cabinet members to go up to the Congress," Kissinger says in rebuttal. "They'd find ways of weaseling out of it . . . the Congress was determined. There was no way to get more money." AI, Henry Kissinger.

overrode the INS and, in tandem with congressional allies, authorized the parole of 125,000 refugees, this on top of several thousand already evacuated to Clark Air Force Base in the Philippines.

Much of this shadowy operation was the work of Ken Quinn, a young staff member of the National Security Council who was himself married to a Vietnamese woman. Convinced that Ambassador Martin, befogged in self-denial, was unlikely to authorize evacuation planning until it was too late, Quinn enlisted the help of State Department officer Lacy Wright and CIA analyst Frank Snepp. Together they would rescue hundreds if not thousands of high-risk Vietnamese in a furtive operation that was kept from Martin, but shared with Dave Kennerly, who in turn passed word on to the president.

On Friday afternoon, April 18, Ford flew to New Hampshire, where he addressed state legislators in Concord before going on to Boston for the unofficial launch of the nation's bicentennial with a moving ceremony in the Old North Church. His emotions fired by schoolboy history, he confessed to feeling "goosebumps" on entering the Georgian-style house of worship, two hundred years to the night since the church sexton hung a pair of lanterns in the steeple to alert Paul Revere and other patriot couriers of the movements of British redcoats embarked on a smash-and-grab foray into the Massachusetts countryside. Climbing a spiral staircase to the high pulpit and lighting a symbolic third lantern to inaugurate the third century of American independence, Ford thought of his stepfather, a devout Episcopalian, and how much it would have meant "for him and me" had he lived to see this day.

As he spoke a discordant chorus of street protesters could be heard through the thick brick walls of Old North, their voices raised on behalf of Cyprus, Native Americans, an end to court-ordered busing—and American involvement in Vietnam. The sounds of dissent were louder still the next day at Concord Bridge, site of the first armed resistance of the American Revolution. An estimated twenty-five thousand protesters, some of them waving yellow banners proclaiming, DON'T TREAD ON ME, nearly drowned out the president's call for "reconciliation, not recrimination."

It was a theme to which he had given considerable thought of late, and one to which he returned in a brainstorming session with White House speechwriters ahead of a visit to New Orleans's Tulane University on April 23. Looking to the future, Ford wanted students to redirect their energies and rekindle a sense of purpose after a decade defined by Vietnam.

"I don't know why we have to spend so much time worrying about a war that's over as far as we are concerned."

"Well, then why don't you just say that?" asked Bob Hartmann.

"Well, I don't think Henry would like it."

Why should Kissinger's feelings matter? Hartmann shot back. "Sometime at some point it has to be said, and you're the one who has to say it, so why not now?"

Ford promised to think it over.

The penultimate emotion in Kübler-Ross's grief cycle is depression. Ford, who liked to characterize himself as "disgustingly sane," was not immune to feelings of grief and humiliation over "getting kicked out of Indochina." If so, the mood was a passing one. In case he had ever doubted it, Vietnam's death throes reminded Ford just how much the presidency is an exercise in multitasking. Under increasing pressure from Congress to accelerate the pace of withdrawal, he must simultaneously arbitrate differences between the Pentagon and the State Department, humor a willful Graham Martin and reassure nervous allies that the United States had no intention of retreating into a new isolationism.

That he could seriously entertain Hartmann's suggestion at all was due in no small measure to the stalwart aviators—commercial, military and more than a few black-ops and Air America pilots answerable to the CIA— who flew over fifty thousand Americans and "endangered Vietnamese" to safety. Renegade pilots from the South Vietnamese air force spirited their fugitive countrymen to Thailand, Singapore and Taiwan. When Philippine president Ferdinand Marcos objected to the waves of refugees crashing on his shores, thousands bivouacked at Clark Air Force Base were rerouted to Tin City, a corrugated barracks compound at Andersen Air Force Base on the island of Guam. In the States, meanwhile, Dean Brown and his task force scouted out a trio of military installations—Camp Pendleton in Northern California, Camp Chafee in Arkansas, and Florida's Eglin Air Force Base—as suitable accommodations (euphemistically dubbed "reception centers").

One night in mid-April, dismayed by Ambassador Martin's ostrichlike denials, Ken Quinn slipped across West Executive Avenue to call on Brent Scowcroft, "always the last person to leave" his office at the end of the workday. Quinn briefed Scowcroft on the secret evacuation plans he had developed, the safe houses and communications networks ready to be activated the moment Martin received back-channel orders to do so. From his meeting with Quinn, Scowcroft in turn went straight to the president, who immediately authorized a message directing his reluctant ambassador in Saigon to assist in evacuating the people on Quinn's list.

On April 19, with time running short and the refugee flow showing no signs of abatement, the US made a diplomatic overture to Hanoi via the Soviet Union. Kissinger told the president there were "two chances in twenty" that Moscow would press Hanoi to grant an unofficial cease-fire of sufficient duration to extract the remaining Americans and an indeterminate number of South Vietnamese—producing what Kissinger called "a controlled takeover." As for Thieu, "we can jettison him as soon as we can get an answer from the Russians." In the event, the Soviets came back with an ambiguously worded response, liberally interpreted by Washington as acquiescence. Thieu announced his resignation in a tearful, rambling speech in which he paid sardonic tribute to "our great ally, leader of the free world." Biting the hand that no longer fed him, Thieu mocked, "you Americans with your 500,000 soldiers in Vietnam. You were not defeated . . . you ran away!"

With Saigon encircled, its purported government crumbling, and the numbers of Americans diminishing daily, Ford flew to New Orleans on April 23. The president's Tulane remarks were entrusted to Milt Friedman, a staunch opponent of the war who feared it might yet be prolonged in a futile attempt to revive the diplomatic option ("Kissinger had been talking about bringing some aircraft carriers back to do air strikes again," according to Friedman). Sacrificing a night's sleep, Friedman crafted a speech around the 1815 Battle of New Orleans, fought *after* diplomats in distant Ghent, Belgium, concluded a peace treaty, in which Andrew Jackson and his ragtag army had restored the pride of a nation still reeling from the destruction of its capital five months earlier. The unobjectionable text Friedman carried with him onto *Air Force One* had been vetted by Kissinger's deputy Brent Scowcroft.* Only when they were in the air did Friedman graft onto Ford's speech the cathartic language for which it would be remembered.

"Today America can regain the sense of pride that existed before Vietnam," he wrote. "But it cannot be achieved by refighting a war that is finished as far as America is concerned." After getting Bob Hartmann's approval, Friedman showed his last-minute insertion to the president.

"Has Henry seen this?"

"No. There wasn't time."

"Well, this is something I want to say. I might as well say it today."

That night a packed fieldhouse exploded in delight and disbelief as Ford delivered the line that formalized his belated acceptance of events beyond

---

* Kissinger himself raised no objections to a second Ford text, scheduled for delivery before the national convention of the Navy League. This was to generate considerable confusion when questions were raised about Kissinger's involvement with "the New Orleans speech."

the control of any American president. Broadly favorable press coverage greeted him the next morning, mingled with the bitter reproach of his secretary of state. Kissinger's anger was directed less at Ford's truth telling about the war than his unhesitating "No" when asked by reporters on the flight back from the Big Easy whether the secretary had been consulted in the speech's preparation. Members of the presidential speechwriting team, summoned to the Oval Office for an impromptu pantomime, found their nemesis energetically pacing the room.

"This we don't need, Mr. President. How is it that I knew nothing about this? How can I explain it all to the people who are calling me to ask what it means?"

It was a fair question. But they had shared the first draft of the Tulane speech with Brent Scowcroft, protested writer Paul Theis. "Of course," Hartmann told the president, "you made some minor changes on the airplane, as you always do." No one in the overworked writers' group had imagined for a moment the fuss that would be engendered by a single throwaway line. No one mentioned its genesis in Ford's response to Hartmann's verbal provocation.

"We *were* a little pressed for time," Ford interjected. Donning his stern face (though Hartmann detected the hint of a wink) he turned to the wordsmiths before him.

"Just be sure it doesn't happen again."

With his Tulane speech, Ford started the clock ticking toward the final exit of his countrymen from Vietnam. In a meeting of the National Security Council the next day, he directed that the number of Americans be reduced from 1,700 to 1,090 within twenty-four hours.

"That is a lot in one day," observed Jim Schlesinger.

"That is what I ordered," said Ford. Moreover, he wanted all nonessential governmental personnel out by Sunday, April 27.

One day before the president's deadline, five days after assuring his people, "I am resigning, but I am not deserting," an inebriated Thieu was driven to the airport by a CIA operative, who put him on a plane bound for Taiwan and permanent exile.

FORD'S SCHEDULE FOR Monday, April 28, maintained an illusion of normality. His daily CIA briefing was followed by a morning address to the US Chamber of Commerce; an Oval Office discussion of aid to parochial schools; meetings with two Cabinet secretaries, the governor of North Carolina and Lisa Lyon, 1974–75 Miss National Teenager. Kissinger kept him abreast of developments in Saigon. At 5:28 that afternoon he notified the president that Communist shelling of Tan Son Nhut Airport had forced at least a temporary halt to fixed-wing flights. Two American marines guarding the adjacent compound of the Defense Attaché Office (DAO—code-named "The Alamo") had been killed by incoming rockets. The shattered remains of an C-130 transport plane littered the tarmac.

Around six p.m. Brent Scowcroft interrupted a Cabinet Room discussion of energy policy to hand the president a note. Conditions at the airport were deteriorating, putting US embassy personnel at risk and raising the possibility that fixed-wing aircraft would have to yield to a risky evacuation by helicopter. Giving nothing away, Ford had resumed the discussion of his planned tariffs on imported oil when Scowcroft returned, this time to receive whispered instructions to convene a seven thirty meeting of the National Security Council. At the appointed hour Bill Colby led off by describing the situation at Tan Son Nhut, ground zero for any mass evacuation, where three South Vietnamese aircraft had been shot down, one by an SA-7 heat-seeking missile.

Ford responded with a flurry of questions. What was the status of the runway? "Do you have air controllers?" Any C-130s presently on the ground? The DAO—how many people did it house?

About four hundred, replied Jim Schlesinger, "including contractors."

"If the C-130s can land, they should," said Ford, mindful that three of the four evacuation scenarios developed by military planners relied on turboprop aircraft used to transport troops and cargo out of Tan Son Nhut. "How many are there?"

According to General George Brown, chairman of the Joint Chiefs of Staff, the plan was for thirty-five aircraft to make two sorties each. Brown reiterated the danger posed by the SA-7, against which choppers and aircraft were defenseless.

"If the risk is too great," Ford replied, "the man on the ground has to judge. We cannot."

From Tan Son Nhut the discussion pivoted to the US embassy and the drastic staff reductions ordered by the president. Kissinger raised the possibility of "suppressive fire" being required to cover the evacuation. Schlesinger dissented. "There is no authority now for suppressive fire, only for the chopper lift." As his dueling Cabinet officers debated the potential use of fighter planes, Ford wondered how long it would take to establish air cover. Even if US aircraft took no offensive action, he reasoned, "they"— meaning the swarming North Vietnamese—"would still have enough radar to pick up our presence." In other words, the mere presence of the warplanes in force might act as a deterrent.

On the other hand, said Ford, if they *did* decide on air cover, "we have to go for the evacuation of Saigon and not just Tan Son Nhut." He turned to General Brown. "Are we ready to go to a helicopter lift?" This was the fourth, and most hazardous, option considered by military planners. Brown told the president he could have choppers in the air within an hour (a prediction that was to prove wildly inaccurate owing to delays in refueling and assembling the needed ground security forces). Schlesinger urged that "subtle priority" be accorded Americans in the airborne exodus. The decision was made: if the airport was closed to the C-130s they would go with option four. In any event, Ford concluded, "Today is the last day of Vietnamese evacuation."

A grim-faced president departed for the residence, hands in his pockets. A few minutes later he joined Betty in the second-floor West Hallway, which served as the social center of the Ford White House for family and visiting friends. The endgame was at hand, he told her. Leaning back on a floral sofa, his feet up on a coffee table, Ford mentioned the two marines killed at the airport: Corporal Charles McMahon of Woburn, Massachusetts, and Lance Corporal—and Eagle Scout—Darwin Lee Judge of Marshalltown, Iowa.

"They were only nineteen and twenty-[one]," said Ford. Betty placed her hand on his shoulder.

"You should write notes to their parents."

Ford promised to do so.

At eight thirty, Kissinger and Scowcroft appeared on the second floor to report continued shelling at Tan Son Nhut. Ford, still reluctant to rule out fixed-wing aircraft, put off a formal authorization. He wanted to hear more from those on the ground. He sipped a martini, had a late dinner of corned beef and cabbage. At 10:25 Kissinger called back with an update. The airport was being overrun, not by enemy soldiers but by South Vietnamese, civilians and military, frantic to be part of the American airlift. "It's getting to be like [Da Nang]," Kissinger told Ford.

"Well, Henry, we've done the best we could and you and I are taking it on the chin and [I] just hope the Good Lord is with us."

Further complicating matters, Ambassador Martin, still convinced that things were not as dire as they appeared, was resisting an evacuation by chopper. Though suffering from walking pneumonia, the heavily medicated diplomat had insisted on seeing Tan Son Nhut for himself, to verify with his own eyes this latest impediment to his personal idée fixe—the clandestine rescue of ten thousand South Vietnamese under the noses of his Washington tormentors (a 180-degree reversal from his earlier foot-dragging).

At the DAO he was met by General Homer Smith, the highest-ranking US military officer in Vietnam. With C-130s in holding patterns sixteen thousand feet above the combatants, Smith saw no alternative to option four.

At the Pentagon, Jim Schlesinger agreed. "For Christ's sake, let's go to the helos," he bellowed at Brent Scowcroft. He gave similar advice to Ford, presumably in more measured language. In Saigon, Martin yielded, but only after being literally shaken by incoming rockets. Back at the embassy a couple thousand civilians waited for buses to take them to the DAO, while a crowd several times as large surrounded the compound, making it impossible for the convoys to operate. At ten forty-five, Kissinger called the president to update him on the rapidly changing evacuation plan. "You tell them to make damn sure every reasonable effort is made to get everybody out," Ford replied. Under current conditions, Kissinger said, "we can't take any more Vietnamese." Reluctantly Ford agreed, while confiding, "it sickens me."

"And then he started crumbling," Kissinger would remember, "and it took about two or three phone conversations . . . 'Can't we do something for these five thousand people that are lining the airfield there?'" As Ford wavered, other refugees self-evacuated. Thirty miles out to sea, sailors and marines on the USS *Blue Ridge*, command ship of evacuation Task Force 76, were startled to see a CH-47 Chinook belonging to the South Vietnamese Air Force execute an emergency landing on the ship's aft deck helipad. The twenty Vietnamese who disembarked from the aircraft figured it was better to chance the South China Sea than wait for a bus to the DAO that might never arrive.

At the embassy an improvised security force of marine guards and CIA agents, some clutching ancient submachine guns or grenade launchers, patrolled the high concrete wall surrounding the compound. "Women were like, 'Take us in, we'll have sex with you,'" one guard remembers. "'Here's gold, I have this money, here's jewels. Please just let me in.'" Mothers tried to pass infant children to GIs manning the gates. In the courtyard, per orders of Ambassador Martin, a much-loved tamarind tree was cut down to make it possible for the big CH-53 choppers to take off from and land there. A rooftop incinerator strained to accommodate the volume of classified documents and US currency belatedly marked for destruction.

It was the twenty-ninth day of April, the last full day of the Republic of Vietnam.

I N WASHINGTON, SEPARATED from Saigon by twelve time zones, it was still the twenty-eighth. After weighing Schlesinger's analysis and Martin's reluctant recommendation, Ford gave the order to launch Operation Frequent Wind at 10:50. "We have no choice but to send in the helicopters," he told Brent Scowcroft, "get our Americans out, and try to save as many friends as we can." In the South Vietnamese capital, radio stations began broadcasting the Mayday signal, a fake weather report and the opening bars of Irving Berlin's "White Christmas." A few minutes later Ford appeared in the basement Situation Room, a hive of activity, where incoming reports were being channeled through Scowcroft, and Kissinger had spread out maps of the Saigon region to illustrate potential evacuation routes.

Upstairs, Ford paused for a brief, wordless communion with Don Rumsfeld, joined by Dick Cheney, Ron Nessen and Dave Kennerly. The patter of April showers against West Wing windows added to the midnight gloom. "Somebody said, 'Sleep well, Mr. President,'" Nessen recalls. "And he walked out and I said, sort of under my breath, 'If you can.'"

Outside the pressroom, Ford encountered Phil Jones.

"Working late?" the reporter asked the president.

"With good reason," said Ford.

Now, the waiting. Figuring he might as well get such sleep as he could, Ford climbed into bed just before twelve thirty. He hadn't turned out the light when Kissinger phoned with news that Duong Van "Big" Minh, South Vietnam's third president in a week, had ordered all Americans out of his country. The secretary called a second time a few minutes later, and again at 1:08 to confirm that the operation was underway. As the president dozed, three dozen marine helicopters lifted off from the deck of the USS *Hancock* and made their way through rain squalls toward the DAO compound on Saigon's outskirts. Two narrow air corridors, one code named Michigan, the other Ohio, had been reserved for incoming and outgoing traffic. The weather played havoc with these plans, forcing pilots to fly at lower altitudes, where they were exposed to small-arms fire and rocket-propelled grenades. Of more immediate concern, the failure to transport embassy personnel and untold numbers of Vietnamese to the DAO and Tan Son Nhut meant the last-minute diversion of choppers—a lot of choppers—to the embassy. One evacuation turned into two.

At the White House lights burned through the night in the West Wing, where aides grabbed slices of pizza in between calls to congressional leaders as stipulated by the War Powers Act. Black humor broke the tension, dark as the small hours of the morning. There was good news, said Dave Kennerly, and there was bad news. "The good news is the war is over. The bad news is that we lost." Kissinger got into the act, describing himself as the only secretary of state in history to have lost two countries in three weeks. As the night stretched out some staffers snatched an hour or two of fitful sleep on office sofas. Around three a.m. came the first verifiable report of a CH-53 "Jolly Green Giant" chopper returning to the USS *Midway* with fifty passengers from the DAO compound. By four thirty the number transported to safety was approaching two thousand.

At Martin's embassy the pace lagged by comparison. It seemed that for every evacuee airlifted out, two took their place. Additional marine guards were brought in to bolster security, some using rifle butts to keep people from scaling the perimeter walls. By the time Ford woke at five thirty Washington time, CH-46 helicopters were landing on the embassy roof and the heavier CH-53s in the courtyard at ten-minute intervals. Other pilots dodged small-arms and sniper fire to rescue stranded Americans and Vietnamese on nearby rooftops. Ford's morning coffee was left to cool as he pumped Kissinger for overnight developments. "The Pentagon wanted to make sure that if anything went wrong, they'd stick the White House with it," Kissinger asserted in a 2009 interview. "So they meticulously tracked every helicopter with us so that they could say they had an order to get out. And then, every hour or so, I'd give Ford a report."

In the Oval Office at 7:35, the president notified Ron Nessen that there were to be no public briefings by any department ahead of the White House. It was critical, Ford said, that the administration speak with one voice (it did not go unnoticed that Secretary Schlesinger had found time to appear on the nation's TV screens at six a.m.). Citing crew fatigue and diminished visibility due to bad weather and fires on the ground, the defense secretary wanted flights suspended overnight. Already one CH-53 had ditched in the South China Sea with the loss of its two pilots. Maintenance issues downed many of the remaining eighty aircraft. An order halting flights to Tan Son Nhut was rescinded after David Kennerly conveyed a report from Ken Quinn contradicting claims that the airport was under bombardment. In the end, Ford accepted Kissinger's argument that any interruption in the airlift would be misinterpreted by the North Vietnamese, who had held their fire, it was assumed, pending a complete American withdrawal.

In the West Wing tempers were beginning to fray. Four hundred Americans remained in the embassy compound. "Can't you tell him to get them out of there?" General Brown fumed to Henry Kissinger.

"Those are his bloody orders, goddammit!" Kissinger exploded. "We can't tell him how to load his helicopters."

Neither man had to identify "him." As darkness enveloped Saigon, Martin said he needed thirty choppers "damn quick." Using every argument he could conjure up to prolong the airlift he had resisted for so long, he messaged the White House: "Perhaps you can tell me how to make some of these Americans abandon their half-Vietnamese children, or how the president would look if he ordered this." What had been intended as a surgical extraction, lasting two and a half hours, stretched into an eighteen-hour marathon. It was still going on shortly before ten a.m. Washington time when the president took his seat at a meeting of the Cabinet. One factor extending the operation was the growing number of Americans flocking to the embassy—"They are coming out of the woodwork." Nevertheless, said Ford, by avoiding unnecessary panic and practicing teamwork when most needed, the US appeared poised to "come out of a tragic situation as well as possible."

His assessment would be roundly disputed on Saigon's anarchic streets.

From the Cabinet Room, Ford returned to the Oval Office for a discussion of the intractable Middle East with Jordan's visiting King Hussein. The king feared the repercussions of Vietnam on US foreign policy. Americans were "sick and tired of war," Ford conceded. "But in a crisis we will stand up."

Noon came and went. Midnight in Saigon. A somber Ford brought congressional leaders up to date in a hastily arranged session unleavened by small talk. Concerns were voiced about the number of Vietnamese who might be resettled in the United States, one lawmaker observing that fifty thousand represented the outer limit of what could be absorbed. No one in the room disputed his assertion.

A one o'clock press briefing was postponed. Forty-five minutes later Schlesinger called to report there were eight choppers in the air, enough to bring out 450 of the 750 reportedly still at the embassy.

"I would say God-damnit this is the end," he snapped.

Ford sympathized with his overwrought Cabinet officer. Yet he was as reluctant to order an end to the operation as he had been to inaugurate it. Not until 3:27 a.m. Saigon time did he authorize a final relay of nineteen choppers to rescue the remaining Americans and US Marine guards and Martin himself. An hour later, illuminated by firefights in the streets below, a CH-46 Sea Knight piloted by Captain Gerald Berry, making his thirty-third flight since the operation began, set down on the embassy roof.

"I'm here to get the Ambassador," said Berry.

"Well, the Ambassador isn't coming."

"Really? I'm supposed to get him."

Berry reluctantly loaded a waiting group of Vietnamese.* He then re-

---

* Kissinger occasionally lost his temper during this most stressful of operations, but not his acerbic wit. He paid mock tribute to Ambassador Martin for getting "five hundred of his last hundred evacuees out."

peated his orders: he wasn't leaving without Martin. Three magical words—
"The President sends . . ."—produced the ambassador with his overnight
bag and black poodle, Nitnoy.

At 4:58 *Lady Ace 09* lifted off. "Tiger, Tiger, Tiger," Berry said into his
radio. "The tiger is out of his cage." Unfortunately, this was misinterpreted
up the chain of command to mean that the evacuation of *all* Americans
was complete. In the White House briefing room, Ron Nessen read a brief,
oddly impersonal statement from the president commending Ambassador
Martin and the crews responsible for the largest helicopter evacuation on
record. "This action closes a chapter in the American experience. I ask all
Americans to close ranks; to avoid recrimination about the past, to look
ahead to the many goals we share and to work together on the great tasks
that remain to be accomplished."

A few minutes later Kissinger appeared with Nessen in the auditorium
of the Old Executive Office Building. "The first thing I said at the briefing was, 'All the Americans are out of Saigon,'" Nessen recalled. Waiting
offstage, Brent Scowcroft had news no one wanted to hear. Nessen had
jumped the gun. A rear guard from the Ninth Marine Amphibious Brigade remained in Saigon. Their plight sparked heated discussion among the
president's aides, Kissinger railing against "the most botched-up, incompe-
tent operation I have ever seen" and Nessen adopting a silence is golden
attitude. Why call attention to an unintentionally delayed rescue mission?
Don Rumsfeld answered the question. "This war has been marked by so
many lies and evasions that it is not right to have the war end with one last
lie." And so Nessen returned to the briefing room to correct the record.

As he spoke hundreds of Vietnamese, frantic at their imminent abandonment, forced their way into the now-vacated Saigon embassy. The handful
of remaining marine guards, protectors in need of protection, withdrew to
the roof, pausing only long enough to drop tear gas and smoke grenades on
the throng surging up the stairwell. Using fire extinguishers and wall lockers to barricade the rooftop door, the marines hoisted themselves aboard a
waiting marine CH-53 chopper just as several of their pursuers scrambled
onto the helipad. One man got close enough to throw himself at the wheels
of *Swift 22*, the last of 682 sorties that evacuated 1,373 Americans and
5,595 Vietnamese and third-country nationals.

The time in Saigon was 7:53 a.m. One minute later Secretary Schlesinger
called the president to convey news that had eluded four of Ford's predecessors. Excusing himself from a stag dinner he was hosting for King Hussein, Ford slipped next door to the office of Chief Usher Rex Scouten. In
this nondescript setting, almost twenty-five years to the day since Harry

Truman created the Defense Attaché Office to oversee US aid to French colonial forces battling the Viet Minh, America's longest war reached its agonizing conclusion. Putting down the phone, Ford returned to the State Dining Room to toast his Jordanian visitor. Before he turned in for the night a Vietcong flag was raised over Saigon's presidential palace.

Meanwhile the unofficial evacuation was gathering momentum. Skies over the South China Sea were crowded as pilots from the now-defunct South Vietnamese Air Force landed their machines on whatever US Navy ship had deck space to accommodate them, and some that did not (at least forty-five Hueys were pushed overboard to make room for incoming aircraft). Other airmen ditched their choppers and were fished from the water. The same Task Force 76 reported taking in twenty-two thousand (a figure shortly raised to forty-five thousand) South Vietnamese refugees in sampans and fishing boats, on high-sterned junks and homemade rafts. Simultaneously a convoy of twenty-six vessels carrying eighteen thousand escapees left Phu Quoc Island under a US Navy escort. Refused permission to enter Philippine territorial waters under the flag of South Vietnam, the armada halted long enough for command of each ship to be transferred to an American officer.

At the White House, reporters demanded to know under what authority Ford had evacuated South Vietnamese nationals. Ron Nessen was reminded of all the news executives and individual journalists who had besieged his office for weeks seeking Washington's help in bringing out their Vietnamese employees. When the questions persisted, Nessen offered a direct presidential quote: "I did it because the people would have been killed, and I'm proud of it," said Ford.*

* Unmentioned was a recent exchange involving Ford and William Small, president of CBS News. Following a nationally broadcast television interview with Walter Cronkite and two of his CBS colleagues, Ford had invited the participants to join him for a drink in the family quarters. Availing himself of the president's hospitality, a distinctly unawed Small turned to Ford and remarked, "Well, at least we got *our* people out." David Kennerly, *Shooter* (New York: Newsweek Books, 1987), 176.

22

STARTING OVER

—◦◦◦—

By resolute and skillful leadership in the Mayaguez crisis,
President Ford has seemingly moved from the doldrums
of Hooverdom to . . . the vigor of Harry Truman.

—C. L. SULZBERGER, *NEW YORK TIMES*, MAY 17, 1975

FORD MAY HAVE steeled himself to the loss of Indochina, but there were limits to his acceptance. These were breached the morning of May 1, when Ron Nessen handed him some AP copy ripped from a clattering teletype machine. The House of Representatives had voted to reject $327 million for the evacuation and eventual resettlement of an estimated seventy thousand South Vietnamese refugees.

"Those sons of bitches," Ford exploded.

"I'd never heard him curse all the time I'd known him," Nessen reflected. At Ford's request he channeled the president's anger into a formal statement to be read before television cameras that afternoon. In it Ford said he was "saddened and disappointed" by the House's action. Contradicting "the values we cherish as a nation of immigrants . . . It reflects fear and misunderstanding rather than charity and compassion." He referenced earlier waves of politically dispossessed immigrants, Hungarians and Cubans who had found sanctuary in the shadow of the Statue of Liberty. "Now, other refugees have fled from the Communist takeover in Vietnam. These refugees chose freedom. They do not ask that we be their keepers, but only, for a time, that we be their helpers."

His position was not a popular one. In the same week that Saigon fell, government statisticians revealed a US unemployment rate of 8.9 percent, a post–World War II record. Amid widespread grumbling about foreigners taking scarce jobs from American workers, Senator Robert Byrd (Democrat, West Virginia) insisted on screening out as "undesirables" any "barmaids, prostitutes and criminals" in the refugee ranks. On the left, Representative John Conyers of Michigan asked whether federal dollars should be spent "on Vietnamese refugees or should we spend them on Detroit 'refugees'?"

George McGovern, as critical of refugee policy as he had been of the war that forced thousands of its victims to seeks asylum on US soil, declared, "The Vietnamese are better off in Vietnam."

According to a Gallup poll, 54 percent of Americans agreed with Byrd, Conyers and McGovern. Outside the refugee camp at Fort Chaffee, Arkansas—referred to by resentful locals as "Gooksville"—one angry protester reminded a reporter that the local climate was much colder than what the newcomers were accustomed to. "With a little luck, maybe they'll take pneumonia and die." Florida's congressional delegation pressured the military to reduce the refugee allotment at Eglin Air Force Base from 25,000 to 2,500. This necessitated the opening of a fourth camp, at Fort Indiantown Gap in Pennsylvania Dutch country.

Offsetting the ugly taunts were spontaneous gestures of solidarity from across the political spectrum. Assuring Ford that "your humanitarian policy represents the best in our constitutional tradition and we support it very strongly," San Francisco mayor Joseph Alioto offered help in negotiating the use of federal properties in his city as group homes for refugees who wished to maintain their sense of community. James Longley, Maine's independent governor, contacted the White House to volunteer his state's summer-camp facilities as temporary housing for displaced children. Ford received vocal support from Pope Paul VI, the American Jewish Congress and the Protestant-based World Church Service. The American Civil Liberties Union applauded the president's leadership, while disputing administration plans to disperse refugees across the United States as a possible infringement of their constitutional rights.

The most politically potent endorsement came on May 3, when the AFL-CIO's George Meany weighed in. "Yes, we have 9 percent unemployment," Meany told reporters. "If this great country can't absorb another 30,000 to 40,000 and find them jobs, we're denying our own heritage." Meany's position stood in contrast to others who held back out of political timidity or rank hypocrisy. "It just burns me up, these great humanitarians," Ford told a group of Republican congressional leaders on May 6. "They just want to turn their backs. We didn't do it to the Hungarians, we didn't do it to the Cubans and, damn it, we're not going to do it now."

His combative mood was on display that evening in a nationally broadcast news conference. He reminded viewers that the United States, itself a nation of immigrants, had traditionally opened its doors to victims of foreign oppression. No doubt the prospect of competing for scarce jobs with tens of thousands of newcomers intensified worker concerns. But a majority of those "either here or on their way" were children. "They ought to be given an opportunity," said Ford. After all, the United States had absorbed far greater numbers of Cubans fleeing Castro's tyranny, and almost as many Hungarians uprooted by their failed revolution in 1956.

"The president's 'damn mad' reaction to what was largely meanspirited racist opposition," concluded the *New York Daily News*, "coupled with his powerful appeal to the essential decency and generosity of Americans . . . appears to have brought the Congress around." Spurred by Ford's public opinion offensive, the rapid exhaustion of funds and a corresponding increase in refugee numbers, lawmakers moved with uncharacteristic speed to enact the Indochina Migration and Refugee Assistance Act of 1975. Besides an appropriation of $405 million, the legislation signed on May 23 created an Inter-Agency Task Force on Indochinese Refugee Resettlement chaired by Julia Taft, a thirty-two-year-old former White House Fellow at the dawn of a remarkable career directing the American response to foreign conflicts, earthquakes, floods—even a plague of locusts in Ethiopia.

At one point Taft confronted a hostile General Leonard Chapman Jr., head of the Immigration and Naturalization Service, over his prediction that "these people are going to be wearing loincloths and carrying spears." Chapman recanted his opposition after greeting the first plane load of carefully clothed and suited newcomers, many from Vietnam's professional and business classes. California governor Jerry Brown was less easily persuaded. "We have enough people in California," the first-term governor informed Taft, "and we don't need any Vietnamese." Taft threatened to go on TV and denounce the governor for his refusal to let any church, synagogue, military family or humanitarian agency in the Golden State provide assistance to the victims of a Communist regime in Asia. She got her way.

In the battle of the political dynasties, score one for Taft over Brown. "She did one fabulous job," said Ford. By the first week of June 23, 742 refugees had been resettled in the United States, and 1,969 released to other countries. Another fifty thousand temporarily housed on Guam were soon reduced to half that number. On a sweltering day in August 1975, Ford received an enthusiastic welcome from twenty thousand refugees lining the main street of Fort Chaffee's resettlement camp. In addition to visiting a hospital ward, he met briefly with Nguyen Van Nhan, a former helicopter pilot who, together with his wife and four children, was preparing to take up residence in the president's hometown of Grand Rapids. By the time the camp at Fort Chaffee closed that December, approximately 130,000 Vietnamese had become Vietnamese-Americans.

Operation New Life foreshadowed even greater waves of Vietnamese boat people, including thousands of ethnic Chinese who fled the repressive government in Hanoi. In 1979 a United Nations conference in Geneva paved the way for several times the original migration of Vietnamese refugees to find new homes in the US and other Western nations. When pressed long afterward to explain his actions, Ford offered a simple rationale: "My conscience would have bothered me the rest of my life if we hadn't done that job."

A MERICANS REELING FROM the fall of Saigon, coupled with the highest unemployment rate since 1941, might be forgiven for believing they had exchanged their long national nightmare for what columnist George Will branded the Unromantic Presidency, a skin graft that had yet to take. Nine months after his East Room inaugural, Gerald Ford was broadly acknowledged for his personal integrity and political moderation—admirable qualities to be sure, especially in contrast to his recent predecessors, though hardly likely to inspire ardent loyalty. To millions of his countrymen, Ford was a character actor attempting a leading man's part. If he had a base, he rarely played to it. Indeed, his avowed aim of restoring public confidence in government institutions shredded by Vietnam and Watergate was incompatible with partisan zeal. To social conservatives in his own party the president appeared a room-temperature centrist stamped indelibly with the consensus-seeking brand of Capitol Hill. At the other end of the political spectrum, lunch pail Democrats, no less than their McGovernite brethren, glimpsed in Ford's economic policies the shade of Herbert Hoover.

The president's self-possession amid the Vietnam debacle had not gone unnoticed, any more than his principled refusal to abandon those orphaned by Hanoi's victory. Yet it was by no means certain these would be enough to compensate for the shock of Uncle Sam's expulsion from Indochina, or to avoid recriminations of the "Who Lost China?" variety that had so poisoned the well of 1950s politics. It wasn't just regional allies like Japan and the Philippines who were eager to distance themselves from Washington. "You sold out Vietnam and Cambodia," Syrian dictator Hafez al-Assad remarked cuttingly to Henry Kissinger. "Why should we not suppose you will also sell out Israel?" Hoping to refute such assumptions, Ford welcomed leaders from Europe, Asia and the Pacific Basin to the White House for some revolving-door diplomacy. Acknowledging the "traumatic experience" his country had been through, he reassured Australian prime minister Gough Whitlam that the US would honor its commitments, "and, to the extent that the President can, I will keep Congress' feet to the fire."

On May 3 the president traveled to Norfolk, Virginia, for the commissioning of the USS *Nimitz*, prototype of the next generation of oceangoing flattops. "We will continue to be strong," he told the crowd assembled in Norfolk's shipyard. "We will keep our commitments, and we will remain a great country." The imposing carrier from whose deck Ford spoke embodied the faith of its builders in high tech as the determining factor in modern warfare, a conviction unshaken by the rank failure of asymmetrical firepower in Vietnam. Within days the theory would be retested in the *Mayaguez* affair, a nautical mismatch pitting the US Navy, Air Force and Marines against Cambodian pirates dressed in black pajamas and clutching AK-47s to their adolescent chests. Little about the hastily improvised operation aimed at freeing the captive vessel and its crew went according to plan. Before it was over, more Americans were to lose their lives than gain their freedom. Still, the result, a facsimile of victory, was sufficient to raise post-Vietnam spirits and, with them, Gerald Ford's standing in the polls.

I DON'T KNOW why it is"—Ford sighed—"but these crises always seem to occur on Monday." It was the twelfth of May, a few minutes before eight a.m. During the course of his regular CIA briefing the president learned of events transpiring overnight in the shipping lanes off Cambodia's southwest coast, where vessels flying the flags of South Korea and Panama had been harassed in recent days by forces answerable to the enigmatic new regime in Phnom Penh. A little after three a.m. Washington time, the American merchant ship *Mayaguez* had come under a hailstorm of machine-gun fire as it plied the oil-rich waters around Poulo Wai, a rocky islet variously claimed by Cambodia, Vietnam and Thailand.

The latter-day tramp steamer was no stranger to military confrontation. Christened *White Falcon* in 1944, the future *Mayaguez* had performed convoy duty in the North Atlantic while Jerry Ford and his *Monterey* crewmates battled Pacific typhoons and Japanese kamikazes. In the three decades since, the onetime navy escort had undergone a shipyard makeover from which emerged a first-of-its-kind cargo carrier. Operating with a crew of thirty-nine, the repurposed *Mayaguez* transported truck-sized containers from Hong Kong to Singapore, with an interim stop at Sattahip, Thailand. It had been traveling this route at a slightly arthritic 12.5 knots when it was attacked by Cambodian gunboats. Discussing the situation with his secretary of state, Ford couldn't know that the *Mayaguez* and its crew were presently anchored off Poulo Wai, a quick-witted Captain Charles T. Miller having persuaded his captors that to proceed sixty-five miles to the mainland port of Kompong Som without functioning radar was to risk disaster.

"I have some bad news for you," the president told Ron Nessen. After relating what little he knew about the ship's disappearance, Ford gave Nessen a taste of the decision-making already weighing on him. "What would you do?" he asked rhetorically. "Would you go in there and bomb the Cambodian boat and take a chance of the Americans being killed? Would you send helicopters in there? Would you mine every harbor in Cambodia?" A noontime meeting of the National Security Council supplied few answers. CIA director Colby guessed that the *Mayaguez* had been taken to Kompong Som (the former Sihanoukville), greatly lessening chances for an early rescue attempt. Speaking for the Pentagon, Jim Schlesinger speculated that the ship's seizure might be "a bureaucratic misjudgment" by a Khmer government whose long-standing hostility toward its South Vietnamese neighbor transcended Communist solidarity.

"We should not look as though we want to pop somebody," observed Henry Kissinger, "but we should give the impression that we are not to be trifled with." Vice President Rockefeller underlined Kissinger's line in the sand, if that is what it was. "I remember the *Pueblo*," said Rockefeller. No one present needed reminding of the US Navy spy ship held for eleven months in 1968 by a North Korean government that had inflicted maximum humiliation on Washington while roundly abusing the ship's crew. "A violent response is in order," the vice president continued. "The world should know that we will act and that we will act quickly . . . If they get any hostages, this can go on forever."*

"I can assure you that, irrespective of the Congress, we will move," Ford promised.

That the *Mayaguez* crisis, first of its kind since passage of the War Powers Act, presented an opportunity to dispel his country's post-Vietnam funk was not lost on Ford. Unfortunately there was little the United States could do immediately to enforce its rights. Two weeks after the evacuation of Saigon, Task Force 76 had long since disbanded, its ships and helicopters repatriated to distant ports and home bases. At the suggestion of Secretary Schlesinger, Ford directed an amphibious brigade from the Third Marine Division, 1,100 strong, to be flown from Okinawa to Utapao, Thailand, over Thai protests. Orders were dispatched to the aircraft carrier *Coral Sea*, en route to Australia, to reverse course and head for the Gulf of Siam. From the US base at Subic Bay in the Philippines, Ford summoned the destroyer

---

* It is quite possible that Rockefeller's aggressive stance was influenced by his searing experience in September 1971, when a prison uprising at New York's Attica State Correctional Institution put at risk the lives of more than forty hostages, and he ordered the facility retaken in a quasi-military operation that resulted in the deaths of thirty-three inmates and ten correctional officers. For many the incident left Rockefeller's reputation permanently scarred.

escort *Harold E. Holt*, the guided missile destroyer *Henry B. Wilson*, and several other ships. At the same time a trio of P-3 naval reconnaissance planes was tasked with locating the *Mayaguez* and keeping it under constant surveillance.

Ford asked the Pentagon to produce a list of military options for overnight review. On the diplomatic front stiffly worded protests were drafted— but to whom should they be delivered? Washington had neither formal nor back-channel means of contacting the hostile regime in Phnom Penh. Playing a long shot, the State Department invited Huang Chen, head of the Chinese Liaison Office in Washington, to convey a message to Cambodian authorities. When he refused, a similar request was routed through its US counterpart in the Chinese capital. This note was also returned, presumably unopened, by Chinese officials.

Shortly after the NSC meeting broke up, Ron Nessen appeared in the White House pressroom to relay the president's denunciation of "an act of piracy." Failure to release the *Mayaguez* and its crew unharmed, warned Nessen, would have "the most serious consequences." Disdaining the theatrics reporters had come to expect from other administrations in times of crisis, Ford made a point of adhering to his regular schedule. Henry Kissinger left town that same afternoon to honor speaking commitments in St. Louis and Kansas City.

T WICE THAT EVENING Brent Scowcroft telephoned the president with conflicting claims as to the whereabouts of the *Mayaguez*. Ford, no stranger to the fog of war, drew on his naval officer's training for perspective. So he wasn't surprised when Scowcroft rang him a third time at 2:23 a.m., seeking authorization for F-4 fighter planes to strafe Cambodian patrol boats escorting the captive *Mayaguez* to Kompong Som. There, it was feared, the American crewmen might vanish into urban captivity and the dreaded *Pueblo* scenario be reenacted. Moments later Scowcroft phoned yet again, in effect instructing his weary chief to forget what he had just been told—the *Mayaguez* was now said to be anchored off Koh Tang Island, a densely jungled dot on the map halfway between Poulo Wei and the mainland.

This was confirmed by Secretary Schlesinger in a lengthy telephone conversation with the president shortly before six o'clock that Tuesday morning. With late reports suggesting at least some of the *Mayaguez* crew were being held there, Ford told Schlesinger that he wanted the island "quarantined." Even now, however, the plight of the *Mayaguez* did not monopolize his attention. That afternoon Ford listened as New York City mayor Abe

Beame and the state's governor, Hugh Carey, pleaded with him for a $1.5 billion federal loan guarantee to avert municipal bankruptcy. Countering their appeal, Vice President Rockefeller insisted that Albany had the resources to provide Beame the financial lifeline he was seeking. Certainly Ford was persuaded. His "Dear Abe" letter, dated May 14, left the state no option but to advance Beame's city hall $200 million from its own depleted coffers. But it was a stopgap solution at best, a tourniquet applied in haste to keep America's greatest city from bleeding out.

Ford had barely seen off Beame and Carey when, a little after six p.m., he learned from Secretary Schlesinger that a helicopter carrying twenty-three air police and crew bound for Utapao had crashed shortly after departing the US Air Force base in Nakhon Phanom, Thailand, near the border with Laos. There were no survivors. Ford was still absorbing the news as members of his staff began phoning leaders of both parties on Capitol Hill to notify them of measures being taken to interdict any craft sailing to or from Koh Tang.* The president's dinner with Betty was interrupted by word that a Cambodian patrol boat had been sunk while trying to reach the mainland from Koh Tang. Three other boats had turned back under withering fire from A-7 combat planes.

An hour later more startling intelligence was received, involving yet another vessel closing in on Kompong Som. "There may be some Caucasians held in the front," Brent Scowcroft told the president. "The pilot thinks he can stop it without sinking it." When targeted fire from an A-7 failed to halt the boat's progress, Ford had to decide whether to put at risk the very crew members he hoped to liberate. The atmosphere was tense, bordering on combative, when the NSC reconvened at 10:40 p.m. Why, Ford inquired, had it taken so long to implement his order to quarantine Koh Tang? Schlesinger assured him that it had been transmitted within thirty minutes.

"I have got to get the word out," Scowcroft pressed. "What should I tell them?"

"Tell them to sink the boats near the island," said Ford. "On the other boat, use riot control agents or other methods, but do not attack it."

In the end the suspect vessel gave its pursuers the slip.

Like a latter-day officer of the deck, Ford inventoried the forces assembling to reclaim the *Mayaguez* and its crew and inflict a punishing blow

---

* Select lawmakers were tipped off to a possible marine assault to free the *Mayaguez*. One early plan for a risky helicopter landing on the ship was abandoned after fears were expressed that the containers on board might not accommodate the choppers' weight.

against the Cambodian regime held accountable for their capture. The loss of the CH-53 transport helicopter earlier in the day left only eleven choppers to fly marines to and from Koh Tang Island where, according to a Cambodian deserter, a mere sixty lightly armed Khmer Rouge fighters would oppose their landing.* (The CIA pegged their number at a hundred or more. Both estimates were to prove disastrously far off the mark.) Further complicating the situation, the US chargé d'affaires in Bangkok, Thailand, had promised that country's rulers no Thai-based American planes would take part in a *Mayaguez* recovery attempt without their approval.

Thai sensitivities, it is safe to say, ranked low on the list of Henry Kissinger's priorities. Just returned from his Midwestern trip, a characteristically expansive Kissinger said he wasn't thinking of Cambodia at the moment, "but of Korea and of the Soviet Union . . ."

"We have the right of self-defense," a stern-faced Phil Buchen interjected, "but only self-defense."

"I think we should do something that will impress the Koreans and the Chinese," Kissinger reiterated.

Jim Schlesinger, less interested in demonstration projects than he was in the narrower mission of recovering the *Mayaguez* and its crew, objected to the use of B-52s against targets on the Cambodian mainland. Citing the controversial part they had played in the skies over Indochina, he said the big Flying Fortresses were "a red flag" on Capitol Hill. Ford postponed a final decision until the morrow, when the *Coral Sea* and its warplanes would draw within range of Koh Tang Island. Meanwhile separate appeals were made to United Nations secretary general Kurt Waldheim and to the UN Security Council. These smacked of a public relations gesture, given their proximity to a final NSC meeting on Wednesday afternoon at which Secretary Schlesinger argued for carrier-based F-4s and A-7s in preference to the Vietnam-tainted B-52s parked on distant Guam. The *Coral Sea* alone boasted eighty-one aircraft, said Schlesinger, some equipped with laser-guided smart bombs. The hawks around the table—Vice President Rockefeller, Secretary Kissinger and Brent Scowcroft—contended for the B-52s as a more potent deterrent to future aggression.

Ford's inclination was to go with the hard-liners. "Henry and I felt that we had to do more," he explained in his memoirs. As for the Cambodians, "We wanted them to know that we meant business." At this point an un-

---

* The actual number of available choppers was eight, the other three being assigned to convey a second marine contingent onto the deck of the USS *Holt* in preparation for their over-the-side assault on the captive American vessel.

likely voice joined the conversation. As David Kennerly later readily conceded, as White House photographer he did not have a seat at the Cabinet Room table. Even so, "I was very concerned about the B-52 part of this . . . they're going to blow the crap out of these people, is what it boils down to." Thus motivated, the irrepressible Kennerly confronted the president and his military planners.

"Has anyone considered that this might be the act of a local Cambodian commander who has just taken it into his own hands to halt any ship that comes by?" he asked. "Has anyone stopped to think that he might not have gotten his orders from Phnom Penh?" Assuming his hunch was correct, Kennerly pointed out, "you can blow the whole place away and it's not gonna make any difference." In any event, said Kennerly, it was a mistake to think of Cambodia, by whatever name its revolutionary overseers might adopt, as a traditional government, like, say, France. "I was in Cambodia just two weeks ago, and it's not that kind of government at all."

Because Kennerly's intervention was edited out of the official transcript of the meeting, we are dependent on his—and Ford's—reconstruction of what happened, and of what happened next. Not only did Ford accept Schlesinger's reservations about the use of B-52s; he also embraced Kennerly's bombing restraint over the more extensive raids urged by Kissinger.

John Hersey had been right to suspect subtleties of character and outlook lurking behind the president's Good Old Jerry persona. Practicing decency as the coefficient of self-interest, Ford was at once a Cold Warrior for whom American bombs offered a justifiable response to foreign piracy *and* David Kennerly's self-restrained statesman who told his young friend, "I'm not going to make innocent Cambodians pay for us losing Vietnam."

That was not all he said.

"Dave, I really appreciated your input there, but next time, don't open your mouth in one of these meetings. Say something to me privately."

A s US forces acting on his orders prepared to retake the *Mayaguez* and rescue its crew, Ford played diplomatic toastmaster at a black-tie working dinner for Dutch prime minister Johannes den Uyl. Den Uyl had already bent the president's ear for an hour that morning contrasting Dutch generosity toward the Third World with the grasping inequities of American-style capitalism. Under the circumstances Ford probably welcomed an excuse to push dinner back to eight thirty, the delay occasioned by a Cabinet Room briefing for congressional leaders. Here Senator Robert Byrd queried the president over his failure to more fully consult the men around the table before deciding to bomb Cambodian targets. "Why didn't you at least give them a chance to express their reservation?" he demanded.

"We have a separation of powers," Ford replied evenly. "The President is the Commander-in-Chief so long as he is within the law. I exercised my power under the law and I complied with the law. I would never forgive myself if the marines had been attacked by 2,400 Cambodians." He concluded the discussion by inviting his guests to join him in prayer.

Before leaving for the residence to change into his tuxedo, Ford was informed by Brent Scowcroft that the USS *Holt* with its forty-eight marine raiders had drawn alongside the captive *Mayaguez* near Koh Tang Island. More distressing reports reached him of a larger marine force encountering unexpectedly fierce resistance from Khmer forces on Koh Tang. Acting on faulty intelligence, US military planners had ruled out any preliminary bombardment of the island lest it endanger *Mayaguez* crew members who might be held there. This allowed a force of perhaps two hundred seasoned Cambodian fighters to fortify positions from which they were able to rake the attackers with machine-gun fire, mortar shells and rocket-propelled grenades. Five of the first eight troop-carrying US helicopters to approach the island were destroyed; three others severely damaged. Fifteen marines died before they could establish a beachhead. Though modestly reinforced by subsequent landings, the small force of leathernecks on Koh Tang would be outmanned and outgunned throughout a savage daylong firefight.

NINE THOUSAND MILES away, Ford was, by his own admission, "totally preoccupied" as he waited at the North Portico for the car bearing den Uyl. He could hardly be anything else, having learned moments before of a Cambodian radio broadcast, hurriedly translated by the CIA, that combined turgid anti-American propaganda with a last-minute offer to release the *Mayaguez*. No mention was made of the ship's crew. In the State Dining Room, den Uyl and the other guests were spooning their cucumber soup when, a few minutes before nine, Ford pushed back from the table to take a phone call in the nearby usher's office. There he learned from Henry Kissinger that a marine boarding party had clambered onto the *Mayaguez*, only to find it a ghost ship devoid of crew or captors.

Still in the dark regarding the crew's whereabouts, Ford instructed Kissinger to proceed with bombing raids on mainland targets. At the same time he authorized an olive branch of his own. With conventional channels of communication blocked, it was decided to enlist the international press corps. Before a hastily summoned group of reporters in the White House pressroom, Ron Nessen raced through the presidential statement as newsmen struggled to keep pace with their pencils.

"As you know, we have seized the ship. As soon as you issue a statement that you are prepared to release the crew members you hold unconditionally and immediately, we will promptly cease military operations."

"Go file!" Nessen practically shouted at the assembled journalists. The TV networks duly interrupted their regular programming to report Ford's offer. The wire services followed suit. In the State Dining Room, a visibly irritated Prime Minister den Uyl was unmollified by Kissinger's appearance in time for the dessert course. The Dutchman's sulky attitude "annoyed the hell out of me," Ford said afterward. It was all he could do to conceal his feelings of relief when, at a quarter to eleven, den Uyl's car at last exited the Pennsylvania Avenue gate and vanished into the night.

About the same time Ford was seeing off his Dutch guest, a reconnaissance plane overflying the Gulf of Siam spotted a Thai fishing boat carrying an undetermined number of Caucasians. Moments later the vessel was intercepted, and its human cargo offloaded by the *Henry B. Wilson*. The transfer occurred too late to avert successive waves of US carrier–based bombers from hitting Cambodian naval facilities, destroying seventeen planes on the ground and wrecking a nearby oil refinery and barracks complex. By this point events were moving too randomly for any truly coordinated response. At 11:08 Jim Schlesinger telephoned the president to report the recovery of at least thirty crew members from the *Mayaguez*. Seven minutes later he called back, raising the number to thirty-nine, the ship's full complement.

"They're all safe," Ford exclaimed, thrusting a fist in the air for emphasis. "We got them all. Thank God."

Outside the Oval Office a handful of presidential aides broke into cheers. Ford ordered a halt to further military operations, excepting those in support of the beleaguered marines on Koh Tang Island.* It was only natural that Ford should want the news announced from the White House. Too late; Pentagon spokesman Joseph Laitin, suspected all week of leaking to the press to make his bosses look good, had already scooped his commander in chief by tipping off the media about the crew's recovery. Laitin's act of lèse-majesté compounded feelings of grievance arising from the unauthorized cancellation of a fourth and final bombing raid, and the belated discovery that scarcely half the eighty-one carrier-based planes promised by Secretary Schlesinger took part in the operation.

---

* A directive modified at Kissinger's urging. "Let's look ferocious," he told Brent Scowcroft. "Otherwise they will attack us as the ship leaves."

Shortly after midnight, having changed out of his crumpled tuxedo into a business suit, Ford appeared in the White House pressroom to express his appreciation, and that of the nation, to the men whose valor and sacrifice he credited with saving their countrymen from *Pueblo*-style captivity. Afterward he signed off on a letter to Congress detailing his compliance with the War Powers Act. At 12:35 a.m. Ford said he was going home and to bed. Outside he paused briefly in the darkness to scan the first bulldozer scratchings for a South Lawn swimming pool on the site of Lyndon Johnson's dog kennels.

"Boy," mused Ford, "I wish that pool was ready now."

Returning to the residence, he was trailed by several staff members. Given the gravity of still-unfolding events, no one wanted to intrude on his thoughts.

Abruptly, Ford halted, looked over his shoulder and inquired, "How did the [Washington] Bullets do tonight?"

Thursday morning he slept an hour beyond his usual five fifteen wake-up call. On Koh Tang Island, besieged marines were gauging their chances for rescue by the handful of remaining helicopters. In the chaotic evacuation that took place mid–Thursday morning, Washington time, three of the marines would be left behind. Their names, along with those of thirty-eight others lost in the *Mayaguez* affair, would be the last added to the Vietnam War Memorial on the National Mall when it was dedicated in 1982.

Already the *Mayaguez* was, literally, yesterday's news. In its place Ford hosted a midmorning welcoming ceremony for the visiting Shah of Iran. In the privacy of the Oval Office His Imperial Majesty voiced admiration for the previous day's show of strength.

"We perhaps overreacted, to show the Koreans and others our resolve," Ford replied. It was the most revealing comment he ever made about the military demonstration in the Gulf of Siam.

THERE SEEMS TO be a feeling of joy that at last we have won one," observed the *Atlanta Journal.* "And indeed we have." Phone calls to the White House ran ten to one in favor of the military action, whose human cost in blood—eighteen Americans dead or missing, and fifty wounded, plus the twenty-three who perished in the Thai helicopter crash—would not be fully known for several days. Among the handful of congratulatory callers who got through to the president were Richard Nixon and Ronald Reagan. On Capitol Hill, Hubert Humphrey spoke for members of both parties in declaring that Ford had no choice but to act as he did when confronted with such flagrant violation of international law. Damn right, said Barry Goldwater. Had the president done nothing, Goldwater predicted, "every little half-assed nation is going to take shots at us."

The first post-*Mayaguez* Gallup poll showed a twelve-point jump in Ford's approval ratings, to 51 percent. Ford himself credited the ship's rescue for imparting "a whole new sense of confidence," though he didn't say to whom—himself or his war-weary countrymen. On the home front what might be labeled the *Mayaguez* effect paid immediate dividends as the House, by a five-vote margin, upheld his rejection of a $5.3 billion jobs package ("too much stimulus, too late"). When the same lawmakers compromised with Ford on farm price supports and restrictions on the strip mining of coal, it prompted demands from freshman Democrats for the overthrow of Speaker Carl Albert and his replacement by a more aggressive party leader.

On May 27, after Albert's majority—large but unwieldy—failed to meet its own deadline for the passage of energy legislation, Ford went on TV to lambast Congress in what one scholar called "the angriest speech of his presidency." Tearing pages from a calendar to dramatize congressional inaction, he reminded viewers that he had suspended his $1-per-barrel hike in fees on imported oil as a good faith gesture toward Capitol Hill. Now Ford said he would resume the monthly tariffs commencing June 1. He also vowed to accelerate the decontrol of domestic oil prices to incentivize conservation and the search for untapped energy reserves.

The next day Ford left for Brussels, and only the third meeting of NATO heads of state since the alliance was formed in 1949. In a last-minute outbreak of bipartisanship, Congress had strengthened his hand by rejecting a seventy-thousand-man reduction in American combat troops deployed overseas. The president got almost all the $27 billion he sought for research, development and purchase of advanced weaponry. This included funds for the proposed B-1 bomber and Trident submarine. Finally, the Senate by one vote reversed its ban on arms shipments to Turkey, thereby raising the prospect of a US-brokered solution to the Cyprus stalemate.

Ford's post-*Mayaguez* windfall boosted his diplomatic credibility no less than his domestic standing. It escaped nobody's notice when French president Valéry Giscard d'Estaing flew to Brussels just to have dinner with his American counterpart. Ford enjoyed less success with Prime Minister Vasco dos Santos Gonçalves of Portugal, where Marxist elements in the armed forces had consolidated their grip on power following the

April 1974 Carnation Revolution. "We are not a trojan horse in NATO," Gonçalves sputtered in response to Ford's insinuation to the contrary. He failed to convince the president, if Ford's remarks before the North Atlantic Council were any indication. Restating his belief in NATO as the "foundation" of US-European relations, Ford maintained that a successful alliance required "unqualified participation" by member nations. He could see no justification for "partial membership or special arrangements" (Portugal having already been evicted from NATO's nuclear planning group).

In the same speech Ford raised the delicate matter of Spanish integration into a Western fraternity that had always shunned the repressive regime of Generalissimo Francisco Franco. The aging dictator showed his gratitude the next day, when Ford arrived in Madrid for a twenty-two-hour stopover en route to Salzburg, Austria, and two days of talks with Egypt's Anwar Sadat aimed at renewing the peace talks broken off in March. Frail and shuffling, Franco brought out tens of thousands of Spaniards to cheer him and his American guest in their open car, flanked by a ceremonial guard of a hundred horsemen dressed in white tunics and silver helmets.

"What a lot of smiling people," Ford observed to the old man beside him.

"Young people are always smiling," Franco mumbled obscurely, "unless they are being poisoned by other people." Their formal discussions confined to thirty minutes on account of El Caudillo's declining health, it fell to Prime Minister Carlos Arias Navarro to press his country's case as an anti-Communist bulwark deserving of full NATO membership and a substantial boost in US recompense for the lease of three air bases and a naval facility at Rota. Before departing the Spanish capital, Ford squeezed in a forty-minute visit with Prince Juan Carlos, Franco's designated successor, who impressed the visitor with his desire to play a constructive role in any transition toward less authoritarian rule.

Ironically the images of Ford's European trip that would linger in public memory were not of Spain's infirm ruler for life, but of the athletic president whose awkward arrival at a rainswept Salzburg airport on June 1 made him a foil for physical comedians like *Saturday Night Live*'s Chevy Chase. "I was with him," Don Rumsfeld remembers, watching closely as Ford, "always a gentleman," took his wife by the arm. "Then an Air Force sergeant handed him an umbrella. He ended up going down a long slippery set of stairs, holding on to an umbrella and Betty instead of the rail." Losing his footing—and the heel from one shoe—on the rain-slick metal stairway, Ford fell the last five steps. He landed on one knee and both hands before quickly righting himself. Making light of the incident, an embarrassed Ford told his Austrian hosts, "I am sorry I tumbled in." But he knew images of his airport fall, broadcast around the world, would overshadow any diplomatic progress he might achieve in his talks with Sadat.*

"Rummy, I am so mad at myself," Ford confided on reaching his Austrian guesthouse. Characteristically he refused to pass the buck when courtiers blamed photographers for capturing his mishap.

"Of course they took the picture," said Ford. "If they hadn't they would have lost their jobs."

CONCEDING THE IMMEDIATE morale boost for a nation awash in self-doubt, the recovery of the *Mayaguez* proved to be less game changer than sugar high. Even before its effects wore off, thoughtful policymakers were speculating about the long-term consequences of the US defeat in Indochina. For Ford, the Brussels conference had offered the ideal forum in which to reassure nervous allies the United States had no intention of withdrawing into the isolationism of his youth. Yet it takes two to make an alliance. To partner with Washington made sense only as other nations perceived it to be in their self-interest.

Enter Anwar Sadat. Almost singlehandedly Sadat would reaffirm American influence in the world by insisting on American leverage in the Middle East, where Ford's country held "at least 99% of the cards." Within minutes of their first encounter in a fifteenth-century hunting lodge on the outskirts of Salzburg, Ford assessed the Egyptian president as "the type of man I could establish a great rapport with." Contributing to the upbeat atmosphere was some maladroit lobbying by the Rabin government and its friends on Capitol Hill. Ten days before he and Sadat dodged rain clouds on the shores of Lake Fuschl, Ford had received a strongly worded appeal to support Israel's latest request for $2.6 billion in military and economic assistance. Its Washington, DC, postmark fooled no one as to the origin of the letter, which bore the signatures of seventy-six members of the United States Senate.

Publicly Ford downplayed his resentment of this unsubtle intervention in domestic American politics. Around Sadat he was more forthcoming. Half the senators whose names were affixed to the letter hadn't bothered to read it, Ford claimed. In any event the senatorial petition would have "negligible" impact on his administration's Middle East policy. Quick to

---

* "Betty tripped me," Ford joked with congressional leaders on his return. "I went flat on my face in the rain and she walked off with the umbrella."

exploit the strategic opening, Sadat offered a full-throated denunciation of the Soviet Union, his chief weapons supplier until he booted fifteen thousand Soviet troops from Egyptian soil following the October 1973 war that left Israel astride the Sinai Peninsula. After lunch the leaders moved to Salzburg's Old Town and the sprawling archiepiscopal palace known as the Residenz. There Sadat asked if Ford had brought with him any fresh proposals from Tel Aviv. The president's negative response could hardly have come as a surprise. Having three-quarters of the United States Senate in their pocket gave the Israelis scant incentive to make concessions, even theoretical ones.

Now as in the failed spring talks, both sides were fixated on the strategically placed Gidi and Mitla passes east of the Suez Canal. Sadat needed to reestablish sovereignty over the passes to demonstrate progress, through peaceful means, toward his goal of regaining the entire Sinai. Almost as critical for his country's battered economy was unfettered access to the Israeli-occupied oil fields at Abu Rudeis on the Red Sea. A more conventional statesman would have played on Ford's emotions, accusing Israel of acting in bad faith and reminding the president that by snapping his fingers he, Anwar Sadat, could send UN peacekeepers packing. As a last resort he could have played the slightly soiled Russian card, implying that his eviction notice could always be rescinded.

In the event, Sadat did nothing of the kind. Hiding his disappointment, for a few pregnant moments he sucked on his pipe while marshaling his thoughts.

"All right," he said at last. "We are willing to go as far as you think we should go. We trust you, and we trust the United States."

At that moment doubts about the US and its post-Vietnam role were effectively stilled. Presenting himself as a risk-taker for peace, Sadat introduced the notion of a buffer zone around the Gidi and Mitla passes, with a handful of US civilian personnel keeping watch on troop movements as a kind of early warning system. Ford liked what he heard. For the proposal to have any chance of adoption, however, it must be publicly divorced from its author, in part to avoid Israeli rejection, but also to shield Sadat himself from the wrath of more militant Arab regimes as well as the Palestine Liberation Organization (PLO), for whom the slightest nod in the direction of the Jewish state was seen as a rank betrayal of the Arab cause.

Henry Kissinger, very much in his element, explained to Ford how they might persuade Rabin to take the bait when he visited Washington later that month: "You could suggest that he and I work out something that

Israel can support. I will then try to move him toward the Egyptian positions without telling him about them. Then you tell him to go home and see if he can sell it to the Cabinet, and if he can, I will go to Egypt to sell it." Questions about who might operate the proposed surveillance stations found Kissinger equally prepared. If "they"—the Israelis—"offer American manning, we can say we have to run it by the Egyptians." Assuming Rabin could get his Cabinet to approve Sadat's concept (masquerading as an exclusively American initiative), "we'd get it done before the Soviet Union and Syria get set for a Geneva conference."

There was no time to waste, for any gathering in Geneva was sure to be dominated by Israel's enemies, not excluding the terrorists of the PLO. Were the US to attend a Geneva conference on the Middle East, it would be to ratify, not negotiate, a peace agreement. This fact, too, would be withheld from Tel Aviv and other cities. There was no reason for Washington to show its diplomatic hand so early in the game.

O N JUNE 11, one week after his return from Europe, Ford welcomed Yitzhak Rabin to the White House for what diplomats euphemistically call a frank and useful exchange of views. The man before him may have lacked Sadat's charisma, but he yielded to no one in his tenacious defense of Israel's territorial integrity. As a negotiator Rabin would contest every kilometer, leaving Ford at times "exasperated . . . but as long as we got results, hard as it was, I respected him as a person who was fighting his battle." With dour eloquence the prime minister decried the emptiness of international guarantees and the risks of "a peace by diplomats and governments and not by people." Sadat's offer of nonbelligerency, for example—it could hardly be confused with any true normalization of relations.

"I don't want the Israelis to be like the Christians in Lebanon," Rabin told the president. "Egypt's the key," he insisted.

More than a phased military disengagement, the right kind of interim agreement with Egypt might herald a regional peace worthy of the name. "The problem is how much longer can the status quo be maintained without political movement," Ford replied. The only alternative to a Geneva conference "with all of its pitfalls as you suggest" was to move rapidly— "Within two or three weeks"—toward some kind of interim settlement. "And it has to be worked out before I get there," Kissinger interjected. At this point Ford and Kissinger executed the scheme, hatched at Salzburg, to introduce as their own Sadat's idea of a US-guaranteed buffer zone around the disputed passes. Kissinger expressed the hope that he might, in his separate discussions with Rabin, conjure up something "practical . . . which the prime minister could then put to his Cabinet and we could then put to the Egyptians."

Ford signaled his assent, but not so fast as to arouse suspicions. Seizing on a remark by Rabin, he reminded his visitor that whatever concerns other American presidents may have raised in Tel Aviv by outlining, even vaguely, the shape of an overall settlement, "for me to talk platitudes is not my style . . . I would intend to be specific in what I would announce." In other words, a Geneva conference might go even worse for Israel than Rabin feared.

"I think you ought to see if you and Henry can come closer."

"Let's talk it over," a wary Rabin concluded.

The bait set out, Ford escorted his visitor to his car.

APPROACHING HIS FIRST anniversary in the Oval Office, there was widespread agreement that the trappings of office had done nothing to inflate Ford's ego. Speechwriter Milton Friedman was meeting with the president one morning when White House secretary Nell Yates interrupted with news that Friedman's mother was on the phone about some emergency. Friedman asked Yates to tell her that he was with the president and would call her back.

"Milton, you'd better go take the call," said Ford. "It's your *mother*." A red-faced Friedman excused himself. A few minutes later he returned, more sheepish than ever.

"Was there anything serious?" Ford asked. Not really, Friedman responded. "My mother's toilet was stopped up."

The president's laughter was quickly replaced by a congressman's concern. "Well, for an older person that's very serious," he reminded Mrs. Friedman's dutiful son. "Did you call a plumber?"

Such incidents fed the popular belief that Gerald Ford was unchanged by the presidency. Yet Ford *had* become more guarded in his comments, his caution born of the biggest security leak of his presidency, in an off-the-record lunch he had hosted for several *New York Times* executives and reporters back in January 1975. It was just after the paper ran Seymour Hersh's front-page indictment of CIA wrongdoing. Ford had responded by establishing the Rockefeller Commission, which he vigorously defended against those around his lunch table who questioned its investigatory zeal. Disputing the idea, popular among journalists and many in Congress, that nothing in the intelligence field should be off-limits to public scrutiny, Ford said that since assuming office he had learned of past CIA misdeeds,

a complete inventory of which could damage the reputations of his predecessors going back to Harry Truman.

"Like what?" snorted *Times* managing editor Abe Rosenthal.

"Like assassinations," Ford blurted out.

Only Ron Nessen's steely enforcement of the ground rules to which they
had agreed, coupled with their own gentlemanly discretion, prevented the
Times men from sharing this revelation with their readers. That said, few
expected Ford's gaffe to stay secret for long. Among those working dili-
gently to get to the bottom of what he called the "Son of Watergate" was
CBS newsman Daniel Schorr. Now, propelled on a wave of rumor and his
own dogged reporting, Schorr made an appointment to see Bill Colby.

"I understand from the president that there's been assassinations going
on in this country," Schorr told the DCI. "Well, no, not in this country,"
a poker-faced Colby replied. Pressed for details, Colby said he couldn't talk
about it.

On the last night of February viewers of the *CBS Evening News* were
informed by Schorr that Ford had warned "associates"—a curious euphemism for *New York Times* editors chafing against the restrictive code of secrecy enforced by the White House—"That if current investigations go too
far, they could uncover several assassinations of foreign officials in which
the CIA was involved." In the resulting furor, Ford reluctantly broadened
the investigatory mandate of the Rockefeller Commission. On March 5 he
warned Frank Church and other senatorial inquisitors against reducing the
CIA "to the level of a newspaper clipping and filing agency." With the fate
of covert operations and, some thought, the agency itself hanging in the
balance, Bill Colby seemed reluctant to be part of White House damage
control. His subsequent cooperation with congressional investigators inevitably raised questions.

"He would come home and I'd say, 'You're spilling the beans. You're
volunteering information they're not even asking you about,'" recalled Colby's son Carl. "He said, 'So they'll learn about these secrets. Place needs a
housecleaning.'"

None of this sat well with Nelson Rockefeller, entrusted with the politically thankless task of investigating CIA misdeeds, and proposing reforms
that would pass congressional muster. Like Ford a product of the World
War II generation, steeped in Cold War polarities and the deference shown
to spymasters like Allen Dulles and John McCone, Rockefeller was even
more reluctant than the president to air the agency's dirty laundry. Dismayed by Colby's latest information dump, he told the DCI, "We realize
there are secrets that you fellows need to keep, and so nobody here is going

to take it amiss if you feel there are some questions you can't answer quite as fully as you seem to feel you have to.'"*

The summer of 1975 was not a happy time for the vice president. In the Senate his support for an extended Voting Rights Act, coupled with a series of parliamentary rulings he made that reduced the number of votes required to break a filibuster, rekindled conservative opposition to Rockefeller and the president who selected him. Fresh from a rancorous dispute with Don Rumsfeld over control of the White House Domestic Policy Council, Rockefeller blamed the chief of staff for the latest hot potato tossed his way.

"This was another way of chopping my head off," he said later, "and of getting me out there where I was the one who was putting the finger on the Kennedys" for their alleged involvement in the deaths of South Vietnam president Ngo Dinh Diem and the Congo's Patrice Lumumba. By no means immune to conspiracy theory, Rockefeller confided to Ford his belief that the Castro regime, seeking revenge for US attempts on the life of the Cuban dictator, was somehow complicit in JFK's assassination. Inevitably the exchange dredged up memories of the Warren Commission a decade in the past. Just as Ford's belief in Oswald's guilt was unswayed by the new evidence, so was his faith in covert action as a necessary component of American intelligence. (This did not prevent him in May 1975 from pulling the plug on Operation Shamrock, initiated by Harry Truman, that subjected practically every international phone call or teletype message to surveillance by the National Security Agency.)

Ford blamed presidential aspirants like Frank Church for sensationalizing the foreign assassination issue. "From what I am told, we made some clumsy attempts," the president conceded. "If he pushes it, it could make Kennedy look bad. But at the same time, it is so clumsy that it makes CIA look bad." Rockefeller didn't help by setting a publication date for his commission's report without first running it by Ford, then in Europe for the NATO meeting and his subsequent talks with Sadat. Presidential counselor Roderick Hills was tasked with reviewing the unpublished draft. On the afternoon of June 4, Ford's first full day back in the Oval Office, Hills warned the president, "This is not a complete report." As evidence, Hills pointed to several allegations that required additional scrutiny before public exposure.

"You're right," said Ford. "This can't come out." Postponing publication by even a few days sparked concerns the entire 299-page document might be suppressed on national security grounds.

---

* This was mild compared to Kissinger's rant against "the psychopath who's running the CIA. You accuse him of a traffic violation and he confesses murder."

Over the weekend Ford read the report for himself. He decided to release it on Wednesday, June 10, minus Chapter Twenty, an eighty-six-page coda examining the issue of foreign assassinations. Exercising the traditional presidential license to classify intelligence, Ford said he would make this information available to Senate and House investigators "under procedures that will serve the national interest." "I am not going to second-guess my predecessors," he told Kissinger. "If Church wants to, let him. The Kennedys will get him."

Ford's decision to hand off the most explosive CIA materials to congressional investigators satisfied almost no one except his vice president. Passing the buck "got the President off the hook," said Rockefeller, "got me off the hook [and] got it right back to where it belonged in the Congress." The price paid for these evasions was measured in public credibility. One early poll found that almost half of those surveyed viewed the commission's findings as another cover-up. That it might have been worse, far worse, was revealed in 1988, when David Belin, Ford's handpicked executive director of the Rockefeller Commission, published his memoir *Final Disclosure*. Belin, a well-regarded deputy counsel on the Warren Commission, had clashed with Rockefeller over evidence linking the CIA and the Mafia in plots to kill Castro. Insisting the commission expand its inquiry beyond strictly domestic activities, Belin won his point when a majority of commission members voted, in effect, to override their chairman. Ford then agreed to extend the group's mandate by sixty days, to June 1.

It was to prove a pyrrhic victory, however, as Kissinger blocked access to National Security Council files and foot-dragging bureaucrats at State and the CIA ran out the clock. Frustrated by official duplicity and the last-minute burial of Chapter Twenty, Belin booked a conference room at the Mayflower Hotel for the morning of June 18. He fully intended to expose the Castro murder plots, the unholy alliance between spy agency and organized crime and the degree to which previous presidents had known of or encouraged these abuses. "If there was no White House direction," Belin wrote, "we have the CIA operating out of control." Yet the very thought of a presidential blessing for such planning "is wholly repugnant to the principles for which this nation stands."

On June 17, one day before his scheduled press conference, Belin shared his draft statement as a courtesy with Phil Buchen, Dick Cheney and Ron Nessen. Buchen and Cheney, citing State Department concerns, advised Belin that however much he styled himself a concerned citizen speaking outside of his official capacity, his comments would inevitably color perceptions of the Rockefeller Commission and of Ford, on whose recommendation he had been made the commission's executive director. In the face of such arguments, Belin yielded. Canceling his hotel booking, the would-be whistleblower counted on Frank Church and his Senate Select Committee to reveal his secrets.

Here, too, he was to be disappointed, as the Church Committee wasted—in Belin's view—five months on the assassination issue, reaping headlines through "sensational reports of botulism pills, plots to discredit Castro, the use of the Mafia, and presidential affairs with a woman with Mafia connections."* At the White House, Ford and Church negotiated a compromise whereby the committee promised to consult the administration before releasing classified material. By and large the agreement held up, thanks in no small part to the committee's ranking Republican, John Tower of Texas. Salty and irascible, Tower knew how to disarm his ambitious chairman.

"When things got really rough," remembers Rod Hills, Ford's point man with the committee, "he would say something that would make Frank mad. Frank was a wonderful man, but it was easy to get him mad." When that happened, Church and Tower recused themselves in favor of their deputies, Walter "Fritz" Mondale and Howard Baker. The senatorial substitutes would then repair to the White House Situation Room, where Hills joined them for a ritualistic discussion of the latest confidential document in a box.

"Rod," said Democrat Mondale, "would you characterize again what's in the box as you did yesterday?" Hills readily complied.

"Howard, do you believe him?"

"Yeah," Republican Baker replied. "I believe him. Fritz, do you want to look at that box?"

"I don't want to look at that box. Howard, do you want to look at that box?"

Their curiosity sated, the senators returned to the Hill.

At no point in the Church inquiry did the White House invoke the doctrine of executive privilege, in sharp contrast with the investigation by the House Select Committee on Intelligence chaired by Otis Pike. A maverick New Yorker from the eastern tip of Long Island, Pike harbored senatorial ambitions that magnified his natural combativeness. On the other hand, the Pike Committee was less easily distracted by bright shiny objects like the preposterously named Nondiscernible Microincubator that Frank Church waved before the cameras like James Bond's favorite dart gun. (The

---

* The latter a reference to JFK mistress Judith Exner Campbell, sometime Oval Office courier to mob boss Sam Giancana.

acerbic Richard Helms said the device would be a more effective weapon if thrown rather than fired at a prospective victim.)

A series of leaks eroded the credibility of the Pike Committee, until its exposure of US eavesdropping on Egyptian communications moved Ford to cut off the flow of classified intelligence (this after Ford told Bill Colby, "We're not going to classify anything to cover it up, whether it is a mistake, an error of judgment, no matter how bad it is."). A full-scale constitutional crisis loomed, as a defiant Pike issued fresh subpoenas to the CIA and the National Security Council. With Kissinger arguing resistance, Colby seeking accommodation, and Attorney General Levi estimating White House chances in the courts as no better than 50–50, Ford assembled all the key players in the Oval Office on September 26, 1975.

The consensus reached in the Oval Office lasted barely a week. Kissinger's refusal to provide the Pike committee internal communications criticizing administration policy on Cyprus raised the specter of a contempt of Congress citation. The New York Times rallied to Kissinger's defense. So did George Kennan, the dean of Cold War diplomacy. The deadlock was eventually broken, but only after the White House dispatched a presidential emissary to read aloud to the committee from contested documents.

Then, two days before Christmas, Richard Welch, CIA station chief in Athens, was murdered by Greek terrorists after his cover was blown by a left-wing magazine called CounterSpy. At Ford's direction Welch was given a hero's burial in Arlington National Cemetery. His assassination contributed to a reversal of popular sentiment toward the CIA. Suddenly it was the investigators who were on the hot seat. In a stunning rebuke, the full House voted not to publish the Pike Committee's final report. Leaked to Daniel Schorr, it was in turn slipped to the Village Voice.* When the Church Committee released its findings in April 1976, Ford took modest satisfaction in the chairman's retraction of his earlier description of the CIA as a rogue elephant.

By then the president had issued Executive Order 11905, prohibiting "political" assassinations as an instrument of American foreign policy. The same directive incorporated most of the administrative and oversight reforms recommended by the Rockefeller Commission. In addition to an enlarged Foreign Intelligence Advisory Board, originally established by

---

* In congressional testimony, Schorr exercised his First Amendment right to keep the identity of his source secret. (Behind the scenes he reportedly sought to deflect attention by spreading false rumors that his CBS colleague Lesley Stahl was behind the leak.) CBS parted company with Schorr in September 1976.

Dwight Eisenhower, a new Intelligence Oversight Board composed of private citizens was created to act as "an independent auditor" of CIA activities and guidelines. For the first time senior intelligence officials had their duties and authority explicitly spelled out. Their involvement with private business or academic institutions was curtailed. The same executive order made anyone who disclosed classified intelligence information liable to prosecution.

The CIA wasn't the only intelligence-gathering agency to be reined in. After considerable internal debate, the Ford administration decided to support legislation that would require a warrant before the government could conduct electronic surveillance for foreign intelligence purposes on American soil. This provision was incorporated in the Foreign Intelligence Surveillance Act of 1978, which also created FISA courts to rule on such requests from the government. Finally, to the dismay of FBI traditionalists, Attorney General Levi crafted stringent written guidelines that all but eliminated warrantless wiretaps of the sort routinely employed by J. Edgar Hoover's bureau. The *New York Times* singled out this reform in eulogizing Levi as "the model of the modern Attorney General."

F ORD'S EXECUTIVE ORDER reflected his shock and disgust over the treatment of Frank Olson. At the time of his death in November 1953, Olson was a forty-two-year-old civilian biochemist assigned to the CIA's Special Operations Division at Maryland's Fort Detrick. Ten days earlier Dr. Olson and several colleagues had been administered the drug lysergic acid diethylamide (LSD) without their knowledge, part of CIA experiments on behavior modification sparked by reports of Soviet trials with drugs and hypnosis. Developing side effects that included sleeplessness and feelings of depression, Olson was treated by a New York City psychiatrist, who recommended that he be hospitalized. Arrangements were made for Olson to enter a facility near his home in the Washington, DC, area.

Before that could happen he spent the night of November 27 at the Hotel Statler across from New York's Penn Station. At approximately two thirty in the morning the doctor's roommate was awakened by the sounds of shattered glass. Olson had crashed through the closed window and fallen ten stories to his death on Seventh Avenue. Compounding the horror of his last minutes, Olson's wife and three children were led to believe their husband and father had, for unknown reasons, decided to end his life in the aging railroad hotel on Manhattan's West Side. Under the Federal Employees Compensation Act, their sole statutory remedy, the Olson family derived benefits totaling $147,513.22 paid out over twenty-two years. In

1975 Olson's widow was receiving a monthly stipend from the government of $792.

Ford learned all this for the first time, as did the Olsons, from the Rockefeller Commission report. "Outraged. Absolutely outraged" by the news, he told Hills, "I want you to make sure that that family is taken care of." Unfortunately the Justice Department, ignorant of the true cause of Olson's suicide, had long ago settled the case. "Let's say it's not settled," Hills theorized to Rex Lee, head of the department's Civil Division. "Tell me what the money is you would pay if it had not been settled, and it's up to the president to get the bill passed to pay that amount."

Later that day Hills received a phone call from Attorney General Levi. Memories at the Justice Department "of having the White House beat up on them" were still fresh," said Levi. "Be calm, I'll take care of it."

With the Olson family contemplating a lawsuit against the United States, Ford was urged to pass the buck to Justice. Charting his own course, on July 11, Ford issued a public apology. He expanded on this ten days later in an emotional Oval Office session with Mrs. Olson and her children. What happened to Frank Olson was "a horrible episode in American history," the president told them. They could all agree that the United States was a great country, "but that does not right the wrong that's been done to you." Besides making available whatever information the government had on the case, Ford said he would ask Ed Levi to meet with the Olson family lawyer, and be prepared to discuss a more generous financial settlement.

When the Justice Department balked at any figure higher than $500,000, Ford assumed responsibility for getting a private bill through Congress. Eventually the family settled for $750,000 and a promise they could go back to Capitol Hill for more in 1977. By then Ford was out of office. Before leaving, however, he had the satisfaction of knowing that no future American would experience the terrors visited on Frank Olson.

23

THE DANGEROUS SUMMER

*It is easy to be a cold warrior in peace time. But it would
be irresponsible for a President to engage in confrontation
when consultation would advance the cause of peace.*

—GERALD R. FORD, AUGUST 19, 1975

A STRAY COMMENT made at the height of the *Mayaguez* crisis captured Ford's state of mind after nine grueling months as president. Trotting between the Executive Office Building and the White House Situation Room he confided to a Secret Service agent, "I really love this job." By the summer of 1975, talk of a placeholder presidency had faded. "Ford in Command," asserted *Time* in a cover story crediting the unelected president for increased sales of pipe tobacco, ski clothes and "the old-fashioned martini." "Ford Is Mr. Right," proclaimed rival *Newsweek*, although its editors could not resist patronizing him as "a kind of anti-hero whose homely virtues of thrift, honesty, hard work and modesty about the capacities of government exactly suit a diminished national mood."

The faint praise would have been inaudible had the nation's economy remained in the basement. At a June 9 press conference Ford claimed credit for substantially reducing inflation. He didn't fail to mention that 450,000 Americans had found jobs during the previous two months. If anyone in Washington was entitled to say *I told you so*, it was Alan Greenspan. Six months earlier, convinced the nation was entering a classic inventory-led recession, Greenspan had advised the president "to do as little as possible; if we could keep our hand off the panic button, the economy would correct itself." Dissatisfied with the lack of real-time data then available to policymakers, Greenspan had cobbled together his own set of economic indicators to calculate GNP in something close to real time. Persuaded by the numbers that consumer demand was holding up even as inventories contracted sharply, Greenspan was vindicated when third-quarter growth topped 13 percent, the highest since 1950. The stock market followed suit, with the Dow soaring 38 percent in 1975 and an additional 18 percent in 1976.

Ford's Capitol Hill allies might bemoan the lack of a more aggressive legislative program coming from the White House. "But I was so convinced of the Greenspan economic philosophy," the president explained, "I finally decided that good economics was good politics." Bolstered in his inclination to stay the course, Ford told a Chicago audience in July 1975 that a president's primary task was to establish "a stable, solid foundation for the years ahead." This meant passing on expedients promising "short-term benefits for some Americans while inflicting long-term damages on all Americans." Above all it demanded that he resist panicky calls to prime the pump through government spending to meet emergencies as defined by Congress, editorial writers—and campaign pollsters.*

Replacing the broken Keynesian consensus did not produce any obvious political chits. To the contrary: giving priority to inflation over the monthly jobs report invited comparisons to apple vendors and Hoovervilles. Ford persisted. Through lavish use of his veto pen, the self-proclaimed "middle of the road conservative" established himself as a prudent manager of taxpayer resources. His knowledge of Capitol Hill enabled him to forestall Keynesian relics like the Humphrey-Hawkins Act, a back-to-the-future panacea that mandated government as the employer of last resort whenever the adult unemployment rate breached 3 percent.

Ford rounded out his Cabinet with like-minded managers whose profile did not always fit the Republican mold.† To replace Caspar Weinberger at the department of Health, Education, and Welfare (HEW), the president sketched his idealized candidate for Don Rumsfeld. "A political moderate with an academic background," he proposed, preferably a youthful Southerner. Ford didn't know it, but he had described, sight unseen, David F. Mathews, the thirty-nine-year-old president of the University of Alabama. Mathews's nomination was made official on June 26, 1975. A political independent, raised in a state "where I didn't see a Republican until I was thirty years old," the new secretary proved to be a deft administrator as

---

* A similar disdain for the quick fix characterized Ford's approach to energy policy. Returning from the Hill one day to report on his lobbying efforts, Frank Zarb threw up his hands in mock despair.

"Mr. President, we have managed to piss off both the Republicans and Democrats."

"That means we have it just right," Ford said, grinning. AI, Frank Zarb.

† Ford spent precious political capital to confirm Stanley Hathaway as secretary of the interior. Hathaway, a two-term Republican governor of Wyoming with the environmental record one might infer from that fact, scraped through to Senate confirmation, only to resign after six weeks on the job, pleading severe depression. His successor was Thomas Kleppe, a former congressman from North Dakota then heading the Small Business Administration.

he grappled with the fallout from court-ordered busing and referenced the work of famed sociologist Margaret Mead in convincing Ford to fund community-based health care directed at underserved Americans.

As Ford hit his stride, so did his concept of Cabinet government. Approximately every three weeks the full Cabinet, officials of Cabinet rank, agency directors and White House staff—perhaps thirty in all—would sit down with the president for ninety minutes that might include a legislative update from Jack Marsh, Ed Levi's planned revision of the Federal Criminal Code, and a warning from Alan Greenspan linking an economic upturn with costly new spending initiatives in Congress. Occasionally the discussion grew heated, as when the pugnacious Treasury Secretary Bill Simon denounced Vice President Rockefeller's proposed Energy Independence Authority (EIA), a $100 billion government-backed effort to speed development of synthetic fuels. EIA, in structure and financing, was not unlike the overleveraged public authorities to build housing or hospitals in New York State, the precarious standing of which threatened the state's credit just as its greatest city flirted with bankruptcy.

Simon, who rarely had an unexpressed thought, especially those directed at perceived enemies, appealed to Ford before the full Cabinet not to let his vice president "do to the United States of America what he did to the state of New York!" In private Ford asked his secretary of the Treasury to "lay off Nelson, I've got to throw him a bone." There was no chance Congress would enact such a massive intrusion by government into the private sector. "We don't have to fight about things that have no prayer," the president told Simon, "and Nelson Rockefeller's idea about utilities has no prayer."*

Saturday afternoons were often set aside as a budget tribunal, a time for Cabinet secretaries to make a direct pitch to the president for some program or activity axed by budget director James Lynn. Bill Coleman successfully argued for an extra $1 billion with which to complete all ninety-eight miles of the Washington Metro subway, the largest public works project since the Interstate Highway System. Carla Hills went to bat for a modest $10 million program counseling public housing residents on how to care for their new homes. Her husband, Rod, was shocked to find not a single economist on the payroll of the Securities and Exchange Commission.

---

* The politician in Ford could live with what the Eagle Scout found morally objectionable. When courting Nelson in the summer of 1974 he had promised him a different kind of vice presidency, with Rockefeller as influential in domestic policymaking as his protégé Henry Kissinger was in foreign affairs. It hadn't worked out that way; nor could it, if Ford hoped to be anything more than a figurehead. But he liked to think of himself as a man of his word, which helps to explain why he gave nominal support to Rockefeller's EIA.

After making some adjustments to the proposed SEC budget, Rod Hills presented his case before the president and budget director Jim Lynn.

"Well, Jim, Rod's trading three lawyers and two automatic typewriters," Ford said to his budget director. "At least we get rid of the lawyers."

Hills got his economist.

Ford made sure to invite Cabinet members to join any discussion that might involve them and another department. He raised no objection when Bill Simon published a plan calling for "radical simplification" of the tax code, with most deductions abolished and the current top individual rate of 70 percent cut in half. Likewise, when Ed Levi floated a proposal to ban all handguns in any metropolitan area whose crime rate was 20 percent above the national average, Ford pronounced the idea "very interesting." Needless to say, he took a harder line in crafting anti-crime legislation for congressional review. His approach was heavily influenced by Harvard sociologist James Q. Wilson, whose most recent book, *Thinking About Crime*, stressed punishment of repeat offenders and mandatory sentencing over the rights of criminals. Rejecting more sweeping handgun registration, Ford did call for a ban on manufacture and sale of the cheap handguns known as Saturday Night Specials. The proposal never made it out of legislative committee.

Ed Levi was a legal legend, but so was Bill Coleman. At a Cabinet meeting in September 1975 the two men mixed it up in front of their colleagues. With the New York City fiscal crisis rapidly worsening, Levi recommended the city file for bankruptcy, entrusting the politically wrenching allocation of scarce resources to a federal judge. As it happened the next item up for discussion was Boston's continuing anguish, now entering its second year, over court-ordered busing of Black and white students into hostile neighborhoods in pursuit of racial balance. At issue was whether the administration should file an amicus curiae brief with the Supreme Court endorsing the petition of Boston's white parents to overturn the lower courts and halt the buses. Levi, like Ford himself, distinguished between legally mandated, or de jure segregation, and de facto separation of the races resulting from historical and demographic patterns. Where the latter held sway, Levi contended, judges shouldn't interfere in the administration of local school systems. Moreover, the Boston plan involved some twenty-one thousand students, a majority from lower grades—too many for Levi's taste.

Then it was Coleman's turn.

"Mr. President," he began, "the last two issues we have discussed present an interesting contrast. The attorney general has suggested that all fiscal, budget and operational problems of New York City be turned over to an unelected public official, a federal district judge . . . On the other hand, when we are discussing the Boston school case, where the local public officials have remained silent for decades on ending racial segregation in the public schools, we are criticizing the federal courts for meddling in local affairs. In my judgment, the courts are simply acting, with great restraint, to enforce the Fourteenth Amendment to the Constitution."

Ford, acknowledging that both issues were "really tough," said he wanted more time to decide. Clinging to his boyhood ideal—walking each day from his Grand Rapids home to a thoroughly integrated school—he declined to file the desired brief in the Boston case, or in a companion suit involving students in Louisville, Kentucky. He opposed a constitutional amendment to outlaw busing for the simple reason it had no practical chance of enactment. Why stoke unrealistic hopes or combustible fears? His stance, portrayed as an ungainly straddle by much of the press, alienated anti-busing forces without winning support from integrationists.

FOR ANYONE WONDERING why the Ford presidency lacked a catchy moniker like JFK's New Frontier or LBJ's Great Society, Bill Seidman had a ready explanation. Reflecting the transactional ethos of Capitol Hill, "Ford just wasn't a theme type of man. He wasn't someone who attempted to sell things by packaging them for the media." Beyond this, claimed Seidman, his Grand Rapids friend "was issue reactive. He wasn't an initiator." Seidman's view is contradicted by what Alan Greenspan calls "the great unsung achievement" of the Ford administration, credit for which has been largely assigned to Jimmy Carter and Ronald Reagan. In fact all three men, each in his own way, were committed to loosening the regulatory straitjacket that frustrated innovation, stymied competition and robbed consumers and taxpayers alike. Ford, however, was first out of the gate, when resistance to change was most entrenched.

In the twenty-first century, when market forces dominate and technology has erased national borders, it is hard to believe that as recently as 1974 Washington regulators determined airfares and airline schedules; dictated which routes truckers could travel before returning with empty rigs to their point of origin; denied crumbling railroads flexibility in setting rates; and kept long-distance phone rates unreasonably high in order to subsidize local service. Uncle Sam fixed the price of natural gas. The US Postal Service enjoyed a monopoly over delivery of first-class mail. Agricultural "marketing orders" inflated consumer costs by restricting crop sales and imports. "Most regulated industries have become federal protectorates," charged Federal

Trade Commission chairman Lewis Engman, "living in a cozy world of cost-plus, safely protected from the ugly specters of competition, efficiency and innovation."

Ford's campaign to change all this began as an offshoot of his war on inflation (by some estimates federally imposed red tape cost American families as much as $2,000 a year). Within days of taking office he established a Council on Wage and Price Stability and charged it with reviewing proposed regulations for their impact on the cost of living and doing business. He codified this in a November 1974 executive order, forerunner of countless Economic Impact Statements measuring the cost of federal mandates on productivity and competition. In the first week of February 1975, Ford proposed legislation to abolish the "inequitable, inefficient and uneconomical" Civil Aeronautics Board (CAB), a New Deal legacy whose chairman, Robert Timm, though a Nixon appointee, was not above price-fixing among major airlines, if that was what it took to keep an ailing Pan Am from being permanently grounded.

Timm's CAB embodied the culture of complacency described by Engman. For example, Continental Airlines waited eight years for board permission to fly between San Diego and Denver. And by denying the request of a start-up cargo carrier to utilize cost-effective large aircraft, the same agency nearly strangled Federal Express in its cradle.

Still, few took Ford's vows of reform seriously. Deregulation was seen as an election-year polemic, a predictable applause line trotted out by Republican candidates before the Chamber of Commerce. Nelson Rockefeller said it had the political sex appeal of a sick alligator. Ford's instincts told him otherwise. He took soundings on Capitol Hill, which convinced him the stars were aligned for meaningful change. He assigned Rod Hills to confirm his hunch.

"I want to find the bipartisan limits," Ford told him, "and I want you to test them."

Fate, or at least coincidence, was on Hills's side. On the Senate Judiciary Committee, Ted Kennedy's recently hired executive assistant, Stephen Breyer, was an old friend ever since he rented the attic of a Cambridge house occupied by Rod and Carla Hills as visiting professors at Harvard Law School. "Steve was interested in airline deregulation," according to Hills. "Howard Cannon"—chairman of the Senate Commerce Committee—"was interested in trucking and train deregulation." With Ford's encouragement, two dozen senators and representatives formed a bipartisan caucus around the issue. They were invited to the White House for a lengthy strategy session with the president. About the same time Yale economist Paul

MacAvoy joined Ford's Council of Economic Advisers, where he became the dominant force on a Review Group on Regulatory Reform, which met every Wednesday in the Roosevelt Room.

They had their work cut out for them. On July 10, 1975, Ford convened an unprecedented summit of independent regulatory commission chairs—independent, that is, of executive branch dictation or oversight. Before this roomful of skeptics, Ford promoted cost-benefit analysis and argued for government's intrusion in the free market "only when well-defined solid objectives can be obtained by such intervention, or when inherent monopoly structures prevent a free, competitive system from operating." Simply put, "Government should foster rather than frustrate competition. It should seek to ensure maximum freedom for private enterprise."

The message was lost on some in his audience. To Richard Wiley, chairman of the Federal Communications Commission, regulatory reform meant fewer pages in the *Federal Register*.

"Richard, I don't think you get my drift," Ford told him. "I'm asking the question: do phone rates have to be regulated?"

He didn't stop there. "I don't understand why we have an Interstate Commerce Commission," Ford mused. He could not have picked a more venerable target. Grover Cleveland was in the White House when Congress established the ICC to guard against railroad monopolies later marked for extinction by Theodore Roosevelt. The robber barons were long in their graves, and by the 1970s it appeared the railroads might soon join them. Increased competition from truck and barge carriers, coupled with stringent limits on pricing and restrictive labor contracts, had left a third of the nation's rail lines on the brink of insolvency. Dismissing talk of railroad nationalization, Ford reluctantly endorsed the consolidation under federal auspices of seventeen money-losing rail lines serving the freight needs of the Northeast and Midwest. (Washington had already created Amtrak to serve the needs of passenger traffic.)

At the same time, however, he let it be known there would be no funding for the new consortium, dubbed Conrail, unless lawmakers passed his Railroad Revitalization and Regulatory Reform Act. The 4R Act, as it was identified in legislative shorthand, provided ailing lines pricing flexibility and the freedom to abandon unprofitable routes. When enacted in February 1976, it became the first law on the books to deregulate transportation. Of course, it was subject to interpretation by the ICC, where a majority of Nixon-era holdovers remained faithful to the status quo. After one of his own appointees to the board labeled deregulation "a prescription for disaster," Ford took greater care in vetting prospective commissioners for their loyalty to his agenda.*

Ford could not fire independent regulators, but he could hold them to account for questionable behavior. A nasty spat over illegal campaign contributions by regulated airlines, and the suicide of a CAB staffer prepared to testify that Robert Timm had tried to shut down the ensuing investigation, culminated in Timm's forced resignation from the board in December 1975.

When Rod Hills left the White House to become chairman of the Securities and Exchange Commission that October, the regulatory portfolio was entrusted to deputy counsel Ed Schmults, who moved quickly to eliminate a president's power to distribute overseas air routes.† Taking the reins of the CAB, Chicago lawyer John Robson likened the airline industry to "a forty year old still living at home with his parents." Under his chairmanship, the agency in April 1976 did the unthinkable, becoming the first regulatory body to support deregulation. Robson himself forged a working partnership with Senator Kennedy, testifying before his committee and introducing pricing flexibility within the limits of existing statute. Lower fare requests were routinely granted, charter airlines were permitted to sell tickets in advance and the CAB reversed its earlier position on air freight carriers. When Jimmy Carter in November 1977 signed legislation deregulating the air freight industry, it was the making of FedEx and other companies offering overnight delivery service.

This was part of a pattern. Three years elapsed between Ford's Aviation Act of 1975 and congressional passage of the Airline Deregulation Act. The CAB was abolished in 1984, a decade before Congress put the ICC out of its misery. The battle over trucking deregulation was even more prolonged. Opposition from organized labor to the Ford Administration's Motor Carrier Reform Act, introduced in November 1975, was ferocious. Bill Coleman recalled visiting Teamsters president Frank Fitzsimmons to brief him

---

* Encouraged by the early signs of progress, Ford was set straight by Paul MacAvoy. "We've accomplished a lot, but you've got to remember that a regulatory agency is like a turtle. It's got a very thick hide. You can put a little lettuce in front of it—it will quicken its pace ever so imperceptibly, but only imperceptibly." So began the Order of the Turtle, complete with crystal ashtray in the shape of its namesake creature on his back, periodically awarded to someone in the administration who advanced the cause of regulatory reform. AI, Rod Hills.

† In the course of which he was personally lobbied on behalf of their clients by Mel Laird, Bill Scranton, and PepsiCo chairman Don Kendall.

on the legislation. "He almost threw me out of his office," said Coleman. Nor were the major trucking companies any more hospitable.

At one point Fitzsimmons and his counterpart from the American Trucking Associations, a Grand Rapids native who persisted in calling the president Jerry, were invited to the White House for some friendly persuasion. "Fitzsimmons could hardly speak, he was so mad," recalls one participant in the meeting. As for the president's friend, he spoke glibly of market complications, alluding to the fact that transcontinental shippers didn't necessarily "want to take milk to the schools of Grand Rapids." Ford was nonplussed. "Are you telling me that the country that walked across the continent in covered wagons can't get milk to Grand Rapids unless we regulate it?" Truckers would be spared the travails of market competition until 1980, and even then firms were still required to file rates with the ICC. That same year Congress enacted the Staggers Act to complete the railroad revival that began with Ford's 4R Act. Conrail showed its first profit in 1981. Six years later the company was privatized for $1.9 billion, at the time the largest initial public offering in the nation's history.

Ford's regulatory reforms were not restricted to trucks, trains and airplanes. In March 1975 he proposed legislation allowing banks to offer interest-paying checking (NOW) accounts and credit unions to market thirty-year mortgage loans. Three months later he signed the Securities Acts Amendments to promote competition among stockbrokers and establish a national stock market system. The repeal of so-called fair-trade laws saved consumers an estimated $2 billion by denying manufacturers the exclusive right to dictate the retail price of their products. Finally, Ford established the Commission on Federal Paperwork to gauge the economic impact of government red tape.

There are many standards by which to judge a president's historical record. The bills he passed, the programs he initiated, the crises he weathered—all are useful criteria. But sometimes a president's most significant achievements bear fruit after he leaves office. It was that way with Harry Truman and Medicare. Ford was similarly prescient with regulatory reform. Taking much more than the first step in the proverbial journey of ten thousand miles, careful to distinguish between regulation of the market and government rules affecting health, safety and the environment, he upended conventions in place since the Great Depression centralized economic management in Washington. Though little recognized at the time, Ford's challenge to the regulatory state would have consequences far more lasting than the rescue of a cargo ship seized by Cambodian pirates in the Gulf of Siam.

F ORD'S STATURE WITHIN the GOP Best Since He Took Office"—Jack Germond's June 20, 1975, column in the *Washington Star* could not have been more encouraging. That same week the rival *Washington Post* reported that Ronald Reagan, having fallen substantially behind the president in post-*Mayaguez* polling, was leaving the door open to replacing Nelson Rockefeller as Ford's 1976 running mate. Such portents encouraged wishful thinking, with Ford resisting the idea that Reagan, chief proponent of the Eleventh Commandment ("Thou shalt not speak ill of any fellow Republican"), would actually challenge an incumbent president of his party. *Post* reporter and Reagan biographer Lou Cannon thought this absurd. "Ford just couldn't get it through his head . . . that anybody could consider himself not conservative enough" to win a Republican nomination for president, said Cannon. "And that was because he was a creature of a Congress in which the words liberal and conservative didn't have that kind of power.""

That Reagan factored into Ford's calculations more than he let on was evidenced in stalled negotiations over the Panama Canal. Successful talks leading to a transfer of US sovereignty to the Panamanians "would give Reagan all he has lost by *Mayaguez*," the president told Henry Kissinger. "We need to find a way skillfully to postpone it for eighteen months." No such restraint inhibited his dealings with Yitzhak Rabin's shaky ministry in Israel. On June 27, the same day he proposed putting the Suez Canal issue on the back burner, Ford sent Rabin a five-page letter outlining a stark set of options for his consideration. His government should enter into serious negotiations with Egypt and the US over the disputed Sinai Desert passes, Ford told the prime minister, or Washington would consent, however reluctantly, to a Geneva conference at which all bets were off.

In the end, however, it wasn't the Panama or Suez canals that ended Ford's post-*Mayaguez* honeymoon, but a speaking invitation from George Meany's AFL-CIO. Given his lifelong hostility to Soviet-style Communism, no one should have been surprised when the labor chieftain asked Aleksandr Solzhenitsyn, Nobel laureate and political activist, to address a

---

* Ford admitted as much to Trevor Armbrister. He kept reading editorial comments describing him as the most conservative president since Herbert Hoover, said Ford. If so, then "damn it, the extreme right wing ought to be satisfied. But the truth is they never are unless they lock you into a little ideological circle that is a minuscule number of voters in the American public." TAFI, 1817.

labor dinner in Washington on June 30. For Solzhenitsyn this would be his first major public statement since publication of *The Gulag Archipelago*, his monumental history of Soviet prison labor camps, got the author banished from his native Russia early in 1974. From his new home in tiny Cavendish, Vermont, Solzhenitsyn produced a jeremiad for 2,500 Washington dinner goers, saving his harshest words for his host country. Flaying American leaders for "one capitulation after another," Solzhenitsyn drew a line from FDR's 1933 recognition of the Soviet state to the upcoming "funeral of Eastern Europe" coinciding with the Conference on Security and Co-Operation in Europe (CSCE) that Ford planned on attending in Helsinki, Finland.

Ford's absence from the dinner had not gone unnoticed.* On July 2 Jack Marsh forwarded separate requests from Senators Jesse Helms and Strom Thurmond for the president to meet with Solzhenitsyn before the Russian exile left town three days later. A rejection would undoubtedly be met with harsh criticism from Senator Helms. So Ford was reminded.

"Tell Jesse Helms to go to hell."

The resulting foul-up would be attributed to logistics: Ford was traveling on the third, and the fourth was the nation's birthday and nominal kickoff to its bicentennial celebration. But the truth is, he had been primed to say no to the senators based on arguments advanced earlier by Kissinger and now restated by Brent Scowcroft in a twenty-minute Oval Office face-off with Marsh. A meeting with the president would poison the well on the eve of a major diplomatic conference, and in the midst of delicate talks to secure a SALT II arms limitation agreement. For his part Marsh cared less about Soviet feelings than he did for the votes of wavering conservatives on Capitol Hill, plus "the bloc groups . . . the Poles, the Slovaks" and other American representatives of oppressed nationalities languishing behind the Iron Curtain. In Congress, Ford had been counted among their staunchest friends. As president, however, he must gauge the cost of a Solzhenitsyn snub against a possible breakthrough on SALT II.

The calendar complicated his decision-making. On July 17, US and Soviet astronauts were due to link up in their Apollo and Soyuz spacecraft and with a simple handshake put paid to the space race launched with

---

* Although the administration was well represented, with head table guests including Secretary of Defense Schlesinger, Labor Secretary John Dunlop and Ambassador to the United Nations Daniel Patrick Moynihan. Ironically, among the dinner attendees to whom Solzhenitsyn was introduced was Simas Kudirka, the Lithuanian sailor Ford rescued from a Soviet jail cell after his failed attempt to gain asylum via the US Coast Guard.

Sputnik in 1957. Literally the high point of détente, what should have been an occasion of bipartisan pride was in danger of being overshadowed by the Solzhenitsyn fracas. Ultimately Ford chose foreign diplomacy over domestic politics. Ron Nessen told reporters it was unclear what the president would gain from a meeting with Solzhenitsyn, leading columnist George F. Will to note that time had recently been found in Ford's schedule for the Cotton Queen and Brazilian soccer star Pelé. Dick Cheney passed along a Herblock cartoon showing Ford cowering under his desk as Kissinger pointed to Solzhenitsyn's receding figure through a White House window. "It's all right to come out now," a watchful secretary tells Ford. "If you had met him, Brezhnev might have disapproved."

In an accompanying memo Cheney reminded the president that any SALT II treaty would require the support of GOP conservatives, a prospect weakened, perhaps fatally, by Ford's standoffish attitude toward Solzhenitsyn. The White House reversed its earlier position, letting it be known that the writer had an "open invitation" to call on the president after his return from Europe. That seemed unlikely in light of Solzhenitsyn's renewed attack on the Helsinki meeting, "when an amicable agreement of diplomatic shovels will bury and pack down still-breathing bodies in a common grave."

As so often in his still-young presidency, Ford confronted multiple foreign brushfires. On July 24, notwithstanding an intensive lobbying effort by the president and members of his Cabinet, the House voted 223–206 against any relaxation of the Turkish arms embargo. An irate Ford judged it "the single most irresponsible, shortsighted foreign policy decision Congress had made in all the years I'd been in Washington." The Turks, angry and humiliated over the denial of access to equipment for which they had already paid, suspended US military activities at twenty-six jointly operated bases and intelligence centers. Ten thousand Americans living in Turkey had their movements restricted.

The next day Helsinki reclaimed Ford's attention. The president whose schedule wouldn't accommodate Solzhenitsyn met with sixteen representatives of Europe's captive nations and their congressional supporters. He offered them renewed assurances that nothing signed at Helsinki would constitute US recognition of Soviet sovereignty over the Baltic states. When the same pledge found its way into Ford's brief remarks on his July 26 departure for Europe, Brent Scowcroft classified it "a disaster" in need of a fast rewrite. Bob Hartmann, more sensitive than the general to the political clout of Latvian- and Polish-Americans, told Scowcroft no typists were available to assist him. The resourceful Kissinger deputy revised Ford's statement on his own, eliminating any reference to the Baltics in favor of

upbeat language about Helsinki as "a forward step for freedom." Unfortunately for Scowcroft, Kissinger and the president for whom they worked,
the advance copy handed out to reporters at the airport retained the offending lines authored by Hartmann's shop. The incident raised fresh doubts
about who was in charge.

N OT SINCE THE Congress of Vienna in 1814–15 reconstituted Europe's balance of power following Napoleon's downfall had so many
heads of state assembled in one place to chart the future course of their continent. Along with the United States and Canada, every European nation
save tiny Maoist Albania was represented in Helsinki's white marble–clad
Finlandia Hall. They were there to ratify a thirty-thousand-word declaration of principles labeled with chilling inexactitude as the Final Act. For the
Soviet Union, which had clamored for this gathering since 1954, Helsinki
was a long-delayed World War II peace conference at which its spoils of
victory would be recognized as "inviolable." Never mind the existing patchwork of treaties and protocols that fixed the postwar boundaries of Europe.
Leonid Brezhnev revealed the Soviet mindset when West German chancellor Willy Brandt, after concluding his own set of agreements recognizing
Soviet-drawn borders, asked why the Russian leader still felt the need to
convene a full-scale European security conference.

"A document," replied Brezhnev. "We want a document."

Desire breeds vulnerability, magnified in Brezhnev's case by a sick man's
increasingly desperate search for a legacy. The general secretary wanted his
document formalized before the 25th Soviet Party Congress scheduled for
February 1976. To get it, he agreed to a series of preconditions advanced
by the United States and its NATO allies. Virtually alone in his administration, Ford sensed the possibilities afforded by a Helsinki-like conference to
wring concessions from the same adversary that had withheld its signature
from the Universal Declaration of Human Rights adopted in 1948. The
Soviets, having slightly relaxed their grip over the divided city of Berlin,
appeared willing to get serious about Mutual Balanced Force Reduction
(MBFR) talks, the grandiloquent name bestowed on negotiations underway
in Vienna aimed at reducing conventional forces on both sides of the Iron
Curtain.

Ford's willingness to trade recognition of "legitimate" postwar boundaries for a Soviet acknowledgment that national borders were changeable
exclusively "by peaceful means" was to have historic repercussions for the
eventual reunification of Germany along democratic, pro-Western lines.
Together with language attesting to the right of Europeans to decide "their

internal and external political status, without external influence," this renunciation of physical force amounted to a repeal of the Brezhnev Doctrine
cited as recently as 1968 to justify the deployment of Soviet tanks to crush
Czechoslovakia's short-lived experiment in political liberalization.

Believing humanitarian issues to be a strictly internal matter, the Soviets
objected strenuously to any inclusion of human rights in the Final Act.
Ford had a different agenda in mind. Accordingly the US conditioned its
attendance at Helsinki on a robust human rights package involving the free
movement of people and ideas, an end to radio and television jamming,
and freedom of conscience, speech, association and belief. And it lent teeth
to the agreement by incorporating follow-up arrangements, including a
conference in Belgrade in 1977 to assess compliance with these guarantees.

Only late in the game did the Soviet leadership awake to the potential
consequences of their trade-off. Anatoly Dobrynin recalled heated debates
within the Politburo: Did the clauses guaranteeing existing frontiers and
nonintervention in a country's internal affairs—both eagerly sought by
Moscow—eclipse the West's introduction of human rights into the conversation? For his part, Ford could take gratification in knowing that for all
the criticism he received for going to Helsinki, Brezhnev had his hands full
convincing himself, his generals and dissenting party colleagues that they
hadn't been snookered by Western diplomats.

H E MIGHT BE a stranger to the dense, semimystical prose of Solzhenitsyn, but like any good politician Ford understood symbolism as a
substitute for words. By including Poland, Yugoslavia and Romania on his
European itinerary, he would call attention to Soviet vulnerabilities while
strengthening political and economic links between his own country and
these increasingly restive satellites. Stopping first in West Germany, he
offered reassurances that nothing about the Vietnam experience had diminished the American commitment to European defense. Ford compared
notes on the state of the global economy with Chancellor Schmidt, who in
turn endorsed the concept of an allied economic conference before the end
of 1975 (this was the genesis of the annual meeting of the G7, or 8, or 20,
although in fact only six nations participated in that fall's original summit
at the Parisian suburb of Rambouillet).

In Warsaw a quarter million people turned out to see Ford's motorcade from the airport to Wilanów Palace, the Polish Versailles, built by a
seventeenth-century monarch and confiscated by the postwar Communist
government. At the conclusion of their private talks, Ford and Communist
party leader Edward Gierek signed a communiqué emphasizing the need

for military as well as political détente, a not-so-subtle appeal to expedite the MBFR talks in Vienna. On July 29 Ford made a grim visit to the Nazi death camp of Auschwitz. Unprepared for the lingering horror of the place, he was visibly moved by the ruins of gas chambers and barracks where life spans had been measured in days or less.

From there he flew to Helsinki. Before the formal conference got underway, Ford renewed his acquaintance with Brezhnev at the residence of the American ambassador. For two hours the leaders sparred over the Middle East and Jewish emigration in the birch-paneled dining room of a faux Virginia plantation house. Brezhnev, thinner than Ford remembered, seemed more dependent than before on Gromyko and others to explain the arcana of arms control. Occasionally he slurred his words. When Brezhnev referenced Jim Schlesinger's recent refusal to rule out the first use of nuclear weapons against the Soviet Union, Ford reminded him that the American president, not his defense secretary, makes and executes nuclear policy.

The oddest exchange of the trip took place outside the ambassador's residence just before the opening session of the conference at Finlandia Hall. As preserved in a Soviet memo, the scraps of which were subsequently retrieved from Brezhnev's ashtray by a quick-witted National Security Council staffer named Jan Lodal, the Soviet leader informed Ford "confidentially and completely frankly" that his government supported the president's bid for a full term in 1976. "And we for our part will do everything we can to make it happen," he added. Ford, masking his surprise, offered perfunctory thanks. He expected to win the election, he told Brezhnev, and he looked forward to improving Soviet-American relations. The incident, quickly forgotten by Ford, was revisited forty-two years later, when Vladimir Putin's Russia placed a thumb on the scale for Donald Trump in his contest with Hillary Clinton. "The difference is that this time the offer was accepted," wrote Lodal, "while in 1975 the United States was led by a man of great experience and absolute integrity, who ignored the offer."

Finnish President Urho Kekkonen's ceremonial welcome inaugurated a parade of national leaders, each putting his own spin on Helsinki's conference and its future implications for the continent. With everyone on their best behavior, old hostilities were muted, if one didn't count Greece and Turkey. In a fractious lunch on July 30, Ford took President Konstantinos Karamanlis to task for the Greek government's heavy-handed lobbying of Congress. The House defeat on the Turkish arms embargo meant "our leverage [over Turkey] is not zero," said Ford, "it is negative." Caught out, Karamanlis minimized the conflict—"It is not like the Middle East"—and indicated a willingness to accommodate Turkish demands for a divided Cyprus, so long as the two zones roughly corresponded in size to the Greek and Turkish populations on the island. The next day, Turkish premier Suleyman Demirel, in no mood for compromise, complained to Ford of the rapacious Greeks: "They have 3,000 islands and they want Cyprus, too."

Ford changed the subject. He had promising news from Capitol Hill, where New York congressman Charles Rangel was dangling the prospect of sixteen additional votes for the administration position *if* the Turks intensified their efforts to limit the poppy crop, whose opium, distilled into heroin, was ravaging his Harlem constituents and other urban neighborhoods represented by Rangel's colleagues in the Congressional Black Caucus. Demirel described for Ford the special coordinating unit he had established to combat the lovely, lethal flower. This, coupled with repeated admonitions to Cabinet colleagues for help in suppressing the drug trade at its source, hardly constituted a dynamic program on the scale envisioned by Rangel.

Still, Ford, so often accused of lacking imagination, could be quite creative where tactics were concerned. "We could say that your Cabinet Committee has been working on this problem and that it has been very helpful," he told Demirel. Hadn't a UN committee singled out the prime minister's efforts for praise? This could be emphasized in any message to Congressman Rangel. "In fact, I could call him on the telephone today. I could also tell him that you have put this in the hands of your Minister of Agriculture." With Congress on the cusp of adjourning for the summer, there was no time to spare. If a little puffery and Rangel's converts could secure a House vote *before* the gavel fell that evening, Ford might yet lure the Turks back to the negotiating table. He duly called Rangel at 6:20 p.m. Helsinki time. What both men underestimated was the ability of embargo supporters to run out the clock. As a result, it wasn't until the first week of October that Rangel and Ford could execute their deal and the House voted to release almost $200 million in Turkish armaments already in the pipeline.*

The Turkish cliffhanger distracted Ford from the main business at Helsinki, an oratorical marathon whose organizers, employing French as the lingua franca of diplomacy, placed the American president twenty-sixth on

---

* The Turks, unimpressed by this grudging action, would maintain their squeeze on US bases and personnel until August 1978, when the Carter White House persuaded lawmakers on Capitol Hill that the embargo had undermined the southern flank of NATO while failing to dislodge Turkish invaders from Cyprus.

the roster of speakers. Competing versions of his remarks reflected continuing tensions between Kissinger's State Department and the White House team led by Bob Hartmann. "Do we need to place East and West in military confrontation?" Ford scribbled after reading Hartmann's rambling and bellicose draft. "Why not amplify HOPE which all want . . ." On the night before he was to speak, Ford recalled, "I got Kissinger and Scowcroft and Hal Sonnenfeldt (the State Department counselor thought of as "Kissinger's Kissinger") and Hartmann together, and we spent about three hours going over the speech, paragraph by paragraph . . . It was not a pleasant evening."

However grueling the process, it produced what the *Los Angeles Times* called the most impressive speech of Ford's presidency. The tone throughout was one of muted optimism. Noting recently relaxed tensions over Berlin, Ford cited the former German capital as "a flashpoint of confrontation in the past [that] can provide an example of peaceful settlement in the future." He acknowledged public skepticism toward the document he was about to sign. "Peace is not a piece of paper . . . but lasting peace is at least possible today because we have learned from the experience of the last 30 years that peace is a process requiring mutual restraint and practical arrangements."

Describing Helsinki's Final Act as "a challenge not a conclusion," Ford addressed the Soviets and their client states with a candor verging on bluntness. It was important, said Ford, "that you recognize the deep devotion of the American people and their government to human rights and fundamental freedoms and thus to the pledges that this conference has made regarding the freer movement of people, ideas, and information." East-West cooperation held out a promise of economic, scientific and technological benefits on an unprecedented scale. But only if both sides honored the commitments codified in Finlandia Hall. "History will judge this conference not by what we say today, but by what we do tomorrow—not by the promises we make, but by the promises we keep."

On the morning of August 2, Ford and Brezhnev met again in an attempt to break the SALT II deadlock that had developed since Vladivostok. At issue were the American cruise missile (a subsonic, highly accurate weapon that eluded detection by flying under the radar screen), and the Soviet Backfire bomber, whose capacities and mission were hotly debated. The Pentagon classified the plane as a strategic weapon capable of reaching US airspace if refueled in the air, or on a one-way mission landing in Cuba. Not so, Brezhnev countered. The Backfire was a medium-range aircraft, designed with Eastern, not Western, targets in mind. Ironically it was their versatility that made both weapons stumbling blocks to any SALT II agreement. Once again technology had outpaced diplomacy. There would be no breakthroughs at Helsinki, no Brezhnev visit to the United States that fall and no East Room treaty signing to bolster Ford's chances in the coming campaign the Soviet leader purportedly hoped he would win. Failure to complete the work begun so auspiciously at Vladivostok would rank near the top of Ford's list of postpresidential regrets.

O N THE OTHER hand, over time the Helsinki Accords more than fulfilled Ford's vision of an international bill of rights—in the words of Anatoly Dobrynin, "the manifesto of the dissident and liberal movement." Publication of the Accords in the Soviet newspaper *Pravda* led to formation of the Moscow Helsinki Group to monitor official compliance with the provisions governing human rights. For daring to insist their government honor its commitments made at Helsinki, dissident leaders including physicist Yuri Orlov and Nobel Peace Prize winner Andrei Sakharov would be imprisoned or exiled by Brezhnev and his successors. These and other punitive actions only confirmed the regime's essential weakness. Beset by a declining economy, the Soviet leadership was squeezed between its need for Western trade and technology and the repressive policies of a crumbling empire.

Powerless to prevent Helsinki Watch groups from springing to life in diverse parts of its far-flung union, Moscow shuddered when Poles in December 1975 petitioned the speaker of that country's parliament demanding the implementation of human rights promises made at Helsinki. Soon the powerful Catholic Church took up the cause. Within five years of Ford's concluding challenge in Finlandia Hall, the Polish government was forced to recognize the non-Communist, Solidarity trade union led by Lech Wałesa. In neighboring Czechoslovakia the Charter 77 movement arraigned the nation's rulers for their neglect of the same human rights agenda. Activists took advantage of follow-up meetings in Belgrade in 1977–78 and Madrid in 1980–83 to focus international attention on the continuing denial of basic freedoms behind the rusting Iron Curtain.

In 1987, general secretary of the Communist Party Mikhail Gorbachev ordered the release of hundreds of Helsinki-inspired activists languishing in Soviet prisons. Thereafter the Soviet Union accepted Western human rights monitors on its soil, their presence like an underground stream eroding the foundations of the state. When Hungary in 1989 opened its borders to vacationing East Germans who sought asylum in neighboring Austria, the action was taken in compliance with the Helsinki Accords. And when the Berlin Wall fell in November 1990, precursor to German reunification and the final collapse of the Soviet Union the following year, President George

H. W. Bush recognized it as part of the Helsinki legacy. By then Ford, out of office for more than a decade, was entitled to crow about what he called "my finest hour" as president.

And yet. If Helsinki redounded to anyone's immediate political benefit domestically, it was Ronald Reagan. In the short term, CSCE gave the Soviets the territorial legitimacy they craved. This allowed Reagan and others on the right to speak of Munich-style appeasement. Rare was the 1976 Republican primary voter in New Hampshire or Florida who confused Ford's post-Helsinki visits to Communist Yugoslavia and Romania with a victory lap. Much of the blame lay with the White House, and Ford himself, for failing to articulate a consistent message on Helsinki. Veering between the hard line urged by his political team and the more hopeful interpretation that had prompted his journey to the Finnish capital in the first place, Ford sounded defensive, at times apologetic, culminating in his March 1976 concession, "I don't use the word détente anymore."

THE HELSINKI BACKLASH wasn't the only threat to Ford's election prospects. The recently established President Ford Committee (PFC) labored under the organizational shortcomings of Howard "Bo" Callaway, a forty-eight-year-old former Georgia congressman and secretary of the army who took the campaign job only after the president's first choice, Mel Laird, had turned him down. Callaway's background in business and politics did not compensate for his reluctance to delegate, his slowness in filling key slots or his habit of poor-mouthing Ford's chances in early primary contests. He hadn't been on the job a hundred days before the committee's finance chairman, David Packard, and political director, Lee Nunn, both quit in frustration.

As director of Richard Nixon's Southern delegate–gathering operation in 1968, Callaway's antennae were most sensitively tuned to his native region, which also happened to be the area of Ronald Reagan's greatest strength. No sooner had he taken the helm of the PFC than Callaway publicly questioned Nelson Rockefeller's appeal in Dixie. His suggestion that Ford might want to choose a younger running mate moved Hugh Scott to inquire whose campaign Callaway was working for—Ford's or Reagan's. Rockefeller's response may easily be imagined. Ford tried to undo the damage by going to bat for Rocky's Energy Independence Authority, loathed by conservatives for its massive expansion of government. In a carefully staged show of solidarity, the president and vice president shared *Marine One* for the ride to Andrews Air Force Base and Ford's European send-off. Finally, on August 9 the Fords marked their first anniversary in the White House by hosting Nelson, Happy and their children at a private dinner and tour of the Oval Office.

Twenty-four hours later Jerry and Betty were in Vail for the start of a two-week working vacation. They arrived just in time to watch Betty's interview with Morley Safer, taped three weeks earlier for the hugely popular CBS broadcast *60 Minutes*. It had taken almost a year for the First Lady and her press secretary, Sheila Weidenfeld, to become comfortable with the idea. "I never knew exactly how I would find her if I scheduled something in advance," said Weidenfeld. "You never knew if it was going to be a good day or a bad day." The European trip had followed this pattern—"Fine in Germany, exhausted in Poland, high in Helsinki, but low in Romania."*

On the morning of the scheduled taping Weidenfeld found her employer close to tears over a renegade social secretary accused of padding guest lists at state dinners with her personal favorites. She was in no condition to go before the cameras, Betty insisted. Her son Jack could take her place. Weidenfeld diagnosed a bad case of stage fright. Granted, butterflies in the stomach seemed out of character for the former Martha Graham dancer, who cheerfully acknowledged being a ham. "She was conflicted—there was Betty Ford versus Mrs. Gerald Ford," according to Weidenfeld. "She was so much more at ease as Betty Ford." Weidenfeld herself was torn, exclusively addressing her employer by her married name, yet determined to showcase Betty's originality and low-key charisma to the vast *60 Minutes* audience. "I think this interview will make you feel good," she pleaded with the First Lady. It would be a happy distraction from whatever else darkened her outlook. Back out now, said Weidenfeld, and "you'll work yourself into a real depression." Having exhausted her store of enticements, Sheila stood in the second-floor West Hall contemplating a disaster in the making when Betty's familiar if not always predictable voice coaxed her back from the brink.

"Okay, you win. I'll do it."

On the night of August 10, twenty million households tuned in to see America's premier news program venture into areas traditionally off-limits to presidential spouses. Commenting on the pressures experienced by political wives, Betty told Morley Safer it all depended on "the type of husband that you have. Whether he's a wanderer or whether he's a homebody." She had "perfect faith" in Jerry. Only when he stopped enjoying the sight of "a pretty girl" would she start to worry. In any event, "he really doesn't

---

* Baffled by such volatility, on at least one occasion Weidenfeld reportedly told White House physician William Lukash to "lay off the pills" prescribed for the First Lady's pinched nerve.

have time for outside entertainment. Because I keep him busy." With equal frankness she described her visits to a psychiatrist ("I was a little beaten down. And he built up my ego") and her support of the Equal Rights Amendment and the Supreme Court ruling in *Roe v. Wade* that brought abortion "out of the backwoods and put it in the hospitals where it belonged." She shared her formula for a successful marriage: "You go into it, both of you, as a seventy-thirty proposition." Of course no marriage was conflict-free. She and Jerry had had their fights.

Over money? asked Safer.

"No. Never had any money to fight over."

Over politics?

That was different. In recent months Betty had successfully argued for a woman in the Cabinet. "And I'm working on another"—if she could get a woman on the Supreme Court, "I think that I'll really . . . have accomplished a great deal."

As the conversation swerved from politics to personal behavior, the First Lady conceded that more and more young people were living together without benefit of a wedding license. Safer saw his opportunity and seized it. What, he asked, if Susan Ford were to declare, "Mother, I'm having an affair"?

"Well, I wouldn't be surprised," said Mrs. Ford, clearly surprised by a question never put to any other White House resident. "I think she's a perfectly normal human being like all young girls." As for herself, she would want to counsel and advise her—"She's pretty young to start affairs."

But old enough, Safer interjected.

"Oh, yes, she's a big girl."

It was at this point that the president in Vail tossed a pillow at his wife with mock lamentations over the ten million votes she had cost him. The number doubled with Mrs. Ford's comments on drug use, specifically her assumption—denied by her children—that each of them had "probably" tried marijuana. She breezily confessed that Betty Bloomer as an adolescent would likely have tried the stuff, "like your first beer or your first cigarette." Near the end of the segment she regained control by mentioning her cancer operation as another example where her willingness to break long-standing taboos had encouraged countless women to get their own checkups. She credited their belief in God for enabling her and the president to come through the ordeal of her surgery and subsequent treatment.

"The morning after the interview," recalled deputy press secretary Patricia "Patti" Matson, "all hell broke loose." Anyone reading the *New York Times* lede could understand why. "Betty Ford said today that she wouldn't s:HE.

THE DANGEROUS SUMMER | 565

be surprised if her daughter Susan, eighteen years old, decided to have an affair," the paper reported. "Mrs. Ford suggested that in general premarital relations with the right partner might lower the divorce rate." Phyllis Schlafly zinged her rival for lowering the moral standards of the White House. It was no defense to praise her honesty, said Schlafly. "The Happy Hooker did not become virtuous when she honestly described her immorality." William F. Buckley went further, accusing Mrs. Ford of trying "to rewrite the operative sexual code of Western Civilization."

Among the letter writers who flooded the White House with their complaints was Maria von Trapp, now a Vermont innkeeper and not sounding at all like Julie Andrews's loving nun–turned–loving governess, wife and scourge of Austrian Nazis. "Do you realize how much harm you have done to the American family and to the American youth?" she demanded. Clearly Mrs. Ford was not one of her favorite things. For good measure, von Trapp wrote separately to the "wishy-washy" president, admonishing him, "You simply must tell your wife to shut up, and keep shut up in public." Even Mike Ford registered a dissent from his mother's views on premarital sex. "I guess I'm more old-fashioned," he said.

From Kansas, where she was working as a summer intern at the *Topeka Capital-Journal*, Mike's sister, Susan, telephoned to say how proud she was of Betty's performance (and to ask, "What is an affair?"). Betty Friedan, the anti-Schlafly, chimed in with praise of the First Lady's "sensitivity and strength." Supporters hoisted placards declaring, JERRY PLAYED FOOTBALL— BUT BETTY KNOWS THE SCORE. They were very much in the minority, at least initially. This didn't surprise Patti Matson. "If a person hadn't seen the interview in person, there was no context, no understanding that her statements were in direct response to specific questions," Matson theorized. "Without seeing it, many wouldn't have heard the tone of her answers, or seen her soft demeanor." As the interview was replayed, in whole or in part, "the reaction began to subside. Her answers made sense once viewers heard the questions." That said, Matson continued, "I won't try to tell you that the political people were happy with the interview."

Advised to speak with his wife about "toning it down," the president found the idea laughable. "You want to tone things down—then YOU talk to her about toning it down." It wouldn't be long before he was basking in the reflected popularity of a First Lady whose candor lent credibility to the Ford White House and its posture of post-Watergate openness. Yet even as buttons appeared declaring "Betty's Husband for President," there were signs Mrs. Ford's forthrightness was unlikely to win many votes for her husband where it counted. In polling its readers, the *Ladies' Home Journal* found a clear discrepancy between the parties, with 59 percent of Democrats approving of her frankness, while just 48 percent of Republicans felt the same.

When the former president published his memoirs in 1979, he included his wife's *60 Minutes* interview with the controversial Helsinki Accords and his failure to meet Solzhenitsyn as factors in his declining poll numbers, and a corresponding increase of support for his conservative challenger, Ronald Reagan. One consequence of the Reagan surge was a decision by strategists to carry the battle, as yet undeclared, to the former governor's own backyard. Campaign planners harbored no illusions about winning the state's June 1976 primary. They planned an aggressive courtship of Golden State voters anyway, according to one, "just to keep Reagan coming back to California to make sure he wasn't losing." With autumn's approach Californians could expect to see a lot of *Air Force One*. The president's schedule had two visits penciled in for September alone.

24

"THE CAT HAS NINE LIVES"

———◆———

It is a little startling to look down to shake hands and you see a hand with
a gun in it . . . Say a little prayer and the Lord will take care of you.

—GERALD R. FORD, SEPTEMBER 5, 1975

O H, THIS IS a nice day. I'll walk."
It was 9:57 a.m. on Friday, September 5, 1975. Gerald Ford stood on the sidewalk outside Sacramento's Senator Hotel, whose location at Twelfth and L Streets, just across from the state capitol, had made it the hub of local political activity since the nine-story, Moorish-style structure opened its doors in 1924. Awash in California sunshine, the president was feeling decidedly upbeat that morning, his mood bolstered by a warm reception from 1,200 guests at a breakfast sponsored by the California Chamber of Commerce. Sacramento was the last stop on a three-state trip that included four Republican fundraisers, a Seattle conference on the regional economy, a speech before thirteen thousand young people attending a bicentennial rally in Portland, Oregon, and the presentation of a bicentennial quilt crafted for their president by the residents of Port Townsend, Washington.

Everywhere he went, Ford had won applause by hailing Henry Kissinger's latest triumph. Capping ten days of shuttle diplomacy intensive even by his well-traveled standards, the secretary of state had nailed down the Sinai II pact between Egypt and Israel that had eluded him in March. Under its terms the Israelis agreed to withdraw east of the Gidi and Mitla passes. The Egyptians promised to open the Suez Canal to nonmilitary Israeli traffic. Both sides accepted a diplomatic formula for coexistence that fell just short of the formal declaration of non-belligerency sought by Tel Aviv. Assuming Congress approved the deal, the United States would station 150 or so civilian technicians to patrol the buffer zone between the former combatants. It would also provide Israel and Egypt military and/or economic assistance totaling $3.5 billion.

Now, hardly twenty-four hours after briefing congressional leaders on the Middle East breakthrough, Ford decided to indulge in a little ego grat-ification. Exiting through his hotel's skylit lobby, he spied several hundred people assembled in the shadow of the capitol dome, hoping for a glimpse of the president in his limousine before he disappeared inside the build-ing to address a joint session of the state legislature. Ford waved away the armored 1972 Lincoln Continental waiting curbside to transport him the short distance to the capitol's east entrance. Walking through the parklike grounds, he told advance man Frank Ursomarso, would be "more fun."

The next few minutes were "just chaos," according to Larry Buendorf, one of a dozen Secret Service agents shielding the president as he bolted across L Street and onto a paved walkaway lined with well-wishers. Buen-dorf stayed close by Ford's shoulder "making sure people didn't grab his watch or hold on too long." Also working the ropeline was presidential ap-pointments secretary Terry O'Donnell. "Because people would constantly give the President things, envelopes, pictures to sign, letters . . . and the agents had to keep their hands free," O'Donnell went ahead of Ford, an executive bag man clutching a big briefcase into which the day's loot was stashed.* They had covered perhaps half the distance between the hotel and the Capitol when, out of the corner of his eye, O'Donnell observed a woman "with eyes that were scary," dressed in "some kind of an orange cape thing." To O'Donnell she seemed less menacing than out of place.

A few feet behind him, shaking hands with excited onlookers in the front row, Ford scanned those behind them with equal parts caution and political calculation. "I was on the Warren Commission," he explained. "In my own way, as I go down a line, I am conscious. I am looking." His attention was drawn to a nearby magnolia tree. Also moving toward it, seemingly paralleling his progress toward the capitol, Ford noted a red-haired woman with a "weather beaten" face and an "unusual" dress of red or orange. Thinking she might want to speak to him, Ford fixed his sights on her right hand as it came up from near the ground to a position "a little above my knee." He flinched at the sight of a .45 Colt semiautomatic pistol pointed in his direction. Barely two feet separated the president from his would-be assassin.

"The country's in a mess," she announced in her juvenile voice. "This man's not your president."

"Gun!" shouted Larry Buendorf. Stepping in front of Ford, he simulta-

---

* "Then we'd take them back to the White House," says O'Donnell, where "every letter would be answered and every request would be dealt with one way or another." Once a congressman, always a congressman. AI, Terry O'Donnell.

neously lunged for the weapon unsheathed from the woman's ankle holster. Before she could pull back on the slide forcing the first of four bullets into the gun chamber, Buendorf managed to insert his hand—actually the thin membrane of skin between his right thumb and forefinger—between the gun's hammer and firing pin. "If she'd had a round chambered, I couldn't have been there in time," the agent said. "It would have gone through me and the President." As it was, Buendorf twisted the diminutive 120-pound assailant away from her intended target and forced her to the ground. Thanks to his lightning-quick reflexes the only blood shed that morning was caused by the hammer puncturing Buendorf's outstretched hand.

"It didn't go off," wailed Lynette Alice "Squeaky" Fromme. "Can you believe it? It didn't go off."

Handing her gun to another agent, Buendorf restrained Ms. Fromme with some handcuffs borrowed from a local cop. Simultaneously Ford's detail seized the president by the tails of his suitcoat—collapsing him, in agent lingo, and rendering him momentarily invisible to any accomplices who might be lurking in the crowd. They formed a flying wedge to rush Ford into the capitol. Eager to avoid additional panic, Ford slowed the pace as he neared the building's doors. He straightened his tie in preparation for his meeting with Governor Jerry Brown. Inside the two men chatted amiably about the California economy and the state's energy needs. Ford's brush with death went unmentioned. "The other thing was past," he'd later explain matter of factly, "and there was no point in my taking his time or my time to talk about it."

After twenty minutes or so a gubernatorial aide entered the room and whispered something in Brown's ear. "You okay, Mr. President?" the governor inquired.

"I'm just fine," Ford replied. Addressing the legislature on his chosen topic of crime, he ignored the morning's events in favor of his now-standard pitch for mandatory jail sentences "for persons found guilty of crimes involving the use of a dangerous weapon." Presumably that would include Squeaky Fromme, a twenty-six-year-old disciple of Charles Manson, hailed by his followers as the Second Coming of Jesus Christ. Law enforcement more accurately labeled Manson a psychopathic killer responsible for at least seven deaths, including the grisly slayings of actress Sharon Tate and grocery store owner Leno LaBianca and his wife in August 1969. With her roommate Sandra Good, Fromme had moved to Sacramento to be near Manson, then incarcerated twenty-five miles away in Folsom Prison. The two women were self-appointed members of the International People's Court of Retribution sworn to kill corporate or elected polluters of the environment.

It was because she feared for the survival of California's redwood trees, Fromme claimed, that she had targeted Ford over his environmental policies. Skeptics thought it more likely she brought a gun to Capitol Park to impress Charlie. As for Ford, before leaving town he all but apologized to the people of California for any negative publicity resulting from the act of a lone individual in no way representative of their state. He praised the Secret Service and other law enforcement for their professionalism. He would express his appreciation to agent Buendorf in private. He also called Betty at the White House to reassure her. At the Sacramento International Airport Ford's instinct was to go over and shake hands with those who had gathered to see him off. This time the Secret Service said no, and made it stick.

On the flight back to Washington, Ford appeared strangely detached from the day's drama. It was like football, recalls one who observed him closely—"There was a play and it wasn't a good play and he got up and went, 'What's the next play?'" In fact his demeanor concealed a heightened sensitivity to the vulnerabilities of his position. That weekend Jack Ford and a friend planned a camping trip through Minnesota's Superior National Forest. Fed up with the constraints of his unsought position, Jack ditched his Secret Service detail for the duration of his time in the woods. "Do you know what it is like to have them shadowing me all the time?" he asked accusingly. "Participating in every aspect of my life? . . . And even if I forget they are there, no one else does."

Jack's father had somewhat greater appreciation for the agents whose participation in his life had extended it, even while risking their own. Taking pains to conceal his involvement, the president encouraged park rangers to watch out for Jack and to keep the White House informed about his son's whereabouts and safety.*

A PLANNED RETURN trip to California two weeks later touched off "a big fight between security and political people." Declaring, "I'm not going to be a captive in the White House," Ford said, "we're going back." Another argument erupted over the subject matter of his speech before the

---

* On November 26, following a trial from which her disruptive antics got her ejected, Fromme was convicted of attempting to kill the president. (Ford made history as the first president to give videotaped testimony in a criminal case.) In mid-December she was sentenced to life imprisonment and sent to the Alderson Federal Correctional Institute in West Virginia. In December 1987, Fromme escaped from the facility and remained at large for two days, an infraction for which she paid with a five-year extension of her sentence. In all, Fromme spent thirty-four years behind bars by the time she was paroled in August 2009.

annual convention of the AFL-CIO building and trades unions meeting in San Francisco. A "devastating" memo from Alan Greenspan disputed the assumptions underlying the Energy Independence Authority, Nelson Rockefeller's crash program to create synthetic fuels under the auspices of the federal government. Questioning the vice president's rosy forecast of job creation through the EIA, Greenspan feared that public money would drive out private investment. Another dissenting voice, Don Rumsfeld, told Ford his vice president was "a serious problem . . . he is enthusiastic and he's imprecise and other people are afraid of him." This alone should make the president think twice before accepting Rockefeller's claims of widespread support for the EIA. Ford, unpersuaded, gave his stamp of approval to a speech broadly endorsing Rocky's moon shot as the centerpiece of his September 22 visit to San Francisco.

Leaving the Hyatt Regency Hotel following his presentation, Ford deferred to Secret Service recommendations and climbed into his limousine for the one-block transit to the St. Francis Hotel. Half a dozen agents surrounded the vehicle as it entered the St. Francis through an underground garage. Upstairs the president addressed the World Affairs Council of Northern California. A question-and-answer session with the group preceded a half-hour interview with a local television station. Responding to a query about gun control, Ford restated his opposition to the registration of weapons or their owners. "I prefer to go after the person who uses the gun for illegal or criminal purposes," he asserted.

It was almost three thirty p.m. By now the crowd in nearby Union Square had grown to perhaps three thousand. Sprinkled among the usual gawkers were protesters espousing multiple causes and chanting sentiments none too friendly. "Get him out of town as fast as you can," an NBC cameraman told Ron Nessen. "These are not good people." Picking up similar vibes, the Secret Service radioed to Ford's agents: when ready to go he should leave the hotel by a rear entrance and head directly to his limousine—no pressing the flesh or tempting fate as in Sacramento. Terry O'Donnell personally delivered this message to the president. "Fine," said Ford. Exiting the St. Francis he covered the short distance to his car in six paces. Waiting for an agent to open the car door he waived briefly to the crowd on the other side of Post Street.

His hand was still in the air when the crack of a gunshot split the air. Fired from a .38-caliber revolver less than forty feet away, the bullet missed the president. Nicking the wall of the hotel behind him, it ricocheted toward the curb to Ford's left before striking an off-duty cabdriver in the groin (he was not seriously injured). Across the street a pair of San Francisco policemen lunged at a front-row spectator named Sara Jane Moore before she could get off a second shot. She might have succeeded but for Oliver Sipple, a thirty-three-year-old disabled marine who, seeing the chrome-plated weapon in Moore's hand, had reached out with both of his to deflect the path of the first bullet. Meanwhile Ford's agents pushed the president roughly to the sidewalk, from which position he "crawled"—his word—into the limousine. Two agents and Don Rumsfeld piled in on top of him, being careful to stay below window level.

Sirens wailing, the motorcade raced toward the airport at speeds approaching ninety miles per hour. After a few minutes Ford broke the tension by complaining, "C'mon Rummy, you guys get off. You're heavy."

Back at the St. Francis, Ford's attacker was carried headfirst across Post Street and into the hotel's Borgia Room. Under interrogation the dowdy forty-five-year-old fantasist with a past that included five husbands and three abandoned children revealed motives as muddled as her politics. Her thwarted attempt on Ford's life, Moore told the *Los Angeles Times*, "was kind of an ultimate protest against the system." Until recently Moore had been plugged into the Bay Area radical underground as a bookkeeper for People in Need, a $2 million food program established as part of the ransom demanded by the kidnappers of newspaper heiress Patty Hearst.

Yet Moore was also a government informant, familiar to the San Francisco Police as well as the Secret Service, both of which had occasion to interview her within the previous twenty-four hours. Ironically, it was her background as a government plant that in all likelihood kept the Secret Service from detaining Moore when they met with her on the night before Ford's visit. Just a few hours earlier local cops had confiscated an unloaded .44-caliber revolver and over one hundred rounds of ammunition from Moore's cluttered apartment in the low-rent Mission district. The next day, armed with a hastily obtained replacement, Moore joined the crowd outside the St. Francis. Luckily for Ford, Oliver Sipple was there as well. Moore herself blamed her failure to kill the president on the cheap Smith & Wesson .38 revolver purchased that morning. As she blithely related to her interrogators in the Borgia Room, "If I had my .44 with me, I would have caught him." Like Squeaky Fromme before her, Moore was quickly tried, convicted and sentenced to life imprisonment for trying to kill the president.*

The ultimate victim of September 22 was Oliver "Billy" Sipple, the Viet-

---

* Also like Fromme, she was released from prison after Ford's death in 2006. On her release she expressed regret for her actions, and gratification that Ford had survived.

nam veteran who had moved to San Francisco following his honorable discharge from the Marine Corps in 1970. Three days after Sipple ruined Sara Jane Moore's line of shot, Ford wrote him expressing "heartfelt appreciation" for his selfless action in averting harm to the president and others in the crowd. Before Ford's letter could reach him, the hero of Union Square was outed as homosexual by the *San Francisco Chronicle*. His parents in Michigan disowned their son. Sipple turned to alcohol to numb the pain, physical and emotional. He papered the walls of his $334-a-month quarters in the Tenderloin District with newspaper stories recalling the day he saved a president's life. The place of honor was reserved for his letter from Ford, framed for display, icon-like. It was still on the wall when Sipple's lifeless body was found in the apartment on February 2, 1989. It was thought he had been dead for two weeks. He was forty-seven.

S YMPATHY YIELDED TO sarcasm with the October 11 debut of NBC's irreverent comedy series *Saturday Night Live*. No Oliver Sipple was on hand to deflect the verbal shots taken by cast member Chevy Chase at the candidate allegedly running on the slogan "If He's So Dumb, How Come He's President?" Publicly Ford played the good sport. "He didn't let stuff like that get under his skin," said Terry O'Donnell. (It was at Ford's insistence that editorial cartoons, few of them flattering to him, were added to the daily White House news summary.) Even after Ron Nessen hosted the show in March 1976, in an ill-advised effort to show the White House was in on the joke, Ford limited himself to the mild observation that he didn't understand that kind of humor.

His accident-prone reputation would keep *SNL*'s writers busy. On October 14, Ford was in Hartford, Connecticut, to promote his latest economic initiative at a GOP fundraising dinner. With the temporary tax cuts enacted earlier in the year due to expire on January 1, no one wanted to enter an election year saddled with responsibility for what was, in effect, a tax increase. Seizing the initiative, Ford, in a nationally televised address on October 6, had proposed a record $28 billion permanent cut, but only if it was matched dollar for dollar by reductions in the FY77 budget. Transparently political, the package was no more likely to pass Congress than the Energy Independence Authority, whose carefully planned San Francisco rollout had been spoiled by Sara Jane Moore.

Ford's Hartford sequel bordered on farce. Leaving the dinner crowd at the Hartford Civic Center, the president's limousine was struck broadside by a yellow 1968 Buick carrying six teenagers and driven by a nineteen-year-old sheet metal worker named James Salemites. As Ford and two other passengers tumbled onto the floor of their bulletproof vehicle, Secret Service agents sprang into action. Weapons drawn, they surrounded the battered Buick and forced the astonished teens out of the car. Salemites took and passed a sobriety test. It subsequently developed that a local motorcycle cop had failed to cover his assigned intersection. "Well, the cat has nine lives," Ford concluded. The next day he phoned the adolescent driver from the White House, just to assure him there were no hard feelings.

Ford's latest near miss coincided with a more ominous, slow-motion collision touched off by New York City's fiscal crisis, now reaching a critical stage. Liable for an estimated $12 billion in debt, half of it short-term, the city had an operating deficit considerably larger than the $600 million it acknowledged. New York's garbage cost twice as much to pick up as that of San Francisco or Boston, while heavily taxed Con Edison charged the highest utility rates in the country—50 percent higher than in neighboring Connecticut. In June 1975 the state of New York established the Municipal Assistance Corporation (MAC), under the chairmanship of investment banker Felix Rohatyn, to meet the city's borrowing needs through the sale of bonds backed by sales tax receipts and stock transfer taxes. In return for this eleventh-hour lifeline, MAC demanded major reductions in the municipal workforce, a wage freeze, increased subway fares and an end to free tuition at the City University of New York.

By the first week of August, with sales of MAC bonds lagging, Nelson Rockefeller raised the possibility of a federal loan guarantee as a "bridge" to reassure jittery markets. In doing so he dissented from the administration line that profligate New Yorkers must not expect Washington to come to their rescue. Spurred on in part by White House criticism of the city's fiscal mismanagement, Governor Hugh Carey pushed through legislation to create the Emergency Financial Control Board. Mayor Abe Beame protested the loss of home rule. It didn't matter. Henceforth the fate of his city was effectively entrusted to Albany—and Washington. "You've had a great time running around Washington seeking help," Ford told Carey when the two men met at the White House the evening of September 2, "but I know that you know that you don't have the votes."

In an effort to prove him wrong the governor turned to their mutual friend Mel Laird. "Here's your problem," Laird informed Carey. "Jerry's been told by somebody that when New York City collapses, Chicago may become the financial center of the U.S." According to Laird this accounted for the meager support New York enjoyed from the pivotal Illinois congressional delegation. When the governor asked him to identify the source of the claims, Laird replied, "Well, Rummy comes from Illinois." Carey hurriedly scheduled a heart-to-heart with Chicago mayor Richard Daley. Laird, meanwhile, cautioned Ford against beating up on New York, city and state, which might yet play a decisive role in the 1976 election. Reinforcing the message were Republican chairmen representing all sixty-two of New York's counties who assembled at Nelson Rockefeller's Westchester estate on September 26 for his annual Governor's Lunch. These party pros warned the vice president of a backlash at the polls if the Ford Administration appeared indifferent to New York City's fate.

Rockefeller shared the group's reservations at his weekly luncheon with Ford on October 2. An emphatic second was registered by West German chancellor Helmut Schmidt, in town for a meeting of the International Monetary Fund. Responding to Ford's inquiry about the state of the German currency, Schmidt snapped, "Never mind the Bundesbank or the mark. If you let New York go broke, the dollar is worth—*scheisse!*" None of this kept Ford, at an October 8 press conference, from restating his opposition to a New York bailout. Only by keeping the pressure on, he felt, could the city's political class be forced to make painful spending cuts demanded by the markets. At the same time his language was deliberately restrained, and he pointedly declined to comment on any of the prospective legislative remedies working their way through Congress.

The resulting vacuum was quickly filled by his vice president. In a Columbus Day speech on his home turf, Rockefeller called on Congress to avert the "catastrophe" of a New York default once Mayor Beame and the Emergency Financial Control Board produced a credible plan to eliminate the city's deficit by 1978. The unusual show of independence earned Rockefeller an equally rare, if muted, rebuke, the president asking him not to make things "more confusing."*

Ford might wish to avoid it, but confusion was exactly the image being conveyed by the widening differences between himself and his vice president over a New York rescue plan; the ongoing Kissinger-Schlesinger feud and the contradictory signals being sent about SALT II; what much of the White House staff were wont to call "the Hartmann problem" and which the president insisted was a problem he could live with; latent tensions between the East and West Wings; Kissinger's distrust of Ron Nessen and Nessen's distrust of his former colleagues in the press; and Bill Simon's

---

* His vice president was "a bit upset," Ford told Kissinger on October 17. Did he think he would be mollified if asked to accompany Anwar Sadat throughout the Egyptian leader's upcoming visit to the US? "Okay," said Kissinger, "but don't saddle me with him. I have Schlesinger."

leaks to favored columnists conveying his dismay over the deficit and energy policy.

During a two-hour meeting of his informal kitchen cabinet on October 16 Ford listened as Bryce Harlow deplored the "internal anarchy" overtaking the Ford presidency, beginning with a national security team notorious for its lack of cohesion. Plainly Kissinger wasn't going anywhere, although the parameters of his authority were subject to reduction. That left Jim Schlesinger. "Ford could get along with just about anybody," noted journalist Fred Barnes, who covered his White House for the *Washington Star*, "but not some professorial, condescending, I-know-better type."

Even this might have been overlooked had Schlesinger demonstrated a surer grasp of congressional relations. On Monday, October 20, for the only time since he became president, Ford was confined to his bedroom with a severe cold and sinus infection. That morning Schlesinger called a Pentagon press conference to assail the House Appropriations Committee and its chairman, George Mahon, for "deep, savage, and arbitrary cuts" to the Pentagon's $104.6 billion budget request for FY77. Schlesinger's outburst scuttled, at least temporarily, Ford's plan to restore some of the money by working quietly with Senate appropriators. This wasn't the first time the acerbic Cabinet officer had crossed swords with Mahon, so valued a Ford intimate that the president resisted inscribing a photo taken for campaign purposes with Mahon's Republican opponent. Perhaps Schlesinger knew and didn't care. Either way, Ford was determined to replace him.

By Tuesday the president was running a temperature of 101 degrees. His condition did not preclude Vice President Rockefeller from strong-arming energy czar Frank Zarb into a sickroom confrontation over funding levels for Rocky's Energy Independence Authority. "So there was poor President Ford in bed, looking tired," Zarb recalls, "and Nelson makes his pitch and I make mine." Solomon-like, Ford split the fiscal baby—"A bone" tossed to his vice president, Zarb concluded. The next day Ford confided to Rumsfeld his fantasy scenario in which Rockefeller replaced Schlesinger at Defense, and George Bush was brought home from Beijing to become the nation's fourth vice president in two years.

Back in the Oval Office on Thursday, the twenty-third, a still-recuperating Ford spent much of the day closeted with Bryce Harlow. That evening Rumsfeld staged the equivalent of an intervention. In league with Dick Cheney, he had prepared for the president's eyes only a sternly worded performance review.* According to its authors, most of the administration's problems could be traced to Hartmann ("He simply seems not to work well with other people"), Rockefeller and Kissinger. Reflecting their privileged status, the Domestic Policy Council, the NSC and the speechwriting operation functioned as independent satraps. Appointments were often made without reference to the personnel office or political factors. A major Russian grain deal negotiated by the State Department, to the exclusion of Agriculture Secretary Earl Butz, had damaged the president among farmers, a key constituency.

Needless to say, all this reflected badly on Ford, weighing down his poll numbers and undercutting his genuine personal appeal. Of Americans generally, Rumsfeld and Cheney wrote, "They like you as an individual, but have doubts about your performance as president." Much of their memo read like a plea for more centralized oversight and coordination—by a fully empowered chief of staff. Other strictures were no doubt familiar through repetition. Ford should not identify himself "with Hoover or the Republican Party," publicize his time on the Vail slopes, remind people he had been in Washington for twenty-six years or "refuse to see Solzhenitsyn." He should cut back on his travel, including the upcoming November Rambouillet economic summit and a planned trip to China.

The energy issue was a political loser, the Rumsfeld-Cheney memo stated. If Ford nevertheless insisted on making it a priority, he at least should not be so quick to embrace compromise—"It gives the appearance of diving for the middle." That was alien territory to Rumsfeld and Cheney, at least so long as Ronald Reagan was in the wings contesting Ford's conservative bona fides. So they urged Ford to adopt an unabashedly right-of-center agenda. That meant resisting efforts by the Justice Department and Rockefeller's Domestic Policy Council to water down a crime package emphasizing mandatory sentences for repeat offenders, until it emerged from the policy meat grinder a gun control measure. Court-ordered busing was another issue that defied nuance. Tackling a few sacred cows like food stamps would steal Reagan's anti-Washington thunder and refute the popular view of Ford as a split-the-difference pragmatist.

---

* Candid to a fault, the authors of the memo noted criticism aimed at them for running a White House deficient in teamwork. Regrettably they could not defend themselves lest the blame be redirected at Ford personally—"Even though we know and feel deeply that things aren't going right and that we have told you so." Thus they appealed to the sympathies of a president they elsewhere faulted for softheartedness where staff performance was concerned.

Three takeaway lines in the memo stood out:

"Be for tax cuts, and be seen as being for tax cuts."

"You will be reelected on how you govern, not your skill as a candidate."

"Fire someone visibly."

Regarding the last point, any changes should be announced *before* Reagan made his candidacy official. Better yet, no later than November 5, the date selected by the authors as a "Congressional confirmation window." If it was any consolation, Rumsfeld assured Ford, the problems he had identified were "solvable."

The president said he would like to meet with the memo's authors on Saturday morning, the prelude to a longer afternoon session involving Rumsfeld and Kissinger. Sensing a Cabinet shuffle in the works, Rumsfeld advised Ford that he might want to reconsider his plans, since he, the president's closest aide, was thinking of leaving the administration. This startling intelligence was confirmed shortly after eleven a.m. Saturday, when an expanded version of the Rumsfeld-Cheney memo, now grown to twenty-eight pages, was handed to the president. Rounding out this edition were draft letters of resignation from both men. For the third time in six weeks, Gerald Ford had a lethal weapon pointed at him.

F OUR HOURS LATER Ford executed *his* plan for rebooting his presidency. In businesslike fashion he outlined for Kissinger and Rumsfeld a series of personnel actions, beginning with Kissinger's replacement as national security advisor by his trusted deputy Brent Scowcroft. Ford wanted Rumsfeld to take over at the Pentagon, not least because he thought the former congressman would do a better sales job on Capitol Hill. To fill Rumsfeld's current place Ford proposed to make Dick Cheney the youngest ever White House chief of staff, in fact if not in title. At the CIA, director Bill Colby, too badly scarred by the ongoing investigations into his agency to restore its battered morale or implement needed reforms, would make way for George Bush, who had grown restless in his exotic, peripheral posting in Beijing.

Rumsfeld questioned Ford's timing. Firing Schlesinger on the cusp of the 1976 campaign risked creating a martyr for the right and a magnet for anyone unhappy over détente or suspicious of Soviet intentions. Declaring his views on arms control substantially the same as Schlesinger's, Rumsfeld asked for time to reflect on the president's offer. With Ford's permission Rumsfeld sought guidance from Paul Nitze, whose long career as a Cold War strategist made him the military-diplomatic equivalent of Bryce Harlow. To Nitze it was self-evident: Rumsfeld had no choice but to defer to Ford's judgment. Schlesinger was his friend, too, the older man remarked, but where the president was concerned, "They don't get along. It's obvious. And Kissinger and Schlesinger don't get along." Ford needed a secretary of defense in whom he had total confidence, someone who was capable of doing the job and could get confirmed by the Senate. "Tell me two other people that fit that template."

There remained the unpleasant task of notifying the losers in this still unfolding round of musical chairs. For Nelson Rockefeller the music stopped on the afternoon of October 28. Ford chose his words carefully, then and later, when describing the scene. Rockefeller was "too active and dynamic a man, too full of new ideas" to be happy as vice president for another four years. Thus Ford convinced himself that he was freeing his restive vice president from the constraints of a position Rockefeller himself had twice rejected as incompatible with his temperament.

Rockefeller, unsurprisingly, remembered things differently. "If you want to get the record straight," the retired vice president told Ford's ghostwriter, Trevor Armbrister, "he asked me to do it." There had been compliments, no doubt sincerely meant, for Rockefeller's contributions and his loyalty to the team. Apparently Bo Callaway's name had come up as one of Ford's political advisers who believed the more liberal New Yorker's presence on the ticket would diminish their chances against Reagan. Hearing this, Rockefeller the good soldier offered to write Ford a letter explaining his decision to retire from the field. The formal announcement was penciled in for Monday, November 3, separate from the Cabinet and related changes, which would be made public a week later.

Ford was scheduled to address the National Press Club on October 29 about New York City's fiscal crisis, and the speech was the subject of heated in-house debate. At a senior staff meeting a few days before the event, Ford asked if anyone present favored legislation to keep New York City from defaulting on its debts. "The answer is not just 'no,'" said Rumsfeld, "It's 'Hell no.'" For once Bob Hartmann was in agreement with his chief antagonist. "If the president wants to win the election, he will *not* give aid to New York City," said Hartmann. "If he wants to lose the election, he *will* give aid to New York City." Bill Seidman disagreed. A political moderate sympathetic to the vice president, Seidman thought he had pruned the most strident rhetoric from Ford's text, only to recoil on hearing it, and worse, when Ford stepped to the Press Club podium.

Likening New York's free spending habits to an "insidious disease," Ford fairly bristled as he promised to veto any legislative bailout for the city.

Instead he proposed to amend existing bankruptcy laws to permit an orderly transition, during which the city's financial affairs would be entrusted to a federal judge, and the federal government would, if necessary, guarantee continuation of essential services. To Ford it was a teachable moment, with implications far beyond the five boroughs of New York. The leader accused of a vision deficiency asked his listeners to imagine the consequences if New York's spendthrift ways and dodgy accounting became the norm. "If we go on spending more than we have, providing more benefits and more services than we can pay for, then a day of reckoning will come to Washington and the whole country just as it has for New York City." When the inevitable occurred, Ford concluded, "who will bail out the United States of America?"

More than what he said, it was the accusatory tenor of Ford's speech, ramped up from earlier drafts containing no explicit veto threat, that infuriated William Brink, managing editor of the *New York Daily News*. That afternoon Brink toyed with several front-page headlines—"Ford Refuses Aid to City," "Ford Says No to City Aid"—before hitting on five words that memorably captured Brink's sense of abandonment:

"Ford to City: Drop Dead."*

At Elaine's, a fashionable nightspot on Manhattan's East Side, an otherwise despondent Hugh Carey took one look at the early edition of the *Daily News* and chortled, "Now we're going to win." White House political advisers were just as certain the president's line in the sand was a domestic *Mayaguez*. Early polling suggested they were right. A newspaper survey in Charleston, South Carolina, recorded 7,604 readers opposed to federal aid to New York City, with just 263 in favor. And it wasn't only conservatives who approved of Ford's stance. "They spent their way in, they should tax their way out," declared Governor Dan Evans of Washington State, a liberal Republican immune to Reagan's appeal.

S ATURDAY, NOVEMBER 1. All Saints' Day. The Day of the Dead in Mexican lore. A little after six p.m. Ford took a panicky phone call from Kissinger. *Newsweek* had the story of his demotion, he told the president.

---

* Brink's anger did not prevent the *Daily News* from endorsing Ford against Jimmy Carter a year later. Still the headline stung Ford, and the wound remained fresh a quarter century later, according to Hugh Carey's biographers, Seymour Lachman and Robert Polner. While attending ceremonies surrounding his receipt of the JFK Library's Profile in Courage Award in May 2001, Ford spotted in the crowd onetime Carey aide David Burke. "I want to get one thing straight," the former president said to Burke. "I never said 'New York City drop dead.' I never said that."

Ford said he would move up the announcement of personnel changes by a week. "Then we don't get people denying this and denying that."* Ford's schedule had him flying to Jacksonville, Florida, the next day to meet with Anwar Sadat, who was on the last leg of a weeklong American tour. Suddenly everything accelerated, as orderly change became hasty improvisation. Calls were placed to Schlesinger and Colby requesting early morning meetings with each. Ford didn't want either man learning his fate from the news media.

At precisely eight o'clock Sunday morning Bill Colby entered the Oval Office. "Good morning, Mr. President. Jack said you wanted to see me."

Ford didn't bother to get up from his desk. "Yes. We are going to do some reorganizing of the national security structure." He thanked Colby for his service under the most challenging circumstances. He offered to make him ambassador to NATO or, if he preferred, Norway, scene of his World War II heroics under the banner of the OSS. A gentleman to the end, Colby said he would get back to the president shortly (as he did, to decline both appointments with thanks). Fifteen minutes after it began, the meeting ended. On his way out Colby encountered Schlesinger.

"What the devil are you doing here at this hour?" the secretary barked.

Colby muttered something about the latest exposure of CIA operations, in an upcoming *60 Minutes* broadcast.

With Schlesinger cooling his heels outside the Oval Office, Dick Cheney, on his first day as de facto chief of staff, went in to smooth the way. "Now, Mr. President," said Cheney, "have you thought about maybe offering Jim something? Making him an ambassador some place, or at least give him the offer to save face here a bit?" Ford, chewing on his pipe, red spots dotting his cheeks—"That's when you could tell when he was pissed off"—said, "Dick, get him in here so I can fire him."

The next hour was, by Ford's recollection, "one of the most disagreeable conversations I have ever had." As with Colby, he began by making the case for change, emphasizing that Schlesinger was not being singled out. Several times he referred to Schlesinger's "resignation."

"I haven't resigned, sir. You're firing me."

"You can put it that way, Jim. But that's not the way I would like to have it understood. I believe that we can find an important place for you in the

---

* Excellent advice, this. Ford would have been well advised to take it at his hastily scheduled press conference on Monday evening, November 3. By his own later admission, he was "less than candid" in denying that personal and policy differences, especially between Kissinger and Schlesinger, were motivating factors in his assembling "my team."

administration . . . one place I think your experience would be helpful is at the Export-Import Bank."

"That is not something that I could accept."

And so it went, with Schlesinger arguing for his job and Ford, politely but firmly, declaring his mind made up. The contentious session ended with Schlesinger's abrupt departure and Ford's conclusion that he hadn't heard the last from his estranged Cabinet officer.

Later that morning the president flew to Jacksonville. There he completed his reshuffle by phoning Elliot Richardson in London with a request that he take over for David Packard as finance chairman of the nascent Ford campaign. Richardson was more interested in replacing Rogers Morton at the Commerce Department. Ford made the offer, though his enthusiasm was minimal. As he explained the move to Kissinger, "With this Nelson thing I had to balance it out with something for the liberals."

In Jacksonville a distracted Ford joined the Sadats for a dinner hosted by Florida governor Reubin Askew. Cementing the bond forged at Salzburg, Sadat told Ford he considered him a brother. Later, on the tarmac with Kissinger as he prepared to board his plane for the flight home, the Egyptian leader said that he looked forward to staging a welcome for Ford twice as big as that accorded Nixon, whenever it would do his friend the most good politically. "I know you are thinking of resigning," said Sadat. "Don't do it. Ford is a good man. And we need you for a while longer in the Middle East."

L ONG AFTER NEW York secured its federal loan guarantees, the critical first step in a dazzling comeback, Bill Seidman lifted the veil on Ford's strategy. "Although we never said it, we both knew that he was going to give them something in the end, but we were going to get what we needed before that," Seidman explained. Washington's help wouldn't come cheap. The state of New York levied $200 million in new income, estate and cigarette taxes. From the city's retirement system Governor Carey exacted the purchase of an additional $860 million in MAC and city-issued securities. Most important, banks agreed to extend the maturities, and reduce interest rates, on more than $1.6 billion in MAC bonds and city notes. In the end, it was this debt restructuring, and the moratorium on payment amounting to default that enabled the White House to claim victory in the long-running battle to impose fiscal discipline on New York City Hall.

That is how Ford described the outcome to the leaders of Britain, France, West Germany, Italy and Japan, with whom he met for two days in mid-November at Rambouillet, just outside Paris. "The only way we have

achieved results is to be difficult," he acknowledged, "a sort of brinksman-
ship by the Administration forcing New York City and New York State to
take responsible action." With stringent financial controls in place, and
bitter-ender Bill Simon assigned to monitor compliance, Uncle Sam agreed
to loan New York up to $2.3 billion a year, for three years, at an interest rate
1 percent above the national average, the full amount to be paid back each
year by June 30. Seasonal loans were preferred to loan guarantees because
they gave Washington greater control over the implementation of promised
reforms.

For all that it was an uphill slog to pass the bailout bill through a skeptical House of Representatives. After the fact Ford pointed to the narrow,
ten-vote margin as proof his uncompromising stance had made it politically possible for just enough Republicans to swallow their doubts, and
their prejudices. According to Senator James Buckley, New Yorkers owed a
great debt of gratitude to Ford, by then a former president after his failure
to win the Empire State and its forty-one electoral votes against Jimmy
Carter. "By maintaining a hard line, he kept the pressure on the city and
state that assured the adoption of measures that I believe will place the city
back on its feet." In fact the city was able to eliminate its short-term debt by
1978. And a new mayor, former congressman Ed Koch, brought a tough-
minded realism to municipal finances.*

At the same Rambouillet summit, Ford turned in the best economic re-
port card of any leader around the table. He rallied the industrial democra-
cies to oppose protectionism—at one point Britain's prime minister Harold
Wilson called for a new round of trade talks, in the tradition of the earlier
Kennedy Round, named for the American president. "It should not be the
Ford Round," piped up Helmut Schmidt, "because that would be unfair to
General Motors." Schmidt's jest assumed that Ford would still be in office
when such negotiations commenced.

Ronald Reagan had a different take on the future. On November 20,
the day after he notified the president of his planned candidacy in a stilted
four-minute phone conversation, the former governor of California made
it official. Denouncing the Washington "buddy system" as insensitive to
the needs of American workers who supported it with their taxes, Reagan
clearly hoped to kindle a populist brushfire even as he embraced Chamber
of Commerce economics and blamed Ford for record deficits. Initial polling brought smiles to the Reagan camp, as a Gallup poll gave the challenger
a 40–32 percent lead over Ford. A stunning reversal from the same pollster's

---

* The US Treasury made a $30 million profit off the deal.

findings less than a month earlier, Gallup pointed to a volatile Republican electorate.

Events were in the saddle, with Ford benefiting from his tough-on–New York City stance, only to fall behind when his botched housecleaning was perceived as an act of desperation. Said New York congressman Barber Conable, a Ford supporter, "The country still views him as the guy who is filling the gap between Watergate and the next election." Others might see him as a caretaker, but Ford felt he had earned the right to pursue an agenda of his own making, one not defined by crisis management. After fifteen months on the job he finally had a Cabinet of his choosing, and a White House staff imbued with the requisite sense of teamwork.

He was especially pleased with Dick Cheney's performance in his new oversight role. More open and less polarizing than Rumsfeld, Cheney proved as good a listener as Ford himself. "I used to talk to Cheney three times a day," said Bob Schieffer of CBS, "and I wasn't the only one . . . he didn't tell us any secrets. He wasn't 'Here's the list of all the things we've done wrong' but he was straight. He'd tell you something, and if he couldn't tell you something, he'd say, 'I can't talk to you about that.'"

Presumably this extended to rumors, long circulating, that an ailing Justice William O. Douglas was thinking of retirement. On November 10, Phil Buchen delivered to the president a confidential letter from Chief Justice Warren Burger alluding to such a possibility and spelling out some criteria for Ford's consideration in filling the likely vacancy. Simultaneously acting on Ford's invitation, Attorney General Levi submitted a lengthy list of potential court nominees. Eschewing politics, Levi reduced his original battery of candidates to three: Dallin Oaks, a Chicago lawyer and law professor then serving as president of Brigham Young University; former solicitor general Robert Bork; and John Paul Stevens, another litigator-scholar from Chicagoland, who since his 1970 appointment by Richard Nixon to the Seventh Circuit Court of Appeals had established a reputation for judicial independence and literary craftsmanship.

After the long-expected resignation letter from Justice Douglas was placed in the president's hands on November 12, Ford joined Levi and Phil Buchen for a preliminary discussion of these and other contenders. Two of Levi's initial favorites were quickly eliminated from consideration— Bork because of his unsought role in the firing of Watergate prosecutor Archibald Cox, and Oaks because his prominence in the Church of Jesus Christ of Latter-day Saints might be a source of controversy.

Ford's idea of a perfect justice was a pragmatic centrist of transparent integrity, young enough to help shape the court long after he had himself vacated the Oval Office, yet with sufficient judicial experience to preclude the need for on-the-job training. Anxious to avoid a prolonged confirmation fight, Ford gave precedence to intellect over ideology in his list of credentials. Restoring a practice discontinued by the Nixon White House, the American Bar Association was invited to assess leading prospects for their legal qualifications. Within a week of Douglas's retirement no fewer than nineteen names had been provided by the administration to the ABA for review, including those of at least two women—US district judge and Michigan native Cornelia Kennedy, and HUD Secretary Carla Hills.

By coincidence the evening of November 24 was set aside for the annual White House dinner honoring the federal judiciary. Ford's guests were just finishing dessert when he pulled a chair up to a table whose occupants included more than one court finalist. The New York City fiscal crisis provided a lively topic of conversation. Judge Stevens came away from his brief encounter with the president convinced of two things—"His command of facts made it obvious that he was a good lawyer; and he was extremely likeable." Publicly, Ford remained tight-lipped about the direction of his thinking. To Betty alone he tipped his hand, confiding to her on the dance floor after dinner that "the top man on the list is here tonight."

"Man?" she replied. "MAN?"

The rest of Thanksgiving week Ford pored over the legal opinions of Judges Stevens and Arlin Adams of Philadelphia, who would be the first of his faith to occupy the Court's unofficial "Jewish seat" since the 1969 resignation of Abe Fortas. By Friday, November 28, his mind was made up. It wasn't yet noon in Chicago when Stevens's secretary alerted him to a call from the White House. The fifty-five-year-old appellate judge wasted no time in accepting Ford's offer to become the 101st justice of the Supreme Court. Praise for Stevens's nomination was bipartisan, excepting the complaint from the National Organization for Women over Ford's failure to name the first female justice. That wasn't enough to keep the Senate from unanimously confirming Stevens nineteen days after Ford submitted his name for consideration.

Thus began a career on the Court whose thirty-four years were second only to the man Stevens replaced. His long tenure was made notable by Stevens's willingness to reconsider earlier positions in support of capital punishment and against affirmative action. His perceived drift to the left prompted second thoughts among some of those involved in his original vetting. But not Ford. In a 2005 letter proudly displayed in Stevens's chambers, the man who put him on the court said he was perfectly willing to let history's judgment on his presidency rest "if necessarily, exclusively" on his selection of John Paul Stevens to fill the vacancy created by the retirement of Justice Douglas. "He has served his nation well," Ford wrote the organizers of a Fordham University symposium on Stevens's Supreme Court career, "at all times carrying out his judicial duties with dignity, intellect and without partisan political concerns." In complimenting the middle-of-the-road jurist, who came to be seen as the court's leading liberal voice, Ford hinted at his own postpresidential reversion to the youthful One Worlder as committed to human rights as he was to fiscal responsibility.

A MONG THE INCONVENIENT truths spotlighted in the Rumsfeld-Cheney blueprint for change none was more politically sensitive than the issue of Ford's foreign travels. When NATO seemed at best a distraction and Helsinki a political bull's-eye, clearly the president's globe-trotting had reached the point of diminishing returns. Now, almost four years after his predecessor mesmerized the world's media by visiting the Great Wall of China, Ford was hard-pressed to justify a return trip in December 1975. With both Party Chairman Mao and Premier Zhou Enlai in deteriorating health, there was little chance of making substantive progress toward full diplomatic recognition, even if Ford, confronting a serious challenge from his right, had been more willing to write off Taiwan as the price of normalized relations. Even Kissinger downplayed prospects for anything significant emerging from the three-day visit. "We are planning only one substantive meeting per day," he told the president. "There isn't that much to say."*

At the welcoming banquet in Beijing's Great Hall of the People on December 1, First Deputy Premier Teng Hsiao-p'ing announced, "The wind sweeping through the tower heralds a rising storm in the mountains." Less elliptically, Zhou Enlai's surrogate lambasted "hegemonism," his government's code word for Soviet aggression. He was equally blunt in decrying US policies of peaceful coexistence with Moscow. "Rhetoric about 'détente' cannot cover up the stark reality of the growing danger of war," Teng lectured his American visitor.

He continued in this vein the next morning, likening the Soviets to Nazi Germany, and détente to '30s-style appeasement. Pushing back, Ford reminded Kissinger's "nasty little man" of ongoing efforts by the United States to strengthen NATO, exclude a Communist Portugal from the West-

---

* There was much to see, however, beginning with an Alaskan stopover and a visit to the future pipeline it was hoped would liberate Americans from "unreliable" sources of foreign oil.

ern alliance and limit grain sales to the Soviets. As for Teng's historical analogies, "It is true that the West made some mistakes against Hitler," the president told him, but the invasion of Poland brought a belated response. In the East no resistance flickered until Hitler's armies invaded their erstwhile partner, the Soviet Union. "So we all made mistakes. Let's not repeat them in the future."

On the afternoon of December 2, the president, First Lady and daughter Susan were notified that Chairman Mao was prepared to receive them at his residence in the Forbidden City. A few minutes after four p.m. the eighty-one-year-old Great Helmsman was helped to his feet to greet his American guests. His waxen features brightened perceptibly as he took eighteen-year-old Susan's hand in his. Much would be made of the hour and fifty minutes that Mao spent with Ford, twice as long as his meeting with Nixon in February 1972. Then again the chairman's bout with Lou Gehrig's disease made communicating twice as onerous, with Mao requiring the services of two interpreters, one to interpret the indistinct sounds coming out of his mouth, and another to translate what she was told into English. Picking up where Teng left off, Mao focused the conversation on "the Socialist Imperialists . . . the one in the North."

Ford offered assurances that Soviet hegemonism could be restrained if "you put pressure from the East, and we will put on pressure from the West."

"Yes. A gentleman's agreement."

The American president broadly hinted at formal recognition by Washington after the 1976 election. "This is just talk," Mao mumbled. More talk ensued, of Yugoslavia after Tito, Spain after Franco, and the recent Sinai Agreement curtailing Soviet influence in the Middle East. More immediate was the threat posed by Soviet-backed troops and Cuban mercenaries battling not one but two independence armies in the former Portuguese colony of Angola.

"You don't seem to have many means," said Mao. "Nor do we."

To the contrary, Ford said, before leaving Washington he had approved $35 million in covert funding to combat the avowedly Marxist MPLA (Movimento Popular de Libertação de Angola). A recent invasion by pro-Western forces aligned with South Africa's apartheid regime had convinced the Chinese to back away from the tangled conflict. For now Mao appeared content to cheer the Americans on. As for the future of US-Sino relations, "We can go at it bit by bit."

"We will work on it, too," said Ford.

After the high drama of Nixon's 1972 breakthrough, Ford's China sequel felt like treading water. The picturesque backdrops had lost their novelty. If the trip had an electrifying moment, it came when Betty Ford kicked off her brown leather pumps and joined the dancers at the Central May Seventh College of Arts in a quasi-spontaneous turn that landed on front pages and newscasts around the world.

Leaving Beijing, a hastily arranged visit to the Indonesian capital of Jakarta was reciprocation for a Camp David meeting earlier in the year between Ford and Suharto, the military strongman who professed to rule by consensus the fractious country of 130 million people, 6,000 islands and 300 ethnic groups. With the fall of Saigon fresh in memory, Washington was looking for Asian allies wherever they could be found, and at this relatively early stage of his thirty-two-year dictatorship, Suharto could boast of restoring order and fostering unprecedented prosperity in the world's largest Muslim nation. That he should wish to extend his authority over the island of Timor, long divided between Dutch and Portuguese colonial rulers, set off no alarms in the West.

East Timor had very different plans following the April 1974 Carnation Revolution that hastened the breakup of Portugal's vestigial empire. Rival nationalist movements competed for popular favor among the colony's eight hundred thousand inhabitants, 98 percent of them Catholics. In July 1975 the Revolutionary Front for an Independent East Timor, or Fretilin, scored a decisive win in local elections. When the victors formally declared independence at the end of November, Jakarta countered with its own proclamation treating East Timor as the island nation's twenty-seventh province. It was against this backdrop that Ford arrived in the Indonesian capital on December 5. His visit was short, barely twenty-four hours, but it was to have historic consequences.

From the outset Ford was operating with diminished leverage, being indebted to the Suharto regime for its refusal to go along with OPEC price increases. A fervent anti-Communist, Suharto dusted off the old domino theory to raise doubts about the future of Thailand, Malaysia and the Philippines. He portrayed pro-independence forces in East Timor as "Communist-inspired." With Congress threatening to end US aid to pro-Western combatants in Angola's complicated civil war, the Ford Administration was content to have one fewer of Portugal's former colonies become a Cold War battleground.

"We want your understanding if we deem it necessary to take rapid or drastic action," Suharto told the American president.

Ford, just as covert in his response, promised that the US "will not press you on the issue. We understand the problem and the intentions you have."*

Indonesia invaded its neighbor before Ford was back in Washington. Civilians were slaughtered in the thousands. Some Fretilin sympathizers were crushed by bulldozers; others were dropped into the sea from helicopters. The ensuing carnage was concealed from the outside world through the simple expedient of killing all foreign reporters. Over the next few years programs of forced sterilization and man-made famine reduced the population of East Timor by 20 percent or more. As recounted by Andrew Downer Crain in his history of the Ford presidency, Ford lived to express remorse over the tragic consequences of his inaction. Having no wish to alienate Indonesia, he confessed to historian Douglas Brinkley, neither he nor Kissinger were inclined to object when Suharto all but announced his aggressive intentions. "I don't want to pass the blame," Ford added. "Given the brutality that Indonesia exhibited in East Timor, our support was wrong . . . We needed allies after Vietnam. Henry—and I'm not exonerating myself—goofed."

It was a mistake ratified by three successive American presidents of both parties for whom a strategic alliance with the aging despot in Jakarta took precedence over any stated commitment to human rights. In May 1998, having enriched himself and his family to the tune of at least $15 billion, and with his country's economy in tatters, Suharto was driven from office by pro-democracy demonstrators. Only his deteriorating health saved him from prosecution for his crimes. In 1999 three-quarters of East Timor voters opted for independence in a UN supervised referendum. Another three years were to pass before the Indonesian occupation came to an end, and Timor-Leste was formally welcomed into the family of nations.

ENTERING THE FINAL weeks of 1975 many Americans, when asked by pollsters to identify a major Ford legislative accomplishment, drew a blank. No doubt the president's extensive use of the veto contributed to this attitude, as did his stay the course management of the economy and stubborn opposition to new initiatives until the national house was reordered to his satisfaction. A still bigger factor was the legislator's outlook Ford brought to the most executive of positions. The muddle over energy and tax policy—ironically two areas in which Ford *had* tried to seize the

* Kissinger offered a feeble caveat, telling Suharto, "You appreciate that the use of U.S.-made arms could create problems." Beyond that, Kissinger noted, "it is important that whatever you do succeeds quickly."

initiative—reflected the crablike movements of congressional dealmaking far more than the heroic presidency celebrated by court historians and Hollywood mythmakers. It was the difference between D-Day and trench warfare, or, to put it in more personal terms, between the USS *Monterey* and *PT 109*.

On December 17, Congress passed a six-month extension of the $18 billion tax reduction approved in the spring as part of Ford's economic recovery package. No spending cuts were included in the bill, which Ford vetoed as soon as it reached his desk. Much to his surprise the House within twenty-four hours upheld the veto. At this point the calendar practically mandated compromise, as no one wanted to go home for the holidays only to be blamed for a rise in withholding rates come January 1. Two days before Christmas, Congress adopted nonbinding language linking a tax rate cut to Ford's demand for equivalent reductions in spending. In the end the president got very nearly the $395 billion spending plan he wanted, given billions in unspent appropriations that, ironically, might have stimulated the economy in time to aid his reelection campaign.

Just as grudging was Ford's endorsement of the Energy Policy and Conservation Act of 1975, following a yearlong standoff between the free marketeers in his White House and a Democratic majority in Congress committed to cheap energy. In place of the immediate decontrol of domestic oil prices sought by the president, lawmakers permitted gradual deregulation over forty months. First, however, they insisted on rolling back existing prices on domestic crude from $8.75 a barrel to $7.66. This MO of one step forward, two steps back guaranteed intense opposition to the bill from Bill Simon and Alan Greenspan. The president took their views seriously. By making a show of vetoing the legislation and daring Congress to give him something better, he might have drawn a line in the sand, impressing conservative primary voters with his fidelity to principle and dispelling the fuzzy aura of incrementalist leadership.

Yet a veto also risked an abrupt lifting of all price controls, with corresponding increases in fuel and home heating costs—hardly a record on which to win votes in New Hampshire's midwinter Republican primary. Moreover, Congress was not the Cambodian navy. A veto would imperil the half loaf Ford had extracted from a hostile majority on Capitol Hill. Phased deregulation over a protracted period was better than no deregulation at all. "He knew we wouldn't get everything we wanted with full price deregulation, but we'd get a start," said Frank Zarb, who strongly advised the president to sign the legislation, with all its imperfections. "Ford's view was, when I'm elected I'll go back for the full plan and with the election behind me I'll get it."

Besides, as Zarb was quick to point out, the bill included much that Ford had asked for. Fulfilling a White House ambition expressed as early as 1944, Congress authorized a Strategic Petroleum Reserve with up to a billion barrels stockpiled against a national emergency; expanded powers for the executive in the event of another oil embargo; incentives for more environmentally friendly coal production; and tough new efficiency mandates for electrical appliances. The same omnibus legislation contained the first CAFE standards requiring greater mileage per gallon in US automobiles, up 50 percent by 1985.*

Like the New York City rescue plan, this first draft of a national energy strategy was a victory that didn't feel like one, the full benefits of which would not be felt until long after Ford was out of office. Concluding it was the best he was likely to get out of a Congress eager to make him a one-term president, Ford signed the legislation on December 22, the same date *Newsweek* featured the president's image on its cover under the heading "Ford in Trouble."

I F NOT THE worst month of Ford's presidency, December 1975 was surely the roughest stretch since April's nightmarish last act played out in Saigon. On Capitol Hill, Senate Democrats conditioned George H. W. Bush's confirmation as CIA director on his Shermanesque withdrawal as a potential Ford running mate in 1976. In a concession to his nominee's bruised feelings, the president himself signed the letter to the Senate Armed Services Committee after Bush told him, "I just can't put it in writing."

An even less palatable choice confronted Ford when he met with his secretary of labor, John Dunlop, to discuss common-situs picketing legislation, a painstakingly crafted compromise growing out of a 1951 Supreme Court ruling that outlawed secondary boycotts of a construction site when the striking union's dispute was limited to a single contractor or subcontractor. Over the years repeated efforts to devise a legislative formula acceptable to both organized labor and the construction industry had fallen short. When Dunlop early in his tenure expressed confidence he could do better, Ford encouraged him to try. Each man had a powerful incentive to succeed: Dunlop as part of a grand deal with the Teamsters Union to achieve labor peace in the construction industry; and Ford in hopes of gaining support from the Teamsters and building trades unions for his reelection campaign.

Twice that spring Ford examined the issue with Robert Georgine,

---

* By the 1990s the CAFE standards alone were credited with conserving as much as three million barrels a day.

president of the AFL-CIO's Building Trades Department. At length Georgine and union president George Meany were won over to a bill that permitted secondary boycotts at construction sites while banning wildcat strikes and requiring local unions to obtain permission from the international before walking off the job. Finally, Dunlop's version of common-situs would entrust oversight of the collective bargaining process to a new labor-management panel, an innovation with implications far beyond the construction industry. With the labor secretary having made good on his vow to the president, the ball was in Ford's court. By all accounts he was prepared to honor his conditional promise to Dunlop and sign the legislation following its passage early in December. Then the Associated General Contractors of America reversed its earlier support of the compromise bill after being bombarded by an angry membership. Seven hundred thousand letters, telegrams and phone calls swamped the White House, the vast majority demanding that Ford veto Dunlop's handiwork. Senator John Tower threatened to resign his chairmanship of Ford's election campaign in Texas if the president approved the bill.

On December 11, Ford told Dunlop frankly that to defy public opinion on this issue risked ceding his party's nomination to a surging Reagan. "John, what good will I do to you if I'm not reelected?"

Dunlop's letter of resignation was dated January 13. For once the ritualistic expressions of mutual regret were, if anything, understated.

As the year wound down, Ford was being squeezed between the Reagan insurgency on his right, and the Watergate class of congressional Democrats sworn to apply the lessons of Vietnam through the War Powers Act. On December 19, following revelations in the *New York Times* of covert US involvement in Angola's messy civil war, the Senate voted 54–22 to cut off military assistance to pro-Western elements in the former Portuguese colony. When the House ratified the Senate ban on US aid, Ford accused lawmakers of having "lost their guts." But he reluctantly went along, the alternative being to put at risk the entire Defense Appropriations act.*

---

* Another casualty of the Angolan conflict was the Ford Administration's halting attempts to improve relations with Castro's Cuba. In July 1975, Kissinger deputy Lawrence Eagleburger met with Cuban diplomats amid great secrecy at New York's Pierre Hotel, after which the US government announced modifications to the economic embargo imposed on Cuban trade since 1961. The diplomatic initiative fell apart when Cuba convened a Puerto Rico Solidarity Conference to promote the island's independence from Washington, and Castro dispatched over thirty thousand combat troops to bolster Marxist forces in Angola. On December 20, ten weeks before a Florida Republican primary whose outcome was likely to be determined by Cuban-American voters in and around Miami, Ford publicly ruled out any further outreach to Havana so long as Cuban forces remained in Africa.

Twenty-five years on Capitol Hill, most spent in the ranks of the minority, had taught Ford not to personalize policy setbacks. They went with the job. Much harder to accept were complaints on the home front about his family's goldfish-bowl existence. Two weeks before Christmas, Steve Ford confessed to his parents that he had been skipping classes at California State Polytechnic University. To the putative rancher the classroom held scant appeal, life in the White House even less. His brother Jack, having missed his father's October 1 deadline for finding a job, was entertaining second thoughts about a travel agency position arranged for him by Sheila Weidenfeld's husband. For Jack, perhaps more than any of his siblings, celebrity was proving to be a double-edged sword. At one point late in 1975 an anonymous tip to law enforcement claimed that he had been seen purchasing cocaine in Salt Lake City. Jack's father, as the nation's chief law enforcement officer, instructed agents to proceed as they would in any such case. Investigators quickly established that Jack had been nowhere near Utah's capital at the time of the alleged transaction.

"If this is what the White House is going to do to my family, then I will get out of the White House," Ford raged. "I just won't run." The storm passed. His 1976 campaign would have no more enthusiastic surrogates than Jack and Steve. In the meantime, Ford's loved ones united in common antipathy toward Bob Hartmann. Betty Ford joined David Kennerly in lobbying her husband for his chief speechwriter's replacement. Kennerly had drafted a memo to the president blaming Hartmann for the administration's seeming inability to communicate a coherent story line.

"And he read it," Kennerly recalled, "and he put it in his pocket and he said, 'Well, thank you for saying that. I really appreciate your honesty and your directness' . . . And nothing really happened."

Dick Cheney enjoyed no greater success in his attempts to separate Ford from Hartmann. "I made an effort at one point to get him to appoint Bob as an ambassador. To get him out of town. And he wouldn't do it." Even though, as Cheney is quick to add, "He knew Bob was a problem." To Don Penny, a former actor turned speechwriter and performance coach hired at the urging of his friend Dave Kennerly, Ford confided matter-of-factly, "You need to talk to Bob before noon."

"Why's that?" said Penny.

"Because he drinks."

After one particularly grueling day on the road, *Air Force One* landed at what was then Andrews Air Force Base. "Bob was booked on the helicopter to fly into the South Lawn with the president and myself and a few others," according to Cheney. "We got on the helicopter—there's no

Hartmann. So I sent somebody back to check, and Bob's passed out on [*Air Force One*]. The president waited and waited, and finally we had to leave him and take off."

Ironically, Hartmann shared many of the same strategic concerns that had found their way into the Rumsfeld-Cheney white paper. While still in Beijing as part of Ford's traveling White House, he had drafted his own memo warning the president that he faced certain defeat in the coming election unless he dramatically altered public perceptions of his leadership. Under the circumstances the 1976 State of the Union address, scheduled for January 19, took on more than the usual significance. As inspirational as a report to the stockholders, these constitutionally mandated presentations typically possessed a shelf life no greater than the average New Year's resolution. This time, however, was different. Coming at the start of America's bicentennial year, the SOTU afforded Ford a unique opportunity to dispense with the usual laundry list of programs targeting poll-tested audiences. By all means break with tradition, said Ford campaign pollster Bob Teeter. A more thoughtful SOTU might help to recast the presidential image while setting out a positive agenda for a second Ford term.

As early as October 1975, Ford had begun sending handwritten notes, relevant quotations and speech outlines for Hartmann's review. In a paperback edition of Revolution-era writings, he discovered Thomas Paine's famous admonition from *The Crisis*: "These are the times that try men's souls. The summer soldier and the sunshine patriot will, in this crisis, shrink from the service of their country; but he that stands it *now*, deserves the love and thanks of man and woman." There had been much in 1975 to try men's souls, the president seemed to be saying. That included the struggle to fashion a State of the Union at once coherent and politically helpful.

Bob Hartmann was under no illusion that he had the field exclusively to himself. Ford had at least three, and probably more, groups working up ideas. "Mostly without knowledge of one another," Hartmann said, "although we did have certain knowledge, spy networks." Jim Cannon and the membership of Rockefeller's Domestic Policy Council were tapped for thematic suggestions. Bill Baroody from the Office of Public Liaison tried his hand at a draft. And there was David Gergen, the Nixon White House speechwriter more recently employed by Bill Simon. One of Dick Cheney's earliest acts as chief of staff was to reclaim Gergen and install him as "our secret speechwriter" (resentful wordsmiths in Hartmann's shop referred to Gergen as Roman Numeral II).

Ford joined the conspiracy, sometimes going to comical lengths, as described by Cheney. "I'd go down at night and the president would give me the latest draft that he'd gotten from Hartmann on a speech . . . I'd have Gergen rework it and rewrite it. I'd take it and give it back to the president. The president would give it to Hartmann, as though it were the president's own words. Hartmann would accept that." But there were limits to what Hartmann would accept.

It was at his insistence that Ford, in the first week of January, told his assembled staff that Hartmann was in charge of preparations for the State of the Union. With Camp David booked up, Hartmann convened a brainstorming session for January 9–11 at Colonial Williamsburg. This produced the opposite of consensus, and a bulky first draft some two hours in length.* On January 15, Hartmann delivered a more polished version for Ford's review. "And it turned out to be a total laundry list," in Gergen's recollection, a dreary, warmed-over legislative stew devoid of vision or political sex appeal. Along with Alan Greenspan, Gergen had been tasked "over Hartmann's objections, but with the president's approval" to write an alternative. "And it was very thematic." Roman Numeral II wanted Ford to acknowledge, by way of background, an overreaching government whose promises had exceeded its performance. "We tried to be a policeman abroad," Ford would tell his countrymen, "and the indulgent parent here at home." Such hubris had upset the delicate equilibrium of Madisonian democracy. "A new realism" was called for, and a new balance between the public and private sectors, between the individual and his government, between Washington and the states, between domestic spending and national defense.

On Thursday night, January 15, Ford took the rival speech drafts and, in an exercise akin to ordering from a Chinese dinner menu, selected the most appetizing passages from each. A Friday-afternoon review was attended by no fewer than thirteen would-be editors. As the debate stretched into its third hour, Ford called for a show of hands. All but two were raised on behalf of the Gergen-Greenspan draft. "The two votes against were Hartmann and Jerry Ford," says Gergen. "And guess which speech he gave." As late as Sunday afternoon Dick Cheney was shuttling between the Oval Office and the speechwriting group, sans Hartmann, who were watching the Dallas Cowboys and Pittsburgh Steelers battle to the last play of Super Bowl X. It was a fitting metaphor for the contest of wills that wasn't decided

---

* At one point in his text, Hartmann blamed a loss of public confidence in government on its being "too big and too bumbling." Ford scratched out "bumbling" and wrote in its place "impersonal." Robert Schlesinger, *White House Ghosts: Presidents and Their Speechwriters from FDR to George W. Bush* (New York: Simon & Schuster, 2008), 255.

until just before halftime, when Cheney appeared in the Roosevelt Room clutching a heavily edited hybrid—part new realism, part laundry list, an Americanized Queen's Speech that did not fail to recommend tax changes to promote industrial plant expansion or claim credit for reform of the international monetary system.

More a missed opportunity than an outright disaster, the speech left White House image makers scrambling for ways to disarm the president's detractors. In fact Gergen and company had a fresh stunt in mind, a high-wire act that hadn't been attempted in a quarter century. With New Hampshire Republicans going to the polls in less than five weeks, the consequences of failure could be irretrievable. But Gerald Ford hadn't spent all those years on the Appropriations Committee for nothing, as a Yankee electorate, and the Washington press corps, were about to discover.

25

THE ROAD TO KANSAS CITY

———◆◆◆———

One actor in the campaign is enough.

—GERALD R. FORD, AUGUST 1976

"THE REAL EXPRESSION of an administration policy is in the budget documents," maintained Paul O'Neill, Ford's OMB deputy director (and future secretary of the Treasury under George W. Bush). "It may be dry and boring to some people, but to people who are insiders, they know that's where the priorities are set." Few presidents of the modern era have shared O'Neill's enthusiasm. Typically the annual budget presentation entails a short opening statement preceding a battery of questions directed at OMB bean counters, the whole exercise lasting barely half an hour. Not since Harry Truman in 1950 had a chief executive submitted to the scrutiny of reporters eager to trip him up on his command of budgetary minutiae.*

O'Neill reminded Ford of this while both men were working late one night to finalize a spending blueprint that would also, inevitably, become a campaign manifesto. He should emulate Truman's example, he told the president, and present the budget personally "because you know this budget so well, you don't need any of the rest of us." No one had to point out to the president the risks he would be taking. Harry Truman's government spent $43 billion annually. The budget for fiscal year 1977 was nine times as large and correspondingly complex. Yet something about the challenge appealed to Ford. At the next budget meeting he said, "I think I want to do that. I know I can handle it." On the morning of January 20, Ford strode onto the stage of the State Department auditorium, where three hundred journalists jostled for space with as many government officials. This year's budget presentation would be a little different, he explained. After brief

---

* It was no accident that both Truman and Ford had served on the Appropriations Committee in their respective legislative branches. Truman would look back on his Senate career as the happiest ten years of his life. Ford had similar feelings for the House.

introductory remarks he opened the floor to questions, fifty-six of them in ninety minutes.*

"It was open season," according to Paul O'Neill. Fluent in the language of government accountants, Ford was equally adept at translating their jargon for the uninitiated. He unraveled the mysteries of health billing, boasted of administrative savings to be achieved through the consolidation of "15 or 16" categorical grant programs and explained—to his own satisfaction at least—how one could recommend a $10 billion tax cut while increasing Social Security benefits by the same amount.

Ford's advisers were understandably delighted with his performance. Their joy was dampened, however, on learning of the decision by the press office to embargo any television broadcast of the briefing until the budget was officially released the next day. This concession to congressional protocol deprived millions of viewers of the opportunity to see and hear Ford at the top of his form. But the exercise hadn't been wholly in vain. "I would swear that was the first time reporters understood that Ford was a bright man," insisted David Gergen. "Much brighter than they imagined."

That said, it was not immediately apparent how Ford's mastery of federal spending habits would translate into votes on primary day in New Hampshire. Enter Stu Spencer, a low-slung, forty-eight-year-old Californian with a leprechaun's grin and the mouth of a stevedore. Sixteen years after he and his professional partner, Bill Roberts, founded their eponymous political consulting firm, Spencer-Roberts was regarded as the gold standard of campaign management. In 1962 the pair had been instrumental in reelecting Senator Thomas Kuchel, a liberal Republican, who returned the favor by recommending them to Nelson Rockefeller, then gearing up for a 1964 White House bid against conservative favorite Barry Goldwater. Hired to manage Rockefeller's faltering campaign in the Golden State, Spencer-Roberts narrowed a 32 percent gap in the polls to just 3 percent on primary day. A badly scarred Goldwater advised Ronald Reagan to hire "the sons of bitches" to run his 1966 campaign for governor.

Spencer's habit of speaking truth to power was demonstrated when Reagan, playing coy about a 1968 presidential candidacy, observed piously that "the office seeks the man." "That's bullshit," Spencer retorted. "If you want

---

* He didn't forget the budget experts, accustomed to toiling in obscurity, who flanked him at a long table onstage. At one point, having comprehensively replied to a question about Medicare funding, Ford turned the spotlight in another direction. "But I think Paul O'Neill down there can give you a more detailed answer." AI, Paul O'Neill.

to be President of the United States you've got to go get it and you've got to fight for it." Yet by the time Reagan was ready to climb in the ring Spencer was missing from his corner. Angered at the palace guard around the governor for undercutting his consulting business, Spencer was ready to listen when Don Rumsfeld reached out to him in the summer of 1975. The two men were deep in conversation in a Sacramento hotel at the moment Squeaky Fromme was aiming her Colt .45-caliber pistol at Gerald Ford a hundred yards away. "End of conversation," says Spencer.

The connection was reestablished a few days later at the party's state convention, where campaign manager Bo Callaway lobbied him hard to join the president's effort. Shut out of the Reagan operation and eager to escape an unhappy marriage, Spencer committed to work on the first two primary contests, in New Hampshire and Florida. Late in September, Ford's new director of political organization walked into a Washington headquarters torn by personal rivalry and strategic conflict. "I almost got a plane and came home," recalled Spencer with a shudder. Adding to his dissatisfaction was a legion of Ford's friends, most veterans of Capitol Hill, headed by Minority Leader John Rhodes and "fuckin' Mel Laird," who were more than generous with advice and criticism.

Spencer poured out his grievances to Bryce Harlow. He feared the White House was underestimating Reagan, while exaggerating Ford's capacity for absorbing early losses. "These people don't realize that if they don't win New Hampshire they are dead meat," he told Harlow. In any event, he said, after Florida on March 9 he was going home.

Two hours later Spencer's phone rang. "Come on over to dinner tonight," Ford told him. "You're not going anywhere."

"Okay, then the ground rules change. Your goddamn friends"—whom Spencer proceeded to name at some length—"they've got to get off my back . . . They can be helpful, but let me go to them to ask for their help. They're taking up four hours a day."

"I'll take care of it," Ford assured him.

"I never heard from them again," says Spencer.

He had his work cut out for him, beginning in New Hampshire, where Congressman Jim Cleveland was too focused on his own reelection to build a statewide organization for the president. The Granite State was in many respects tailor-made for Reagan's insurgency. With a population under 750,000, it provided an ideal setting for old-fashioned retail politicking, at which the Californian excelled. Reagan's celebrity drew crowds, and his unassuming manner disarmed skeptics. Even Ford conceded his opponent's mastery of the one-liner. Recycling his trademark four-by-six-inch cards,

Reagan offered a snappy put-down of détente, whose chief benefit for the US was "the right to sell Pepsi-Cola in Siberia."

Reagan's anti-Washington message had strong appeal to a Yankee (and French Canadian) electorate wedded to tradition and suspicious of activist—read: expensive—government. Residents of the Granite State prided themselves on having neither an income nor a sales tax. Republican governor Meldrim Thomson Jr., elected in 1972 on a promise to "Ax the Tax," was determined to keep it that way. Thomson's preference for Reagan over Ford had never been in doubt, although the White House was probably just as glad to keep its distance from the polarizing governor, who had once proposed to arm the state's National Guard with tactical nuclear weapons.

Thomson was a mere irritant compared to the *Manchester Union Leader*, New Hampshire's only statewide newspaper and the personal megaphone of publisher William Loeb, a bullet-headed crank whose national reputation rested on his abusive treatment of "Dopey Dwight" Eisenhower, "Moscow [Ed] Muskie" and "Kissinger the Kike." Ford received double the scorn accorded his predecessors. Originally designated Jerry the Jerk, he was rechristened "Devious Gerald" during the primary season, when he visited the state accompanied by his "stupid" and "immoral" wife.

"Pay no attention to him," Ford instructed his managers. Setting aside Loeb and the *Union Leader*, the worst wounds suffered by the Ford campaign were mostly self-inflicted. To a reporter who asked if the president would be taking to the local ski slopes before the March primary, Ron Nessen replied, truthfully, that Ford preferred Colorado powder to the icy terrain and unpredictable snows of New Hampshire. The resulting uproar quieted only after Susan Ford was dispatched to ski the competition course of Mount Cranmore in the state's far north. Such damage control wasn't Stu Spencer's idea of how to win an election. "I'm a guy who swings for the homerun," he vowed. Or, as in the case of his former client-turned-adversary, the glass jaw. "I know one thing about Reagan," Spencer said. "I know that he was a rhythm candidate . . . and my whole goal was to get him out of rhythm in New Hampshire every chance I could get." That meant confronting him with questions he couldn't handle, or something he may have said in the past "that is kind of stupid now." Disrupt his rhythm, Spencer implied, and Reagan might take a week to recover.

Within days of his arrival in Washington, Spencer received a phone call from a Chicago-based reporter for the *Washington Post*, angry because his editors had buried his story on page seven. Had Spencer heard about a recent Reagan speech delivered in the Windy City? What Reagan biographer Lou Cannon calls "the speech that arguably prevented Reagan from winning the 1976 Republican presidential nomination" was the product of Jeffrey Bell, a former Nixon campaign staffer and Capitol Hill director for the American Conservative Union. In promoting a smaller, decentralized government, the September 24 address—formally titled "Let the People Rule"—echoed many of Ford's priorities as expressed through programs like revenue sharing and block grants to the states. Bell, however, went much further, arraigning the federal government as the culprit behind rampant inflation, soaring deficits, crippling unemployment and a choke hold on once-abundant energy supplies.

Reagan's answer was the New Deal in reverse—"Nothing less than a systematic transfer of authority and resources to the states—a program of creative federalism for America's third century." Handing off to state and local officials responsibility for "welfare, education, housing, food stamps, Medicaid, community and regional development, and revenue sharing," Reagan said, would save $90 billion—enough for Washington to balance its budget, make a token $5 billion payment on the national debt, and reduce federal taxes on individuals by 23 percent. It all added up to the most sweeping blueprint for change in four decades. Yet, bafflingly, the national press corps had virtually ignored Reagan's proposal.

Which explained the telephoned tip to Stu Spencer. Admitting "I'm not a big policy wonk," Spencer nevertheless concluded that "this doesn't look good. So I turn it over to some research guys and said, 'I want you to take this and extrapolate it. I want to know how much this would cost the people of New Hampshire, of Florida and every other state.'" Meanwhile Ford on his own asked Jim Lynn at OMB to analyze Reagan's numbers. Only drastic cuts in program spending, Lynn reported back to Ford—reductions of 70 percent or more—could avoid major tax increases resulting from what became notorious as Reagan's "$90 billion speech."

Stu Spencer put it more trenchantly. "It was going to cost everybody in New Hampshire twelve grand or so," he observed. He assembled a press kit in league with Peter Kaye, a straight-talking former *San Diego Union* reporter. The two men notified their journalistic contacts to expect something big. "The day we hit it," Spencer remembered, "it was on the seat of every press guy on the Reagan plane, as well as every press guy who was traveling with Ford."*

---

* Candid to a fault when interviewed over two days in 2010, Spencer clammed up when asked how he had managed to infiltrate the Reagan campaign plane with Ford propaganda. "You have ways," he finally allowed.

R EAGAN'S RESPONSE TO this manufactured controversy was all Spencer could have wished for. "He stumbled all over New Hampshire for
three days. Couldn't answer the damn question . . . It bought us a week's
time." Ford piled on in his State of the Union address. "Complex welfare
programs cannot be reformed overnight," he told a national television audience. "Surely we cannot simply dump welfare into the laps of 50 states,
their local taxpayers, or their private charities, and just walk away from it."

On February 5, just as he was beginning to close the polling gap with
Reagan, it was Ford's turn to be surprised when the Chinese government
revealed that Richard Nixon had accepted an invitation to revisit the People's Republic—the same week New Hampshire Republicans went to the
polls. "Nixon is a shit," muttered the straitlaced Brent Scowcroft, whose
reaction pretty much summed up the feelings of the Ford White House.* In
a telephone conversation initiated by Ford, his predecessor assured him of
his support, stressed the utility of his contacts with a Chinese government
in flux following the recent death of Premier Zhou Enlai and reiterated that
he was in no way impinging on official diplomacy. Publicly Ford downplayed the trip, even expressing delight that the former president's health
would permit him to make so arduous a journey.

None of this deterred a thousand Ford supporters from showing up to
greet the president and First Lady at a reception in the New Hampshire
state capital of Concord on February 8. Before a capacity crowd of 3,500
at the University of New Hampshire in Durham, Ford cited a battery of
economic numbers to prove that his policies were working. The nation's
unemployment rate, for so long hovering at or above the 8 percent level,
had slipped to 7.7 percent. Virtually all of the 2.1 million jobs lost in the
recession had been recovered. Personal income for 1975 was up 9.2 percent, well above the rate of inflation. "That means real earnings, real purchasing power is climbing."

Reprising his recent budget presentation, Ford answered a litany of student questions despite persistent heckling from representatives of the People's Bicentennial Commission. On the environmental front, Ford took
credit for $6.5 billion set aside to help local communities build wastewater
treatment plants, "a 65% increase over the current fiscal year and a 90%
increase over the past fiscal year." Lake Erie, once given up for dead, was

---

* Ford had been assured by at least one Nixon surrogate that the former president wouldn't
be returning any time soon to the scene of his 1972 triumph.

now swimmable—a fact cheerfully confirmed by a student questioner from Buffalo. To counter suggestions the US was falling behind the Soviet Union in naval power, Ford touted seventeen new capital ships, costing almost $7 billion, that were on the drawing board. He didn't forget to single out for praise the state's Portsmouth Naval Shipyard—"Which we are not going to close, incidentally."*

On the same trip the Ford campaign deployed its not-so-secret weapon. No one questioned Betty Ford's political value now. If anything she was in danger of being overworked by heedless schedulers. The New Hampshire trip was "enjoyable but strenuous," the First Lady told a friend from Grand Rapids. She hoped to accompany Jerry on most of his campaign swings, while taking a pass on his current trip to Florida. "Five cities in two days is a bit much for me." Hoping to build on the gains produced by his in-person campaigning, the White House released Ford's most recent tax returns, showing that he and Betty had paid nearly half their 1974 income in federal, state and local taxes—a none-too-subtle jab at Reagan, whose failure to pay any income tax in 1970 owing to unspecified "business reverses" was public knowledge only because persistent journalists had dug out the facts.

Ford got another lucky break when Reagan told a Florida audience that Social Security funds would earn a higher rate of return if invested in the stock market. By implying support for privatization of the system, however faintly, Reagan demonstrated anew why Social Security was considered the third rail of American politics. By the time Ford returned to New Hampshire for a second visit on the eve of the primary, the challenger's own polls showed his lead over the president slipping. None of which kept Governor Thomson from going on *Meet the Press* and predicting a Reagan win by ten points or more. Reagan's New Hampshire chairman, former governor Hugh Gregg, believed the closing hours of any campaign were best reserved for grassroots mobilization. The candidate only got in the way of this. Deferring to Gregg's wishes, Reagan left the state for Illinois, whose primary was scheduled for March 16. Before leaving, however, the challenger revealed Ford's earlier offers to him of two separate Cabinet positions, a well-timed rejoinder to criticism from the Ford camp of Reagan's readiness to govern.

Stu Spencer had a markedly different game plan for the final stretch of a

---

* This was a reference to over 150 military installations his administration marked for closure or consolidation. Political advisers pleaded with the president to scale back or cancel these plans. Ford was adamant. For him it was 1953 all over again, when a young congressman turned a deaf ear to pleas from local interests dependent on the economic boast provided by military installations, obsolete or not.

contest he judged too close to call. "That's why we drove our phone banks to the end," completing an estimated sixty thousand calls to Ford supporters, and flying *Air Force One* with the president on board "into some burg town" for a last minute—and largely uncontested—bid for attention. As Spencer put it, "Their polling data showed them winning by eight points or something and they pulled out and left us a 24 hour gap with all the accoutrements of a president—the airplane, the Secret Service—all the trappings that you can take into a little state like New Hampshire."

Primary day, February 24, would reveal which campaign had made the right strategic call. Good weather promoted only a moderate turnout, adding to the anxiety at the White House, which was hoping for more. Reagan, sufficiently confident of victory to return to the state for the final count, appeared vindicated when the polls closed at seven thirty p.m. and he jumped out to an early but consistent lead of around four percentage points. Ford supporters took heart from polls showing late-deciding voters breaking decisively for the president. Their patience was rewarded as ballot boxes were tallied in friendly areas like Concord and Portsmouth. By midnight Reagan's margin had dwindled to three hundred votes. Late returns from the northern part of the state—ski country—gave Ford a photo-finish victory by 1,311 out of 110,000 votes. In separate balloting he won seventeen of the state's twenty-one convention delegates.

White House aides seemed as surprised by the result as their counterparts on the Reagan campaign. Ford himself took the news in stride, though not without a gruffly worded claim of vindication.

"I hope that's the last God damn time I hear that the only thing I ever won was the Fifth District in Michigan," he remarked.

H OWEVER NARROW, HIS New Hampshire win supplied rocket fuel to the Ford campaign. Moreover, the calendar was on his side, at least until Texas voted on May 1. Reagan barely contested the next two primaries, in Massachusetts and Vermont, leaving Ford to harvest most of their sixty-one delegates.

On March 9 it was Florida's turn. A Reagan loss there would put a serious crimp in the challenger's fundraising; if large enough, it might even cause the onetime actor to leave the stage. At least that was the hope in Ford's camp. With its theme parks, launchpads and distinctive Latin flavor—1976 was the first year primary ballots in Dade County were printed in Spanish as well as English—Florida would appear to have little in common with the snowcapped Granite State. In fact, both states topped the nation in their concentration of residents sixty-five and older. In trailer parks and gated communities, golf resorts and oceanfront condominiums, retirees, many from New York and the Midwest, were expected to comprise as much as 40 percent of Republican primary voters. This alone guaranteed a heavy emphasis by the Ford campaign on the Social Security issue.

Reagan was not without advantages of his own. In the words of conservative author Richard Whalen, he was "unsullied by Watergate, untainted by Vietnam, and uncorrupted by a Washington system that isn't working." In Florida he could count on a warm reception from the state's large Cuban-American population. Refugees from Castro's tyranny, residents of Miami's Little Havana were uniquely receptive to Reagan's criticism of Henry Kissinger and his alleged willingness to bargain away US military superiority. The challenger also had significant strength in the state's Panhandle region, popularly dubbed the Redneck Riviera. This was George Wallace country, where tourists escaping the Northern winter mingled with yellow dog Democrats in crab shacks and on sugar-white beaches. Yet even here Reagan was disadvantaged by the state's election laws, which prohibited crossover voting in another party's primary.

On paper the president's campaign in the Sunshine State was entrusted to Congressman Lou Frey, another Ford crony from Capitol Hill, whose gubernatorial aspirations made him reluctant to offend Reagan partisans. It didn't help when Frey turned day-to-day campaign management over to a former administrative assistant who was plainly out of his depth. On a visit to the state, Spencer found a campaign on life support. "I'd go in the office one day and open a bottom drawer and there's about twenty thousand dollars' worth of unpaid bills sitting in it." No phone banks had been installed, or volunteers recruited to man them. Buttons and posters were scarce. Spencer's response was to call up his former partner, Bill Roberts, then ailing and restless in semiretirement. "How'd you like to go to Florida and do the Ford thing for us?" he asked.

Roberts jumped at the offer. Known to be lavish in deploying other people's money, Roberts would have virtually free rein in Florida, thanks to a revitalized fundraising effort spearheaded by Robert Mosbacher and ably assisted by Ford's longtime Detroit ally Max Fisher. Described as "a blunt spoken Texas oilman" from Houston, Mosbacher exhibited in the finance job the same fiercely competitive instincts that had made him a gold-medal yachtsman. By the first week of March the campaign was more than halfway to its goal of $10 million, with a pending request for $500,000 in federal matching funds. On his first visit to the state, a two-day swing in mid-February, Ford was seen in person by one hundred thousand Floridians. Many more than that benefited from presidential largesse. In Orlando he announced the city's selection to host a convention of the International Chamber of Commerce, an economic coup secured with help from his own Commerce Department. Dipping deeper into the executive goody bag, Ford pulled out defense contracts and a mass transit system for Miami. St. Petersburg received a new veterans' hospital. In Fort Lauderdale, Ford stressed the benefits of federal revenue sharing ($12 million thus far to the Venice of America—aka Spring Break Central—over $900 million to Florida as a whole).

As the president exploited incumbency for all it was worth, Bill Roberts used direct mail to pound Reagan on Social Security. Ford commercials, some dubbed in Spanish, were running in all major media markets. A radio spot featuring the First Lady got heavy play. The campaign reported nearly two hundred thousand phone calls made by the time Ford returned to the state on February 28. Addressing a naturalization ceremony for 1,200 new citizens in Miami, he denounced Fidel Castro as "an international outlaw" with whom his administration would have no dealings. This was followed by an eight-hour motorcade threading forty miles of shopping malls and mobile home parks amid tropical downpours. Ford spoke to rain-soaked crowds in ten separate locations.

"There's a saying that aristocracy is a matter of the soul, not of the cloth," a waterlogged Ford told supporters gathered in the Royal Park Shopping Plaza near Boca Raton. "I don't look very good, but I think I'm a darned good president." On March 9 enough primary voters agreed with his assessment to give Ford a convincing 53–47 percent victory over Reagan. He did better still in the race for delegates, taking forty-three of the sixty-six up for grabs. At the White House the president pronounced himself "overjoyed" at the result. The reaction among his campaign strategists was more equivocal. Although grateful for the win, they had hoped to land a knockout blow in Florida.

That they fell short of their goal was attributed to Reagan's nonstop campaigning—the exact reverse of New Hampshire—and, more ominously, to the challenger's escalating attacks on Ford-Kissinger foreign policies. Skillfully exploiting a post-Vietnam malaise, Reagan blamed the current administration for a loss of American will. Wherever he went Reagan bristled with indignation over the possible turnover of the Panama Canal to a "tinhorn dictator" like Panamanian strongman Omar Torrijos. Nowhere did he explicitly mention the last days of Saigon. He didn't have to.

FORD WAS FEELING upbeat as the primary spotlight moved to Illinois, his mood a reflection of paternal pride as much as poll-driven optimism. Jack Ford, preceding his father to the Land of Lincoln, had drawn large and enthusiastic crowds as he crammed twenty-nine appearances and twelve press conferences into a four-day blitz of the state.* An even more ambitious foray into Texas led *New York Times* reporter James Naughton to write a story describing Jack as a better campaigner than his old man. The next time he boarded *Air Force One*, Naughton was told the president would like to see him. "Uh-oh," thought Naughton. "He didn't like that Texas story." To the contrary, Ford invited the reporter to join him for a drink. Speaking for the First Lady as well as himself, he thanked Naughton for what he had written. Adjusting to life in the White House had been harder for Jack than for other members of the Ford family. Naughton's piece in the *Times* confirmed to Jack's parents that their son had overcome his problems; that he had, in fact, turned a corner in his life.†

On the eve of Illinois's primary, reports surfaced that Bo Callaway, on his last day as army secretary, had met in his Pentagon office with officials from the Agriculture Department, the parent agency of the US Forest Service, as part of his ongoing effort to win government approval to expand his family's Crested Butte ski resort in Colorado. Conceding he had been "naïve" in his choice of a meeting place, Callaway told Cheney, "Hey, give me a week, I'll get it cleared up." The story persisted. Callaway did not. His leave of absence was made permanent as of March 30.‡

On March 16, Ford swept 59 percent of the vote in Illinois and, with it, seventy of the state's ninety-six delegates. Stu Spencer, now the campaign's acting chairman, publicly called for Reagan's withdrawal "the sooner the better." Similar advice came from a more surprising source. "Ronnie has to get out," Nancy Reagan told campaign press secretary Lyn Nofziger. "He's going to embarrass himself if he doesn't." Reagan campaign manager John

---

* The cost of Jack's political travel was personally paid by his father. Dick Cheney to Phil Buchen, July 8, 1976. GRFL.

† "That was the real Jerry Ford," Naughton observed in notes accompanying records of his work placed at the Ford Library.

‡ There was abundant evidence of partisan politics being played, especially by Colorado senator Floyd Haskell, whose Senate Interior subcommittee heard from seventeen witnesses, none of whom claimed impropriety or inappropriate pressure applied on Callaway's behalf. Callaway was subsequently cleared of any wrongdoing by the Justice Department. More convincingly, he was absolved by *Harper's Magazine* in a July 1977 article that concluded the former army secretary was scapegoated by Haskell's committee, drawing on accusations made by a local newspaper publisher hostile to Callaway's ski resort. In March 1978 the Forest Service granted Crested Butte its desired expansion. That fall, much to Ford's satisfaction, Haskell lost his Senate seat to Republican William Armstrong.

Sears dropped hints to Ford operatives that the governor might be willing to shut down his insurgency after North Carolina voted on March 23—*if* the White House backed off its harshest criticisms of Social Security and the "$90 billion speech." This was fine with Ford, who was keenly aware that his chances of winning in November depended on a united GOP.

Taking his foot off the pedal, Ford addressed the annual convention of North Carolina's Future Homemakers of America on March 20, where he lamented the fact that some Americans "have disparaged and demeaned" the traditional role assumed by members of his audience. "I say—and say it with emphasis and conviction—that homemaking is good for America." Reagan, meanwhile, broadened his attacks to include Ford's budget deficits and "disastrous" energy legislation. Above all Reagan laced into "Dr. Kissinger and Mr. Ford" for détente and their proposed Panama Canal giveaway. "We bought it," Reagan thundered, "we paid for it, it's ours, and we're going to keep it." It might not supplant "Fifty-four forty or fight" in the lexicon of political sloganeering, but Reagan's cri de coeur beat Ford's bromides about homemaking hands down.

An invigorated Reagan campaign didn't rely exclusively on crowd-pleasing rhetoric to regain its footing. The president had North Carolina governor James Holshouser in his corner. But Reagan had Senator Jesse Helms, and a formidable statewide organization led by Helms's chief political operative, Tom Ellis.* In the face of skepticism from senior campaign officials, Ellis obtained a grainy, half-hour Reagan TV speech first broadcast in Florida. With local references scrubbed in favor of an appeal for contributions, the speech was broadcast by fifteen North Carolina stations in the closing days of the race. The ten thousand people who turned out to see Ford in Winston-Salem for a bicentennial-themed event constituted one of the largest crowds of the campaign, but they were a fraction of Reagan's TV audience.

Neither candidate stuck around to wait for the verdict. On the evening of March 23, Reagan was in Wisconsin. At the White House, Ford's schedule included a congratulatory phone call to Governor Holshauser between nine thirty and ten p.m. Instead he learned from Dick Cheney, a little after nine o'clock, that Reagan had grabbed an early lead. Hoping for a repeat of his dramatic New Hampshire comeback, Ford instead made history as the first incumbent president to lose a primary since 1952, when Democrats in

---

* Ellis, a virtuoso of direct mail, was not averse to hardball tactics. To Reagan's credit he refused to approve an Ellis-orchestrated statewide mailing contending that Massachusetts senator Edward Brooke, the only African-American member of that body, was on Ford's short list of vice presidential candidates.

the Granite State chose Estes Kefauver over Harry Truman. Within days of his embarrassing loss, Truman had announced his retirement. It was one parallel with the man from Missouri that Ford had no intention of emulating.*

NORTH CAROLINA WAS "a shocker," said Ford, who blamed his own campaign's overconfidence for the result. Meanwhile the demands of office did not let up even as Ford sought to extend them beyond January 20, 1977. Here incumbency had its disadvantages—witness the twenty days Reagan had spent barnstorming across Florida, five times what Ford could spare. Scheduling aside, in an election year virtually every presidential action, not excepting matters of life and death, is viewed through the prism of partisan self-interest. Almost half a century before scientists identified a lethal strain of coronavirus called SARS-CoV-2, the Ford White House confronted its own epidemiological reckoning.

Ground zero for the great swine flu scare of 1976 was New Jersey's Fort Dix. In January several soldiers on the base experienced flu-like symptoms. Eventually a dozen were hospitalized. Serologic, or antibody, testing revealed an additional two hundred recruits had been infected. By itself this was hardly cause for alarm. After all, influenza is an annual occurrence in the United States. In the mid-1970s it was blamed for around seventeen thousand deaths a year. Unusually severe outbreaks of the disease were thought to occur at approximately ten-year intervals. In 1957 an especially virulent strain was blamed for sixty thousand deaths. More recently still, the so-called Hong Kong flu killed thirty-three thousand Americans during the fall and winter of 1968–69.

Bad as these were, nothing in modern times compared to the worldwide pandemic of 1918–19, which claimed 675,000 American lives, the first on a Kansas army base much like Fort Dix, where, on February 4, 1976, a nineteen-year-old recruit named David F. Lewis collapsed and died after taking part in a five-mile forced march. Blood samples collected from him and others at the fort were sent to the Centers for Disease Control in Atlanta. Most came back labeled A/Victoria/3/75, the most common flu strain at the time. But Lewis and one other soldier were found to be infected with a form of swine flu (immediately dubbed A/New Jersey/76) reminiscent of the deadly World War I–era plague misnamed the Spanish Flu.

For Gerald Ford's generation, the words conjured childhood memories

---

* To be sure, Lyndon Johnson lost to Eugene McCarthy in Wisconsin's 1968 Democratic primary, but he had already withdrawn from the race two days before votes were cast in the Badger State.

of black-draped neighborhoods where, aided by remedies no more potent than turpentine and beef tea, lives flickered out in a matter of hours. Grand Rapids' first reported case, at the end of September 1918, was none other than Arthur Vandenberg, editor and publisher of the *Grand Rapids Herald*. He recovered, but on October 3 the *Herald* reported the death of a thirty-year-old salt worker. Michigan governor Albert Sleeper declared a statewide lockdown affecting theaters, movie houses, pool halls and all places of worship. Under heavy pressure, Sleeper lifted his order after just three weeks, only to see the number of new cases skyrocket. Residents were cautioned to wear masks and avoid crowded settings. By the end of 1918, Grand Rapids counted 295 deaths.

Did the random horrors of those days resurface in Ford's mind as he learned of the outbreak at Fort Dix? Certainly the town in which he grew up hadn't forgotten the wartime pestilence that exacted a heavy toll among children five—Ford's age in 1918—or younger. In 1918 there was no CDC to coordinate Washington's response and no miracle drugs to combat the raging fevers or bluish tint that discolored flu victims as a bloody froth, expelling oxygen from the lungs, caused patients to suffocate in their own fluids. Echoes of the earlier pandemic did surface at the CDC. After scientists there confirmed the first documented transmission of swine flu to humans since 1930, agency director David J. Sencer drafted a memo for HEW Secretary David Mathews in which he called for a mass immunization program. "They said we just can't afford to take any chances with the health of the American people," recalled Mathews. "I think it was the right call. But at HEW, we checked with people who were not at CDC to see if there were any dissenting views in the medical community. The White House did the same with its contacts."

On March 10 the Advisory Committee on Immunization Practices of the U.S. Public Health Service weighed in. It agreed with the CDC. All the ingredients for a pandemic were present. Persons under fifty years of age had no antibodies to resist the new strain traced to Fort Dix. Fortunately early detection of the New Jersey virus meant there was still time to produce a vaccine, made from fertilized hens' eggs, before the onset of the 1976–77 influenza season running from late fall through February. With that the National Influenza Immunization Program (NIIP) was born. Vaccination on the scale recommended by the CDC was unprecedented. It could only happen if the federal government acted quickly to contract with private pharmaceutical companies to produce the quantities of vaccine needed to safeguard two hundred million Americans. An action plan was devised, with Washington channeling $135 million ($500 million in 2021 dollars) to state health departments and private medical practices. They would bear responsibility for administering the vaccine, after large-scale trials were conducted to ensure its safety and effectiveness. All of this would, needless to say, require the approval of Congress.

On March 24, Ford met with experts from HEW, public health officials, virologists and representatives of the pharmaceutical industry. Special attention was paid to a pair of medical legends, Dr. Jonas Salk, his name synonymous with the first successful polio vaccine; and his rival Dr. Albert Sabin, originator of the oral vaccine that, beginning in 1962, supplanted Salk's breakthrough. The two men had cordially detested each other throughout their careers, making it all the more desirable that they reach a consensus on the looming danger. Everyone around the Cabinet Room table was sensitive to the calendar. "We went over the data and the recommendations and asked if there were any second thoughts," Mathews recalled. Ford let it be known that he would be in his office for half an hour, available to meet privately with anyone who held views "you are not willing . . . to express in front of this larger group." No one took the president up on his invitation. Mathews cautioned Ford, "This is a no-win situation. If the virus is virulent and comes, we're going to be accused of not being prepared enough. If it doesn't come, we're going to be accused of squandering millions of dollars and God only knows what else. I'll be happy to make this announcement and you can stay presidential."

Ford declined Mathews's offer. Moments later he appeared in the pressroom to alert the public to the threat and his administration's plans to contain it. The response was typified by American Red Cross president Frank Stanton, who told Ford within twenty-four hours that the organization's three thousand chapters and one million volunteers awaited his directions. Congress was almost as quick to vote the necessary funds, enabling four pharmaceutical companies to embark on the 1976 version of Operation Warp Speed, the Trump Administration's crash program to develop an effective anti-COVID vaccine in 2020. Then the insurance industry notified the Office of Management and Budget of its refusal to cover liability unless the government indemnified vaccine makers, one of which, Parke-Davis, manufactured three million doses using the wrong seed virus before HEW discovered the error.

As spring turned to summer, a chorus of state and local officials complained that federal funding for the NIIP was inadequate. In July Dr. Sabin retracted his support for universal inoculation. His reasoning mirrored the data-driven caution of the World Health Organization. Historically, local outbreaks of flu had preceded a global pandemic. Yet, with the single exception of Fort Dix, no cases of swine flu had been reported. A Gallup poll at the end of August found that while 93 percent of the public was aware of the swine flu program, only 53 percent intended to get a shot themselves. And the liability issue had yet to be resolved. "The federal government was going to have to assume some of the risk," David Mathews later said. "The issue was, how much? Once that was determined, we could negotiate with the insurers to get protection for vaccine manufacturers."

On July 19, Ford received assurances from HEW assistant secretary Theodore Cooper that a pandemic remained "possible," with probabilities "unknown." Writing long afterward in *BBC Future*, Richard Fisher noted a "language gap" separating scientific experts from public policymakers: "To a scientist, possible can mean a one in a million chance; to a politician, it necessitates action." On July 25, Ford appealed to Congress to insure the insurers. His message was reiterated in a flurry of phone calls he made to friends on the Hill. The result was a tort claims bill protecting vaccine makers, except when they were found to be negligent. Ford signed the legislation as soon as it cleared both houses on August 12.

Yet no amount of presidential arm-twisting could erase the fearsome headlines when, in the first week of August, a deadly outbreak of pneumonia at an American Legion convention in Philadelphia killed twenty-nine Legionnaires—immortalized in the belatedly diagnosed disease that bears their name. "Greatly relieved" as he was to learn that the tragic deaths in Philadelphia were not the result of swine flu, Ford said, "let us remember one thing: they could have been." Vaccination of high-risk candidates began on October 1. Within days news outlets ignorant of the nuances of epidemiological science ran scare stories linking the vaccine to unrelated fatalities. After three elderly Pittsburgh residents died from heart attacks on October 12, the *New York Post* described a fictitious scene at "the Pennsylvania Death Clinic" where an old woman, wincing at the hypodermic needle in her arm, managed a few "feeble steps" before dropping dead "right in front of their eyes."

Such distortions prompted a rare on-air rebuke from CBS News icon Walter Cronkite, who urged his journalistic brethren to be more factual and less speculative in their reporting. It didn't matter. Within twenty-four hours, nine states halted their inoculation programs. Hoping to stem the flow of negative publicity, Ford submitted to a televised vaccination on October 14. (His opponent in the fall campaign, former Georgia governor Jimmy Carter, pointedly refused to have the shot.) This would later be cited as evidence the Ford White House deliberately inflated the swine flu menace for political gain. The accusation feels unfair; even more, it feels uncharacteristic. Ford's advisers, having experienced the fierce backlash from his Nixon pardon and a score of unpopular vetoes since, wished he were *more* politically minded. How many times had they heard him repeat his mantra: "Good government is good politics"?

For all its shortcomings, the program successfully vaccinated some forty million people by mid-December, when fresh media reports insinuated a link between the vaccine and Guillain-Barré syndrome (GBS), a paralyzing nerve condition that can be fatal. History argued for caution, given the record of prior public health campaigns jeopardized by medical errors. In 1954 a single tainted batch of the Salk vaccine had left ten youngsters dead and nearly two hundred paralyzed. Then–HEW Secretary Oveta Culp Hobby quietly resigned, and President Eisenhower had committed his prestige to the restoration of popular trust. In the 1960s, Sabin's oral vaccine, promoted as a safer alternative to Salk's injections, was shown through subsequent genetic sequencing to have caused a handful of paralysis cases. Testing in 1976 calculated the odds of contracting GBS from the swine flu vaccine at one in 105,000. (The extra death rate from GBS as measured in a 2009 swine flu outbreak was two in a million.)

Already suffering the erosion of public confidence and a corresponding drop in the vaccination rate, the NIIP was suspended, with Ford's pained assent, on December 16. Five days later a *New York Times* editorial condemned the program as a "debacle" and "fiasco." What David Sencer called the paper's "simple and sinister innuendoes" lodged in the journalistic memory bank, to be recycled in 2020—sometimes verbatim—by journalists and anti-vaxxers skeptical of government mandates to combat COVID-19.[*]

Otherwise the ghost epidemic that was swine flu had little in common with the global devastation of COVID-19. One president was accused of overreacting to a theoretical danger, the other of minimizing a mass killer lapping at 1600 Pennsylvania Avenue. We know things now that we didn't know in 1976: for example, that the 1918–19 pandemic originated with birds, not pigs, and that the ten-year cycle for mass flu outbreaks is spurious. In the end it fell to Dr. Sabin, reinforcing David Mathews's early warning to the president, to put things in context. "You are preparing for a potential attack that might never happen," Sabin told White House reporters. "This is what the Defense Department is doing, too. We are spending a lot of money and a lot of effort against something that may never happen. But should we do it?"

Ford thought the question answered itself. This was not without irony,

---

[*] Sencer himself lost his job in February 1977, on the same day the new Carter Administration resumed distribution of the controversial swine flu vaccine on a limited basis.

for the NIIB snugly fit the profile of other government programs, taxpayer-funded altruism launched with better intentions than results, that Ford had made a career out of opposing in Congress.

EIGHT YEARS AFTER impressing a Gridiron Dinner audience with an unexpected display of wit and comic timing, Ford confronted an even tougher audience at the annual Radio and Television Correspondents' Association dinner the night of March 25. Emceeing the event was none other than his *Saturday Night Live* tormentor Chevy Chase. Never inclined to take himself too seriously, Ford laughed heartily at Chase's pratfalls and other barbed humor at his expense. "He's a funny guy," he remarked during a lull in the program. When introduced by Chase, Ford put on his own display of physical comedy. He became entangled in the tablecloth, sending silverware flying in all directions. He placed his speech, actually a bundle of blank pages, on the podium, only to scatter its pages across the head table guests. Turning to his nemesis, Ford allowed that "Mr. Chevy Chase" was "a very, very funny suburb." By the time he lifted Chase's signature line, declaring, "I'm Gerald R. Ford and you're not," there was no doubt who had won the evening's competition for belly laughs.

Ford's carefully rehearsed routine was the inspiration of a short, shaggy-haired former actor–turned–gag writer, speech coach and $150 a day consultant named Don Penny. Recruited by David Kennerly, a longtime friend with his own Hollywood connections, the forty-three-year-old Penny was charged with imparting a little of the media savvy demanded by the modern presidency. As with Kennerly, irreverence was Penny's stock in trade. At the outset he told Ford that he was a humfer—someone who hums at the same time he talks.

Penny sensed a deliberate attempt by his client to appear calm, a by-product of his difficult childhood ("not much better than an orphan," Ford confided) and a corresponding lifelong struggle to hold his temper in check. Unfortunately, said Penny, "That ain't what attracts people." Ford must learn to take a written piece and turn it into a spoken piece, "which means you have to be an actor."

"Oh, God, I'll never be an actor."*

Penny's makeover wasn't limited to Ford's speechmaking. If the sartorial

---

* Sometimes, his competitive instincts turned inward, Ford could get down on himself. "I know I'm not very good," he once confessed to Penny.

"Well, Mr. President, you have an incredible strength."

"Yeah, what's that?"

"Honesty."

look of the Ford presidency became more polished at the start of 1976, chalk it up to a conspiracy between Penny and the First Lady. "Mr. President, I'm a natty guy from Brooklyn, you know," the gagster informed his presidential client. "And I'm a little disappointed in your outfits with the white belts."

"So?"

Penny arranged for a tailor from Britches, a popular men's clothing store in Georgetown, to come to the White House with a selection of suits bearing the Ralph Lauren label. Saul, the tailor in question, was a Holocaust survivor, whose story Penny related to the president before the nervous outfitter arrived. The two men shook hands, then repaired to Ford's nearby study where Saul could take measurements and his client identify the outfits most to his liking. Passing by the room, Penny caught a glimpse of Ford talking earnestly to Saul, who then stood and embraced the president. Moments later they reappeared in the second-floor hall with reddened eyes. Once outside, Saul told Penny what had happened, how Ford had brought up Auschwitz before complimenting the tailor, a naturalized citizen whose appreciation of his adopted homeland and its democratic values made him one of the best Americans.

Returning to the second floor, Penny tracked Ford to his study, glued to a television set.

"Mr. President? I just want to tell you I think you are a terrific guy and I love you."

"Don, I think you're a terrific guy, too, and I love you. Now can I watch the ballgame?"*

E VEN DON PENNY would be impressed by the roster of celebrities who signed on to the Ford campaign. By April 1976 the president boasted endorsements from legendary entertainers Cary Grant, Ella Fitzgerald, Zsa Zsa Gabor and Howard Keel; football great Terry Bradshaw and Olympic medalists Peggy Fleming and Cathy Rigby. Eleven former chairmen of the Republican National Committee added their support. North Carolina looked like an aberration when, on April 6, Ford swept all forty-five delegates in Wisconsin's primary. After Reagan skipped the April 27 "beauty contest" in Pennsylvania, the president was positioned to claim most of that state's 103 convention votes as well. Ohio's primary deadline passed with a majority of the state's ninety-seven delegate slots uncontested by Ford's

---

* Ford made sure to invite Saul and his wife to the state dinner for Israeli prime minister Yitzhak Rabin on January 20, 1976. There Betty Ford joined the visiting Rabin and the Georgetown tailor in dancing the hora.

rival. Under the circumstances, Ford readily waived his rights under the equal time rule administered by the Federal Communications Commission so that Reagan's old movies could be broadcast on television without political repercussions.

Not that Reagan had backed off his challenge. In a nationwide broadcast on March 31, he renewed his attacks on Ford and Kissinger for allegedly yielding American military supremacy to the Soviet Union. The harshly worded speech generated $1.5 million to replenish Reagan's campaign coffers as the primary battleground shifted to the south and west, terrain considered more favorable to the challenger. On April 1, John Connally came to dinner at the White House. He told the president that Reagan's popularity and organization in Texas made him a prohibitive favorite in the state's May 1 primary. Hoping to disprove Connally's pessimistic forecast, the Ford campaign committed $800,000 and several days of in-person campaigning to what Stu Spencer calls the biggest mistake of his career.

Waxing indignant over Reagan's charge that he intended to recognize North Vietnam, Ford left it to his state chairman, Senator John Tower, to fend off the challenger's attacks on US military preparedness. The crowds greeting the president on a multi-city tour of the state in mid-April were friendly, if not wildly enthusiastic. "I can remember helping set up an event in Lubbock, Texas," said George W. Bush, then a Midland-based oilman who had agreed to help out when asked to run West Texas for the Ford campaign. "He came out there and there were a handful of people that came to see him." It was clear to Bush from the beginning that Ford's cause was hopeless—"Tamale or no tamale"—a reference to the notorious incident where the president offended Texan sensibilities at the Alamo of all places. Offered a plate of tamales, Ford, clearly unfamiliar with Tex-Mex cuisine, bit into the corn husk without bothering to remove it. Broadcast repeatedly, the gastronomic faux pas drew far more attention than Ford's upbeat message contrasting the current American economy and global standing with the summer of 1974.

Other elements were working against him in the deeply conservative state. "Ronald Reagan was, you know, riding a wave . . . a political tsunami," explained Bush. The Panama Canal was a galvanizing issue. State election law allowed crossover votes, and the Reagan campaign was not subtle in appealing to conservative Democrats seeking a champion following the virtual collapse of George Wallace's candidacy and the rise of former Georgia governor Jimmy Carter, until recently a mere asterisk in most polls. Then there was the Kissinger factor. Nationwide polls showed Kissinger remained broadly popular, with a 58 percent approval rating. Not so in Texas, whose critical primary coincided with the secretary's headline-making tour of seven African nations. The continent was a powder keg, with the CIA predicting Cuban troops in Angola would join forces with Black nationalists against the white minority regime of Rhodesian prime minister Ian Smith.

On March 13, Ford declared the United States "totally dedicated to seeing to it that the majority becomes the ruling power in Rhodesia." The implications of his statement went far beyond the borders of Rhodesia, to include the apartheid regime of South Africa, long allied with the United States as a bulwark of anti-Communism. On April 27, four days before Texans voted in their GOP primary, Kissinger spelled out details of the most dramatic volte-face in US foreign policy since Nixon's trailblazing China trip of February 1972. Speaking in the Zambian capital of Lusaka, he listed American objectives for a postcolonial, pro-democratic Africa. These included a negotiated end to the Smith regime in Rhodesia; an early deadline for South Africa to grant self-determination to its former trust territory of Namibia; explicit condemnation of apartheid coupled with a call for demonstrable progress toward racial equality and basic human rights; and guarantees to protect the rights of the white minority. Finally, Kissinger offered American economic assistance to a continent throwing off its former dependence and suspicious of Western capitalism.

Inevitably the Ford-Kissinger initiative became a campaign issue, with Reagan accusing the secretary of inflaming racial tensions. Texas primary voters appeared to agree. Smashing all turnout records, Texas Republicans, and Republicans for a day, gave Reagan a clean sweep of the ninety-six delegates chosen that Saturday. Indicative of the bitter feelings souring the contest, Senator Tower, the first Republican to win statewide election in Texas since Reconstruction, was denied a place on his state's delegation to the upcoming GOP convention in Kansas City. "This is just one primary," Ford told Cheney and other disappointed staffers. "We are going to win more primaries and we are going to win the nomination."

His optimism was put to the test on May 4, when Reagan reaped an even bigger delegate harvest after scoring victories in Georgia, Alabama and Indiana. A week later he upset Ford in Nebraska, an embarrassment only partially offset by a convincing Ford win in West Virginia. That same day the president convened a National Security Council meeting at which he reiterated his commitment to Black majority rule in Africa. To be sure, he allowed, "we got a little political flak" as a result of Kissinger's recent trip. So be it. American foreign policy could not be held hostage to the political calendar. "We will continue to do what is right regardless of the primaries."

ALL EYES NOW turned to Michigan. A loss in his home state's May 18 primary, or an unimpressive win, would likely condemn Ford to the dismal fraternity of incumbent presidents (Chester Arthur, Franklin Pierce, Millard Fillmore) denied their party's nomination for a second term. Behind the scenes of the Ford campaign cracks were showing. Much of the criticism was directed at Rogers Morton, the six-foot-seven-inch gentle giant, former congressman and Cabinet officer who had reluctantly taken the helm after Bo Callaway resigned and Stu Spencer declined the position of campaign director. A disheveled Morton let himself be pictured, on the night of Ford's triple defeat in the South and Midwest, behind a row of empty liquor bottles. Worse, Morton told reporters that no strategic changes were planned. As he put it, all too quotably, "I'm not going to rearrange the furniture on the deck of the *Titanic*."

This was met with a rare display of presidential temper. "Rog," Ford growled the morning after, "we had a real bad night and your comments didn't help one damn bit."

Ford's presidency hung in the balance when *Air Force One* landed at Detroit's Wayne County Metropolitan Airport on the morning of May 12. Michigan was another open primary state in which George Wallace Democrats could emulate their Texas brethren by supporting Reagan. Rejecting the advice of Michigan Republican officials who frowned on any appeal to independents and Democratic crossovers, Ford had decided to contest Reagan for every voter from Motown to Sault Ste. Marie. Why should Republicans—and, equally important, non-Republicans—vote to keep him in the White House? he asked the first of many audiences he would address that rainy Wednesday. "Because I have done a good job and I am proud of it. Because I have turned a lot of things around and we are going in the right direction. Because I want a mandate from Michigan and the American people to finish that job."

The message was simple, and pitched to the pragmatists in both parties whose memories reached back at least as far as August 1974. What made it compelling was the means of its delivery—from the rear platform of the *Presidential Express*, a seven-car Amtrak train reminiscent of whistle-stop campaigns last conducted by Harry Truman and Dwight Eisenhower. Campaign officials had objected to the train's cost. Ford overruled them, in his best decision since returning at the last minute to campaign in the snows of New Hampshire. Energized by Dixieland bands and trackside crowds with their signs and flags, Michigan's first president made a blatant appeal to

Wolverine pride. In Durand, population 3,628, he told at least that many people assembled at the train station, "I need your help on Tuesday— I won't let Michigan down. Don't you let me down."

In between stations the president visited with state officeholders in the dark-wood-paneled observation car, or chatted up campaign workers and reporters as he prowled the length of the train, his mood brightening the closer he got to his old congressional district. Back at the White House, Ford called old friends in the Fifth District seeking their support, while veterans from his Capitol Hill staff implemented their own telephone tree to promote a big turnout on primary day. Their efforts paid off as a million voters, double the previous high for a Republican primary, gave Ford the Texas-sized mandate he was seeking. Watching Reagan on TV declaring himself "pleased" with his 34 percent showing in Michigan, Ford said he was glad to hear it. "I hope we can keep him that pleased in the other primaries."

As a bonus Ford won an additional forty-three delegates that night in Maryland's uncontested primary. His double victories couldn't have been better timed given the next round of contests in six states. Of Arkansas, Tennessee, Kentucky, Nevada, Idaho and Oregon, only the last, a West Coast bastion of moderate Republicans like Senators Mark Hatfield and Bob Packwood, was thought to be leaning in Ford's direction. Hoping to muffle the impact of expected bad news on May 25, Richard Rosenbaum, Nelson Rockefeller's handpicked chairman of the New York GOP, persuaded the vice president that the time had come to execute what Rosenbaum aptly dubbed "Operation Rescue." Between them the two men twisted enough arms to guarantee Ford at least 119 previously uncommitted delegates. Rockefeller also had a hand in persuading eighty-eight members of Pennsylvania's pivotal delegation to announce their support for the president.

On May 25 Reagan won easily in Arkansas, Nevada and Idaho. He fared better than expected in Oregon. But he tripped himself up in Kentucky and, especially, Tennessee, after a casual last-minute comment reminiscent of Barry Goldwater's 1964 call to sell the government-backed Tennessee Valley Authority, on which millions relied for cheap energy and thousands more for their livelihood. The Ford campaign, quick to pounce on Reagan's misstep, enlisted Tennessee senators Howard Baker and Bill Brock to warn of the consequences of a privatized TVA. Even with their help the race was nip and tuck, with Ford ahead at the wire by 1,688 votes out of 250,000 cast. By winning Tennessee and neighboring Kentucky, the president had bested his rival in the expectations game, while setting himself up for a continent-spanning finale on June 1, when New Jersey, Ohio and Reagan's own California combined to offer the richest delegate harvest of the primary season.

The last was hardly a contest. "I knew we'd have our tail handed to us," Stu Spencer said of the Golden State. "It was just a matter of—you get beat by ten points or twenty points." To date the Ford campaign, its eyes fixed on November, had shied away from going negative in its advertising. But as late polls showed California out of reach, leaving Reagan free to raid Ford's Ohio bailiwick, an incautious remark suggesting the challenger's willingness to provide a "token" show of US force to avert further bloodshed in Rhodesia triggered a media assault reminiscent of 1964's infamous "daisy spot" in which Lyndon Johnson equated the election of Barry Goldwater with a license for nuclear war. Only this time, in place of a mushroom cloud was Spencer's equally provocative tagline: "When you vote Tuesday, remember: Governor Ronald Reagan couldn't start a war. President Ronald Reagan could."

An enraged Reagan reportedly punched a fist through the bulkhead of his campaign plane. As in New Hampshire, the rhythm candidate was thrown off his game. Predictably, when Reagan ventured into Ohio for some last-minute campaigning, all the press wanted to talk about was Rhodesia. "That was my whole goal," Spencer claimed long afterward. "Always trying to keep him screwed up." On June 1, Reagan swamped Ford in winner-take-all California. But Ford had his own shutout in New Jersey (sixty-seven delegates) and a near thing in Ohio, where the state party organization under Governor James Rhodes limited Reagan to just nine of the state's ninety-seven delegates. At the end of the primaries Ford owed his overall lead in the popular vote to Pennsylvania, where his had been the only name on the ballot.

ATTENTION SHIFTED TO ten states that chose delegates by caucus or convention. Here Ford operated at an organizational disadvantage. "The strategy was knock them out early in the primary states," recalled Jim Baker, the former Commerce Department official recruited by Ford "to help Rog" by heading up the campaign's delegate pursuit, "and we won't have to mess around with the convention states." The resulting vacuum was filled by ideologically zealous Reaganites eager to evict the GOP establishment in a replay of Barry Goldwater's hostile takeover of 1964, beginning, ironically, in Goldwater's Arizona. Goldwater himself favored the president, and his legendary delegate counter, F. Clifton White, was working for Ford. But Reagan captured twenty-seven of the state's twenty-nine delegates anyway. Likewise Missouri's GOP organization, led by Governor Kit Bond, threw in with Ford early. At Bond's urging, Ford made an in-person appeal at the state convention on June 12. He might as well have spent the weekend at Camp David. Reagan's precinct organizers snared all but one of the nineteen delegates awarded that day.

Iowa governor Bob Ray, having no desire to be similarly embarrassed, was counting on a presidential visit to offset Reagan's appearance at the state convention in Des Moines. Three days before delegates met on June 19, the campaign script was rewritten when Lebanese kidnappers thought to be aligned with a Palestinian splinter group seized US ambassador Francis E. Meloy Jr., his economic counselor and their Lebanese driver as Meloy navigated the chaos of war-torn Beirut on his way to see President-elect Elias Sarkis. The bodies of all three men were subsequently discovered at a seaside garbage dump.

At the White House, Ford ordered a voluntary evacuation of Americans from the war zone contested by Christian militia and Palestinian militants. Plans to move US citizens and other Westerners by road convoy to the Syrian capital of Damascus were twice postponed due to firefights along the route. Eventually the convoy was abandoned in favor of a US-led naval operation. Ford stayed up most of the next two nights to monitor events. On June 19 he passed up Iowa's convention to be at Andrews Air Force Base when the bodies of the slain American diplomats arrived home. In Des Moines Betty Ford substituted for her husband. Nothing she said could prevent a Trojan horse on the state nominating committee from reneging on his pledge to the Ford camp and voting for an even split of the state's half dozen at-large delegates. More frustrating still were the dueling accusations hurled at the absent chief executive: that he was ducking a confrontation with Reagan while conducting a military operation the *New York Times* labeled "Operation Iowa Primary."

The next few weeks offered the political equivalent of trench warfare. Both sides exploited party rules wherever possible to pad their delegate totals. Reagan forces played hardball in Montana. Ford supporters reciprocated in Minnesota. Not all delegates were model citizens. Jim Baker kept a logbook of improper or illegal requests made of the Ford campaign, like the Colorado congressman who offered to trade his vote for a place on the Federal Communications Commission, or the delegate from the Virgin Islands who wanted a new federal building to bear his name. White House state dinners saw an influx of unfamiliar faces, their ethnic backgrounds carefully matched to the nationality of a visiting head of state. The guest list for a July dinner with Britain's visiting Queen Elizabeth and Prince Philip was said to include more than one Republican delegate from Wyoming, resplendent in white tie and cowboy boots.

Some delegates, relishing their time in the spotlight, deliberately strung out the courtship. "The West Virginians were the worst," in Baker's estimation. Leaving a White House lunch for his state's delegation, one grizzled veteran of Mountain State politics whispered to Baker, "I'm really for you but I can't do it yet." Five weeks went by. On the Kansas City convention floor, the same delegate told Baker, "If you can get the NBC cameraman over here, I'm ready to commit to the president." Fat chance. With Ford assured by then of the necessary majority, the publicity-seeking West Virginian had overplayed his hand.

Other delegates, with a surer sense of timing, landed prime seats aboard the USS *Forrestal* from which to observe the July 4 Tall Ships extravaganza in New York harbor. Insensible to such enticements, one New Yorker wanted a federal judgeship for his brother (he didn't get it). A party leader from Long Island's Suffolk County used his fifteen minutes in the Oval Office to pitch the needs of his county's sewer district. "We had this woman from Brooklyn who would announce she was for Reagan, and then she would announce she was for Ford, and so forth," Dick Cheney recalled. "Finally, we got hold of her, and I asked her, 'What do we have to do to get you to commit to the president?' She said, well, she wanted to bring her family down, and have them meet with the president in the Oval Office. So, she did, and on the appointed day they showed up in the West Wing lobby—the scruffiest bunch of people you ever saw in your life . . . I didn't even go in, I couldn't stand to watch. They came out about thirty minutes later, the president made the sale, and she committed, and she stayed right with us."

One July evening the Fords had retired for the night when a thirty-year-old Washington taxi driver named Chester Plummer scaled the White House fence armed with a three-foot piece of metal pipe. Refusing multiple orders to stop in his tracks, the intruder ignored a warning shot as well. He was finally brought down by a rookie member of the executive protective force. Plummer's death left the young guard "really shaken up," according to Cheney. "So they got him down in the Secret Service command post underneath the West Wing. It got very quiet for just a moment and one of the older agents leaning over against the wall announced, 'Gentlemen, if that fellow we just shot was an uncommitted delegate, we're in deep shit.'"

THE FORD WHITE House had special reason to celebrate July 4, 1976. With Watergate "finally interred" and the economy in recovery mode, *Time* magazine said the nation was in better shape than at any time since November 22, 1963. Federal planners had originally envisioned a grand world's fair in Boston or Philadelphia to serve as the epicenter of the nation's bicentennial observance. When both cities came up short on popular enthusiasm and credible financing, Washington made a virtue out of necessity. The bicentennial would be decentralized and, hopefully, diversified. The American people themselves would be on exhibit, and their experiment in self-government acknowledged as a work in progress.

Corporate America moved swiftly to fill the vacuum. McDonald's took the occasion to hawk its red, white and blue milkshakes. Coca-Cola enlisted Paul Revere and Betsy Ross to remind flag-waving consumers that "Coke Adds Life." Registering their dissent from such commercialized patriotism, the People's Bicentennial Commission observed the Glorious Fourth with a march from the Jefferson Memorial to the National Mall, punctuated by chants of "Mobil, Exxon, ITT—Down with corporate tyranny."

Ford missed the show, having decamped after an early morning church service for Valley Forge and the first stop on a whirlwind tour of Revolutionary landmarks. Responding to the streamlined management style of Dick Cheney, the White House speechwriting operation had outdone itself by producing half a dozen texts around the theme of "The American Adventure." The first of these was delivered on July 1, when the president dedicated the Smithsonian's new National Air and Space Museum on the Mall. Saluting a nation of "explorers and inventors, pilgrims and pioneers . . . whose imagination and determination could not be confined," Ford looked on as a signal from the *Viking 1* spacecraft orbiting Mars commanded some industrial-strength shears to slice through a red, white and blue ribbon.

The next day he stood before America's founding charters, housed at the National Archives, and encouraged his fellow citizens to read "the dull part, the negatives" of Jefferson's Declaration, "because the injuries and invasions of individual rights listed there are the very excesses of government power which the Constitution, the Bill of Rights, and subsequent amendments were designed to prevent." He acknowledged the long-delayed promise of equality scratched out by the pen of a Virginia slaveholder; how, two centuries on, it remained unfulfilled in the nation for which Jefferson had supplied both justification and birth certificate. Here, too, Ford sounded a hopeful note. "The Declaration is the promise of freedom," he said. "The Constitution continually seeks the fulfillment of freedom."

At Valley Forge on the Fourth, Ford greeted a wagon train of latter-day pioneers completing their transcontinental journey of discovery. Sitting at a desk once used by the first president, the thirty-eighth signed legislation making the site of Washington's fabled winter encampment a national historical landmark. From there Ford choppered the short distance to Philadelphia's Center City to address a vast assemblage in front of Independence Hall. Paying tribute to "this union of corrected wrongs and expanded rights," he declared the American Revolution unique because it had been fought in the name of law no less than liberty. "Our Founding Fathers knew their Bibles as well as their Blackstone. They boldly reversed the age old political theory that Kings derive their powers from God, and asserted that both powers and unalienable rights belong to the people as direct endowments from their Creator."

As millions watched on television, Ford hopscotched the Eastern Seaboard, alighting on the deck of the USS *Forrestal* long enough to initiate a nationwide bell ringing at the hour of the Declaration's formal adoption, then reviewing from a nearby ship the grand nautical procession of over two hundred vessels representing over fifty nations, the world under sail, as it glided up the Hudson River as far as the George Washington Bridge. Blasé New Yorkers massed around the Battery broke into "The Star-Spangled Banner." They were a fraction of the seven million celebrants in the city so recently flirting with bankruptcy. In St. Louis, a million revelers were drawn to the city's iconic arch beside the Mississippi River. Millions more lined parade routes in Los Angeles, Atlanta and Florida's Walt Disney World.

After dark, four hundred thousand Bostonians, the largest audience ever for a classical music performance, thrilled to the *1812 Overture* conducted by Arthur Fiedler. By the time Maestro Fiedler picked up his baton, Ford was back at the White House, watching from the Truman Balcony as the nation's capital launched thirty-three tons of pyrotechnics into the skies. The next morning, Washingtonians were still rubbing sleep from their eyes when the president helicoptered to Virginia's Albemarle County, site of Thomas Jefferson's mountaintop home. He had reserved his best speech of the long weekend for 106 naturalized Americans on the front lawn of Monticello. Shunning bombast, Ford explained the United States as "a new kind of nation," one in which geographical or blood ties mattered less than "the most revolutionary idea in the world."

He quoted Jefferson: "Men may be trusted to govern themselves without a master." Over time this idea, however imperfectly executed, had served as a magnet for untold immigrants who cleared the wilderness, plowed the prairie and transfused their own traditions into the American bloodstream. For all their astonishing variety, they yet belonged to a community of shared values. "To be an American," said Ford, "is to subscribe to these principles which the Declaration of Independence proclaims and the Constitution protects: the political values of self-government, liberty and justice, equal rights and equal opportunity."

Unity in diversity—in Ford's judgment the greatest single achievement in two hundred years of American independence—called to mind the long-ago Sunday-school teacher who had regaled him with the story of Joseph's many-colored coat. Applying the youthful lesson to contemporary America produced some of the most moving words of Ford's presidency. Written and delivered with posterity in mind, they warrant extensive quotation. "'Black is beautiful' was a motto of genius which uplifted us far above its first intention," Ford told his audience.

> *Once Americans had thought about it and perceived its truth, we began to realize that so are brown, white, red and yellow beautiful . . . I believe Americans are beautiful—individually, in communities, and freely joined together by dedication to the United States of America.*
>
> *I see a growing danger to this country in conformity of thought and taste and behavior. We need more encouragement and protection for individuality. The wealth we have of cultural, ethnic, religious and racial traditions are valuable counterbalances to the overpowering sameness and subordination of totalitarian societies.*

"You came as strangers among us," Ford reminded the newest Americans assembled on Jefferson's lawn,

> *and you leave here citizens, equal in fundamental rights, equal before the law, with an equal share in the promise of the future. Jefferson did not define what the pursuit of happiness means for you or for me. Our Constitution does not guarantee that any of us will find it. But we are free to try.*
>
> *Remember that none of us are more than caretakers of this great country. Remember that the more freedom you give to others, the more you will have for yourself. Remember that without law, there can be no liberty. And remember, as well, the rich treasures you brought with you from whence you came, and let us share your pride in them. This is the way*

we keep our Independence as exciting as the day it was declared, and the United States of America even more beautiful than Joseph's coat.

He spoke for ten minutes. When he finished, Ford took time to individually greet the naturalized citizens and their families. Just after noon he was driven the short distance down Jefferson's little mountain to the family graveyard where the third president was laid to rest on July 5, 1826, one hundred fifty years to the day before his distant successor came to pay him homage. The official party stood back as Ford climbed the steps leading to the wrought-iron fence and gate, on the other side of which stood the obelisk erected by the United States in 1883 to replace the original, smaller marker eroded by time and souvenir hunters. Later that day, Ford would argue his case for reelection in phone calls to uncommitted delegates from Colorado, Connecticut, North Dakota and Hawaii. But for now, alone with his thoughts, he drew inspiration from his proximity to the draftsman of American independence, the embodiment of his country's contradictions.

26

I'M FEELING GOOD ABOUT AMERICA

∎◈∎

There's a change that's come over America,
A change that's great to see
We're living here in peace again
We're going back to work again
It's better than it used to be . . .
I'm feeling good about America
I'm feeling good about me.

—FORD CAMPAIGN SONG

SHORTLY AFTER NOON on July 26, 1976, Dick Cheney and Jack Marsh presented themselves in the Oval Office.

"We just got the best news we've had in months," said Cheney. Ronald Reagan, his campaign stalled a hundred votes shy of a convention majority, had stunned even his own partisans by announcing three weeks ahead of the Kansas City showdown his choice of a running mate—Pennsylvania senator Richard Schweiker.

"Oh, come on," Ford replied. "You guys are pulling my leg."

His reaction was understandable. "If the Reagan-Schweiker ticket is a political coalition," columnist George Will observed, "then sauerkraut ice cream is a culinary coalition." As recently as July 12, Reagan had called it "a foolish mistake" for the president to invite a Northern liberal to run with him in November. Yet in Schweiker, a sixteen-year congressional veteran with Pennsylvania Dutch antecedents, Reagan had settled on the one member of the Senate with a 100 percent pro-labor voting record as measured by the AFL-CIO. Teaming Reagan and this original sponsor of the Humphrey-Hawkins full employment bill was akin to the Hatfields inviting the McCoys over to soak in the hot tub.

Against this was the logic of desperation. Campaign manager John Sears "had been conning the political media for weeks, claiming that Reagan had more delegates than in fact he had," Lyn Nofziger acknowledged. "He had to find more delegates somewhere." Sears was soon making cow eyes at the big Pennsylvania delegation, most of whose 103 members had signaled an informal preference for Ford. Would adding Dick Schweiker, one of their own, to the Reagan ticket materially change the arithmetic? The answer to that question would go a long way toward determining the Republican nominee for president. Deny Ford an early win on the convention floor, so the theory went, and delegates released from their first ballot obligations would flock to the candidate whose outsider status gave him an advantage in tapping the sour anti-Washington mood gripping the country.

This story line nourished the Reagan candidacy during the fallow weeks after the last state convention had allocated its delegates. Sustaining it all the way to Kansas City would not be easy, especially if, as rumored, CBS News were to declare Ford over the top in the delegate count before the opening gavel fell on August 16. With Harry Dent, coauthor of Nixon's Southern strategy, now working for his successor, both sides concentrated on Mississippi's thirty delegates, and as many alternates, who would vote together as a unit. State party chairman Clarke Reed, a forty-eight-year-old camera-courting businessman from the Delta community of Greenville, had led both camps to believe they had his backing. Reed expressed anger when Ford ally Gil Carmichael, the party's most recent candidate for governor, took to lobbying Mississippi's alternate delegates. He wasn't too angry to accept an invitation to the White House dinner honoring Queen Elizabeth.

The hospitality was returned when Dick Cheney visited Jackson on Sunday, July 25, to meet with delegates and scope out a possible Ford appearance later in the week. No one wanted a repeat of Missouri, where the president had come up empty-handed after making a personal appeal to the state convention. Returning to Washington Monday morning, Cheney was as surprised as anyone to learn of the Schweiker bombshell.* Exploiting media desire for a prolonged contest, Sears had succeeded in scrambling the race before CBS or other news organizations could declare it effectively over. Whether he had bought more than time depended on Schweiker's ability to flip a couple dozen Ford-leaning delegates in his home state, including Drew Lewis, the delegation chairman who also happened to be Schweiker's lifelong friend and campaign manager.

From the airport, Cheney recalled, "I went to the White House. I got

---

* He might have been equally astonished had Sears succeeded in recruiting his initial choice: former EPA director and deputy attorney general William Ruckelshaus. The idea fizzled when the hero of the Saturday Night Massacre told Sears of his support for the president.

hold of Drew Lewis . . . Drew, to his credit, said, 'Tell the president I'm with him. There's nothing to worry about in Pennsylvania, it doesn't matter what he [Reagan] does with Schweiker. We're Ford people.'" Yet another incoming call indicated which way the wind was blowing. John Connally, eager to be with the winner, told Ford he was now ready to shed his previous neutrality. Arrangements were hurriedly made for Connally to fly to Washington the next day for a public announcement. That Tuesday, July 27, began with Ford phoning the sage of Greenville, Mississippi. His call was well timed. "Clarke was pissed at Reagan and what he'd done picking Schweiker," Cheney remembered. "And in the anger of the moment, he committed to Ford."

He was not alone. One poll showed fully a third of self-identified conservatives thought less of Reagan because of his pairing with Schweiker. This was forcefully communicated by a group of Alabama Reaganites whose chairman, a confirmed teetotaler, informed Reagan *and* Schweiker that on learning of the latter's selection, he went home and drank a pitcher of whiskey sours. "Governor," he added earnestly, "I'd rather that my doctor called me and told me that my wife had the clap."

COMING OFF A string of impressive showings in state conventions, one might logically expect public opinion to swing in Reagan's direction. Yet the final Harris poll of Republicans and independents taken before Kansas City showed Ford leading his challenger 63–31 percent. The discrepancy reflected more than unhappiness over the Schweiker gambit. Beginning in North Carolina, Reagan had stormed back into contention by voicing angry, sometimes strident, opposition to the SALT talks, the Helsinki Accords and any loss of US sovereignty over the Panama Canal. His nationalist agenda struck a powerful chord with conservatives of the Jesse Helms and Phyllis Schlafly school. Yet the more fervor he aroused on the hard right, the more Reagan undercut his appeal among self-described moderate, independent, affluent and/or college-educated voters.

Protesting that he didn't want history to repeat itself with an irretrievably divided GOP suffering a Goldwater-sized rout in November, Reagan demonstrated his sincerity in party platform deliberations. Unlike Helms, who threatened an ideological bloodbath over as many as twenty-two planks, the Reagan and Ford campaigns labored to find compromise language wherever possible. So they softened Reagan's demands regarding the Panama Canal while accommodating his views in support of Taiwan and a "superior national defense." To be sure, the White House worked overtime to ensure inclusion of the Equal Rights Amendment in the party manifesto. And Reagan forces drafted an omnibus "morality in foreign policy" amendment designed to embarrass the administration and Henry Kissinger in particular. In Kansas City, Kissinger was to serve as a lightning rod for True Believers eager to seize the reins from the Republican Establishment. The resulting clash of cultures set the GOP on an increasingly conservative course for the next forty years. In their drive for organizational control, their populist rhetoric and their equation of pragmatism with surrender, these heirs to the Goldwater uprising of 1964 foreshadowed Pat Buchanan's culture wars, the Obama-era Tea Party and Donald Trump's shambolic America First.

A S IT HAPPENED, Reagan was not the only candidate willing to take risks. Regarding the upcoming convention as a preliminary to the main event, Ford that spring had pondered an even larger contingency. His electoral prospects in the fall depended on a continuation of the economic recovery that registered a sizzling 9.3 percent growth during the year's first quarter. Almost four million new jobs had been created since March 1975. The unemployment rate had drifted down to 7.3 percent before bouncing back close to 8 percent that summer as formerly discouraged workers, many of them women or young people, reentered the workforce. None of this surprised Alan Greenspan. "Because a modern economy involves so many moving parts," he wrote, "it rarely accelerates or decelerates smoothly." Greenspan alerted the president to expect a pause of several months in the current expansion, a slowdown that could last through the election.

"It's pretty hard to explain that a pause is beneficial to the guy who is out of work," Ford reflected. He didn't lack for quick fixes. Time-honored practice dictated that any election-year weakness in the economy could be corrected, or at least disguised, through a heavy dose of government spending and/or a well-timed surge in the money supply. Most recently Richard Nixon had turned on the spigots in advance of his 1972 bid for a second term, bequeathing to his successor a legacy of runaway inflation and budgetary red ink to match. Now, confronting his own day of reckoning at the polls, Ford rejected short-term stimulus or election-year jobs programs. Having done the heavy lifting to reduce inflation below 5 percent by the spring of 1976, he figured "it would have been one hell of a job to turn it around again." Accordingly, Ford said, "I would rather gamble that we could win the election even though there was a pause, and then have a healthy economy to start out the new term."

T HE $350-A-DAY ROYAL Suite of Kansas City's Crown Center Hotel had been given a presidential makeover. On the walls a collection of oversized images, identical to the big framed photos in the hallways of the West Wing, showed Gerald Ford as commander in chief inspecting troops, climbing on a tank, boarding a naval vessel. "So we had the suite looking good and conservative," Terry O'Donnell remembered, "and then Baker and the group would bring up the delegates, onesies, twosies, and threesies at a time." On Monday morning, August 16, Ford breakfasted with the Hawaiian delegation while across town Betty and three of the Ford children attended the convention's opening session in Kemper Arena, a windowless structure built on the former site of the Kansas City stockyards. Their arrival evoked the week's first display of party disunity as members of Texas's solidly pro-Reagan delegation stood in an effort to block television cameras from recording the first family's entrance.

The Texans, resentful that the Ford-controlled Republican National Committee had assigned them hotel accommodations fifty miles away, were not above hurling rolls of toilet paper over the railing that separated them from the Fords in their box. This was disputed by Ernie Angelo, a Reagan stalwart from Midland who accused Jack Ford of dumping "two big baskets of trash" on the offending Longhorns. "It wasn't trash," Jack insisted. "It was red, white and blue confetti . . . My mom was sitting beside me, and she wouldn't let me dump trash." Hoping to camouflage such animosities, convention organizers instructed the house orchestra to play "God Bless America" over warring chants of "We Want Reagan" and "We Want Ford." On Monday evening Barry Goldwater and Nelson Rockefeller, aging combatants in an earlier GOP schism, spoke back to back in a mutual show of reconciliation. Delegates paid more attention to Nancy Reagan, elegant in her trademark red, whose appearance set off a fifteen-minute ovation. It was still going on when Betty Ford entered her front-row box at the opposite end of the hall. First round of the battle between "the queen of the north galleries" and "the queen of the south galleries" was awarded to the Red Queen with the California tan.

It would be a different story on night two, when decisive votes on Reagan initiatives affecting party rules and the platform roiled emotions beyond Irving Berlin's powers to pacify. First up: Rule 16-C, a classic John Sears sleight of hand that would have required Ford to name his running mate *before* the presidential balloting on Wednesday night. Whomever he selected, Sears reasoned, Ford was sure to offend some other candidate or bloc of delegates—especially if, with November in mind, the president should veer back to the center and choose an Elliot Richardson or William Ruckelshaus. (Ironically, one of the most potent arguments Ford supporters used against the rules change was its elimination of Reagan as a vice presidential prospect.)

Because no state laws bound delegates to support the Ford or Reagan position on 16-C, this left an unknown number of fifth columnists—Reagan sympathizers in Ford delegations, and vice versa—to vote their conscience on what the White House had taken to calling, in a not-so-subtle jab at the backlash generated by Schweiker's selection, the "misery loves company" amendment. As late as five that afternoon, according to Peter McPherson, Jim Baker's deputy in the hunt for delegates, "the vote count showed an undecided convention, with Mississippi being the swing." Even as Clarke Reed's delegation caucused, Reagan's Californians taunted the enemy with chants of "Who's Our Veep? Who's Our Veep?"

Ford operatives, determined to prevent a repetition of Monday night's upstaging of the First Lady, had smuggled hundreds of signs into the hall, where they could be stashed under delegate seats until Betty's entry generated a foot-stomping display of organized spontaneity. The nearly simultaneous arrival of Mrs. Reagan ignited a rival show of enthusiasm. Judged by decibel level, the contest was very nearly a draw. Then the orchestra struck up "Tie a Yellow Ribbon 'Round the Old Oak Tree," a musical earworm whose originator, Tony Orlando, just happened to be in the Ford box.[*]

"That's your song, Tony," said Susan Ford. "Come on, you and Mom get up and dance." The ensuing performance had the desired result, as television cameras remained fixed on Betty Ford, and the crowd—half of it, anyway—chanted its delight. Speaking for the other half, Texas delegate Robert Monaghan groused, "She's like a chorus girl, wiggling her bottom."

John Connally was at the podium addressing an unruly crowd of nineteen thousand when copies of the afternoon *Birmingham News* began circulating among Clarke Reed and his brethren. The paper's banner headline— "Ford Writing Off Cotton South?"—was a garbled version of Rogers Morton's latest press gaffe. The inflammatory front page gave Reed the excuse he needed to abandon Ford. If even a handful of Mississippi delegates or alternates followed suit, then all thirty of the state's votes could revert to Reagan on 16-C. Only a hastily arranged phone call from the president

---

[*] Orlando had previously volunteered to spend "a couple of hours" instructing the president in the use of body language. "Oh, Maria," Betty Ford responded to the courier of the entertainer's offer, "he'd kill us both." AI, Maria Downs.

to Reed averted this last-minute calamity. Indicative of the continuing turmoil within its ranks, Mississippi passed the first time it was called on by the convention secretary.

By then it was clear that Sears's claims of hidden support were exaggerated. To the contrary, Ford showed surprising strength in delegations like Georgia and Indiana, where the lingering desire for a Ford-Reagan ticket overrode the governor's explicit denials of interest. Did every vote count? Certainly the Ford managers thought so. When one of their delegates fell and broke her leg, her handlers fashioned a splint out of convention programs rather than send the woman to an emergency room. Their reasoning was simple enough—her alternate was a Reagan supporter. In the end her vote wasn't needed any more than the thirty cast by Mississippi under the unit rule against 16-C. The Ford position prevailed by just over a hundred votes.*

But John Sears wasn't through. "The thing that I feared the most in Kansas City was the platform," recalled Stu Spencer. "It's like you got a matchbook and you got gasoline sitting there. If you throw a match on it, you've got emotions that just start going crazy." Sharing Spencer's concerns, Jim Baker puzzled then, and for decades afterward, at Sears's strategic choices. Rather than "a technical, procedural thing"—Baker's dismissive characterization of 16-C—"why didn't they come up with a platform that said, 'Fire Kissinger'?"

They very nearly did, as Dick Cheney remembered it. Besides explicit praise for Aleksandr Solzhenitsyn, the substitute plank denounced "secret agreements" in American diplomacy and took aim at the Helsinki Accords for normalizing Soviet domination of Eastern Europe. Without explicitly naming Kissinger, noted Cheney waggishly, "you knew who that querulous bastard was who was leading Ford astray."

How vigorously should the White House contest the issue? Kissinger, backed by Rockefeller and Scowcroft, argued that to accept the Reagan plank was to repudiate a record of historic accomplishment and hand Democratic nominee Jimmy Carter a stick with which to beat the Ford campaign every day until November 2. Where was the morality in that? At one point Kissinger threatened to resign if the president, in effect, took a dive. "Henry, will you do it now and do it loudly? We need the votes," said Tom Korologos, the administration's sharp-tongued envoy to Capitol Hill.

---

* Though Clarke Reed, playing both sides of the street to the last, waited until he was sure Ford had thirty-one of Mississippi's sixty delegates and alternates, and only then voted in favor of the Reagan position on 16-C.

Cheney, so often portrayed in later years as a rigid ideologue, was the voice of pragmatism as he gauged the odds against success. "This is late at night, we've already won on the 16-C, our people are in bars all over Kansas City thinking we won the fight. We'll never get them back," he concluded.

It was a line of reasoning endorsed by Jim Baker. "We knew we had the nomination if we didn't lose any other votes, and we also knew that the platform is something written at the convention and then forgotten." Ford was persuaded. Shortly before two a.m. the Reagan plan was adopted on a voice vote. "We have made political history out there," Kissinger observed acidly. "The first party that condemns itself in the name of morality." With freakish timing the president accused of coddling Communists would, in a matter of hours, wake to confront the world's most fanatical Communist regime over the brutal murder of two American soldiers. Their deaths were unlikely to be avenged by platform rhetoric.

WHAT HISTORIAN GORDON F. Sander calls "Ford's finest forgotten hour" overlapped with the business of getting nominated in Kansas City. Flush with his victory on 16-C, Ford learned Wednesday morning— hours before the climactic roll call to determine the party's nominee—of the overnight assault by ax-wielding North Korean troops on a handful of US and South Korean soldiers trimming a poplar tree that blocked the allies' view of the Joint Security Area within the demilitarized zone (DMZ) separating the two Koreas. The North Koreans claimed the tree in question had been planted by the Great Leader, Kim Il-Sung, himself. When this failed to halt the pruning operation, reinforcements from the Korean People's Army (KPA) arrived on the scene. They quickly surrounded the work party ultimately answerable to the United Nations. In the ensuing melee, Americans Captain Arthur Bonifas and First Lieutenant Mark Barrett were beaten with clubs and iron pipes, then finished off with knives and axes.

In Kansas City, Ford asked for military options that would punish the North Koreans without escalating the crisis into something far worse. Before leaving town on Friday morning he gave the go-ahead for Operation Paul Bunyan, under which UN forces were to complete what Captain Bonifas and his company had started earlier in the week. In support of their mission Ford ordered the largest demonstration of US military might on the Korean peninsula since the 1953 armistice. The aircraft carrier *Midway* was redirected to the Korea Strait, along with a squadron of F-4 Phantom fighters from Okinawa and twenty F-111s summoned from their Idaho home base. In addition, a contingent of B-52 bombers from Guam would take a flight path calculated, wrote Robert McFarlane in his history of the

Korean crisis, "to make their presence visible, but sufficiently far from the DMZ to remain nonprovocative."

Henry Kissinger wanted more. Merely cutting down the disputed poplar would look "ridiculous," he contended. Better to deter future aggression from the North by eliminating the tree *and* shelling KPA barracks on the other side of the zone. Ford preferred to hold such reprisals in reserve. "To gamble with an overkill might broaden very quickly into a full military conflict," he explained later, "but responding with an appropriate amount of force would be effective in demonstrating U.S. resolve." The president was in Vail for a postconvention review of campaign strategy when a US engineering team carrying chain saws and axes entered the disputed area, backed by nearly one hundred security and South Korean Special Forces. Massed on the other side of the so-called Bridge of No Return was a larger group of KPA soldiers, most clutching machine guns.

Any North Korean urge to renew hostilities was effectively stifled by the sudden appearance on the horizon of a fleet of US Cobra helicopters. It took just forty minutes for the tree trimmers to complete their task. Later that day Kim Il-Sung issued the nearest thing to an apology in the history of his dynasty. It was "regretful," said the Great Leader, "that an incident occurred in the Joint Security Area." Efforts must be made—on both sides—to prevent a recurrence of such incidents in the future. Within days the North Koreans dismantled a quartet of illegally stationed guard posts on the UN side of the DMZ.

Ford's carefully modulated use of American power showed how much he had gained in self-confidence since the *Mayaguez* affair. The latter may have generated more immediate political rewards. But history has been kinder in assessing the military confrontation whose very obscurity testifies to the finesse with which Ford, exercising his own judgment, averted full-scale combat while putting North Korea's paranoid regime back in its cage.

So MUCH EMOTION had been generated by Tuesday night's test votes on 16-C and the platform that Wednesday's roll call for the nomination felt anticlimactic. Reagan supporters in the hall, unreconciled to defeat, vented their frustrations by blasting plastic horns and pushing Ford's moment of triumph out of television's prime time in the East. Not until twelve thirty Thursday morning was West Virginia governor Arch Moore able to announce the infusion of delegates that sealed Ford's victory. The final tally was Ford 1,187, Reagan 1,070. A standing agreement between the rival camps dictated that the winner should without delay call on the loser to begin the arduous task of reuniting the party. "You have to promise in advance that you will not offer the vice presidency," John Sears told Bill Timmons, the Ford convention manager, who related the message to Cheney on Wednesday morning. Sears repeated himself that night after the balloting. Timmons went to the president's hotel suite, where he found Ford and Cheney about to depart for Reagan's quarters at the Alameda Plaza Hotel.

"Well, Mr. President, I have some very bad news for you."

"What's that?"

"The condition is that you not offer Reagan the vice presidency.

"Boy, he smiled and puffed about six puffs" on his ever present pipe, Timmons recalled, "and he was just so happy about that because he was worried that Reagan wanted the vice presidency."*

It was after one a.m. Thursday when Ford and Reagan met alone in the governor's tenth-floor suite ("no wives" being another precondition of their meeting). Acknowledging "we are way behind" in the latest polls, Ford asked for his rival's assistance in closing the gap. Reagan pledged his cooperation, though he declined Ford's invitation to address the convention on its final night, telling the president, "It's your show." Ford mentioned half a dozen potential running mates under consideration. A subdued Reagan indicated that Bob Dole would be acceptable. Neither man felt comfortable in the other's company, nor did they make much effort to pretend otherwise. "The tension of the whole contest was reflected in that fifteen minutes," Ford would remember. "We had never been close and these circumstances certainly didn't bring us any closer."

Returning to the Crown Center, Ford was joined by Rockefeller, Senators Bob Griffin and John Tower and half a dozen key advisers in a two-hour discussion of vice presidential possibilities that didn't begin until 3:15 a.m. Pollster Bob Teeter's latest voter surveys indicated significant resistance to John Connally and, to a lesser degree, Reagan. Only in the Deep South did the California governor surpass the modest advantage Howard Baker

---

* An alternate narrative involves a second back channel linking Ford's close friend Leon Parma and William French Smith, Reagan's personal lawyer and a mainstay of his Sacramento kitchen cabinet. At the start of the week, Smith approached Parma with a request from the governor. Would Ford be willing to designate someone to handle sensitive communications "other than through Sears" or some other campaign operative? With Ford's concurrence Smith and Parma, shielded from curious reporters, met several times in Parma's hotel room. "Nothing substantive came from it until the last day when Bill said, 'The Governor wants to be sure that the President doesn't ask him to be the vice president . . . He doesn't want to say no to the President.'" Parma relayed this message to Ford, who replied, "Tell him not to worry." AI, Leon Parma.

brought to the ticket nationwide. Ford kept his counsel, though some attendees thought they detected a mild preference for Baker. Had the Democrats nominated anyone but Jimmy Carter, the Tennessee senator might
well have gotten the nod. But at this stage of the campaign the Ford team
was inclined to write off most of the South, the region in which Baker
would presumably have his greatest appeal.

Bill Ruckelshaus was Teeter's candidate. A Catholic moderate with a
strong environmental record and a feminist wife, his role in the Saturday
Night Massacre would help to inoculate Ford on the pardon issue. But
Ruckelshaus added little to the ticket geographically, his home state of Indiana being next door to Michigan. Moreover, he had never won a statewide race, calling into question his prowess as a campaigner in a year when
the vice presidential candidate would be expected to compensate for Ford's
deficiencies on the stump. As the discussion droned on inconclusively, Stu
Spencer introduced the name of Anne Armstrong, the smart, personable
US ambassador to the Court of St. James's, Texas cattle rancher and first
woman to keynote a national political convention. "Hey," Spencer told his
skeptical colleagues. "We're thirty-two points behind today. What have you
got to lose?" Ford was intrigued, but Teeter's polling showed voters in both
parties skittish about the idea of a woman on the ticket.

The looming stalemate fostered the candidacy of Kansas senator Robert
Dole. Darkly handsome as a sort of rustic Bogart, Dole was that rare Republican officeholder who had managed to stay friends with both camps
even while making no secret of his preference for Ford. Now in his second
Senate term, the fifty-three-year-old Dole had been Republican national
chairman at the time of Watergate, but few held it against him, especially
after his vaunted independence got him fired from the job by an unappre-
ciative Richard Nixon. Besides farmers, Dole numbered veterans and the
handicapped among his personal constituencies. His sharp tongue and biting wit gave pause to some, but no one questioned Dole's ability to fill the
role of attack dog while Ford donned the statesman's mantle and executed
a Rose Garden strategy.

The senator had divorced his first wife, an occupational therapist who
aided his remarkable comeback from near-fatal wounds inflicted by German machine guns on an Italian hillside in April 1945. Dole's second wife,
Elizabeth, shared his political interests. Part trailblazer—as a Harvard Law
School graduate appointed to the Federal Trade Commission by Nixon—
she was thoroughly traditionalist in her North Carolina upbringing and
bonny Southern charm. Not her regional appeal but her husband's figured
in Stu Spencer's rationale for choosing Dole over Baker or Ruckelshaus.

"You've got a problem in the farm states," he reminded Ford. "We're losing Republican territory." True enough. His three-month embargo on grain sales to the Soviet Union in the summer of 1975 had damaged Ford in Republican strongholds like Dole's Kansas. Simply put, his base needed shoring up.

Still, nothing had been decided when, shortly after five o'clock, the group broke up, to reconvene after everyone grabbed a few hours' sleep. Overnight, word reached the Ford team that Southern delegates who wanted him to choose Reagan considered Dole an acceptable substitute. Clarke Reed confirmed this in a telephone conversation with Dick Cheney.

A last-minute effort by Republican state party chairmen to impose a Ford-Reagan ticket lent a note of urgency to the second round of vice presidential talks early that Thursday morning. Across town a group of Reagan die-hards roused the governor and his wife, who were sleeping in after their late night. Before they could hear yet another plea for Reagan to accept a job he didn't want and hadn't been offered, the phone rang. "Excuse me, Governor," said Mike Deaver, "but it is President Ford." The visitors to the suite eavesdropped, their spirits visibly sagging as Reagan put down the receiver and told them, "Fellas, I know why you're here, but that was the President. And he just told me that he picked Bob Dole." That evening over a hundred Reagan delegates abstained on the single vice presidential ballot; as many more cast a protest vote for Jesse Helms. Responding to cries from the floor, convention chairman John Rhodes twice invited Reagan to leave his skybox and address the convention. Both times he demurred.

At the Crowne Plaza, Ford watched as precious television minutes were being lost and his acceptance speech receding beyond prime time in the east. "Dammit," he shouted at Cheney. "Get hold of Rhodes and tell him to get this thing under control. NOW." The continuing disarray only magnified the pressure to deliver the speech of his lifetime, one that would simultaneously dispel hard feelings, banish defeatism within Republican ranks and reach beyond the Kemper Arena to reintroduce himself to millions of television viewers accustomed to the caricature Ford of *Saturday Night Live*. For two weeks Don Penny had rehearsed his client to the point of exhaustion. Reviewing his performance on videotape, Ford had corrected garbled words and exaggerated gestures. When Betty said the speech was much too long, Penny cut thirteen pages. At the last minute, telling no one but Betty and Bob Hartmann, Ford wrote out a challenge to Carter to debate him face-to-face. ("We were sitting on pins and needles," he revealed afterward, "hoping Carter would not challenge me before I made my challenge.") The insert went into the president's reading copy and no one else's.

Further delayed by an introductory campaign film, it was nearing midnight on the East Coast when the convention orchestra sounded the opening notes of "Hail to the Chief." Watching from his hotel room, campaign ad maker Doug Bailey muttered, "This is either the end of the campaign or the beginning." Bailey was in for a surprise. From his opening lines, Ford sounded a Trumanesque defiance of the polls, vowing to wage "a winning campaign . . . from the snowy banks of Minnesota to the sandy plains of Georgia." Invoking Dwight Eisenhower, he reminded his countrymen that he was the first president since Ike "who can tell the American people America is at peace."

When Ford in his next sentence declared the nation "on the march to full economic recovery," the audience response mirrored Doug Bailey's.

"My God," Bailey whispered. "He sounds *good*."

Ford's unexpected challenge to debate Carter brought the crowd to its feet. There were cheers for his middle-class manifesto directed to the people "who pay the taxes and obey the laws . . . You are the people who make America what it is. It is from your ranks that I come and on your side that I stand." Recalling the unique circumstances by which he had acceded to the nation's highest offices—"Not as prizes to be won, but a duty to be done"—Ford contrasted the American condition in August 1974 with the "incredible comeback" achieved in spite of "a vote-hungry, free-spending congressional majority on Capitol Hill." His vetoes of extravagant legislation had saved American taxpayers billions, he said, giving credibility to his crowd-pleasing claim to be "against the big tax spender and for the little taxpayer." A litany of conservative initiatives to lower taxes, restrict court-ordered busing and crack down on crime and drugs had been stymied by congressional inaction, Ford complained. Where Congress did take action, it was to slash the defense budget.

Running on his record "of specifics, not smiles," Ford would test the axiom that elections are about the future rather than the past. Aside from promising to submit a balanced budget by 1978, the vision section of the speech was mostly generalities—Social Security would be protected, the quality of life improved, taxes reformed. He would see to it that "the party of Lincoln remains the party of equal rights." In a blatant mea culpa to disaffected farmers, he pledged, "There will be no embargoes." Ford's upbeat narrative was meant to drive home his steady, unflashy leadership of a nation in recovery. And to begin to sow doubts about his untested opponent.

"An electrifying performance," concluded William F. Buckley. "As if Joe Palooka had appeared in the Roman Senate and outshone Cicero." "We may have lost a Vice-President," remarked Doug Bailey, a Ruckelshaus booster, "but we have gained a President." And it wasn't over. Before the balloons dropped, Ford invited the Reagans to join him onstage. "Ronnie, don't go," Nancy told her husband, but Ford would not be put off. Five minutes later the adversaries were sharing the spotlight. "This convention is full of things we have never seen before," marveled NBC anchorman John Chancellor. Stepping to the microphone as Ford requested, Reagan improvised a tribute to the hard-line platform—"A banner of bold unmistakable colors with no pale pastel shades"—and an imagined letter addressed to Americans of 2076 reflecting on the "erosion of freedom" that their own generation was duty bound to resist.

Kansas City, journalist David Broder would write, was the only convention in history to witness two acceptance speeches. That Reagan failed to mention the Ford-Dole ticket in his nonvaledictory was lost in the euphoria generated by the semblance of unity. Afterward the president isolated himself, Bob Hartmann and Don Penny in a room away from the celebratory hubbub.

"All right, now I want you two to shake hands and cut it out."

Hartmann allowed as how Penny had done "a fine job" coaching Ford on the speech. "You did a pretty good job on writing it," said Penny. The two men embraced. Ford could only hope his party was as easily reconciled.

BEFORE LEAVING THE Alameda Plaza, Ford had publicly complimented Reagan by telling him, "You really got us in shape." Others in his entourage were less generous. "Ford became a better *candidate* because of some of the things that Reagan did," Doug Bailey conceded shortly before his death in 2013. "But his *candidacy* was all the more difficult because of what Reagan did." By this Bailey meant that Reagan's challenge made Ford a better speaker and a more focused executive. Securing his party's endorsement over so formidable a competitor imparted to the Ford presidency a luster missing from Richard Nixon's imprimatur. Yet Bailey also had cause to lament the resulting diversion of time and energy into "how to get the nomination, not how to win the election."

Ford, too, reflected on his administration's postponed accomplishments—SALT II being the most obvious—and politically questionable positions forced on him as the price of victory in Kansas City. His plight was summarized in a 120-page campaign battle plan drafted by White House aide Michael Raoul-Duval. "If past is indeed prologue," Ford read, "you will lose on November 2—because to win you must do what has never been done: close a gap of about 20 points in seventy-three days from the base of a minority party while spending approximately the same amount of money as your opponent." (Post-Watergate campaign finance laws guaranteed each candidate $21.8 million to spend between the conventions and election day.)

Ford and his campaign planners were roundly criticized for their failure to include Reagan or his representatives in their Vail strategy sessions. This was no oversight. On the heels of the governor's vice presidential ultimatum, Leon Parma had received fresh instructions from his convention-week collaborator William French Smith.

"Please tell the President not to call Reagan."

Never, Smith told a stunned Parma, had he seen two people so hurt as Ron and Nancy Reagan in the aftermath of Kansas City.*

"Well, if that's what he wants," Ford said on learning of his rival's wounded feelings. Not until September 2, two weeks after they shared the stage in Kansas City, did the two men speak. The ice was barely dented in their seven-minute phone conversation. Reagan agreed to introduce Ford at a fundraising dinner in Los Angeles on October 7, but only after Lyn Nofziger made certain that Paul Haerle, the unabashedly pro-Ford chairman of the California Republican Party, would be barred from attending. Reagan pleaded scheduling conflicts to avoid campaigning with the president in the conservative heartland of Southern California. Eleventh-hour visits to battleground states Mississippi and Texas were ruled out for the same reason.

Ford turned his attention to his Democratic opponent. Tapping a level of regional pride reminiscent of Catholic fervor for John F. Kennedy in 1960, Jimmy Carter, the Bible-quoting populist from tiny Plains, Georgia, was ideally placed to exploit the bipartisan distrust of Washington bred by Vietnam, Watergate and the Nixon pardon. Railing against "the bloated bureaucracy," Carter's vow to reduce 1,900 existing federal agencies to 200 was the Democratic equivalent of Reagan's $90 billion reshuffle of government responsibilities. "You've got to remember there was a romance about Carter," David Gergen asserted. "He wasn't Nixon, he wasn't Ford, he wasn't an insider, and that was working powerfully in his direction . . . there was something almost Obamaesque about the Carter campaign. It was well choreographed."

---

* It bears noting that Lyn Nofziger in his eponymous memoir goes even further than Smith in describing an uncharacteristic bitterness Reagan felt toward the man he believed had stolen their party's nomination. John Connally, disappointed at being left off the ticket, turned down Ford's request to become Republican National Committee chairman. But he at least showed his face in Vail.

Obscured by contrasting styles, Ford and Carter had more in common than their partisans acknowledged. Both men were Horatio Alger strivers, imbued with a sense of place and the small-town values it imparted. Both had had their horizons broadened by service in the navy. Devout in his faith, each was the happily married father of three sons and a younger daughter on whom he doted. If Ford's affability was sometimes mistaken for weakness, Carter's steely self-confidence bred charges of ruthlessness. "Jimmy is like a beautiful cat with sharp steel claws," according to his mother, the formidable Miss Lillian. Rarely was it said of Carter that what you see is what you get. Novelist turned Carter speechwriter Patrick Anderson thought the Democratic nominee "a combination of Machiavelli and Mr. Rogers." More detached observers attributed Carter's complex character to the dueling influences of Old South and New—of a sternly traditionalist father; a book-loving, racially sensitive mother; and the perfectionism of his naval mentor, Admiral Hyman Rickover.

With inspired symbolism Carter launched his homestretch drive on Labor Day, against the backdrop of FDR's Little White House in Warm Springs, Georgia. That he should draw parallels between himself and Roosevelt's 1932 campaign to defeat an allegedly callous Herbert Hoover was not surprising. Yet even here, before the altar of modern liberalism, Carter balanced his support of new social programs with a pledge of "tough management and careful planning, leading to a balanced budget." In many ways he was ahead of his time.* By 1976 the old Roosevelt coalition of Northern Blacks and Southern racists, farmers and union men, big-city bosses and Ivy League academics had fractured as traditional economic concerns vied for attention with more divisive social issues.

The Supreme Court's 1972 decision in *Roe v. Wade* introduced abortion to the national debate. Overnight both parties were forced to recalibrate their messages, especially those aimed at some thirty million Catholic voters concentrated in a dozen large states. A reliable Democratic voting bloc, by the summer of 1976 they were thought to be in play owing to Carter's Southern Baptist heritage and his relatively liberal stance on *Roe v. Wade*. On September 10, Ford welcomed several high-ranking members of the church hierarchy to the White House. Afterward the bishops offered qualified praise for their host's qualified support of a constitutional amendment that would let states determine for themselves the legal status of abortion.

---

* It was no accident that, until Barack Obama's election in 2008, the two Democrats to reach the White House during the period of conservative ascendancy that began in 1968 were centrist Southern governors, Jimmy Carter and Bill Clinton.

For his formal campaign kickoff on September 15, Ford returned to Ann Arbor and a mixed reception from his alma mater. Following dinner with the Wolverines football team, Ford braved student hecklers and an exploding cherry bomb that unnerved the Secret Service and fifteen thousand spectators in Crisler Arena. In his speech the president talked of revising the federal tax code to make college more affordable and assist parents enrolling children in parochial and other private schools. Mostly, however, he drew bright lines of credibility between himself and his opponent. "The American people are ready for the truth, simply spoken, about what government can do for them and what it cannot," he declared. "It's not enough for anyone to say, 'Trust me.' Trust must be earned. Trust is not having to guess what a candidate means. Trust is leveling with the people before the election about what you are going to do after the election."*

Doug Bailey conveyed his theory of the race. "If you focus on Ford, you like Carter. If you focused on Carter, Ford had a chance." Put another way, if, on election day, "the question in voters' minds was, do I know enough about Jimmy Carter . . . then Gerald Ford would be elected president. If the question was, do I think that Gerald Ford ought to be reelected, then Jimmy Carter would win." Unscripted events would drive the campaign narrative, with the news cycle itself exercising more control than the candidates. "Somewhere about '76," Stu Spencer has observed, "the media decided . . . process [is] more important than the ideas on issues." Blame it on Theodore H. White and his landmark *Making of the President* series. Beginning in 1960, with novelistic flair, the self-described "storyteller of elections" had converted a quadrennial civics lesson into nail-biting stagecraft, a horse race without horses.

White's journalistic successors were more inclined to insert themselves into the narrative. Besides Woodward and Bernstein there was Joe McGinniss's *The Selling of the President 1968*, Timothy Crouse's seminal *The Boys on the Bus*, and Hunter S. Thompson's scathing *Fear and Loathing: On the Campaign Trail '72*. The Watergate drama accelerated the synergy between print journalism and television, with its sound-bite sensibility and emphasis on controversy over nuance. The multiplier effect of such a collaboration was demonstrated on September 21, when advance copies of the November issue of *Playboy* magazine produced the campaign's first media feeding frenzy. Buried within a long, reflective interview in which

---

* Ford's words of warning were not crafted in a vacuum. Pollster Lou Harris had recently reported that fully two-thirds of those he sampled expressed doubts "about a politician who says he'll never lie," as Carter had done.

the Democratic nominee for president sought to explain himself and his faith-based values to a skeptical electorate was Carter's acknowledgment that "Christ set some impossible standards for us." He cited a passage from the Book of Matthew in which the Lord declares anyone who looks on a woman with lustful intent to be guilty of adultery.

"I have looked on a lot of women with lust," he volunteered. "I've committed adultery in my heart many times." Carter expected God to absolve him. Indeed, he continued, Christian forgiveness was extended equally to the sinner who "screws a whole bunch of women" and to the man who remains faithful to his marital vows. Former presidents, on the other hand, made poor candidates for redemption. Guided by his faith, Carter expressed confidence he would never emulate his predecessors Nixon or Johnson in "lying, cheating and distorting the truth."

In the swing state of Texas, Johnson's widow let it be known that she was "distressed, hurt and perplexed" by Carter's comments. Others took offense that Carter should air his intimate thoughts in a journal many of his neighbors held to be little better than pornography. The next day, in a previously scheduled meeting with a group of prominent evangelical leaders led by "the Protestant pope" W. A. Criswell of Dallas's First Baptist Church, Ford revealed that he had been approached for an interview in *Playboy* "and I declined with an emphatic 'No'!"*

Two factors aided the Carter campaign in containing the firestorm. One was an accident of timing, as eighty-five million Americans on September 23 tuned in to the first debate ever between a sitting president and his challenger. The second was a change of subject thanks to columnist Jack Anderson, who informed viewers of ABC's *Good Morning America* that Ford while in Congress had accepted free golf outings at a New Jersey golf club from U.S. Steel vice president and lobbyist William Whyte. "I get so tired of this horseshit," Ford erupted on hearing it said that his conduct violated a 1968 House code of ethics precluding members from accepting a "gift of substantial value . . . from any person, organization or corporation having a direct interest in legislation before the Congress." Ford disputed the "substantial value" of a golf game, played with a family friend of twenty years standing, one close enough to host the initial meeting of Ford's transition

---

* Ironically, many of the religious spokesmen voicing outrage over Carter's indecorous language had been just as displeased by Betty Ford's candid treatment of premarital sex in her *60 Minutes* interview. By the time Ford graced the pulpit of Criswell's church in October, Carter had supplied the forces of traditional morality with fresh grounds for outrage by employing the F-word in a *New York Times Magazine* interview with Norman Mailer.

team in August 1974, and reciprocated when he and Whyte traded gimmes at the Burning Tree Country Club, where both were members.

Anxious to get out in front of the story, the White House let it be known that Ford had been the recipient of corporate hospitality from at least three other companies prior to becoming vice president. Blunting the impact somewhat were revelations that Carter, as governor, had stayed at vacation lodges owned by companies doing business in Georgia. Polls showed public indifference to the issue. It went unmentioned, as did the *Playboy* kerfuffle, in the September 23 debate at Philadelphia's Walnut Street Theatre. Quickly shedding his initial nervousness, Carter accused his rival of economic mismanagement and insensitivity toward those who were out of work. Playing the populist card, he denounced the nation's tax laws as "welfare for the rich." Ford noted that those same laws had been written by Democrats in control of the Congress since 1955.

If he bore responsibility for the Democratic Congress, Carter rebutted, then Ford was culpable for the Nixon Administration, "of which he was a part." Ford defended his use of the veto pen, which had saved taxpayers $9 billion and had been employed more frequently by FDR as president and Carter himself as governor. Ford also sought to make an issue of Carter's reorganization of Georgia state government, resulting, he claimed, in increased work rolls and administrative costs.

Near the end of the program the failure of an ABC audio amplifier plunged the stage into silence. For twenty-seven minutes the combatants remained frozen behind their podiums. By the time sound was restored, much of the audience had drifted away, dissatisfied with a format that amounted to dueling press conferences.

Snap polls showed Ford winning on points; 40 percent to 31 percent in a Harris survey. A postdebate sampling of opinion by the same pollster showed a narrowing of the race overall, with the president cutting his deficit from thirteen to nine points. Yet Carter had more than held his own, exhibiting impressive command of facts and reassuring Democrats fretful over the *Playboy* interview. The *Boston Globe* rendered a split decision, assessing Ford as the more skillful debater and Carter "the more attractive personality."

ANY MOMENTUM GENERATED by Ford's debate performance was short-lived, as a series of distractions put him on the defensive through mid-October. His vow to keep politics out of the Justice Department boomeranged when an unnamed informant (subsequently identified as Helen Delich Bentley, a Reaganite from Baltimore whom Ford had failed to reappoint as chair of the Federal Maritime Commission) went to the FBI asserting that Congressman Ford had pocketed maritime union funds laundered through the Kent County (Michigan) Republican organization. Unknown to the White House, in July Attorney General Levi had authorized Charles Ruff, Leon Jaworski's successor as Watergate Special Prosecutor, to investigate these claims. That they had already been examined by the FBI in conjunction with Ford's vice presidential selection was confirmed by Jaworski on September 28. The former special prosecutor said his own inquiry into the Kent County GOP had turned up nothing improper.*

With the election barely six weeks off, Ford knew only what he read in the press or watched, with growing anger, as, night after night, the three television networks led their evening newscasts with the shadowy story. "I couldn't talk to Levi about any of this because I had given him my word that I would never interfere with an investigation," he wrote in his memoirs. To publicly challenge Ruff, whose teaching obligations at Georgetown University made him, in effect, a part-time special prosecutor, "would smack of Nixon and the Saturday Night Massacre all over again." Besides complying with Ruff's request for personal financial information, Ford authorized the IRS to provide the special prosecutor with the complete work product from its 1973 audit of his tax returns for the previous six years.

Adding to his troubles were reports of a slowing economy and the scabrous indiscretions of his agriculture secretary Earl Butz. "Earl lives by the tongue and he's going to die by the tongue," his wife, Mary Emma, had predicted. Her prophecy was fulfilled on a postconvention flight from Kansas City to the West Coast. Working the aisle of the plane, Butz had stopped to talk with Pat Boone, a Reagan delegate worried about Republican chances in the fall. Seated beside Boone was John Dean, the former Nixon White House counselor fresh from his gig covering the GOP convention for *Rolling Stone*. Discussing with Boone their party's failure to attract more Black voters, Butz, a master of barnyard humor, outdid himself. Why didn't more "coloreds" vote for the party of Lincoln? As Butz told it, they wanted only

---

* The reader will recall from Chapter Twenty how Ford in December 1974 had reluctantly vetoed cargo preference legislation requiring 20 percent of all US oil imports to be carried in American vessels. Jesse Calhoon, president of the Marine Engineers' Beneficial Association (MEBA), swore revenge on his former ally. Twice in May 1975 Calhoon met with Jimmy Carter and, after obtaining written assurances of the governor's sympathy for his industry, he sponsored a $1,000-per-person fundraising dinner that raised $150,000 for the Carter primary campaign.

three things: "first, a tight pussy; second, loose shoes; and third, a warm place to shit."

This obscenity found its way into Dean's *Rolling Stone* reporting, in which he attributed the story to an unnamed Cabinet official. Butz's cover was blown when an enterprising *New Times* editor reviewed the travel schedules of several Cabinet members. On Friday, October 1, the day after his involvement became public, Butz was dressed down in the Oval Office. Ford told him that his remarks were equally offensive to the American people and their president. Butz's apology did nothing to calm the waters. By the time he returned three days later to tender his resignation, Ford was taking heat from farm organizations for cutting the popular secretary loose, and from virtually everyone else for not firing him at the earliest opportunity.

That Monday afternoon a distracted Ford left Washington for San Francisco, scene of his second televised debate with Jimmy Carter. Assailing the Ford record from both ends of the political spectrum, Carter vowed to withdraw US troops from the Korean peninsula *and* maintain US control of the Panama Canal. To employ a historical parallel suitable to these two naval veterans, it was like watching the Civil War ironclads *Monitor* and *Merrimac*: the former vessel boasting a revolving turret and sufficient agility to evade the lethal ram of its heavily plated, less maneuverable adversary. What halted the Ford campaign dead in the water was the president's response to the debate inquiry of *New York Times* associate editor Max Frankel about US-Soviet relations; specifically, whether the United States had not bailed out the Soviet economy through huge grain sales and, by signing on to the Helsinki Accords, all but conceded Russian dominance in Eastern Europe.

To the contrary, Ford replied, the recent arms control talks at Vladivostok had been conducted from a position of American strength. Grain sales contributed to American prosperity. As for Helsinki, the agreement formalized there had won the approval of thirty-five nations on both sides of the Iron Curtain. Among the signatories was the Vatican secretary of state on behalf of His Holiness the Pope, someone not generally viewed as a Communist sympathizer. The same pact mandated advance notice of any planned military maneuvers by NATO and the Warsaw Pact. No, Ford concluded, "There is no Soviet domination of Eastern Europe, and there never will be under a Ford Administration."

Caught off guard by this startling fallacy, Frankel sought clarification. "I don't believe, Mr. Frankel," said Ford, "that the Yugoslavians consider themselves dominated by the Soviet Union." Likewise the Romanians and Poles. "And the United States does not concede that those countries are under the domination of the Soviet Union." As evidence he mentioned his own visits to the three countries in question.

In the greenroom where the Ford staff watched the debate on TV screens set up for that purpose, Brent Scowcroft was the first to speak up. "You've got a problem," he told Stu Spencer.

"What do you mean, I've got a problem?"

Scowcroft recited the number and firepower of Russian divisions currently on Polish soil.

"Holy Toledo, we've got a problem."

Dick Cheney later claimed that "Ford thought Frankel was asking about the Sonnenfeldt Doctrine," this a reference to State Department counselor Helmut Sonnenfeldt and his clumsily phrased diplomatic formula envisioning an "organic" relationship between the Soviet Union and its Eastern European satellites that many took to mean US acquiescence in the status quo.* The resulting controversy had been skillfully exploited by Reagan during the primaries to bolster his critique of Helsinki-style diplomacy. Ford, haunted yet by the bitter fight for his party's nomination, fell back on a programmed response to a question he misread. "What he'd pulled up was an answer that he'd been primed to give on the Sonnenfeldt doc," said Cheney. "Basically, it came out that Poland is not dominated by the Soviet Union."

"If I had simply said the Russians have six divisions in Poland, but regardless of the number of Russian troops, they will never dominate the Polish spirit," Ford would later reflect. "That's all I had to say. I just forgot there were six Russian divisions. Well, what the hell." Even so it did not, at first, feel like a turning point in the campaign. Bob Teeter's initial postdebate poll had Ford winning the encounter by eleven points. An Associated Press survey awarded the victory to Carter 38.2 percent to 34.6 percent. Even Scowcroft may have been reassured when stopped by reporters as he left the building. "We didn't get a single question on it, not a one. By the time we had gone in the motorcade back to the hotel, there was a furor."

According to Ron Nessen, before briefing the press, Scowcroft sug-

---

* Specifically, his staff had prepped Ford to refute Carter's assertion that "at Helsinki you endorsed Soviet domination of Eastern Europe and he cites the Sonnenfeldt Doctrine as proof of this." Robert D. Schulzinger, "The Decline of Détente," in *Gerald Ford and the Politics of Post-Watergate America*, volume 2, eds. Bernard J. Firestone and Alexej Ugrinsky (Westport, CT: Greenwood Press, 1993), 416.

gested to his colleagues, "let's make sure we understand, the president made a mistake, and we ought to acknowledge that. Tell them 'it was a slip of the tongue. Of course he knows . . .' Well, we didn't do that. And so overnight this was a huge thing." As Cheney mounted the stage of the briefing room, Lou Cannon's voice could be heard from the back of the room.

"Hey, Cheney, how many Soviet divisions are there in Poland?"

The atmosphere in the presidential suite was more celebratory. "I walk in the door," Stu Spencer recalled. "I hear"—in unsubtle Germanic accent— "Mr. President, you were wonderful tonight." Kissinger's shameless flattery initiated an uneven contest between the foreign policy maestro complimenting his brightest pupil, and "the hack, Spencer, and the staff," whose contrasting message could be boiled down to: "You were shitty, we've got to straighten this out." Ford wasn't taken wholly by surprise. Earlier he had asked that a television set be turned on, confident he would hear Kissinger's fawning assessment repeated by network pundits. "So, I turn the TV on and he is getting hammered," remembered one aide, "and he can't believe what he's hearing."

"He was living in a bubble up there," said David Gergen. "We saw the polls change. Right after the debate was over, people said he won the debate. And then the media started replaying the Polish thing and by noon the next day we'd lost the debate and we knew we had to reverse it." This unenviable task fell to Cheney. "The next day we flew down to Southern California. I went up and talked to him in the cabin of [*Air Force One*], said I really thought this was out of hand. Carter was having a field day, the press was going bananas, we're getting phone calls from the Washington office saying what the hell are you guys doing out there? I said to the president that we need to go down and make it clear that you understand that Poland is still under Soviet purview. And he threw me out. So I got Stu and took Stu back up, and he threw both of us out."

At least Ford saw his chief of staff and most trusted campaign strategist. "He wouldn't talk to me," Brent Scowcroft confided long after the fact. "I mean literally he wouldn't talk to me." A halfhearted correction by Ford at a business breakfast on the morning of October 7 only prolonged the agony. It took three days for Ford to change his mind. "Spencer and Cheney finally convinced me," Ford confessed. "I was stubborn as hell because I was mad at myself first. I was mad because I thought the point I made was distorted. And I was stubborn about it."

Finally a makeshift press conference was arranged in Glendale, Califor-

nia. Before it could begin, an apprehensive Cheney asked the president if
he had formulated with precision what he intended to say.

"Yeah," said Ford, turning with mock menace on his youthful assistant.
"I'm going to say Poland is not dominated by the Soviet Union." His roguish grin told Cheney his boss was ready to put the worst week of the campaign behind him.

F ROM CALIFORNIA, FORD flew to Oklahoma, and then two days in
Texas. In Dallas he made another attempt to exorcise ghosts by riding
past the Texas School Book Depository, John Connally at his side. To the
dismay of the Secret Service, Ford poked his head through the hatch of the
presidential limousine before halting his motorcade and wading into an
enthusiastic crowd of sidewalk spectators. Better than words, the gesture
evoked the image of a battle-scarred nation coming to terms with its turbulent recent past. Another page was turned on October 13. Arriving late at
the White House following a strenuous day of campaigning in New York
and New Jersey, Ford found waiting for him the long-sought letter from
Charles Ruff, formally clearing the president of any wrongdoing in the
handling of congressional campaign funds.

He was not entirely out of the woods. The same day he received Ruff's
good news, John Dean made the first of three appearances on NBC's *Today* show to promote his memoir *Blind Ambition*, its publication date advanced three months to ride the wave of campaign-related publicity. Dean
garnered headlines with his charge that Ford, at Richard Nixon's instigation, had personally intervened to thwart a congressional investigation of
the Watergate break-in. This was not the first time Dean had described
a campaign, originating in the White House, to pressure House Banking
Committee members of both parties into denying their chairman, Wright
Patman, the subpoena power he sought to compel witness testimony about
possible money laundering by the Nixon reelection campaign. (The original episode involving the House Banking and Currency Committee is covered in detail in Chapter Thirteen.)

Going beyond his June 1973 testimony before the Senate Watergate
Committee, Dean said he had raised with Nixon congressional liaison Bill
Timmons the possibility of blackmailing committee holdouts over questionable campaign contributions. Timmons had thrown cold water on
the idea, mentioning Ford's own potential vulnerability as a reason not to
proceed with the scheme. Timmons flatly denied any such exchange took
place. His White House assistant, Dick Cook, dismissed as "vicious lies"

Dean's assertion that he, Cook, had repeatedly discussed the Patman investigation with Ford.

Leon Jaworski questioned the timing of Dean's return to the headlines. "What bothers me is why hold a matter of this kind for several years," Jaworski told an interviewer, "especially when the man is nominated for vice president, succeeds to the presidency and it's still withheld. And then, here shortly before the election, it comes out in connection with the sale of a book." Ford himself was inclined to say nothing, until the night his son Steve called from the road, where he was campaigning through the rural West in a Winnebago.

"Dad, I'm getting all these questions about you and John Dean. What's going on?"

The next day, October 14, reporters crowded into the East Room for Ford's first televised press conference since February. More important to him than any public office, Ford said, was his personal reputation for integrity. Again he referred questioners to his testimony before the Senate committee examining his qualification to be vice president. He expressed hope that the remainder of the campaign might rise "to a level befitting the American people." Afterward he and Betty danced until one in the morning following a White House ceremony honoring dance pioneer Martha Graham, and in celebration of their twenty-eighth wedding anniversary.

Charles Ruff declined to investigate Dean's latest allegations and the news cycle moved on. NBC's credibility took a further hit when Ford learned, from CBS's Walter Cronkite no less, that the peacock network had paid Dean a reported $7,500 for an option on his unpublished book, with the prospect of considerably more should the option be picked up for filmmaking or documentary purposes. In *We Almost Made It*, his account of the 1976 campaign, Ford adman Malcolm MacDougall wrote that Dean's book plugs on *Today* coincided with the auctioning of the paperback rights.

On October 15, Ford campaigned in the farm belt, where a bumper wheat crop had sent prices tumbling. Standing in "the Iowa version of the Rose Garden" he signed legislation extending credit to hard-pressed livestock producers. That night he watched the vice presidential debate between Bob Dole and Walter Mondale staged in Houston. Unsurprisingly, he disagreed with the consensus opinion that Dole's maladroit reference to "Democrat wars" in the twentieth century had tipped the scales in his opponent's favor. The next day farm issues again dominated the agenda

as Ford, replicating his successful train trip across Michigan ahead of that
state's primary, rode the rails through downstate Illinois. He had more on
his mind than the price of soybeans. Angered by Carter's comment that he
had been "brainwashed" on Eastern Europe, Ford lashed out at his opponent as someone who would say anything, anywhere, to advance his presidential ambitions.

From his New York hospital bed Hubert Humphrey advised the president to spend less time answering Carter's charges. Judging from the latest
batch of polls, the Polish gaffe that so dominated campaign reporting mattered less to voters than it did to the national press corps. A Harris survey
taken after the second Ford-Carter debate measured the challenger's lead
at four points, down from seven just two weeks earlier. An NBC poll had
the race even closer. Speculation turned to the potential impact of Eugene
McCarthy's independent candidacy. The former senator and antiwar activist best known for his challenge to LBJ in 1968 would be on the ballot in
at least thirty states. Multiple polls pegged his support at around 5 percent
of likely voters, a showing attributed to disenchantment with the two major party contenders rather than any surge of enthusiasm for the quirky
McCarthy, whose reluctance to quit the stage threatened to make him the
thinking man's Harold Stassen.[*]

H OPING TO DRAMATIZE the virtues of their self-effacing client, the
political consulting firm of Bailey, Deardourff sent a camera crew
to Grand Rapids to film interviews with Ford's onetime neighbors, teachers and teammates. A comment from the president's high school football
coach, Cliff Gettings, stood out. Describing Ford's aggressive style of play,
Gettings said "he seemed to be able to run faster when he was chasing
somebody than he could if they were chasing him." The last ten days of
the campaign would test Coach Gettings's theory, as Ford scheduled visits
to all eight of the nation's most populous states, at least five of which he
must carry if he was to defy oddsmakers, who made Carter a three-to-one
favorite. The president also planned on barnstorming at least half a dozen
smaller states where polls showed him closing fast.

Having hoarded its cash while taking full advantage of the Rose Garden strategy, the Ford campaign entered the homestretch with close to
$18 million on hand, roughly twice what the Carter forces had to spend.
"Nobody knows Jerry Ford," Dick Cheney told admen Mal MacDougall

---

* The onetime "boy governor" of Minnesota who, in his later years, became something of a
national joke for his quixotic White House campaigns, nine in all between 1948 and 1992.

and John Deardourff. Grand Rapids reserve made for less-than-compelling television. Worse, said Cheney, "President Ford freezes on camera." Thus was born *The Joe and Jerry Show*, half a dozen thirty-minute broadcasts targeting key states thought to be within reach by the Ford campaign. The format was unvarying: freshly edited footage of Ford's visit to a local media market set the scene for softball queries lobbed at him by journeyman St. Louis Cardinals catcher (lifetime batting average: .257) and current *Today* show personality Joe Garagiola. From their first chance encounter at that summer's All-Star game in Philadelphia, the two old jocks had established instant chemistry. Their subsequent pairing, often in tandem with former Democratic congresswoman Edith Green of Oregon, showed Ford to advantage in a medium that was rarely his friend.

Yankees legend Mickey Mantle and Mean Joe Greene of the Pittsburgh Steelers were among the "Jocks for Jerry" who volunteered their services to the campaign. So did Ohio State's irascible Woody Hayes, as long as he didn't have to mention Michigan. For succinct advocacy, no one topped Alabama coach Paul "Bear" Bryant. "Let me tell you why I'm for my friend President Ford," said Bryant. "If I had a quarterback who took the team from his own one yard line past the fifty yard line, I wouldn't take him out. I would keep him in the ballgame because I think he will continue to be successful." On the other hand, Telly Savalas, then at the height of his popularity as the streetwise New York City detective Theo Kojak, reneged on a promise to do an ad crediting Ford with the Big Apple's recovery from near bankruptcy.

Five-minute biographical spots emphasized Ford as family man. Tying it all together, a San Francisco adman named Robert Gardner contributed a song, as brilliantly evocative of national revival as MacDougall's campaign slogan ("He's Making Us Proud Again"). "I'm Feeling Good About America" might have been written for a high school marching band at halftime. But it was just as infectiously stirring in country-western tempo, hard rock or salsa beat. That fall it became ubiquitous, the most effective campaign jingle since Irving Berlin's "I Like Ike" in 1952. Bailey, Deardourff did not limit themselves to feel-good messaging. Hoping to cut Carter down to size, the Ford campaign ran sixty-second spots featuring man- and woman-on-the-street interviews with Georgians disparaging their "just average . . . run of the mill governor." In the same genre, a cross-country montage of citizens employed the words "wishy-washy" to convey their doubts about the peanut farmer, Georgia state senator and one-term governor few had heard of at the start of the year.

By any comparison with the character assassination routinely vomited by modern campaigns, Ford's negative ads—like their counterparts from Carter image maker Gerald Rafshoon—were mild as milk. At one point Mal MacDougall crafted an ad citing eleven issues, from right-to-work laws to capital punishment, to make the case for Carter as flip-flopper. Out of the question, said the campaign's lawyers. Mere newspaper articles, the source of many of MacDougall's quotes, were not considered reliable enough to justify such attacks.

Some of the lawyers' caution seeped into preparations for the final debate of the campaign, set for October 22 in Williamsburg, Virginia. To Ford the encounter felt "anticlimactic." Viewers disagreed. Most found the discussion more substantive and less confrontational than its predecessors. Carter in particular dialed back the prosecutorial mien Eric Sevareid described as an "instinct for the deliberate insult, the loaded phrase and the broad innuendo." At one point in the evening Carter said Ford should be "ashamed" of an answer he gave in defense of his economic record. But he also acknowledged his own error in granting *Playboy* an interview.

For his part, Ford took Carter to task for ruling out the use of American forces to counter a Soviet attack on post-Tito Yugoslavia. The president drew a parallel with Dean Acheson's public exclusion of the Korean peninsula from the US defense perimeter months before the Communist North invaded its southern neighbor. Yet Ford, too, conceded his shortcomings as a campaigner, acknowledging that "like most others in the political arena" he sometimes resorted to "rather graphic language" to distinguish himself from his opponent.

THE DEBATES BEHIND him, a sprint to the finish turned into a grueling marathon. Before election day Ford would cover nearly sixteen thousand miles, visiting twenty-six cities in sixteen states for what weary aides dubbed the Bataan Death March. Somewhere in this improvised blur of factory visits and airport rallies, the indifferent campaigner found his voice even as he overtaxed his vocal cords. And the crowds responded, beginning in the blue-collar Los Angeles suburb of Fountain Valley, where twenty-five thousand people welcomed Ford and John Wayne on Sunday afternoon, October 24. Semifrantic efforts by the campaign to entice Ronald Reagan to the same event produced nothing more than a telegram expressing "sincere regrets that I will not be able to join you."

From San Diego, Ford flew to the Pacific Northwest, making stops in Seattle and Portland before overnighting Monday in Pittsburgh. Tuesday morning featured remarks before that city's Economic Club. By lunchtime Ford was in Chicagoland taping another *Joe and Jerry* outing for statewide broadcast that evening, at a time when he would be addressing suburban audiences critical to Republican success. Wednesday the twenty-seventh found the president, joined by the reigning Miss America, appealing to Garden State voters on the boardwalk in Atlantic City.* The balance of the day was spent in Pennsylvania, which was rapidly becoming ground zero (as it would be in 2016 and 2020) in the quest for 270 electoral votes. In Philadelphia, Ford met with Cardinal John Krol and squeezed in another *Joe and Jerry* taping for Keystone State viewers. He concluded the day at a suburban theater in the round, with what was universally ranked among his best performances of the campaign.

Speaking in hushed tones through a handheld microphone, Ford told the audience he would like to hear a pin drop. He asked them to remember what things were like in August 1974. "America was in very deep trouble. Faith and confidence in the White House had been lost." Twelve percent inflation. The worst economic recession since the 1930s. The unfinished agony of Vietnam. Since then, Ford asserted, trust in the presidency had been restored. Confidence had returned to their country. The war on inflation was being won. New jobs were being created. And overseas the United States enjoyed a respect few could have foreseen in April 1975.

His efforts were rewarded the next morning, when someone handed Ford an advance copy of the final Gallup poll, to be released the Sunday before election day. To his astonishment it showed him leading Carter for the first time since March. The difference was a microscopic 1 percent, well within the polltaker's margin of error. But it was more than enough for an excited Ford to call his running mate, two time zones away, in New Mexico.

"Morning, Bob. What are you doing?"

"Well, Mr. President, it's only five o'clock out here. I'm still sleeping."

Gallup's numbers, precarious as they were, convinced Ford that he was headed for the biggest political upset in the nation's history. Another set of figures, also released that eventful Thursday by the Commerce Department,

---

* Thomas Kean, a future governor of the Garden State, was a lowly campaign operative in 1976. Taking an instant shine to the president who was so eager to have his input, Kean briefed Ford on a controversial ballot referendum to legalize casino gambling in Atlantic City. Better, said Kean, to take no position on an issue that could only cost him votes in the closely contested state. Ford nodded in agreement. "And as soon as he was asked about the issue by a reporter," Kean remembers, "he replied without hesitation—'I've never been a fan of casino gambling. If I were a New Jersey voter I'd be voting no.'" Ford's candor presumably did nothing to aid his candidacy in Atlantic County, which he lost to Carter 52 to 46 percent. Statewide he fared better, scoring heavily in Republican suburbs to outpace Carter by sixty-five thousand votes overall. AI, Tom Kean.

might have tempered his optimism. The government's monthly index of leading economic indicators was down, for the second consecutive month, by seven-tenths of 1 percent. Ordinarily such news would merit at most a passing reference from network anchors. But in the closing days of an exceedingly close presidential contest, any fragment of relevant information takes on exaggerated significance. Coupled with a recent stock market pullback and a slowdown in GNP growth for the latest quarter, the report gave Ford's critics ammunition with which to dispute his claims of recovery.

From Philly, Ford hopscotched to events in Indianapolis, Cincinnati and Cleveland. In a rare display of temperament, he voiced dissatisfaction on being assigned to a hotel's grandly named Emperor's Suite. Within minutes the suite's nameplate was obscured by a piece of cardboard on which someone inscribed, JERRY FORD'S ROOM. In Washington, Betty Ford worried about her husband's physical condition more than his political health. "He's doing too much," she complained. "He's going to have a heart attack!" Ford's response was to pile on still more shopping mall rallies and radio interviews. On the twenty-ninth he was in Milwaukee, meeting with representatives of ethnic groups, a transparent attempt to repair damage caused by his premature liberation of Poland and other Soviet satellites. "It's like the last quarter of the big game," he exhorted a crowd at the Marc Plaza Hotel, "and boy, this is big."

Missouri was another bellwether state that could go either way. Ford could spare but four hours for St. Louis and its suburbs—less time than he spent in traveling to, attending and returning from a Friday-night high school football game outside Houston, described as "a little bit of a break" for the bone-weary campaigner. Wherever he went, Ford traveled in a car equipped with a microphone and sound system, enabling him to talk to spectators without ever leaving the secure confines of his Secret Service–approved vehicle. The talking car became a favorite toy, employed to sometimes startling effect in urban motorcades and more sparsely attended airport runs. It was put to good use on Saturday the thirtieth as Ford returned yet again to Pennsylvania before blitzing upstate New York. At Buffalo's St. Stanislaus Catholic Church he was greeted with signs proclaiming POLONIA WITA FORD (roughly translated as "Polish People with Ford") and a strongly worded antiabortion letter from the local bishop.

In the afternoon the president canvassed Long Island, a traditional Republican stronghold, where polls suggested he was falling short of the numbers needed to overcome Carter's overwhelming advantage in the five boroughs of New York. The local GOP organization, operating with its usual efficiency, packed the Nassau County Veterans Memorial Coliseum with fifteen thousand cheering supporters. The festive mood was only slightly marred by sketchy reports out of Plains involving the First Baptist Church favored as a place of worship by the Carters. Earlier that morning the church had canceled its weekly service rather than admit the Reverend Clennon King and three other African-American applicants to church membership. A publicity-seeking cleric activist, King had run for president of the United States in 1960 under the auspices of the Independent Afro-American Party. Since then he had applied unsuccessfully for political asylum in Cuba, Jamaica and Mexico. He had also served time in a California jail for failure to pay court-ordered support for his six children. Nothing King did should have come as a surprise. Still, the timing of his confrontation with the white power structure in Plains instantly roused suspicions of a Republican dirty trick.

"I hope none of our people are involved in this," Ford told Dick Cheney on *Air Force One*. Cheney assured him they were not. "I hope that's true," Ford repeated. "I'm telling you—if any of our people are involved with any of this, I will deal with them personally." Unknown to him, the President Ford Committee, under the signature of Jim Baker, was sending telegrams to some four hundred Black ministers calling attention to the incident. (Baker later said this was a mistake.) Whatever its origins, Clennon King's stunt shone a spotlight on Andrew Young, Jesse Jackson and Coretta Scott King as they rallied to Carter's side. The candidate himself denounced the segregationist practices of his church's deacons.*

The finish line beckoned. Almost thirty years after a young naval officer defied the odds by challenging an entrenched congressman over his isolationist voting record, Ford's last day as a candidate for office included stops in Akron and Columbus, Ohio (a state without which no Republican had ever won the presidency), and a rally in a suburban Detroit mall, culminating in a Grand Rapids homecoming. Fifty thousand people welcomed Jerry and Betty back with a parade that featured sixteen marching bands, thirty antique autos and one talking car. The last two years had been challenging ones, Ford told the vast crowd with what remained of his speaking voice. "But we kept the ship of state on the right course and I tried to keep a firm, commonsense hand on the tiller." As a result, "things have turned around . . . But we are going to get better. We are going to make America

---

* As it happened, Carter had long opposed the exclusionary policies of the deacons. In time the Plains congregation would splinter over issues of race and church administration and Carter as an ex-president would leave the denomination because of its hard-shell stance on the role of women.

what our forefathers said it would be." Once he had asked his countrymen to pray for him. "But tomorrow I ask that you confirm me with your votes."

TUESDAY MORNING, FORD breakfasted on blueberry pancakes at Granny's Kitchen, an election-day tradition, before attending the un-veiling of an airport mural honoring him and his parents. The president was not the only attendee moved to tears. "You did the impossible," Ron Nessen told Ford aboard *Air Force One*. "You made Helen Thomas cry." Outside the plane's windows the weather was picture perfect, distressingly so for those who equated sunny skies with a heavy turnout in Democratic bastions like Detroit, Cleveland, Philadelphia and New York.* In his Fifth Avenue triplex, Nelson Rockefeller likened the contest to a football punted away and bouncing around the field. "You don't know where it's going to land."

At the White House, Henry Kissinger upbraided Brent Scowcroft over the time of Ford's arrival. "If I had been told it was 1:30 I would have moved my lunch back. I was told 1:50." Informed that the campaign's latest "rock bottom estimate" of electoral votes for the president was 271 (Ford himself predicted 278), Kissinger snorted. "Nonsense . . . my gut tells me Carter will win." He shared his doubts with Richard Nixon. "I don't know," the former president mused. "I think Ford should get it." Of course, "You have got to watch what labor does in the last two weeks." John Connally expected Ford to carry Texas with 150,000 votes to spare. Nixon felt equally confident about California, though by a smaller margin than re-cent polls suggested. That said, he was worried about Ohio and Michigan.

At the White House, steam inhalation, tea and honey were prescribed for Ford's ravaged voice. A little after six o'clock, Cheney, Spencer, Jim Baker and Bob Teeter appeared in the Oval Office to review early exit polling data. The numbers were not encouraging. Ford received the news impassively. He offered his guests cigars as they left. By six thirty he was up-stairs welcoming a diverse guest list that included all four Ford children and several of their friends; Clara Powell, the indispensable Ford family house-keeper from Crown View Drive; David Kennerly; Senator Jacob Javits and his wife; Bob and Elizabeth Dole; Pearl Bailey; Edith Green; and family friends from Grand Rapids and Southern California.

---

* Sure enough, fair skies and seasonal temperatures led to early-day reports of a heavy, perhaps even record, turnout. As the day progressed the numbers slacked off. In the end the Ford-Carter contest attracted 54 percent of the electorate to the polls, the lowest figure since the Truman-Dewey race in 1948.

Joe Garagiola, invited to spend the night with his wife and four kids, played cheerleader, minimizing early losses by muttering, "Big deal. It isn't even the second inning yet." To more practiced observers, the early trends bred anxiety, as Carter swept the South while Ford had to content himself with a handful of Republican strongholds like Indiana and Dole's Kansas. The first family and their guests, perhaps three dozen in all, mingled in front of television sets in the second-floor center hall or the nearby den/exercise room. From time to time Dick Cheney or Ron Nessen would come over from the West Wing to update results. Around nine fifteen, Greg Willard, a junior staffer who had accompanied Steve in his campaign Winnebago, joined the party at Steve's invitation. The electoral map in his copy of *Time* was treated like a Delphic oracle, with the president frequently comparing its predictions against the network calls (including, most embarrassingly, the state of Oregon, originally assigned to Carter, then retracted as more votes came in, and finally won by Ford in a squeaker).

"It's all right, Prez," chortled Garagiola. "We've given up a couple of runs, but the ballgame is only in the top of [the] fourth. We got a long way to go." And sure enough, the raw vote count *was* an improvement over the exit polls. Ford carried Michigan, Connecticut and, in an upset, New Jersey. Ohio was too close to call, but John Connally telephoned the president to say things looked good in Texas. Senator Dole discerned more troubling trends. Close to midnight he commented, almost to himself, "We just aren't going to be able to pull it off." An hour later, boosted by near-record margins in New York City, Carter was declared the winner in New York with its forty-one electoral votes. Pennsylvania was looking out of reach.

Now it was Ford's turn to insist, "We're still in this thing."

At 1:40 a.m., seeking reassurance, he put in a call to Connally for a Texas update. After a few minutes he returned to his seat in the West Hall, confiding to Steve and Greg Willard that Connally had turned pessimistic. Twenty minutes later the networks awarded the Lone Star State to Jimmy Carter. "That one hurts," said Ford. "That one really hurts." Victory in North Dakota (four electoral votes) hardly compensated for the shock of losing Texas. Ford willed himself to appear upbeat. "It's still a helluva ballgame!" he declared, punching his left fist into his right hand. Most of America south of the Mason-Dixon Line and east of the Mississippi River was colored Carter red on the network graphics, with practically everything west of the Mississippi declared or leaning Ford blue.

"Go blue," Joe Garagiola shouted with every state to join the Ford column. Iowa came through, and the Rocky Mountain states without exception. Illinois appeared promising. The first returns from California were in line with Nixon's forecast. The problem was, by two o'clock they were running out of states, with Carter stalled just shy of the necessary 270 electors.

A little after three Ford asked everyone to assemble in the West Hall. "No, this isn't a concession speech," he told them. "I think Bob Teeter has some interesting information for us." The president's pollster explained that Hawaii's four electoral votes were still up for grabs. He thought the networks were mistaken in calling Wisconsin for Carter. Either of those states, coupled with Ohio, where the race was breathtakingly close, would give Ford a bare majority in the Electoral College.

Seemingly the calmest man in the room, Ford told his guests that nothing would be certain for several hours, and he was going to bed. Meanwhile, in the press office, Ron Nessen was scanning the latest bulletin ripped from his UPI machine:

FLASH
Washington–Carter Wins Presidency

The call was made on the basis of Mississippi's seven electors. By fewer than fifteen thousand votes, Mississippi, the brutally contested epicenter of the Ford-Reagan slugfest, would supply Jimmy Carter with the key to the White House. Nessen, encountering Ford on his way to bed, didn't have the heart to share the news with him. This didn't prevent Chief Usher Rex Scouten from trying to put things in perspective. Offering praise for Ford's thirty-five years of service to his country, in the navy, on Capitol Hill and in his unexpected presidency, Scouten remarked tentatively, "Maybe, Mr. President, it's just time for you to take a well-earned rest."

"I don't think so," Ford croaked.

Another half hour went by before the remaining guests, now all congregated in the West Hall, heard John Chancellor make it official on NBC. Several minutes of silence ensued, broken at last by Mrs. Ford. "Do we dare wake him?" she asked. It was decided not to interrupt the president's sleep. A few minutes later Jimmy Carter's broadly smiling visage came on the television screen. Betty Ford chuckled at the sight.

"Governor, you have no idea what you're in for in here."

Around four thirty, the little group of die-hards decided to call it a night. As he was leaving, Greg Willard apologized to the First Lady. She took his hands in hers. "Now listen! We're going to leave here in January with no regrets and many wonderful memories. And remember, when we leave, we'll have our heads up with lots of pride."

It was after eight when Dick Cheney broke the bad news to the president: "Even if we win Ohio, no other state would give us the additional four votes we need." A little after eleven, Ford called Carter. "I can't talk," he rasped, before putting Cheney on the phone to read the telegram congratulating the president-elect.

A despondent-looking Ford family posed for Kennerly's camera. Trooping over to the pressroom, they heard Mrs. Ford read her husband's concession, in which he pledged a smooth transition and appealed to all Americans to "put the divisions behind us and unite the country once more in the common pursuit of peace and prosperity." The president, Betty and the children shook hands and swapped stories with members of the press. Back in the Oval Office Ford put an arm around Terry O'Donnell, thanked him for all his hard work. "If there is anything I can do for you let me know." With that, O'Donnell and Kennerly lost it. Tears flowed again when Joe Garagiola appeared in the doorway for an emotional embrace with his partner in the *Joe and Jerry Show*.

"Damn it, we should have won," Garagiola sobbed. "We should have won."

"Hey," said Ford, "there are more important things to worry about than what's going to happen to Jerry Ford."

"Not today, damn it."

Ford's thoughts were already turning to the future.

"I've got to give him the White House in better shape than I got it," he told Garagiola.

Part V

WHEN IS THIS RETIREMENT GOING TO START?

1976–2006

———◆———

27

DO WHAT YOU CAN

❦

My whole lifetime, whether in athletics or politics, my philosophy
always was: do what you can to win. If you lose, don't sit around
and worry about it. There's another day, another opportunity down
the road. So even though we lost, I was ready to find something
else that would be constructive and interesting. And I did.

—GERALD R. FORD, NOVEMBER 1996

D EMOCRATIC LEADER OF the Senate Mike Mansfield, just back from
China, was slightly mystified. "I don't know what to say," he told Ford
on the Friday after the election. "You got the states but not the votes." This
was no mere courtesy. Late returns gave Ford twenty-seven states, still the record for any losing presidential candidate. A switch of fewer than nine thousand votes in Ohio and Hawaii would have reversed the outcome. Lingering
resentment over the Nixon pardon was the most widely cited factor in Ford's
defeat. To this one might add economic uncertainties rekindled by eleventh-
hour reports of a pause in growth that was anything but refreshing. The
president's veto of common situs legislation may have preserved his nomination in Kansas City, but it also spurred organized labor to unprecedented
voter registration and turnout efforts on behalf of the Carter-Mondale ticket.
Henry Kissinger got no argument from Ford when he cursed the Pentagon
for undermining a SALT II agreement and letting billions in budgeted funds
go unspent at a time when the sluggish economy craved stimulus. (Left unmentioned was Ford's own refusal to rethink a long list of military base
closings in pivotal states like Texas and Missouri.)*

---

* For those who accompanied Ford on his last frenzied dash to the finish line, it became
an article of faith: given a few more days to campaign, he could have supplanted Harry
Truman as the patron saint of political underdogs. One can argue just as forcefully that
Gallup's final poll showing Ford edging into the lead actually worked to his disadvantage
among uncommitted or hesitant voters. "As long as the polls seem to say Carter's going to
win, Carter's going to win," Doug Bailey explained, "then my thought as one of those voters

Ford's mail provided a measure of solace. "There is no man who in four years could do half the job you have done for us in two," tennis great Jimmy Connors assured him. "You walked into an impossible situation at a very difficult time," wrote Notre Dame president Father Theodore Hesburgh, "and somehow brought order out of chaos and dignity out of disgrace." Similar sentiments were voiced by New York Cardinal Terence Cooke, Barbara Stanwyck, Willie Mays, Ansel Adams, John Paul Stevens, Barbara Walters and Mo Udall. One of the most touching tributes, because one of the most unexpected, came from Clarence Mitchell, chief Washington lobbyist and spokesman for the NAACP. After praising Ford's "sincerity and fundamental decency," Mitchell defined the philosophical differences between the parties as he saw them; one representing the many Americans "well off, better situated and more enlightened than ever before in our history"; the other sensitive to the inequitable distribution of these blessings and to the plight of those "caught in what appear to be alleys from which there is no escape." Despite their political differences, Mitchell added, "nothing will ever change the fact that you entered the office at a time when the Nation was in great travail and you left it with the respect and, in many instances, the affection of your supporters and many of your opponents as well."

M ORE SHAKEN BY the outcome than he let on, Ford privately groaned, "I can't believe I lost to a peanut farmer." Resolved that Carter would have the orderly transition denied to himself, he directed his senior staff to work closely with their counterparts in the Carter camp. Thirty-five transition officers in as many departments and agencies set to work preparing issue books outlining problems likely to confront the new administration. On November 5, Ford embarked on an eight-day Palm Springs holiday. Renting the fourteen-room villa of his friend Leonard Firestone on the grounds of the exclusive Thunderbird Country Club, the president got in plenty of golf and socializing, interspersed with discussions about the final budget he planned on submitting.*

---

in the middle is, 'God, who is this guy?'" Unfortunately for the incumbent, the same lesser-of-two-evils logic worked in reverse. By refocusing attention away from Carter, Ford's goal line rally raised doubts in the electorate as well as hopes. Supporting this theory were exit polls showing that late-deciding voters broke 2–1 for Carter.

* "He did a lovely thing," said Strobe Talbott, then *Time*'s White House correspondent, of inviting Talbott and his professional colleagues to accompany him to Palm Springs "just as long as you leave me alone out there. We're going to just charge the hell out of your companies for the seats" on the press plane, Ford explained, "and you can bring your families for $5 apiece." AI, Strobe Talbott.

The old athlete's natural resilience gradually reasserted itself. Yet even now his emotions were raw, as Sheila Weidenfeld discovered when she mentioned press interest in Ford's postelection phone conversation with Richard Nixon. "He is a politician, an ex-President. I wanted to hear his analysis of the election. I just wanted to call him," Ford snapped. "And God dammit, I can call anyone I want!" Deputy Press Secretary John Carlson got the same treatment after forwarding a request from the paper of record for an interview to update Ford's obituary.

"John, I'm not ready to die. Tell that to the *New York Times*."

Back in Washington, Ford threw himself into a series of diplomatic meetings aimed at holding down potentially crippling price increases by the OPEC cartel. Appealing directly to Saudi Arabia's King Khalid, he raised the specter of Western unrest leading to Communist gains in NATO mainstays Italy and France. The Saudis agreed to step up their production, thereby undercutting the cartel's nominal price hikes. One evening, Ford invited Hubert and Muriel Humphrey, another couple acquainted with electoral disappointment, for dinner and conversation that stretched late into the night. The weekend before Thanksgiving he and Betty were guests at the Rockefeller estate in Tarrytown, on the Hudson, where the president played golf with Nelson and his brothers John and Laurance. Afterward Ford participated in a ceremony designating the mansion originally built for the first John D. Rockefeller a national historic landmark.

On November 22, the Fords welcomed the president-elect and Mrs. Carter to the White House. Expecting a short, ritualized meeting in the Oval Office, Ford spent over an hour with his successor. Their talk was almost exclusively of foreign policy. Carter said he hoped that the Panama Canal issue might be settled before inauguration day. Ford said it was "very doubtful." An awkward moment ensued when he led Carter on a tour of the Oval Office and surrounding work spaces. In her nearby office, Dorothy Downton heard the president's familiar voice outside her door. "Here's the head," Ford told his visitor, "and here's the pantry and here's my personal secretary."

Carter introduced himself, asked where Dorothy came from. When she replied, "Michigan," he seemed momentarily at a loss for words.

"You don't have to worry about her," Ford intervened. "She's going with me."*

His plans were otherwise vague. Pepperdine University wanted Ford for

---

* Ford proposed that Carter use the office space as his own during the transition period, an offer the president-elect declined with thanks.

its new chancellor, Metropolitan Life Insurance for its board of directors. Sensitive to potential conflicts of interest, Ford designated Washington lawyer Dean Burch to act as his representative, assisted by the president's military aide, army major Robert Barrett, who also accepted an invitation to organize Ford's postpresidential office after January 20. In the wings, agent Norman Brokaw of the William Morris Agency was conjuring lucrative deals involving books and television contracts for both the president and First Lady. For the moment these were strictly hypothetical. "He didn't know anything about that until he left the presidency," insisted Barrett, who admittedly bent the rules by confessing to Ford his knowledge of the Pepperdine overture. "Sir . . . you just can't do it. You have to trust me . . . There will be better offers than this."

Ford took satisfaction in government reports pegging the inflation rate for 1976 at 4.8 percent, the lowest figure in four years. He asked Secret Service agent Dick Keiser to bring his family in so he could meet them. "I took you away on a lot of holidays and I'd like to say thank you." He declined a request from Bill Simon that he pardon former Nixon attorney general John Mitchell for his part in the Watergate scandal. On January 12, Ford returned to Capitol Hill for his last—"Maybe"—State of the Union address. In light of more recent history, his tribute to the orderly transfer of power is worth quoting: "There are no soldiers marching in the streets except in the Inaugural Parade; no public demonstrations except for some of the dancers at the Inaugural Ball; the opposition party doesn't go underground but goes on functioning vigorously in the Congress and in the country; and our vigilant press goes right on probing and publishing our faults and our follies, confirming the wisdom of the framers of the First Amendment."

Even as he became reconciled to his newfound status, Ford hatched an elaborate charade to lift Betty's spirits. It began with a party at David Kennerly's Georgetown pad. Plainly enjoying herself, Betty was reluctant to leave the festivities, yielding only after being told that the Marine Band was waiting to have a final picture taken with the president and First Lady. "The South Portico of the White House was dark as our limousine pulled up to the entrance shortly after ten o'clock that Tuesday night," Ford recalled. "We took the family elevator to the first floor. All was quiet, only a few lights were on, but I could see the members of the band waiting for us by the foot of the grand staircase."

As long as they were there, he said to Betty, why not enjoy a last dance. The strains of "Thanks for the Memory" echoed through the house as, suddenly, with her back turned to them, "other couples began gliding [onto] the floor—the Buchens, the Harlows, the Kissingers, the Lairds—more than a hundred people in all, and they were beaming with delight." So, the moment she turned to face her unexpected guests, was Betty Ford. "Seeing her so happy," wrote Ford, "was one of the greatest joys of my life." The next day, January 19, he placed calls to Leonid Brezhnev, Jim Callaghan, Yitzhak Rabin and Pierre Trudeau. He announced an end to price controls on gasoline, a symbolic gesture sure to be overturned by Congress within the requisite fifteen days.*

An administration defined for many by the Nixon pardon concluded with another act of executive clemency as Ford pardoned Iva Toguri D'Aquino, one of several women dubbed Tokyo Rose by GIs exposed to their propaganda broadcasts during World War II. In the evening the Fords welcomed as their final overnight guests Nelson and Happy Rockefeller and their two young sons. After dinner the Rockefeller men joined the president on an improvised driving range in the third-floor hallway. A Gallup poll showed Ford leaving office with a 53 percent approval rating. Sixty percent of those polled said that history would regard him as an "average" president.

J ANUARY 20 DAWNED cold and cloudless in the nation's capital. A little after eight, Ford entered the State Dining Room, where seventy-five members of the Cabinet and senior staff had gathered to say goodbye. Going around the room, he paused to shake hands and thank each individual for contributing to an administration "which I think was good and which history will treat kindly." In the Oval Office the last of his pictures were stripped from the walls, as a cleaning crew emptied Ford's desk of presidential cuff links and a drawer full of undistributed WHIP INFLATION NOW buttons. At ten thirty, the outgoing president reappeared in the State Dining Room, accompanied by Betty, who tearfully hugged maids and butlers she had come to regard as a second family. Presently the Carters and Mondales arrived for coffee and strained conversation in the Blue Room. Ford assured his successor that he would be available should the new president ever desire his help.

James Monroe's French bronze mantel clock chimed eleven, signaling it was time to leave for the traditional ride to the Capitol, a journey Ford remembered as "very congenial." Neither president wore a top coat in the subfreezing cold (a final bit of friendly competition?). Ford was unprepared

---

* Jimmy Carter saved lawmakers the trouble by rescinding Ford's order on his fourth day as president.

for Carter's generous acknowledgment in his inaugural address, thanking his predecessor for "all he has done to heal our land." As the applause built, Ford rose from his seat, a bit hesitantly, shook hands with the new president, and sat down—only to stand a second time, close to tears, as 150,000 onlookers signaled their approval. After the ceremony the Fords and the Rockefellers boarded an army helicopter for the short flight to Andrews Air Force Base. The pilot readily complied with a request that he circle the Capitol dome to afford the now-former president a parting glimpse.

"That's my real home," Ford said.

Several hundred spectators were on hand at Andrews to see them off. At Ford's request there were no speeches. A band played "God Bless America." Happy Rockefeller presented Betty a bouquet of red roses. The Fords hugged the Rockefellers and the Kissingers before boarding their plane, a backup to *Air Force One*. A festive mood prevailed during the first half of the flight to California, where Ford would spend his first forty-eight hours as ex-president playing in the Bing Crosby Pro-Am golf tournament. By midcontinent, the atmosphere turned more somber. At one point Ford wandered back into the press cabin.

"Do you have anything to tell us, Mr. President?" someone asked.

"Well," Ford replied, the hint of a smile creasing his face, "anything I say isn't very important any longer."

Any inclination to self-pity ran into the buzz saw of spousal wit, as demonstrated later in the week, when the Fords flew to Houston for a dinner benefiting the American Cancer Society. Nearing their destination, a brooding Ford noted that he had been in office when the group invited him to be their speaker. He hoped his audience wouldn't feel let down by hearing from a mere former president.

"Don't worry, honey," said Betty. "It's me they're coming to see anyway."

THE FORDS HAD originally planned on retiring from the White House to their old Alexandria home. The Secret Service pretty much ruled that out on security grounds. Besides, the president told a friend inquiring about his plans, "None of the kids' friends are left in Virginia." As for Grand Rapids, said Ford, his family hadn't lived there since Mike's infancy. Taking into consideration Betty's health and his own preference for outdoor exercise, "we'll probably float between California and Vail." Early in 1977, 514 Crown View Drive was put on the market. It quickly sold for the asking price of $137,000. After checking out a couple properties on the Monterey Peninsula, the Fords decided the dry heat of Palm Springs was better suited to Betty's arthritis. The Fords would be in distinguished company, as every president since Truman had vacationed in the mountain-fringed desert resort one hundred miles east of Los Angeles.

Still more exclusive was Rancho Mirage, a recently incorporated enclave described as a place where people want to live when they have the money to live anywhere they want. The town's first billionaire, publishing tycoon Walter Annenberg, played host to England's royal family on the 160-acre Sunnylands. A different breed of royalty imparted Hollywood glamour to Rancho Mirage, symbolized by the weekend retreat Lucille Ball and Desi Arnaz built in 1954 on the seventeenth fairway of the Thunderbird Country Club.* By the mid-1970s, local streets bore the names of such resident luminaries as Frank Sinatra, Bob Hope, Bing Crosby, Dinah Shore, Dean Martin, Danny Kaye, Claudette Colbert and Greer Garson.

In time Gerald Ford would join this starry pantheon, though his first few months in Rancho Mirage gave little hint of the special position he and Betty were to occupy. Until they built a place of their own on two lots sold them by Leonard Firestone, the Fords rented a home in a gated community overlooking the Thunderbird Golf Course. The former president's official business was initially transacted in the kitchenette of a nearby bungalow shared by four staffers. After a few weeks the office was relocated to another rental property belonging to actress Pat Priest, best known as Marilyn Munster on the short-lived television series.

"I remember those early weeks," said Greg Willard, "how lonely it was for the Fords." More so for Betty, he might have added, as her husband's schedule replayed the near-constant travel of his minority leader days. "Jerry's retirement was a fraud," Betty would write. "He might as well have been campaigning." She had a point. In his first three years out of office Ford covered 1.2 million miles, visiting ten countries and fifty-six college campuses. Besides serving as president of the Eisenhower Fellowships Exchange, Ford was a board member of the Boys Clubs, and honorary vice president of the Boy Scouts of America. He barnstormed for dozens of GOP candidates, delivered almost as many paid speeches, wrote a book and raised $9.3 million for his presidential library and museum.

David Kennerly put it all in perspective. "Even heroin addicts get time to withdraw," he concluded. "But they don't know nothing until they look at someone trying to kick the presidency. It's real cold turkey, man." Ford's

---

* While Lucy was invited to join the exclusive club, Desi was not—no minorities being admitted to its select precincts. The situation repeated itself in the 1960s, when the reigning queen of comedy divorced Desi and married Gary Morton, a Borscht Belt comic who was Jewish.

position on leaving the White House was as singular as it had been on August 9, 1974. Not a father figure like Eisenhower, nor a politically spent force like Truman, LBJ and Nixon before him, Ford was sixty-three, in rugged health, with undiminished energies and no desire to be put on the political shelf. Even the possibility of a rematch with Jimmy Carter made him—for now—shy away from corporate directorships or other potential sources of controversy.

Ford's biggest single commitment took him to a dozen or more college campuses a year under the aegis of the American Enterprise Institute, a conservative think tank in Washington with which he was to enjoy a close relationship for the rest of his life. The former president relished living in student housing and taking meals with undergraduates. He proved a popular instructor, illustrating his lectures with real-world examples and providing candid answers like his response to the Yalie who asked about his handling of the Solzhenitsyn affair. "You know," said Ford, "I wish I had that one to do over." Ultimately he would visit over two hundred campuses, public and private, including several historically Black schools.

On the evening of March 21, 1977, the former president and First Lady received a standing ovation from a full house of New Yorkers waiting for the curtain to rise on the musical smash *A Chorus Line*. For Betty, the action onstage evoked memories of long-dormant aspirations. She and Jerry laughed heartily at the dancer whose career was rejuvenated through plastic surgery on her "tits and ass." They adopted the show's eleven o'clock number "What I Did for Love" as their personal anthem. Later that week Ford returned to Washington for the first time since January 20. At the White House, Carter hosted him for a ninety-minute discussion of domestic and foreign concerns. Ford begged off when asked by a reporter to grade Carter's first months in office: "He has enough trouble without me nit-picking or speaking out." In return Carter credited Ford's lobbying of GOP senators with providing the margin of victory by which the Senate in April 1978 approved an end-of-century handover of the Panama Canal.

Eyebrows were raised when Jerry and Betty signed his-and-hers book contracts with Harper & Row/Reader's Digest for a combined sum of $1 million. Norman Brokaw, the über-agent who counted Marilyn Monroe and Clint Eastwood among his clients, negotiated an even larger deal with NBC that would have both Fords doing commentary and occasional programming for the network. Critics accused Ford of cashing in on his brief presidency. This overlooked a long history of presidential profiteering. No one complained in 1953, when Harry Truman contracted to write his memoirs for $670,000 ($1,460,000 in 1977 dollars). Earlier still, Dwight

Eisenhower got a favorable tax ruling from the IRS that allowed him to keep an additional $400,000 generated by his wartime chronicle *Crusade in Europe*. As late as 1964, Ike was reportedly "handsomely paid" by ABC for on-camera analysis at that year's Republican National Convention. Lyndon Johnson received $1 million for his White House memoir and another $300,000 for several television interviews with Walter Cronkite. Topping them all, Richard Nixon took home over $2 million for his autobiography *RN*, to which television personality David Frost added $600,000 plus 20 percent of the profits from a series of conversations, which drew a huge viewership when broadcast in May 1977.

Ford left the White House with government pensions totaling $104,000 and a net worth under $400,000. All his years of party building, and the countless trips he had made in support of Republican candidates, had come at the expense of his family's long-term financial security. Now he resolved to make up for lost time. Brent Scowcroft understood his motivations, even if he questioned some of Ford's dealmaking. "First of all he had a tough childhood. I think he really worried about his children and his children's ability to make it. I think he felt that he had to provide for them. It's an obsession, I think—still is," Scowcroft remarked in 1990. Ford admitted as much to Bob Barrett. Having grown up in the Depression, he told Barrett, "I've just always worried about not having money." Memories of youthful privation explain Ford's reluctance to part with a jacket woefully out of fashion—"Well, it's still good. It's not worn out"—and the almost comical reprimand his children could expect after leaving the lights on in an empty room: "What do you think, you own stock in the electric company?"

His phobia about debt was the stuff of family lore. "Once a year you'd get called into the office," according to Susan Ford Bales, "and it was like, 'How are your finances? Do you have a retirement fund? Are you putting enough money away every year to take care of you? How are the children? Are you paying for their college?'" If she or her siblings were financially hard-pressed, "he would do short-term loans . . . And you signed a document that you were going to pay it back." To be able to move into a $650,000 home in Rancho Mirage and a $2 million ski chalet near Vail, Colorado— mortgage-free—was for Ford a source of unabashed pride. (His faith in the free enterprise system was more easily defended after Tip O'Neill decided to pad *his* retirement income by becoming a well-compensated pitchman for Miller Lite and Comfort Inns.)

Moreover, no one had prepared Ford for the chief financial obligation that went with his new status. *Someone* had to raise sufficient private funds to build his presidential library before it was turned over to the National

Archives to administer. Ford had deeded his papers to the University of Michigan a month before leaving office. But friends and family in Grand Rapids wanted a piece of the action. So Ford decided, Solomon- (or congressman-) like, to split the facility into two components, with a research library in Ann Arbor and a tourist-friendly museum in his old hometown. Henceforth many of his speaking fees were donated to the foundation entrusted with building and programming both institutions.

In April 1977 the Fords returned to Grand Rapids for the Easter holiday. For Jerry it was an opportunity to scout out potential museum sites.* With her husband's assistance, Betty was able to navigate the steps leading to Grace Episcopal Church. But she was forced to cancel the rest of her public schedule due to crippling neck and arm pain compounded by spinal arthritis. On the flight back to California, Greg Willard observed Betty clasping her hands tightly. Just how tightly became clear when they got home and he noticed a "huge bruised welt," self-inflicted by a woman who internalized her suffering. That evening the president asked Willard to come to the Fords' residence. There he found Mrs. Ford in agony. "I can't live like this anymore," she gasped. "I know the risks. I want to reconsider the possibility of surgery."

Doctors scheduled a myelogram, a test in which colored dye is injected into the spine before X-rays are taken. Betty's negative reaction to the procedure caused her to be admitted to the nearby Eisenhower Medical Center. As her condition rapidly deteriorated, brain scans and blood samples were ordered to rule out the possibility her cancer had returned. The results came back negative, and she was discharged after a few days, her illness blamed on the myelogram. Except she continued to languish at home. "No one ever connected the dots," says Willard, marveling at the failure of Betty's doctors, family and staff to detect symptoms of withdrawal from any of her painkillers, tranquilizers, sleeping medications or other drugs she took to counteract their side effects. By her own estimation she was swallowing twenty-five pills a day. Chris Chase, her coauthor on the manuscript that became *The Times of My Life*, suspected more than pharmaceutical agents were encroaching on their work time. "She'd wake up and she'd have her breakfast and everything," Chase remembers. "By lunch she'd start drinking and so . . . right after lunch we'd just give it up."

---

* Offered a building site on the outskirts of town, Ford opted instead for a downtown location in a riverfront area then cluttered with warehouses and other structures suggestive of urban decay. Together with the revitalized and expanded Pantlind/Amway Grand Hotel just across the Grand River, the Ford museum sparked a building boom that has transformed the onetime Furniture City into a popular convention spot and medical research hub.

Jerry, more disciplined if no more reflective, compiled nearly four thousand pages of taped interviews with writer Trevor Armbrister. Ford's work on the book was sandwiched into a schedule that had him on the road for twenty-seven days in the month of May 1977. At the University of Kentucky on May 8, he delivered a major speech critical of the War Powers Act. Quoting de Gaulle's formula that "Parliaments can paralyze policy; they cannot initiate it," Richard Nixon complimented his successor for making the same point "powerfully and profoundly." A few weeks later Ford, an eighteen-handicap golfer, scored his first and only hole in one at the Danny Thomas Memphis Classic Pro-Am before twenty thousand cheering spectators. "It's the best day I've had since November," he boasted.

He had always fantasized about living on a golf course. His ambition was realized in March 1978, when the Fords moved into their new house on the thirteenth fairway of the Thunderbird Golf Course. The 6,300-square-foot, six-bedroom residence was designed by Welton Becket & Associates, whose namesake was responsible for LA's Capitol Records Building. Betty worked closely with interior designer Laura Mako, a favorite of Bob Hope and a prime exponent of the "desert modern aesthetic," with its open floor plans and expansive glass walls showcasing the rugged landscape of the Coachella Valley. A seven-foot-tall portrait of Betty that had hung in the White House welcomed visitors to a sunken living room with a white brick fireplace. Wicker and chintz predominated, along with Betty's favorite colors of green and blue. The master bedroom suite included a sitting room, study, exercise room and two dressing rooms. A butcher-block kitchen opened onto the sun patio and twenty-by-forty-five-foot swimming pool.

The Fords and their guests would share a tennis court with Leonard Firestone. A few steps in the other direction, a house formerly inhabited by Ginger Rogers's mother was leased by the General Services Administration to provide working space for the former president's office staff, as well as his Secret Service detail. A stone's throw from the security gate on Sand Dune Road, the Fords' new neighbors included Spiro Agnew and his wife, Judy. The two men studiously avoided each other except on rare occasions in a golf locker room or a party at Frank Sinatra's, when they made polite chitchat.

H E MIGHT BE out of office, but Ford remained very much in the public eye, as he demonstrated on New Year's Day 1978 by serving as grand marshal of the Tournament of Roses Parade. Two weeks later he was in Washington for a ceremony in the Capitol Rotunda honoring Hubert Humphrey, who had succumbed to cancer on January 11. The occasion marked Richard Nixon's first appearance in the capital since his resignation. Persuaded against his wishes to attend a reception in Howard Baker's office, the former president retreated to one corner of the room. Ford sought him out, his hand extended in greeting.

"Good to see you."*

The ice broken, the two men compared their golf scores. Soon they were joined by Henry Kissinger, Lady Bird Johnson and a plainly uncomfortable Jimmy Carter. "Everybody made a maximum effort to be congenial," Ford said afterward, "but it was not relaxed." He might have said as much about conditions under his own roof. A recent visitor to the Ford home confided to *Time* magazine that Betty Ford was a virtual zombie, addicted to painkilling drugs yet "effectively abandoned" by her family, all but Susan. Christmas in Vail had been nightmarish, with Betty seemingly oblivious to everything except the attempted watering of her drinks. Casting a further cloud over the holiday season was her embarrassing performance as host and narrator of NBC's broadcast of *The Nutcracker* by Moscow's famed Bolshoi Ballet. Millions of viewers saw a vacant-eyed woman who took forever to get dressed, eat a meal, finish a sentence. Friends stopped asking Betty to lunch. Jack and Steve, each of whom had taken up residence in Southern California, stopped visiting their parents' home.

"We've sent her to psychiatrists, we've sent her to doctors, massage therapists, acupuncture. There's nothing left to do," the former president told Caroline Coventry, his wife's well-meaning assistant. A classic enabler, adept at making excuses for his tardy or absent wife, Ford had become resigned to living with this sadly diminished vestige of the woman to whom he had been married for almost thirty years. One night while he was out of town, Betty fell in the bathroom, cracked her ribs and chipped a tooth. Susan, "fed up" with filling in for her mother at all the events from which the former First Lady withdrew, turned to Dr. Joseph Cruse, himself a recovering alcoholic, and a gynecologist whose patient list included both Ford women. To assist in their hastily arranged intervention, Susan reached out to Caroline and to Clara Powell, the former Ford family housekeeper, who was ostensibly recruited to help Betty get settled into the new house.

---

* Only a few months earlier, Ford had confirmed reports that he had guest lists at upcoming events checked to guard against crossing paths with Nixon. "Everybody recognizes it would be difficult for this to happen at this particular time," he told Marjorie Hunter of the *New York Times*. Marjorie Hunter, "Ford Careful to Avoid Meetings with Nixon," *New York Times*, August 15, 1977.

On Thursday, March 30, they assembled amid packing boxes in her green-and-blue living room to hear Dr. Cruse relate his personal recovery story to Betty. It was not well received. "You are all a bunch of monsters," Mrs. Ford shrieked. "Get out of here and never come back." For good measure she threatened to get the doctor's license to practice revoked if he bothered her again. Jerry was in Rochester, New York, when he received a phone call from a furious spouse. "She told me she wants half of our house, and she's going to New York," he confided to Dr. Cruse. Apparently it was not the first time she had made such threats. Fearing failure and having no wish to inflict additional pain on someone in torment, Jerry was reluctant to take the emotional plunge of a full-scale intervention.

"Susan was the one who said we just could not go on like this," according to Mike Ford. "She was the one who told Dad, get your rear end home, Mother's in a bad way." Shocked into action, Ford rearranged his schedule. Henry Kissinger agreed on the spot to pinch hit for the former president at a speaking date in Virginia. Flying through the night, Ford walked into his office a little after 7 o'clock on Saturday morning, the first of April. "Are you sure this needs to be done? And that she can be helped?" he said to Dr. Joseph Pursch, the navy psychiatrist recruited by Dr. Cruse to lead the intervention. Assured on both counts, Ford swallowed his doubts. "In that case, we'll go."

The next two hours were agony. Jerry took the lead. "Betty, the reason we're here is because we love you," he told her, adding that she had become chemically dependent, and the doctors in attendance wished to talk to her. In fact, practically all the talking was done by her family. As the oldest child, Mike went first, conveying sympathy for the pressures she had endured while in Washington. But her current behavior was putting her relationships and her very life at risk. His wife, Gayle, landed an emotional punch by telling Betty how much she and Mike looked forward to starting a family, and how important it would be for their children to know their grandmother "as a healthy, loving person." Jack explained his reluctance to bring friends to the house, and Steve recounted the time he had gone to the store, shopped for and prepared an elaborate dinner for his mother, only to have Betty say she didn't feel like eating. Susan, struggling to get the words out amid her sobs, contrasted the graceful dancer she had admired with the clumsy woman who repeatedly fell down.

As emotional barriers were breached, discretion was drowned in long-repressed anger and resentment. Intimidated by the force of their love, and their ruthless determination that she get well, Betty agreed to accept treatment. A harrowing week of at-home detoxification—the president wanted to know from her nurse companion why Betty kept throwing up—was followed by a soul-baring month at Long Beach Naval Hospital, where the former First Lady was shamed into sharing a room with three other women alcoholics. Gradually Betty weaned herself off her "celebrity hang up" that made group therapy acutely painful for someone clinging to imagined status in lieu of genuine self-esteem. Ten days into her hospitalization, Dr. Pursch threw her a curveball by confronting Betty with her dependence on alcohol as well as pills. She protested vehemently, rejecting a public statement of the kind Pursch wanted her to put out because, she said, "I don't want to embarrass my husband." Only after Jerry assured her, "You won't embarrass me," was Betty able to write out the words she had for so long resisted speaking to herself, let alone to millions of strangers.

Her publisher insisted that in light of recent developments she add a chapter to a manuscript ready to go to the printer. When *The Times of My Life* was published that fall, most reviewers echoed what Jane Howard wrote in the *New York Times*. "As American as a suburban shopping mall," she said of the author. Reading her book "is like going to the mall to buy just one thing and ending up staying all afternoon." Howard had only one complaint. The order of chapters should have been reversed. "I'm Betty and I'm an alcoholic" would have made "a smashing opening sentence." Not long after his wife's story was published, Jerry told a friend, "You know, it doesn't work when one's drinking and one isn't drinking." Thereafter he contented himself with a nightly "cocktail" of club soda and lime.

For his sixty-fourth birthday, celebrated in Vail, Betty gave her husband a T-shirt bearing the message: BET MY BOOK OUTSELLS YOURS. (It did.) Two months before the scheduled publication of *A Time to Heal* in May 1979, someone slipped a copy of the two-hundred-thousand-word manuscript to Victor Navasky, editor of *The Nation* magazine. The magazine's April 7 issue featured a 2,250-word article incorporating roughly 300 words taken verbatim from Ford's narrative. *Time* magazine, which had contracted to print a 7,500-word excerpt from the book, asked the publisher if it could advance publication by a week. When the answer came back no, *Time* canceled the deal and withheld its second payment of $12,500. Ford's publishers sued *The Nation*, which claimed its scoop fell under the "fair use" clause of the Copyright Act. In a 6–3 ruling handed down in March 1985, the Supreme Court disagreed. By lifting Ford's own words to expose the most newsworthy parts of his story, Justice Sandra Day O'Connor wrote, the magazine had "effectively arrogated to itself the right of first publication."

Reviewers were less inclined to side with the former president. Lacking in major revelations or personal introspection, his book was judged "sincere but dull" by the *Detroit News*. A dissenting view was expressed by Harry McPherson, Lyndon Johnson's White House counsel. He praised Ford's candor and thought he made a persuasive case for the Nixon pardon. Ultimately Ford had been a successful president, wrote McPherson, not because he was clever, articulate, or a man of vision, "but because he was honest, straightforward, forgiving and possessed of sound judgment. He was all that his predecessor was not, and so he was exactly what we needed."

WHEN NOT ON the road, Ford still liked to begin each day with a half-mile swim in the backyard pool of his Rancho Mirage compound. He avoided golf carts and begged off post-tournament socializing with other celebrity players, explaining, "I've still got some office work to do." Six days out of seven he was at his desk by nine a.m., reviewing his mail, conferring with staff and weighing a daily stack of twenty to thirty invitations to future events. Once a week, navy personnel delivered a top secret update of current events courtesy of the Carter White House. Besides reading half a dozen newspapers Ford stayed current by phone with political cronies like Tip O'Neill. They had "great rapport," according to Ford. "He gives me hell and I give him hell."

Technology had never been his friend. The office telephone system was beyond his comprehension, although he did learn to play Solitaire on a computer and to receive, if not send, email. For several years in the 1980s Ford would stop whatever he was doing at eleven to watch his son Steve, who had landed a regular part on the TV soap opera *The Young and Restless*. After lunch with Betty he returned to the office to catch up on paperwork and greet visiting friends or, occasionally, reporters. Another hour was set aside for autographing not just letters or photos, but footballs, baseballs, "everything you could imagine, they would send in," according to Shelli Archibald, who replaced Dorothy Downton as Ford's personal secretary in 1981. Typically Ford signed and sent out over a thousand Eagle Scout letters a month.

He was less accommodating of the repeater's list containing the names of autograph hounds not above claiming they had cancer, or a dying child, to secure a Ford signature that could be sold for a quick sawbuck online. One afternoon the former president was signing autographs in the large conference room with its picture window framing the adjacent golf course when a gardener's lawn mower dislodged a rock and sent it flying against the plate glass. Instantly Ford went down, ducking for cover as his agents had taught him in Sacramento and San Francisco. Old habits die hard. Ford's work ethic was reminiscent of the freshman congressman insistent on answering constituent letters within twenty-four hours. Saturdays were no different. "Do we really have to come in tomorrow?" one staffer inquired at the end of the normal workweek.

"Well, sure. There's mail."*

Many of Ford's correspondents expressed dissatisfaction with the Carter Administration and hoped the former president might run again in 1980. An October 1979 Harris poll showed Ford leading Carter 55–42 percent, while Ronald Reagan trailed the incumbent by seven points. Egged on by governors anxious to prevent a Reagan takeover of the GOP, Ford commissioned a private poll from Bob Teeter. It showed less enthusiasm for his prospective candidacy. There were two groups opposed to the idea, said Ford, tongue in cheek. One objected to his running "because I pardoned Richard Nixon." A second demurred "because they're afraid I might pardon Jimmy Carter." By then any window of opportunity for a 1980 bid was closing fast. Stu Spencer said as much at a meeting of Ford loyalists called to assess the former president's prospects. Try as they might, Ford's delegate counters were unable to devise a winning strategy. He couldn't defeat Reagan, only bloody him.

On March 15, 1980, Ford called a press conference at which he effectively conceded the nomination. "America needs a new President," he told reporters. For him to embark on a campaign so late in the day would further divide the party, dampening its prospects for victory in November. Unknown to Ford, Reagan pollster Dick Wirthlin had identified three potential running mates—George Bush, Howard Baker and Ford himself—who would boost Republican chances against Carter.

On June 5 Reagan paid a call on Ford in Rancho Mirage. The presumptive presidential nominee had come to ask the appointed vice president and unelected president to run for vice president with the man he held most accountable for his defeat in 1976. Ford suggested George Bush as a better pick. But Reagan wasn't so easily put off.

What happened next remains, four decades later, tangled in conflicting memories and debatable motives. "Let's test the waters was kind of Reagan's idea," insisted Mike Deaver. "Well, testing the waters got out of control like a prairie fire." Monday, July 14, was the convention's opening day. It was also Ford's sixty-seventh birthday. Ron and Nancy Reagan dropped by the

---

* Reminded that a federal holiday was looming on the calendar, Ford responded, "Well, I didn't vote for it." AI, Lee Simmons.

former president's suite on the seventieth floor of the Detroit Plaza Hotel to toast the occasion with champagne and present Ford an antique Crow Indian "peace pipe." The gift was a hit, though it didn't change Ford's skepticism toward the so-called Dream Ticket being bandied about the Motor City. That no one knew how to address a former president reborn as a vice president was the least of impediments. How could Ford sign on to a ticket opposed to the Equal Rights Amendment? Was he really prepared to side with Phyllis Schlafly in opposition to Betty? For that matter, how would Reagan supporters deal with the inevitable sneers that their hero required the services of a former president to help him do the job?

One can hardly blame Reagan for thinking his onetime adversary might be persuadable. In his barn burner of a speech to the convention on Monday night, Ford rejected the label of "elder statesman," telling the cheering delegates, "I've never spent much time on the sidelines . . . This country means too much for me to comfortably park on the bench. So, when this convention fields the team for Governor Reagan, count me in!" The next forty-eight hours would be consumed in divining the meaning of this delphic utterance. Feeling an obligation to the party and its nominee to at least explore the possibility, Ford designated Alan Greenspan, Jack Marsh, Bob Barrett and Henry Kissinger to represent him in discussions with the Reagan campaign.

No doubt savoring the spotlight, Ford continued to send mixed signals in public, telling NBC's *Today* show on Wednesday morning that "if all the other questions could be resolved," pride would not stand in the way of his acceptance. "What in the world is going on?" an anxious Dorothy Downton asked her boss when he called his Rancho Mirage office. "I don't like what I'm hearing."

"Well, what are you hearing?"

"That you're going to be his running mate?"

Ford started laughing. "Dorothy, don't believe everything you hear."

That afternoon, hours before the convention was to nominate Reagan, the candidate's aides produced a list of ten talking points to address the former president's concerns. Reprising his familiar role as convention manager for the winning side, Bill Timmons backstopped the discussions from start to finish. "I was the only one who could type, and I typed the agreement," he later explained. Timmons said he would deny "to my dying day" that anything he typed and retyped resembled a copresidency. It was, in fact, a fairly conventional attempt to codify Ford's role as vice president—"Meeting with the president often, having a voice in policy, comment on policy, personnel, special attention to Congress . . . Ford wanted some active role in the defense department."*

Reagan's national security advisor, Richard Allen, arriving at the governor's hotel suite one floor below the Fords around five thirty, was surprised to encounter the former president leaving the premises, accompanied by his Secret Service detail.

"What do you think of the Ford deal?" Reagan asked Allen.

"What deal?"

As Reagan explained it, Ford was pressing for Kissinger as secretary of state and Greenspan at the Treasury. "That is the craziest deal I have ever heard of," said Allen. There was more. Borrowing an idea of Nelson Rockefeller's that he had rejected while in office, Ford wanted the vice president to double as White House chief of staff. He likened the presidency to a well-run business "where you have a chief executive officer and a chief operating officer." Ford's presidential CEO "makes all the decisions. He signs all the actions. He is the final authority." As for the vice president, "He manages the White House . . . he is in the job of making the Administration work."

Not surprisingly the Ford-Reagan talks became a media sensation, especially after Ford honored his commitment that evening to appear on camera with America's most trusted newsman. "Oh my God, I've never seen anything like it," reflected Bob Schieffer of CBS, "and if ever there was a scoop, I mean, getting Ford up and into that booth with Walter Cronkite. Everything at the convention came to a dead stop while people were looking up there, watching and trying to figure out what was going on."†

In response to Cronkite's questioning, Ford played coy, restating his earlier refusal to accept a draft while speculating on what "meaningful role" it would take for him to be something other than "a figurehead vice president." Cronkite—"Old Walter," Schieffer chuckles, "I mean, he knew when he had something, he wasn't going to let it go"—then confronted Ford with the phrase that set George H. W. Bush on the road to the White

---

* Unfortunately the document in question did not survive the aborted talks. "When it kind of fell apart, I took the last copy," Timmons recalled, and with Reagan's future attorney general, Ed Meese, as a witness "folded it up and flushed it down the toilet." AI, Bill Timmons.

† Rival Barbara Walters of ABC "went up and started banging on the door of our anchor booth, 'Let me in! Let me in!'" said Schieffer. "And we told her, 'No, you can't come in here.' And we always wondered what she would've done if she'd have gotten in there. Would she have gone in there and sat down with Walter Cronkite and joined in the interview which was being broadcast on CBS or would she have tried to grab Ford and drag him off the set?" AI, Bob Schieffer.

House. "It's got to be something like a co-presidency?" Ford replied that this was something Reagan "really ought to consider."

Ronald Reagan happened to be among Cronkite's viewers. According to Reagan aide Mike Deaver, at the mention of a copresidency he "almost [choked] on whatever he was eating." Another aide describes a stunned Reagan muttering, "That's it. No way." Deaver was sent to retrieve Ed Meese and pollster Dick Wirthlin from the Ford suite upstairs. They, in turn, told Deaver to mind his own business. "We're just about ready to make a deal, and you're going to screw this whole thing up."

"I'm telling you, you better not make a deal," Deaver shot back, "because Reagan wants to talk to you."

With the delegates in Joe Louis Arena becoming visibly restive, and rumors sweeping the hall that a Reagan-Ford ticket was a done deal, Meese and campaign manager William Casey described for Reagan and his team a Rube Goldberg scenario in which the vice president would have veto power over the secretaries of state and Treasury, while the president would exercise the same authority over OMB and the National Security Council.

Meanwhile, sitting with Ford in his darkened living room, Jack Marsh remarked, "I think you'd better pull the plug on it."

"I do too," said Ford. "That's what I felt all along." He had been thinking, a visibly weary Ford confessed to Bob Barrett. "It just wouldn't be fair to Betty."*

At eleven thirty he made a final call on Reagan. The two men were seen to be "laughing and joking" in a corner of the suite out of earshot from everyone around them. After perhaps five minutes they shook hands and Ford left. "It's off," Reagan told his entourage. "Jerry said it's a bad idea. It won't work." Before leaving, however, Ford had vowed to do everything in his power to help Reagan get elected.

Moments later Reagan was on the phone to George H. W. Bush.

Some thought Ford had overplayed his hand as payback for past grievances. Dick Cheney took a different view. "I had the feeling that Ford didn't really want the job and one way he could get out of it was ask for the moon, and he did." Ford himself appeared highly pleased with the week's results as he boarded a plane that would take him back to Vail for the rest of the summer. Kicking off his shoes, he leaned back in his seat and observed

---

* In discussing with Betty Ford her husband's imminent return to the political wars, Stu Spencer heard her say jokingly, "I'll divorce him if he does that." "Okay," Spencer concluded, "I've got the answer to this one; they're all playing games up there on the top floor." AI, Stu Spencer.

to Bob Barrett, "Well, Robert, not a bad convention. I gave a good speech, and I got Bush as vice president."

Barrett's suspicions of a mutual head fake were rekindled.

"They were conniving," he said to Ford. "Were you conniving too?" He never got a satisfactory answer.

28

LIGHTS IN A TREE

———◆·◆———

I feel OK, but I'm old!

—GERALD FORD, TO WELL-WISHERS INQUIRING AFTER HIS HEALTH

RONALD REAGAN WAS nine months into his presidency and still re-cuperating from a near-fatal assassination attempt when, in October 1981, Islamic fundamentalists murdered Egyptian president Anwar Sadat. Precluded from attending Sadat's funeral on security grounds, Reagan welcomed Secretary of State Al Haig's proposal that all three of Reagan's living predecessors accompany Haig to the Cairo services. Ford's family, recalling his violent encounters with Squeaky Fromme and Sara Jane Moore, were reluctant to see him go. But Ford was of a generation for whom a presidential request, whether from Lyndon Johnson after Dallas or Ronald Reagan in response to Anwar Sadat's martyrdom, had the force of command.

It would be a long trip to Cairo, and not just as measured in flight time. Less than three weeks had passed since Ford, accompanied by Reagan and several other world leaders, dedicated his presidential museum in Grand Rapids. Neither Nixon nor Carter had graced that occasion. Carter, still nursing emotional bruises inflicted by his loss of the White House, and resentful of Ford's harsh campaign rhetoric aimed at avenging *his* defeat in 1976, characterized their relationship as "oil and water." The Ford-Nixon connection was even more battle-scarred. Still, resigned to spending the next thirty-six hours in their company, Ford suggested to his fellow travelers, "Look, for the trip, at least, why don't we make it just Dick, Jerry, and Jimmy?"

Relegated to the economy section of the plane by Haig and his minions, the presidents talked of Sadat and the impact his death would have on the volatile Middle East. They compared notes on memoir writing and library building (or more accurately, library fundraising). Here Ford was the voice of experience, a defeated candidate with no capacity to reward donors, who had nevertheless overcome his lifelong distaste for money raising to build not one but two facilities entrusted to the National Archives.

Arriving in Cairo, the presidential trio paid a sympathy call on Sadat's successor, Hosni Mubarak. That evening there was a dinner at which Carter spoke first, followed by Ford. When it was his turn, Nixon used the occasion to extol the virtues of Foreign Service personnel as well as the chauffeurs and waiters who had eased his global travels since 1953. As Nixon posed for pictures with the hotel waitstaff, Ford whispered to Henry Kissinger, "Sometimes I wish I had never pardoned that son of a bitch."

The next day the three presidents, cocooned in a massive security operation that held ordinary Egyptians at bay, donned bulletproof vests before walking half a mile to the deserted reviewing stand where Sadat had been gunned down. There Ford assured Jehan Sadat that the American people shared her grief. Back at the airport Carter and Ford learned that Nixon would not be returning with them, having carved out a separate itinerary through Saudi Arabia and several other Middle Eastern countries. It was just as well. With their mission successfully completed, the two men were free to relax and get better acquainted. They passed the long flight across the Atlantic talking about their families and backgrounds, swapping stories from the campaign trail and agreeing in principle to appear at each other's libraries. They also made some news, telling reporters on board that it was only a question of time before the US recognized the Palestine Liberation Organization.

By the time they reached Andrews Air Force Base, both men felt a page had been turned. It was the start of a friendship unmatched since the autumnal reconciliation of John Adams and Thomas Jefferson. To those who expressed surprise at his sudden change of heart, Ford said it was simple: "I misjudged him." True to their airborne pact, Ford and Carter would jointly sponsor conferences in Atlanta and Grand Rapids addressing subjects like arms control, the Middle East, the former Soviet Union and Europe's emerging democracies in the aftermath of the Cold War. Betty Ford and Rosalynn Carter would stage their own program on the First Lady's changing role. The two women made a formidable team testifying before Congress on behalf of their mutual interest in treating mental health and alcohol/substance abuse.

In 1988 Ford and Carter cochaired the American Agenda, an ambitious effort by academic experts and former administration officials to compile a précis of the country's needs to help guide the winner of that year's presidential election. In 2001 their National Commission on Federal Election Reform proposed sweeping changes in election law: making election day a national holiday; allowing voters challenged by poll workers to cast provisional ballots; and restoring to felons their voting rights. Twenty years later their recommendations remain mired in partisan bickering. Ford no doubt smiled over an August 1982 letter from Ted Kennedy, whose 1980 insurgency held a place in Carter's memory akin to Ford's resentment of the Reagan challenge four years earlier. "I read that you and President Carter are in the process of reconciling," Kennedy wrote. "As soon as you finish, would you mind sending me a few hints on how to do it?"

In his January 2007 eulogy of Ford—the two men had agreed that whichever one survived would speak at the other's funeral—Carter recalled a *New Yorker* cartoon enjoyed by both. In it a small boy, looking up at his father, earnestly declares, "Daddy, when I grow up, I want to be a former president." The job had its perquisites, to be sure, although these were sometimes hard to distinguish from penalties. In November 1985 Ronald Reagan asked Ford to represent the United States at ceremonies commemorating the fifteenth anniversary of Oman's Sultanate. In 1999 he joined Bill Clinton, Jimmy Carter and George H. W. Bush at the funeral of Jordan's King Hussein.* Wherever he went, Ford was invariably surprised at the recognition he stirred. On a private visit to Athens, onlookers broke into spontaneous applause as he scaled the Acropolis. In Jerusalem he and Betty blended into the crowds walking the Via Dolorosa.

Not all his receptions were so friendly. An Adriatic cruise was marred by a confrontation between Ford and a badly scarred German war veteran near the Croatian city of Dubrovnik.

"You did this to me!" the old soldier shouted at the American visitor.

"I wonder if he ever heard of Hitler," Ford said afterward. "He started this thing, not us."

At an audience with Pope John Paul II, the pontiff was visibly surprised to encounter female Secret Service agents accompanying the former president.

"Security? Women?" he asked Ford.

"Yes, Your Holiness. They're very, very good."

"Women. Security."

FORD'S RELATIONS WITH the Reagan White House were collegial if less than intimate. Shrewdly, Reagan insisted that Ford on return visits to the Oval Office should occupy the "presidential" chair before the fireplace. "Little things like that . . . did not go unnoticed," said David Kennerly. Such courtesies hardly made up for the administration's embrace of supply-

---

* In 1993 he had performed similar duties in Brussels at ceremonies honoring Belgium's late King Baudouin.

side economics ("Don't use that term with me," Ford erupted when asked to assess policies at odds with his more orthodox approach to debt and taxes) or the growing imbalance between moderates and what he called the hard right in his beloved GOP. Opposing elements in the party that he considered "anti-Negro, anti-Semitic, anti-minorities, or anti-women's rights," Ford lent his name to Norman Lear's People for the American Way, an organization conservatives frowned upon as a haven for Hollywood liberals unnerved by the Reagan Revolution.

Ford was far more supportive of abortion rights than he had ever been while in the White House. In 1987 he and Betty took part in Palm Springs's first Desert AIDS Walk to raise awareness and money with which to fight the disease. When Mary Fisher, daughter of their close friend Max Fisher and herself the first White House advance woman, was diagnosed in 1991, Ford was red-faced with anger at Fisher's husband, from whom she contracted HIV. The next year he was in the presidential box at Houston's Astrodome wiping away tears as Fisher shared her story with Republican convention delegates. To a *Detroit News* columnist inquiring about his views on same-sex couples, Ford replied, "I think they should be treated equally. Period." Equal treatment extended to Social Security, tax and other federal benefits. In joining the board of the Republican Unity Coalition, he became the first former or current American president to align himself with a gay rights advocacy organization. Several years before his death in 2006, Ford endorsed same-sex marriage over halfway measures like civil unions. As he put it, "We need to short-circuit all the damage that's being done to people and just let it happen."

Ford was "liberated" after 1980, asserted Tom DeFrank, whose extensive private conversations with the former president formed the basis for his 2007 volume *Write It When I'm Gone*. "He didn't care what people thought. And if they disagreed with him in the party, too bad." Ironically, the same journalist earned Ford's ire by questioning his decision to join half a dozen corporate boards (a number that would, in time, swell to ten) and become what *Newsweek* labeled "Jerry Ford, Incorporated." Ford's business interests, combined with his speaking fees, earned him an estimated $1.3 million in 1985 alone. By the end of his life an informed guess as to his net worth would be in the range of $10–15 million.

Having no wish to be remembered as the president who opened the door for his successors to land vastly larger book, media and speaking deals, Ford protested that he turned down ten offers for every one he accepted, and kept his distance from defense contractors. His sensitivity on the subject was not unlike his wounded feelings at the time of the Nixon pardon. Because he felt his motives to be honorable and his scruples intact, the Eagle Scout in Ford assumed that others would reach the same conclusion.

Banker-financier Sandy Weill recruited Ford for Shearson Loeb Rhoades, an investment banking firm that later merged with American Express. Weill subsequently founded Commercial Credit, forerunner of Citigroup, on whose board Ford served as an "honorary director" until his death. Weill thought the criticism aimed at his friend's plunge into the private sector misplaced. Shunning the role of lobbyist or rainmaker, Ford never missed board meetings, asked pertinent questions, did his homework. "Here was somebody that had a unique experience that could be of great advantage to how a company operates," says Weill. "You're talking about a person that had the highest ethical standards . . . and an enormous amount of common sense."

Accustomed to dealing with outsized egos, Ford brought to the boardroom the consensus-building skills honed on Capitol Hill. Rod Hills, his former White House counsel, served with Ford on the board of the Santa Fe International Drilling Company. After it was sold to Kuwait Oil in 1981, the Kuwaiti oil minister announced that board meetings would be reduced from twelve to four a year. In addition the minister said that current directors' fees of $25,000 were inadequate. "'We'd like to make the fees $100,000 a year.'" Ford cut in. "Before anybody else says something, Ali, let me say that if we're going to cut the meetings from twelve to four, it's hardly the time to raise our income. Probably we should reduce it, but I understand you want us to come to Kuwait City for one of our meetings, and it takes a little more time, so if my colleagues don't object"—here he looked sharply at Hills, the chairman of the audit committee—"we will accept $25,000."

That said, a former president's connections were undoubtedly useful for his corporate patrons (in Singapore Sandy Weill had a seat at the head table when President Lee Kuan Yew hosted a dinner for Jerry and Betty). Weill saw another side to the utilitarian Ford on the golf course at Pebble Beach. "The President, he shanked it on this Par 3 and the ball went up in the air and into the woods and we heard a clunk. I was praying it was a tree, but it wasn't a tree, it was a lady's head. And he went over to apologize, and I thought it was a little funny. And he said to me as we walked up to the green with the ball in his pocket . . . 'Sandy, you're not going to think this is so funny. You are my insurance carrier.'"

Ford's charitable calendar was nearly as crowded as his travel schedule. He and Betty were regulars on the Palm Springs charity circuit, often attending two or three events weekly. In the 1980s he chaired the Campaign for Michigan, which raised $178 million for his alma mater. For twenty years beginning in 1977 the former president hosted an eponymous golf tournament in Vail that was heavily attended by pro players and celebrities alike. In the winter the Jerry Ford Celebrity Cup raised Vail's profile in the ski world as it raised funds for a local hospital. Ford was also instrumental in the creation of a handsome interfaith chapel set amid the balsam trees of Beaver Creek, the Vail offshoot originally planned to host the 1976 Winter Olympics.

With the 1982 dedication of the Betty Ford Center, an addiction treatment facility on the campus of the Eisenhower Medical Center in Rancho Mirage, a proud husband willingly ceded the spotlight to his wife. At annual reunions of center alumni, Jerry could be found grilling hot dogs. "She's the president," he told people who addressed him by that title. "I'm just the former president." Asked if he wasn't even a little bit jealous of Betty's success, he replied, "Hell no, I think it's wonderful." Ford beamed with pride when she received from George H. W. Bush a Presidential Medal of Freedom a decade before he did. And he told people that when the history books were written, her contributions would loom larger than his own.

As fate would have it, one of the celebrity patients treated at the Betty Ford Center was Chevy Chase, the breakout star of *Saturday Night Live* famous for his sendups of Ford as president. Chase's subsequent movie career was checkered, and by 1985 he was seriously addicted to painkillers prescribed for his back. Before sitting down to lunch at the center, Chase and the former president looked on as their wives, on all fours, tried to figure out the wiring of a videotape recorder. At one point Chase got up from his chair and suggested to Ford that they might lend a helping hand.

"No, no, Chevy," Ford said, gently pulling him back to his seat. "Don't even think about it. I'll probably get electrocuted and you'll be picked up and arrested for murder."

T HE PRESIDENCY OF George H. W. Bush (1989–93) felt like something of a restoration. Not only did the new administration recruit Ford alumni (Dick Cheney, Jim Baker, Brent Scowcroft and Carla Hills) for key positions; it prioritized traditional Republican concerns like budget discipline and the environment. At Bush's request secure telephones were installed at Rancho Mirage and Beaver Creek. Early in his presidency Bush called to apologize for "gossipy and ugly little press whispers" that found their way into print when *Time* magazine erroneously reported that Ford had asked for the use of Blair House and rides on *Air Force One*. In fact it was the Bush White House that sought Ford's help in corralling votes for a controversial deficit reduction plan, only to see it shot down in the House when Republican insurgent Newt Gingrich held Bush to his "read my lips, no new taxes" pledge made at their party's 1988 convention.*

Ford admired Bush for his political courage in tackling a runaway deficit and his diplomatic skill in assembling a thirty-five-nation coalition to expel Iraqi dictator Saddam Hussein from Kuwait, five months after Saddam's August 1990 invasion and occupation of the oil-rich emirate. In the afterglow of Operation Desert Storm, Bush's approval numbers soared to record heights. His reelection appeared a cinch. Yet military glory proved short-lived amid economic recession and a perception that Bush's historical mission had been fulfilled through his adroit management of the implosion of the old Soviet empire, much as Ford's claims of restoring public trust in the aftermath of Watergate were deemed insufficient to overcome a grassroots hunger for change in 1976.

Ford enjoyed a closer relationship with Bush's successor than the public knew, or Ford himself anticipated in January 1993. It began with a thirty-minute phone call in which he described for Bill Clinton a recent trip through Ukraine made in tandem with former British prime minister James Callaghan. Stressing the importance of the breakaway republic with nuclear weapons and a democratically elected president, Leonid Kravchuk, Ford advised Clinton against putting all his eggs in the basket of Russian leader Boris Yeltsin. In April 1993, Betty Ford wrote to Hillary Clinton urging that any legislation crafted by the First Lady's health-care task force emphasize education and treatment of drug abuse over interdiction and prosecution. Invited to meet with Mrs. Clinton at the White House, Betty was impressed by the First Lady's hospitality and detailed knowledge of health care, a notoriously complex subject.

The two families became better acquainted that August, when the Clintons briefly occupied Leonard Firestone's house next door to the Fords' in Beaver Creek. Presidential daughter Chelsea was interested in ballet, and Vail each summer played host to Moscow's Bolshoi Academy and its legendary director Sofia Golovkina. One evening the Fords and Clintons attended a performance by the dance company. Mrs. Clinton gave Ford a photo recalling the summer of 1968, when she had worked as an intern

* Gingrich's confrontational style rubbed Ford as counterproductive, although the two men established a personal relationship that was perfectly friendly. Hearing Ford complain about the future Speaker's aggressive partisanship, a mutual friend inquired, "President Ford, am I wrong in thinking that when you were a young congressman, you led a movement to impeach William O. Douglas?" Ford thought a moment before removing the pipe from his mouth and laughed out loud. AI, Chris DeMuth.

for the House Republican Conference. For the golfaholic Bill Clinton, the Vail respite offered a chance to partner with Jack Nicklaus against Ford and Clinton's friend Ken Lay, the Houston businessman later convicted of fraud and conspiracy in the collapse of energy giant Enron.

"We have a few things where we have similar views," Ford told reporters at the Country Club of the Rockies. He promised to be "as helpful as I possibly can" in securing passage of the hemispheric free-trade package known as NAFTA. In response to press inquiries about his game, Clinton reported a score of eighty. Jack Nicklaus, out of earshot, muttered to Ford, "Eighty with fifty floating mulligans." Whatever he might think of Clinton's ethics on the golf course, Ford decided the new president was "nicer than I thought" and "a helluva salesman." Employing his own persuasive skills, Ford pitched Republican lawmakers on the virtues of the North American Free Trade Agreement. Following its enactment in December 1993, Clinton was suitably appreciative. He thanked Ford again for helping secure House passage of a ban on semiautomatic assault weapons. Though the two men differed over Clinton's diplomatic outreach to North Korea, the White House took Ford's objections seriously, dispatching officials from the National Security Council to brief the former president on the latest developments. In 1995, Ford returned to the spotlight, briefly, as he led the US delegation commemorating the twentieth anniversary of the Helsinki Accords.

A grimmer reminder of time's passage was the Boston funeral of Tip O'Neill in January 1994. Three months later Richard Nixon suffered a massive stroke at his home in Park Ridge, New Jersey. He died on April 22. At his funeral the Fords joined four other presidents and as many first ladies in a holding room prior to the outdoor service and burial on the Nixon Library grounds in Yorba Linda, California. Ford made a point of thanking Clinton for his attendance. Finding the normally gregarious Ronald Reagan somewhat distant—"You had the uneasy feeling that he wasn't quite sure who you were," said Ford—his doubts were confirmed a few months later when Reagan wrote his poignant letter to the American people revealing his diagnosis of Alzheimer's disease.

It may have been coincidental, but around this time Ford began to scale back his outside activities in order to spend more time with Betty and their extended family. None of the Ford children showed interest in a political career. Eldest son Michael was associate dean of campus life at Wake Forest University, his brother Jack a San Diego businessman. Steve Ford juggled an acting career with his first love, ranching. Susan teamed up with her mother to help launch National Breast Cancer Awareness Day (eventually a month). Divorced from Secret Service agent Chuck Vance in 1988, she remarried the next year, to Tulsa lawyer Vaden Bales.

As of 1994 there were five Ford granddaughters with their own attic loft in the three-story, eleven-thousand-square-foot house in Beaver Creek, the mountain resort to which the former president and First Lady retreated from May through October. Here the children rode fire trucks in the Fourth of July parade and assisted their grandfather in lighting the big Christmas tree near Vail's covered bridge. Ford made sure all his grandchildren could swim ("OK, show me your laps"), and somehow kept his temper in check when five-year-old Tyne Vance pushed him into the pool, in business suit and tie.

Just when it looked as if there would be no male heirs to the Ford name, Jack Ford and his wife, Juliann, produced two sons, Christian (1997) and Jonathan (1999). The boys were introduced to golf at an early age. Football was another shared interest. If the Michigan Wolverines were playing on TV and things weren't going their way, Ford did not hesitate to verbalize his unhappiness. "Jerry," Betty interrupted, "you probably used some words that . . . the grandchildren didn't need to hear." After her father hurled his glasses at the offending tube once too often, Susan gave him a sponge brick in the Michigan colors of maize and blue. "Here, Dad, use this when you get mad."

"He was a fantastic bridge player," she recalled. "He and Mother played Gin their entire life. They played a penny a point, and . . . they totaled up who won at the end of the year." Fans of old movies, the History Channel and *60 Minutes*, the couple were occasionally seen visiting the nearest cineplex. They enjoyed *Apollo 13*, though Betty walked out of *Titanic* after the iceberg made its appearance. The former president amused the females on his staff by recommending *First Wives Club* as "a good chick flick." A virtual member of the family was Lorraine Ornelas. Formerly chef at a Palm Springs–area Marriott, Lorraine was twenty-eight years old and nine months sober after going through the Betty Ford Center when she got a call from its namesake asking if she might be interested in working for the Fords as their personal chef. That meant seasonal travel between the desert and the mountains of Colorado. To conquer her fear of flying, Ford had their pilot explain in detail how the plane worked and why she should feel perfectly safe. Mrs. Ford provided verbal reassurance to the nervous traveler throughout the ninety-minute flight to Vail. Unprepared for the sight of December snow, Lorraine was given driving lessons by Dick Garbarino, the Fords' caretaker and the president's occasional traveling companion after Betty and the children decided he should not be on the road without someone besides the Secret Service.

The former president personally enrolled Lorraine in the local ski school, admonishing the instructor on duty, "Make sure she doesn't get hurt." Lorraine marveled at the discipline of the elderly gentleman who called her Pal, sharing stories of his successful friends who, like Lorraine, were dyslexic or had never finished high school. "I don't know too many people who could have a big juicy piece of pie with ice cream and stop themselves at half," she later said.* Finding she could talk to Mrs. Ford about anything, Lorraine nevertheless expected to be fired when she briefly relapsed into her old habits. To her surprise, Betty related her own temptations, and the meetings she attended as part of a lifelong recovery program. Lorraine's job was safe. Motivated by the example of her employers, she went on to earn her high school diploma, discovering her own sense of purpose in emulation of the former president who "showed me that you set a goal and you do it day after day, in and out."

I T WAS NO secret around the Ford household that Betty had a more ribald sense of humor than her husband. When one of their granddaughters introduced her fiancé at a family dinner, she took some ribbing about the couple living together before marriage. This went on for a while, to the obvious discomfort of the prospective bride and groom, until Mrs. Ford shut down the conversation by declaring, "Well, you've got to try the shoe on before you buy it." It was in that same spirit of irreverence that the former First Lady in January 1998 returned from her weekly appointment at the beauty parlor in Rancho Mirage. Sketchy news reports of a sexual affair involving President Clinton and a twenty-one-year-old White House intern named Monica Lewinsky had caused a media feeding frenzy, with some observers questioning Clinton's ability to weather the storm.

"Did you hear?" asked Betty mischievously. "Al Gore is one orgasm away from the presidency."

The former president, though personally appalled by Clinton's Oval Office dalliance with Lewinsky, kept his silence for most of a year dominated by hairsplitting defenses and naked partisanship. Boosted by the salaciously detailed findings of independent counsel Kenneth W. Starr, Republicans on the House Judiciary Committee set a vote for Monday, October 5, on

---

* She didn't know the half of it. One day Susan, annoyed by her father's lecture on the perils of cigarettes, vowed to quit the habit the day he gave up pipe smoking. Accepting the challenge, Ford retrieved a box from his office and went through the house scooping up every pipe in the place for shipment to the Ford library. He quit tobacco as abruptly, and thoroughly, as he had sworn off alcohol.

whether to recommend to the full House a formal impeachment inquiry to examine charges of perjury—in effect, lying about sex—and obstruction of justice. This despite polls indicating strong opposition to removing a president whose most recent approval rating stood at 64 percent. The committee leadership was not happy, on the eve of its momentum-generating vote, to find a *New York Times* op-ed piece with Gerald Ford's byline. Entitled "The Path to Dignity," Ford's public intervention reflected what he had told friends seeking his opinion of the most serious White House scandal since Watergate: "I'm going to say the same thing I said when I forgave Nixon. We need to forgive Clinton. Sure he was wrong in what he did, but we need to get on and remember what's good for the country."

His *Times* article echoed these sentiments. "Whether or not President Clinton has broken any laws," Ford wrote, "he has broken faith with those who elected him." Under the circumstances "a simple apology is inadequate, and a fine would trivialize his misconduct by treating it as a mere question of monetary restitution." Presenting himself as a man of the House once called on to restore popular confidence in the nation's governing institutions, Ford invited his former colleagues to consider the long-term consequences "of removing this President from office based on the evidence at hand."

Instead of impeachment, with all its prolonged and divisive distraction from the nation's business—and, though he didn't say it, a foreordained acquittal by a Senate unable to muster the necessary two-thirds vote to convict—Ford called for an improvised ceremony in which the president appeared in the well of the House to accept a formal rebuke from the people's representatives. "No spinning, no semantics, no evasiveness or blaming others for his plight," Ford wrote of the famously slippery Clinton. "Let all this be done without partisan exploitation or mean-spiritedness. Let it be dignified, honest and, above all, cleansing. The result, I believe, would be the first moment of majesty in an otherwise squalid year."

The element of surprise magnified the impact of Ford's proposal, as well as the anger with which it was received by Clinton's would-be prosecutors. One member of the House Republican leadership wrote the former president a lengthy rebuke of his own, citing all the constitutional barriers to what Ford was suggesting. (The same congressman followed this up a short time later with a request for Ford's help in getting him into an exclusive country club for golf. Neither letter received an answer.) At the White House, special counsel Greg Craig welcomed the former president's initiative without committing his client to anything. Clinton's defense team instead sought to enlist Ford as its sole witness before the House

Judiciary Committee, a stunning proposition conveyed by former Democratic national chairman Bob Strauss. Ford told Strauss that he was asking too much. He had written the op-ed piece, and he informally committed to making phone calls "when the time comes."

Still not satisfied, after the House voted to impeach, and Ford and Jimmy Carter collaborated on a second appeal to lawmakers, White House counsel Charles Ruff asked Ford if he could do something more to forestall a Senate trial. By this time the Clinton camp, having defied the odds and gained seats in the November congressional elections, was feeling more confident than ever of prevailing in any Senate vote. Ford told Ruff that the president would have to acknowledge lying under oath about his relationship with Lewinsky. He repeated this to Clinton when the former president called him on December 30. Clinton said that was out of the question. Buoyed by his party's recent triumph at the polls, Clinton adopted the pre-election arguments of his Republican tormentors who—so long as they thought they had the upper hand—rejected Ford's proposed solution as unconstitutional. Simply put, Clinton argued, Congress had no right to impose such a penalty on him or any other chief executive.

"Bill," Ford reportedly answered, "I spent twenty-five years up there, and in my experience they can do pretty much anything they want."

The drama played out to its predictable conclusion, with the Senate falling well short of the sixty-seven votes required to undo the mandate of two presidential elections.

FAR FROM HARBORING any hard feelings, on August 15, 1999, Clinton presented Ford with the Presidential Medal of Freedom in an East Room ceremony twenty-five years after Ford's inauguration there. Two months later Clinton joined lawmakers of both parties as they awarded the Congressional Gold Medal to the Fords as a couple. The former president couldn't let the occasion pass without commenting on the nation's polarized politics. "Some people equate civility with weakness and compromise with surrender," he said. "I disagree." In July 2000, almost exactly sixty years after he and thousands of youthful insurgents flooded Philadelphia to demand that Republicans nominate Wendell Willkie, Ford returned to the city to applaud the George W. Bush–Dick Cheney ticket.

Disregarding a painful abscess on his tongue, Ford agreed to several press interviews in advance of a Tuesday-evening tribute by the full convention. They did not go well. Asked for his views on Iran, the former president told C-SPAN that he enjoyed roast beef and onions. Clearly something was amiss. Ford assumed his garbled speech was caused by the abscess. Only after the evening convention session concluded did he visit the emergency room of Philadelphia's Hahnemann Hospital complaining of pain in his face. He stayed just long enough for doctors to administer some routine pain medications. After a sleepless night Ford went back to the hospital, faint-headed and slurring his words, at Betty's insistence. This time a more thorough exam, including CAT scans, revealed the source of his persistent headache: Ford had suffered one, possibly two strokes, traced to the brain stem at the back of his head. He was admitted to the hospital, where he recovered fast enough to be discharged on August 9, the twenty-sixth anniversary of his East Room inauguration.

Soon he was back to his pre-stroke routine, swimming four laps before breakfast and as many again before dinner. He became an advocate for early stroke detection and treatment. Citing doctors' orders, he was initially reluctant to make the trip to Boston in May 2001 to receive the prestigious Profile in Courage Award from the John F. Kennedy Library. Betty helped change his mind, and the May 21 ceremony helped change his life. Acknowledging his contemporary opposition to the Nixon pardon, Ted Kennedy said "time has a way of clarifying past events, and now we can see that President Ford was right." For a quarter century, Ford said afterward, wherever he went people had asked him the same questions regarding his role in the pardon. Once the Kennedy Library gave its imprimatur to his decision, they stopped asking.*

Following the September 11 terrorist attacks in New York, at the Pentagon and in the skies over Western Pennsylvania, the Fords joined four other presidential families beneath the vaulted ceiling of Washington National Cathedral for a televised prayer service. Stirring fears of domestic terrorism, a series of anonymous letters laced with deadly anthrax spores were delivered to multiple media and congressional offices. Half a century after answering his first constituent letter, mail remained Ford's drug of choice. With postal deliveries severely curtailed, the president's secretary Shelli Archibald and her Secret Serviceman husband were forced to don the equivalent of hazmat suits before sorting and opening Ford's mail in a sealed-off garage. The former president wouldn't have it any other way.

Ford welcomed US leadership in toppling Afghanistan's Taliban regime that had sheltered Osama bin Laden and other instigators of the 9/11 attacks. He harbored doubts about the 2003 invasion of Iraq. "I can understand the theory of wanting to free people," he told Bob Woodward in a

---

* In a grim coincidence, Phil Buchen, a leading player in the drama of Nixon's pardon, died on the day Ford was in Boston accepting his award from Ted and Caroline Kennedy.

conversation embargoed until Ford's death. At the same time he was skeptical "whether you can detach that from the obligation number one, of what's
in our national interest . . . I just don't think we should go hellfire and dam-
nation around the globe freeing people, unless it is directly related to our
own national security." Privately he questioned the Bush Administration's
rationale for taking action against Saddam Hussein, who was widely believed to possess and be willing to use weapons of mass destruction against
his enemies. "He felt uncomfortable about it," Bush acknowledged in a
2010 interview, "particularly after we didn't find any. I felt uncomfortable
about it, too," but as president when "everybody's saying he's got weapons
of mass destruction" and he "chose to kick the inspectors out and not show
the weapons, I concluded, as did a lot of other people, that he was hiding
something."

Bush and Vice President Cheney were both "surprised" to learn subsequently of Ford's reservations, "because he never brought it up." The situation was especially awkward for Ford. A Republican Party loyalist and
longtime friend of the Bush family, he was even closer to Cheney and Defense Secretary Donald Rumsfeld, to name only the most prominent of his
protégés behind the invasion and occupation of Iraq.

On July 16, 2003, President and Mrs. Bush celebrated Ford's ninetieth
birthday with a party at the White House. Bush thanked his predecessor
for the "most interesting idea" of awarding Eugene McCarthy the Presidential Medal of Freedom. Their paths again crossed in June 2004 at funeral services for Ronald Reagan. Such ritualized farewells were inescapable
for a man in his tenth decade. 2005 marked seventy-five years since the
Thanksgiving Day football game that had cinched the state championship
for young Junie Ford and his South High teammates. When one of the few
surviving players wrote the former president seeking permission to disband
the 30–30 Club, which had convened every year since to commemorate the
big game, Ford voiced no objection.

S TEVE FORD HAD never cared for golf. Now, living closer to Rancho
Mirage than his siblings, and without their family responsibilities, he
fell in love with the game as something that he and his father could do
together. "There was a friend of his that owned a private golf course, and
we got to play on it," Steve recalled. "Every once in a while he'd catch his
drive pretty good, and it would go 175 yards. This was a ninety-two-year-
old man. And you could just see him kind of . . . bristle up like a big old
rooster." Proud and fastidious, the former president was reluctant to ask
for help in tying his necktie, or getting up from a chair. He shunned the

support of a cane as long as possible, and refused to use a wheelchair in public. Betty was equally resistant to the idea of live-in nursing care. "Let's try it part-time," the kids told her, initially from ten at night until ten in the morning.

No outsiders were present on the Sunday afternoon in April 2006 when President Bush dropped by 40365 Sand Dune Road. "We solved all the problems, didn't we?" Ford quipped at the end of their hour-long visit. In a handwritten follow-up, Bush promised to report back to Ford's many friends in DC "that you looked great and were strong of mind and spirit." In fact this was to be the former president's last public appearance. Soon after, he stopped going to the office, content to transact a dwindling work-load from home.

Strongly advised by his doctors to avoid the alpine setting of Beaver Creek (altitude 8,100 feet), Ford dug in his heels, insisting that nothing would deny him and Betty another Colorado summer. His pleasures there were simple and, for the most part, physically undemanding: filling the birdfeeder outside the big bay window where he read his newspapers; daily swims—even if Ford complained they had been reduced to mere wading; and, most enjoyable of all, savoring the view of the Beaver Creek ski moun-tain from his bedroom window.

For his ninety-third birthday on July 14, Jimmy and Rosalynn Carter sent love and best wishes. "One of my most prized possessions has been the close relationship you and I have shared during our own post-presidential years," wrote the thirty-ninth president to the thirty-eighth. That same month Ford was briefly admitted to a Vail hospital for shortness of breath. At a subsequent meeting attended by his doctors and family members, with the Mayo Clinic participating by telephone connection, Ford considered his medical options. These were necessarily limited. He could undergo heart bypass surgery to replace a bad valve and, if successful, enjoy an en-hanced quality of life for another year or two. The alternative was perhaps six months during which he could "look out the window and see the kids come by," as the doctors put it.

Ford, sitting in a wheelchair in the middle of the room, polled those around him. "Vaden, what do you think?" he asked Susan's husband. The same question was put to Susan and Mike, and Betty, and his doctor Jack Eck and cardiologist Larry Gaul. After everyone had said their piece, Ford sat silent for several minutes. "We all just stared straight ahead," Eck re-members. "And then I saw him rise up . . . and I thought, God he looks like an athlete. I mean, you could see determination."

"Let's go to Mayo."

In the event, surgery was ruled out because Ford could not tolerate the anesthetics. Doctors did perform an angioplasty to increase the blood flow in two coronary arteries. In addition they inserted a pacemaker to try to regulate his failing heart. As it happened, the only Ford to have surgery at Mayo was Betty, unexpectedly operated on for blood clots in her legs. Occupying separate rooms, the former president and First Lady kept visitors busy answering their questions about how the other was doing.

On August 29, the Fords left Mayo and returned to Rancho Mirage. For months Ford had declared his intention to attend the October 16 dedication in Ann Arbor of Weill Hall, the newly constructed home of the Gerald R. Ford School of Public Policy. "We raised $33 million in six months, because we were in a hurry to get it done," said Paul O'Neill, who cochaired the campaign. Unfortunately a fresh round of tests at the Eisenhower Medical Center prevented Ford from making the trip.* On departing the hospital, his Secret Service detail assumed that he wanted to go straight home. Think again. Pulling up at the nearest In-N-Out Burger, Ford got out of the vehicle, walked inside on his own power and stood in line for his burger.

On November 12 he surpassed Ronald Reagan's 93 years, 120 days, to become the nation's longest-lived president. A few days later he watched the Michigan-Ohio State game on television with his friend Leon Parma. Though confined to a hospital bed in his den, Ford offered a running commentary on the so-called Game of the Century, finally won by Ohio State 42–39. Ford's spirits were raised shortly after Thanksgiving when Don Rumsfeld appeared at his front door carrying an artist's renderings of the navy's newest class of nuclear aircraft carrier—the USS *Gerald R. Ford*. "Rummy!" Ford called out from the living room. He wasted no time in donning the gift hat bearing the ship's name.

The Ford children came in shifts now, hoping to relieve the strain on Betty of caring for her husband. Just before Christmas, Susan left to spend the holiday with her family. This left her three brothers in the den turned hospital room, where Mike read to his father from the Gospel of John and the Twenty-Third Psalm. Holding her hand tightly, the dying man made Jan Hart, Mrs. Ford's personal assistant, promise to stay on and look out for Betty in his absence. On Christmas Eve, Steve slipped out of the house to buy the last set of lights at a nearby drugstore, then strung them around

---

* "My brother Mike had to break the news to him," Susan Ford Bales told the student paper. "It wasn't pleasant, and I cannot repeat some of the things that were said." *Michigan Daily*, October 17, 2006.

the TV set near his father's bed to give the room a touch of holiday cheer. Ford smiled and murmured his thanks.

His family was convinced he willed himself not to die on Christmas. His voice silenced, Ford squeezed each boy's hand to acknowledge the presence of Mike, Jack and Steve. By Tuesday morning, the twenty-sixth, he no longer recognized Betty. An abbreviated winter's afternoon faded to dusk, and the white Christmas bulbs draping the olive tree beside the front door flickered to life in the early evening darkness when Jerry Ford died at six forty-five p.m., surrounded by his wife of fifty-eight years and their three sons.* Betty was to leave those lights in place so that Jerry, seeing them from above, would know she was okay. With his death she retreated from public view. "There's a time when you know you don't have to do everything anymore," she explained. Her time lasted until one night in July 2011 when her oft-expressed desire to rejoin "my boyfriend" was realized at last.

---

* Doctors listed the cause of death as arteriosclerotic cerebrovascular disease and diffuse arteriosclerosis, elaborate terms for what Ford's generation called hardening of the arteries.

"GOD HELP THE COUNTRY"

———❖———

The ultimate test of leadership is not the polls you take, but the risks you take. In the short run, some risks prove overwhelming. Political courage can be self-defeating. But the greatest defeat of all would be to live without courage, for that would hardly be living at all.

—GERALD R. FORD, MAY 20, 2001

FORD WANTED HIS Washington funeral more Trumanesque than Reaganesque. So he dispensed with the riderless horse and caisson-led procession to the Capitol with which presidents were traditionally memorialized.

It was well after dark on a drizzly Saturday night, December 30, when the hearse carrying his remains traversed Alexandria, Virginia, neighborhoods familiar to the young Congressman Ford and his family. Entering the District, the motorcade paused at the World War II Memorial, where a contingent of female alumnae of US military academies saluted Ford as the president who signed legislation opening those institutions to women. Ford had never been a television president, and many of the week's most poignant moments occurred off camera—Jimmy Carter pacing the aisles of *Air Force One* carrying Gerald Ford's six-month-old granddaughter, Joy Vance Berlanga, on his shoulder; the big plane flying low over Ann Arbor's football stadium and dipping its wings in tribute to Number 48.

The story of Secret Service agent Stacy Bauerschmidt likewise went unreported. Together with her Washington-based colleagues, Bauerschmidt had seen her protective responsibilities greatly expanded as part of a post-9/11 reorganization entrusting the agency with security planning for large-scale public events like presidential inaugurations and state funerals. For Stacy this meant giving up her holidays to work the Ford obsequies. When friends commiserated over her sacrifice, she brushed them off for reasons going back half a century.

Stacy hadn't been born when her parents early in their marriage faced eviction from the tiny trailer in West Michigan they called home. Her father, Leo Bauerschmidt, was a World War II veteran of the Army Air Corps dependent on his disability checks from the military. For reasons unknown, these were being held up. As the young couple's plight became desperate, Mr. Bauerschmidt finally called the local office of his congressman with a plea for help. Within days the bureaucratic logjam was broken, and Bauerschmidt's check was personally delivered to him by Congressman Gerald Ford. Coordinating security for Ford's funeral was a chance for Stacy to show appreciation.

No doubt Ford had her family and countless others like it in mind when someone asked him in 1981 how he would like to be remembered. "I hope people remember me as a person who was trying to make government work to the benefit of individuals," he replied. "I want them to think of me as an honest, forthright, hardworking friend who chose public service as a career." This was both his strength and his shortcoming, this president with a congressman's instincts, his deceptively modest public agenda attached to a list of Eagle Scout virtues.

More than forty years have passed since Ford left the White House. Today he is recalled chiefly for how he came to live there rather than for anything in particular he did while in residence, with one overriding exception—his pardon of Richard Nixon. It would take a quarter century and the virtual paralysis of American government caused by Bill Clinton's affair with a White House intern, and a fiercely partisan attempt by congressional Republicans to remove him from office, to put the Nixon pardon in a fresh light, and validate Ford's original rationale—his desire to refocus the nation's attention on more pressing matters of state. That is not, however, how it appeared at the time. Hoping to clear the national palate of Watergate's foul aftertaste, Ford instead fostered a view of himself as Nixon Lite, the stolid surrogate handpicked to be a steward of existing policies.

This is ironic, since it was memories of Nixon's graceless demand for wholesale staff resignations following his landslide reelection that prevented any similar housecleaning during Ford's first months in office. As an exercise in public policymaking, and public relations, that fall's White House Summit on Inflation was staged in deliberate contrast to the cloistered decision-making of the Nixon White House. Even Ford's rejection of an all-powerful chief of staff could be traced to his unhappy experience with Nixon's Prussian Guard.

Had he spent his first weeks in the Oval Office mentally inquiring *What Would Nixon Do?* it's a safe bet Ford would not have arranged an early meeting with the Congressional Black Caucus, replaced Al Haig with Don Rumsfeld or announced a Vietnam amnesty, even a conditional one. Nor, in all likelihood, would he have chosen Nelson Rockefeller as his vice president, notwithstanding Nixon's parting advice to do so, a year after he had himself spurned the patrician New Yorker whose name was synonymous with his party's fading Eastern Establishment. In downscaling the role of White House staff and practicing the closest thing in modern times to Cabinet government; in depoliticizing the selection process for the Supreme Court and fumigating a Justice Department in disrepute; in tearing up the White House enemies list and welcoming to the East Room some of his severest critics in Congress and the press, Ford might in fact be called the Un-Nixon.*

A second assumption about his presidency is that it was essentially reactive: the first half spent cleaning up inherited messes, the second warding off Ronald Reagan's conservative uprising. While this has more than a grain of truth, it significantly understates the range of issues competing for Ford's attention. Other chief executives before him had been tasked with maintaining public confidence in times of economic distress. Some, plagued by scandal, had left it to their successors to scour institutions muddied by corruption or administrative betrayal. Still others were driven to defend their constitutional franchise against legislators fearing the encroachment of presidential claims. Deprived of the legitimacy uniquely bestowed by the ballot box, Ford was to confront all these and other tests during his 895 days as president. Not to mention the challenges posed by a lost war in Indochina and the need to rebuild American credibility in its aftermath.

Like his somewhat improbable role model, Harry Truman, Ford was better at making decisions than communicating them to the public. Asked by a reporter late in life if he had regrets, Ford replied frankly, "I wish I were a better public speaker. I would have liked to be able to communicate more effectively."† His verbal deficiencies may have undermined public confidence in Ford's leadership abilities, yet faith in his personal integrity was

---

* Asked to name Ford's biggest accomplishment in the Oval Office, Bob Hartmann offered a characteristically blunt response. "Well, the President of the United States ceased to be a dirty word." JCI, Robert Hartmann, June 19, 1991.

† Even here Ford was capable of surprising those around him. David Gergen got a call one day from the former president's office, asking him if he would review a speech in draft. "It was beautifully written, but very complex; lots of three or four syllable words, and I thought, somebody has written an elegant speech, but it's not Ford. And what he wants me to do is to 'dumb it down,' to put it into simpler language." When Ford called to get his opinion, Gergen said it was a wonderful speech. "Really elegant, but it's not quite you. Did you want me to work on it or whatever? And I could hear him chuckling in the background" as the former White House speechwriter talked himself further and further out on a limb. "For a long, long time," Ford responded, "I've never had the space in my life to write my own speech. This is the first time I've done it . . . but I wanted you to read it." AI, David Gergen.

undiminished. If the polls were to be believed, Americans tended to view Jerry and Betty as the ideal next-door neighbors. This was not an unmixed blessing. Dissatisfaction with the imperial presidency did not manifest itself in a clamor for the return of Ozzie and Harriet, as the 1976 campaign demonstrated. A Harris survey of voters taken on election day showed 62 percent of them "alienated or disenchanted" with their government, more than double the number recorded ten years earlier.

But if we overstate Ford's success in healing wounds inflicted by a turbulent decade of assassinations, political scandal, official mendacity and an unwinnable war, we have overlooked his role in resetting the course of domestic and foreign policy. No one grasped this better than John Osborne, the *New Republic*'s resident oracle, who delivered a split decision on Ford's twenty-nine months in the White House. Interpreting the outcome of the 1976 election as a majority judgment "that goodness and decency in the Presidency are not enough," Osborne nevertheless concluded that where foreign policy was concerned, "there was not a single substantive problem area in which [succeeding administrations] could do more than follow or at best improve upon the course bequeathed to them by President Ford and Henry Kissinger."

Certainly nothing Ford did as president had more lasting impact than signing on to the Helsinki Accords, mistakenly embraced by Moscow as an affirmation of Soviet dominance in Eastern Europe, but seen in perspective as a trumpet blast for human rights and a critical milestone on the road to Western victory in the Cold War. The framework negotiated by Ford and Brezhnev at Vladivostok in November 1974 was to guide US arms control policy until Ronald Reagan and Mikhail Gorbachev made history in 1987 by eliminating an entire class of nuclear weaponry. Ford's 1976 U-turn on Africa committed the United States to the end of white minority rule in Rhodesia, and seeded the ground for the end of South African apartheid. In league with Kissinger, the ultimate Nixon holdover, Ford restored mutual trust to US alliances in Europe and Japan. His personal relationships with leaders as prickly as Canada's Trudeau and France's Giscard d'Estaing helped to strengthen NATO, while his measured military response to a brutal, though isolated, act of aggression on the Korean peninsula extracted something very close to an apology from the future Eternal General Secretary of North Korea Kim Il-Sung.

In the Middle East, Ford practiced some of the tough love he reserved at home for spendthrift New Yorkers. Neither instance was appreciated at the time. Both look good in retrospect. The Sinai II Agreement negotiated by Kissinger in 1975 broke a diplomatic stalemate, paving the way for Jimmy

Carter's historic peace initiative in 1978. (It was no accident that the first person Carter and Anwar Sadat called while flying back to Washington from Camp David was Gerald Ford.) Ford reversed the erosion of the defense budget, preserving from the budgetary ax the cruise missile, the B-1 bomber, and the AWACS (Airborne Warning and Control System) reconnaissance plane that would serve as NATO's "eyes in the sky" following Iraq's 1990 invasion of Kuwait and persistent border tensions involving Russia and Ukraine.*

Even Ford admirers conceded, in the words of one, "The President is not a visionary. But he took office when we didn't need to look into the future but assure ourselves we had one." Ford thought this a bum rap. "Ever since FDR, that word 'vision' has been equated by the media with new federal attempts to solve new problems," he argued. "The more costly these attempts, the more 'vision' their backers possess—and never mind that these 'visionaries' are spending money that doesn't belong to them." Ford's Cabinet secretary, James Connor, offered his own definition of the term. Vision, said Connor, entails "understanding trends in a society that are not immediately apparent and don't make much in the way of headlines."

In the Ford White House, this meant recognizing the long-term consequences of a public sector growing faster than the private economy that sustained it. No new program Ford might unveil could be as radical as his stated determination to undertake *no new programs* until the economy was righted and the deficit brought under control. That didn't make him hostile to innovation. The first piece of legislation to which he affixed his presidential signature rewrote federal housing policies in place since the late 1940s. Henceforth mayors weren't limited in their choices to seven spending programs designed in Washington. Regulations were slashed from 2,600 to 30 pages, review time for simplified applications cut from twenty-six months to an average of forty-five days. By entrusting local officials, with mandated public input, to allocate federal housing and community development funds, the new approach helped revitalize existing housing stock, while aiding the cause of neighborhood preservation.

Against a backdrop of runaway inflation and deficit spending, Ford proposed a rethink of the landmark Employment Act of 1946, which for thirty

---

* Another controversial weapons system was the Trident submarine, officially categorized as the Ohio class after Ford turned down an offer from the navy to name it for his home state of Michigan. It wasn't Ford's modesty so much as his political antennae that made him decline the honor. Ohio was a key swing state. It made no sense to offend untold numbers of Buckeye voters, especially if the alternative was to exalt their Wolverine enemy.

years had enshrined Keynesian theory as the password to "maximum employment." In its place, Ford and Alan Greenspan envisioned capital formation policies sufficient to generate ten million private-sector jobs by 1980. "The key to [the Ford Administration]," explained Bill Seidman, "was to go back to fundamentals. You deregulate the economy so you fight high inflation with competition. You reduce the government's take so the private sector can provide growth. And by doing that you can reduce interest rates." Ford could boast of reducing double-digit inflation below 5 percent. Unemployment was slower to respond to his policies, but it was at least headed in the right direction by the end of his term.*

Still more transformational was the change wrought by economic deregulation, which, along with the Helsinki Accords, stakes Ford's claim to being a significant president in his own right. For better *and* worse, he ushered in the entrepreneurial society with its Individual Retirement Accounts and cross-state banking; its competitive fares and heightened productivity; its lowered barriers to market entry and increased choice for consumers, who replaced government planners as the ultimate arbiters in an economy where there was less red tape and more R&D. Nor was it any surprise that, as a former president, Ford should champion NAFTA and other global trade deals.

Brief as it was, Ford's presidency incubated a remarkable group of future policymakers. Besides the most prominent—the elder George Bush, Baker, Rumsfeld, Cheney, Scowcroft, Hills, O'Neill, Gates, Gergen, Seidman and Zarb—are other, less instantly recognizable names—Stephen Hadley, Richard Parsons, Bob Hormats, Julia Taft, Bob Teeter, James Cannon and Russell Train, to name just a few, who would apply into the twenty-first-century lessons first learned in the Ford White House. Of the many condolence letters Betty Ford received at the time of her husband's death, that from UN ambassador Bill Scranton stood out. Written on behalf of "those of us who worked with him," Scranton recalled a boss "so down to earth, so sound, so open and, most of all in my book, a President who was always thinking of and working for the USA and not of and for himself."

Scranton might have been referring to Helsinki, or Ford's handling of the New York City fiscal crisis, or his Vietnam amnesty program. He may have been thinking of the administration's outreach to Black Africa at the height of the Republican primary season, or Ford's unpopular campaign to

---

* The 7.5 percent jobless rate in Ford's last month in office was to recur around election day 1980, when Jimmy Carter's presidency fell victim to the "misery index" with which Carter had arraigned Ford four years earlier.

resettle tens of thousands of Vietnamese refugees in a country eager to forget the humiliation of military defeat. Politics are grounded in expediency, but Ford as a rule paid little attention to the polls. In accepting the Profile in Courage Award from the JFK Library, the former president reminded his audience that "in the age old contest between popularity and principle, only those willing to lose for their convictions are deserving of posterity's approval."

Bryce Harlow thought Ford's quarter century on Capitol Hill made him "a splendid behind the scenes operator . . . very familiar with which buttons would need pushing in a particular situation." Yet Harlow hadn't forgotten Dwight Eisenhower's view of Congress as "the worst recruiting ground for presidential candidates." Legislators were taught to value the art of compromise, said Ike, no doubt a useful political skill, but incompatible with the decisiveness required of any executive. Bob Hartmann thought in some ways Ford never stopped being a congressman "in the sense that he was always seeking, if not a golden at least a doable mean. He was not planting a standard and rallying the troops around the standard. He was trying to find a place where he could get the most troops to come."

For all his admirable personal qualities, journalist David Broder said of Ford, "there was a reason, perhaps, why he had never been into the normal path to the presidency . . . He may have needed a larger ego and larger ambition-slash-vision in that office. You may have needed to be a little more ruthless . . . than I think he was by nature." Would another four years in the White House have added significantly to Ford's historical legacy? Probably not, according to Tom DeFrank. "I think he did what history will decide was important for him to have done. How do you top becoming president after the first president in history resigns under pain of impeachment, and bringing the country through that, getting it back on an even keel?"*

Post-bicentennial America was entering a more ideological phase than her problem-solving president was accustomed to. "Maybe Ford was the last true pragmatist—someone who concentrated on making the political system work," suggested David Mathews, one of the more thoughtful of his Cabinet appointees. A classic Eisenhower Republican, fiscally conser-

---

* DeFrank covered Ford longer and more perceptively than just about anyone, but here, I think, he underestimates the likelihood of a popularly elected Ford achieving treaties on arms control, the Panama Canal and normalization of US-China relations, not to mention a lessening of tensions in the Middle East—in short, the ambitious foreign policy agenda pursued by the Carter Administration.

vative and socially moderate to liberal, Ford was virtually marooned in his later years, when he and Betty offered shelter to the dwindling band of like-minded Republicans who supported the Equal Rights Amendment, freedom of choice for women exercising their reproductive rights and the freedom to marry for same-sex couples.

We can only speculate as to Ford's feelings about our contemporary scorched-earth politics, in which consensus yields to conspiracy, Facebook masquerades as a news source, and demagoguery is delivered round the clock in 140-character doses. But he left a clue in an unusually thoughtful and far-ranging interview he gave Neil MacNeil of *Time* in October 1980. After setting out the necessary experience and qualifications for future presidents, Ford was asked to identify any qualities that were "especially disabling" for the job. Weighing his words carefully, he told MacNeil, "The one that came to mind was arrogance. That's a terrible characteristic for a President. Fortunately an arrogant person will most unlikely ever be elected, but if you had an arrogant President—and I mean in a vicious way—God help the country."

I BEGAN THIS book by describing George McGovern's White House visit as Ford's guest in the spring of 1975. It seems only fitting to conclude with another act of official hospitality designed to compensate for past exclusions.

On August 5, 1976, a year after he approved Title IX regulations providing equal access to the playing field for women as well as men, the sports-loving president invited the US Olympic team to a Rose Garden celebration of its impressive showing at the recently concluded Montreal games. To lend additional luster to the occasion, Ford asked an earlier Olympian, the legendary track-and-field star Jesse Owens, to accept the recognition denied him after his unprecedented feat of winning four gold medals at the Berlin Olympics of 1936. Returning home from Hitler's games, Owens had been accorded a ticker tape parade through the canyons of Manhattan. But at the Waldorf-Astoria hotel, where a dinner in his honor was being staged, Owens was forced to use the service elevator, the sole means of access for African-Americans like himself. Moreover, the Olympic sensation who had single-handedly exploded Nazi theories of racial supremacy was pointedly left off the White House invitation extended to his white teammates by President Franklin D. Roosevelt.

Forty years on, Ford hoped to rectify this injustice by presenting Owens the Presidential Medal of Freedom, the highest civilian honor his country could bestow. For Ford the occasion represented much more than unfinished business. This was personal. Setting aside the long-standing rivalry between their schools, the Wolverine jock paid homage to his contemporary from Ohio State. "I saw Jesse Owens at a Big Ten track meet in Ann Arbor," Ford reminisced about the jaw-dropping day in May 1935, when Owens broke three records and tied a fourth.

In accepting his medal from the president, Owens reciprocated by recalling their mutual friend Willis Ward, the college football phenom benched, to Jerry Ford's dismay, at the insistence of a segregationist rival school. Owens didn't mention how he, the celebrated Olympian, had struggled to support his family working as a gas station attendant and playground janitor until Willis Ward, his competitor on the track, offered him employment as an assistant personnel director at the Ford Motor Company. And now he was shaking hands with the president of the United States.

Less than three years had passed since Spiro Agnew's replacement as vice president introduced himself to the country as a Ford, not a Lincoln. Few Americans at the time would have disputed his modest self-appraisal. On this muggy August afternoon in 1976, however, Ford was at least Lincolnesque. For what else was his belated recognition of Jesse Owens but an appeal to the better angels of our nature?*

---

* It bears noting that Jesse Owens was singled out that day as the first in a distinguished company of Medal of Freedom recipients whose contributions Ford would recognize before he left office. The list included Alexander Calder, Georgia O'Keeffe, Norman Rockwell, Lowell Thomas, General Omar Bradley, Irving Berlin, Martha Graham and historians Will and Ariel Durant.

ACKNOWLEDGMENTS

———◆———

IN THE SUMMER of 1975, fresh out of college, I joined the latest class of summer interns to work at the Ford White House, an experience I was to describe for the *Washington Post's Potomac* magazine in an act of postgraduate smart-assery that would surely have engendered a lifetime ban from any other presidential alumni group. That it didn't prevent me from serving nearly six years as director of the Gerald R. Ford Presidential Library & Museum in Ann Arbor and Grand Rapids, respectively, is evidence of Ford's aversion to grudge holding. In addition, the former president invited me to work with him on a number of speeches and articles in the last decade of his life. In January 2007, at his request, I delivered a final eulogy at his Grand Rapids funeral (an honor extended to me a second time in July 2011, when Mrs. Ford was granted her wish to be reunited with her husband of fifty-eight years).

Such personal involvement made me think twice when the subject of a Ford biography came up. To be sure, other presidential biographers have maintained objectivity while gaining valuable insight through prior knowledge of their subjects; Jon Meacham and Doris Kearns Goodwin come readily to mind. August Heckscher's status as president of the Woodrow Wilson Foundation did nothing to lessen the scholarly detachment that made his 1991 life of Wilson a model of its kind. Precisely because I knew the Fords as I did, I felt confident they shared the sentiments expressed to me by David Kennerly, who knew them far better. As he put it, "If you're not critical, you have no credibility."

Any lingering doubts I might have had were effectively dispelled by David McCullough over a memorable lunch just across Lafayette Park from what Gerald Ford insisted on calling "the residence." "Write what you know," said David, thereby inaugurating an eight-year process of discovering just how much I didn't know. Throughout I drew inspiration from the example of my friend Jim Cannon, himself a Ford intimate, whose 1994 biography *Time and Chance* brilliantly re-created events leading to August 9, 1974. Indeed, it was Cannon's publishing experience that did as much as anything to prompt the present volume. As Jim described it to me, his original editors were chiefly interested in seeing Ford through the prism of Watergate and the Nixon pardon that followed. By squeezing his entire postpardon presidency—twenty-eight out of twenty-nine months, to be precise—into a single chapter, they inadvertently reduced Ford to a supporting player in his own story. And they planted the seeds of *An Honorable Life*, a biographical sequel Jim hoped would compensate for what was left out of his earlier account.

Besides Jim, I owe a debt of gratitude to many other Ford scholars who have enlarged our knowledge and understanding of the thirty-eighth president. Heading the list: John Robert Greene, Tom DeFrank, Douglas Brinkley, Bud Vestal, Jerald terHorst, Bob Hartmann, Barry Werth, Yanek Mieczkowski, Hendrik Booraem, Andrew Downer Crain, Mark Rozell, Scott Kauffman and Edward and Frederick Schapsmeier.

The work you hold in your hands was made possible by the generosity and encouragement of Hank Meijer, esteemed biographer of Arthur Vandenberg, and a cherished friend of almost thirty years. Hank earmarked part of a substantial gift to the Gerald R. Ford Foundation to underwrite the production of a new and comprehensive Ford biography. Recognizing that any such work must be wholly independent of the foundation, Hank routed funding for the project through Grand Valley State University in Grand Rapids and Allendale, Michigan. I am deeply grateful to Scott Blinkhorn and his associates at GVSU for taking on this responsibility. The same holds true for Hank and his fellow manuscript readers, Martin J. Allen and David Frye—friends of the Ford family whose commitment to objective scholarship made them at once intellectual overseers as well as literary cheerleaders.

Gleaves Whitney, the Ford Foundation's current executive director, showed similar enthusiasm for the project, as did Ford library and museum director Brooke Clement, and her immediate predecessor, Elaine Didier. I may be biased, but based on personal experience, I regard the archival staff at Ann Arbor as the best in the business. Especially helpful in guiding me through the library's voluminous holdings were Geir Gunderson, John O'Connell and Stacy Davis. Just as Stacy uncovered many a needle/footnote in the mountainous haystack that is a modern presidential library, so her colleague Elizabeth Druga employed her detective skills in uncovering illustrations, many published here for the first time. Brenda Thacker of the St. Louis Public Library tracked down the scandalous death notice of George R. Ford, with accompanying exposé of Ford's bigamist relationship with Emma Tutton, as reported by the *St. Louis Post-Dispatch*.

Don Holloway, much more than a museum curator during and after my years as Ford Library director, shared his own original research, not to mention a quarter century's worth of deeply informed hunches, observations, and more than a few provocative theories. Former library director Frank Mackaman read every page of the manuscript, improving many through questions of fact, interpretation or style. It was Frank who alerted me to an untapped gold mine in the papers of Neil MacNeil, Capitol Hill correspondent for *Time* from 1958 until his retirement in 1987. It would be hard to find a more trustworthy source of Washington reporting, analysis and irresistible gossip as compiled by one of his generation's premier journalists.

The helpful staff of the Harry Ransom Center at the University of Texas in Austin made for a successful foray into the Woodward/Bernstein Watergate collection housed there. In a class by himself, David Kennerly supplied numerous images from his personal archive. He was equally generous with his memories, being especially helpful in re-creating the tragic final days of Saigon. David shared an early draft of the manuscript with Ken Quinn, then a conscientious member of the National Security Council, more recently president emeritus of the Iowa-based World Food Prize. Ambassador Quinn in turn was kind enough to recount the harrowing month of April 1975 as he experienced it, when the United States lost a war but not, thanks to himself and other bighearted diplomats, its moral compass.

Next to Hank Meijer, my greatest debt is to over 170 individuals who spoke with me as part of a multiyear Ford Oral History Project carried out under the auspices of the Gerald R. Ford Foundation. (Their transcripts can be found on the foundation's website.) I am grateful to each of them, and saddened by how many did not live to see their contributions reflected in these pages. Special appreciation is owed Lawrence Meyer and Jack Hushen, who disclosed for the first time their part in events leading to Spiro Agnew's resignation and the defining early weeks of the Ford presidency. Brian Lamb read the evolving narrative and spoke candidly about his own involvement with an ad hoc transition group so secretive its existence was hidden from the vice president whose future it was charting. Laurence Lynn Jr., the group's other surviving member, kindly answered my queries by e-mail. Derek Leebaert, James Strock, Steve Chapman, Mark Updegrove, Ed Cheffy, Jonathan Martin and John Garnett read all or parts of the manuscript, which benefited greatly from their comments. Needless to say, the responsibility for any remaining errors is mine alone.

Gordon Olson's encyclopedic knowledge of Grand Rapids history was tapped for my early chapters. A former city historian, Gordon was also project manager and editor of *Flight to Freedom: The Story of the Vietnamese of West Michigan*. I am indebted to him and to Barbara McGregor Packer for providing me a copy of the book and its companion CD, *From Saigon to Sanctuary*, produced by Huan Le and Frank Jamison. Katherine Smith Kennedy and Melissa Rabidoux gave me a memorable tour of Frank McKay's old haunts in the Grand Rapids tower that bears his name. The Honorable David M. Murkowski, chief probate judge for Kent County, helped to clarify a critical incident in Ford's early political career. Erik Nelson organized and videotaped nearly all the Ford oral history project interviews. He also sifted through box after box of research materials essential to the footnoting process. Linda Kay Provo has typed and retyped books for me since 2001. Her professionalism is on a level with her patience, higher than the cornflower-blue skies over Lawrence, Kansas. Gail Ross excelled as agent and advocate. Roger Labrie worked his editorial magic on what must have appeared at first glance a very long book about a very short presidency. In league with Jonathan Jao and David Howe at HarperCollins, his thoughtful editing brought into sharper focus Ford's personal and political development. The subsequent copyediting of the manuscript by Janet Robbins Rosenberg was both meticulous and (in the best sense) unsparing.

This is the fifth and final volume in a series crafted over forty years, beginning with the 1982 publication of *Thomas E. Dewey and His Times*. Subsequent biographies of Herbert Hoover, Colonel Robert McCormick of *Chicago Tribune* fame, Nelson Rockefeller and now Gerald Ford collectively trace the evolution of American politics, and the Republican Party in particular, from Theodore Roosevelt to George W. Bush. Along the way I have had the immense good fortune to work with, and learn from, legendary editors, including Alice Mayhew, Robert Loomis, and Christopher Carduff, as well as the above mentioned team of Jao and Labrie.

The name on my dedication page is, formally, a stranger to me. And yet I feel as if I know Billy Sipple, like so many others who have shared in his struggle for dignity and self-respect. I only wish he could have lived to see how much history he was part of, and not just on September 22, 1975.

NOTES

⬥

THIS BOOK IS based primarily on the vast documentary riches of the Gerald R. Ford Library at the University of Michigan in Ann Arbor. Researching and writing Ford's life has only increased my regard for the dedication and sheer persistence of anyone who sets out to describe in detail a modern presidency. I say this as someone who spent seventeen years as director of four presidential libraries (Hoover, Eisenhower, Reagan and Ford) operated by the National Archives and Records Administration, as well as the Abraham Lincoln Presidential Library and Museum sponsored by the state of Illinois. Complicating my task, the Ford Library was for all practical purposes closed to researchers for nearly two years due to the COVID-19 pandemic. One practical consequence of this was to preclude my reviewing the last eleven of sixty-six massive scrapbooks, the first assembled a century ago by Dorothy Gardner Ford as part of a series documenting her son's life from his earliest high school football heroics until his final days in the California desert. This deficiency was more than compensated for by what I extracted from the library's holdings long before COVID made itself felt: the first fifty-five scrapbook volumes, plus thousands of pages of letters, memoranda, diary entries, telephone and other private conversations, National Security Council and Cabinet minutes, newspaper clippings and articles from journals both popular and scholarly. An immensely useful virtual archive was accessed through the library's voluminous digitized holdings. The National Security Archive was equally helpful, not just for its unparalleled collection of declassified documents but also for the interpretive context within which they are presented.

As useful as the written word are the recorded memories—over 250 interviews in all, the majority conducted by the author under the auspices of the Gerald R. Foundation during the years 2008–11. Of special significance are some four thousand pages of interviews Ford compiled with Trevor Armbrister, his collaborator/ghostwriter on the former president's White House memoir, *A Time to Heal*. Though they were closed to researchers since they were originally deposited at the Ford Library, I was nevertheless able to see anything I requested, as will be apparent to anyone who examines the endnotes. Ford tended to pull his punches in print; for example, agreeing with Armbrister's characterization that Richard Nixon's conduct toward his vice president amounted to a "betrayal," while ruling out the use of such a loaded word in print. His more candid recollections, coupled with those of dozens of his associates, which have, until now, been embargoed by the library, will hopefully contribute to a fuller understanding of the thirty-eighth president.

Abbreviations of Sources

AI	Author Interview
APP	American Presidency Project
DGDP	Dorothy Gardner Divorce Papers
DRA	Donald Rumsfeld Archive
FCN	Ford Congressional Newsletter
FCP	Ford Congressional Papers
FD	Ford Diary (more precisely, his desk calendars with handwritten comments)
FLOHP	Ford Library Oral History Project
FRUS	Foreign Relations of the United States, US State Department
GFNR	Gerald Ford Naval Records
GRFL	Gerald R. Ford Library
GROHP	Grand Rapids Oral History Project
HUC	Hofstra University Conference
JCI	James Cannon Interview
NSA	National Security Archive
NMP	Neil MacNeil Papers (Dirksen Congressional Center)
NOHP	Nixon Oral History Project
PP	Presidential Papers
PSF	Press Secretary and Speech File
ROD	Roberts Oral Diary (maintained by Bill Roberts, Ford's assistant press secretary, July 24–September 12, 1974).
TAFI	Trevor Armbrister—Ford Interview
TAI	Trevor Armbrister Interview
TML	*Times of My Life*
TTH	*A Time to Heal*
WBP	Woodward-Bernstein Watergate Collection, Harry Ransom Center, University of Texas (Austin)

Introduction: A Capacity for Surprise

1   "Ford was hidden": Bob Woodward interview with Robert Hartmann, December 10, 1997, WBP, Harry Ransom Center, University of Texas.

1   "Lyndon Johnson never invited me to dinner": AI, George McGovern.

1   "the house belongs to everyone": Memo, Bob Barrett to Ron Nessen, January 8, 1976, Folder: Bob Barrett, Box 126, Ron Nessen Papers, GRFL.

1   "Nixon really fucked up the pea patch": AI, Peter Secchia.

3   "That's the difference between a peacock": Letter from Ted Morgan to author.

3   "the man who saved the country from Nixon": John Robert Greene, "'A Nice Person Who Worked at the Job': The Dilemma of the Ford Image," *Contributions in Political Science* 300 (1993): 645.

3   "Nobody leaves the plane without a parachute": FLOHP, Paul Theis, September 1, 1998.

3   "My whole philosophy of life is": TAFI, 3572.

4   "Do you know who would make a great Supreme Court justice?": AI, Ken Lazarus.

7   "I think he ought to tell the American people": AI, James Baker.

7   "I'm never quite who I want to be": AI, Don Penny.

7   "Forgive me, Mr. President": AI, Dick Cheney.

Prologue: Seven Days in June

9   "I was standing this far away": TAFI, 325; AI, Vern Ehlers.

9   "I wish I could do this more often": AI, Leon Parma.

10   "She holds down the fort": Myra MacPherson, "Gerald Ford: The Crusader," *Washington Post*, July 5, 1970.

10   "*Don't*," Ford advises potential candidates: "Do's and Don'ts for a young man going into politics," GRF handwritten reflections. Scanned from collection Gerald R. Ford: Materials from the writing of *A Time to Heal*, GRFL.

10   no one knew how she was going to get back to Michigan: *Grand Rapids Herald*, May 5, 1957.

10   He duly complied: The Presidents UFO Website, Gerald Ford UFO Talk, March 25, 1966, FCP, Press secretary and speech file (hereafter PSF), Box D37.

11   "The strongest weapon in a political campaign": "How You Can Enjoy Politics," Ford speech to Junior Chamber of Commerce, undated, 1951; FCP, Box 14, PSF, GRFL.

11   "I made everyone else's problems my problems": AI, Greg Ford.

11   "I keep reading that I'm a plodder": "The 38th President: Growing into Power," *Grand Rapids Press*, September 13, 1981. Nothing contributed more to Ford's image problems than his droning speeches and lack of platform command. He never used a teleprompter, "for one particular reason. I'm nearsighted, and if I'm going to be certain of following the teleprompter, I have to wear glasses. And I don't like to wear glasses. The other option was to go to contact lenses, and they got them for me, but I always had trouble getting the damn things in and out, so I said forget it . . . I was too old to learn." TAFI, 1918.

11   "a kind of young Eisenhower": TAI, Paul Miltich.

12   His executive secretary, Mildred Leonard: *Grand Rapids Press*, November 28, 1965.

12   "You have to be nice": AI, Anne Holkeboer.

12   "98% teddy bear and 2% grizzly:" AI, Bob Barrett.

12   "But I'll tell you what": AI, Steve Ford.

13   "Jerry really thinks it's a great idea": AI, Vern Ehlers.

13   "Well, Vern, you've got to recognize one thing": Ibid.

14   "I did not know what a Calder was": Curt Wozniak, "Hometown Values Shared with a Nation," *Grand Rapids Magazine*, March 2007, 18.

16   "It is so important": George H. W. Bush, *All the Best, George Bush: My Life in Letters and Other Writings* (New York: Scribner, 2013), 97–98.

16   "Mother is merely a shoplifter": *Congressional Record*, June 19, 1972, 21410.

17   "Jerry, I know this fellow McCord": AI, Jack Marsh.

17   "Dick Nixon is much too smart a politician": James Cannon, *Gerald R. Ford: An Honorable Life* (Ann Arbor: University of Michigan Press, 2013), 99.

18   "Maybe you ought to have a little fire": Jeb Stuart Magruder, *An American Life: One Man's Road to Watergate* (New York: Atheneum, 1974), 247. It bears noting that Magruder's account differs significantly from his earlier testimony before the Senate Watergate committee and what he told DC prosecutors. James Rosen, John Mitchell's sympathetic biographer, questions Magruder's veracity on this and much else.

18 "It is not far from the truth": Woodrow Wilson, *Congressional Government* (Baltimore: Johns Hopkins University Press, 1981), 69.

19 "He never cajoles": *Baltimore Sun*, September 24, 1974.

19 "There are three parties in the House": Bart Barnes, "Former Iowa Rep. H.R. Gross Dies at 88," *Washington Post*, September 24, 1987.

19 "I'll be with you sled's length": TAI, Jack Marsh.

19 Good Old Jerry can play hardball: AI, Brian Lamb.

21 "I said sure, I'd see what I could do": Bob Hartmann, *Palace Politics: An Insider's Account of the Ford Years* (New York: McGraw-Hill, 1980), 56. Ford's calendar shows him meeting with Liddy and two representatives of newly elected congressman Hamilton Fish at eleven a.m. on December 18, 1968.

21 He had intended to write earlier: G. Gordon Liddy to GRF, May 28, 1969, FCP, Box A 189, Folder: General information- L (Folder 48).

22 "I need three": A. James Reichley interview of William Timmons, November 29, 1977, Nixon White House interviews, Box 1, GRFL.

22 "Keep your feet shuffling all the time": John Hersey, *The President: A Minute-by-Minute Account of a Week in the Life of Gerald Ford* (New York: Knopf Doubleday, 2019), 126.

23 "You have to have some people way out in front": TAFI, 1760.

23 "the most decent man I ever knew": Paul N. McCloskey to GRF, February 2, 2005, GRFL.

25 "He's nuts!": Bob Woodward, "Nixon and Ford Discuss a Democrat's Attack on FBI Director Hoover," *Washington Post*, December 29, 2006.

25 "their investigation is now leading into some productive areas": Transcript of a Recording of a Meeting Between the President and H. R. Haldeman in the Oval Office on June 23, 1972, 10:04–11:39 a.m., https://www.nixonlibrary.gov/sites/default/files/forresearchers/find/tapes/watergate/trial/exhibit_01.pdf.

1: Secrets

29 "So few people knew": JCI, Gerald Ford, November 13, 1990, GRFL.

30 "The Dutch had a thing": AI, Maury DeJonge. An excellent source for early Grand Rapids history is *The Story of Grand Rapids*, ed. Z. Z. Lydens (Grand Rapids, MI: Kregel Publications, 1966), 1–47. The same volume treats the great water scandal of 1900–04, 56–68. Also useful is Gordon Olson's *A Grand Rapids Sampler* (Grand Rapids Historical Commission, 1992) and Norma Lewis's *Grand Rapids: Furniture City* (Mount Pleasant, SC: Arcadia Publishing, 2008).

33 "a strong, self-made and very inner-directed man": Bud Vestal, *Jerry Ford, Up Close: An Investigative Biography* (New York: Coward, McCann & Geoghegan, 1974), 44.

34 Dorothy was introduced to Leslie Lynch King: "President's Mother Schooled in Knoxville," *Galesburg Register-Mail*, May 20, 2006.

34 "I don't want to speak to her": "Mrs. King Tells Story of Abuse by her Husband" (undated), GRFL Vertical File, Folder: Dorothy Gardner, 1892–1967. Hereafter referred to as DGDP.

35 During the night she woke: "King Divorce Case in Court," *Omaha Evening Bee*, December 1, 1913.

35 "it didn't work out": "President Ford's Parents Wed in Harvard Church," *Sunday Courier-News*, Elgin, Illinois, November 28, 1976.

35 It is altogether possible: "President Ford's Early Life in Omaha Clouded by Parents' Stormy Marriage," *Omaha World-Herald*, August 16, 1974.

36 "Bye, bye," uttered at ten months: Baby Book of Leslie King, Junior, May 9,

1914, "Cunning Things Done by Baby," GRFL Vertical File, Folder: Ford, Gerald Rudolph—Biography: Early Years (1), GRFL.

37 "He was quite a man": Lloyd Shearer, "President Ford's 'Other' Family," *Parade*, September 9, 1974.

38 Junior one day tossed his grandfather's shoes: Baby Book, op. cit.

38 "If I inherited anything": JCI, Gerald Ford, November 13, 1990, GRFL.

38 "A very headstrong little boy": Terence Sheridan, "Portrait of the Next President as a Young Man," *New Times*, June 14, 1974.

38 "She gave me unshirted hell": JCI, Gerald Ford, November 13, 1990, GRFL.

38 "I tend to get more angry at things than people": JCI, Gerald Ford, April 26, 1990 (part 1), GRFL.

40 "With my present salary": Leslie King to A. C. Pancoast, February 8, 1921, GRFL Vertical File, Folder: Ford, Gerald Rudolph—Genealogy—King, Leslie Lynch.

41 "Left handed when I've been sitting down": TTH, 45.

42 "She was a doer": JCI, Gerald Ford, April 25, 1990, GRFL.

42 "She was strong physically": JCI, Betty Ford, July 20, 1990, GRFL.

42 "Mother had a lot of guts": JCI, Gerald Ford, November 13, 1990, GRFL.

43 "Some of my friends, especially in the political arena, believe I overdid it": "Revenge or Forgiveness—Have I Attempted Revenge or Extended Forgiveness," scanned from collection Gerald R. Ford: materials from the writings of *A Time to Heal*, GRFL.

43 "like an old-fashioned Christmas card come true": Vestal, *Jerry Ford, Up Close*, 49.

44 an "abnormal": TAFI, 273.

44 "very stern, stoic woman": AI, Linda Burba.

44 "a very busy woman . . . very direct": AI, Susan Ford.

44 "the most selfless person I have ever known": TTH, 44.

44 "character and integrity were the most important things in the whole world": JCI, Richard Ford, July 2, 1990, GRFL.

44 "There was always politics in the house": JCI, Janet Ford, February 27, 1991, GRFL.

44 "He went in with a true salesman's philosophy": AI, Richard Ford.

45 "an unnecessary commodity": TAFI, 289.

45 "Jerry is not demonstrative": JCI, Janet Ford, February 27, 1991.

2: Luck and Pluck

46 "that's where he will learn more about living": JCI, Gerald Ford, April 24, 1990, GRFL.

46 "You're like me in a way": AI, Don Penny.

47 "I never felt jealous": JCI, Gerald Ford, April 26, 1990 (part 2), GRFL.

47 "quiet and attentive, yet always prepared": "She Taught the President," *Logansport Indiana Paras Tribune*, undated, GRFL. Another Ford teacher remembered him as "not a real outgoing type"—always cooperative and popular because of football, yet "pensive at times." "Resident Recalls Ford . . . ," *Petoskey News Review*, August 16, 1974.

47 "Betty says that I can't even listen in tune": TAFI, 2435.

49 "Hey, Whitey," Ford quoted Gettings, "you're a center": Hendrik Booraem V, *Young Jerry Ford: Athlete and Citizen* (Grand Rapids, MI: William B. Eerdmans 2013), 88.

49 "He couldn't stand to lose": AI, Jim Trimpe.

50 "I would give three of you Hollanders": "Portrait of the Next President," *New Times*, June 14, 1974.

50 "He doesn't like to think of himself as a loser": Art Brown interview, Box E41, Campaign '76 Office: General Election, John Deardourff Files, GRFL.

50 "It made no difference": GROHP, Art Brown, January 26, 1980, 4.

51 "we'd go down and drain the shellac drum": JCI, Tom Ford, February 27, 1991, GRFL.

52 "Are you Leslie King?": Booraem, *Young Jerry Ford*, 16.

53 "Apparently he didn't have a good watch as a kid": AI, Tom DeFrank.

53 "I don't want to say Dad hid it": AI, Susan Ford. Ford himself told James Cannon that while he knew Gerald Ford Senior was not his real father, "99 and 9 tenths of the people in Grand Rapids thought I was Jerry Ford's child."

54 "Here he was": Cliff Gettings interview, Box E41, Campaign '76 Office: General Election, John Deardourff Files, 13, GRFL.

55 "You can go up to your knees": AI, Maury DeJonge.

55 "I don't dance—but I can neck": Hendrik Booraem notes (undated).

56 "Just like that Ford had a bunch of us": GROHP, Art Brown, January 26, 1980, 15; Hendrik Booraem notes (undated).

56 "It's the only time he ever stuffed the ballot box": JCI, Tom Ford, February 27, 1991, GRFL.

57 "Everybody is after this boy Ford": Paul Goebel to Harry Kipke, May 4, 1931, Board in Control of Inter-Collegiate Athletics, Box 16, Bentley Historical Library, University of Michigan.

57 "I've always wanted to be a student at the University": GRF to Harry Kipke, June 3, 1931, GRFL.

58 "I'm Jerry Ford. Aren't you Willis Ward from Detroit?": David Pollock, "Jerry Ford at Michigan," *Independent Times*, September 2006.

58 "We *always* kicked off": Gerald Ford with John Underwood, "In Defense of the Competitive Urge," *Sports Illustrated*, July 8, 1974.

58 "the ideal player for any coach to work with": "Kipke Counts Heavily Upon Ford as M Center This Fall," *Grand Rapids Press* (undated, fall 1934).

58 "Have a fine time": GRF to Dorothy Ford, May 14, 1932, Thomas G. Ford Papers, 1900–1980, Box 1, Folder: Gerald R. Ford Jr. Correspondence, GRFL.

60 "You should have seen the campus this week-end": GRF to Frederica Pantlind, October 5, 1933, Frederica Pantlind Papers, Box 1, GRFL.

60 "The Feds are making the rounds": Hal Flynn to GRF, June 30, 2006, as shared with Elaine Didier, GRFL.

61 "From now until two weeks from now": GRF to Frederica Pantlind, January 25, 1934, Frederica Pantlind Papers, Box 1, GRFL.

61 "Saw all of the fair in 25 minutes": GRF to Frederica Pantlind, November 2, 1933, Frederica Pantlind Papers, Box 1, GRFL.

61 "I wanted to play": Ford revealed perhaps more than he intended when he described for Jim Cannon the frustrations of being Chuck Bernard's backup. "I learned a lot sitting on the bench," Ford said. "I learned that there was the potential always that somebody could be better than you." James Cannon, *Of Time and Chance* (New York: HarperCollins, 1994), 19.

61 "He was one guy who would stay": Geoff Larcom, *Ann Arbor News*, December 27, 2006.

62 "I will tell all of the conference teams": "Jerry Ford at Michigan," *Independent Times*, September 2004.

62 "The administration wasn't eager to have Blacks": Vestal, *Jerry Ford, Up Close*, 60.

62 "I never dreamed there would be so much agitation": Fielding Yost to W. A. Alexander, December 5, 1934, Board in Control of Inter-Collegiate Athletics, Box 20, Bentley Historical Library, University of Michigan.

63 "He was color blind": AI, David Kennerly.

63 "Where's your ni**er player?": Vestal, *Jerry Ford, Up Close*, 60. Vestal substituted "black" for the N word, unlike Hendrik Booraem V in *The Education of Gerald Ford* (Grand Rapids, MI: William B. Eerdmans, 2016). Ford's subsequent comment about the vicious hit on Georgia Tech's Charlie Prescott—"We did that for you!"—is on page 172 of Tyron Kai Steward's 2013 Ohio State dissertation, "In the Shadow of Jim Crow: The Benching and Betrayal of Willis Ward."

63 "That Georgia Tech game killed me": Jon Zemke, "Black Athlete Not Allowed to Compete Against Georgia Tech," *Michigan Daily*, April 9, 1999.

64 "Son, do you know what respect is?": Brian Kruger and Buddy Moorehouse, *Black and Blue: The Story of Gerald Ford, Willis Ward and the 1934 Michigan-Georgia Tech Game*, Stunt3 Multimedia, 2011.

66 "probably . . . the best game I ever played in college": Within days of becoming president, Ford received a letter from Rip Whalen, in which he recounted the true story of the Northwestern–Michigan game that led to Ford's being asked to play in the East-West Shrine game. According to Whalen, his coach Dick Hanley grabbed him by his jersey at halftime. "You Irish *SOB*, you ought to pay that *G.D.* Jerry Ford bus fare. He's been driving you all over the field." Edward "Rip" Whalen to GRF, Box 3C, Folder: Presidential Handwriting, 9/15/1974, GRFL.

66 "You know why you're here?": JCI, Gerald R. Ford, April 24, 1990, GRFL.

67 "We pay in full after each contest": E. Curly Lambeau to GRF, February 11, 1935, *The Wearing of the Green (and Gold)*, June 1, 2012, https://packersuniforms.blogspot.com/2012/06.

67 "He was definitely interested in politics": Phil Buchen's assessment is in "My Years at Michigan Made Me a Better Person," *Grand Rapids Magazine*, October 1974, 41.

67 "Of boxing I knew next to nothing": Ford and Underwood, "In Defense of the Competitive Urge."

68 "Such recognition means all the more": Walter Nowinski, "No Ford, but Dedication Goes On," *Michigan Daily*, October 18, 2006.

3: Yale

71 it was both "undemocratic and un-American": John Heilprin, "Ford Only President to Serve as Park Ranger," *Montana Standard*, December 27, 2006.

72 As he fed out the lifeline: The successful rescue attempt is described in "President Ford's Heroic Days as a Yellowstone Park Ranger," *National Enquirer*, April 29, 1975.

72 He judged Ford to be: James M. Naughton and Adam Clymer, "President Gerald R. Ford, Who Led U.S. out of Watergate Era, Dies at 93," *New York Times*, December 28, 2006.

73 A friend from Corby Court: Robert D. Stuart to EBF, December 28, 2006, Betty Ford Condolence Letters, Box 27, GRFL. See also Gaddis Smith, "Life at Yale During the Great Depression," *Yale Alumni Magazine*, November/December 2009.

73 "The most beautiful girl I ever knew": JCI, Gerald Ford, April 24, 1990, GRFL.

73 "She was really a classy girl": "She was such a glamour girl," Phil Buchen remembered. "So striking that on first impression, you thought, well, here is a very beautiful girl. Yet she had a good personality as well; it wasn't just looks. I just think she had a lot of spark, a lot of life, and was always laughing, always joking." GROHP, Phil Buchen, January 18, 1980, 12.

74 "a pretty torrid romance": GRF, April 24, 1990, op. cit. Tom Ford told James Cannon that within the Ford family, it was "a foregone conclusion" that Jerry and Phyllis were to be married. "But the story given to me," he added, "was that she wanted him to stay in New York." JCI, Tom Ford.

74 "some deb thing": JCI, Phyllis (Brown) Haller, April 28, 1990.

75 "I don't think I have ever thought": Ibid.

75 "I wouldn't have liked to be the little congressman's wife": Ibid.

75 "I still think about him": AI, James Cannon.

76 "one of the most flagrant cases": Investigation Report, January 24, 1942, US Navy Department of Naval Intelligence, Box 1, Folder: GRF Investigatory Information, GRFL.

77 "Arrange with your mother": "Dad" to GRF, October 17, 1938, DGDP.

77 "the blonde Venus": B. B. Kelly to GRF, December 17, 1938, DGDP.

77 Once that was resolved: "Frankly I'm rather tired of being the go between [sic] you and Mother over such a disagreeable situation," Ford wrote his birth father. King refused his appeal to post collateral as demanded by Dorothy Ford's lawyers. "I wish you would write her and explain to her what an imposition she is putting me too [sic]," he told his son. GRF to Leslie King, April 1, 1939; Leslie King to GRF, April 21, 1939, DGDP.

77 "Nobody is trying to cheat you Gerald": Leslie King to GRF, April 21, 1939, DGDP.

77 "I am very much interested in you and Phil": Leslie King to GRF, April 27, 1938, DGDP.

77 "Swell," replied Ford: "Jerry" to Leslie King, April 1939 (undated), DGDP.

78 "I am going to fight this to a finish": James Cannon, Gerald R. Ford: An Honorable Life, 25.

78 "We found a way we could enjoy": Gerald Ford/Phil Buchen interview by Jay Foonberg and Walter Russell, 1995. ABA Special Committee on Solo and Small Firm Practitioners, Accession #96-NLV-025, GRFL. Also see James Podgers, ABA Journal, February 1997.

79 "When you really think he is not even paying attention": JCI, Phyllis (Brown) Haller.

79 "She just wouldn't sacrifice the nice things": TAFI, 1309.

79 "You know, when I was a youngster": AI, Don Penny.

80 "I damn near cried. At the same time I was exhilarated": TAFI, 1215.

81 "I'm a businessman first, politician second": William A. Syers, "The Political Beginnings of Gerald R. Ford: Anti-Bossism, Internationalism, and the Congressional Campaign of 1948," Presidential Studies Quarterly 20, no. 1(Winter 1990): 128. Also see "Frank D. McKay: Teflon Boss," The Downfall Dictionary, http://downfalldictionary.blogspot.com/2010/08/.

82 "He sold surety bonds": GROHP, Willard VerMeulen, January 26, 1980.

82 "Frank made a lot of friends": AI, Maury DeJonge.

83 "Tell McKay he can have them if we get the votes": Steve Neal, Dark Horse: A Biography of Wendell Willkie (Garden City, NY: Doubleday, 1984), 115.

84 What could account for his abrupt departure: Undated letters from GRF to R. D. Stuart Jr., Box 66, America First Committee Papers, Hoover Institution, cited in Justin Doeneck, In Danger Undaunted: The Anti-Interventionist Movement of 1940–1941 as Revealed in the Papers of the America First Committee (Stanford, CA: Hoover Institution Press, 1990), 7–8.

85 "I enjoyed all aspects of the law": David Horrocks, "Gerald R. Ford: All American Counsel," American Lawyer-Presidents, ed. Norman Gross (Evanston, IL: Northwestern University Press, 2004), 282.

86 "a superior type of young man": M. Wilson McFarlin–J. Edgar Hoover, July 3, 1941, Box 1, Special Agent (FBI) Application Case File, GRFL.

86 "one of the best applicants": M. Wilson McFarlin–J. Edgar Hoover, December 22, 1941, Box 1, Special Agent (FBI) Application Case File, GRFL.

87 "Grand Rapids," muttered one visiting physician: GROHP, VerMeulen, 18.

87 "If I could get ten guys": Ibid., 19.

87 "I haven't got any clients": Ibid., 20.

88 "They had to own their own business": Ibid, 21.

88 "He's such a hard worker": Ford/Buchen interview, ABA Special Committee.

4: The View from the Bridge

89 "I didn't know a gun from a bow and arrow": JCI, Gerald R. Ford, April 26, 1990 (part 3), GRFL.

90 "as easy to wear as an old shoe": Roland Giduz, "Gerald Ford '38 (Law): As His Classmates Remember Him," *Carolina Alumni Review*, January 1975. An excellent account of the "Tunny Fish Program" is Donald W. Rominger Jr., "From Playing Field to Battleground: The United States Navy V-5 Preflight Program in World War II," *Journal of Sport History* 12, no. 3 (Winter 1983).

90 "The prettiest girl here": Emily Irby Short to GRF, August 26, 1975; GRF to Mrs. W. D. Short, TR3/ST26/76, Central Files, GRFL.

90 At one point in the evening: David E. Brown, "The War Years," *Carolina Alumni Review*, September/October 1995.

90 "Ford crashed the plane": "Gerald Ford Took a Liking to UNC, But to Horace Williams Airport, Not So Much," UNC General Alumni Association, December 29, 2006. See also Giduz, "Gerald Ford '38 (Law): As His Classmates Remember Him."

90 "In a war, single, no responsibility": JCI, Gerald Ford, April 22, 1992 (part 1).

91 "When we were fully loaded": TAFI, 303.

91 "We were fast": AI, Frank Weston.

91 "where you all were drinking and having a good time": TAFI, 297.

92 "Don't shoot at this guy": Richard Brooks, "Memories of Flying the Panama Canal with Gerald Ford," September 27, 1943, GFNR, Box 2, GRFL.

92 "you never saw so many people ill": TAFI, 297.

93 "The landing signal office is my shepherd": *Flat Top Flashes*, "Psalm of a Carrier Pilot," August 16, 1943, GFNR, Box 3, GRFL.

94 "Get their engine or their wings": James F. Hall, "Recollections of Life on the Monterey," GFNR, Box 1, Folder 1, 1, GRFL.

94 "One time we didn't meet our tanker": AI, Frank Weston.

95 "as much action as I'd ever hoped to see": Bob Drury and Tom Clavin, *Halsey's Typhoon* (New York: Grove Press, 2007), 153.

95 "Which meant I was on the bridge": JCI, Gerald Ford, April 26, 1990 (part 3), GRFL.

96 "a damn good drink": GRF to "Dear Margaret," September 1, 1944, GFNR, Box 1, GRFL.

97 On the other end was Mr. Ford: GROHP, VerMeulen, 29.

97 "He turned out to be a nice guy": JCI, Gerald Ford, April 26, 1990 (part 3), GRFL.

98 On the lower decks: Hall, Recollections, "Sometime around the middle of December 1944 . . . ," Box 1, Folder 2, 1, GFNR, GRFL.

98 "That night was hell": Vivid contemporary descriptions of the December 1944 typhoon can be found in John Cadwalader's memoir of his service on the *Monterey*, 85–87, as well as an eleven-page account contained in James F. Hall's "Recollections." Both documents are in Box 1, GFNR, GRFL.

99 "We had a chief petty officer": AI, Frank Weston.

99 "The minute I woke up I could smell smoke": TAFI, 301.

100 "I spread myself out": TAFI, 302. Ford succinctly described his near miss for author Robert Drury in September 2004: "I had about a two second slide down across the desk that scared the hell out of me."

101 "We can fix this": Drury and Clavin, *Halsey's Typhoon*, 157.

101 "It was a lucky ship": TAFI, 298.

103 "They said if he is called and opens up": GROHP, VerMeulen, 31. Stephan P. Dresch, "Three Bullets Sealed His Lips—A Timely Reminder of Capital Sleaze," *Traverse City Record Eagle*, May 2, 1993. This review of the book of the same title by Bruce A. Rubenstein and Lawrence E. Ziewacz is an excellent introduction to the strange case of Warren Hooper.

103 "Nobody is going to bother me": Ibid., 32.

104 "I'm no Roosevelt man": GRF to John Cadwalader April 23, 1945, GFNR, Box 1, GRFL.

5: Betty

107 "a brilliant man": See Carrie Sharlow, "Michigan Lawyers in History: Julius H. Amberg," *Michigan Bar Journal*, June 2015.

107 "I want to train you": JCI, Gerald R. Ford, April 24, 1990, GRFL.

108 "That guy would get down here": GROHP, Niel A. Weathers, January 28, 1980, 1, GRFL.

109 "When are you going to start dating again?": TTH, 62.

109 "there were a lot of girls that were chasing him": TAI, Betty Ford, January 18, 1978.

109 "You're doing what you want to do": TAFI, 293.

109 "Who's around that a bachelor my age can date?": TAFI, 3358.

110 "Oh I can't do that": TAI, Betty Ford.

110 "I won't tell you what she said": AI, Lillian Fisher.

110 "It was a very peculiar romance": TAI, Betty Ford.

111 "Oh great," thought Betty: Ibid.

111 "I certainly had had my fill of it": Ibid.

111 "That didn't seem like a long way in the summer": JCI, GRF, April 30, 1990.

111 "the strongest man I have ever known": Barbara Howar, "Spotlight on Betty Ford, a New Breed of Wife in the Nation's Capital," *Family Circle*, November 1974.

111 "I was very immature": JCI, Gerald R. Ford, April 21, 1993.

112 "Oh, you're Jerry's new girl!": Howar, op. cit., November 1974.

112 "I gather you didn't care much for them": TML, 53.

114 "There were so many things missing in her life": John Robert Greene, *Betty Ford: Candor and Courage in the White House* (Lawrence: University Press of Kansas, 2004), 5.

114 "If you have to do it": Lynn Minton, "Betty Ford Talks About Her Mother," *McCall's*, May 1976.

114 "A very dignified lady": AI, Lillian Fisher.

115 PLEASE DO NOT FEED THIS CHILD: TML, 7.

115 "Dance was my happiness": TML, 17.

115 "Oh my, she's pretty": AI, Lillian Fisher.

116 "I couldn't get my knees straight enough": TML, 17.

116 "In pioneer days": 1931 *Helios Yearbook*, Central High School, GRFL.

116 "We were always telling the girls": AI, Lillian Fisher.

116 "If you don't do it well": Minton, "Betty Ford Talks . . ."

117 "I worshipped her as a goddess": Greene, *Betty Ford*, 9.

118 "those great padding steps": Martha Graham, *Blood Memory* (New York: Doubleday, 1991), 103–4. See also Angelica Gibbs, "The Absolute Frontier: Martha Graham's Mode of Dance," *New Yorker*, December 19, 1947.

118 "I did not want it to be beautiful or fluid": "Martha Graham Biography," PBS, *American Masters*, September 16, 2005, www.pbs.org/wnet/americanmasters/martha -graham-about-the-dancer.

118 "motion pictures for the sophisticated": "The Blessed Damsel Comes Down from Rochester," *Dance Magazine*, July 1926.

118 "Ugly girl makes ugly movements onstage": Susan Ware, *Letter to the World: Seven Women Who Shaped the American Century* (New York: W. W. Norton, 1998), 213.

118 "An American dance": Martha Graham, "Platform for the American Dance," *I See America Dancing: Selected Readings, 1685–2000*, ed. Maureen Needham (Chicago: University of Illinois Press, 2002), 203–4.

119 "Tunics are worn over leotards": Janet Mansfield Soares, *Martha Hill and the Making of American Dance* (Middletown, CT: Wesleyan University Press, 2009), 63; Susan Green, "How Modern Dance Took Root in Vermont," *Burlington Free Press*, March 5, 2012.

119 "If you got her knee in your back": Katherine A. S. Sibley, *A Companion to First Ladies* (Hoboken, NJ: John Wiley & Sons, 2016), 553.

120 "I wanted to learn something": Betty Ford Notebook, Bennington School of Dance, Vertical File, Folder: Ford, Betty (1918–2011): 1924–1959, 63–72, GRFL.

121 "You can't carouse and be a dancer, too": TML, 30.

122 "Bill Warren was one of the best looking men": In later years Mrs. Ford professed not to remember the date of her first wedding. What did she take away from her "five year misunderstanding," as she labeled her marriage to Warren? "It taught me that there are some people you just can't change." Jane Howard, "The 38th First Lady: Not a Robot at All," *New York Times*, December 8, 1974.

122 "I was young and stupid": Minton, "Betty Ford Talks . . ."

6: Time for a Change

124 "I like your style": Vestal, *Jerry Ford, Up Close*, 92.

124 "once a probate judge always a probate judge": TAFI, 312. Earlier still, Michigan secretary of state Fred Alger invited Ford to be his top deputy, an offer Ford spurned. "Even in those days I had no desire to get involved in state politics." TAFI, 3354.

125 "I remember, it was right after the war": AI, Ralph Hauenstein.

125 "You probably can't win": JCI, Gerald R. Ford, April 30, 1990.

126 "There were those who said": Ibid.

126 "If you were divorced": AI, Maury DeJonge.

126 "They never went to the movies": AI, Frank Weston.

127 "They knelt before me and prayed for my soul": AI, Werner Veit.

127 "I thought, oh well, he'll never beat Jonkman": TAI, Betty Ford.

127 "There was no pretense there": GROHP, Kay Clark, 5. Ford was not above playing the age card, pointing out that "Mr. Jonkman is over 64 years old. I am 35." Holland Radio Interview, September 9, 1948, FCP, Box D37, PSF, GRFL.

128 "You've got to remember that": AI, Seymour Padnos.

128 "We cannot convincingly carry the torch": James Cannon, *Gerald R. Ford: An Honorable Life*, 69.

129 "Jonkman lost his temper": GROHP, Dorothy Judd, May 21, 1980, 14.

129 "There are no issues in this campaign": "Ford Criticizes Jonkman Hideout," *Star and Alliance*, September 1948 (undated).

129 "Got seven farms to do": Jerald F. terHorst, *Gerald Ford and the Future of the Presidency* (New York: The Third Press, 1974), 103. Ford proved to be a quick study, as his friend Jack Stiles observed after the election, when Congressman Ford encountered a pig farmer, "a Dutchman . . . mad as hell about a quota on raising pigs. I mean this guy was a real clod hopper." As for Ford, "he doesn't know his ass from third base about pigs. So this guy goes back there and talks to Jerry for 30 minutes, and when he comes out he's *smiling*. He says, 'I vote for dat Jerry Ford every time. He knows all about pigs!' Hell, all Jerry did was *listen* to him." Sheridan, "Portrait of the Next President . . ."

131 "This is very embarrassing to the firm": JCI, Gerald Ford, April 30, 1990, GRFL.

131 "And that teed off Lee Woodruff": JCI, Gerald R. Ford, April 25, 1990, GRFL.

131 Voters should be on their guard: Syers, "The Political Beginnings of Gerald R. Ford," 134.

132 "I had breakfast for two weeks": "Barn Razing Erases Vintage Landmark," *Grand Rapids Press*, January 3, 2007.

132 "We had a small wedding and a big reception": Minton, "Betty Ford Talks . . ." The tale of Ford's mismatched shoes is related by Paul Goebel in his National Archives oral history of January 25, 1980, 7.

133 "Do you suppose you could fix me a bowl of soup": TML, 61.

133 "Jerry's mistress will not be a woman": TAI, Betty Ford, 27; terHorst, *Gerald Ford and the Future of the Presidency*, 50.

133 "Like every woman I thought": JCI, Betty Ford, April 30, 1990.

133 "Well, I got nine kids and I need the job": JCI, Gerald R. Ford, April 30, 1990.

134 Only members of Congress: *Grand Rapids Herald*, January 6, 1949.

135 "He was sort of a conscience": TAFI, 721.

135 "I cannot forget": "Coopersville Speech, ca, 1948," FCP, PSF Sub-Group 1948–73, Box D 13, GRFL.

137 "One type of letter, however": FCN, August 14, 1952.

138 Recognizing that only a handful of voters: The best account of Ford's mobile office and his aggressive approach to constituent service is Ralph Truex and Gerald Elliott's unpublished article, "Congressman on Main Street," Composite Grand Rapids Accessions, Box 4, GRFL.

138 "Is there any hope for a free Poland?": September 1, 1950, PSF, Box D 14, GRFL.

138 "We are just worms here": Michael O'Brien, *John F. Kennedy: A Biography* (New York: Thomas Dunne Books, 2005), 222.

139 "We talked about how nice it would be": Douglas Brinkley, *Gerald Ford* (New York: Henry Holt, 2007), 15.

139 "Pick something on which you want to be expert": JCI, Gerald Ford, April 25, 1990, GRFL.

140 "We never ate chowder and we never marched": GRF interview with Dale Van Atta, August 18, 1937, Mel Laird Papers, Box D 98, GRFL.

140 "I had to grow with him": JCI, Betty Ford, April 30, 1990, GRFL.

7: A Modern Republican

145 "Do you need Papa's parakeet?": Dorothy Ford telegram to GRF, July 10, 1952, GRFL.

146 "the leader in the crimson clique": Lincoln Day Banquet—Introducing Richard Nixon, February 10, 1951; FCP, PSF Sub-Group 1948–73, Box D 14, GRFL.

146 "joyous and very happy": TAFI, 3428.

147 Vandenberg had gone his own way: "How You Can Enjoy Politics," Ford speech to Junior Chamber of Commerce, undated, 1951, FCP, GRFL.

148 "He reminds you a lot of Scrooge": Richard P. Fenno Jr. interview with GRF, June 8, 1958, NARA Center for Legislative Archives.

148 "They have to be willing to work": Richard P. Fenno Jr. interview with Charles W. Halleck, March 1963, NARA Center for Legislative Archives.

149 "We discussed that for a good many nights": TAI, Betty Ford.

149 "the greatest lobbyists in the world": *Grand Rapids Press*, January 2, 1950.

149 "nothing less than a 210 horsepower sedan": FCN, July 2, 1953, GRFL.

150 "I took a copy home to Betty": FCN, April 12, 1951.

151 "They want you to come home": Jim Newton, *Eisenhower: The White House Years* (New York: Random House, 2011), 55.

151 "Art always was suspicious of me": Summerfield was an only slightly more sophisticated version of the typical Republican voter in the Midwest described by Illinois's
Bob Michel, who was elected to the first of his nineteen terms in the second Eisenhower landslide of 1956. "Well Bob," he summarized the feelings of his constituents, "you just go down to Washington and cut the cost of government and lower
my taxes. We don't want anything. Just get the government off my back." AI, Robert
Michel.

152 Excluded from union halls: TAFI, 1223.

152 "Eisenhower believes": FCN, September 11, 1952.

153 "I felt as though we were up against a firing line": Ibid., November 6, 1952.

154 "There's no question about it": Ibid., November 13, 1952.

154 "Try, sometime, to put yourself": Ibid., November 20, 1952.

154 For Ike to have gone head-to-head: Ford's criticism of the Wisconsin senator was
muted. He called McCarthy's feud with army secretary Robert Stevens "regrettable,"
adding that congressional investigators had a responsibility to concentrate on the
facts—to "keep their eye on the ball." *Grand Rapids Herald*, March 3, 1954. Betty
Ford attended at least one session of the climactic hearings that spring leading to
McCarthy's downfall and eventual censure.

155 "All these damn Admirals": JCI, Gerald R. Ford, April 25, 1990, GRFL.

155 "Give the Army only what it needs": *Grand Rapids Press*, August 9, 1953.

155 "Why, that's minor league": Richard F. Fenno Jr. interview with John McCormack,
January 1961, NARA Center for Legislative Archives.

156 "I have the most nonpartisan committee in Congress": Chuck Lanehart, "George
Mahon: Greatest of West Texas Statemen," *Lubbock Avalanche-Journal*, January 11,
2020.

157 "an almost fanatical interest in golf": Dale Van Atta, *With Honor: Melvin Laird in War,
Peace, and Politics* (Madison: University of Wisconsin Press, 2008), 39.

157 "He doesn't scheme for any sinister reason": Ibid., 113.

157 "the kind of guy who would put poison in the river": AI, Bob Dole.

157 "You had to kick him in the ass": Van Atta, *With Honor*, 113.

157 "If Uncle Sam doesn't spend the money": FCN, July 23, 1953.

158 As it was, said Ford: Ibid., July 30, 1953.

159 "It is no easy task": Ibid., September 10, 1953.

160 "Speeches, pictures, bally-hoo": James Cannon, *Gerald R. Ford: An Honorable Life*, 63.

160 "an exceedingly interesting hour": FCN, September 3, 1953.

161 "If the Japanese had known": Ibid., September 24, 1953.

161 "The loss of Indo-China to the Communists": Ibid., April 29, 1954.

8: The Middle of the Road

163 "I always wake up in the morning": Nick Carbone and Ishaan Tharoor, "Ford's Man:
(Most) Modest Proposals," *Time*, October 14, 1974.

163 *"Viva Puerto Rico libre"*: Ford's account of the attack on the House is in FCN, March
4, 1954.

164 "I called it the Knife and Ford": AI, Bill Gill.

165 "The Republican Party is a humane party": Lincoln Day speech, February 12, 1955,
FCP, PSF, Box D 14, GRFL.

165 "No longer can Communists holler and scream": FCN, June 10, 1954.

166 "The longer you stay on the committee": Richard F. Fenno Jr. interview with Ben F.
Jensen, May 28, 1959, NARA Center for Legislative Archives.

166 "God, what I would give": Robert Caro, *The Years of Lyndon Johnson: Master of the
Senate* (New York: Alfred A. Knopf, 2013), 157.

167 "Way way below even half price": Dorothy Ford to Betty Ford (undated), Ford Special Materials (1941–2007), Box B2, Betty Ford Special Letters, GRFL.

167 "There was some talk": JCI, Betty Ford, April 30, 1990.

167 "He was away—even then": Ibid.

168 "to remind me of atomic energy's menace": *Current Biography*, March 1961, 21. That there was more legislative oversight of the CIA during this period than previously suspected has been documented by Cold War historians like Tim Weiner, David Barrett and L. Britt Snyder. To be sure, much of this involved the agency's budget. During the Republican-controlled 83rd Congress (1953–54), the House subcommittee under Chairman John Taber convened more than a dozen oversight meetings with CIA representatives. Unwilling to rely on their numbers, Taber hired sufficient staff to conduct an independent review of agency spending.

168 on the floor of the House: Before returning the chairman's gavel to Clarence Cannon at the start of 1955, Taber had succeeded in imposing the first limits on the CIA's fifteen-thousand-member workforce, as well as a 15 percent reduction in authorized expenditures.

168 "a real comer": "Walter Pforzheimer Reminisces," *CIA Studies in Intelligence* 42, no. 4 (Fall 1998).

169 "the worldwide atheistic communist conspiracy": FCN, April 12, 1961.

169 "Dulles is a legendary figure": Peter Grose, *Gentleman Spy: The Life of Allen Dulles* (Boston: Houghton Mifflin, 1994), 531.

169 "What was that meeting about?": Ibid., 415.

169 "Because this business inevitably involves": FCN, May 18, 1960.

170 "I must admit": "Notes from McNamara Lemnitzer Appearances Before Defense Subcommittee of House Appropriations Committee," February 1, 1962, www.cia/gov/readingroom/document/cia.

170 "It was a mistake to stop them": Ibid.

170 "my closest political and personal friends in this world": Richard Nixon to GRF, January 12, 1956, GRFL.

172 community of Grand Haven, Michigan: AI, Gordon Vander Till.

172 "A very wonderful lady": TAFI, 570.

173 "There goes the guest room": TML, 91.

174 "the glue that kept us all together": AI, Susan Ford Bales.

174 "There was just no discussion": Ibid.

174 With a candor that would do his parents proud: AI, Mike Ford.

174 "He liked people": AI, Susan Ford Bales.

174 "An old timer in Congress": Paul Miltich, "No 8-Hour Day for Rep. Ford," *Grand Rapids Press* (undated).

175 "he would always stick us in front of a typewriter": AI, Mike Ford.

175 "And you didn't really talk much at breakfast": AI, Steve Ford.

175 After telling the children how much he missed them: AI, Mike Ford.

176 "Some of the constituents are convinced": *Detroit News*, May 3, 1961.

176 "Sometimes I think": *Detroit Free Press*, July 1, 1962.

176 "You are one of the guys": Sherman Adams to GRF, November 13, 1956, Ford Scrapbooks Unbound Items, Box 4, Folder: June 1960, Eisenhower; Spivak.

177 "From all sides I had reports": Dwight Eisenhower to GRF, September 11, 1957, Gerald Ford Scrapbooks, Volume 7, "January–December 1957."

177 "but I won't pay to find out this year": Stephen E. Ambrose, *Eisenhower: The President* (New York: Simon & Schuster, 1984), 433.

177 "and boy, did he operate": "Any compromise that Lyndon got, he got better than 50 percent," Ford remembered. "He knew exactly what he wanted. He would play one

person against another. He wouldn't shout but he would be very firm. He would be very hardline, and then he would give enough to get what he wanted without selling out what he didn't want." TAFI, 1279.

178 "an able and courageous legislator": John W. McCormack to GRF, August 5, 1958, Gerald R. Ford Scrapbooks, Box 3 (unbound), Folder: August 1958 [S1–36].

178 an "Eisenhower echo": *Grand Rapids Press*, October 16, 1958.

178 "You could always tell what they had for dinner": AI, Leon Parma. Martin was dubbed Mr. Malaprop for his mishandling of the English language on formal occasions, like the time he welcomed the prime minister of Indonesia to the House chamber by introducing him to his colleagues as the PM of Indiana. AI, John Dingle.

179 "I am a gut fighter": "The Congress: The Gut Fighter," *Time*, June 8, 1959.

179 "I'm going to an execution": John L. Steele, "GOP Tactics That Toppled a Veteran Leader," *Life*, January 19, 1959.

180 "the most able": *Grand Rapids Herald*, June 1, 1958.

9: The Congressman's Congressman

182 "Too many people try to move too fast in politics": *Detroit Times*, December 29, 1959.

182 "young, attractive and conservative": *National Review*, January 5, 1960.

182 "a smashing candidate": *Wall Street Journal*, April 21, 1960.

183 "Some of my friends are talking me up": "Open Contest Predicted over Vice Presidency," *Los Angeles Times*, December 17, 1959.

184 "I have heard many nice things about him": Lawrence E. Spivak to GRF, June 20, 1960, Ford Scrapbooks Unbound Items, Box 4, Folder: June 1960, Eisenhower; Spivak.

185 "Ford's main difficulty is that no one knows him": Walter R. Beardsley to Montgomery Shepherd, July 14, 1960, Ford Scrapbooks, Unbound Items, Box 4, Folder: Republican National Convention 1960.

185 "Caliber-ability-fresh approach": Stephen E. Ambrose, *Nixon: The Education of a Politician, 1913–1962* (New York: Simon & Schuster, 1988), 553.

185 "I am completely realistic": Edward L. and Frederick H. Schapsmeier, *Gerald R. Ford's Date with Destiny: A Political Biography* (New York: Peter Lang, 1989), 68.

185 Betty Ford shared her husband's fatalistic outlook: *Grand Rapids Press*, July 17, 1960.

185 "You must be mighty proud of them!": Pat Nixon to Betty Ford, June 29, 1960, Gerald R. Ford Scrapbooks, Unbound Items, Box 4, Folder: June 1960.

185 "As always, I am deeply grateful": Richard Nixon to GRF, July 5, 1960, GRFL.

185 "If I feel I have a chance": *Grand Rapids Press*, July 19, 1960.

186 "Forget whether I was qualified or not": TAFI, 2177.

186 "A moderate conservative": James Cannon, *Gerald R. Ford: An Honorable Life*, 70–71.

187 "In light of additional experience": FCN, January 11, 1961.

188 "If I had three minutes to live": Billy Hathorn, "Otto Passman, Jerry Huckaby, and Frank Spooner: The Louisiana Fifth Congressional District Election of 1976," *Louisiana History: The Journal of the Louisiana Historical Association* LIV, no. 3 (Summer 2013): 336.

188 "The only way I can square this with my kids": "Mrs. Ford Brings 'New' GOP Look," *Washington Evening Star*, January 24, 1965.

189 "What right does this generation of Americans have": *Holland (Michigan) Evening Sentinel*, December 11, 1961.

189 "His was a good life": Phillip W. Buchen to GRF, February 1, 1962; GR Ford Scrapbooks Unbound Items, Number 5, Folder: Jan.–Mar. 1962.

189 "My wife disagrees with me": *Grand Rapids Press*, November 21, 1961.

191 "avoid awkward sentences": Ford Scrapbooks, Jan.–July 1962, Carl Vinson to GRF, March 22, 1962, GRFL. The fullest account of the B-70 controversy is in Gordon F. Sander, "JFK's Forgotten Constitutional Crisis," *Politico*, May 29, 2017.

192 McCone thought otherwise: terHorst, *Gerald Ford and the Future of the Presidency*, 79–80.

192 Over a nonsecure phone line: TAFI, 355–56.

192 "They're Jupiter missiles, made in Detroit": *Grand Rapids Press*, October 27, 1962.

193 "As I leave the political arena": This was, in fact, a political form letter, sent to Ford and other victorious candidates in that year's elections. Donald Rumsfeld, *Known and Unknown: A Memoir* (New York: Penguin Group USA, 2011), 87.

194 "This party isn't divided": Meg Greenfield, "Charles A. Halleck and the Restless Republicans," *Reporter*, March 29, 1962.

195 "I happened to be in the line of fire": *Roll Call*, January 13, 1963.

195 Ford's response was conciliatory: *Michigan Chronicle*, June 14, 1963.

196 "Congratulations on making General Eisenhower's Hit Parade": Robert Novak to GRF, August 21, 1963, GRFL. Ike's list is spelled out in Eisenhower to William Knox, August 22, 1961, Ford Special Materials, Box A4, Folder: Dwight David Eisenhower, GRFL.

198 "There wasn't another soul around": TML, 104.

198 It seems a curious appointment: Halleck was, in fact, "mad as hell" at Johnson for passing over senior ranking Republicans to choose Ford. TAFI, 1282.

10: The Warren Commission

201 According to Baker: Todd S. Purdum, "Sex in the Senate," *Vanity Fair*, November 19, 2013. Purdum's article is based on Bobby Baker's June 1, 2009, oral history with Senate historian Richard Baker.

202 "It is always extremely gratifying": Max Holland, "The Ford Files," *Washington Decoded*, August 11, 2008.

202 "a religious man": Louis Nichols interview, October 18, 1973, FBI Investigatory Report on Gerald R. Ford, Nominee for Vice President (Internet Archive), https://archive.org/details/Gerald Ford FBI/1118463–002–161-HQ-9896-Section 3/page 212-13(of 248). Nichols served as Nixon's campaign security chief in 1968. Following J. Edgar Hoover's death in May 1972, and his replacement by Nixon loyalist L. Patrick Gray, Nichols would tell the president that Gray's appointment was "a tragic mistake." This was a view shared by FBI official Mark Felt, soon to be immortalized by Woodward and Bernstein as Deep Throat. Beverly Gage, "Deep Throat, Watergate, and the Bureaucratic Politics of the FBI," *Journal of Policy History* 24, no. 2 (2012): 157–83.

202 "I was never intrigued": TAFI, 1571.

202 "You have to think of the ten bad things": Thomas DeFrank, *Write It When I'm Gone: Remarkable Off-the-Record Conversations with Gerald R. Ford* (New York: G.P. Putnam's Sons, 2007), 205.

202 "You know, his wife had a serious drug problem": Office of the Senate Historian, Bobby Baker Oral History, 134.

204 "I don't like that man": Walter Pincus, "Transcripts Show LBJ's Maneuvers on Setting Up Warren Commission," *Washington Post*, September 23, 1993.

204 "Speculation about Oswald's motivation": Warren Commission, Staff Report of the Select Committee on Assassinations, US House of Representatives, March 1979, 4.

205 Speaking in "the strictest of confidence": George Lardner Jr., "Documents Show Ford Promised FBI Data—Secretly—About Warren Probe," *Washington Post*, January 20, 1978.

205 "Ford indicated he would keep me thoroughly advised": C. D. DeLoach to Mr. (John) Moher, December 12, 1963, quoted in Holland, "The Ford Files." The same document concludes by noting, "We have had excellent relations with Congressman Ford for many years," effectively demolishing claims the bureau had to blackmail the congressman into serving as an FBI informer.

206 "Typical of Hoover": Arlen Specter, *Passion for Truth* (New York: William Morrow, 2000), 94.

206 To a friend he later confided: Howard P. Willens, *History Will Prove Us Right: Inside the Warren Commission Report on the Assassination of John F. Kennedy* (New York: Overlook Press, 2013), 58.

206 Warren had publicly attributed Kennedy's murder: Holland, "The Ford Files."

206 Wrenched out of context: Michael J. Sniffen, "Ford Told FBI About Panel's Doubts on JFK Murder," *Washington Post*, August 9, 2008.

207 "They asked me questions from time to time": Interview of Cartha D. "Deke" De-Loach, May 1, 2007, Society of Former Special Agents of the FBI, Inc., 149.

208 "Knew his answers": FD, May 5, 1964.

208 "If that doesn't work": *Grand Rapids Press*, May 27, 1964.

209 "he is only President of the United States": *Time*, February 14, 1964.

210 "Not too bright or articulate": FD, March 26, 1964.

210 "Seems to be telling truth": Ibid., April 22, 1964.

210 Ford sought interviews: GRF to J. Lee Rankin, March 20, 1964, Box E1, Folder: Warren Commission Working Papers, 3/11–20/64, GRFL.

210 "Could shots have come from that": GRF to J. Lee Rankin, March 23, 1964, Box E1, Folder: Warren Commission Working Papers, 3/21–3/31, 1964.

211 "Our record seems pretty sketchy": GRF to J. Lee Rankin, May 22, 1964, Box E1, Folder: Warren Commission Working Papers, 5/64.

211 "There were three or four people": Specter, *Passion for Truth*, 48.

212 anyone else on the commission staff: Vincent Bugliosi, *Reclaiming History* (New York: W. W. Norton, 2007), 341.

212 "I like him": Warren Commission Executive Session Transcript, May 19, 1964, History-matters.com, 6624. Ford may well have leaked his concerns about Redlich to Jerry terHorst, his future White House press secretary, who authored an article in the May 18, 1964, *Washington Star*: "Warren Probe. Redlich May Go in Cutback."

212 "He is a born crusader": Warren Commission Executive Session Transcript, May 1964, History-matters.com, 6633.

212 "Jerry," Warren interjected: Ibid., 6647.

213 "I don't like being quoted": Warren Commission Executive Session, June 4, 1964, 2–10, History-matters.com.

213 "I'll give you Sunday": Specter, *Passion for Truth*, 106. Ford wrote a separate, three-page diary entry titled "Dallas Trip, June 7, 1964," which is by far the most detailed account he ever provided of his visit to the assassination site and his subsequent encounter with Jack Ruby.

215 "Everything seemed to go well for about 45 minutes": FD, June 7, 1964.

215 "Chief, you've got to get me to Washington": Specter, op. cit., 113.

215 "You see. I told you he was crazy": Ibid., 114.

216 "Reaction to Ronald Reagan television show excellent": GRF telegram to Dean Burch, October 28, 1964, GRFL.

216 "Are State Department procedures sufficient": GRF to J. Lee Rankin, July 24, 1964, Box E2, Folder: Warren Commission Working Papers, 7/22–30/64, GRFL.

216 "This is not a satisfactory way to summarize evidence": GRF to J. Lee Rankin,

August 13, 1964, Press and Policy in Dallas, Box E2, Folder: Warren Commission Working Papers, 8/64, GRFL.

217 "A significant question that occurs to me": Ibid., 3, "General Comments."

217 "the one time I saw him get irritated": Specter, *Passion for Truth*, 58.

217 Soviet embassy in the Mexican capital: TAFI, 361.

218 "Senator Russell and Representative Ford were not completely convinced": William T. Coleman, *Counsel for the Situation: Shaping the Law to Realize America's Promise* (Washington, DC: Brookings Institution Press, 2010), 176.

218 "very high and way off the side": The opposing view, backed by the diaries and testimony of John and Nellie Connally, is argued by Don Folsom, "Gerald Ford's Role in the JFK Assassination Cover-Up," *Crime Magazine*, October 3, 2009.

220 "a narrative as well as a portrait": *New Yorker*, July 3, 1964.

220 "Since it adds nothing new": Douglas Kiker, *Washington Post Book Week*, July 4, 1965.

220 "an extremely partisan politician": FBI interview with Earl Warren, FBI Investigatory Report on Gerald R. Ford (Internet Archive), https://archive.org/details/Gerald FordFBI/1118463-002-161-HQ-9896-Section 3/page86-87(of 248). Warren was interviewed by the FBI on October 17, 1973. The Warren interview is part of an FBI file on Gerald Ford obtained in 2017 by independent scholar Ivan Greenberg, who has written extensively on the bureau, government surveillance and civil liberties issues.

220 "But I think you have to weigh that": Bill Kurtis interview with GRF, February 11, 1992, for *Investigative Reports*. David Belin Papers, Box 42, Folder: Ford, Gerald, 1989–1994 (2), GRFL. Ford's request that any remaining assassination-related files be opened is in his letter to Chairman Louis Stokes, January 21, 1992. David Belin Papers, Box 42, Folder: Ford, Gerald, 1989–1994 (1).

221 "Lee Oswald was that kind of a person": Gerald Ford with Jack Stiles, *Portrait of an Assassin* (New York: Simon & Schuster, 1964), 496.

11: Holding the Line

222 "I went to Charlie Halleck": William A. Syers interview with Mel Laird, July 24, 1985, GRFL.

222 "He was a heavy drinker": AI, Leon Parma.

222 "Charlie's drinking room": *Time*, June 8, 1959.

222 "Halleck doesn't read books": Meg Greenfield, "Charles A. Halleck and the Restless Republicans," *The Reporter*, March 29, 1962.

223 "He was constantly trying to set up coalitions": William A. Syers interview with Hugh Scott, June 3, 1985, GRFL.

223 "Charlie Halleck was like Everett Dirksen": William A. Syers interview with Barber Conable, February 1, 1985, GRFL.

223 "the affirmative . . . aggressive approach": TAFI, 367.

224 Infused with Rumsfeld's sense of urgency: Henry Z. Scheele, "Prelude to the Presidency: An Examination of the Gerald R. Ford-Charles A. Halleck House Minority Leadership Contest," *Presidential Studies Quarterly* 25, no. 4 (Fall 1995): 77.

225 "This is no time to hunt for scapegoats": *New York Times*, November 11, 1964.

225 "If Jerry doesn't go against Halleck, I will": To Dick Clurman, The Washington Report—Revolt in the Ranks, December 7, 1964, NMP, DCC.

225 "Everyone seems to think that Mel Laird was one of the plotters": William A. Syers interview with Charles Goodell, February 27, 1985, GRFL.

226 "That way I didn't have to get involved": *Grand Rapids Press*, January 12, 1965.

226 "It seems to me this is the time": From: MacNeil To: Parker, House Republicans (nation), December 19, 1964, NMP, DCC.

227 "the high middle road of moderation": James Robinson, "Ford Urges Split with Dixie," *Detroit Free Press*, December 18, 1964.

227 "We're over the 50–50 mark": From: MacNeil To: Parker, "Congress I" (nation), December 23, 1964, NMP, DCC.

228 "A few minutes after three": William A. Syers interview with Charles Goodell, February 27, 1965, GRFL.

228 "He promised to take me out to dinner": Marie Smith, "Betty Ford Paces as Husband Wins," *Washington Post*, January 5, 1965.

228 "I'm from Indiana": *Detroit Free Press*, January 4, 1965.

229 "I don't think the party can sit back": *Globe Gazette (Mason City, Iowa)*, May 22, 1965.

229 "I don't need to tell you": Richard Nixon to GRF, January 20, 1965, Scrapbook, 18 A, Ford Scrapbooks Collection, GRFL.

229 "Jerry's first three or four months": Charles Goodell interview, William A. Syers Papers, February 27, 1985, GRFL.

230 "As a very young member of Congress": AI, Lee Hamilton.

230 "Republican members would not simply be against": Eric F. Goldman, *The Tragedy of Lyndon Johnson* (New York: Alfred A. Knopf, 1949), 283.

231 "The patience of fair-minded people is wearing thin": *Grand Rapids Press*, February 28, 1965.

232 "the look of a man betrayed": Goldman, op cit., 330.

232 That fall another symbolic gesture: Larry L. King, "Republican Quarterback in Search of an Offense," *True*, September 1966.

233 "He has conviction and courage": Interview with Don Rumsfeld, August 5, 1966, Box 2, Robert L. Peabody Research Interview Notes 1965–67, GRFL.

234 "Things are going to get better, honey": Paul Miltich, "Ford's Busy in Capitol," *Grand Rapids Press*, March 14, 1965.

234 "I realize Jerry has all these people": Ibid.

234 "I remember one evening": AI, Max Friedersdorf.

234 "It allowed her to have some control": AI, Mike Ford.

235 Her mother was "very sick": TML, 124.

235 "I wanted them to worry about me": Ibid.

236 Willard Edwards wrote of Ford: *Chicago Tribune*, June 27, 1965.

236 "No man in Congress": Walter Trohan, "Ford Scores as Coach," *Chicago Tribune* (undated).

236 "That isn't the reason we're supporting you, Mr. President": *Chicago Tribune*, May 13, 1965.

236 Ford claimed that Democratic lawmakers: King, "Republican Quarterback in Search of an Offense."

237 "Once in a while": Samuel Shaffer, *On and Off the Floor: Thirty Years as a Correspondent on Capitol Hill* (New York: Newsweek Books, 1980), 267.

237 "For three days, I wandered around in a daze": Ibid., 269.

238 "The flute and drum": Mary McGrory, *Washington Evening Star*, January 11, 1965.

238 "He went pretty far, didn't he?": terHorst, *Gerald Ford and the Future of the Presidency*, 100.

239 "That had a tremendous effect on Dirksen": AI, Mel Laird.

239 "nasty, vindictive and loud": Richard Reeves, *A Ford, Not a Lincoln* (New York: Harcourt Brace Jovanovich, 1975), 47.

239 "Bob is a hell of a nice guy": TAFI, 2323. "Made everybody mad," Ford acknowledged of Hartmann, "but at the same time . . . kept you from becoming complacent, self-satisfied . . . A prodigious worker when he puts his mind to it, but the tone of his voice, the words he will say, are like daggers in people."

239 "We all said, who is that?": AI, Anne Holkeboer.

240 "He can write a hell of a speech when he wants to": TAI, Paul Miltich.

240 "He had Ford's back": AI, Gail Raiman.

240 "This is coming down the track": AI, Bob Hynes.

241 "We're going to win forty seats":" FLOHP, David Broder.

241 "I feel a lot more like a leader": *Newsweek*, January 23, 1967.

242 "to drive southern Democrats": "At Home," *National Review Bulletin*, August 29, 1967.

242 "I'm not going to coalesce with some s.o.b.": Don Oberdorfer, "Where Are You Going, Jerry Ford?," *Detroit Free Press*, August 27, 1967.

242 "You can leave the troops just so often": terHorst, *Gerald Ford and the Future of the Presidency*, 107. Adam Clayton Powell's slurs on his fellow spokesmen for Black Americans are in "Adam Cast Out," *Newsweek*, January 23, 1967.

243 "Why are we pulling our best punches in Vietnam?": A. James Reichley, *Conservatives in an Age of Change: The Nixon and Ford Administrations* (Washington, DC: Brookings Institution, 1981), 277.

12: The Good Soldier

245 "On the gridiron I always wear my helmet": Gridiron Club Speech, March 9, 1968; Ford Scrapbooks, Unbound Items, July–August 1967, Box 8, Folder: March 1968.

247 "A proper way to call attention to grievances": *Grand Rapids Press*, May 1, 1968. Ford's tribute to Dr. King, and his description of a US Capitol under siege, are in his April 20, 1968, radio script, PSF, GRFL.

247 "I know that in the past, Jerry": TTH, 85. Indicative of just how commodious was Ford's big tent, one might compare the ideological rankings the minority leader of the House earned next to those of his preferred vice presidential candidate. Measured at the start of John Lindsay's final full term in Congress, the New Yorker enjoyed an 87 out of 100 as bestowed on the future mayor by the liberal interest group Americans for Democratic Action. ADA awarded Ford a measly 12. The numbers were reversed in performance appraisals by the conservative Americans for Constitutional Action, with Ford scoring an 82 and Lindsay a 35.

247 Hidden from the public: AI, Bill Timmons.

248 "Ford's raucous whinny shattered the silence": Robert T. Hartmann Papers, Box 19, Folder: Post–White House Subject Files, *Palace Politics*, Draft 1, 5–6, GRFL.

248 "we want all of them to support the national ticket if they can": *Charlotte Observer*, September 24, 1968; Herb Klein's rejoinder is in *Chicago Tribune*, September 25, 1968.

249 "An absolute trust": *U.S. News & World Report*, November 18, 1968.

249 "When we get to Washington": AI, Hal Bruno.

249 "Jerry, you and I have had": TTH, 87.

250 "the official alibi": Ford Scrapbooks, Unbound Items, January 1969, Box 10, Folder: October–December 1960, "White House, December 22, 1969."

250 According to White House Counselor John Ehrlichman: John Ehrlichman, *Witness to Power* (New York: Simon & Schuster, 1982), 204.

251 "He was the kind of man": Ford statement on the death of Senator Everett Dirksen, September 8, 1969, FCP, Box D4, Folder: Ford Press Releases, August–December 1969.

251 "Hugh, I understand the problem": AI, Tom Korologos.

252 "I must build a wall around me": Reeves, *President Nixon*, 29.

252 "the door to door salesman": AI, Aram Bakshian.

252 Scheduling White House Christmas: Robert Schmuhl, "A Revealing Look at the Ultimate Loner," *Chicago Tribune*, September 23, 2001.

252 "it would be goddamned easy to run this office": Haldeman Diary, May 19, 1971.

252 "What Nixon wanted": TAI, Bob Hartmann.

253 "That was the end": AI, Tom Korologos.

253 "I don't believe in muzzling demonstrators": *Hartford Times*, November 9, 1969.

253 "Remember, he is the President of the United States": Ford statement on death of Dwight Eisenhower, March 31, 1968, Box D55, "Individuals," FCP, Statements, PSF.

254 The most damaging opposition: Rick Perlstein, *Nixonland: The Rise of a President and the Fracturing of America* (New York: Scribner, 2008), 286–87.

254 "to save Bill Douglas": Laura Kalman, *Abe Fortas: A Biography* (New Haven, CT: Yale University Press, 1990), 334.

255 In truth Ford's interest in Douglas: AI, Lawrence Meyer. An excellent source on Ford's campaign against Douglas is the Library of Congress monograph (Legislative Reference Service), "Role of Vice-President Designate Gerald Ford in the Attempt to Impeach Associate Supreme Court Justice William O. Douglas," Box 229, Ford Vice Presidential Papers, GRFL.

256 "He didn't have any place to peddle this story": TAI, Bob Hartmann.

256 In a late-night phone call: Hartmann, *Palace Politics*, 64.

257 "Five or six sheets of no letterhead": TAI, Bob Hartmann.

257 "I gave him the files": Will R. Wilson Sr., *A Fool for a Client: Richard Nixon's Free Fall Toward Impeachment* (Austin: Eakin Press, 2000), 21.

257 "I've got a little more": TAI, Bob Hartmann.

258 "An impeachable [offense]": "Impeach Justice Douglas," April 15, 1970, PSF, Box D29, GRFL.

258 "irresponsible and malicious criticism": *Buffalo News*, May 11, 1971.

259 "Ford will appoint some bastard": Bruce Allen Murphy, *Wild Bill: The Legend and Life of William O. Douglas* (New York: Random House, 2003), 491.

259 "It is my sincere hope": GRF to William O. Douglas, November 12, 1975, Box 62, Folder: Supreme Court—Justice Douglas Resignation, Philip Buchen Files, GRFL.

260 "We've got to get together more often": Murphy, op. cit., 497.

261 "They weren't voting against the SST": Evans and Novak, *Seattle Times*, March 20, 1971. Ford told Bob Woodward in 1997 that he "detested" Nixon's officious White House surrogates. "They would come up to Capitol Hill and have a meeting with me and my Republican associates and say the President wants this, the President wants that . . . Fortunately I knew Dick Nixon better than they did and I knew he didn't operate that way."

262 "You know, I'm not tough enough": Confidential source.

262 "Isn't that the god damnedest thing?": Richard Nixon and Gerald R. Ford, June 17, 1971, 8:31–8:34 p.m., Nixon Telephone Tapes, Miller Center, Presidential Recordings Digital Edition, University of Virginia.

263 "Nixon continually looked for people": AI, Pat Buchanan.

263 "Bill, you know I like you": AI, Bill Timmons.

264 "the wisdom of these decisions": *Washington Evening Star*, December 27, 1971.

264 A month later it was the turn of: By far the best source for Ford's first China visit (June 26–July 5, 1972) is the exhaustive diary of the trip compiled by Frank Meyer, housed at the Ford Library. See also the Ford/Boggs Report to the President, July 20, 1972, Betty Ford Papers, Mrs. Ford Personal Files, Box 1, GRFL; and FCN, July 17, 1972.

265 "Well, a small Ping-Pong ball has brought us together": TTH, 97.

13: Ahead of the Curve

269 "In politics": *Baltimore Sun*, October 13, 1973.

269 "McGovern really thought the war would win the election": NOHP, Joseph Califano, April 3, 2007.

269 "I don't know why any Republican": *Raleigh News*, August 20, 1972.

270 Ford's failure to take "an active interest": Stanley Kutler, *Abuse of Power: The New Nixon Tapes* (New York: Free Press, 1997), 150. Contemporary news coverage emphasized Patman's failure to secure enough support from his fellow Democrats to obtain the subpoena authority he sought. E. W. Kenworthy, "Patman Balked on Watergate," *New York Times*, October 13, 1972.

270 "A real conniver": TAFI, 707.

271 "At the meeting Ford mostly sat": Bill Stanton memo [Oct. 1976], "President-Wright Patman Investigation Memoranda," Phil Buchen Files, Box 55, GRFL.

272 "checking on everyone": AI, Cokie Roberts.

272 "You think things are bad now": TTH, 101.

273 A Democrat of the old school: "Dale Anderson Convicted on 32 Counts," Anderson obituary, *Baltimore Sun*, July 28, 1996. The investigation actually began in October 1972. According to Agnew, he first learned of his own vulnerability in a phone call and visit from his lawyer George White early in February 1973. For the author's purposes, the smoking gun in the case is *Newsweek*'s acknowledgment in August 1973 that Attorney General Richard Kleindienst had been alerted to the probe as early as February—thus confirming Lawrence Meyer's February 6 lunch with Kleindienst's press spokesman Jack Hushen, who immediately told the attorney general what he had learned from Meyer. This set in motion a string of events culminating in Agnew's resignation eight months later.

274 "because I have been paying off the vice president": Richard M. Cohen and Jules Witcover, *A Heartbeat Away: The Investigation and Resignation of Vice President Spiro T. Agnew* (New York: Viking, 1974), 5.

275 Ford asked Meyer to repeat it: Lawrence Meyer to author, July 10, 2016.

275 "Jack, you're not going to believe what I'm about to tell you": AI, Lawrence Meyer.

276 "Smells to me like your Democratic law firm": AI, Jack Hushen.

276 "If he's done what you say he's done": Ibid.

277 "the best damned birthday present I've ever gotten": James Cannon, *Gerald R. Ford: An Honorable Life*, 114.

277 "They were the ones who blew open the cover up": Earl Silbert Oral History, Historical Society for the District of Columbia Circuit, 103.

277 "Go before the Senate committee": "Watergate Unit Studying Rules," *New York Times*, April 17, 1973.

277 "and have another life": James Cannon, *Gerald R. Ford: An Honorable Life*; "Watergate Unit Studying Rules," *New York Times*, April 17, 1973.

278 "If you have so many enemies": *(Independence, Missouri) Examiner*, August 13, 1974.

278 "an open and shut, cut and dried case": NOHP, Al Haig, November 30, 2007.

279 "The vice president wanted to consult with you": TTH, 102.

279 "Then I'm going after Richardson": NOHP, William Ruckelshaus.

279 "I'm caught between number one and number two": NMP, Gerald Ford, Take III, October 13, 1973, DCC.

279 "along with Senators Percy and Baker": GRF to William Timmons, October 4, 1973, GRFL.

279 "I'm interested in the job for myself": Rowland Evans Jr. and Robert D. Novak, "Jerry Ford: The Eisenhower of the Seventies?," *Atlantic*, August 1974, 26.

280 "I'd like to be in the shape": *Time*, October 22, 1973.

280 He only did it when alone: Ibid., 102.

280 "Jerry, Agnew just resigned": FLOHP, Al Cederberg, September 16, 1998.

280 "Dear Jerry": Agnew to GRF, October 10, 1973, Special Materials, GRF Special Letters, Box A1, GRFL.

281 "What are the procedures?": NMP, Vice presidential selection, Take 4, October 11, 1973, DCC.

282 "I'm going to clear the decks": NOHP, Al Haig.

282 "Have you got any recommendations?": NMP, Vice President Successor, Take 5, October 12, 1973, DCC.

282 "I'll do whatever the president wants me to do": TTH, 104.

283 "Jerry I want you to be my nominee": Nixon came closest to expressing his true feelings about Ford as a potential vice president in a phone conversation with Henry Kissinger the day after Agnew's resignation. After both men minimized Ford's 1976 prospects, Nixon declared, "He is basically a safe Truman." Telecon Nixon/Kissinger, October 11, 1973, 6:35 p.m., NSA.

283 "I've lived with it all my life": *Grand Rapids Press*, December 6, 1973.

283 "I know it's Jerry": *Birmingham News*, October 16, 1973.

283 "on cloud nine": TAI, Bob Hartmann.

284 "Can you deny that Nixon has picked you?": TAI, Paul Miltich.

284 "You will know tonight at 9 pm": Nancy L. Ross, "God, Country and Jerry Ford," *Washington Post*, October 15, 1973.

284 "It's not going to happen": A. James Reichley interview with Gerald R. Ford, March 8, 1978.

284 "The president wants you to be his nominee": NMP, Bonnie Angelo to *Time* Nation, October 13, 1973, DCC.

285 "Mother's been so good to us": "Betty: The New First Lady," *Newsweek*, August 19, 1974.

285 "vice presidents don't really do anything": AI, Steve Ford.

285 "I said I have no intention of running": NMP, Bonnie Angelo, op. cit., DCC.

285 "It's good for the country": NMP, Neil MacNeil, Gerald Ford—Take III, October 13, 1973, DCC.

285 "I'm as happy for him": *Chicago Tribune*, October 14, 1973.

285 "for this Gerald Ford may well be the right man": Donald W. Riegle Jr., "Ford: Nixon's Better Idea," *Washington Post*, December 9, 1973.

285 "a good lineman for the quarterback": *Time*, October 22, 1973.

286 "Here I have been trying": Andrew Glass, "Gerald Ford Favored to Become First Unelected Vice President, October 12, 1973," *Politico*, October 12, 2015.

286 "They were all looking for junk": AI, Maury DeJonge.

286 "Anything you want you'll find it there": Bob Woodward interview with Bob Hartmann, op. cit., WBP, Harry Ransom Center, University of Texas (Austin).

287 "We've got a worse problem than Agnew": NOHP, William Ruckelshaus.

287 "that fucking Harvard professor": Barry Werth, *31 Days: Gerald Ford, the Nixon Pardon, and a Government in Crisis* (New York: Anchor Books, 2006), 30.

288 "No more tapes": Ray Locker, *Haig's Coup: How Richard Nixon's Closest Aide Drove Him from Office* (Lincoln: University of Nebraska Press, 2019), 182.

288 "Elliot Richardson was a consummate politician": AI, Tom Kuyper.

289 "The President had no other choice": *Grand Rapids Press*, October 21, 1973.

289 "Get off your goddamned ass": James Cannon, *Of Time and Chance*, 219.

290 "As sure as I'm alive, General": Bob Woodward interview with Benton Becker, WBP, Harry Ransom Center, University of Texas (Austin).

290 "I do not think a vice president should just rubber stamp": TAFI, 585.

290 "If a play has been made": Carl Leubsdorf, "Ford Shows He's His Own Man," *Honolulu Star-Bulletin*, November 2, 1973.

291 "Can Richard Nixon save his presidency?": Marjorie Hunter, "Ford Favors Role for Congress," *New York Times*, November 6, 1973.

293 Finally, there was Dr. Hutschnecker: "Ford Accuser Faces Charge of Perjury," *Washington Post*, November 8, 1973.

293 "You have a reputation for openness and honesty": *Detroit Free Press*, November 6, 1973.

293 "He puts one in mind of a big sloppy dog": Harriet Van Horne, *New York Post*, November 12, 1973.

294 "The voting record reflects Grand Rapids": "The New President: A Man for This Season," *Time*, August 19, 1974.

294 "I have never before voted for a Republican vice president": *Washington Post*, November 27, 1973.

294 "I believe that Mr. Ford is a decent man": NMP, Neil MacNeil, December 8, 1973, DCC.

295 "It was a very impressive ceremony": Saul Friedman, "In Praise of Honest Ignorance," *Harper's Magazine*, August 1974.

14: "The Worst Job I Ever Had"

296 "He was almost pathological": AI, David Kennerly.

297 "What I have to watch out for": Mark J. Rozell, *The Press and the Ford Presidency* (Ann Arbor: University of Michigan Press, 1992), 22. Ford always maintained a public optimism about Nixon's chances to complete his second term. That he privately harbored more complex emotions was spelled out for Trevor Armbrister, his ghostwriter for Ford's 1979 memoir, *A Time to Heal*. Asked for his reaction on learning of Nixon's White House taping, he was unsparing. "Well, it struck me that the ballgame was in the bottom of the ninth, and we didn't have anybody on base, and we didn't have much chance of prevailing." TAFI, 642.

297 "Jerry, this is scary": AI, Maury DeJonge.

297 "You'll want an advance man": Hartmann, *Palace Politics,* 39–40.

297 "Thanks," said the former minority leader: Ibid., 88.

298 "What we need is three more Henry Kissingers": Memcon, Blair House, March 13, 1974, 11:00–11:45 a.m., Ford, King Hussein, etc., Box 65, Vice Presidential Papers, GRFL.

298 "I want my own team": Hartmann, *Palace Politics*, 35–36. Ford told Hartmann that he didn't want any former members of Congress working for him. "A lot of them are volunteering and most are good guys, but they wouldn't be able to adjust to the new situation. They'd expect things to be just like before."

299 "No, he's from Utah": AI, Roger Porter.

299 "It was a twin engine prop plane": AI, Phil Jones.

299 "Why do you fly this piddling little airplane?": AI, Tom DeFrank.

301 "You will undergo a change of residence": AI, David Kennerly.

301 "the worst job I ever had": AI, Dick Cheney. See also Benton Becker, "Adequacy of Current Succession Law in Light of the Constitution and Policy Considerations," *Fordham Law Review* 79, no. 3 (2011): 897–903.

301 "Ford had to walk this terrible tightrope": AI, Tom DeFrank.

301 "The way to save the Presidency": David Belin, "Jerry Ford at the Crosswords," January 28, 1974, Box 224, Robert Hartmann Subject File, Ford Vice Presidential Papers, GRFL.

301 "You won the hearts of congressional wives": Pat Nixon to Betty Ford, November 28, 1973, Betty Ford Special Letters, Box B 3, GRFL.

302 "They're the kind of people you can trust": David Kennerly, *Shooter* (New York: Newsweek Books, 1979), 155.

302 "Drink this": Ibid., 116.

302 "I sympathize with the press": *San Francisco Chronicle*, April 13, 1974.

302 "you can tell she's on something": Kandy Stroud, "She Likes Being Second Lady," *Women's Wear Daily*, April 19, 1974.

302 "We had an understanding": AI, Jack Marsh. Ford said he limited any speculation about his becoming president to Betty. "In talking with Hartmann and Marsh, they would come in one day and say: Well it looks closer and closer . . . And I would always sort of pooh-pooh it because regardless of what I said, in jest or otherwise, the minute I told anybody, even those two close intimates, the story would somehow get out that, well, Ford is really beginning to undermine Nixon." TAFI, 937.

303 "I am sure the President feels": *Washington Star*, January 11, 1974.

303 "I told 'em I wanted a real tough speech": Hartmann, *Palace Politics*, 104.

304 "It was amazing how many people he knew": AI, Ron Nessen.

304 "We don't think he should get involved in Watergate": *San Francisco Examiner*, January 18, 1974.

305 "You can't mean that": "Michigan Democrat Wins in Voting for Ford's Seat," *New York Times*, February 19, 1974.

305 "No, Mr. President, it is Watergate": TAFI, 508.

305 "I don't happen to believe": Marjorie Hunter, "Ford Bars Plea That Nixon Quit," *New York Times*, March 12, 1974.

306 "The fatal defect of CREEP": Remarks before Midwest Republican Leadership Conference, March 30, 1974, Box 131, Ford Vice Presidential Papers, GRFL.

306 "Will you dance in the White House?": *Grand Rapids Press*, April 4, 1974.

306 "I think he went as far as he felt he could": AI, Roger Porter.

307 "I'm not so sure my father disagrees with me": Congressional Quarterly, *Presidency 1974* (Washington, DC: Congressional Quarterly, 1975), 63.

308 "But when the pages of history are written": DeFrank, *Write It When I'm Gone*, 14.

309 "God knows what the unexpurgated tapes would show": "The Nation: Nixon Has Gone Too Far," *Time*, May 20, 1974.

309 "a continuous series of revelations": "Alone," *New York Times*, May 12, 1974.

309 "Jerry, you're doing too much": TAFI, 795.

309 "If they are, they are doing so": James Naughton, "Ford Friends Had Secret Transition Plan," *Detroit Free Press*, August 26, 1974.

310 "Jerry Ford wanted Phil Buchen in this town": Brian Lamb remarks at Service of Thanksgiving, July 31, 2008, Box 71, Clay Thomas Whitehead Papers, Library of Congress.

310 "I love the act of creating things": Margaret Whitehead remarks, ibid.

310 "We have to do some planning for Jerry": "Clay T. Whitehead, Guide of Policy That Helped Cable TV, Is Dead at 69," *New York Times*, July 31, 2008.

311 "My understanding is that it was Buchen": AI, Roger Porter.

311 "My clear impression is that Ford was aware": Lawrence Lynn Jr. to author, August 24, 2018.

311 "We decided the statement was an implicit one": James Naughton, "The Change in Presidents: Plans Began Months Ago," *New York Times*, August 26, 1974.

312 "There will be great pressure from within": Undated, 018 Transition Team—CTW Initial Notes, Pre-Swearing-In, 1974 Aug. pdf, Box 048, The Papers of Clay T. Whitehead, 61, GRFL.

312 "I had a chat with the Secret Service": "The Zigzagging Missionary," *Time*, June 10, 1974.

313 "Bring the car around": AI, Richard Keiser.

313 "He's got a congressional point of view": NMP, Bonnie Angelo to *Time* Nation, May 30, 1974, DCC.

313 "It's a bad call, Mr. President": TAFI.

314 "small gauge, very partisan": Richard Reeves, "Is Jerry Ford Big Enough?," *New York Magazine*, May 6, 1974.

314 "The vice president should go underground": Mary McGrory, *Washington Star-News*, May 13, 1974.

314 "Can you see Jerry Ford sitting in this chair?": Admitting he felt "damn mad" on reading this reported put-down of the vice president in *Newsweek*, Ford asked Nixon to his face whether he had in fact made the derisory comment. "He said that this was one of those stories that was made up by our enemies." TAFI, 806. Only much later did Jim Cannon confirm the story, which he heard from Rockefeller personally.

314 "I hope you don't mind": JCI, William Whyte.

315 "Ford said to me": AI, Jack Marsh.

315 "Gee Mom, you've gotten good-looking": *(New Orleans) Times-Picayune*, March 19, 1974.

315 "Do you ever become accustomed to this?": Stroud, "She Likes Being Second Lady."

316 "Go on down": TML, 154.

317 "I have a feeling that he's going to be impeached": AI, Bill Gill.

317 "Larry, I think you're making a mistake": TAFI, 838.

317 "The decision was not what we expected": AI, Jerry Jones.

318 "We're thinking about not obeying the order": AI, Robert Bork.

318 "Before I leave": FLOHP, David Broder, February 25, 1997.

319 "Fred usually kept the tapes": AI, Jerry Jones.

15: "My God, This Is Going to Change Our Whole Life"

320 "the major problem confronting America": "Excerpts from Nixon's Address on the Economy," *New York Times*, July 26, 1974.

320 "we have no government at present": Arthur F. Burns Journal, July 25, 1974, Accession 2006-NLF-057, 122, GRFL.

322 "It's getting close": AI, Leon Parma.

322 "End career as a fighter": Jonathan Aitken, *Nixon: A Life* (Washington, DC: Regnery, 2015), 612.

323 Things were "deteriorating": TTH, 2.

323 "How's the President holding up?": Hartmann, *Palace Politics*, 127.

323 "Never mind": Ibid., 128.

324 "I've been lied to for the last time": Bob Woodward interview with Charles Wardell, March 19, 1998, WBP, Harry Ransom Center, University of Texas (Austin).

324 Finally Nixon could resign his office: TTH, 4.

324 "One thing is sure": Alexander Haig with Charles McCarry, *Inner Circles: How America Changed the World* (New York: Warner Books, 1992), 483.

325 "I'll have to think about all these things you've told me": JCI, Gerald R. Ford, April 22, 1993, Part 1A. "I think I acted in this case like I had acted in most instances where I had a big issue presented to me and I had to make a decision," Ford told Cannon. "I think I acquired over the years a capability of listening to the pros and cons and then saying, 'Well, I want to think about it.'"

325 "a very, very accurate summary": AI, Jack Marsh.

325 "What did you tell him?": Bob Woodward, *Shadow: Five Presidents and the Legacy of Watergate* (New York: Simon & Schuster, 1999), 7.

326 "before this goes any further": James Cannon, *Of Time and Chance*, 296.

326 the vice president seemed unusually quiet: Betty Beale, *Power at Play: A Memoir of Parties, Politicians and the Presidents in My Bedroom* (Washington, DC: Regnery, 1993), 220.

326 "My God, this is going to change our whole life": Andrew Downer Crain, *The Ford Presidency: A History* (Jefferson, NC: McFarland, 2009), 26.

326 "Mrs. Ford was far more eager than he": Bob Woodward interview with Phil Buchen, April 8, 1998, WBP, Harry Ransom Center, University of Texas (Austin).

326 "He came home with the word from General Haig": TAI, Betty Ford.

327 "I told Al I had talked to Betty": TTH, 10.

327 "I hope you understand there was no agreement": Bob Woodward, "Closing the Chapter on Watergate Wasn't Done Lightly," *Washington Post*, December 28, 2006.

327 "Goddamnit, what did you do to me": Woodward, *Shadow*, 8.

328 "He just wanted to get it all over with": Hartmann, *Palace Politics*, 135.

328 "What are your plans?": TAI, Hugh Scott.

329 "You're all we've got now": Ibid.

329 "There is just no misunderstanding": "Seven Days in August," *Newsweek*, August 19, 1974.

329 "I cannot for a moment believe": Anthony Lewis, "Abroad at Home; Questions for the General," *New York Times*, January 8, 1981.

330 "Let them impeach me": Haig, *Inner Circles*, 487.

330 "Bob, do what you think is right": Charles E. Harmon, "Our National Nightmare: Michigan's Watergate Connections," *Michigan History Magazine*, July/August 1999, 75.

331 "Pompous little jackass": Haig, *Inner Circles*, 489.

331 "Damn it, Al, this is not what I asked for": Ibid., 491.

332 "we don't want the old man resigning right now": AI, Pat Buchanan.

332 "Jesus Christ," said Burch: Ibid.

332 "The public interest is no longer served": John Herbers, "Ford Seeks to Stand Aloof in Impeachment Debate," *New York Times*, August 6, 1974.

332 "It's over," said Elfin: Tom DeFrank, "Five Days in August: What It Was Like to Report Watergate," *Atlantic*, August 2014.

333 "This is unreal": TAI, George H. W. Bush.

333 "If the man had made up his mind": Bob Woodward interview with James Lynn, undated (file added October 15, 2011), WBP, Harry Ransom Center, University of Texas (Austin). Woodward relied on Lynn's notes, the most complete contemporary account of Nixon's final, somewhat surreal, Cabinet meeting.

334 "Everyone here recognizes the difficult position I'm in": Ford's statement, underlined for emphasis, is in Box 27 of the Philip Buchen Files at the Ford Library.

334 "Is that *all* the President had to say?": *Newsweek*, August 19, 1974.

335 "I screwed it up good, real good, didn't I?": James Cannon, *Gerald R. Ford: An Honorable Life*, 198.

335 "Now for Christ's sake": TAI, Barry Goldwater.

335 "Do you want me to put it back?": AI, Dorothy Downton.

335 "Even in times of worst crisis": TAI, Phil Buchen.

335 "you'd better tell me what's going on": Bob Woodward and Carl Bernstein, *The Final Days* (New York: Simon & Schuster, 1976), 400.

336 "It is happening; we'll rise to it": Ibid., 401.

336 *The Making of the President in 72 Hours*: James Naughton, "The Change in Presidents," *New York Times*, August 26, 1974.

337 "Look after the President, Al": Haig, *Inner Circles*, 498.

337 "Your friend wants you and Barbara to be back here": AI, Leon Parma.

337 "Please give him a straight story": *Newsweek*, August 19, 1974.

338 "Some were turkeys": *Time*, August 19, 1974.

338 "The bar is open": Naughton, "The Change in Presidents."

338 "We can sit here and speculate": Phil Buchen interview, WBP, Harry Ransom Center, University of Texas (Austin).

338 "How long do we have?": Naughton, "The Change in Presidents."

338 "We have to hurry up[,] guys": JCI, William and Peggy Whyte, February 26, 1991.

339 "I really hate to leave this pool": "Enter Ford," *Time*, August 19, 1974.

339 "The President wants to see you": TAFI, 1103.

340 "I have made the decision to resign": Ibid., 1106.

340 "I will not call you unless you call *me*": TAFI, 3455.

341 "This is the last time I'll call you Jerry": Christopher Klein, "The Last Hours of the Nixon Presidency," History Stories, History.com, August 8, 2014, www.history.com /news/the-last-hours-of-the-nixon-presidency-40-years-ago.

341 "I'll see you tomorrow": TTH, 30.

341 "Al said that he came away from the meeting": Bob Woodward interview with J. Fred Buzhardt, March 1975, Container 4.16, WBP, Harry Ransom Center, University of Texas (Austin).

342 "Better than I am": Hartmann, *Palace Politics*, 159.

342 "You're the President of the United States": AI, Jack Marsh.

342 "Oh, I've got to be there": TTH, 31.

343 "Christ, Jerry[,] isn't this a wonderful country?": TAFI, 1132.

343 it flowed from the typewriter of speechwriter Milton Friedman: FLOHP, Milton Friedman. Interestingly, Hartmann declined to take personal credit for the phrase in the days immediately after Ford's inauguration, telling reporters, "the function of a ghost is to be invisible." *Washington Star-News*, August 26, 1974.

344 "Yes, of course": TTH, 33.

344 "The Michigan mafia is taking over": ROD, August 1974.

16: Change and Continuity

346 "We have a new coach": "Bing, Bing: Game Over," *States-Item (New Orleans)*, August 10, 1974.

346 "It would have had a better ring": TAFI, 1136.

347 "Like VE Day if you were German": AI, Aram Bakshian.

347 "A great weight has been lifted": "The People Take It in Stride," *Time*, August 19, 1974.

347 "The prisons of Georgia": *Time*, August 19, 1974.

347 "Not after this speech, Al": Werth, *31 Days*, 30–31.

348 Presented with three names to choose from: The three names in question were Rumsfeld, Frank Carlucci and Bill Clements.

349 "the saddest day of my life": "Woman of the Year," *Newsweek*, December 29, 1975.

349 "I will keep a muzzle on my mouth": Ted Roelofs, "At Her Funeral in Grand Rapids, Betty Ford Is Remembered as a 'Trailblazer,'" *Grand Rapids Press*, July 14, 2011.

349 "These are the President's kids": "The First Family's First Days," *Time*, August 19, 1974.

349 "You'll find the Lincoln Sitting Room": "Washington Without Nixon," *Grand Rapids Press*, August 18, 1974.

349 "Well, Betty, you'll see so many of these": "The Nixon departure was difficult," Ford wrote James Cannon on April 22, 1993. "We strained to say the right words. It wasn't easy. We waved as the helicopter took off for Andrews Air Force Base. We turned and walked into the White House, holding hands. We were joined in determination to meet our new responsibilities."

349 "Drop us a line if you get the chance": Robert Sam Anson, *Exile: The Unquiet Oblivion of Richard M. Nixon* (New York: Simon & Schuster, 1984), 20.

350 "We can do it": Werth, *31 Days*, 11.

350 "To the President": Hartmann, *Palace Politics*, 170.

350 "I don't think that's on the schedule": ROD, August 11, 1974.

351 "And then the whole mood changed": Bush, *All the Best*, 195.

351 "I wanted to appear strong": TAFI, 1150.

352 "Authentic Jerry Ford": *Time*, August 19, 1974.

352 "All the Nixon people left": AI, Leon Parma.

352 "I was standing there with Al": AI, Gerald Warren.

353 "A tiny little guy": AI, Bob Schieffer.

353 "do not commit yourself": Werth, *31 Days*, 13.

353 "He has our blessing": Memcon, Briefing for Arab Ambassadors, August 9, 1974, GRFL.

354 "he wanted his good friends to give him hell": "Meeting in the Cabinet Room," August 9, 1974, 5:30–6:30 p.m., DRA.

354 "I already sent you my letter of resignation": AI, Leon Parma.

354 "He said not to mess": AI, Dick Cheney.

354 "people could speak their minds honestly": "Meeting in the Cabinet Room," August 9, 1974, 5:30–6:30 p.m., DRA.

354 "Sit down, sit down": AI, Leon Parma.

355 "Just wear your golf shirt": Ibid.

355 "My friends, we've got a lot of work to do": Hartmann, *Palace Politics*, 179.

355 "Jerry, there's something wrong with this picture": AI, Steve Ford.

356 "Total access all the time": David Kennerly, *Extraordinary Circumstances: The Presidency of Gerald R. Ford* (Austin: Center for American History, University of Texas, 2007), 14.

356 "Tell him to call back, I'm busy": Ibid., 15.

356 "Well, you better get over here then": Ibid.

357 "The first time I tried to leave this office": David Priess, *The President's Book of Secrets* (New York: PublicAffairs, 2016), 80.

357 "I really think that you're better off": FLOHP, William Simon, November 11, 1998.

357 "[A] great president": Memcon, Cabinet Meeting—August 10, 1974, 1. At this same meeting Attorney General Saxbe expressed a hope that the clamor for Nixon's scalp "might just sort of die out"—precisely the kind of wishful thinking to which Ford himself would cling until his eye-opening press conference of August 28. GRFL.

357 "I don't think anybody wants": Barry Koek, "Jaworski Cautious on Nixon Change," *Washington Star-News*, August 11, 1974.

357 "Please be affirmative with them": Cabinet Meeting, August 10, 1974.

357 At least US intelligence agents thought so: The story of the *Glomar Explorer* is told in "Project Azorian," *CIA Studies in Intelligence* (Fall 1985), made public by the National Security Archive on February 12, 2010.

358 "the whole gut-fighting attack group": George H. W. Bush Diary, "September the 10th," https://www.archives.gov/exhibits/eyewitness/html.php?section=12.

358 Soon it was noticed: ROD, September 12, 1974.

358 "He's on a bus full of people": J. Brooks Flippen, *Conservative Conservationist: Russell E. Train and the Emergence of American Environmentalism* (Baton Rouge: Louisiana State University Press, 2006), 163.

358 "Milt Friedman . . . spent the whole day": ROD, August 11, 1974.

359 "There continued to be": AI, David Gergen.

359 "Mr. Ford said we had to try and get along": AI, Anne Holkeboer.

359 "We've got people coming in": AI, Terry O'Donnell.

359 "Boy, I'll tell you, Bob": AI, Bob Barrett.

360 Obscured from all three men: Bill Gully with Mary Ellen Reese, *Breaking Cover* (New York: Simon & Schuster, 1980), 228–30.

360 "That truck does not leave here": AI, Benton Becker.

361 "All my children have spoken for themselves": TML, 167–68.

361 "I can't imagine anything worse": Hugh Sidey, *Portrait of a President* (New York: Harper & Row, 1975), 119.

362 "Hi, I am Jerry Ford": Vernon Ehlers, *Gerald R. Ford, Memorial Tributes Delivered in Congress* (Washington, DC: U.S. Government Printing Office, 2007), 21.

362 "In Nixon's day you'd go over there": AI, Tom Kauper.

362 "Secretary Coleman thought it was constitutional": AI, Ken Lazarus.

363 "Hi, Charlie!": *Time*, August 26, 1974.

363 "What in hell are you doing?": TAI, Dean Burch.

364 "Today we have a new president": Cost of Living Task Force, Hearings Before the Committee on Banking, Housing, and Urban Affairs, United States Senate, August 14 and 15, 1974, 83.

364 "Turkey is more important to us": Memcon, August 13, 1974, 9:00 a.m., Ford, Kissinger, Scowcroft. GRFL.

365 "Lists have ends to them": AI, Tom Korologos.

365 "He just goes on and on and on": "The Once and Future Ford," *Newsweek*, August 19, 1974.

368 "This is not just for the Jewish emigrants": Memcon, August 15, 1974, 8:00 a.m., Ford, Kissinger, Scowcroft, Timmons; Senators Jackson, Javits, Ribicoff. GRFL.

368 "Jerry, we can't do this": AI, Jerry Jones.

369 "Do you feel good, executing a sick man?": Werth, *31 Days*, 103.

369 "We need to be strong to negotiate withdrawal": Memcon, White House, 11:30–12:30 p.m., August 16, 1974, Ford, King Hussein, Kissinger, etc. GRFL.

370 "This house has been like a grave": "Chatting with Betty and Susan," *Time*, August 26, 1974.

370 "I'm just going to tell the guests": *Time*, August 26, 1974.

370 "My God," Henry Kissinger joshed: At one early Ford White House dinner, attended by guests considered persona non grata by the previous administration, *Washington Post* owner Katharine Graham declared, "It's marvelous—all the underground is here." "Betty and Jerry Are at Home," *Time*, December 30, 1974.

17: The Pardon

372 "You can't pull a bandage off slowly": Rhondi Jeffrey-Jones, *The CIA and American Democracy* (New Haven, CT: Yale University Press, 2003), 196.

372 "I really want to bind up the wounds": Hugh Sidey, "So Like the Rest of America," *Time*, September 2, 1974. In its August 20 report, Ford's transition team included an "earned reentry" program among its recommendations for the new president. The complete report is in Box 63, Phil Buchen Files, GRFL.

373 "Steve hasn't registered for the draft": AI, Jack Marsh.

374 "The purpose of amnesty": "The Amnesty Issue," *Time*, September 9, 1974. See also Alan Jaroslovsky, "Gerald Ford's Clemency Board Reconsidered," *Federal History Journal*, no. 13 (2021): 127–50.

374 From the island of Maui: "I didn't want to come right out and campaign for the Vice Presidency," Richardson told Henry Kissinger, even as he did exactly that. Telecon Elliot Richardson/Secretary Kissinger, August 13, 1974, 12:55 Virtual Reading Room, https://foia.state.gov/Search/Results.aspx?collection=KISSINGER.

375 "Let's say that I narrowed it down to six": Hartmann, *Palace Politics*, 229.

375 "This is the closest I'm ever going to get": Richard Norton Smith, *On His Own Terms: A Life of Nelson Rockefeller* (New York: Random House, 2014), 643.

376 "Amnesty and Rockefeller in the same 24 hour period": "You can kiss the Republican Party goodbye," said Goldwater. "He left the party three times and now he gets the cream and sugar." Box C1, Folder: Presidential Handwriting, 8/22/1974 (2), GRFL.

376 "Any way we can get Nixon back?": Ibid.

376 "It was the impression and not the reality that counted": William A. Syers interview with Mel Laird, June 6, 1985, GRFL.

376 "Rumsfeld stopped me in the hall": AI, Jerry Jones.

377 Ford had "put a lock on his records": Bob Woodward interview with Al Haig, Subseries C, Container 72.7, WBP, Harry Ransom Center, University of Texas (Austin).

377 "had the feeling everybody was working for the cause": TAI, Bob Hartmann. Hartmann wasn't the only accomplished White House leaker, or so he told Dave Horrocks of the Ford Library staff on September 2, 1998. "This is something nobody will believe but it's true. I never leaked a story that Ford hadn't told me to leak or that I hadn't told him that I intended to leak if it was all right with him. Then I waited for the stories to come back blaming everyone—not only mine but all the others as well—because I'm a goddamned former journalist and I take care of my friends, you know. But Ford just laughed at these stories. The people that were telling him didn't know why he was laughing so hard." FLOHP, Bob Hartmann.

378 "I didn't see the Oval Office": AI, Benton Becker.

378 "Those damn things are not leaving here": AI, Benton Becker.

378 "You have made a great beginning": President's Mail—August 30, 1974, President's Handwriting File, Box C2, GRFL.

378 George McGovern promised his help: Ibid., August 31, 1974.

379 A different kind of threat: "Police, CIA Thwarted 1974 Plot to Kill Ford," *Atlanta Constitution*, December 19, 1984.

379 "Those characters have finally learned to count": AI, Hal Bruno.

379 "Where in the hell's my pardon?": AI, Jerry Jones.

380 "Dammit!" Ford said on learning the news: Hartmann, *Palace Politics*, 199.

380 "Fuck Haig. I work for the President": Richard Reeves, "Power Failure in the White House," *New York Magazine*, October 20, 1975.

380 "It's distracting": AI, Jerry Jones.

381 "They're lawyering it now": Werth, *31 Days*, 214.

381 "You don't want to read this thing": JCI, Phil Buchen, February 15, 1991, GRFL.

382 "Well, that's the first one": "Ford: Plain Words Before an Open Door," *Time*, September 9, 1974.

382 "Yet for me to say": TAFI, 1990.

382 "just the right balance": *New York Times*, August 29, 1974.

382 "Is this the way it's going to be": AI, Tom DeFrank.

383 Reluctant to saddle Jaworski: Bonnie Buchen Diary notes, August 28, 1974, Philip W. Buchen Papers, 1973–1992, Box 2, GRFL.

383 "Look, if you're going to do this": TTH, 164. Asked long after to identify Ford's biggest mistake as president, Hartmann replied, "that he didn't wait a month to pardon Richard Nixon."

383 "I don't need anything in writing": FLOHP, John Logie, March 4, 2014.

383 "it was just like any football game": Werth, *31 Days*, 224.

384 "Will there *ever* be a right time?": Hartmann, *Palace Politics*, 259.

384 "I don't need the polls": TTH, 162.

385 "If he resigns, we can drop all this": Robert McClory, "Was the Fix In?," *National Review*, October 14, 1973.

385 "I don't like the idea of amnesty": "Next, a Vietnam Amnesty," *Newsweek*, September 16, 1974.

385 "Amnesty isn't supposed to be punishment": "Limited Program, Limited Response," *Time*, September 30, 1974.

386 "There are many angles to it": Nancy Gibbs and Michael Duffy, "The Other Born-Again President?," *Time*, January 2, 2007. See Part II, "The Pardon of Richard Nixon," HUC, 15–75. This examination of Ford's decision to pardon his predecessor includes remarks by Stephen Ambrose, Stanley Kutler and Mark Rozell, with responses from Benton Becker, Bob Hartmann, Jerry terHorst, William Hungate, Bob Woodward and Elizabeth Holtzman—a stellar cast of historical participants whose memories remain fresh, even if, over time, their views of the pardon itself may have changed.

386 Jaworski appeared a lonely: Bonnie Buchen Diary notes, September 5, 1974, op. cit., GRFL.

387 "It sounded like he was saying": Laura Kalman, *Right Star Rising: A New Politics, 1974–1980* (New York: W. W. Norton, 2010), 54.

387 "Can't tell you why I'm asking it": TAI, Phil Buchen.

387 Harlow "told me and I agreed": AI, Henry Kissinger.

387 "If you know a man is about to walk a plank": *U.S. News & World Report*, September 9, 1974, Box 18, Robert Hartmann Files, GRFL.

387 "I have never, even slightly": Littlefair, Duncan (1912–2004), Harvard Square Library, A Digital Library of Unitarian Universalist Biographies, History, Books, and Media. Also see, Robert Sherrill, "What Grand Rapids Did for Jerry Ford—and Vice Versa," *New York Times*, October 20, 1974. In a newly disclosed interview with Trevor Armbrister, Buchen was explicit: "Course I was influenced by Littlefair" on emphasizing mercy over practical considerations—Ford's considerations—of changing the national conversation.

389 Americans in their hearts: John Logie, who had Phil Buchen for his godfather, claimed that Littlefair met with Ford as well as "Uncle Phil" for a discussion of the Nixon pardon. It is an assertion no one else has made. That the visiting pastor *did* spend time in the White House complex on the afternoon of September 4 is attested to by Secret Service log-in records, which show him being directed to Buchen's office around two o'clock.

   Here special mention should be made of Michael W. Grass, an enterprising Grand Rapids writer and filmmaker, who discovered the Littlefair-Buchen connection to the Nixon pardon, an action Littlefair publicly defended (while saying nothing of his personal involvement) in a September 15, 1974, sermon at Fountain Street Church. In October 2003, Grass donated to the Ford Library Buchen's January 10, 2001, letter describing Littlefair's involvement in the days preceding Ford's Sunday-morning bombshell. It says nothing of any Ford-Littlefair meeting. My own narrative is based largely on contemporaneous notes by Buchen's wife, Bunny, for a projected diary that seems to have gotten no further than January 1, 1975.

389 The lawyers agreed: Benton Becker, Memorandum: History and Background of Nixon Pardon, September 9, 1974, Box 2, Benton Becker Papers, GRFL.

389 According to Al Haig: Bunny Buchen Diary Notes, September 6, 1974, op. cit., GRFL.

390 "Some of my staff people are too": AI, Benton Becker.

390 "Hello, Mr. Becker": Benton Becker Memorandum, September 9, 1974, GRFL.

390 "Because I'm going home": Ibid.

391 "We didn't tell Jerry": TAFI, 2031–2032.

391 "He was looking for a way out": AI, John Hushen; this view is seconded by John Carl-

son, who went on to serve as Ron Nessen's deputy, as well as the late Bonnie Angelo of *Time*, who told the author that terHorst acknowledged as much to her. In another twist, speechwriter Milton Friedman revealed to a Ford Library interviewer in 1998, "Jerry really wanted to bring out a book about Ford, which he wouldn't have been able to do had he remained." The book in question duly appeared before the end of 1974.

391 "The job was overwhelming": AI, Ron Nessen.

391 "No statement is better than that statement": TAI, Benton Becker.

392 "Ron, he's right": Miller's sixteen-page memo to Jaworski, in effect a plea for leniency on the grounds that pretrial publicity made it impossible for Nixon to receive a fair trial, can be found in Box 16, Folder: Richard Nixon-Pardon: General, Ron Nessen Papers, GRFL.

393 "Am I going to like it?": JCI, John Rhodes, December 4, 1990.

393 "I don't mean '76": TTH, 172.

394 "We can't tolerate any weakened statement": Ibid., 173.

394 Laird remembered things differently: AI, Mel Laird. Judge Robert Bork spoke for many in dismissing Laird's latest scheme. "I think if I were a congressman and were asked to participate in that thing, I'd say, 'I'm terribly busy. I'll see you around!" AI, Robert Bork.

394 On second thought, Richardson: FLOHP, Elliot Richardson, April 25, 1997.

395 "You know what this means?": In his Ford book, terHorst does not recount how he first learned of the impending pardon. In his White House memoir, *Palace Politics*, Bob Hartmann says terHorst was told about the pardon by the president on Friday night (Ford was returning to Washington after a bicentennial event in Philadelphia). Pat Buchanan recalls a conversation he had with terHorst in late August, during which the Nixon loyalist protested Ford's conditional amnesty for Vietnam draft resisters. "Well, we've got to do this one before we do the other one," terHorst replied. AI, Pat Buchanan.

395 how refreshing it was: Walter Heller, *Washington Post*, September 8, 1974.

395 "You'll find out soon enough": Barry Werth, "The Pardon," *Smithsonian Magazine*, February 2007.

396 "What are you pardoning him for?": TAI, Barry Goldwater.

396 "If I had known on Friday": *Piqua (Ohio) Daily Call*, September 10, 1974.

396 "Oh no," said one member of the group: AI, Jack Hushen.

396 "Mr. President, I have something": terHorst, *Gerald Ford and the Future of the Presidency*, 226.

397 "I hope you would reconsider": TTH, 176.

397 "Don't hurt the President this way": Werth, *31 Days*, 319.

397 "I am sworn to uphold our laws": APP, "Remarks on Signing a Proclamation Granting Pardon to Richard Nixon," September 8, 1974.

397 "I can see nothing": Jerald terHorst to GRF, September 8, 1974, 11:15 a.m., GRFL.

398 "You just don't understand": John Osborne, "The Pardon," *New Republic*, September 28, 1974.

398 "Why must it be tomorrow?": Hartmann, *Palace Politics*, 264.

399 "At times the father of a family": AI, Steve Ford.

399 "Do you know that Jerry terHorst is thinking of resigning?": ROD, September 10, 1974.

18: Clearing the Decks

401 "I was driving a car": AI, Robert Bork. No one was more shocked by Ford's action than the president's brother Dick. "My first reaction was the same as everyone else's—how could he pardon that crook?" AI, Richard Ford.

401 Recanting their earlier support: Paul Newman and Joanne Woodward to GRF, September 11, 1974. White House Central Files Subject File on Judicial and Legal Matters, File Unit: JL1/Nixon: Amnesties-Clemencies-Pardons/Nixon.

401 "you can't do the will of God": Johnny Cash to GRF, September 10, 1974, GRFL.

401 "Two days ago": Mary McGrory, *Washington Star-News*, September 10, 1974.

402 "The phone calls": TAI, Benton Becker.

402 "Hey, you Ford people": AI, Tom Korologos.

403 "Leon Jaworski called me": Ford Special Materials Collection, Buchen to GRF, 9/17/74, Box A1, Folder: Phil Buchen, GRFL. "Watergate Grand Jury Tried to Indict President Richard Nixon," June 17, 1982, UPI Archives.

403 That night Ford took a call: Handwritten notes from Nixon-Ford telephone call, September 17, 1974, Ford Library Vertical Files, Folder: Nixon, Richard Milhouse (1913–1994)—Pardon.

403 "We will do our best": Memcon, September 13, 1974, 12:15 p.m., Ford, Rabin, Kissinger, Scowcroft, Ambassador Simcha Dinitz. GRFL.

403 "We were crushed": AI, Bob Schieffer. One of the more perceptive commentaries was that of *Time*'s Hugh Sidey. "In over simplified terms, the pardon came down to a compromise," Sidey wrote, "the kind he had negotiated as minority leader. The nation's problems needed attention, but they would not have it if everyone was preoccupied with the pardon and the tapes." Hugh Sidey, "Second Sight on the Pardon," *Time*, October 7, 1974.

404 "As long as Nixon was in": *Newsweek*, September 30, 1974.

405 "Totally different job skills": AI, William Brock.

405 "Kind of holding your cards close to your vest": FLOHP, Donald Rumsfeld.

405 "to articulate a vision": Rozell, *The Press and the Ford Presidency*, 15.

406 "Headlines can only be written": TAI, James Cannon.

406 Keiser immediately sought out Ford: AI, Richard Keiser.

407 "Is this Mr. Whyte?": "Ford as Mr. Right," *Newsweek*, August 11, 1975.

407 "He always accepts responsibility": TAI, Jack Marsh.

407 As minority leader: AI, Bob Schieffer.

408 "I didn't know whether to stand up or salute": AI, Gary Walters.

409 "I couldn't stand to look at soldiers": TML, 174.

409 "Ford's probably the only man in history": TAI, James Cannon.

409 "Marsh was the kind of a guy": Ibid.

409 but she begged off: AI, Bonnie Angelo.

410 "always standing on the sidelines": Ron Nessen, *It Sure Looks Different from the Inside* (New York: Simon & Schuster, 1978), 11.

410 "If I can't sell the program": Ibid., 13.

410 "You are a White House reporter": AI, Ron Nessen.

411 "You're the president of the United States": Woodward, *Shadow*, 29.

411 "I was trying to do too many things": TAFI, 3534.

411 "a sad looking Barca Lounger": TAI, Clem Conger.

412 "A terrible concept": AI, Don Rumsfeld.

413 Ford had unwittingly yielded leadership: FLOHP, Don Rumsfeld. See also Stephen J. Wayne, "Running the White House: The Ford Experience," *Presidential Studies Quarterly* 7, no. 2/3 (Spring–Summer 1977): 95–101.

413 "a Haldeman who smiles": AI, H. P. Goldfield.

413 "One of the smartest political operators": AI, James Cannon.

413 "If we keep getting consumed": FLOHP, Don Rumsfeld.

413 "The second time they said": AI, Dick Cheney.

414 "If they don't like it": Lisa McCubbin, *Betty Ford* (New York: Simon & Schuster, 2018), 151.

414 "I don't know": AI, Sheila Weidenfeld.

414 "Nobody ever speaks to me": TAI, Betty Ford.

415 "He was *never* home before": AI, Anne Holkeboer.

415 "Baloney," she wrote: Nessen, *It Sure Looks Different*, 28. Mrs. Ford looked askance at political staff apt to heap praise on her husband as part of their job, "whereas I think the wife is a more critical person. At least *I* am my husband's strictest critic. He comes to me for it." Betty Ford, "How I Influence Jerry," *Family Circle*, November 1975.

415 "If he doesn't get it in the office": Maurine H. Beasley, *First Ladies and the Press: The Unfinished Partnership of the Media Age* (Chicago: Northwestern University Press, 2006), 136.

415 "The new Dolly Madison has arrived!": Peggy Stanton, "Dear Betty and Nancy," August 18, 1974, GRFL.

416 "The president always kept his little notebook": AI, Maria Downs. Ms. Downs's unpublished memoir, *Mostly Wine and Roses*, is an invaluable record of White House entertaining during the Ford years. GRFL.

416 "You are," said Ford: AI, Don Penny.

417 "The members of my trio": *Washington Star-News*, February 6, 1975.

417 "The Shah likes attractive young ladies": The lavish party for the Shah and his Shabanou is described in the June 2, 1975, issue of the *New Yorker*. The Shah wasn't alone in his appreciation of the opposite sex. David Kennerly remembers a state dinner for Pakistan's visiting prime minister Ali Bhutto, which he attended with his date Candice Bergen. "We left early. The next morning—I'm always in there early—the buzzer rings and it's the President. 'Come up here,' he said. 'Where did you go last night??' I said, 'Ah, well, we got tired.' He said, 'I wanted to dance with Candice. Edgar Bergen is my old friend.' I said, 'You didn't want to dance with her because of Edgar Bergen, don't give me that.' But it was funny. He was great. It was so fun to be around him." AI, David Kennerly.

417 "No tinsel, no sequins": "Woman of the Year," *Newsweek*, December 29, 1975.

418 "This does a lot for the ego": "Betty Ford: Balm for the Ego," *Time*, October 7, 1974.

418 "I'm not so good on the small things": FLOHP, Donna Lehman interview with Patricia Mattson, GRFL.

419 "Sometimes I wish": *New York Times*, September 28, 1974.

419 "I see you are having a party": *Newsweek*, October 7, 1974.

419 "the loneliest of my life": TTH, 191.

419 "No written words": McCubbin, *Betty Ford*, 171–72.

419 "Good night, Sweet Prince": Sidey, *Portrait of a President*, 125.

420 "That was Dr. Lukash": Hartmann, *Palace Politics*, 294.

420 "Dr. Lukash assured me": PP, August–December 1974. "Remarks Concluding the Summit Conference on Inflation," September 28, 1974, 205.

421 "Mother and Dad could have said": AI, Susan Ford Bales.

421 "She sent word": AI, Ron Nessen.

421 "You know, I don't have any business": AI, William Archer.

421 "You have been through a rough ordeal": Richard M. Nixon to Betty Ford, October 1, 1974, GRFL.

422 "Everybody sat in the hospital waiting room": Mary McGrory, *Washington Star*, October 1, 1974.

422 "With all the drains": McCubbin, *Betty Ford*, 175.

422 "Don't be silly": Ibid.

422 "Not bad," she concluded: "Susan Ford: From Student to Stand-In First Lady," *People*, October 21, 1974.

423 "Oh my God, we're saved": AI, Benton Becker.

424 "Labor leaders led by George Meany": Reeves, *A Ford, Not a Lincoln*, 157.

424 "Let us not be beguiled": *Newsweek*, September 16, 1974. An excellent source on Ford's break with Keynesian orthodoxy is Andrew D. Moran, "More Than a Caretaker: The Economic Policy of Gerald R. Ford," *Presidential Studies Quarterly* 41, no. 1 (March 2011): 39–63.

424 "We must stop kidding ourselves": Milton Friedman, "Statement at the Conference on Inflation," September 27–28, 1974, *The Collected Works of Milton Friedman*, compiled and edited by Robert Leeson and Charles Palm, miltonfriedman.hoover.org.

425 "We discussed it in the Economic Policy Group": TAFI, 1608.

427 "the power of the atom": Joint Session of Congress, October 7, 1974, Box 1, PSF: Reading Copies, GRFL.

427 "to recommend the uncomfortable": TTH, 195.

19: "Run Over by History"

431 "We were talking about something": AI, Jack Marsh.

432 "He isn't going to get hurt": Woodward, *Shadow*, 25. "Jaworski has submitted his resignation," wrote Bunny Buchen. "He was strongly in favor of Jerry's Pardon but never could admit it (or felt he couldn't)—because his whole Prosecution Staff threatened to resign in protest if he stated approval. His deputy *did* resign." As for her husband, Bunny said he felt "partly betrayed" due to Ford's failure to divulge his earlier meetings with Haig, St. Clair, etc., as well as Jaworski's unwillingness to admit "that he favored the pardon." Bunny Buchen Diary Notes, October 12, 1974, GRFL.

432 "The whole world knows": Charles W. Sandman Jr. to GRF, October 1, 1974, PHF, Box 4, GRFL.

432 "I am here not to make history": APP, "Statement and Responses to Questions from Members of the House Judiciary Committee Concerning the Pardon of Richard Nixon," October 17, 1974.

433 "They never asked for one document": NOHP, Elizabeth Holtzman.

433 "pretty anticlimactic": *Washington Post*, October 18, 1974. Bunny Buchen Diary Notes, October 12, 1974, GRFL.

434 "So I'll tell you what": AI, Pat Buchanan.

434 "Claude Brinegar was Nixon's secretary of transportation": AI, Dick Cheney.

435 "Look, these guys aren't ready for the meeting yet": Patricia Dennis Witherspoon, *Within These Walls: A Study of Communication Between Presidents and Their Senior Staffs* (Westport, CT: Praeger, 1991), 156.

435 "There is something fishy about the word 'intellectual'": Tevi Troy, "Robert Goldwin, R.I.P.," *National Review*, January 13, 2010.

435 "the first conservative": Ibid.

435 After one such seminar: A. James Reichley interview with Robert Goldwin, February 27, 1978, Box 1, Ford White House Series, GRFL.

435 "Ford always brought the talk down": Robert Goldwin, March 31, 1994, Yanek Mieczkowski Research Interviews, 1994–2007, GRFL.

436 "Stick to a shopping list": APP, "Remarks to the Annual Convention of the Future Farmers of America," Kansas City, MO, October 15, 1974.

436 Gerald Ford's Bay of Pigs: A. James Reichley interview with Richard Cheney, May 20, 1977, GRFL.

436 "You cannot talk your way out of inflation": Tracy Lucht, *Sylvia Porter: America's Original Personal Finance Columnist* (Syracuse, NY: Syracuse University Press, 2013), 142.

437 "It's going to be an avalanche": Yanek Mieczkowski, *Gerald Ford and the Challenges of the 1970s* (Lexington: University of Kentucky Press, 2005), 61.

437 "He didn't want to leave the White House": AI, Terry O'Donnell.

438 "If there's no place in politics": Stephen E. Ambrose, *Nixon: Ruin and Recovery, 1973–1990* (New York: Simon & Schuster, 1991), 473.

438 "I don't want to push it": *New York Times*, November 1, 1974.

438 "Oh, Mr. President": Kasey S. Pipes, *After the Fall: The Remarkable Comeback of Richard Nixon* (Washington, DC: Regnery, 2019), 37. On *Air Force One* after the Ford visit to his predecessor's hospital room speechwriter Bob Orben was approached by presidential physician Bill Lukash. "You know, Bob, you might be thinking of a statement from President Ford if Nixon doesn't make it through the night," said Lukash. "It's that close?" Orben replied. "It's that close."

439 "The people have spoken": APP, "Statement on the Results of the 1974 Elections," November 5, 1974.

439 "Goddam it, Jerry": Bob Woodward interview with James Cannon, October 14, 1997, WBP, Harry Ransom Center, University of Texas (Austin).

440 "First limitation": Anatoly Dobrynin, *In Confidence* (Seattle: University of Washington Press, 2001), 253.

441 "just a hardline Japanese politician": TAFI, 1407.

443 "a tough and shrewd union boss": Memorandum for the President, From: Henry Kissinger, "Leonid Brezhnev: The Man and His Style," NSA Outside the System Chronological File (Box 1–11/1/74–11/12/74), GRFL.

444 "I understand you're quite an expert on soccer": Pool Report, Airport Comments by General Secretary and Ford, November 23, 1974, White House Press Release Unit (Box 4B–Nov. 23–26, 1974), GRFL.

444 "How much snow do you get in Moscow?": Pool Report, Riding the Rails with Ford and Brezhnev, November 23, 1974, ibid. See also Adam J. Stone, "Gerald R. Ford and Vladivostok: A Study in Foreign Policy Formation," bachelor of arts thesis, Department of History, University of Michigan, March 31, 2017.

445 "Let us speak not as diplomats": Memcon, On Board Train Between Vozdvizhenka Airport and Okeanskaya Sanitorium, November 23, 1974, 2:30 p.m., Ford, Brezhnev, etc., GRFL.

445 "an abandoned YMCA camp in the Catskills": TTH, 214.

446 "I'm a sheep in wolf's clothing": Pool Report, Start of Formal Talks at White Stucco Conference Hall, November 23, 1974, White House Press Release Unit (Box 4B–Nov. 23–26, 1974), GRFL.

447 "I can't talk or think with these things on": Kennerly, *Extraordinary Circumstances*, 110.

447 "And you can imagine whom I am talking about": Kalman, *Right Star Rising*, 94.

448 "I don't think there are any holy people in the military": Memcon, November 23, 1974, Okeanskaya Sanitorium, Ford, Brezhnev, Kissinger, Gromyko, etc., 15, GRFL.

448 Ford said he had never been to Mongolia: Ibid., 23.

448 "How did the game turn out?": TTH, 217–18.

449 "An interesting suggestion": Memcon, November 24, 1974, 10:10 A.M., Okeanskaya Sanitorium, Ford, Brezhnev, Kissinger, Gromyko, Dobrynin, etc., GRFL.

450 "With Watergate aside": Memcon, Wednesday, December 4, 1975, 3:01–4:25 p.m., Ford, Pierre Elliott Trudeau, Kissinger, etc., GRFL.

450 "Does it mean strategic nuclear attack": Memcon, November 14, 1974, 1:40–2:05 p.m., Okeanskaya Sanitorium, Ford, Brezhnev, Kissinger, Gromyko, etc., GRFL.

451 "I do not want to inflict that upon my people again": Odd Arne Westad, *The Cold War: A World History* (New York: Basic Books, 2017), 360.

451 "We have averted an arms race": "The 'Breakthrough' on SALT," *Time*, December 9, 1974.

451 "The best that could be done at the time": Ibid.

452 "A leader has two important characteristics": *Burlington (VT) Free Press*, December 18, 1970.

453 "It's a bad bill": AI, Frank Carlucci.

453 "The economy is now in the midst": Andrew D. Moran, "Gerald R. Ford and the 1975 Tax Cut," *Presidential Studies Quarterly* 26, no. 3 (Summer 1996): 743.

454 "I've been called worse things": Gil Troy and L. Ian MacDonald, "U.S. Presidents and Canadian Prime Ministers: Good Vibes, or Not," *Policy Options, Politiques*, March 1, 2011.

454 "For a middle power like Canada": Ibid.

454 "one hell of a strong man": Helmut Schmidt, *Men and Powers: A Political Retrospective* (New York: Random House, 1989), 166.

455 "You do not need to reassure us": Memcon, Washington, December 5, 1971, 11 a.m., Ford, Helmut Schmidt, Kissinger, etc., GRFL.

455 "We don't plan to go to a producer meeting": Memcon, Martinique, December 15, 1974, 10:52 a.m.–12:26 p.m., Ford, Kissinger, Giscard d'Estaing, GRFL.

456 "I have to admit": TTH, 222.

457 "We're going to be in serious economic trouble": A. James Reichley interview with Alan Greenspan, January 28, 1978, GRFL.

457 "The 75 freshmen Democrats are a different breed": Donald W. Riegle to GRF, December 14, 1975, Seidman Files, Box 63, Folder: Economy: Background Material (1).

457 "You just can't do that to the country": TTH, 224.

458 "The Soviets have to keep détente": Memcon, January 7, 1975, 9:18–10:18 a.m., Ford, Kissinger, Scowcroft, GRFL.

458 "It looks like we'll be pretty busy at Vail": "Meeting with the President," December 20, 1974, 8:30–9:04 a.m., DRA.

458 "That's all we need": Ibid., December 20, 1974, 5:55 to 6:50 p.m.

458 Not likely, said Mel Laird: Meeting with the President, December 20, 1974, 8:30–9:04 a.m., Oval Office, DRA.

460 "He's been run over by history": "Sizing Up Gerald Ford," *Newsweek*, December 9, 1974. "Reforming the Intelligence Community," HUC, 463–502, provides an excellent overview of the executive-congressional tug of war over reform of the nation's intelligence agencies.

20: A President in the Making

461 "boundless powder, open slopes, and open sky": David O. Williams, "50 Years of Vail in Words and Pictures," *Vail-Beaver Creek Magazine*, holiday 2012/2013 issue.

462 "faster than the agents who were supposed to be protecting him": AI, Billy Kidd. Observant, too, according to Secret Service agent Richard Keiser, who advised the president to change his ski outfits every day so as to present a less obvious target to would-be assailants. Meanwhile, accompanying Ford on his daily visits to Vail's slopes was agent Larry Buendorf.

"Very good skier," recalls Keiser, "and he [Buendorf] had bought a brand-new ski outfit. I'll never forget, it was a bright yellow jacket—canary—and he wore that every day. So on the way back, we were flying home after vacation and President Ford called me up to the compartment and he said, 'Dick, I really, really enjoyed skiing. I felt very safe. I thought the guys did a great job. I really appreciate that. It was just great.' He said, 'I do have one question.' And I said, 'Yes, sir.' And he said, 'Well, it was a good idea that I change mine periodically, but all you had to do was see the guy in the yellow jacket and you'd know I was the guy next to him.' And I said, 'Yes, sir.'"

462 "Every skier takes a fall once in a while": AI, Ron Nessen.

462 "Those reporters": Pool Report, Thursday Morning, December 26, 1974, Box 6, White House Press Releases, GRFL.

462 "Is that still a live option?": "Ford's Risky Plan Against Slumpflation," *Time*, January 27, 1975.

463 "The more progress you make": Alan Greenspan, *The Age of Turbulence* (New York: Penguin Press, 2007), 68.

464 "As long as it's a one-shot deal and doesn't become permanent": Memcon, January 7, 1975, 9:18–10:18 a.m., Ford, Kissinger, Scowcroft, GRFL.

464 "The people who need it the most": "Ford's Risky Plan," op. cit.

464 "I want to talk to you": Telecon, Rumsfeld to Sec. Kissinger, December 23, 1974, 9:35 a.m., NSA.

465 his agency's congressional overseers: "Intelligence—The Colby Report, 12/27/74," Box 5, Richard Cheney Files. Cheney draft memo outlining options for the Ford Administration to deal with the aftermath of Seymour Hersh's December 22 story in the *New York Times* alleging domestic intelligence activities in violation of the CIA's charter. See also Kenneth Kitts, "Commission Politics and National Security: Gerald Ford's Response to the CIA Controversy of 1975," *Presidential Studies Quarterly* 26, no. 4 (Fall 1996): 1081–98.

465 "It is curious that you don't think": William Colby Oral History, March 15, 1988, CIA Oral History Archives, Oral History: Reflections of DCI Colby and Helms on the CIA's "Time of Troubles"—Central Intelligence Agency.

465 Almost as an afterthought: Tim Weiner, *Legacy of Ashes: The History of the CIA* (New York: Doubleday, 2007), 391; "Memorandum for the President" from Lawrence H. Silberman, January 3, 1975; "Memorandum for the File," James A. Wilderotter, Associate Deputy Attorney General, January 3, 1975, Box 7, Folder: Intelligence—President's Meeting with Richard Helms, Richard B. Cheney Files, GRFL.

466 "Was the CIA involved in Watergate?": Memcon, January 3, 1975, 11:10 a.m.–12:18 p.m., Ford Schlesinger, Marsh, Buchen, Scowcroft, GRFL.

466 "A character straight out of a John le [Carré] spy novel": Henry Kissinger, *Years of Renewal* (New York: Simon & Schuster, 1999), 325.

466 "we have four to Jane Fonda": Memcon, January 3, 1975, 5:30 p.m., Ford, Colby, Buchen, Marsh, Scowcroft, GRFL.

467 "We don't want to destroy": Ibid.

467 Beyond naming Colby as his source: Bunny Bucher Diary Notes, January 2 and 3, 1975, GRFL.

467 "What is happening is worse": Memcon, January 4, 1975, 9:40 a.m.–12:20 p.m., Ford, Kissinger, Scowcroft, GRFL.

467 Before he could keep his appointment with Ford: Lawrence G. Meyer to the author, May 14, 2020.

467 "At the base is Congressional oversight": Memcon, January 4, 1975, 12:37–1:00 p.m., Ford, Helms, Marsh, Buchen, Scowcroft, GRFL.

468 If its members restricted their activities: Memcon, January 4, 1975, 5:25–7:05 p.m., Ford, Rockefeller, Kissinger, Rumsfeld, Buchen, Marsh, Scowcroft, GRFL.

469 "this stuff will be all over town soon": Ibid.

469 Undercutting positive reviews: *Wall Street Journal*, January 14, 1975.

470 "Hartmann is like a lot of newspaper guys": TAI, Paul Miltich.

470 "Go back and give me one speech": AI, Bill Seidman.

470 "We sat down in a room": Ibid.

470 "I must say to you": State of the Union Address, January 15, 1975, Box 4, PSF: Reading Copies (Third Draft).

472 "Your conclusions were great, Mr. President": "Ford's Risky Plan," op. cit.

472 "his most presidential speech": *Wall Street Journal*, January 16, 1975.

472 "He stepped forward": *Time*, January 27, 1975, op. cit.

472 Drunk on more than power: "A Firecracker Explodes," *Newsweek*, October 21, 1974.

473 Ford telephoned Mills at the hospital: Memo, Bob Barrett to Ron Nessen, January 8, 1976, Folder: Bob Barrett, Box 126, Ron Nessen Papers, GRFL.

473 "The American people have a right": APP, "Remarks at the White House Conference on Domestic and Economic Affairs," Cincinnati, Ohio, July 3, 1975.

474 "Ford's greatest strength": AI, Jack Marsh.

474 "I see all those things you said about me": Ibid.

475 "thus stimulating the economy": Antonio Fins, "Falling for Ford," *South Florida Sun Sentinel*, August 1, 1999.

475 "the quality and stability of his leadership": It was in this same speech, delivered before the second annual gathering of the Conservative Political Action Conference, that Reagan issued his call for a revitalized GOP, one that would "march under a banner of bold colors rather than pale pastels."

476 "This is as far as we dare to go": American Presidency Project, Ford Address to the Nation upon Signing the Tax Reduction Act of 1975, March 29, 1975.

476 "repugnant to American principles": APP, "The President's News Conference," Hollywood, Florida, February 26, 1975.

477 "Can we do that?": AI, Rod Hills.

477 "My daily meetings with Henry": John Hersey, *The President* (New York: Alfred A. Knopf, 1975), ix.

477 "glacial caution": Ibid., 76.

478 "a separation of powers": John Hersey, *Aspects of the Presidency: Truman and Ford in Office* (Boston: Ticknor & Fields, 1980), 9.

478 "It's just a welfare thing": Ibid., 41.

479 "What is it in him?": Ibid., 176.

479 "a first rate college campus": *South Bend (IN) Tribune*, March 18, 1975.

479 Bill Saxbe's replacement: Gerald Ford remarks at Edward Levi Memorial Service, Chicago, April 6, 2000. See Nancy V. Baker, "Rebuilding Confidence: Ford's Choice of an Attorney General," HUC, 79–96; and Howard Ball, "Confronting Institutional Schizophrenia: The Effort to DePoliticize the U.S. Department of Justice, 1974–76," HUC, 97–109.

480 The next thing Levi knew: FLOHP, Edward Levi.

480 "Then I sprang my trap": Levi Memorial, op cit.

481 "I know about that": TAI, Levi.

481 "They didn't tell me": AI, Tom Korologos.

481 "Bowtie": Ibid.

482 "Do I have to eat these things?": AI, Ken Lazarus.

482 Together with his fellow clerk and best friend: AI, William Coleman.

483 "I'm not an urbanologist": AI, Carla Hills.

483 BETTY FORD GET OFF THE PHONE: The Nation: "A Fighting First Lady," *Time*, March 3, 1975. See also Lisa Tobin, "Betty Ford as First Lady," *Presidential Studies Quarterly* 20, no. 4 (Fall 1990): 761–67.

484 "I'm going to stick to my guns": Marlene Cimons, "First Lady Sticks to Her Guns," *Los Angeles Times*, September 18, 1975.

484 "He just smiles": "Betty Ford-ERA Lead," Associated Press, February 14, 1975, Sheila Weidenfeld Files, Box 47, GRFL.

484 No one in the counsel's office was smiling: Memorandum from Philip Buchen to Nancy Howe, etc., March 4, 1975, Philip Buchen Files, Box 49, Folder: President-Personal Family-Betty Ford (1), GRFL.

484 "the only woman in the White House": *Lansing (MI) State Journal*, February 29, 1976.

485 "I know Nancy is a constant problem": Sheila Rabb Weidenfeld, *First Lady's Lady: With the Fords at the White House* (New York: Putnam, 1979), 80.

486 "It's bad enough on a first date": https://www.brainyquote.com/authors/steven-ford-quotes.

486 "My career! My career!": "Life in 'Museum' Spurred Ford Son to Cowboy Career," *Los Angeles Daily News*, January 7, 1993.

486 "I wish the whole thing with my father": "Young Jack Ford," *Florida Today*, February 9, 1975.

486 "that the President's son is smoking dope": Weidenfeld, *First Lady's Lady*, 154.

486 "have a hell of a good time": "Woman of the Year," *Newsweek*, December 29, 1975.

487 "This is the President": AI, Frank Ursomarso.

487 With Susan's parents out of the country: Jim Windorf, "Remembering a Wild Night at the 1975 White House Prom," *Vanity Fair*, May 29, 2015.

488 "A job is all right": Kandy Stroud, "Susan Ford: On the Go in the White House," *Ladies' Home Journal*, January 1975.

21: The Cruelest Month

489 Politely but firmly: "An Anecdote About President Ford at St. John's Church," Powell Hutton to Don Holloway, February 24, 2018; Holloway to author, February 28, 2020.

489 "Reagan doesn't feel that I am a candidate": Donald Rumsfeld, *When the Center Held: Gerald Ford and the Rescue of the American Presidency* (New York: Free Press, 2018), 138.

490 "So the negotiation was": Memcon, March 24, 1975, 8:00 a.m., Ford, Kissinger, Bipartisan Congressional Leadership, GRFL.

491 Ford wrote confidentially to Rabin: Ford to Rabin, March 21, 1975, FRUS, 1969–1976, Vol. xxvi, *Arab-Israeli Dispute, 1974–1976*.

491 "This is no reflection on you": Memcon, Washington, March 24, 1975, Ford, Kissinger, Scowcroft, GRFL.

491 "Nothing has hit me so hard": Memcon, Ford, Kissinger, Max Fisher, Brent Scowcroft, March 27, 1975, 3:10 p.m., the Oval Office, GRFL.

491 a stunning proposal to supply South Vietnam: Frank Snepp, *Decent Interval* (Lawrence: University Press of Kansas, 2002), 216.

492 "I will do everything I can": Memcon, March 25, 1975, 11:00 a.m., the Oval Office, Ford, South Vietnamese Parliamentarians, Ambassador Tran Kim Phuong, etc., GRFL. Also see Jerrold L. Schecter, "The Final Days: The Political Struggle to End the Vietnam War," HUC, 539–51.

492 "I've been in this business for forty years": Thurston Clarke, *Honorable Exit* (New York: Doubleday, 2019), 84. Martin's faith in the Thieu government never wavered. During a February 15, 1975, visit to the White House, he predicted the South Vietnamese president would get 80 percent of the vote in a fair and honest election. At the end of April, with North Vietnamese forces encroaching on Saigon, the US ambassador predicted it would take three months to form a coalition government, with the Communists on the short end of a 60–40 stick.

492 "You are not going over to lose": Memcon, March 25, 1975, 9:22–10:25 a.m., Ford, Kissinger, General Frederick Weyand, Graham Martin, Scowcroft, GRFL.

493 "You know, I would really like to go": AI, David Kennerly. Ford told Jim Cannon that the idea of sending Fred Weyand on his Vietnam fact-finding mission came from Jack Marsh. JCI, April 27, 1990 (part 2).

493 "You be careful": TTH, 251.

493 "Of course, we've told Thieu": Douglas Brinkley and Luke A. Nichter, eds., *The Nixon Tapes: 1973* (Boston: Houghton Mifflin, 2015), 255.

494 "You thought the last Congress was bad": Memcon, March 27, 1975, Ford, Haig, Rumsfeld, Scowcroft, GRFL.

495 Thieu considered: GRF interview by Jerrold Schecter, February 10, 1986, Box 1, Composite Oral History, Accessions, GRFL.

495 "wake up the country": Memcon, March 28, 1975, 9:25–10:10 a.m., Ford, Rockefeller, Kissinger, Scowcroft, GRFL.

495 "It's important to get her out of here": Rumsfeld, *When the Center Held*, 144.

496 "He is going off on vacation": Geoffrey C. Ward, *The Vietnam War: An Intimate History* (New York: Alfred A. Knopf, 2017), 534.

496 "Because my bet": AI, Phil Jones.

497 "the gutsy Cambodians": Memcon, April 14, 1975, 3:30 p.m., Ford, Kissinger, Schlesinger, Senate Committee on Foreign Relations, FRUS, Volume X, 817, GRFL.

497 "There's not a damn thing you can do here": Telecon, April 1, 1975, 7:15 p.m., Ford/Kissinger, NSA.

497 "He didn't bring anything up": AI, Leon Parma.

498 "but I'm no gentleman": "The CIA: The Big Shake-Up in a Gentlemen's Club," *Time*, April 30, 1973.

498 "It's a pity": AI, Henry Kissinger.

498 "Schlesinger was always pressing me": TAFI, 2481. Ford made his feelings clear at a Cabinet meeting on April 16, 1975, telling his associates, "I want no one here to talk about evacuation. That is a code word in Saigon for a bugout."

498 "Probably we will not get a significant amount": Memcon, April 17, 1975, 4:30 p.m., Ford, Kissinger, Schlesinger, Rumsfeld, Marsh, Scowcroft, Max Friedersdorf, GRFL.

498 "In the last two months of Vietnam": AI, Henry Kissinger.

498 "We were sending planes": "Turning Off the Last Lights," *Time*, May 5, 1975.

498 "if we had a sane Secretary of Defense": Kissinger, Telecon with Phillip Habib, April 11, 1975, 9:07 a.m., NSA.

498 "A very difficult person": TAFI, 2901.

499 "I can make an even greater contribution": Back Channel Message from Martin to Scowcroft, April 4, 1975, FRUS, Volume X, *Vietnam, January 1973–July 1975*, 683.

499 "and not run from the press": "Memorandum for Jack Marsh from Charles Lippert, Jr.," April 2, 1975, Box 26, Loen and Lippert Files, GRFL.

499 "This is the least we can do": PSF, Box 7, Folder: 4/3/75. Opening Statement at Press Conference at San Diego, California.

499 "the brink of a total military defeat": NSC Meeting, April 9, 1975, 11:25 a.m.–1:25 p.m., Box 1, National Security Advisor's NSC Meeting File, GRFL.

500 Vietnam couldn't last more than a month: David Kennerly, "In the Room—The Final Days of Vietnam," *David Hume Kennerly Blog*, April 27, 2015, https://kennerly.com/blog.

500 "hundreds of babies": AI, Ron Nessen.

501 "Until now it is not 'Ford's war'": Memo, Ron Nessen to Rumsfeld, April 8, 1975, Vietnam-General, 3/25/75, Box 13, Richard Cheney Papers, GRFL.

501 "That is not the way I am": Memcon, April 8, 1975, 9:00 a.m., Ford, Kissinger, Scowcroft, GRFL.

501 "Leave them up": AI, David Kennerly. Also see Kennerly's "Inside the Final Days of Vietnam," *Politico*, April 29, 2015.

501 "those who wanted an early exit": GRF Remarks at the Opening of the Ford Museum's Saigon Staircase Exhibit, April 10, 1999.

501 "It's all over, Mr. President. We lost": Bob Woodward interview with James Schlesinger, July 31, 1989, WBP, Harry C. Ransom Center, University of Texas (Austin).

502 "There are two kinds of people": Minutes of NSC Meeting, Washington, DC, April 9, 1975, 11:25 a.m.–1:15 p.m., FRUS, Volume X, *Vietnam, January 1973–July 1975*, 777.

502 "The president is aware": Memcon, April 10, 1975, 6:00 p.m., Kissinger, Marvin Kalb, Tom Jarriel, Ted Koppel, Richard Valeriani, Scowcroft, GRFL.

502 "If they think we have given up": Rumsfeld, *When the Center Held*, 156.

502 "to point the finger of blame": APP, "Address Before a Joint Session of the Congress Reporting on United States Foreign Policy," April 10, 1975.

503 Before leaving the White House: "Husband of Aide to 1st Lady Kills Self During Probe," *Detroit Free Press*, April 12, 1975. Also see "Fords Express Sympathy to the Howes," *New York Times*, April 12, 1975.

503 "Maxine Cheshire killed Jimmy Howe": Weidenfeld, *First Lady's Lady*, 120.

504 "It wasn't permitted": TML, 235.

504 "Four thousand on one ship would pull the plug": Memorandum of Conversation, Washington, April 14, 1975, 3:30 p.m., FRUS, Volume X, *Vietnam, January 1973–July 1975*, 819.

504 "I will give you large sums for evacuation": TTH, 255.

505 "It has the same latitude": FRUS, April 14, 1975, op. cit.

506 Much of this shadowy operation: Ken Quinn to David Kennerly (shared with author).

506 he confessed to feeling "goosebumps": TTH, 264–65; John J. Casserly, *The Ford White House: The Diary of a Speechwriter* (Boulder: Colorado Associated University Press, 1977), 77–80. Casserly was deeply frustrated at the disconnect between Ford's formal texts and his unscripted comments. Example: the president's Old North Church speech of April 19, 1975, in which the president appealed for "reconciliation—not recrimination," only to undercut his words by publicly blaming Congress for its failure to approve his latest request for military assistance for Saigon.

506 "I don't know why": Hartmann, *Palace Politics*, 321.

507 "always the last person to leave": Quinn to Kennerly, op. cit.

508 "two chances in twenty": Telecon, Ford/Kissinger, April 18, 1975, 10:10 p.m., NSA.

508 "you Americans": "Nguyen Van Thieu," *Telegraph*, October 1, 2001.

508 "Kissinger had been talking": FLOHP, Milton Friedman.

508 "Today America can regain": Richard L. Madden, "Ford Says Indochina War Is Finished for America," *New York Times*, April 24, 1975.

508 "Has Henry seen this?": FLOHP, Milton Friedman.

509 "This we don't need, Mr. President": Hartmann, *Palace Politics*, 323.

509 "That is a lot in one day": NSC Meeting, April 24, 1975, 4:35 p.m., Box 1, National Security Advisor's NSC Meeting File, GRFL.

509 "I am resigning": "Preparing to Deal in Peace," *Time*, May 5, 1975.

510 What was the status of the runway?: NSC Meeting, April 28, 1975, 7:23–8:08 p.m., Box 1, National Security Advisor's NSC Meeting File, GRFL.

511 "Are we ready to go to a helicopter lift?": Ibid. See also Walter J. Boyne, "The Fall of Saigon," *Air Force Magazine*, April 1, 2000.

511 "They were only nineteen and twenty-[one]": McCubbin, *Betty Ford*, 197.

511 "It's getting to be like [Da Nang]": Henry Kissinger, *Crisis: The Anatomy of Two Foreign Policy Crises* (New York: Simon & Schuster, 2003), 512.

512 "For Christ's sake, let's go to the helos": Bartholomew Sparrow, "Inside America's Massive, Messy Evacuation from Saigon," *New Republic*, April 29, 2015.

512 "You tell them to make damn sure": Kissinger, *Years of Renewal*, 1107.

512 "And then he started crumbling": AI, Henry Kissinger.

512 "Women were like": Margie Mason, "US Marines Recall Being Forgotten in Saigon at War's End," *Morning Call*, April 30, 2015.

513 "We have no choice": Charles Henderson, *Goodnight Saigon: The True Story of the U.S. Marines' Last Days in Vietnam* (New York: Penguin, 2008), 328.

513 "Somebody said, 'Sleep well, Mr. President'": Nessen, *It Sure Looks Different*, 110.

513 "Working late?": "The Last Helicopter: Evacuating Saigon," *Newsweek*, May 12, 1975.

513 "The good news is the war is over": Rumsfeld, *Known and Unknown*, 208.

514 "The Pentagon wanted to make sure": AI, Henry Kissinger.

514 "Can't you tell him to get them out of there?": Minutes of Washington Special Actions Group Meeting, April 29, 1975, 9:04–9:20 a.m., FRUS, Volume X, *Vietnam, January 1973–July 1975*, 933.

514 "Perhaps you can tell me": Ibid.; Backchannel message from Amb. Martin to Scowcroft, April 29, 1975, 1414Z, 939.

515 "They are coming out of the woodwork": Memcon, the President's Meeting with the Cabinet on Indochina, April 29, 1975, 9:45 a.m. GRFL.

515 Americans were "sick and tired of war": Memcon, April 29, 1975, 10:30–11:30 a.m., Ford, King Hussein, Prime Minister Zaid al-Rifai, Kissinger, Scowcroft, Ambassador Thomas R. Pickering, GRFL.

515 "I would say God-damnit this is the end": Telecon, Kissinger/Schlesinger, April 29, 1975, 1:45 p.m., NSA.

515 "I'm here to get the Ambassador": Theresa Kay Albertson, "How One Iowan Rescued Thousands from the U.S. Embassy in Saigon and Ended the Vietnam War," *Des Moines Register*, March 11, 2019.

516 "This action closes a chapter": Public Papers, 1975, Book I, 605.

516 "The first thing I said at the briefing": AI, Ron Nessen.

516 "the most botched-up": Nessen, *It Sure Looks Different*, 112.

516 "This war has been marked by so many lies": Rumsfeld, *When the Center Held*, 162.

517 "I did it because the people would have been killed": Snepp, *Decent Interval*, 563.

22: Starting Over

518 "Those sons of bitches": AI, Ron Nessen.

518 Senator Robert Byrd (Democrat, West Virginia) insisted: David Binder, "Ford Asks Nation to Open Its Doors to the Refugees," *New York Times*, May 7, 1975.

518 On the left, Representative John Conyers: Robert Marsh, "Socioeconomic Status of Indochinese Refugees in the United States: Progress and Problems," *Social Security Bulletin*, October 1980.

519 "The Vietnamese are better off": "The Orange Grove: Vietnam Refugees Had Friend in Ford," *Orange County Register*, January 2, 2007.

519 "With a little luck": Nessen, *It Sure Looks Different*, 115.

519 "your humanitarian policy": Joseph Alioto to GRF, May 5, 1975, GRFL. Mayor Alioto would join sixteen others, among them George Meany, Washington State governor Dan Evans and AMA president Dr. Malcolm Todd, on the President's Advisory Committee on Refugees chaired by John S. D. Eisenhower, son of the late president.

519 "Yes, we have 9 percent unemployment": *New York Times*, May 6, 1975.

519 "It just burns me up": Ibid.

519 "They ought to be given an opportunity": Binder, "Ford Asks Nation to Open Its Doors," op. cit.

520 "The president's 'damn mad' reaction": Jim Wieghart, *New York Daily News*, May 9, 1975.

520 "these people are going to be wearing loincloths": Ann Blackmun, *Off to Save the*

World: How Julia Taft Made a Difference (Rockland: Maine Authors Publishing, 2011), 55.

520 "We have enough people in California": Ibid., 57.

521 "My conscience would have bothered me": TAFI, 2928.

521 "You sold out Vietnam": Steven F. Hayward, *The Age of Reagan: The Fall of the Old Liberal Order, 1964–1980* (Roseville, CA: Prima Publishing, 2001), 412.

521 Acknowledging the "traumatic experience": Memcon, May 7, 1975, 3:00 p.m., Ford, Prime Minister Gough Whitlam, Kissinger, etc., GRFL.

522 "We will continue to be strong": Remarks of the President at the Commissioning of the USS *Nimitz*, May 3, 1975, Box 10, White House Press Releases, GRFL.

522 "I don't know why it is": Roy Rowan, *The Four Days of Mayaguez* (New York: W. W. Norton, 1975), 66.

522 "I have some bad news for you": Nessen, *It Sure Looks Different*, 118.

523 "a bureaucratic misjudgment": NSC Meeting Minutes, May 12, 1975, 12:05–12:50 p.m., Ford, Rockefeller, Kissinger, Schlesinger, etc., FRUS X, *Vietnam, January 1973–July 1975*, 977.

523 "We should not look": Ibid., 983.

523 "I remember the *Pueblo*": Ibid., 981.

524 Ford told Schlesinger: Christopher Jon Lamb, *The* Mayaguez *Crisis, Mission Command, and Civil-Military Relations* (Joint History Office of the Chairman of the Joint Chiefs of Staff, 2016), 19. See also Rowan, *Four Days*, 88–93.

525 "There may be some Caucasians": Lucien S. Vandenbroucke, *Perilous Options as an Instrument of U.S. Foreign Policy* (New York: Oxford University Press, 1993), 81.

525 "Tell them to sink the boats near the island": Box 1, Folder: NSC Meeting, 5/13/75 (evening); National Security Advisor, National Security Council Meeting File, GRFL.

526 "We have the right of self-defense": Box 1, Folder: NSC Meeting, 5/14/1975, 3:52–5:42 p.m., National Security Advisor, National Security Council Meeting Files, GRFL.

526 "Henry and I felt": TTH, 279.

527 "I was very concerned": NOHP, David Kennerly.

527 "Has anyone considered": AI, David Kennerly.

527 "I'm not going to make": Ibid.

527 "Dave, I really appreciated your input there": NOHP, Kennerly, op. cit.

528 "We have a separation of powers": Memcon, May 14, 1975, 6:30 p.m., Ford, Bipartisan Congressional Leadership, Kissinger, Schlesinger, Scowcroft, GRFL.

529 The Dutchman's sulky attitude: TTH, 282.

529 "They're all safe": TTH, 283.

530 "Boy," mused Ford: *Time*, May 26, 1973.

530 "How did the [Washington] Bullets do tonight?": FLOHP, Bob Barrett, September 14, 1998.

530 "We perhaps overreacted": Memcon, May 15, 1975, 11:00 a.m., Ford, Shah Mohammed Reza Pahlavi, Kissinger, Scowcroft, GRFL.

531 "every little half-assed nation": *Washington Post*, May 16, 1975.

531 "the angriest speech of his presidency": Mieczkowski, *Gerald Ford and the Challenges of the 1970s*, 235.

532 "We are not a trojan horse": Memcon, Brussels, Belgium, May 29, 1975, 3:30–4:30 p.m., Ford, Kissinger, Portuguese prime minister Gonçalves, etc., GRFL.

532 "unqualified participation": APP, "Address Before the Council of the North Atlantic Treaty Organization," May 29, 1975.

532 "What a lot of smiling people": Stanley G. Payne and Jesús Palacios, *Franco: A Personal and Political Biography* (Madison: University of Wisconsin Press, 2014), 485.

532 "I was with him": AI, Don Rumsfeld.

533 "I am sorry I tumbled in": "Ford Falls but Is Unhurt," *New York Times*, June 2, 1975.

533 "Rummy, I am so mad at myself": Rumsfeld, op. cit.

533 "Of course they took the picture": Ibid.

533 "at least 99% of the cards": Jewish Telegraphic Agency, March 28, 1977.

533 "the type of man": Memcon, Schloss Klessheim, June 2, 1975, Ford, Scowcroft, Egyptian journalists, GRFL.

534 "All right," he said at last: TTH, 290.

534 "You could suggest": Memcon, the Oval Office, June 11, 1975, 9:34–10:06 a.m., Ford, Kissinger, Scowcroft, GRFL.

535 "a peace by diplomats": Memcon, June 11, 1975, 10:00–12:00 Noon, the Oval Office, Ford, Rabin, Kissinger, Scowcroft, etc., GRFL.

535 "I don't want the Israelis": Ibid.

536 "I think you ought to see": Ibid.

536 "Milton, you'd better go take the call": FLOHP, Milton Friedman.

537 "Like what?": Crain, *The Ford Presidency*, 118.

537 "I understand from the president": Scott C. Monje, *The Central Intelligence Agency: A Documentary History* (Westport, CT: Greenwood Press, 2008), 72.

537 On March 5 he warned Frank Church: Presidential Meeting with Senators Frank Church and John Tower, March 5, 1975, President's Handwriting File, National Security Series, b.30.f (2), NSA.

537 "He would come home and I'd say": Martin Edwin Andersen, "How Late DCI William Colby Saved the CIA, and What That Can Teach Us Today," *Just Security*, January 16, 2020, www.justsecurity.org/how-late-dci-william-colby-saved-the-cia-and-what-that-can-teach-us-today.

537 "We realize there are secrets": William Colby and Peter Forbath, *Honorable Men: My Life in the CIA* (New York: Simon & Schuster, 1978), 400.

538 "This was another way of chopping my head off": Smith, *On His Own Terms*, 668.

538 "From what I am told": Memcon, the Oval Office, June 5, 1975, 2:00 p.m., Ford, Kissinger, Scowcroft, GRFL.

538 "This is not a complete report": AI, Rod Hills.

539 "I am not going to second-guess": Memcon, June 5, 1975, op. cit.

539 Passing the buck: Smith, *On His Own Terms*, 668.

539 "If there was no White House direction": David W. Belin, *Final Disclosure: The Full Truth About the Assassination of President Kennedy* (New York: Scribner, 1988), 169.

540 "sensational reports": Belin's reference is to Taylor Branch's article "How the CIA Won," in the *New York Sunday Times Magazine*, September 12, 1976.

540 "When things got really rough": AI, Rod Hills.

541 "We're not going to classify anything": Rumsfeld, *When the Center Held*, 223.

543 "Outraged. Absolutely outraged": AI, Rod Hills.

543 "a horrible episode in American history": *Washington Star*, July 22, 1975.

23: The Dangerous Summer

544 "I really love this job": David Alsobrook to author, January 23, 2021.

544 "a kind of anti-hero": "Ford as Mr. Right," *Newsweek*, August 11, 1973.

544 "to do as little as possible": Few things annoyed Ford more than to be told he and his administration were lacking in "vision." As he put it privately, "When you're trying to heal a country you better get it healed before you start coming up with a platter full of visionary proposals." TAFI, 3000.

545 "But I was so convinced": TAFI, 2310.

545 "A political moderate with an academic background": TTH, 261.

545 "where I didn't see a Republican": AI, David Mathews.

546 "do to the United States of America": AI, William Simon.

547 "Well, Jim, Rod's trading three lawyers": AI, Rod Hills.

547 "the last two issues we have discussed": Coleman, *Counsel for the Situation*, 223. See also Lawrence J. McAndrews, "Gerald Ford and School Desegregation," *Presidential Studies Quarterly* 27, no. 4 (Fall 1997): 791–864.

548 "Ford just wasn't a theme type of man": TAI, Bill Seidman.

548 "the great unsung achievement": Greenspan, *Age of Turbulence*, 71. An excellent overview is "Moving Ahead" by Thomas Gale Moore, senior fellow at the Hoover Institution, at https://web.stanford.edu/~moore/MovingAhead.html. Also see Andrew Downer Crain, "Ford, Carter and the Deregulation in the 1970s," *Journal on Telecommunications and High Technology Law* 5, no. 2 (Winter 2007): 413–48.

548 "Most regulated industries": "How to Regulate the Regulators," *Time*, October 21, 1974; David Burnham, "Ford Administration Deregulation Proposals Bring Increasing Opposition from Industry," *New York Times*, January 2, 1976.

549 "I want to find the bipartisan limits": AI, Rod Hills.

550 "Government should foster": APP, "Remarks at a Meeting to Discuss Regulatory Reform," July 10, 1975.

550 "Richard, I don't think you get my drift": AI, Rod Hills.

550 "a prescription for disaster": Ford's greatest contribution to the causes of deregulation may, in fact, have occurred after he left office and took up residence at the American Enterprise Institute. "He brought with him a team of people," says AFI's Chris DeMuth, who in turn produced several volumes of research that would inform Jimmy Carter's efforts to relax government controls on the aviation and trucking industries. AI, Chris DeMuth.

551 "a forty year old still": John E. Robson, "Airline Deregulation: Twenty Years of Success and Counting," *Regulation* 21, no. 2 (Spring 1998): 18.

552 "He almost threw me out of his office": Coleman, *Counsel for the Situation*, 251.

552 "Are you telling me": AI, Rod Hills.

553 "Ford just couldn't get it through his head": AI, Lou Cannon.

553 "We need to find a way skillfully": Memcon, the Oval Office, June 27, 1975, 9:30–9:58 a.m., Ford, Kissinger, Scowcroft. GRFL.

554 "one capitulation after another": Aleksandr Solzhenitsyn, "Words of Warning to America," *Imprimis* 4, no. 9 (September 1975).

554 "Tell Jesse Helms to go to hell": Meeting with the President, July 9, 1975, 1725–55, DRA.

554 "the block groups": AI, Max Friedersdorf.

555 "It's all right to come out now": *Washington Post*, July 8, 1975; see also July 8, 1975, memorandum from Dick Cheney to Don Rumsfeld, Subject: Solzhenitsyn, Box 10, Richard B. Cheney Files, GRFL.

555 "when an amicable agreement": W. Paul Kangor, *The Crusader: Ronald Reagan and the Fall of Communism* (New York: HarperCollins, 2007), 44.

555 "the single most irresponsible": Mieczkowski, *Gerald Ford and the Challenges of the 1970s*, 277.

555 "a disaster": Telecon, General Scowcroft/The Secretary, July 25, 1975, 9:40 p.m., NSA.

556 "A document": "A Star-Studded Summit Spectacular," *Time*, August 4, 1975. An excellent source for what follows is Christopher R. Kline's honors thesis, "The Promises We Keep: President Gerald R. Ford's Leadership at the Helsinki Conference," University of Michigan, College of Literature, Science and the Arts, History Department, March 31, 2014.

558 On July 29 Ford made a grim visit: Polish authorities were initially reluctant to have him set foot in the Auschwitz death camp. "I said I was going—period." What he saw—"It was just a terrible torture chamber from the beginning to end"—was personalized by the presence of Henry Kissinger, who lost members of his family at the notorious slaughterhouse. TAFI, 3016–19.

558 "confidentially and completely frankly": Jan Lodal, "Brezhnev's Secret Pledge to 'Do Everything We Can' to Re-elect Gerald Ford," *Politico*, July 26, 1975.

558 "our leverage [over Turkey] is not zero": Memcon, July 30, 1975, 1 p.m., Ford, Caramanlis, Kissinger, etc., GRFL.

559 "They have 3,000 islands": Memcon, July 31, 1975, 8 a.m., Ford, Demirel, Kissinger, etc., GRFL.

559 "We could say": Ibid.

560 "I got Kissinger and Scowcroft": TAFI, 3630.

560 "Peace is not a piece of paper": APP, "Address in Helsinki Before the Conference on Security and Co-operation on Europe," August 1, 1975.

560 The Backfire was a medium-range aircraft: William G. Hyland, *Mortal Rivals: Superpower Relations from Nixon to Reagan* (New York: Random House, 1987), 101–2.

561 "the manifesto of the dissident": Anatoly Dobrynin, *In Confidence* (Seattle: University of Washington Press, 2001), 346.

562 "my finest hour": Gerald R. Ford, "The Ford Presidency: How It Looks Twelve Years Later," HUC, 671. See also Suzanne Nossel, "Once Upon a Time, Helsinki Meant Human Rights," *Foreign Policy*, July 11, 2018; Craig R. Whitney, "Summit in Europe: The Legacy of Helsinki," *New York Times*, November 19, 1990.

562 "I don't use the word détente anymore": Murray Marden, "Ford Stirs Flurry on Détente," *Washington Post*, March 3, 1976. Also see Julian E. Zelizer, "Détente and Domestic Politics," *Diplomatic History* 33, no. 4 (September 2009): 653–70. Plus Robert D. Schulzinger, "The Decline of Détente," HUC, 407–20.

562 His suggestion that Ford might want: *Washington Post*, July 28, 1975.

563 "I never knew exactly how I would find her": AI, Weidenfeld.

563 "Fine in Germany": Weidenfeld, *First Lady's Lady*, 165.

563 On the night of August 10: According to Sheila Weidenfeld, the only editing of her interview Mrs. Ford requested was the removal of her verbal opposition to gun control. AI, Sheila Weidenfeld.

564 the president in Vail tossed a pillow at his wife: Betty Ford with Chris Chase, *Betty: A Glad Awakening* (New York: Doubleday, 1987), 17.

564 "Betty Ford said today": *New York Times*, August 11, 1975.

565 It was no defense: Phyllis Schlafly, "Mrs. Ford and the Affair of the Daughter," *Ladies' Home Journal*, November 1975.

565 William F. Buckley went further: "First Lady Voices Act of Aggression," quoted in Gil Troy, "Betty Ford: A New Kind of First Lady," Gilder Lehrman Institute of American History.

565 "Do you realize how much harm you have done": Caroline Dickey, "Free Spirit Meets Gilded Cage: Betty Ford as Second Lady 1973–74," Liberty University, December 12, 2019, 42.

565 "You simply must tell your wife to shut up": Maria von Trapp to GRF, August 12, 1975, FSM, Box A10, Folder: Maria von Trapp, GRFL.

565 "I guess I'm more old-fashioned": "Susan Ford Likes Mother's Interview, But Others Criticize View on 'Affair,'" *New York Times*, August 12, 1975.

565 "What is an affair?": John Whitcomb and Claire Whitcomb, *Real Life at the White House: Two Hundred Years of Daily Life at America's Most Famous Residence* (New York: Routledge, 2000), 408.

565 Betty Friedan, the anti-Schlafly: Betty Friedan to Betty Ford, October 8, 1975, White House Social Files, Box 62, Folder: HU2 Equality, GRFL.

565 "If a person hadn't seen the interview in person": FLOHP, Patti Matson.

565 "You want to tone things down": Ibid.

24: "The Cat Has Nine Lives"

567 "It is a little startling to look down": Telecon, President Ford/Sec. Kissinger, September 5, 1975, 7:25 p.m., U.S. Department of State Doc #C18090326.

567 "Oh, this is a nice day": AI, Larry Buendorf.

568 "more fun": AI, Frank Ursomarso.

568 The next few minutes: AI, Larry Buendorf.

568 "Because people would constantly give": AI, Terry O'Donnell.

568 "I was on the Warren Commission": "Recollections of President Gerald R. Ford Relayed to Donald Rumsfeld Aboard Air Force One, September 5, 1975," DRA.

568 "The country's in a mess": "The Girl Who Almost Killed Ford," *Time*, September 15, 1975.

569 "If she'd had a round chambered": AI, Larry Buendorf.

569 "It didn't go off": Rumsfeld, *When the Center Held*, 200.

569 "The other thing was past": TAFI, 1480.

569 "You okay, Mr. President?": AI, Terry O'Donnell. As Ford once explained his reaction to the author—"Well, I didn't think it would be very nice to go in and say some lady just tried to shoot me while I was coming to see you."

570 It was like football: AI, Frank Ursomarso.

570 "Do you know what it is like": Weidenfeld, *First Lady's Lady*, 190.

570 A planned return trip: AI, Terry O'Donnell.

571 "a serious problem": Meeting with the President, August 27, 1975, 8:40, DRA. "What we had was a veto strategy in terms of dealing with Congress," according to Dick Cheney. "We put the No New Starts marker out there for good and legitimate policy reasons. Rockefeller just refused to abide by it. He would come in for these weekly sessions with the president and he always had a new proposal. I'd go down that evening, or sometimes the president would even call me right after Rockefeller left, and I'd go down and he'd hand me the new proposal and he'd say, 'Dick, what do we do with this?' And I'd say, 'Well, Mr. President, we'll staff it out.' And so it would go to OMB and then get fed out through the system, and the answer would always come back, 'This is not consistent with our basic fundamental policy, and [got] shot down. Rockefeller became convinced I was out to do him in. It was another one of the charges against me." AI, Dick Cheney.

571 "Get him out of town": Ron Nessen, *Making the News, Taking the News* (Middletown, CT: Wesleyan University Press, 2011), 181.

571 "Fine," said Ford: AI, Terry O'Donnell.

572 Her thwarted attempt on Ford's life: "Making of a Misfit," *Time*, October 6, 1975.

572 "If I had my .44 with me": "Can the Risk Be Cut?," *Newsweek*, October 6, 1975.

573 Before Ford's letter could reach him: "Remember Oliver Sipple" (1941–1989), Archived February 13, 2007, at the Wayback Machine. Also see "The Man Who Grabbed the Gun," *Time*, October 6, 1975; and Lynn Duke, "Caught in Fate's Trajectory, Along with Gerald Ford," *Washington Post*, December 31, 2006.

573 "If He's So Dumb": Todd S. Purdum, "SNL: Skewering Pols for 35 Years," *Politico*, April 29, 2011.

573 "He didn't let stuff like that": AI, Terry O'Donnell.

574 "Well, the cat has nine lives": Nessen, *Making the News*, 184.

574 "You've had a great time": Seymour P. Lachman and Robert Polner, *The Man Who*

Saved New York: Hugh Carey and the Great Fiscal Crisis of 1975 (Albany: State University of New York Press, 2010) 150. See Charles J. Orlebeke, "Saving New York: The Ford Administration and the New York City Fiscal Crisis," HUC, 359–85, followed by discussants Abe Beame, Hugh Carey, Jim Cannon, Ken Auletta and Richard Ravitch.

574 "Here's your problem": Ibid., 151.

575 "Never mind the Bundesbank": Ibid., 152.

575 The unusual show of independence: Smith, *On His Own Terms*, 675.

576 "Ford could get along with just about anybody": AI, Fred Barnes.

576 so valued a Ford intimate: TAI, James Cavanaugh.

576 "So there was poor President Ford": AI, Frank Zarb.

576 The next day Ford confided: Rumsfeld, *When the Center Held*, 213–14.

576 In league with Dick Cheney: Rumsfeld-Cheney Memorandum for the President, October 24, 1975, DRA.

578 If it was any consolation: Ibid. In his memoirs Ford makes no reference to the Rumsfeld-Cheney memo, though it's hard to dismiss in light of the sweeping shake-up revealed within twenty-four hours of the memo's receipt—with the attached resignation letters of its authors.

579 "They don't get along. It's obvious": AI, Don Rumsfeld.

579 "too active and dynamic a man": TTH, 328.

579 "If you want to get the record straight": TAI, Nelson Rockefeller.

579 "The answer is not just 'no'": Meeting with the President, October 24, 1975, 1430 to 1552, DRA.

579 "If the president wants to win": Nessen, *Making the News*, 186.

580 "If we go on spending more than we have": APP, Remarks and a Q and A Session at the National Press Club on the Subject of Financial Assistance to New York City, October 29, 1975.

580 More than what he said: Jen Chung, "Ford's 'Drop Dead' Tactics Actually Helped the City," *Gothamist*, December 28, 2006.

580 "Now we're going to win": Lachman and Polner, *The Man Who Saved New York*, 157.

580 "They spent their way in": "The Birth of an Issue," *Newsweek*, November 10, 1975.

581 "Then we don't get people denying this and denying that": Telecon, President Ford/The Secretary, November 1, 1975, 6:05 p.m., State Department Doc No C18090768, Date 7/24/2015.

581 "Now, Mr. President": AI, Dick Cheney. "Someone," recalls Bob Schieffer, "and it's not my story, but somebody told me Henry Kissinger would go in to see the president and he would say, 'Mr. President as you are aware . . .' and he would outline the problem. Schlesinger would go in and say, 'Mr. President, you don't know this . . .' And Ford didn't like him. Ford didn't like him because he thought Schlesinger thought he was dumb. That's about it. And everybody knew it and when he got the chance, he fired him."

582 "That is not something that I could accept": TAFI, 3551.

582 "With this Nelson thing": Memcon, November 3, 1975, 9:20–10:13 a.m., Ford, Kissinger, Scowcroft, GRFL.

582 "I know you are thinking of resigning": Kissinger, *Years of Renewal*, 843.

582 "Although we never said it": AI, Bill Seidman.

582 "The only way we have achieved results": Memcon, November 15, 1975, 4:00 p.m., Rambouillet Economic Summit, Ford, Helmut Schmidt, Harold Wilson, Valery Giscard d'Estaing, GRFL.

583 "By maintaining a hard line": Alex J. Pollock, *Finance and Philosophy: Why We're Always Surprised* (Philadelphia: Paul Dry Books, 2018), 121.

584 "The country still views him": "Ford in Trouble," *Newsweek*, December 22, 1975.

584 "I used to talk to Cheney three times a day": AI, Bob Schieffer.

585 "His command of facts": John Paul Stevens, *The Making of a Justice: Reflections on My First 94 Years* (New York: Little, Brown, 2014), 125. See also David M. O'Brien, "The Politics of Professionalism: President Gerald R. Ford's Appointment of Justice John Paul Stevens," HUC, 111–36.

586 "He has served his nation well": GRF to Dean William Michael Treanor, Fordham University School of Law, September 21, 2005, quoted in the *New York Times*, April 10, 2010. See also I. Matthew Miller, "Sound Judgment: The Nomination of John Paul Stevens to the Supreme Court of the United States," December 12, 1989, Dallenbach, History 396.004, GRFL.

586 "There isn't that much to say": Memcon, the Oval Office, November 19, 1975, 9:15 a.m., Ford, Kissinger, Scowcroft, GRFL.

586 "The wind sweeping through the tower": Vice Premier Teng Hsiao-p'ing's Toast at the Banquet in Honor of President Gerald R. Ford, December 1, 1975, Gerald Ford's China Visits, GRFL, Presidential Digital Library.

586 Kissinger's "nasty little man": "Ford in China: Warm Hosts," *Time*, December 15, 1975.

587 "So we all made mistakes": Memcon, December 2, 1975, 10:10 a.m.–12:30 p.m., Ford, Teng Hsiao-p'ing, Kissinger, etc., GRFL.

587 "the Socialist Imperialists": Memcon, December 2, 1975, 4:10–6:00 p.m., Ford, Mao Tse Tung, Kissinger, Ambassador George H. W. Bush, etc., GRFL.

587 "You don't seem to have many means": Ibid. Ford's efforts to rebuild the American military, including his proposed six-hundred-ship navy, by the end of the century, are documented in Lawrence J. Korb, "Gerald R. Ford and the Defense Budget: A Man Ahead of His Time," HUC, 421–30.

588 East Timor had very different plans: "East Timor Revisited, Ford, Kissinger and the Indonesian Invasion, 1975–76," December 6, 2001, NSA. In later years Ford expressed remorse for looking the other way as Suharto brutally annexed East Timor. See Brinkley, *Gerald Ford*, 130–32.

589 "I don't want to pass the blame": Brinkley, *Gerald Ford*, 132.

590 "He knew we wouldn't get everything": AI, Frank Zarb.

591 "I just can't put it in writing": Bush almost joined the Ford Cabinet six months earlier. Confronted with a vacancy at the Interior Department after his original nominee, Stan Hathaway, withdrew following treatment for depression, Ford said he seriously considered picking up the phone and asking Bush, still in Beijing, to serve as Hathaway's replacement. TAI, George H. W. Bush.

592 "John, what good will I do": AI, Rod Hills.

593 At one point late in 1975: Weidenfeld, *First Lady's Lady*, 238–39.

593 "If this is what the White House is going to do": Ibid.; "Reference Jack Anderson Story on Jack Ford," November 25, 1975, Box 11, Ron Nessen Papers, GRFL.

593 "And he read it": AI, David Kennerly.

593 "I made an effort at one point": AI, Dick Cheney.

593 "You need to talk to Bob before noon": AI, Don Penny.

593 After one particularly grueling day: AI, Dick Cheney.

594 "I'd go down at night and": Ibid.

595 "And it turned out to be a total laundry list": AI, David Gergen.

25: The Road to Kansas City

597 "One actor in the campaign is enough": Nessen, *Making the News*, 212.

597 "The real expression of an administration policy": AI, Paul O'Neill.

598 "I would swear": AI, David Gergen.

598 "the office seeks the man": AI, Stu Spencer.

599 "Come on over to dinner tonight": Ibid.

600 "the right to sell Pepsi-Cola in Siberia": "Reagan's Longest-Running Act," *Time*, February 23, 1976.

600 "Pay no attention to him": TAFI, 2448. Also, Myra MacPherson, "Who William Loeb Is and Why He's Saying All Those Things About . . . ," *Washington Post*, February 24, 1980.

600 Setting aside Loeb and the *Union Leader*: Nessen, *It Sure Looks Different*, 196–97.

600 "I'm a guy who swings for the homerun": AI, Stu Spencer.

600 "the speech that arguably prevented": Lou Cannon, *Governor Reagan: His Rise to Power* (New York: PublicAffairs, 2003), 406.

601 Only drastic cuts in program spending: Memorandum for the President from James T. Lynn, November 4, Subject: Reagan Speech and Buchanan Analysis, Richard B. Cheney Files, Box 19, Folder: Ronald Reagan, GRFL.

602 "It bought us a week's time": AI, Stu Spencer.

602 "Complex welfare programs": APP, "Address Before a Joint Session of the Congress on the State of the Union," January 19, 1976.

602 "Nixon is a shit": Bartholomew Sparrow, *The Strategist: Brent Scowcroft and the Call of National Security*, Illustrated edition (New York: PublicAffairs, 2015), 210. On October 27, 1975, Rumsfeld told Ford of a recent call from Herb Klein, passing along the news that Nixon, Klein's former boss, would not be going to China before the 1976 election. DRA.

602 In a telephone conversation: Handwritten notes of Nixon-Ford telephone call, February 6, 1976, Ford Library Vertical File, Folder: Nixon, Richard Milhouse (1913–1994)—Pardon.

602 "That means real earnings": APP, "Remarks and a Question and Answer Session at the University of New Hampshire in Durham," February 8, 1976.

603 "Five cities in two days is a bit much for me": Betty Ford to Kay Clark, February 10, 1976; Composite Grand Rapids Accession, Box 5, Folder: Kay Clark Correspondence 1976, GRFL.

604 "That's why we drove our phone banks to the end": AI, Stu Spencer.

604 "I hope that's the last God damn time": AI, Bob Orben.

605 "unsullied by Watergate": Lee Edwards, "Ronald Reagan vs. Gerald Ford: The 1976 GOP Convention Battle Royal," *National Interest*, April 16, 2016.

605 "I'd go in the office one day": AI, Stu Spencer.

606 "There's a saying": Jules Witcover, *Marathon: The Pursuit of the Presidency, 1972–1976* (New York: Viking, 1977), 400.

607 An even more ambitious foray into Texas: James M. Naughton, "A Hidden Asset for Ford: His Son Jack," *New York Times,* April 30, 1976. For the story of Naughton's *Air Force One* encounter with the president, see Ford Library Vertical File, Folder: Naughton, James M. (1932–2012).

607 "Hey, give me a week": A Conversation with Bo Callaway, Georgia Public Broadcasting, www.gpb.org7bo_callaway_transcript (undated). Also, see "U.S. Clears Callaway in Resort Case," *Washington Post*, January 12, 1977.

607 "the sooner the better": William E. Farrell, "Ford Decisively Defeats Reagan in Illinois Voting," *New York Times*, March 17, 1976.

607 "Ronnie has to get out": Lyn Nofziger, *Nofziger* (Washington, DC: Regnery Gateway, 1992), 179.

608 "Dr. Kissinger and Mr. Ford": Hayward, *The Age of Reagan*, 468.

609 "a shocker": TAFI, 3239.

609 Grand Rapids' first reported case: *Grand Rapids Herald*, September 29, 1918.

610 "They said we just can't afford to take any chances": AI, David Mathews.

611 "This is a no-win situation": Ibid. See Mathews's March 24 press conference with Drs. Salk and Sabin. Gerald R. Ford Presidential Digital Library, "The Swine Flu Immunization Program of 1976." The immediate response was typified by Frank Stanton to GRF, March 26, 1975. White House Central Files (sp 2.3–94-Exec), GRFL.

612 "To a scientist": Richard Fisher, "The Fiasco of the 1976 'Swine' Flu Affair," BBC, September 21, 2020, https://www.bbc.com/future/article/20200918-the-fiasco-of-the-us-swine-flu-affair-of-1976.

612 "let us remember one thing": "The President's Remarks Urging Congressional Enactment of the Program," August 6, 1976, Weekly Compilation of Presidential Documents, Office of the Federal Register, NARA, General Services Administration. See also "The Swine-Flu Snafu," *Newsweek*, July 12, 1976. There is, not surprisingly, a huge storehouse of literature relating to the swine flu scare. The standard history is Richard Neustadt and Harvey Fineberg, *The Swine Flu Affair: Decision-Making on a Slippery Slope* (Washington, DC: National Academies Press, 1978). More recent articles include Mary Halford, "Swine Flu in '76: Lessons from an Almost-Epidemic," *New Yorker*, April 28, 2009; Kat Eschner, "The Long Shadow of the 1976 Swine Flu Vaccine 'Fiasco,'" *Smithsonian*, February 6, 2017; Rebecca Creston, "The Public Health Legacy of the 1976 Swine Flu Outbreak," *Discover*, September 30, 2013; Patrick DiJusto, "The 'Great' Swine Flu Epidemic of 1976," *Salon*, April 28, 2009.

613 "simple and sinister innuendoes": David J. Sencer and J. Donald Millar, "Reflections on the 1976 Swine Flu Vaccination Program," Centers for Disease Control and Prevention, *Emerging Infectious Diseases* 12, no. 1 (January 2006), Theme Issue, *Influenza*.

614 "He's a funny guy": AI, Don Penny.

615 "Mr. President? I just want to tell you": Ibid.

616 "I can remember": TAFI, 3227; Elliot Kleinberg, "President Gerald Ford Visited Briny Breezes in 1976," *Palm Beach Post*, February 1, 2017.

617 On March 13, Ford declared: *Department of State Bulletin*, Volume 74, Office of Public Communication, Bureau of Public Affairs, 497.

617 "This is just one primary": Nessen, *It Sure Looks Different*, 209. Also see Sean P. Cunningham, "The 1976 GOP Primary: Ford, Reagan, and the Battle That Transformed Political Campaigns in Texas," *East Texas Historical Journal* 41, no. 2 (2003). Ford got some cautionary advice from Barry Goldwater. "You are not going to get the Reagan vote," the 1964 nominee told Ford in the aftermath of his Texas humiliation. "These are the same people who got me the nomination and they will never swerve, but ninety per cent of them will vote for you for President, so get after middle America. They have never had it so good. They are making more money and they are not at war and, for God's sake, get off of Panama, but don't let Reagan off that hook." Barry Goldwater to GRF, May 7, 1976, Folder: PL(Exec) 6/1–30/16, Box 3, White House Central Files, GRFL.

617 "we got a little political flak": National Security Council Meeting, May 11, 1976, 6:15–7:15, Box 2, National Security Advisor's NSC Meeting File, GRFL.

618 "we had a real bad night": TTH, 382.

618 "Because I have done a good job": TTH, 385.

619 "I hope we can keep him that pleased": Nessen, *It Sure Looks Different*, 217.

620 "I knew we'd have our tail handed to us": AI, Stu Spencer.

620 "That was my whole goal": Ibid.

620 "The strategy was knock them out early": TAI, James Baker.

621 Jim Baker kept a logbook: James Baker with Steve Fiffer, *Work Hard, Study . . . and Keep Out of Politics* (New York: G.P. Putnam's Sons, 2006), 43.

622   "The West Virginians were the worst": TAI, James Baker.

622   "We had this woman from Brooklyn": AI, Dick Cheney.

622   One July evening: "Man Killed at White House; Scaled Fence and Went On," *New York Times*, July 26, 1976.

622   'Gentlemen, if that fellow we just shot': AI, Dick Cheney.

623   "explorers and inventors": "Remarks at the Dedication Ceremony for the National Air and Space Museum," July 1, 1976, PSF, Box 35.

624   "The Declaration is the promise of freedom": "Remarks at a Bicentennial Ceremony at the National Archives," July 2, 1976, PSF, Box 35.

624   "this union of corrected wrongs": "Remarks at Independence Hall, Philadelphia, Pa," July 4, 1976, PSF, Box 36.

624   "a new kind of nation": Remarks at Naturalization Ceremony at Monticello, Charlottesville, Va, July 5, 1976, PSF, Box 36. Ford's visit to Jefferson's tomb is described in Pool Report #1, July 5, 1976, Box 26, White House Press Releases, GRFL.

26: I'm Feeling Good About America

627   "We just got the best news we've had in months": TTH, 394.

627   "If the Reagan-Schweiker ticket is a political coalition": Hayward, *The Age of Reagan*, 476. The candidate himself asked Sears on first being told the name of his prospective running mate, "Are you out of your mind?" His hearing impaired by a long-ago accident on a Hollywood movie set, Reagan thought Sears was pitching Senator *Lowell Weicker* of Connecticut, an even more improbable candidate than Pennsylvania's Schweiker. JCI, Mike Deaver.

627   Campaign manager John Sears: Nofziger, *Nofziger*, 196.

628   "I went to the White House": AI, Dick Cheney.

629   "Governor," he added earnestly: JCI, Mike Deaver.

629   So they softened Reagan's demands: "Showdown in Kansas City," *Time*, August 23, 1976; Martha Wagner Weinberg, "Writing the Republican Platform," *Political Science Quarterly* 95, no. 4 (Winter 1977–1978): 655–62.

630   "Because a modern economy involves so many moving parts": Greenspan, *Age of Turbulence*, 75.

630   "It's pretty hard to explain": TAFI, 3292. Ford's initial reaction to Greenspan's analysis was a bit more pungent. "That's a hell of a time to have a pause," he reflected. TAFI, 3612.

630   "I would rather gamble": TAFI, 1663.

631   "So we had the suite looking good and conservative": AI, Terry O'Donnell; AI, Red Cavaney.

631   The Texans: AI, Bill Archer.

631   "It wasn't trash": Adam Wren, "'It Was Riotous': An Oral History of the GOP's Last Open Convention," *Politico*, April 5, 2016. Also see "Contest of the Queens," *Time*, August 30, 1976.

632   As late as five that afternoon: AI, Peter McPherson.

632   "That's your song, Tony": David Greenberg, "The Last Great Republican Rupture," *Wall Street Journal*, April 30, 2016.

633   "The thing that I feared the most in Kansas City": AI, Stu Spencer. Also see Craig Shirley, "Lessons from the 1976 Republican Convention: Why Ronald Reagan Lost the Nomination," Retro Report, July 18, 2016, www.retroreport.org/transcript/lessons-from-the-1976-republican-convention-why-ronald-reagan-lost-the-nomination.

633   Jim Baker puzzled then: AI, James Baker.

633   "you knew who that querulous bastard was": AI, Dick Cheney.

633   "Henry, will you do it now": AI, Tom Korologos.

634   "This is late at night": AI, Dick Cheney.

634   "We knew we had the nomination": TAI, James Baker.

634   "We have made political history out there": Telecon, Senator Matthias/Secretary Kissinger, August 20, 1976, 8:45 p.m., Virtual Reading Room, foia.state.gov/search.

634   "Ford's finest forgotten hour": Gordon F. Sander, "When the U.S. Almost Went to War with North Korea," *Politico*, September 14, 2017.

635   "to make their presence visible": Richard G. Head, Frisco W. Short and Robert C. McFarlane, *Crisis Resolution: Presidential Decision Making in the Mayaguez and Korean Confrontations* (Boulder, CO: Westview Press, 1978), 194.

635   "To gamble with an overkill": Don Oberdorfer and Robert Carlin, *The Two Koreas: A Contemporary History* (New York: Basic Books, 2014), 63.

635   "You have to promise in advance": AI, Bill Timmons.

636   "we are way behind": TTH, 400.

636   "It's your show": Wren, "'It Was Riotous.'"

636   "We had never been close": TAFI, 3460.

637   "What have you got to lose?": AI, Stu Spencer.

638   "You've got a problem in the farm states": Ibid.

638   "Excuse me, Governor": Michael K. Deaver, *Behind the Scenes* (New York: William Morrow, 1987), 71–72.

638   "Dammit," he shouted at Cheney: Nessen, *Making the News*, 216.

638   "We were sitting on pins and needles": TAFI, 3349.

639   "This is either the end of the campaign or the beginning": Malcolm D. MacDougall, *We Almost Made It* (New York: Crown Publishing, 1977), 88.

639   "You are the people who make America what it is": Transcript of Ford's acceptance speech, *New York Times*, August 20, 1976.

639   "An electrifying performance": William F. Buckley, "The President Comes Alive with the Conservative Spirit," *Los Angeles Times*, August 26, 1976.

639   "We may have lost a Vice-President": MacDougall, *We Almost Made It*, 90.

640   "Ronnie, don't go": Confidential source.

640   "All right, now": J. Glasser interview with Don Penny, June 2, 1998, WBP, Harry Ransom Center, University of Texas (Austin), 22.

640   "You really got us in shape": PP, vol. 3, August 19, 1976, 730.

640   "Ford became a better *candidate*": AI, Doug Bailey.

640   "If past is indeed prologue": Michael Raoul-Dwal, Campaign Strategy for President Ford, 1976, Box 1, Folder: Presidential Campaign—Campaign Strategy (1), Dorothy Downton Files, GRFL.

641   "Please tell the President not to call Reagan": AI, Leon Parma.

641   "You've got to remember": AI, David Gergen.

642   "Jimmy is like a beautiful cat": *Detroit Free Press*, May 12, 1976.

642   "a combination of Machiavelli and Mr. Rogers": Hayward, *The Age of Reagan*, 489.

642   Carter balanced his support: "On Abortion, the Bishops v. the Deacon," *Time*, September 20, 1974.

643   "The American people are ready for the truth": PP, vol. 3, September 15, 1976, 784.

643   "If you focus on Ford": AI, Doug Bailey.

643   "Somewhere about '76": AI, Stu Spencer.

644   "Christ set some impossible standards for us": Kai Bird, *The Unfinished Presidency of Jimmy Carter* (New York: Crown, 2021), 117.

644   In the swing state of Texas: "Trying to Be One of the Boys," *Time*, October 4, 1976.

644   "I get so tired of this horseshit": Nessen, *It Sure Looks Different*, 284.

645   "the more attractive personality": *Boston Globe*, September 24, 1976.

645   subsequently identified as Helen Delich Bentley: TAFI, 3377.

646   "I couldn't talk to Levi": TTH, 418.

646 "Earl lives by the tongue": AI, John Knebel.

648 "You've got a problem": AI, Stu Spencer. Don Rumsfeld, though frustrated like the rest of Ford's inner circle, understood his friend's comments about Eastern Europe, "because all of the Congressmen from ethnic districts like I was and he was—every year on Captive Nations Week, we'd get up there and say, 'We do not concede that the Poles are permanently subjugated . . .' He just kept leaving out the word permanently. He had it in his mind." AI, Don Rumsfeld. For a sample of what Rumsfeld referred to, there is this golden oldie from the September 29, 1958, *Holland Sentinel*: "Poland today," said Ford, "is probably the country behind the Iron Curtain that would be most effective in defeating the Soviet Union if it had a choice."

648 "Ford thought Frankel was asking": AI, Dick Cheney.

648 "If I had simply said": JCI, Gerald Ford, April 27, 1990 (part 2).

648 "We didn't get a single question on it": JCI, Brent Scowcroft.

648 before briefing the press: AI, Ron Nessen.

649 "I walk in the door": AI, Stu Spencer.

649 "So, I turn the TV on": AI, Terry O'Donnell.

649 "He was living in a bubble up there": AI, David Gergen.

649 "The next day we flew down": AI, Dick Cheney.

649 "He wouldn't talk to me": JCI, Brent Scowcroft.

649 "Spencer and Cheney finally convinced me": TAFI, 3414.

650 His White House assistant, Dick Cook: "Ex-Nixon Aide: Never Contacted Ford," *Washington Star*, October 21, 1976. See also "Lifting the Cloud over the President," *Time*, October 25, 1976. Far less attention has been paid to a second Dean television appearance, on ABC's *Good Morning America* (October 21, 1976), in which he appeared to moderate some of the claims he made earlier on the rival *Today*, conceding that Ford in his subsequent actions "was doing what he thought he should do as Minority Leader." In his lengthy dissection of "The Pardon" in the August 1983 issue of the *Atlantic*, Seymour Hersh quotes Charles Colson saying, "I talked to Ford directly" about Nixon's desire to halt the Patman investigation in its tracks. Given his track record and the hyperpartisan speech drafts from his White House shop routinely disregarded by Ford, Colson would appear to carry little weight in the office of the minority leader.

In the same article Hersh contends that Nixon called Ford on September 7, 1974, and essentially blackmailed the president by threatening to expose their "deal" unless the pardon was forthcoming. Ford called the charge "a damn lie." No evidence of such a call has ever been produced. Hersh declined Ford's offer to examine White House telephone logs, insisting they were incomplete and subject to alteration.

651 "What bothers me": Nessen, *It Sure Looks Different*, 297.

651 "Dad, I'm getting all these questions": Weidenfeld, *First Lady's Lady*, 375.

651 NBC's credibility took a further hit: MacDougall, *We Almost Made It*, 170.

652 From his New York hospital bed: Memcon, October 21, 1976, 8:32–9:14 a.m., Ford, Kissinger, Scowcroft, GRFL.

652 "he seemed to be able to run faster": Cliff Gettings interview, Box E41, Campaign '76 Office: General Election, John Deardourff Files, GRFL.

652 "Nobody knows Jerry Ford": Malcolm D. MacDougall, "How Madison Avenue Didn't Put a Ford in Your Future," *New York Magazine*, February 21, 1977.

653 "Let me tell you": TAFI, 1483.

654 Out of the question: MacDougall, *We Almost Made It*, 128.

654 "instinct for the deliberate insult": "Bitter, Not Better, Down the Stretch," *Time*, October 25, 1976.

654 "like most others in the political arena": *Time*, November 1, 1976.

654 Semifrantic efforts by the campaign: AI, Frank Ursomarso; James M. Naughton, "Ford's Final Drive Reflects Optimism," *New York Times*, October 25, 1976.

655 "America was in very deep trouble": APP, "Remarks in Devon, Pennsylvania," October 27, 1976.

655 "Morning, Bob. What are you doing?": TTH, 432.

656 JERRY FORD'S ROOM: AI, Mary Fisher.

656 "It's like the last quarter of the big game": Pool report, "Milwaukee . . . and trip to St. Louis," October 30, 1976, Box 33, White House Press Release, GRFL.

657 Earlier that morning: Witcover, *Marathon*, 632–35.

657 "I hope none of our people are involved in this": TAI, Joe Garagiola.

657 "But we kept the ship of state": APP, "Remarks at a Rally on Arrival at Grand Rapids, Michigan," November 1, 1976.

658 "You did the impossible": Nessen, *It Sure Looks Different*, 311.

658 "You don't know where it's going to land": Telecon, Rockefeller/ Kissinger, November 2, 1976, 11:00 a.m., Virtual Reading Room, https:/foia.state.gov.

658 "If I had been told it": Telecon, General Scowcroft/Secretary Kissinger, November 2, 1976, 11:40 a.m., Virtual Reading Room.

658 "I think Ford should get it": Telecon, Kissinger/Nixon, November 2, 1976, 4:55 p.m., NSA.

659 "It's all right, Prez": AI, Greg Willard.

659 Close to midnight: Ibid.

659 "We're still in this thing": Ibid.

659 "That one hurts": Ibid.

660 A little after three Ford: Ibid.

660 "Maybe, Mr. President": AI, Rex Scouten.

660 "Do we dare wake him?": AI, Greg Willard.

660 "Governor, you have no idea": Ibid.

661 "If there is anything I can do for you": AI, David Kennerly.

661 "Damn it, we should have won": TAI, Joe Garagiola.

661 "I've got to give him the White House": Ibid.

27: Do What You Can

665 "My whole lifetime": John F. Kennedy Jr., "Ford Lately," *George*, November 1996.

665 "I don't know what to say": Memcon, November 5, 1976, 3:25 p.m., Ford, Mike Mansfield, Scowcroft, GRFL.

666 "There is no man": Jimmy Connors to GRF, November 8, 1976, Ford Special Materials, 1941–2007, Box C1, GRFL.

666 "You walked into an impossible situation": Reverend Theodore Hesburgh to GRF, November 8, 1976, Ford Special Materials, 1941–2007, Box C2, GRFL.

666 Despite their political differences: Clarence Mitchell to GRF, November 4, 1976, Ford Special Materials, 1941–2007, Box C3, GRFL.

666 "I can't believe": AI, Bob Barrett.

667 "He is a politician": Weidenfeld, *First Lady's Lady*, 394.

667 "John, I'm not ready to die": AI, Carlson.

667 Ford said it was "very doubtful": Memcon, November 23, 1976, 9:10–10:30 a.m., Ford, Kissinger, Scowcroft, GRFL.

667 "Here's the head": AI, Dorothy Downton.

668 "He didn't know anything about that": FLOHP, Bob Barrett, September 14, 1998.

668 "I took you away on a lot": AI, Dick Keiser.

668 "There are no soldiers marching": APP, State of the Union Address, January 12, 1977.

668 "The South Portico of the White House": TTH, 440.

669 Going around the room: Philip Shabecoff, "Farewells and Tears at White House," *New York Times*, January 20, 1977.

670 "That's my real home": TTH, 442.

670 "Do you have anything to tell us": Marjorie Hunter, "Ford, Arriving on Coast, Declares He Now Has 'Best of Both Worlds,'" *New York Times*, January 21, 1977.

670 "Don't worry, honey": Sue Thomas, *Michigan Live*, July 14, 2011.

670 "None of the kids' friends": Memcon, November 5, 1976, 3:25 p.m., Ford, Mike Mansfield, Scowcroft, GRFL. Besides the local climate and Betty's health, Ford had another reason for avoiding wintry West Michigan. "If we had gone back there, every Rotary Club, every Lions club, every Exchange Club, would expect me to come back [and] keep on speaking . . . Going to California, I didn't have any obligation to do that."

671 "I remember those early weeks": AI, Greg Willard. See also "The Sunny World of Palm Springs," *Newsweek*, March 28, 1975; "Why Be President, When You Can Live in Palm Springs?," *Detroit Free Press*, March 28, 1980.

671 "Jerry's retirement was a fraud": Ford, *Betty: A Glad Awakening*, 47.

671 "Even heroin addicts get time to withdraw": Mary Murphy, "What Do You Do for an Encore?," *New West*, November 21, 1977.

672 "You know": FLOHP, Bob Barrett.

672 "He has enough trouble": Marjorie Hunter, "Ford Careful to Avoid Meetings with Nixon," *New York Times*, August 15, 1977.

672 Eyebrows were raised: Herbert Mitgang, "Ford and Wife Sign Pact for Memoirs," *New York Times*, March 9, 1977; Carey Winfrey, "Fords' Memoirs, TV Shows: Complex Deals and Friendly Chats," *New York Times*, March 23, 1977.

673 "I've just always worried about not having money": FLOHP, Bob Barrett. Typical of Ford's continuing clout as a fundraiser is a January 5, 1998, letter from Notre Dame's former president, Reverend Theodore Hesburgh, citing a report from his brother Jim of a Ford appeal for a day care center that brought in $1.5 million as a first response.

673 "Well, it's still good": AI, Susan Ford Bales.

673 "What do you think": Ibid.

673 "Once a year": Ibid.

674 "I can't live like this anymore": AI, Greg Willard.

674 "She'd wake up": AI, Chris Chase.

675 "Parliaments can paralyze policy": Richard Nixon to GRF, May 8, 1977, Gerald Ford Special Materials, Box A8, GRFL.

675 "It's the best day I've had since November": Murphy, "What Do You Do for an Encore?"

675 He had always fantasized: Rob Haskell, "Gerald and Betty Ford's Restoration Politics," *New York Times Style Magazine*, March 27, 2015. See also Matthew Link, "The Lime Grows on You," *Palm Springs Life*, March 30, 2012.

676 "Good to see you": Anson, "I Gave Them the Sword," 185.

676 "Everybody made a maximum effort": TAFI, 3368.

676 "We've sent her to psychiatrists": McCubbin, *Betty Ford*, 268.

677 "You are all a bunch of monsters": Ibid., 280.

677 "She told me she wants half of our house": Ibid.

677 "Susan was the one": Ford, *Betty: A Glad Awakening*, 17.

677 "Are you sure this needs to be done?": Ibid., 21.

677 "Betty, the reason we're here": Liz McNeil, "How Betty Ford's Daughter Staged an Intervention to Save Her Mom's Life," *People*, September 5, 2018.

677 "as a healthy, loving person": Ford, *Betty: A Glad Awakening*, 22.

678 "I don't want to embarrass my husband": Ibid., 54.

678 "As American as a suburban shopping mall": Jane Howard, "The First Lady Next Door," *New York Times*, November 26, 1978.

678 "You know, it doesn't work": FLOHP, Bob Barrett.

678 In a 6–3 ruling: Linda Greenhouse, "High Court Sustains Ford Memoir Copyright," *New York Times*, May 21, 1985. Also see Ron Ostroff, "Can News Be Private Property? Gerald Ford's Publishers Cry Foul Before the Supreme Court," *Washington Post*, February 10, 1985.

679 A dissenting view: Harry McPherson, "Ford over Troubled Waters," *Washington Post*, May 20, 1979.

679 "I've still got some office work to do": Maury DeJonge, "Happy Retiree Ford 'Keeps Options Open,'" *Ann Arbor News*, January 21, 1979.

679 "He gives me hell": DeFrank, *Write It When I'm Gone*, 76.

679 "everything you could imagine": AI, Shelli Archibald.

680 "Do we really have to come in tomorrow?": AI, Dorothy Downton.

680 There were two groups opposed: AI, Rod Hills.

680 "Let's test the waters": Michael Deaver interview, September 12, 2002, Ronald Reagan Oral History Project, Miller Center of Public Affairs, University of Virginia, 32.

681 "I've never spent much time on the sidelines": Scott Kaufman, ed., *A Companion to Gerald R. Ford and Jimmy Carter* (Oxford: John Wiley & Sons, 2016), 523.

681 "What in the world is going on?": AI, Dorothy Downton.

681 "I was the only one who could type": AI, Bill Timmons.

682 "What do you think of the Ford deal?": Richard V. Allen, "How the Bush Dynasty Almost Wasn't," *New York Times Magazine*, July 30, 2000.

682 "Oh my God, I've never seen anything like it": AI, Bob Schieffer.

683 "We're just about ready to make a deal": Deaver interview, Miller Center, 33.

683 "I think you'd better pull the plug on it": AI, Jack Marsh.

683 "It just wouldn't be fair to Betty": AI, Bob Barrett.

683 "It's off": Deborah Hart and Gerald S. Strober, *The Reagan Presidency: An Oral History of the Era* (Dulles, VA: Potomac Books, 2003), 22.

683 "I had the feeling": AI, Dick Cheney.

684 "Well, Robert, not a bad convention": AI, Bob Barrett.

28: Lights in a Tree

685 "I feel OK, but I'm old!": AI, Paul Jenkins.

685 "oil and water": Douglas Brinkley, *The Unfinished Presidency: Jimmy Carter's Journey to the Nobel Peace Prize* (New York: Penguin Books, 1999), 67.

685 "Look, for the trip, at least": Mark D. Updegrove, *Second Acts, Presidential Lives and Legacies After the White House* (Guilford, CT: Lyons Press, 2006), 133.

686 "Sometimes I wish": Yehuda Avner, "Dick, Jimmy and Jerry Fly Coach to Cairo," *Jerusalem Post*, January 29, 2010.

686 "I misjudged him": AI, Penny Circle.

687 "I read that you and President Carter": Ted Kennedy to GRF, August 20, 1982, Gerald Ford Special Materials, Box A6, GRFL.

687 "You did this to me!": AI, Leon Parma.

687 "Security? Women?": Ibid.

687 "Little things like that": AI, David Kennerly.

688 Opposing elements in the party: Dotson Rader, "What I Would Have Done," *Parade*, April 4, 1982.

688 "I think they should be treated equally": Deb Price, "Gerald Ford: Treat Gay Couples Equally," *Detroit News*, October 21, 2001.

688 "We need to short-circuit": AI, Vaden Bales.

688 "He didn't care": AI, Tom DeFrank.

689 "Here was somebody": AI, Sandy Weill.

689 "We'd like to make the fees": AI, Rod Hills.

689 "The President, he shanked it": AI, Sandy Weill.

690 "She's the president": GRF to author.

690 "Hell no, I think it's wonderful": JCI, Gerald Ford, April 22, 1992 (part 2), GRFL.

690 "No, no, Chevy": Chevy Chase, "Mr. Ford Gets the Last Laugh," *New York Times*, January 6, 2007.

690 Early in his presidency: George H. W. Bush to GRF, December 19, 1989, Gerald Ford Special Materials, Box A1, Folder: George H. W. Bush. Following the 1992 election, which Bush lost to Bill Clinton, Ford got a letter from the attorney representing former secretary of defense Caspar Weinberger, asking the former president to intervene with the Bush White House and endorse a pardon for Weinberger's role in the Iran-Contra scandal. He knew quite a bit about pardons, Ford replied. "And I know from experience it is a very personal decision. Only the President can make that decision. And I feel confident that President Bush will make the right decision. And I'm not going to contact him, urging him to make it one way or another." JCI, Gerald Ford, April 22, 1993 (Part 1B).

692 "We have a few things": Thomas Ferraro, "Clinton, Ford Engage in Bipartisan Golf," August 14, 1993, UPI Archives.

692 "Eighty with fifty floating mulligans": Bob Woodward interview with Gerald Ford, September 22, 1997, 18, WBP, Harry Ransom Center for the Humanities, University of Texas (Austin).

692 "You had the uneasy feeling": Ibid. Bob Woodward interview with Ford.

693 "Jerry," Betty interrupted: AI, Susan Ford Bales.

693 "Here, Dad, use this": Ibid.

693 "a good chick flick": AI, Shelli Archibald.

694 "Make sure she doesn't get hurt": AI, Lorraine Ornelas.

694 "you set a goal": Ibid.

694 "Well, you've got to try the shoe": AI, Tyne Berlanga.

694 "Did you hear?": Confidential source.

695 "I'm going to say the same thing": AI, Jim Greenbaum.

695 "Whether or not": Gerald R. Ford, "The Path to Dignity," *New York Times*, October 2, 1998.

695 The same congressman: AI, Penny Circle.

696 "Bill," Ford reportedly answered: Confidential source.

696 "Some people equate civility with weakness": "Ford Urges Healing of Partisan Rancor," *Los Angeles Times*, October 28, 1999.

697 For a quarter century: GRF to author.

697 "I can understand the theory": Bob Woodward, "Ford Disagreed with Bush About Invading Iraq," *Washington Post*, December 28, 2006.

698 "He felt uncomfortable about it": AI, George W. Bush.

698 "most interesting idea": "Dear Mr. President," March 6, 2001, Gerald Ford Special Materials, Box A2, Folder: George W. Bush, GRFL.

698 "There was a friend of his": AI, Steve Ford.

699 "Let's try it part-time": AI, Vaden Bales.

699 "We solved all the problems, didn't we?": Amy DiPierro, "Presidents' Playground: George W. Bush Came to the Desert with Money and Mountains on His Mind," *Palm Springs Desert Sun*, November 1, 2017.

699 In a handwritten follow-up: "Dear President and Mrs. Ford," April 24, 2006, Gerald Ford Special Materials, Box A2, Folder: George W. Bush, GRFL.

699   "One of my most prized possessions": "To Jerry Ford," July 14, 2006, Gerald Ford Special Materials, Box A3, Folder: Jimmy Carter.

699   The alternative was perhaps six months: AI, Jack Eck.

699   "Let's go to Mayo": Ibid.

700   "We raised $33 million": AI, Paul O'Neill.

700   Pulling up at the nearest In-N-Out Burger: AI, Todd Matanick.

700   "Rummy!": Rumsfeld, *When the Center Held*, 268.

700   Holding her hand tightly: AI, Jan Hart.

700   On Christmas Eve: AI, Steve Ford.

701   By Tuesday morning: AI, Lillian Fisher.

701   Betty was to leave those lights in place: AI, Marty Allen.

701   "There's a time when you know": AI, Jan Hart.

Epilogue: "God Help the Country"

702   Stacy hadn't been born: AI, Mike Wagner.

703   "I hope people remember me": *Grand Rapids Press*, September 13, 1981.

704   "I wish I were a better public speaker": Trude Feldman, "Truth Is the Glue," *Tampa Bay Times*, July 30, 2002.

705   "there was not a single substantive problem area": Nelson W. Polsby, *Congress and the Presidency* (Hoboken, NJ: Prentice-Hall, 1976), 55.

706   "The President is not a visionary": James Naughton, "The Unique and Necessary Presidency of Gerald Ford," *New York Times*, November 14, 1976.

706   "Ever since FDR": TTH, 263.

706   Vision, said Connor: TAI, James Connor.

707   "The key to [the Ford Administration]": Fins, "Falling for Ford," op. cit.

707   Of the many condolence letters: Bill Scranton to Betty Ford, December 29, 2006, Betty Ford Special Materials, Box D1, GRFL.

708   "a splendid behind the scenes operator": William A. Syers interview with Bryce Harlow, July 27, 1985, GRFL.

708   Bob Hartmann thought: JCI, Bob Hartmann, June 19, 1991, GRFL.

708   "there was a reason, perhaps": FLOHP, David Broder.

708   "I think he did what history will decide": AI, Tom DeFrank.

708   "Maybe Ford was the last": AI, David Mathews.

709   Weighing his words carefully: NMP, MacNeil interview with GRF, 37, op. cit.

710   "I saw Jesse Owens": APP, "Remarks to Members of the U.S. Olympic Team and Presentation of the Presidential Medal of Freedom to Jesse Owens," August 5, 1976.

SELECTED BIBLIOGRAPHY

Agnew, Spiro T. *Go Quietly or Else*. New York: William Morrow, 1980.

Aitken, Jonathan. *Nixon: A Life*. Washington, DC: Regnery, 2015.

Ambrose, Stephen E. *Eisenhower: The President*. New York: Simon & Schuster, 1984.

———. *Nixon: The Education of a Politician 1913–1962*. New York: Simon & Schuster, 1988.

———. *Nixon: Ruin and Recovery, 1973–1990*. New York: Simon & Schuster, 1991.

Anderson, Patrick. *Electing Jimmy Carter: The Campaign of 1976*. Baton Rouge: Louisiana State University Press, 1994.

Anson, Robert Sam. *Exile: The Unquiet Oblivion of Richard M. Nixon*. New York: Simon & Schuster, 1984.

Baker, James, with Steve Fiffer. *Work Hard, Study . . . and Keep Out of Politics*. Evanston, IL: Northwestern University Press, 2006.

Baker, Peter, and Susan Glassner. *The Man Who Ran Washington: The Life and Times of James A. Baker III*. New York: Doubleday, 2020.

Barrett, David M. *The CIA and Congress: The Untold Story from Truman to Kennedy*. Lawrence: University Press of Kansas, 2006.

Beale, Betty. *Power at Play: A Memoir of Parties, Politicians and the Presidents in My Bedroom*. Washington, DC: Regnery, 1993.

Beasley, Maurine H. *First Ladies and the Press: The Unfinished Partnership of the Media Age*. Chicago: Northwestern University Press, 2006.

Belin, David W. *Final Disclosure: The Full Truth About the Assassination of President Kennedy*. New York: Scribner, 1988.

Bird, Kai. *The Unfinished Presidency of Jimmy Carter*. New York: Crown, 2021.

Blackmun, Ann. *Off to Save the World: How Julia Taft Made a Difference*. Rockland: Maine Authors Publishing, 2011.

Booraem, Hendrik, V. *The Education of Gerald Ford: Gerald Ford's Early Years*. Grand Rapids, MI: William B. Eerdmans, 2016.

———. *Young Jerry Ford: Athlete and Citizen*. Grand Rapids, MI: William B. Eerdmans, 2013.

Brinkley, Douglas. *Gerald R. Ford*. New York: Henry Holt, 2007.

———. *The Unfinished Presidency: Jimmy Carter's Journey to the Nobel Prize*. New York: Penguin Books, 1999.

Brinkley, Douglas, and Luke A. Nichter, eds. *The Nixon Tapes: 1973*. Boston: Houghton Mifflin, 2015.

Brower, Kate Anderson. *First Women: The Grace and Power of America's Modern First Ladies*. New York: HarperCollins, 2016.

Bugliosi, Vincent. *Reclaiming History*. New York: W. W. Norton, 2007.

Bush, George H. W. *All the Best, George Bush: My Life in Letters and Other Writings*. New York: Scribner, 2013.

Cannon, James M. *Gerald R. Ford: An Honorable Life*. Ann Arbor: University of Michigan Press, 2013.

———. *Time and Chance: Gerald Ford's Appointment with History*. New York: Harper-Collins, 1993.

Cannon, Lou. *Governor Reagan*. New York: PublicAffairs, 2003.

Caro, Robert. *The Years of Lyndon Johnson: Master of the Senate*. New York: Alfred Knopf, 2013.

Casserly, John J. *The Ford White House: Diary of a Speechwriter*. Boulder: Colorado Associated University Press, 1977.

Cheney, Dick, with Liz Cheney. *In My Time: A Personal and Political Memoir*. New York: Simon & Schuster, 2011.

Clarke, Thurston. *Honorable Exit*. New York: Doubleday, 2019.

Cohen, Richard M., and Jules Witcover. *A Heartbeat Away: The Investigation and Resignation of Spiro T. Agnew*. New York: Viking Press, 1974.

Colby, William, and Peter Forbath. *Honorable Men: My Life in the CIA*. New York: Simon & Schuster, 1978.

Coleman, William T. *Counsel for the Situation: Shaping the Law to Realize America's Promise*. Washington, DC: Brookings Institution Press, 2010.

Congressional Quarterly. *Presidency*. Washington, DC: Congressional Quarterly, 1974–1976.

———. *President Ford: The Man and His Record*. Washington, DC: Congressional Quarterly, 1974.

Coyne, John R. *Fall In and Cheer*. New York: Doubleday, 1979.

Crain, Andrew Downer. *The Ford Presidency: A History*. Jefferson, NC: McFarland 2009.

DeFrank, Thomas M. *Write It When I'm Gone: Remarkable Off-the-Record Conversations with Gerald R. Ford*. New York: G.P. Putnam's Sons, 2007.

Dobbs, Michael. *King Richard: Nixon and Watergate—An American Tragedy*. New York: Alfred A. Knopf, 2021.

Dobrynin, Anatoly. *In Confidence*. Seattle: University of Washington Press, 2001.

Doeneck, Justin. *In Danger Undaunted: The Anti-Interventionist Movement of 1940–1941, as Revealed in the Papers of the America First Committee*. Stanford, CA: Hoover Institution Press, 1990.

Drew, Elizabeth. *Washington Journal: Reporting Watergate and Richard Nixon's Downfall*. New York: Overlook Duckworth, 2014.

Drury, Bob, and Tom Clavin. *Halsey's Typhoon*. New York: Grove Press, 2007.

———. *Last Men Out*. New York: Free Press, 2011.

Ehlers, Vernon. *Gerald R. Ford: Memorial Tributes Delivered in Congress*. Washington, DC: US Government Printing Office, 2007.

Ehrlichman, John. *Witness to Power*. New York: Simon & Schuster, 1982.

Firestone, Bernard J., and Alexej Ugrinsky, eds. *Gerald R. Ford and the Politics of Post-Watergate America*. Westport, CT: Greenwood Press, 1993.

Flippen, J. Brooks. *Conservative Conservationist: Russell E. Train and the Emergence of American Environmentalism*. Baton Rouge: Louisiana State University Press, 2006.

Ford, Betty. *Betty: A Glad Awakening*. New York: Doubleday, 1987.

———. *The Times of My Life*. New York: Harper & Row, 1978.

Ford, Gerald R. *A Time to Heal: The Autobiography of Gerald R. Ford*. New York: Harper & Row, 1979.

———. *Portrait of the Assassin*. New York: Simon & Schuster, 1965.

Garment, Len. *Crazy Rhythm*. New York: Times Books, 1997.

Gergen, David. *Eyewitness to Power: The Essence of Leadership, Nixon to Clinton*. New York: Simon & Schuster, 2000.

Gibbs, Nancy, and Michael Duffy. *The Presidents' Club*. New York: Simon & Schuster, 2012.

Goldman, Eric F. *The Tragedy of Lyndon Johnson*. New York: Alfred A. Knopf, 1949.

Graham, Martha. *Blood Money*. New York: Doubleday, 1991.

———. "Platform for the American Dance." In *I See America Dancing: Selected Readings, 1685–2000*. Edited by Maureen Needham. Chicago: University of Illinois Press, 2002.

Graff, Garrett. *Watergate: A New History*. New York: Simon & Schuster, 2022.

Greene, John Robert. *Betty Ford: Candor and Courage in the White House*. Lawrence: University Press of Kansas, 2004.

———. *The Presidency of Gerald R. Ford*. Lawrence: University Press of Kansas, 1995.

Greenspan, Alan. *The Age of Turbulence: Adventures in a New World*. New York: Penguin Books, 2008.

Grose, Peter. *Gentleman Spy: The Life of Allen Dulles*. Boston: Houghton Mifflin, 1994.

Gully, Bill, with Mary Ellen Reese. *Breaking Cover*. New York: Simon & Schuster, 1980.

Haas, Lawrence J. *Harry and Arthur: Truman, Vandenberg and the Partnership That Created the Free World*. Dulles, VA: Potomac Books, 2016.

Haig, Alexander. *Inner Circles: How America Changed the World*. New York: Warner Books, 1992.

Hamilton, Nigel. *American Caesars: Lives of the Presidents from Franklin D. Roosevelt to George W. Bush*. New Haven, CT: Yale University Press, 2010.

Hanhimäki, Jussi M. *The Rise and Fall of Détente: American Foreign Policy and the Transformation of the Cold War*. Washington, DC: Potomac Books, 2013.

Hart, Deborah, and Gerald S. Strober. *The Reagan Presidency: An Oral History of the Era*. Dulles, VA: Potomac Books, 2003.

Hartmann, Robert T. *Palace Politics: An Insider's Account of the Ford Years*. New York: McGraw-Hill, 1980.

Hayward, Steven F. *The Age of Reagan: The Fall of the Old Liberal Order, 1964–1980*. Roseville, CA: Prima Publishing, 2001.

Head, Richard G., Frisco W. Short and Robert C. McFarlane. *Crisis Resolution: Presidential Decision Making in the Mayaguez and Korean Confrontations*. Boulder, CO: Westview Press, 1978.

Henderson, Charles. *Goodnight Saigon: The True Story of the U.S. Marines' Last Days in Vietnam*. New York: Penguin Publishing Group, 2008.

Hersey, John. *Aspects of the Presidency: Truman and Ford in Office*. New Haven, CT: Ticknor & Fields, 1980.

———. *The President: A Minute-by-Minute Account of a Week in the Life of Gerald Ford*. New York: Alfred A. Knopf, 1975.

Holzer, Harold. *The Presidents vs. the Press*. New York: Dutton, 2021.

Horrocks, David A. "Gerald Ford: All-American Counsel." In *America's Lawyer-Presidents*. Chicago: Northwestern University Press, 2004.

Hyland, William. *Mortal Rivals: Superpower Relations from Nixon to Reagan*. New York: Random House, 1987.

Isaacson, Walter. *Kissinger: A Biography*. New York: Simon & Schuster, 1992.

Jeffrey-Jones, Rhondi. *The CIA and American Democracy*. New Haven, CT: Yale University Press, 2003.

Johnson, Loch K. *A Season of Inquiry Revisited: The Church Committee Confronts America's Spy Agencies*. Lawrence: University Press of Kansas, 2015.

Jones, Sidney L. *Greenspan Counsel*. Lanham, MD: University Press of America, 2008.

Kalman, Laura. *Abe Fortas: A Biography*. New Haven, CT: Yale University Press, 1990.

———. *Right Star Rising: A New Politics, 1974–1980*. New York: W. W. Norton, 2010.

Kangor, W. Paul. *The Crusader: Ronald Reagan and the Fall of Communism*. New York: HarperCollins, 2007.

Kaufman, Scott, ed. *A Companion to Gerald R. Ford and Jimmy Carter*. Hoboken, NJ: John Wiley & Sons, 2016.

Kennerly, David Hume. *Extraordinary Circumstances*. Austin: Center for American History, University of Texas, 2007.

———. *Shooter*. New York: Newsweek Books, 1979.

Kissinger, Henry A. *Crisis: The Anatomy of Two Foreign Policy Crises*. New York: Simon & Schuster, 2003.

———. *Years of Renewal*. New York: Simon & Schuster, 1999.

Kunhardt Jr., Philip. "Gerald R. Ford: Healing the Nation." In *The American Presidency*. New York: Riverhead Books, 1999.

Kutler, Stanley. *Abuse of Power: The New Nixon Tapes*. New York: Free Press, 1997.

———. *The Wars of Watergate: The Last Crisis of Richard Nixon*. New York: Alfred A. Knopf, 2013.

Lachman, Seymour, and Robert Polner. *The Man Who Saved New York: Hugh Carey and the Great Fiscal Crisis of 1975*. Albany: State University of New York Press, 2010.

Lamb, Christopher Jon. *The Mayaguez Crisis, Mission Command, and Civilian-Military Relations*. Joint History Office of the Chairman of the Joint Chiefs of Staff, 2016.

Langguth, A. J. *Our Vietnam: The War 1954–1975*. New York: Simon & Schuster, 2002.

Lankevich, George. *Gerald R. Ford, 1913–: Chronology-documents-bibliographical AIDS*. Dobbs Ferry, NY: Oceana Publications, 1977.

Lewis, Norma. *Grand Rapids: Furniture City*. Charleston, SC: Arcadia Publishing, 2008.

Lifset, Robert, ed. *American Energy Policy in the 1970s*. Norman: University of Oklahoma Press, 2014.

Locker, Ray. *Haig's Coup: How Richard Nixon's Closest Aide Drove Him from Office*. Lincoln: University of Nebraska Press, 2019.

Lucas, J. Anthony. *Nightmare: The Underside of the Nixon Years*. New York: Viking, 1976.

Lucht, Tracy. *Sylvia Porter: America's Original Personal Finance Columnist*. Syracuse, NY: Syracuse University Press, 2013.

Lyden, Z. Z., ed. *The Story of Grand Rapids*. Grand Rapids, MI: Kregel Publishers, 1967.

MacDougall, Malcolm D. *We Almost Made It*. New York: Crown Publishers, 1977.

Magruder, Jeb Stuart. *An American Life: One Man's Road to Watergate*. New York: Atheneum, 1974.

Mallaby, Sebastian. *The Man Who Knew: The Life and Times of Alan Greenspan*. New York: Penguin Press, 2016.

McCubbin, Lisa. *Betty Ford*. New York: Simon & Schuster, 2018.

Meijer, Hendrik. *Arthur Vandenberg: The Man in the Middle of the American Century*. Chicago: University of Chicago Press, 2019.

Mieczkowski, Yanek. *Gerald Ford and the Challenges of the 1970s*. Lexington: University of Kentucky Press, 2005.

Mollenhoff, Clark R. *The Man Who Pardoned Nixon*. New York: St. Martin's Press, 1976.

Monje, Scott. *The Central Intelligence Agency: A Documentary History*. Westport, CT: Greenwood Press, 2008.

Morgan, Michael Cotey. *The Final Act: The Helsinki Accords and the Transformation of the Cold War*. Princeton, NJ: Princeton University Press, 2018.

Mount, Graeme. *895 Days That Changed the World: The Presidency of Gerald R. Ford*. New York: Black Rose Books, 2006.

Murphy, Bruce Allen. *Wild Bill: The Legend and Life of William O. Douglas*. New York: Random House, 2003.

Neal, Steve. *Dark Horse: A Biography of Wendell Willkie*. Garden City, NY: Doubleday, 1984.

Nelson, W. Dale. *Who Speaks for the President? The White House Press Secretary from Cleveland to Clinton*. Syracuse, NY: Syracuse University Press, 1998.

Nessen, Ron. *It Sure Looks Different from the Inside.* New York: Simon & Schuster, 1978.

———. *Making the News, Taking the News: From NBC to the Ford White House.* Middletown, CT: Wesleyan University Press, 2011.

Newton, Jim. *Eisenhower: The White House Years.* New York: Random House, 2011.

Nofziger, Lyn. *Nofziger.* Washington, DC: Regnery, 1992.

O'Brien, Michael. *John F. Kennedy: A Biography.* New York: Thomas Dunne Books, 2005.

Oberdorfer, Don, and Robert Carlin. *The Two Koreas: A Contemporary History.* New York: Basic Books, 2014.

Olson, Gordon L. *A Grand Rapids Sampler.* Grand Rapids Historical Commission, 1992.

Osborne, John. *White House Watch: The Ford Years.* Washington, DC: New Republic Books, 1977.

Payne, Stanley G., and Jesús Palacios. *Franco: A Personal and Political Biography.* Madison: University of Wisconsin Press, 2014.

Peabody, Robert L. *Leadership in Congress: Stability, Succession, and Change.* Boston: Little, Brown, 1976.

Perlstein, Rick. *Nixonland: The Rise of a President and the Fracturing of America.* New York: Scribner, 2008.

Phillips-Fein, Kim. *Fear City: New York's Fiscal Crisis and the Rise of Austerity Politics.* New York: Metropolitan Books, 2017.

Pipes, Kasey S. *After the Fall: The Remarkable Comeback of Richard Nixon.* Washington, DC: Regnery History, 2019.

Pollock, Alex J. *Finance and Philosophy: Why We're Always Surprised.* Philadelphia: Paul Dry Books, 2018.

Porter, Roger B. *Presidential Decision Making: The Economic Policy Board.* New York: Cambridge University Press, 1980.

Priess, David. *The President's Book of Secrets.* New York: PublicAffairs, 2016.

Reeves, Richard. *A Ford, Not a Lincoln.* New York: Harcourt Brace Jovanovich, 1975.

Reichley, A. James. *Conservatives in an Age of Change: The Nixon and Ford Administrations.* Washington, DC: Brookings Institution Press, 1981.

Rowan, Roy. *The Four Days of Mayaguez.* New York: W. W. Norton, 1975.

Rozell, Mark J. *The Press and the Ford Presidency.* Ann Arbor: University of Michigan Press, 1992.

Rubenstein, Bruce A., and Lawrence E. Ziewacz. *Three Bullets Sealed His Lips.* Lansing: Michigan State University Press, 1987.

Rumsfeld, Donald. *Known and Unknown: A Memoir.* New York: Penguin Group USA, 2011.

———. *When the Center Held: Gerald Ford and the Rescue of the American Presidency.* New York: Free Press, 2018.

Sargent, Daniel J. *A Superpower Transformed: The Remaking of American Foreign Relations in the 1970s.* New York: Oxford University Press, 2015.

Schapsmeier, Edward L., and H. Frederick. *Gerald R. Ford's Date with Destiny: A Political Biography.* New York: Peter Lang, 1989.

Schmidt, Helmut. *Men and Powers: A Political Retrospective.* New York: Random House, 1990.

Schwartz, Thomas A. *Henry Kissinger and American Power: A Political Biography.* New York: Hill and Wang, 2020.

Shaffer, Samuel. *On and Off the Floor: Thirty Years as a Correspondent on Capitol Hill.* New York: Newsweek Books, 1980.

Sibley, Katherine A. S., ed. *A Companion to First Ladies.* Hoboken, NJ: John Wiley & Sons, 2016.

Sidey, Hugh. *Portrait of a President.* New York: Harper & Row, 1975.

Simon, William E. *A Time for Truth*. New York: Reader's Digest Press, 1978.

Smith, Richard Norton. *On His Own Terms: A Life of Nelson Rockefeller*. New York: Random House, 2014.

Snepp, Frank. *Decent Interval*. New York: Random House, 1977.

Soares, Janet Mansfield. *Martha Hill and the Making of American Dance*. Middletown, CT: Wesleyan University Press, 2009.

Sparrow, Bartholomew. *The Strategist: Brent Scowcroft and the Call of National Security*. Illustrated edition. New York: PublicAffairs, 2015.

Specter, Arlen. *Passion for Truth*. New York: William Morrow, 2000.

Stevens, John Paul. *The Making of a Justice: Reflections on My First 94 Years*. New York: Little, Brown, 2014.

Storing, Herbert J. *The Ford White House: A Miller Center Conference Chaired by Herbert J. Storing*. Edited by Miller Center of Public Affairs at the University of Virginia. Lanham, MD: University Press of America, 1986.

terHorst, Jerald F. *Gerald Ford and the Future of the Presidency*. New York: Third Press, 1974.

Thompson, Kenneth W., ed. *The Ford Presidency: Twenty-Two Intimate Perspectives of Gerald Ford*. Portraits of American Presidents, VII. Lanham, MD: University Press of America, 1988.

Townley, Dafydd. *The Year of Intelligence in the United States: Public Opinion, National Security, and the 1975 Church Committee*. Cham, Switzerland: Palgrave Macmillan, 2021.

Turner, Michael. *The Vice President as Policy Maker: Rockefeller in the Ford White House*. Westport, CT: Greenwood Press, 1982.

Updegrove, Mark. *Second Acts: Presidential Lives and Legacies After the White House*. Guilford, CT: Lyons Press, 2006.

US Congress, House, Committee on the Judiciary. *Nomination of Gerald R. Ford to Be Vice President of the United States: Hearings*. Washington, DC: Government Printing Office, 1973.

US Congress, Senate, Committee on Rules and Administration. *Nomination of Gerald R. Ford of Michigan to Be Vice President of the United States: Hearings*. Washington, DC: Government Printing Office, 1973.

Van Atta, Dale. *With Honor: Melvin Laird in War, Peace, and Politics*. Madison: University of Wisconsin Press, 2008.

Vandenbroucke, Lucien S. *Perilous Options as an Instrument of U.S. Foreign Policy*. New York: Oxford University Press, 1993.

Vestal, Bud. *Jerry Ford, Up Close: An Investigative Biography*. New York: Coward, McCann & Geoghegan, 1974.

Ward, Geoffrey C. *The Vietnam War: An Intimate History*. New York: Alfred A. Knopf, 2017.

Ware, Susan. *Letter to the World: Seven Women Who Shaped the American Century*. New York: W. W. Norton 1998.

Weidenfeld, Sheila Rabb. *First Lady's Lady: With the Fords at the White House*. New York: Putnam, 1979.

Weiner, Tim. *Legacy of Ashes: The History of the CIA*. New York: Doubleday, 2007.

Werth, Barry. *31 Days: The Crisis That Gave Us the Government We Have Today*. New York: Doubleday, 2006.

Westal, Odd Arne. *The Cold War: A World History*. New York: Basic Books, 2017.

Wetterhahn, Ralph. *The Last Battle: The Mayaguez Incident and the End of the Vietnam War*. New York: Carroll & Graf, 2001.

Whitcomb, John, and Claire Whitcomb. *Real Life at the White House: Two Hundred Years of Daily Life at America's Most Famous Residence*. New York: Routledge, 2000.

Willens, Howard P. *History Will Prove Us Right: Inside the Warren Commission Report on the Assassination of John F. Kennedy*. New York: Overlook Press, 2013.

Williams, Daniel K. *The Election of the Evangelical: Jimmy Carter, Gerald Ford, and the Presidential Contest of 1976*. Lawrence: University Press of Kansas, 2020.

Wilson, Will R., Sr. *A Fool for a Client: Richard Nixon's Free Fall Toward Impeachment*. Austin: Eakin Press, 2000.

Witcover, Jules. *Marathon: The Pursuit of the Presidency, 1972–76*. New York: Viking Press, 1977.

———. *Very Strange Bedfellows*. New York: PublicAffairs, 2007.

Witherspoon, Patricia Dennis. *Within These Walls: A Study of Communications Between Presidents and Their Senior Staffs*. Westport, CT: Praeger, 1991.

Woodward, Bob. "Gerald R. Ford." In *Profiles in Courage for Our Time*, by Caroline Kennedy. New York: Hyperion Press, 2002.

———. *Shadow: Five Presidents and the Legacy of Watergate*. New York: Simon & Schuster, 1999.

Woodward, Bob, and Carl Bernstein. *The Final Days*. New York: Simon & Schuster, 1976.

INDEX

ABOUT THE AUTHOR

RICHARD NORTON SMITH is the director of five presidential libraries, a familiar face to viewers of C-SPAN and *PBS NewsHour*, and the author of, among other works, *Thomas E. Dewey and His Times*, which was a finalist for the Pulitzer Prize; *Patriarch: George Washington and the New American Nation*; and *On His Own Terms: A Life of Nelson Rockefeller*.